Springer Series in Computational Mathematics

Volume 48

More information about this series at http://www.springer.com/series/797

Vít Dolejší · Miloslav Feistauer

Discontinuous Galerkin Method

Analysis and Applications to Compressible Flow

Springer

Vít Dolejší
Faculty of Mathematics and Physics
Charles University in Prague
Praha 8
Czech Republic

Miloslav Feistauer
Faculty of Mathematics and Physics
Charles University in Prague
Praha 8
Czech Republic

ISSN 0179-3632 ISSN 2198-3712 (electronic)
Springer Series in Computational Mathematics
ISBN 978-3-319-37123-8 ISBN 978-3-319-19267-3 (eBook)
DOI 10.1007/978-3-319-19267-3

Mathematics Subject Classification (2010): 65M60, 65M15, 65M20, 65M08, 65N30, 65N15, 76M10, 76M12, 35D30

Springer Cham Heidelberg New York Dordrecht London

Printed on acid-free paper

Springer International Publishing AG Switzerland is part of Springer Science+Business Media (www.springer.com)

Preface

Many real-world problems are described by partial differential equations whose numerical solution represents an important part of numerical mathematics. There are several techniques for their solution: the finite difference method, the finite element method, spectral methods and the finite volume method. All these methods have advantages as well as disadvantages. The first three techniques are suitable particularly for problems in which the exact solution is sufficiently regular. The presence of interior and boundary layers appearing in solutions of singularly perturbed problems (e.g., convection-diffusion problems with dominating convection) or discontinuities in solutions of nonlinear hyperbolic equations lead to some difficulties. On the other hand, finite volume techniques based on discontinuous, piecewise constant approximations are very useful in solving convection-diffusion problems, but their disadvantage is their low order of accuracy.

The most recent technique for the numerical solution of partial differential equations is the *discontinuous Galerkin method* (DGM), which uses ideas of both the finite element and finite volume methods. The DGM is based on piecewise polynomial but discontinuous approximations, which provides robust numerical processes and high-order accurate solutions.

During the past two decades the DGM has become very popular and a number of works has been concerned with its analysis and applications. It appeared that the DGM is suitable for the numerical solution of a number of problems for which other techniques fail or have difficulties. We can mention singularly perturbed problems with boundary and internal layers, which exist in solutions of convection-diffusion equations with dominating convection.

Another possibility represents problems with solutions containing discontinuities and steep gradients, as in the case of nonlinear hyperbolic problems and compressible flow. This means that the DGM is suitable for the numerical solution of problems appearing particularly in fluid dynamics, hydrology, heat and mass transfer and environmental protection on the one hand, but also financial mathematics and image processing on the other hand. Moreover, the DGM offers considerable flexibility in the choice of the mesh design; indeed, the DGM easily handles non-matching and non-uniform grids, even anisotropic, with different polynomial approximation degrees on different elements. This allows for a simple treatment of *hp*-variants of adaptive techniques. Finally, the DGM can easily be parallelized, which is demanding in complex numerical simulations.

This book is devoted to the theory and applications of the discontinuous Galerkin method. The first part of this book deals with theoretical aspects of the discontinuous Galerkin (DG) method applied to the numerical solution of scalar nonlinear convection-diffusion problems. Scalar equations serve as models for several applications treated in the second part of the book. Our aim is to present the DG discretization of model problems and to derive (a priori) error estimates. Theoretical results are supported by numerical experiments demonstrating the accuracy of the DG methods.

In order to better understand the basic principles of the discontinuous Galerkin method, we start from a numerical solution of the simple Poisson problem having mixed Dirichlet–Neumann boundary conditions. Hence, in Chaps. 2 and 3, we describe the DG discretization and the derivation of error estimates in detail in order to familiarize non-specialist readers with theoretical tools used in the DGM. We tried to have material self-contained as much as possible. Therefore, these chapters contain similar material on the DGM as other monographs.

In Chaps. 4–6 the main attention is paid to the analysis of discontinuous Galerkin techniques for solving nonstationary, nonlinear convection-diffusion problems. Chapter 7 is devoted to some generalizations of the DGM: the *hp*-version of the DGM, the use of general polygonal elements and the effect of numerical integration. Theoretical results are demonstrated by the solution of numerous test problems.

The second part (formed by Chaps. 8–10) deals with applications of the DGM to solving gas dynamics problems. The numerical schemes, proposed and analyzed in the first part of the book, are extended to solving the system of equations describing compressible flow, namely, in Chap. 8, the compressible Euler equations are solved, Chap. 9 is devoted to the solution of viscous flow described by the compressible Navier–Stokes equations, and in Chap. 10 the DGM is applied to simulating compressible flow in time-dependent domains and to the interaction of compressible flow with elastic structures. We also discuss the numerical solution of the resulting systems of algebraic equations which is a fundamental aspect in the practical use of the DGM for solving industrial problems. The treatment in the last three chapters is accompanied by test problems and also technically relevant applications proving the flexibility, accuracy and robustness of the described discontinuous Galerkin schemes.

We hope that the book will be useful to specialists—namely, pure and applied mathematicians, aerodynamists, engineers, physicists and natural scientists. We also expect that the book will be suitable for graduate and postgraduate students in mathematics and in the technical sciences.

As for references, there is a rapidly increasing amount of literature on theoretical aspects and applications of the DGM. We tried to quote the works relevant to the topics of the book, but it is clear that many significant references have been unintentionally omitted. We apologize in advance to those authors whose contributions are not mentioned or do not receive the attention they deserve. We have tried to avoid errors, but some may remain. Readers are welcome to send any correction electronically to the address dolejsi@karlin.mff.cuni.cz or feist@karlin.mff.cuni.cz.

We are grateful to Profs. I. Babuška, F. Bassi, B. Cockburn, M. Dumbser, A. Ern, R. Hartmann, J. Horáček, P. Houston, M. Křížek, D. Kröner, M. Lukáčová, C.-D. Munz, R. Rannacher, S. Rebay, H.-G. Roos, A. Sändig, C. Schwab, C.-W. Shu, V. Sobotková, E. Süli, P. Sváček, F. Toro, M. Vohralík and W. Wendland for valuable information, advice, comments, inspiring suggestions and stimulating discussions which helped us during our work in the area of the DGM and in the preparation of the manuscript. We also appreciate our cooperation with our colleagues V. Kučera and M. Vlasák and Ph.D. students J. Česenek, J. Hasnedlová-Prokopová, O. Havle, M. Holík and J. Hozman in the DGM. Further, we are grateful to our colleagues V. Kučera, M. Vlasák and Ph.D. students M. Balázsová, M. Hadrava, A. Kosík, I. Soukup, I. Šebestová, P. Šimánek and A. Živčák for reading parts of the manuscript. Particularly, we are gratefully indebted to our colleague K. Najzar, who carefully read the whole book and provided us with a number of helpful suggestions.

Last but not least, we would like to thank Prof. W. Jäger, who recommended that we publish the book at Springer and who managed our contacts with this publishing house. We highly appreciate the cooperation with Springer staff, particularly with project coordinator Ms. Thanh-Ha Le Thi and copyeditor Ms. Ann Konstant.

The work on the book was partially supported by the Czech Science Foundation, projects No. 201/08/0012 and 13-00522S and by the 3rd Call of the 6th European Framework Programme, project ADIGMA, No. AST5-CT-2006-030719. We gratefully acknowledge these supports. We also acknowledge our membership in the Nečas Center for Mathematical Modeling (http://ncmm.karlin.mff.cuni.cz).

Our families gave us considerable support during work on our book. We wish to express our gratitude for their patience and understanding.

Prague
May 2015

Vít Dolejší
Miloslav Feistauer

Contents

Chapter 1
Introduction

The investigation of *convection-diffusion problems* is a very topical subject in theoretical as well as applied research. On the one hand, these problems play an important role in fluid dynamics, hydrology, heat and mass transfer, environmental protection, water transfer in soils, porous media flow, but also on the other hand, in financial mathematics or image processing. The complexity of these problems prevent from obtaining their exact solution. Therefore, developing a sufficiently *robust*, *accurate* and *efficient numerical method* for computing *approximate solutions* of (nonlinear) convection-diffusion equations is a challenging problem.

There is an extensive literature devoted mainly to linear convection-diffusion problems, represented by monographs [187, 226, 245, 250] and references therein. The main difficulty which has to be overcome in the numerical solution of convection-diffusion problems is the precise resolution of the *boundary layers*. In physics and fluid mechanics, a boundary layer is the layer of fluid in the immediate vicinity of a bounding surface in which the effects of viscosity are significant.

If the equation under consideration represents a nonlinear conservation law with a small dissipation, then besides boundary layers, also *shock waves* appear (slightly smeared due to the dissipation), which represent the *interior layers*. This is particularly the case of the Navier–Stokes equations describing viscous compressible flow treated in the second part of this book.

The *discontinuous Galerkin method* (DGM) appears suitable for the numerical solution of problems with solutions containing discontinuities and/or steep gradients, see Sect. 1.1. For the discretization of convective terms, the DGM uses the concept of a numerical flux, which is an important tool for the finite volume method (FVM), using piecewise constant (and hence, discontinuous) approximations. In contrast to the FVM, similarly, as in the finite element method (FEM), the DGM uses polynomial approximations of higher-degree, which lead to higher-order schemes in a natural way. From this point of view, the DGM can be considered as a generalization of the FVM and FEM.

V. Dolejší and M. Feistauer, *Discontinuous Galerkin Method*,
Springer Series in Computational Mathematics 48,
DOI 10.1007/978-3-319-19267-3_1

In this chapter we present a motivating example demonstrating a potential of the DGM in comparison to the finite volume and finite element methods. Moreover, we give a historical overview of the development of the discontinuous Galerkin discretization of elliptic, parabolic and hyperbolic problems and also an application of the DGM to the numerical solution of compressible flow. Finally, this chapter also contains a survey of some mathematical concepts and results that are important for understanding the subsequent treatment.

1.1 DGM Versus Finite Volume and Finite Element Methods

Among several fundamental techniques developed for the approximate solution of *partial differential equations*, two methods figure the principal role:

- *finite element method* (FEM), which is based on piecewise polynomial approximations and applied mainly to elliptic and parabolic, i. e., diffusion problems,
- *finite volume method* (FVM), which is based on a piecewise constant approximations and applied mainly to convection or hyperbolic problems and problems of fluid flow.

In the finite element method, the question arises: should we prefer to use *conforming* or *nonconforming* finite elements? Conforming (i.e., continuous) finite element approximations are suitable for problems with sufficiently regular solutions. However, in the solution of some special problems, using the conforming finite elements leads to nonphysical solutions. As an example we can mention the approximation of viscous incompressible flow. In this case, one often applies noncomforming finite elements, for which the continuity is relaxed to some discrete points on interfaces between neighbouring elements. This is the case of the well-known Crouzeix–Raviart piecewise linear finite elements [68] that are continuous at midpoints of sides of triangular elements.

On the other hand, singularly perturbed problems, nonlinear conservation laws and compressible flow have solutions with steep gradients or discontinuities, and their approximations by the FEM usually suffer from the Gibbs phenomenon, i.e., nonphysical oscillations, called *spurious oscillations*, in the approximate solution. One way to avoid this drawback is to use a suitable stabilization as, e.g., the streamline diffusion method or Galerkin least squares method and shock capturing stabilization, see, e.g., [191]. These techniques are applied with success to the solution of scalar equations describing heat and mass transfer or to the solution of the incompressible Navier–Stokes equations. However, there are difficulties with the application of the FEM to the solution of compressible flow. The extension of the stabilization methods to the compressible Navier–Stokes equations (described, e.g., in [127, Chap. 4]) is rather complicated and other suitable methods have been sought.

Approximating discontinuous solutions (or solutions with steep gradients) by continuous functions (or functions that are continuous at some prescribed nodes) does not appear quite natural. Therefore, the FVM using a piecewise constant approximation could be more suitable, because the finite volume approximations are discontinuous on interelement interfaces. This allows for a better resolution of shock waves and contact discontinuities. On the other hand, the finite volume schemes are of a low-order of accuracy. The construction of higher-order finite volume schemes is connected with various obstacles. We can mention the treatment of boundary conditions. Theoretical analysis of higher-order FVM is not developed.

A combination of ideas and techniques of the FVM and the FEM methods yields the *discontinuous Galerkin method* (DGM) using the advantages of both approaches and allowing us to obtain schemes with a higher-order accuracy in a natural way. The DGM is based on the idea of approximating the solution of a given problem by a piecewise polynomial function over a finite element mesh without any requirement on interelement continuity. Moreover, it is not necessary to construct subsets of finite element spaces in order to approximate the Dirichlet boundary conditions. This is replaced by the application of the *interior and boundary penalty*. This approach comes from theideas of Babuška, Zlámal and Nitsche [17, 228]. Hence, the DGM can be considered as a fully nonconforming finite element technique. The use of high polynomial degree approximations lead to a sufficiently accurate numerical solution and the discontinuous approximations cause greater flexibility of the method, which allows us a better resolution of problems having solutions with discontinuities or steep gradients.

In order to illustrate some of the effects mentioned above, we present a numerical solution of the 1D convection-diffusion problem

$$-\varepsilon u'' + u' = 1 \ \text{ in } \Omega := (0, 1), \quad u(0) = u(1) = 0,$$

where u' and u'' denote the first- and second derivative with respect to $x \in \Omega$, respectively, and $\varepsilon > 0$ is the diffusion coefficient. For $\varepsilon \ll 1$, the problem is convection dominanted and the exact solution has a steep gradient near $x = 1$ in the boundary layer. We solve this problem numerically by three techniques:

- FEM: finite element method using the continuous piecewise linear approximation over a uniform partition of Ω with spacing $h = 1/32$,
- FVM: finite volume method using the piecewise constant approximation over a uniform partition of Ω with spacing $h = 1/32$,
- DGM: discontinuous Galerkin method using the discontinuous piecewise quartic approximation over a uniform partition of Ω with spacing $h = 1/8$.

Let us note that in all three cases, we use approximately 32 degrees of freedom. Moreover, no additional stabilization technique is applied.

Figure 1.1 shows approximate solutions obtained by the FEM, FVM and DGM for $\varepsilon = 10^{-2}$ and $\varepsilon = 10^{-3}$ in comparison with the exact solution. We observe that the approximate solutions obtained by the FEM suffer from nonphysical spurious oscillations, whose amplitude is increasing with decreasing ε. It is possible to avoid them by using a suitable stabilization as, e.g., the streamline diffusion method or Galerkin least squares method and shock capturing stabilization. Further, the FVM does not give a sufficiently accurate approximation of the exact solution, namely

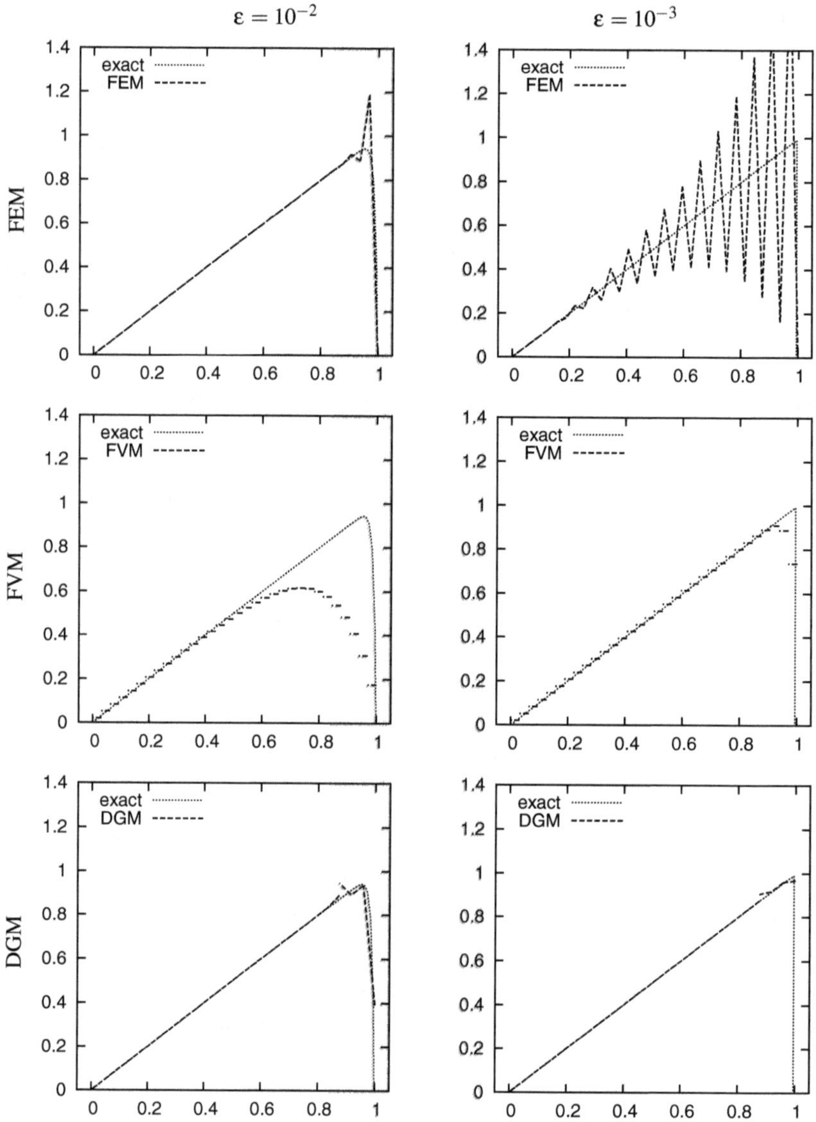

Fig. 1.1 A comparison of the approximate solution obtained by FEM, FVM and DGM and the exact one

in the case $\varepsilon = 10^{-2}$. The steep gradient is smeared. It is possible to overcome this drawback partly by, e.g., a higher-order reconstruction technique. Finally, the approximate solutions obtained by the DGM give the best approximation for both cases. The approximation of the steep gradient near $x = 1$ is not smeared and the solution does not suffer from nonphysical oscillations. Only a small inaccuracy represented by a small wiggle appears near the boundary layer.

We can conclude that discontinuous piecewise polynomial approximations have a potential for constructing efficient, accurate and robust methods for the numerical solution of convection-diffusion problems.

1.2 A Short Historical Overview of the DGM

The numerical solution of partial differential equations with the aid of discontinuous piecewise polynomial approximations started to appear in the 1970s. This approach, later called the *discontinuous Galerkin method* (DGM), was developed almost independently for hyperbolic problems and for elliptic and parabolic problems. Within 30 years, more than 10 principal approaches were developed. In the following, we briefly describe the development of the DGM in both communities.

1.2.1 DGM for Hyperbolic and Singularly Perturbed Problems

The *discontinuous Galerkin method* was used for the first time by Reed and Hill in 1973 [234] for solving the neutron transport equation. The first numerical analysis of the DGM was carried out by Le Saint and Raviart in 1974 [214], who derived suboptimal error estimates. After more than 10 years, in 1986, Johnson and Pitkäranta presented in [192] improved error estimates. Further, Richter in 1988 [236] proved the optimal order of convergence for a linear first-order hyperbolic equation in $\mathbb{R}^2$, and in 1992 [237] he extended the analysis to the case with a constant linear diffusion.

In the period 1989–1998 a rapid development of the DGM started by the series of papers by Cockburn, Shu and coworkers [56, 57, 62, 63, 65], where the DGM was applied to nonlinear conservation laws. In their approach, the space DG discretization is combined with the Runge–Kutta time discretization. Therefore, this method is called the *Runge–Kutta discontinuous Galerkin* (RKDG) method. Due to the local character of discontinuous Galerkin schemes and the explicit time discretization, the RKDG method is highly parallelizable. It was further developed for hyperbolic conservation laws, e.g., by Biswas, Devine and Flaherty in 1994 [34], deCougny et al. in 1994 [75], Bassi Rebay in 1997 [24], Atkins and Shu in 1998 [9]. The review of the development of the RKDG method can be found in [66].

In 1997, Bassi and Rebay [23] adapted the RKDG method to the system of the compressible Navier–Stokes equations in such a way that the solution itself and its gradient were considered as independent variables. This approach is often called the BR method. It recalls the well-known mixed method used in the conforming FEM. However, a counterexample presented by Brezzi et al. in 1997 [39] showed that the application of the BR method to the Poisson problem is rather problematic, because the existence of the approximate solution is not guaranteed. Therefore, an additional stabilization term was proposed and the resulting scheme was analyzed. Then Bassi et al. proposed in [27] a modified method, which is usually referred to as the BR2 method, see [203]. The BR2 method (together with a further variant of the DGM) applied to the Poisson problem was analyzed in 2000 [40].

In 1998, Cockburn and Shu introduced in [64] a generalization of the BR method called the *local discontinuous Galerkin* (LDG) method, which is analogous to mixed methods in the conforming FEM. In 1999, Cockburn and Dawson extended the LDG method to multidimensional equations in [55]. A summary of the development of the RKDG and LDG methods for convection-diffusion problems was given by Cockburn in 1998 [54]. After that, the LDG method became very popular and was extended to various kinds of partial differential equations. Castillo et al. in 2000 [43] proved convergence of the LDG method for elliptic problems. Castillo et al. in 2002 [44] derived optimal error estimates for the hp-version of the LDG method for convection–diffusion problems. In 2001, Cockburn et al. proved in [58] the super-convergence of the LDG method for elliptic problems on Cartesian grids. Elliptic problems were further analyzed by Dawson in 2002 [73]. Moreover, the LDG method was applied by Cockburn et al. in 2002 [61] to the Stokes problem, by Cockburn, Kanschat and Schötzau in 2005 [59] and by Kanschat in 2005 [193] to the linearized incompressible fluid flow problems and by Cockburn, Kanschat and Schötzau in 2005 [60] to the incompressible Navier–Stokes equations. Today there exist more than 160 papers dealing with the LDG method. Reviews on the LDG methods applied to elliptic problems and to incompressible fluid flow are contained in [42, 59], respectively.

1.2.2 DGM for Elliptic and Parabolic Problems

Simultaneously, but quite independently, the DGM was developed for the numerical solution of second-order elliptic and parabolic equations. Several variants of discontinuous approximation were proposed and studied by Douglas and Dupont in 1976 [106], by Baker in 1977 [18], by Wheeler in 1978 [283], and by Arnold in 1982 [7]. These techniques were frequently called the *interior penalty Galerkin* (IPG) method. Later they were called in the literature the *symmetric interior penalty Galerkin* (SIPG) method. The development of these techniques remained independent of the development of the DG methods for hyperbolic equations.

In the IPG methods the requirement of *continuity* for approximate conforming finite element solutions is replaced by the *interior penalty*. The Dirichlet boundary condition is embodied in the DG scheme with the aid of the *boundary penalty*. It was inspired by the concept of penalty-imposed Dirichlet boundary conditions already mentioned by Courant in [67]. It gained interest in the late 1960s, due namely to the paper by J.-L. Lions [216], in which boundary data with very low regularity are treated. In 1970, the Lions' approach was used by Aubin [10] in the framework of finite difference approximations of nonlinear problems. In the finite element method, this technique was used in 1973 by Babuška [11] for the solution of the Poisson problem with Dirichlet boundary condition and by Babuška and Zlámal

[17] for a fourth-order problem. A similar technique, ensuring the consistency of the formulation, was developed in 1971 by Nitsche [228] for the Poisson problem with general boundary conditions.

The interior penalty method was further applied to flow in porous media in 1978 [105] and in 1984 [107]. Since the early 1980s, the IPG methods became less attractive, probably due to the fact that they were never proven to be more advantageous or efficient than classical conforming finite element methods.

Since the 1990s, an interest in this class of techniques has been renewed. It was stimulated by the computational convenience of the DGM, namely its robustness, high-order accurate approximations, the flexibility in the choice of the mesh design, a simple treatment of hp-adaptation techniques and its easy parallelization.

A new type of the DGM, the *Baumann–Oden* (BO) method was introduced in the works 1997 [28], 1998 [230], 1999 [12, 29]. The BO method is very similar to the *global element method* (GEM) introduced and studied by Delves et al. in the papers [76, 77, 172, 173] in 1979–1980. Although the BO method and GEM differ only in the sign in certain terms, the BO method is more stable and satisfies a local conservation property.

In 1999, Rivière, Wheeler and Girault proposed in [239] the *nonsymmetric interior penalty Galerkin* (NIPG) method that is an extension of the BO method, involving the interior and boundary penalty. The NIPG technique was further analyzed by Rivière, Wheeler and Girault in [240].

In 2003, [243] Romkes, Prudhomme and Oden introduced the *stabilized discontinuous Galerkin* method which is related to the BO method, but involves an extra stabilization term with jumps of the normal fluxes across element interfaces.

A simplified version of the IPG method is the *incomplete interior penalty Galerkin* (IIPG) method proposed and analyzed by Sun in 2003 [263] and Dawson, Sun and Wheeler in 2004 [74]. This approach was further developed by Sun and Wheeler in 2005 [264, 265] for transport in porous media.

The interior penalty Galerkin techniques were developed and studied by many authors. For example, the paper of Houston, Schwab and Süli in 2000 [182] contains error estimates for an hp-variant of the DGM for first-order linear hyperbolic equation. In 2002 [183] error estimates for the hp-DGM for a linear advection-diffusion-reaction problem were derived. Furthermore, Wihler, Frauenfelder and Schwab in 2003 [285] proved the exponential convergence of the hp-DGM for diffusion problems. In the same year Schötzau and Wihler in [252] proved the exponential convergence of mixed hp DGM for the Stokes problem. Wihler and Schwab in 2000 [286] established the exponential convergence of the hp-DGM that was uniform with respect to the diffusion coefficient for linear one-dimensional convection-diffusion problems. The interior penalty Galerkin methods were also applied in the framework of optimal control problems, as e.g., in [2, 3].

Finally, let us mention the fundamental paper [8] by Arnold, Brezzi, Cockburn and Marini from 2002, introducing a unified analysis of nine different variants of the DGM, namely the BR method [23], BR2 method [27], Brezzi et al. method [39], Brezzi et al. method [40], SIPG method [106], NIPG method [240], Babuška, Zlámal method [17], LDG method [64] and BO method [29].

It is necessary to mention that the DGM has also been applied to solving ordinary differential equations, e.g., by Johnson [115], and to the time discretization of parabolic problems discretized in space by conforming finite elements by Jamet in 1978 [189], Eriksson, Johnson, Thomée in 1985 [117], Eriksson and Johnson in 1991 [116] and later by Makridakis and Babuška in 1997 [219] and Schötzau and C. Schwab in 2000 [251].

A series of papers published in the period 2002–2013 by Dolejší, Feistauer et al. [46, 94, 95, 96, 97, 99, 125, 129, 134, 139, 168, 257] have been devoted to analyzing the DGM methods for the numerical solution of nonstationary convection-diffusion linear or nonlinear problems.

1.2.3 DGM for the Numerical Solution of Compressible Flow

A history of the numerical solution of the compressible Navier–Stokes equations by discontinuous piecewise polynomial approximations starts at the end of the 1990s. In 1997, Bassi, Rebay and co-authors [23, 27] solved the Navier–Stokes equations with the aid of the BR and BR2 methods. In 1998, Lomtev, Quillen and Karniadakis [218] used DG-space discretization to deal with the convective part of the compressible Navier–Stokes equations and used a mixed method for approximating the diffusion part of the Navier–Stokes system. In 1999, Baumann and Oden [30] applied the BO method to the compressible Euler equations and mentioned the possibility of extending this method to the solution of compressible viscous flow.

From the beginning of the new millennium DGM starts to be intensively applied to the solution of compressible inviscid as well as viscous flows. The space-time DGM with dynamic grid motion was proposed by van der Vegt and van der Ven and co-authors since 2002 in [196, 197, 272, 273]. The SIPG technique was applied by Hartmann and Houston since 2002 in the papers [163, 164, 165, 166]. The BR2 method was further developed by the group of Bassi, Rebay in papers [20, 21, 22, 25, 26]. A high-order DG-based scheme with respect to space and time for flow problems was developed by Munz and co-authors in [110, 145, 146, 176] and further by Dumbser [109]. Dolejší, Feistauer et al. developed accurate, efficient methods that are robust with respect to the Mach number and Reynolds number, allowing for the numerical simulation of compressible flow from the high-speed transonic and hypersonic regimes up to the incompressible limit flow. (See, [86, 88, 92, 93, 98, 123, 124, 133, 135].) The works [47, 48, 130, 131, 135, 167] are concerned with the DGM simulation of compressible flow in time-dependent domains and applications to fluid-structure interaction (FSI). Progress in the development and applications of DGM for compressible flow simulations until the year 2010 can be found in [203].

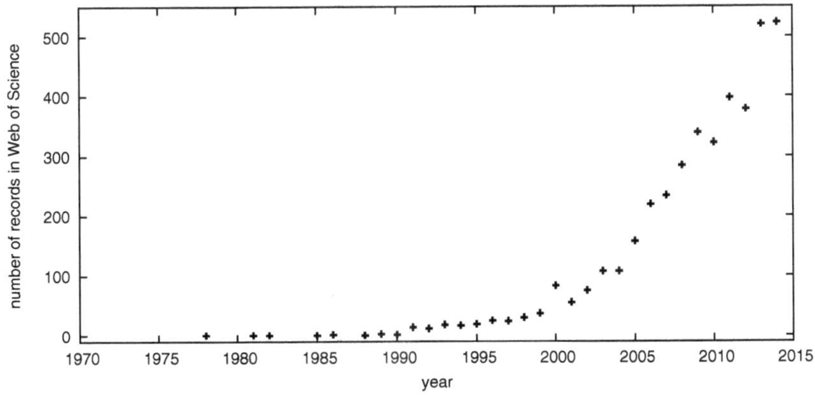

Fig. 1.2 Number of papers in the Web of Science database appearing under the key words *discontinuous Galerkin* for each year

1.2.4 Monographs Dealing with the DGM

To our knowledge there exist four monographs dealing with the DGM. The book by G. Kanschat (2007) [194] deals with the discontinuous Galerkin methods for incompressible Stokes and Navier–Stokes flows.

The book by B. Rivière (2008) [238] introduces the DGM for the linear elliptic and parabolic equations and gives applications of the DGM to the linear elasticity, the incompressible Stokes and Navier–Stokes equations and flow in porous media.

Moreover, the monograph by Hesthaven and Warburton (2008) [174] is concerned with the nodal discontinuous Galerkin method with applications to scalar problems, the Maxwell equations, the Euler equations, the incompressible Navier–Stokes equations, and the Poisson and Helmholtz equations.

Finally, the monograph by Di Petro and Ern (2012) [80] deals with first and second-order scalar equations, the incompressible Navier–Stokes equations, and Friedrichs' systems. Implementation issues are also addressed.

The DGM is a progressively developing technique for the numerical solution of partial differential equations. In the Web of Science database, there are 3,846 records (until the year 2014) appearing under the key words *discontinuous Galerkin*. Figure 1.2 shows the number of these papers for each year separately. The increase of the number of works is obvious.

1.3 Some Mathematical Concepts

In this section for the reader's convenience, we recall some basic tools of mathematical analysis, which are frequently used in the book. We assume that the reader is familiar with mathematical analysis, including the theory of the Lebesgue integral, and elements of functional analysis, see, for example, [247].

If X is a set or space and $n > 0$ is an integer, then the symbol $X^n = (X)^n$ denotes the Cartesian product $X \times \cdots \times X$ (n-times). This means that

$$X^n = (X)^n = \{(x_1, \dots, x_n);\ x_1, \dots, x_n \in X\}. \tag{1.1}$$

By $\mathbb{R}$ and $\mathbb{N}$ we denote the set of all real numbers and the set of all positive integers, respectively. In the Euclidean space $\mathbb{R}^d$ ($d \in \mathbb{N}$) we use a Cartesian coordinate system with axes denoted by $x_1, \dots, x_d$. Points from $\mathbb{R}^d$ will usually be denoted by $x = (x_1, \dots, x_d)$, $y = (y_1, \dots, y_d)$, etc. By $|\cdot|$ we denote the Euclidean norm in $\mathbb{R}^d$. Thus, $|x| = \left(\sum_{i=1}^d |x_i|^2\right)^{1/2}$.

Now we introduce some function spaces and their properties, which will be used in the sequel. For deeper results and proofs, we refer the reader to the monographs [1, 208, 287].

1.3.1 Spaces of Continuous Functions

Let us assume that $d \in \mathbb{N}$ and $M \subset \mathbb{R}^d$ is a domain (i.e., an open connected set). By ∂M and $\overline{M}$ we denote its boundary and closure, respectively. By $C(M)$ (or $C^0(M)$) we denote the linear space of all functions continuous in M. For $k \in \mathbb{N}$ and a domain $M \subset \mathbb{R}^d$, $C^k(M)$ denotes the linear space of all functions which have continuous partial derivatives up to the order k in M. The space $C^k(\overline{M})$ is formed by all functions from $C^k(M)$ whose all derivatives up to the order k can be continuously extended onto $\overline{M}$.

Let $M \subset \mathbb{R}^d$. A function $f : M \to \mathbb{R}$ is *μ-Hölder-continuous* with $\mu \in (0, 1]$, if there exists a constant L such that

$$|f(x) - f(y)| \le L|x - y|^\mu \qquad \forall x,\ y \in M. \tag{1.2}$$

If $\mu = 1$, we speak of a *Lipschitz-continuous* (or simply *Lipschitz*) function. If $M \subset \mathbb{R}^d$ is a domain, then $C^{k,\mu}(\overline{M})$ denotes the set of all functions whose derivatives of order k are μ-Hölder-continuous in $\overline{M}$.

Let us put

$$C^\infty(M) = \bigcap_{k=1}^{\infty} C^k(M) \quad \text{and} \quad C^\infty(\overline{M}) = \bigcap_{k=1}^{\infty} C^k(\overline{M}). \tag{1.3}$$

By $C_0^\infty(M)$ we denote the linear space of all functions $v \in C^\infty(\overline{M})$, whose *support*

$$\operatorname{supp} v = \overline{\{x \in M;\ v(x) \neq 0\}} \tag{1.4}$$

is a compact (i.e. bounded and closed) subset of the domain M.

If $\alpha_i \geq 0$, $i = 1, \ldots, d$, are integers, then we call $\alpha = (\alpha_1, \ldots, \alpha_d)$ a multi-index, and define its length as $|\alpha| = \sum_{i=1}^{d} \alpha_i$. By D^α we denote the multidimensional derivative of order $|\alpha|$:

$$D^\alpha = \frac{\partial^{|\alpha|}}{\partial x_1^{\alpha_1} \ldots \partial x_d^{\alpha_d}}. \tag{1.5}$$

The linear space $C^k(\overline{M})$, $k = 0, 1, \ldots$, equipped with the norm

$$\|u\|_{C^k(\overline{M})} = \sum_{|\alpha| \leq k} \sup_{x \in M} |D^\alpha u(x)| \tag{1.6}$$

is a Banach space. This space is separable but not reflexive.

The linear space $C^{k,\mu}(\overline{M})$, where $k = 0, 1, \ldots$, and $\mu \in (0, 1]$, equipped with the norm

$$\|u\|_{C^{k,\mu}(\overline{M})} = \|u\|_{C^k(\overline{M})} + \sum_{|\alpha| - k} \sup_{x, y \in M,\, x \neq y} \frac{|(D^\alpha u)(x) - (D^\alpha u)(y)|}{|x - y|^\mu} \tag{1.7}$$

is a Banach space. It is called the *Hölder space*. This space is neither separable nor reflexive.

Finally, the symbols ∇ and $\nabla \cdot$ mean the gradient and divergence operators, respectively, i.e.,

$$\nabla u = \text{grad}\, u = \left(\frac{\partial u}{\partial x_1}, \ldots, \frac{\partial u}{\partial x_d} \right)^{\mathrm{T}} \in \mathbb{R}^d \quad \text{for } u : M \to \mathbb{R} \tag{1.8}$$

and

$$\nabla \cdot \boldsymbol{u} = \text{div}\, \boldsymbol{u} = \sum_{i=1}^{d} \frac{\partial u_i}{\partial x_i} \in \mathbb{R}, \quad \text{for } \boldsymbol{u} = (u_1, \ldots, u_d) : M \to \mathbb{R}^d, \tag{1.9}$$

where the superscript ^{T} denotes the transposed vector.

The symbols D^α, ∇ and $\nabla \cdot$ are also used for the distributional derivatives; see Sect. 1.3.3.

1.3.2 Lebesgue Spaces

First we recall some standard notation and results from the Lebesgue theory of measure and integral, see, e.g., [247]. Let $M \subset \mathbb{R}^d$, $d = 1, 2, \ldots$, be a Lebesgue-measurable set. Its d-dimensional Lebesgue measure will be denoted by meas(M) or for short $|M|$. We recall that two measurable functions are *equivalent* if they differ at most on a set of zero Lebesgue measure. Then we say that these functions are equal almost everywhere (a.e.) in M.

For $s \in [1, \infty)$ the *Lebesgue space* $L^s(M)$ is the linear space of all functions measurable on M (more precisely, of classes of equivalent measurable functions) such that

$$\int_M |u|^s \,\mathrm{d}x < +\infty. \tag{1.10}$$

The space $L^s(M)$ is equipped with the norm

$$\|u\|_{L^s(M)} = \left(\int_M |u|^s \,\mathrm{d}x \right)^{1/s}. \tag{1.11}$$

In case that $s = \infty$, the space $L^\infty(M)$ consists of such measurable functions on M for which the norm

$$\|u\|_{L^\infty(M)} = \operatorname*{ess\,sup}_M |u| = \inf \left\{ \sup_{x \in M \setminus Z} |u(x)|;\ Z \subset M, \mathrm{meas}(Z) = 0 \right\} \tag{1.12}$$

is finite. The space $L^s(M)$ is a Banach space for $1 \le s \le \infty$. Moreover, it is separable if and only if $1 \le s < \infty$ and reflexive if and only if $1 < s < \infty$. The space $L^2(M)$ is a Hilbert space with the scalar product

$$(u, v)_{L^2(M)} = \int_M uv \,\mathrm{d}x. \tag{1.13}$$

The *Cauchy inequality* holds in $L^2(M)$:

$$|(u, v)_{L^2(M)}| \le \|u\|_{L^2(M)} \|v\|_{L^2(M)}, \quad u, v \in L^2(M). \tag{1.14}$$

1.3.3 Sobolev Spaces

Let $M \subset \mathbb{R}^d$, $d = 1, 2, \ldots$, be a domain, let $k \ge 0$ be an arbitrary integer and $1 \le s \le \infty$. We define the *Sobolev space* $W^{k,s}(M)$ as the space of all functions from the space $L^s(M)$ whose distributional derivatives $D^\alpha u$, up to the order k, also belong to $L^s(M)$, i.e.,

$$W^{k,s}(M) = \left\{ u \in L^s(M);\ D^\alpha u \in L^s(M)\ \forall \alpha,\ |\alpha| \le k \right\}, \tag{1.15}$$

(See e.g. [122, 208, 213].)

The Sobolev space is equipped with the norm

$$\|u\|_{W^{k,s}(M)} = \left(\sum_{|\alpha|\le k} \|D^\alpha u\|^s_{L^s(M)}\right)^{1/s} \quad \text{for } 1 \le s < \infty, \tag{1.16}$$
$$\|u\|_{W^{k,\infty}(M)} = \max_{|\alpha|\le k} \left\{\|D^\alpha u\|_{L^\infty(M)}\right\} \quad \text{for } s = \infty,$$

and the seminorm

$$|u|_{W^{k,s}(M)} = \left(\sum_{|\alpha|= k} \|D^\alpha u\|^s_{L^s(M)}\right)^{1/s} \quad \text{for } 1 \le s < \infty, \tag{1.17}$$
$$|u|_{W^{k,\infty}(M)} = \max_{|\alpha|= k} \left\{\|D^\alpha u\|_{L^\infty(M)}\right\} \quad \text{for } s = \infty.$$

For $1 \le s \le \infty$, the space $W^{k,s}(M)$ is a Banach space; it is separable if and only if $1 \le s < \infty$ and reflexive if and only if $1 < s < \infty$. For $s = 2$, the space $W^{k,2}(M)$ is a Hilbert space and we denote it by $H^k(M)$. Moreover, we put

$$\|u\|_{H^k(M)} = \|u\|_{W^{k,2}(M)} \quad \text{and} \quad |u|_{H^k(M)} = |u|_{W^{k,2}(M)}. \tag{1.18}$$

If $k = 0$, then we set $W^{0,s}(M) = L^s(M)$, $H^0(M) = L^2(M)$ and

$$|\cdot|_{W^{0,s}(M)} = \|\cdot\|_{W^{0,s}(M)} = \|\cdot\|_{L^s(M)}. \tag{1.19}$$

For vector-valued functions $\boldsymbol{v} = (v_1, \dots, v_n) \in (H^s(\Omega))^n$, we put

$$\|\boldsymbol{v}\|_{H^k(M)} = \left(\sum_{i=1}^n \|v_i\|^2_{H^k(M)}\right)^{1/2}. \tag{1.20}$$

Moreover, with respect to (1.8), (1.17), (1.18) and (1.20), we write

$$\|\nabla v\|_{L^2(M)} = |v|_{H^1(M)},\ v \in H^1(M), \qquad |\nabla v|_{H^1(M)} = |v|_{H^2(M)},\ v \in H^2(M). \tag{1.21}$$

1.3.4 Theorems on Traces and Embeddings

In the modern theory of partial differential equations the concept of a bounded domain $M \subset \mathbb{R}^d$ with Lipschitz boundary ∂M plays an important role. For the definition of a Lipschitz boundary, see, e.g., [122, 208, 287] or Sect. 4.3.2. It is possible to say that such a boundary ∂M is formed by a finite number of parts expressed as graphs of Lipschitz-continuous functions in local Cartesian coordinate systems. On this boundary, the $(d-1)$-dimensional Lebesgue measure meas_{d-1} and integral are defined and also an outer unit normal vector exists at a.e. point $x \in \partial M$. Moreover, Lebesgue spaces $L^s(\partial M)$ are defined over ∂M.

Theorem 1.1 (Theorem on traces) *Let $1 \le s \le \infty$ and let $M \subset \mathbb{R}^d$ be a domain with Lipschitz boundary. Then there exists a uniquely determined continuous linear mapping $\gamma_0^M : W^{1,s}(M) \to L^s(\partial M)$ such that*

$$\gamma_0^M(u) = u|_{\partial M} \quad \text{for all } u \in C^\infty(\overline{M}). \tag{1.22}$$

Moreover, if $1 \le s \le \infty$, then Green's formula

$$\int_M \left(u \frac{\partial v}{\partial x_i} + v \frac{\partial u}{\partial x_i} \right) \mathrm{d}x = \int_{\partial M} \gamma_0^M(u) \gamma_0^M(v) n_i \, \mathrm{d}S, \tag{1.23}$$
$$u \in W^{1,s}(M),\ v \in W^{1,s'}(M),\ i = 1, \ldots, d,$$

holds, where $s' = s/(s-1)$ and $\boldsymbol{n} = (n_1, \ldots, n_d)$ denotes the outer unit normal to ∂M.

The function $\gamma_0^M(u) \in L^s(\partial M)$ is called the *trace* of the function $u \in W^{1,s}(M)$ on the boundary ∂M. For simplicity, when there is no confusion, the notation $u|_{\partial M} = \gamma_0^M(u)$ is used not only for $u \in C^\infty(\overline{M})$ but also for $u \in W^{1,s}(M)$. The continuity of the mapping γ_0^M is equivalent to the existence of a constant $c > 0$ such that

$$\|u|_{\partial M}\|_{L^s(\partial M)} = \|\gamma_0^M(u)\|_{L^s(\partial M)} \le c\|u\|_{W^{1,s}(M)}, \quad u \in W^{1,s}(M). \tag{1.24}$$

Let $k \ge 1$ be an integer and $1 \le s < \infty$. We define the Sobolev space $W_0^{k,s}(M)$ as the closure of the space $C_0^\infty(M)$ in the topology of the space $W^{k,s}(M)$. If M is a domain with Lipschitz boundary, then $W_0^{1,s}(M) = \{v \in W^{1,s}(M);\, v|_{\partial M} = 0\}$.

The space of traces on $\partial\Omega$ of all functions $u \in H^1(\Omega)$ is denoted by $H^{1/2}(\partial\Omega)$. Hence, we can write

$$H^{1/2}(\partial\Omega) = \{\gamma_0^\Omega u;\, u \in H^1(\Omega)\}. \tag{1.25}$$

If $k \in \mathbb{N}$, we define the space

$$H^{k-1/2}(\partial\Omega) = \{\gamma_0^\Omega u;\, u \in H^k(\Omega)\}. \tag{1.26}$$

We speak of Sobolev–Slobodetskii spaces on $\partial\Omega$. (See e. g., [127, Sect. 1.3.3].)

Note that the symbols c and C will often denote a positive *generic constant*, attaining, in general, different values in different places.

1.3.4.1 Embedding Theorems

Definition 1.2 Let X, Y be Banach spaces. We say that X is continuously embedded into Y (we write $X \hookrightarrow Y$), if X is a subspace of Y and the identity operator $I : X \to Y$ defined by $Ix = x$ for all $x \in X$ is continuous, i.e., there exists $C > 0$ such that

$$\|Iv\|_Y \leq C\|v\|_X \quad \forall v \in X.$$

We say that X is compactly embedded into Y ($X \hookrightarrow\hookrightarrow Y$) if the embedding operator I is compact.

Theorem 1.3 *The following properties are valid:*
(i) Let $k \geq 0$, $1 \leq s \leq \infty$ and let $M \subset \mathbb{R}^d$ be a bounded domain with Lipschitz boundary. Then

$$W^{k,s}(M) \hookrightarrow L^q(M) \text{ where } \frac{1}{q} = \frac{1}{s} - \frac{k}{d}, \text{ if } k < \frac{d}{s}, \tag{1.27}$$

$$W^{k,s}(M) \hookrightarrow L^q(M) \text{ for all } q \in [1, \infty), \text{ if } k = \frac{d}{s},$$

$$W^{k,s}(M) \hookrightarrow C^{0,k-d/s}(\overline{M}), \text{ if } \frac{d}{s} < k < \frac{d}{s} + 1,$$

$$W^{k,s}(M) \hookrightarrow C^{0,\alpha}(\overline{M}) \text{ for all } \alpha \in (0, 1), \text{ if } k = \frac{d}{s} + 1,$$

$$W^{k,s}(M) \hookrightarrow C^{0,1}(\overline{M}), \text{ if } k > \frac{d}{s} + 1.$$

(ii) Let $k > 0$, $1 \leq s \leq \infty$. Then

$$W^{k,s}(M) \hookrightarrow\hookrightarrow L^q(M) \text{ for all } q \in [1, s^*) \text{ with } \frac{1}{s^*} = \frac{1}{s} - \frac{k}{d}, \quad \text{if } k < \frac{d}{s},$$

$$W^{k,s}(M) \hookrightarrow\hookrightarrow L^q(M) \text{ for all } q \in [1, \infty), \text{ if } k = \frac{d}{s},$$

$$W^{k,s}(M) \hookrightarrow\hookrightarrow C(\overline{M}), \text{ if } k > \frac{d}{s}.$$

(We set $1/\infty := 0$.)
(iii) Let $1 \leq s < \infty$. Then $C^\infty(\overline{M})$ is dense in $W^{k,s}(M)$ and $C_0^\infty(\overline{M})$ is dense in $W_0^{k,s}(M)$.
(iv) By [213, Exercise 1146, p. 342], if the domain M is bounded, then the space $W^{1,\infty}(M)$ can be identified with the space $C^{0,1}(\overline{M})$.

Remark 1.4 In some cases, it is suitable to use the concept of the domain with boundary having the *cone property*. This is more general than the concept of the Lipschitz boundary, but the above definitions and results remain valid. See [1].

1.3.5 Bochner Spaces

In the investigation of nonstationary problems we work with functions which depend on time and have values in a Banach space. Such functions are elements of the so-called Bochner spaces. If $u(x, t)$ is a function of the space variable x and time t, then it is sometimes suitable to separate these variables and consider u as a function $u(t) = u(\cdot, t)$, which, for each t under consideration, attains a value $u(t)$ that is a function of x and belongs to a suitable space of functions depending on x. This means that $u(t)$ represents the mapping "$x \to (u(t))\,(x) = u(x, t)$".

Let $a,\ b \in \mathbb{R},\ a < b$, and let X be a Banach space with norm $\|\cdot\|$. By a *function defined in the interval* $[a, b]$ *with its values in the space* X we understand any mapping $u : [a, b] \to X$.

We say that a function $u : [a, b] \to X$ is *continuous at a point* $t_0 \in [a, b]$, if

$$\lim_{\substack{t \to t_0 \\ t \in [a,b]}} \|u(t) - u(t_0)\| = 0. \tag{1.28}$$

By the symbol $C([a, b];\ X)$ we denote the space of all functions continuous in the interval $[a, b]$ (i.e., continuous at each $t \in [a, b]$) with values in X. The space $C([a, b];\ X)$ equipped with the norm

$$\|u\|_{C([a,b];\,X)} = \max_{t \in [a,b]} \|u(t)\| \tag{1.29}$$

is a Banach space.

For $s \in [1, \infty]$, we denote by $L^s(a, b; X)$ the space of (classes of equivalent) strongly measurable functions $u : (a, b) \to X$ such that

$$\|u\|_{L^s(a,b;X)} = \left[\int_a^b \|u(t)\|_X^s \,\mathrm{d}t\right]^{1/s} < \infty, \quad \text{if } 1 \le s < \infty, \tag{1.30}$$

and

$$\begin{aligned} \|u\|_{L^\infty(a,b;X)} &= \operatorname{ess} \sup_{t \in (a,b)} \|u(t)\|_X \qquad (1.31) \\ &= \inf\left\{ \sup_{t \in (a,b)\setminus N} \|u(t)\|_X;\ N \subset (a, b),\ \mathrm{meas}(N) = 0\right\} < +\infty, \quad \text{if } s = \infty. \end{aligned}$$

We speak of Bochner spaces. It can be proved that $L^s(a, b; X)$ is a Banach space. (The definition of a strongly measurable function $u : (a, b) \to X$ can be found in [208] or [122, Chap. 8].)

If the space X is reflexive, so is $L^s(a, b; X)$ for $s \in (1, \infty)$. Let $1 \le s < \infty$. Then the dual of $L^s(a, b; X)$ is $L^q(a, b; X^*)$, where $1/s + 1/q = 1$ and X^* is the dual of X (for $s = 1$ we set $q = \infty$). The duality between $L^q(a, b; X^*)$ and $L^s(a, b; X)$ becomes

$$\langle f, v\rangle = \int_a^b \langle f(t), v(t)\rangle_{X^*,X}\,\mathrm{d}t, \quad f \in L^q(a,b;X^*),\ v \in L^s(a,b;X). \tag{1.32}$$

The symbol $\langle f(t), v(t)\rangle_{X^*,X}$ denotes the value of the functional $f(t) \in X^*$ at $v(t) \in X$.

If X is a separable Banach space, then $L^s(a,b;X)$ is also separable, provided $s \in [1,\infty)$. (See, for example, [112, Sect. 8.18.1].)

Let $|\cdot|_X$ denote a seminorm in the space X. Then a seminorm in $L^s(a,b;X)$ is defined as

$$|f|_{L^s(a,b;X)} = \left(\int_a^b |f(t)|_X^s\,\mathrm{d}t\right)^{1/s} \quad \text{for } 1 \le s < +\infty, \tag{1.33}$$

and

$$|f|_{L^\infty(a,b;X)} = \operatorname{ess\,sup}_{t\in(a,b)} |f(t)|_X\,. \tag{1.34}$$

Similarly we define Sobolev spaces of functions with values in X:

$$W^{k,s}(a,b;X) = \left\{f \in L^s(a,b;X);\ \frac{\mathrm{d}^j f}{\mathrm{d}t^j} \in L^s(a,b;X),\ j = 1,\ldots,k\right\}, \tag{1.35}$$

where $k \in \mathbb{N}$, $s \in [1,\infty]$ and $\frac{\mathrm{d}^j f}{\mathrm{d}t^j}$ are distributional derivatives. The norm of $f \in W^{k,s}(a,b;X)$ is defined by

$$\|f\|_{W^{k,s}(a,b;X)} = \left(\sum_{j=0}^{k}\left\|\frac{\mathrm{d}^j f}{\mathrm{d}t^j}\right\|^s_{L^s(a,b;X)}\right)^{1/s} \tag{1.36}$$

for $s \in [1,\infty)$ and

$$\|f\|_{W^{k,\infty}(a,b;X)} = \max_{j=0,\ldots k}\left\|\frac{\mathrm{d}^j f}{\mathrm{d}t^j}\right\|_{L^\infty(a,b;X)}. \tag{1.37}$$

If $s = 2$, we often use the notation $H^k(a,b;X) = W^{k,2}(a,b;X)$.

Let $|\cdot|_X$ denote a seminorm in the space X. Then a seminorm in $W^{k,s}(a,b;X)$ is defined as

$$|f|_{W^{k,s}(a,b;X)} = \left(\int_a^b \left|\frac{\mathrm{d}^k f}{\mathrm{d}t^k}(t)\right|_X^s\,\mathrm{d}t\right)^{1/s} \quad \text{for } 1 \le s < +\infty, \tag{1.38}$$

and

$$|f|_{W^{k,\infty}(a,b;X)} = \text{ess sup}_{t\in(a,b)} \left|\frac{\mathrm{d}^k f}{\mathrm{d}t^k}(t)\right|_X. \tag{1.39}$$

For example,

$$|f|_{H^k(a,b;H^1(M))} = \left(\int_a^b \left|\frac{\mathrm{d}^k f}{\mathrm{d}t^k}(t)\right|^2_{H^1(M)} \mathrm{d}t\right)^{1/2}. \tag{1.40}$$

We also define spaces of continuously differentiable functions on an interval $I = [a, b]$ with values in X:

$$C^k(I; X) = \left\{f \in C(I; X);\ \frac{\mathrm{d}^j f}{\mathrm{d}t^j} \in C(I; X) \text{ for all } j = 1, \dots, k\right\}. \tag{1.41}$$

The norm of $f \in C^k(I; X)$, $k = 0, 1, \dots$, is defined by

$$\|f\|_{C^k(I;X)} = \max\left\{\left\|\frac{\mathrm{d}^j f}{\mathrm{d}t^j}\right\|_{C(I;X)};\ j = 0, \dots, k\right\}. \tag{1.42}$$

These spaces are nonreflexive Banach spaces. They are separable if X is separable.

If X is a Banach space with norm $\|\cdot\|_X$, then by X^* we denote its dual space (simply dual), i.e., the space of all continuous linear functionals on X. The space X^* is also a Banach space with norm

$$\|f\|_{X^*} = \sup_{v\in X} \frac{|f(v)|}{\|v\|_X} \quad \forall f \in X^*. \tag{1.43}$$

Finally, if $p \geq 0$ is an integer and $\omega \subset \mathbb{R}^n$, then by $P_p(\omega)$ we denote the space of the restrictions on ω of all polynomials of degree $\leq p$ depending on $x \in \mathbb{R}^n$. We simply speak of polynomials of degree $\leq p$ on ω.

For nonstationary problems, we use spaces of polynomial functions with respect to time. Let $-\infty < a < b < \infty$. If X is a Banach space, then we put

$$P_q(a, b; X) = \left\{v \in C(a, b; X);\ v(t) = \sum_{i=0}^{q} t^i \varphi_i, \varphi_i \in X, i = 0, \dots, q,\ t \in [a, b]\right\}. \tag{1.44}$$

1.3.6 Useful Theorems and Inequalities

Lemma 1.5 (Young inequality) *If* $s, q \in (1, +\infty)$, $1/s + 1/q = 1$ *and* $a, b \geq 0$, *then*

$$ab \leq \frac{a^s}{s} + \frac{b^q}{q}. \tag{1.45}$$

In particular, if $s = q = 2$ *and* $\lambda > 0$, *then*

$$ab \leq \frac{1}{2\lambda}a^2 + \frac{\lambda}{2}b^2. \tag{1.46}$$

Proof See, e.g., [120, Lemma 1.11.]

Lemma 1.6 (Lax–Milgram) *Let* V *be a Hilbert space with norm* $\|\cdot\|$, *let* $f : V \to \mathbb{R}$ *be a continuous linear functional on* V, *and let* $a : V \times V \to \mathbb{R}$ *be a bilinear form on* $V \times V$ *that is coercive, i.e., there exists a constant* $\alpha > 0$ *such that*

$$a(u, u) \geq \alpha\|u\|^2 \qquad \forall u \in V, \tag{1.47}$$

and continuous (also called bounded) and, hence, there exists a constant $C_B > 0$ *such that*

$$|a(u, v)| \leq C_B\|u\|\,\|v\| \qquad \forall u, v \in V. \tag{1.48}$$

Then there exists a unique solution $u_0 \in V$ *of the problem*

$$a(u_0, v) = f(v) \qquad \forall v \in V. \tag{1.49}$$

Proof See [52, Theorem 1.1.3].

Corollary 1.7 *Let* V_N *be a finite-dimensional Hilbert space with norm* $\|\cdot\|$, *let* $f : V_N \to \mathbb{R}$ *be a linear functional on* V_N, *and let* $a : V_N \times V_N \to \mathbb{R}$ *be a bilinear form on* $V_N \times V_N$ *which is coercive, i.e., there exists a constant* $\alpha > 0$ *such that*

$$a(u, u) \geq \alpha\|u\|^2 \qquad \forall u \in V_N. \tag{1.50}$$

Then there exists a unique solution $u_0 \in V_N$ *of the problem*

$$a(u_0, v) = f(v) \qquad \forall v \in V_N. \tag{1.51}$$

Proof Since the space V_N is finite dimensional, the bilinear form a and the functional f are continuous. Then the application of the Lax–Milgram Lemma 1.6 gives the assertion. Let us note that all norms on the finite-dimensional space are equivalent. □

Lemma 1.8 (Discrete Cauchy inequality) *Let* $\{a_i\}_{i=1}^n$ *and* $\{b_i\}_{i=1}^n$ *be two sequences of real numbers. Then*

$$\left|\sum_{i=1}^n a_i b_i\right| \le \left(\sum_{i=1}^n a_i^2\right)^{1/2} \left(\sum_{i=1}^n b_i^2\right)^{1/2}. \tag{1.52}$$

In the analysis of nonstationary problems, the following versions of the Gronwall lemma will be applied.

Lemma 1.9 (Gronwall lemma) *Let* $y,\ q,\ z,\ r \in C([0,T])$, $r \ge 0$, *and let*

$$y(t) + q(t) \le z(t) + \int_0^t r(s)\, y(s)\, \mathrm{d}s, \quad t \in [0,T]. \tag{1.53}$$

Then

$$\begin{aligned} &y(t) + q(t) + \int_0^t r(\vartheta)\, q(\vartheta) \exp\left(\int_\vartheta^t r(s)\, \mathrm{d}s\right) \mathrm{d}\vartheta \\ \le\, &z(t) + \int_0^t r(\vartheta)\, z(\vartheta) \exp\left(\int_\vartheta^t r(s)\, \mathrm{d}s\right) \mathrm{d}\vartheta, \quad t \in [0,T]. \end{aligned} \tag{1.54}$$

Proof Inequality (1.53) can be written in the form

$$y(t) \le h(t) + \int_0^t r(s)\, y(s)\, \mathrm{d}s, \tag{1.55}$$

where

$$h(t) = z(t) - q(t). \tag{1.56}$$

Let us set

$$z_1(t) = \int_0^t r(s)\, y(s)\, \mathrm{d}s. \tag{1.57}$$

Then $z_1'(t) = r(t)\, y(t)$, $z_1(0) = 0$. Since $r(t) \ge 0$, it follows from (1.55) that

$$z_1'(t) \le h(t)\, r(t) + r(t)\, z_1(t). \tag{1.58}$$

If we set

$$w(t) = z_1(t) \exp\left(-\int_0^t r(s)\, \mathrm{d}s\right), \tag{1.59}$$

then, by (1.58),

$$\begin{aligned} w'(t) &= z_1'(t)\exp\left(-\int_0^t r(s)\,\mathrm{d}s\right) - z_1(t)\,r(t)\exp\left(-\int_0^t r(s)\,\mathrm{d}s\right) \\ &\le (h(t)\,r(t) + r(t)\,z_1(t))\exp\left(-\int_0^t r(s)\,\mathrm{d}s\right) - r(t)\,z_1(t)\exp\left(-\int_0^t r(s)\,\mathrm{d}s\right) \\ &= h(t)\,r(t)\exp\left(-\int_0^t r(s)\,\mathrm{d}s\right). \end{aligned} \tag{1.60}$$

Taking into account that $w(0) = 0$ and integrating (1.60) from 0 to t, we get

$$w(t) \le \int_0^t h(\vartheta)\,r(\vartheta)\exp\left(-\int_0^\vartheta r(s)\,\mathrm{d}s\right)\mathrm{d}\vartheta.$$

This and (1.59) imply that

$$\begin{aligned} z_1(t) &\le \exp\left(\int_0^t r(s)\,\mathrm{d}s\right)\int_0^t h(\vartheta)\,r(\vartheta)\exp\left(-\int_0^\vartheta r(s)\,\mathrm{d}s\right)\mathrm{d}\vartheta \\ &= \int_0^t h(\vartheta)\,r(\vartheta)\exp\left(\int_\vartheta^t r(s)\,\mathrm{d}s\right)\mathrm{d}\vartheta. \end{aligned} \tag{1.61}$$

Hence, by (1.53), (1.55), (1.61) and (1.56), we have

$$\begin{aligned} y(t) + q(t) \le z(t) + z_1(t) \;&\le\; z(t) + \int_0^t h(\vartheta)\,r(\vartheta)\exp\left(\int_\vartheta^t r(s)\,\mathrm{d}s\right)\mathrm{d}\vartheta \\ &= z(t) + \int_0^t z(\vartheta)\,r(\vartheta)\exp\left(\int_\vartheta^t r(s)\,\mathrm{d}s\right)\mathrm{d}\vartheta \\ &\quad - \int_0^t q(\vartheta)\,r(\vartheta)\exp\left(\int_\vartheta^t r(s)\,\mathrm{d}s\right), \end{aligned}$$

which immediately yields inequality (1.54). □

Lemma 1.10 (Gronwall modified lemma) *Suppose that for all $t \in [0, T]$ we have*

$$\chi^2(t) + R(t) \le A(t) + 2\int_0^t B(\vartheta)\chi(\vartheta)\,\mathrm{d}\vartheta, \tag{1.62}$$

where $R, A, B, \chi \in C([0, T])$ are nonnegative functions. Then for any $t \in [0, T]$

$$\sqrt{\chi^2(t) + R(t)} \le \max_{\vartheta\in[0,t]} \sqrt{A(\vartheta)} + \int_0^t B(\vartheta)\,\mathrm{d}\vartheta. \tag{1.63}$$

Proof For any $\vartheta \in [0, T]$ we set

$$\varphi(\vartheta) = 2\int_0^{\vartheta} B(s)\,\chi(s)\,\mathrm{d}s.$$

Then $\varphi(0) = 0$ and

$$\varphi'(\vartheta) = 2B(\vartheta)\,\chi(\vartheta). \tag{1.64}$$

Let us consider an arbitrary fixed $t \in [0, T]$ and denote

$$S_t = \max_{s\in[0,t]} A(s).$$

It is clear that if $S_t = 0$ for some $t \in [0, T]$, then $S_\tau = 0$ for all $\tau \in [0, t]$. Similarly, the condition $\varphi(\vartheta) = 0$ for some $\vartheta \in [0, T]$ implies that $\varphi(\tau) = 0$ for all $\tau \in [0, \vartheta]$. Let us set $t_1 = 0$, provided $S_t \neq 0$ for all $t \in [0, T]$, and

$$t_1 = \max\{t \in [0, T];\ S_t = 0\},\ \ t_2 = \max\{\vartheta \in [0, T];\ \varphi(\vartheta) = 0\},\ \ t_3 = \min(t_1, t_2).$$

By (1.64) and (1.62),

$$\varphi'(\vartheta) \le 2B(\vartheta)\sqrt{S_t + \varphi(\vartheta)}.$$

Then for $t \in (t_3, T]$ we have

$$\int_{t_3}^{t} \frac{\varphi'(\vartheta)\,\mathrm{d}\vartheta}{2\sqrt{S_t + \varphi(\vartheta)}} \le \int_0^t B(\vartheta)\,\mathrm{d}\vartheta$$

and thus,

$$\sqrt{S_t + \varphi(\vartheta)}\Big|_{\vartheta=t_3}^{t} = \sqrt{S_t + \varphi(t)} - \sqrt{S_t} \le \int_0^t B(\vartheta)\,\mathrm{d}\vartheta.$$

This implies that

$$\sqrt{S_t + \varphi(t)} \le \sqrt{S_t} + \int_0^t B(\vartheta)\,\mathrm{d}\vartheta. \tag{1.65}$$

Now, by virtue of (1.62) and (1.65),

$$\sqrt{\chi^2(t) + R(t)} \le \sqrt{S_t + \varphi(t)} \le \sqrt{S_t} + \int_0^t B(\vartheta)\,\mathrm{d}\vartheta. \tag{1.66}$$

Taking into account that

$$\sqrt{S_t} = \sqrt{\max_{s\in[0,t]} A(s)} = \max_{s\in[0,t]} \sqrt{A(s)},$$

from (1.66) we immediately get (1.63). Finally, it is obvious that (1.63) also holds for all $t \in [0, t_3]$. □

Lemma 1.11 (Gronwall discrete lemma) *Let $x_m,\ b_m,\ c_m \geq 0$ and $a_m > 0$ for $m = 0, 1, 2, \ldots$, and let the sequence a_m be nondecreasing. Then, if*

$$\begin{aligned} &x_0 + c_0 \leq a_0, \\ &x_m + c_m \leq a_m + \sum_{j=0}^{m-1} b_j x_j \quad \text{for } m \geq 1, \end{aligned} \tag{1.67}$$

we have

$$x_m + c_m \leq a_m \prod_{j=0}^{m-1} (1 + b_j) \quad \text{for } m \geq 0. \tag{1.68}$$

Proof We start from inequality (1.67), divided by a_m, and use the assumption that the sequence a_m is nondecreasing. We get

$$\frac{x_m}{a_m} + \frac{c_m}{a_m} \leq 1 + \sum_{j=0}^{m-1} b_j \frac{x_j}{a_m} \leq 1 + \sum_{j=0}^{m-1} b_j \frac{x_j}{a_j}. \tag{1.69}$$

Let us set $v_0 = 1$ and $v_m = 1 + \sum_{j=0}^{m-1} b_j \frac{x_j}{a_j}$ for $m \geq 1$. Then by (1.67) and the inequality $c_{m-1}/a_{m-1} \geq 0$, we have

$$v_m - v_{m-1} = b_{m-1} \frac{x_{m-1}}{a_{m-1}} \leq b_{m-1} \left(\frac{x_{m-1}}{a_{m-1}} + \frac{c_{m-1}}{a_{m-1}} \right) \leq b_{m-1} v_{m-1}, \quad m \geq 1.$$

This implies that

$$v_m \leq (1 + b_{m-1}) v_{m-1} \leq v_0 \prod_{j=0}^{m-1} (1 + b_j) = \prod_{j=0}^{m-1} (1 + b_j).$$

Now from (1.69) we get (1.68). □

Part I
Analysis of the Discontinuous Galerkin Method

Chapter 2
DGM for Elliptic Problems

This chapter concerns in basic aspects of the discontinuous Galerkin method (DGM), which will be treated in an example of a simple problem for the Poisson equation with mixed Dirichlet–Neumann boundary conditions. We introduce the discretization of this problem with the aid of several variants of the DGM. Further, we prove the existence of the approximate solution and derive error estimates. Finally, several numerical examples are presented.

The book contains a detailed analysis of qualitative properties of DG techniques. It is based on a number of estimates with various constants. We denote by $C_A, C_B, C_C, \ldots, C_a, C_b, C_c, \ldots$ positive constants arising in the formulation of results that can be simply named (e.g., $_A$ corresponds to approximation properties, $_B$—boundedness, $_C$—coercivity, etc.) Otherwise, we use symbols C, C_1, $C_2, \ldots$. These constants are always independent of the parameters of the discretization (i.e., the space mesh-size h, time step τ in the case of nonstationary problems, and also the degree p of polynomial approximation in the case of the hp-methods), but they may depend on the data in problems. They are often "autonomous" in individual chapters or sections. Some constants are sometimes defined in a complicated way on the basis of a number of constants appearing in previous considerations. For an example, see Remark 4.13.

2.1 Model Problem

Let Ω be a bounded domain in $\mathbb{R}^d$, $d = 2, 3$, with Lipschitz boundary $\partial\Omega$. We denote by $\partial\Omega_D$ and $\partial\Omega_N$ parts of the boundary $\partial\Omega$ such that $\partial\Omega = \partial\Omega_D \cup \partial\Omega_N$, $\partial\Omega_D \cap \partial\Omega_N = \emptyset$ and $\partial\Omega_D \neq \emptyset$.

We consider the following model problem for the Poisson equation: Find a function $u : \Omega \to \mathbb{R}$ such that

V. Dolejší and M. Feistauer, *Discontinuous Galerkin Method*,
Springer Series in Computational Mathematics 48,
DOI 10.1007/978-3-319-19267-3_2

$$-\Delta u = f \quad \text{in } \Omega, \tag{2.1a}$$

$$u = u_D \quad \text{on } \partial\Omega_D, \tag{2.1b}$$

$$\boldsymbol{n} \cdot \nabla u = g_N \quad \text{on } \partial\Omega_N, \tag{2.1c}$$

where f, u_D and g_N are given functions. Let us note that $\boldsymbol{n} \cdot \nabla u = \frac{\partial u}{\partial \boldsymbol{n}}$ is the derivative of the function u in the direction $\boldsymbol{n}$, which is the outer unit normal to $\partial\Omega$. A function $u \in C^2(\overline{\Omega})$ satisfying (2.1) pointwise is called a *classical solution*. It is suitable to introduce a weak formulation of the above problem. Let us define the space

$$V = \{v \in H^1(\Omega);\ v|_{\partial\Omega_D} = 0\}.$$

Assuming that u is a classical solution, we multiply (2.1a) by any function $v \in V$, integrate over Ω and use Green's theorem. Taking into account the boundary condition (2.1c), we obtain the identity

$$\int_\Omega \nabla u \cdot \nabla v \,\mathrm{d}x = \int_\Omega f\, v \,\mathrm{d}x + \int_{\partial\Omega_N} g_N\, v \,\mathrm{d}S \quad \forall v \in V. \tag{2.2}$$

We can introduce the following definition.

Definition 2.1 Let us assume the existence of $u^* \in H^1(\Omega)$ such that $u^*|_{\partial\Omega_D} = u_D$ and let $f \in L^2(\Omega)$, $g_N \in L^2(\partial\Omega_N)$. Now we say that a function u is a *weak solution* of problem (2.1), if
(a) $u - u^* \in V$,
(b) u satisfies identity (2.2).

Using the Lax–Milgram Lemma 1.6, we can prove that there exists a unique weak solution of (2.1), see, e.g., [233, Sect. 6.1.2]. In the following, we deal with numerical solution of problem (2.1) with the aid of discontinuous piecewise polynomial approximations.

2.2 Abstract Numerical Method and Its Theoretical Analysis

In order to better understand theoretical foundations of the DGM, we describe a possible general approach to deriving error estimates. (Readers familiar with concepts of a priori error estimates in the finite element method can skip this section.)

Let $u \in V$ be a weak solution of a given problem. Let V_h denote a *finite-dimensional space*, where an *approximate solution* u_h is sought. The subscript $h > 0$ (usually chosen as $h \in (0, \overline{h})$ with $\overline{h} > 0$) denotes the parameter of the discretization. Further, we introduce an infinitely dimensional function space W_h such that $V \subset W_h$ and $V_h \subset W_h$. (If $V_h \subset V$, then we usually put $W_h := V$ and thus, W_h is independent of h.) Finally, let $\|\cdot\|_{W_h}$ be a suitable norm in W_h. As we see later, the spaces V_h and W_h will be constructed over a suitable mesh in the computational domain, and hence the norm $\|\cdot\|_{W_h}$ may be mesh-dependent.

An *abstract numerical method* reads: Find $u_h \in V_h$ such that

$$A_h(u_h, v_h) = F(v_h) \quad \forall v_h \in V_h, \tag{2.3}$$

where $A_h : W_h \times W_h \to \mathbb{R}$ is a bilinear form and $F : W_h \to \mathbb{R}$ is a linear functional.

In the numerical analysis, we want to reach the following goals:

- the approximate solution u_h of (2.3) *exists* and is *unique*,
- the approximate solution u_h *converges* to the exact solution u in the $\|\cdot\|_{W_h}$-norm as $h \to 0$, i.e.,

$$\lim_{h\to 0} \|u - u_h\|_{W_h} = 0, \tag{2.4}$$

- *a priori error estimate*, i.e., we seek $\alpha > 0$ independent of h such that

$$\|u - u_h\|_{W_h} \le Ch^\alpha, \quad h \in (0, \overline{h}), \tag{2.5}$$

  where $C > 0$ is a constant, independent of h (but may depend on u), and α is the *order of convergence*.

Obviously, an a priori error estimate implies the convergence.

The existence and uniqueness of the approximate solution is a consequence of the *coercivity* of A_h, i.e., there exists $C_c > 0$ such that

$$A_h(v_h, v_h) \ge C_c\|v_h\|^2_{W_h} \quad \forall v_h \in V_h. \tag{2.6}$$

Then Corollary 1.7 implies the existence and uniqueness of the approximate solution u_h.

In order to derive a priori error estimates, we prove the *consistency* of the method,

$$A_h(u, v_h) = F(v_h) \quad \forall v_h \in V_h \tag{2.7}$$

which, together with (2.3), immediately gives the *Galerkin orthogonality* of the error $e_h = u_h - u$ to the space V_h:

$$A_h(e_h, v_h) = 0 \quad \forall v_h \in V_h. \tag{2.8}$$

Further, we introduce an *interpolation operator* (usually defined as a suitable *projection*) $\Pi_h : V \to V_h$ and prove its *approximation property*, namely existence of a constant $\alpha > 0$ such that

$$\|v - \Pi_h v\|_{W_h} \le \tilde{C}(v)h^\alpha \quad \forall v \in V, \quad h \in (0, \overline{h}), \tag{2.9}$$

where $\tilde{C}(v) > 0$ is a constant independent of h but dependent on v. A further step is the derivation of the inequality

$$A_h(u - \Pi_h u, v_h) \le R(u - \Pi_h u)\|v_h\|_{W_h} \quad \forall v_h \in V_h, \tag{2.10}$$

where R depends on suitable norms of the interpolation error $u - \Pi_h u$.

Finally, the *error estimate* is derived in the following way: for each $h \in (0, \bar{h})$ we decompose the error e_h by

$$e_h = u_h - u = \xi + \eta, \tag{2.11}$$

where $\xi := u_h - \Pi_h u \in V_h$ and $\eta := \Pi_h u - u \in W_h$. Putting $v_h := \xi$ in (2.8), we get

$$A_h(e_h, \xi) = A_h(\xi, \xi) + A_h(\eta, \xi) = 0. \tag{2.12}$$

It follows from the coercivity (2.6) and estimate (2.10) that

$$C_c\|\xi\|^2_{W_h} \le A_h(\xi, \xi) = -A_h(\eta, \xi) \le R(\eta)\|\xi\|_{W_h}, \tag{2.13}$$

which immediately implies the inequality

$$\|\xi\|_{W_h} \le \frac{R(\eta)}{C_c}. \tag{2.14}$$

Now, the triangle inequality, relations (2.11) and (2.14) give the error estimate in the form

$$\|e_h\|_{W_h} \le \|\xi\|_{W_h} + \|\eta\|_{W_h} \le \frac{R(\eta)}{C_c} + \|\eta\|_{W_h}. \tag{2.15}$$

This is often called the *abstract error estimate*, which represents an error bound in terms of the interpolation error η.

The last aim is to use the approximation property (2.9) of the operator Π_h and to estimate the expression $R(\eta)$ in terms of the mesh-size h in the form

$$R(\eta) \le \tilde{C}_1(u)h^\alpha, \tag{2.16}$$

which together with (2.15) immediately imply the *error estimate*

$$\|e_h\|_{W_h} \le \left(C_c^{-1}\tilde{C}_1(u) + \tilde{C}(u)\right) h^\alpha, \tag{2.17}$$

valid for all $h \in (0, \bar{h})$. We say that the numerical scheme has the *order of convergence* in the norm $\|\cdot\|_{W_h}$ equal to α.

This concept of numerical analysis is applied in this chapter. (Among other, we specify there the spaces W_h and V_h.) For time dependent problems, treated in Chaps. 4–6, the analysis is more complicated and the previous technique has to be modified. However, in some parts of the book, error estimates are derived in a different way.

Remark 2.2 As was mentioned above, we are interested here in deriving of *a priori error estimates* (simply called *error estimates*). We do not deal with *a posteriori error estimates*, when the error is bounded in a suitable norm in terms of the approximate solution and data of the problem. The subject of a posteriori error estimates plays an important role in practical computations, but is out of the scope of this book. For some results in this direction for the DGM we can refer, e.g., to the papers [5, 91, 118, 166, 185, 190] and the references cited therein.

2.3 Spaces of Discontinuous Functions

The subject of this section is the construction of DG space partitions of the bounded computational domain Ω and the specification of their properties which are used in the theoretical analysis. Further, function spaces over these meshes are defined.

2.3.1 Partition of the Domain

Let $\mathscr{T}_h$ ($h > 0$ is a parameter) be a partition of the closure $\overline{\Omega}$ of the domain Ω into a finite number of closed d-dimensional simplexes K with mutually disjoint interiors such that

$$\overline{\Omega} = \bigcup_{K \in \mathscr{T}_h} K. \tag{2.18}$$

This assumption means that the domain Ω is polygonal (if $d = 2$) or polyhedral (if $d = 3$). The case of a 2D nonpolygonal domain is considered, e.g., in [256], where curved elements are used. See also Chap. 8, where curved elements are treated from the implementation point of view. We call $\mathscr{T}_h$ a *triangulation* of Ω and do not require the standard conforming properties from the finite element method, introduced e.g., in [37, 52, 115, 254] or [287]. In two-dimensional problems ($d = 2$) we choose $K \in \mathscr{T}_h$ as triangles and in three-dimensional problems ($d = 3$) the elements $K \in \mathscr{T}_h$ are tetrahedra. As we see, we admit that in the finite element mesh the so-called *hanging nodes* (and in 3D also *hanging edges*) appear; see Fig. 2.1.

In general, the discontinuous Galerkin method can handle with more general elements as quadrilaterals and convex or even nonconvex star-shaped polygons in 2D and hexahedra, pyramids and convex or nonconvex star-shaped polyhedra in 3D.

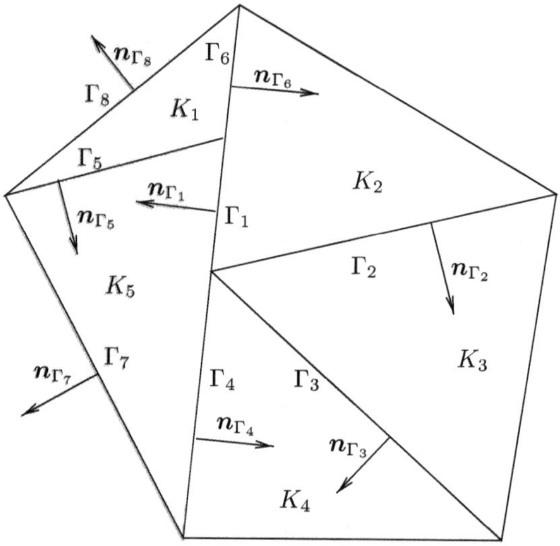

Fig. 2.1 Example of elements $K_l,\ l = 1, \ldots, 5$, and faces $\Gamma_l,\ l = 1, \ldots, 8$, with the corresponding normals $\boldsymbol{n}_{\Gamma_l}$. The triangle K_5 has a hanging node. Its boundary is formed by four edges: $\partial K_5 = \Gamma_1 \cup \Gamma_4 \cup \Gamma_7 \cup \Gamma_5$

As an example, we can consider the so-called dual finite volumes constructed over triangular ($d = 2$) or tetrahedral ($d = 3$) meshes (cf., e.g., [126]). A use of such elements will be discussed in Sect. 7.2.

In our further considerations we use the following notation. By ∂K we denote the boundary of an element $K \in \mathscr{T}_h$ and set $h_K = \text{diam}(K) =$ diameter of K, $h = \max_{K \in \mathscr{T}_h} h_K$. By ρ_K we denote the radius of the largest d-dimensional ball inscribed into K and by $|K|$ we denote the d-dimensional Lebesgue measure of K.

Let $K, K' \in \mathscr{T}_h$. We say that K and K' are *neighbouring elements* (or simply *neighbours*) if the set $\partial K \cap \partial K'$ has positive $(d-1)$-dimensional measure. We say that $\Gamma \subset K$ is a *face* of K, if it is a maximal connected open subset of either $\partial K \cap \partial K'$, where K' is a neighbour of K, or $\partial K \cap \partial \Omega_D$ or $\partial K \cap \partial \Omega_N$. The symbol $|\Gamma|$ will denote the $(d-1)$-dimensional Lebesgue measure of Γ. Hence, if $d = 2$, then $|\Gamma|$ is the length of Γ and for $d = 3$, $|\Gamma|$ denotes the area of Γ. By $\mathscr{F}_h$ we denote the system of all faces of all elements $K \in \mathscr{T}_h$. Further, we define the set of all boundary faces by

$$\mathscr{F}_h^B = \{\Gamma \in \mathscr{F}_h;\ \Gamma \subset \partial\Omega\},$$

the set of all "Dirichlet" boundary faces by

$$\mathscr{F}_h^D = \{\Gamma \in \mathscr{F}_h;\ \Gamma \subset \partial\Omega_D\},$$

the set of all "Neumann" boundary faces by

$$\mathscr{F}_h^N = \{\Gamma \in \mathscr{F}_h,\ \Gamma \subset \partial\Omega_N\}$$

and the set of all inner faces

$$\mathscr{F}_h^I = \mathscr{F}_h \setminus \mathscr{F}_h^B.$$

Obviously, $\mathscr{F}_h = \mathscr{F}_h^I \cup \mathscr{F}_h^D \cup \mathscr{F}_h^N$ and $\mathscr{F}_h^B = \mathscr{F}_h^D \cup \mathscr{F}_h^N$. For a shorter notation we put

$$\mathscr{F}_h^{ID} = \mathscr{F}_h^I \cup \mathscr{F}_h^D.$$

For each $\Gamma \in \mathscr{F}_h$ we define a unit normal vector $\boldsymbol{n}_\Gamma$. We assume that for $\Gamma \in \mathscr{F}_h^B$ the normal $\boldsymbol{n}_\Gamma$ has the same orientation as the outer normal to $\partial\Omega$. For each face $\Gamma \in \mathscr{F}_h^I$ the orientation of $\boldsymbol{n}_\Gamma$ is arbitrary but fixed. See Fig. 2.1.

For each $\Gamma \in \mathscr{F}_h^I$ there exist two neighbouring elements $K_\Gamma^{(L)}, K_\Gamma^{(R)} \in \mathscr{T}_h$ such that $\Gamma \subset \partial K_\Gamma^{(L)} \cap \partial K_\Gamma^{(R)}$. (This means that the elements $K_\Gamma^{(L)}, K_\Gamma^{(R)}$ are adjacent to Γ and they share this face.) We use the convention that $\boldsymbol{n}_\Gamma$ is the outer normal to $\partial K_\Gamma^{(L)}$ and the inner normal to $\partial K_\Gamma^{(R)}$; see Fig. 2.2.

Moreover, if $\Gamma \in \mathscr{F}_h^B$, then there exists an element $K_\Gamma^{(L)} \in \mathscr{T}_h$ such that $\Gamma \subset K_\Gamma^{(L)} \cap \partial\Omega$.

2.3.2 Assumptions on Meshes

Let us consider a system $\{\mathscr{T}_h\}_{h\in(0,\bar{h})}$, $\bar{h} > 0$, of triangulations of the domain Ω ($\mathscr{T}_h = \{K\}_{K\in\mathscr{T}_h}$). In our further considerations we meet various assumptions on triangulations. The first is usual in the theory of the finite element method:

- The system $\{\mathscr{T}_h\}_{h\in(0,\bar{h})}$ of triangulations is *shape-regular*: there exists a positive constant C_R such that

$$\frac{h_K}{\rho_K} \le C_R \quad \forall K \in \mathscr{T}_h \, \forall h \in (0, \bar{h}). \tag{2.19}$$

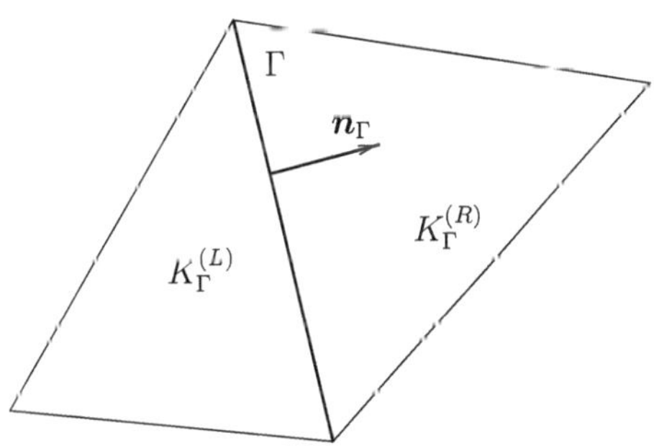

Fig. 2.2 Interior face Γ, elements $K_\Gamma^{(L)}$ and $K_\Gamma^{(R)}$ and the orientation of $\boldsymbol{n}_\Gamma$

Moreover, for each face $\Gamma \in \mathscr{F}_h$, $h \in (0, \bar{h})$, we need to introduce a quantity $h_\Gamma > 0$, which represents a "one-dimensional" size of the face Γ. We require that

- the quantity h_Γ satisfies the *equivalence condition* with h_K, i.e., there exist constants $C_T, C_G > 0$ independent of h, K and Γ such that

$$C_T h_K \le h_\Gamma \le C_G h_K, \quad \forall K \in \mathscr{T}_h, \ \forall \Gamma \in \mathscr{F}_h, \ \Gamma \subset \partial K, \ \forall h \in (0, \bar{h}). \tag{2.20}$$

The equivalence condition can be fulfilled by additional assumptions on the system of triangulations $\{\mathscr{T}_h\}_{h\in(0,\bar{h})}$ and by a suitable choice of the quantity h_Γ, $\Gamma \in \mathscr{F}_h$, $h \in (0, \bar{h})$. We introduce some assumptions on triangulations and several choices of the quantity h_Γ. Then we discuss how the equivalence condition (2.20) is satisfied.

In literature we can find the following assumptions on the system of triangulations:

(MA1) The system $\{\mathscr{T}_h\}_{h\in(0,\bar{h})}$ is *locally quasi-uniform*: there exists a constant $C_Q > 0$ such that

$$h_K \le C_Q h_{K'} \quad \forall K, K' \in \mathscr{T}_h, \ K, K' \text{ are neighbours}, \ \forall h \in (0, \bar{h}). \tag{2.21}$$

(MA2) The faces $\Gamma \subset \partial K$ do not degenerate with respect to the diameter of K if $h \to 0$: there exists a constant $C_d > 0$ such that

$$h_K \le C_d \mathrm{diam}(\Gamma) \quad \forall K \in \mathscr{T}_h \ \forall \Gamma \in \mathscr{F}_h, \ \Gamma \subset \partial K, \ \forall h \in (0, \bar{h}). \tag{2.22}$$

(MA3) The system $\{\mathscr{T}_h\}_{h\in(0,\bar{h})}$ is *quasi-uniform*: there exists a constant $C_U > 0$ such that

$$h \le C_U h_K \quad \forall K \in \mathscr{T}_h \ \forall h \in (0, \bar{h}). \tag{2.23}$$

(MA4) The triangulations $\mathscr{T}_h$, $h \in (0, \bar{h})$, are *conforming*. This means that for two elements $K, K' \in \mathscr{T}_h$, $K \ne K'$, either $K \cap K' = \emptyset$ or $K \cap K'$ is a common vertex or $K \cap K'$ is a common face (or for $d = 3$, when $K \cap K'$ is a common edge) of K and K'.

If condition (MA4) is not satisfied, then the triangulations $\mathscr{T}_h$ are called *nonconforming*.

Remark 2.3 There are some relations among the mesh assumptions (MA1)–(MA4) mentioned above. Obviously, (MA3) $\Rightarrow$ (MA1). Moreover, if the system of triangulation is shape-regular (i.e., (2.19) is fulfilled) then (MA4) $\Rightarrow$ (MA1) & (MA2).

Exercises 2.4 Prove the implications in Remark 2.3.

Concerning the choice of the quantity h_Γ, $\Gamma \in \mathscr{F}_h$, $h \in (0, \bar{h})$, in literature we can find the following basic possibilities:

$$\text{(i)} \qquad h_\Gamma = \operatorname{diam}(\Gamma), \quad \Gamma \in \mathscr{F}_h, \tag{2.24}$$

$$\text{(ii)} \qquad h_\Gamma = \begin{cases} \frac{1}{2}\left(h_{K_\Gamma^{(L)}} + h_{K_\Gamma^{(R)}}\right) & \text{for } \Gamma \in \mathscr{F}_h^I \\ h_{K_\Gamma^{(L)}} & \text{for } \Gamma \in \mathscr{F}_h^B, \end{cases} \tag{2.25}$$

$$\text{(iii)} \qquad h_\Gamma = \begin{cases} \max\left(h_{K_\Gamma^{(L)}}, \, h_{K_\Gamma^{(R)}}\right) & \text{for } \Gamma \in \mathscr{F}_h^I \\ h_{K_\Gamma^{(L)}} & \text{for } \Gamma \in \mathscr{F}_h^B, \end{cases} \tag{2.26}$$

$$\text{(iv)} \qquad h_\Gamma = \begin{cases} \min\left(h_{K_\Gamma^{(L)}}, \, h_{K_\Gamma^{(R)}}\right) & \text{for } \Gamma \in \mathscr{F}_h^I \\ h_{K_\Gamma^{(L)}} & \text{for } \Gamma \in \mathscr{F}_h^B, \end{cases} \tag{2.27}$$

where $K_\Gamma^{(L)}, K_\Gamma^{(R)} \in \mathscr{T}_h$ are the elements adjacent to $\Gamma \in \mathscr{F}_h^I$, see Fig. 2.2, and $K_\Gamma^{(L)} \in \mathscr{T}_h$ is the element adjacent to $\Gamma \in \mathscr{F}_h^B$.

The following lemma characterizes assumptions on computational grids and the choice of h_Γ, which guarantee the equivalence condition (2.20).

Lemma 2.5 *Let $\{\mathscr{T}_h\}_{h\in(0,\bar{h})}$ be a system of triangulations of the domain Ω satisfying the shape-regularity assumption (2.19). Then the equivalence condition (2.20) is satisfied in the following cases:*

(i) The triangulations $\mathscr{T}_h$, $h \in (0, \bar{h})$, are conforming (i.e., assumption (MA4) is satisfied) and h_Γ are defined by (2.24) or (2.25) or (2.26) or (2.27).
(ii) The triangulations $\mathscr{T}_h$, $h \in (0, \bar{h})$, are, in general, nonconforming; assumption (MA2) (i.e., (2.22)) is satisfied and h_Γ are defined by (2.24).
(iii) The triangulations $\mathscr{T}_h$, $h \in (0, \bar{h})$, are, in general, nonconforming; assumption (MA1) is satisfied (i.e., the system $\{\mathscr{T}_h\}_{h\in(0,\bar{h})}$ is locally quasi-uniform) and h_Γ are defined by (2.25) or (2.26) or (2.27).

Exercises 2.6 Prove the above lemma and find the constants C_T and C_G. For example, in the case (iii), when h_Γ is given by (2.25), we have

$$C_T = (1 + C_Q^{-1})/2, \quad C_G = (1 + C_Q)/2, \tag{2.28}$$

where C_Q is the constant from the local quasi-uniformity condition (2.21).

2.3.3 Broken Sobolev Spaces

The discontinuous Galerkin method is based on the use of discontinuous approximations. This is the reason that over a triangulation $\mathscr{T}_h$, for any $k \in \mathbb{N}$, we define the so-called *broken Sobolev space*

$$H^k(\Omega, \mathscr{T}_h) = \{v \in L^2(\Omega); v|_K \in H^k(K) \; \forall \; K \in \mathscr{T}_h\}, \tag{2.29}$$

which consists of functions, whose restrictions on $K \in \mathscr{T}_h$ belong to the Sobolev space $H^k(K)$. On the other hand, functions from $H^k(\Omega, \mathscr{T}_h)$ are, in general, discontinuous on inner faces of elements $K \in \mathscr{T}_h$. For $v \in H^k(\Omega, \mathscr{T}_h)$, we define the norm

$$\|v\|_{H^k(\Omega,\mathscr{T}_h)} = \Big(\sum_{K\in\mathscr{T}_h} \|v\|^2_{H^k(K)} \Big)^{1/2} \tag{2.30}$$

and the seminorm

$$|v|_{H^k(\Omega,\mathscr{T}_h)} = \Big(\sum_{K\in\mathscr{T}_h} |v|^2_{H^k(K)} \Big)^{1/2}. \tag{2.31}$$

Let $\Gamma \in \mathscr{F}_h^I$ and let $K_\Gamma^{(L)}, K_\Gamma^{(R)} \in \mathscr{T}_h$ be elements adjacent to Γ. For $v \in H^1(\Omega, \mathscr{T}_h)$ we introduce the following notation:

$$\begin{aligned} v_\Gamma^{(L)} &= \text{the trace of } v|_{K_\Gamma^{(L)}} \text{ on } \Gamma, \qquad (2.32)\\ v_\Gamma^{(R)} &= \text{the trace of } v|_{K_\Gamma^{(R)}} \text{ on } \Gamma,\\ \langle v\rangle_\Gamma &= \frac{1}{2}\left(v_\Gamma^{(L)} + v_\Gamma^{(R)}\right) \quad \text{(mean value of the traces of } v \text{ on } \Gamma),\\ [v]_\Gamma &= v_\Gamma^{(L)} - v_\Gamma^{(R)} \quad \text{(jump of } v \text{ on } \Gamma). \end{aligned}$$

The value $[v]_\Gamma$ depends on the orientation of $\boldsymbol{n}_\Gamma$, but $[v]_\Gamma \boldsymbol{n}_\Gamma$ is independent of this orientation.

Moreover, let $\Gamma \in \mathscr{F}_h^B$ and $K_\Gamma^{(L)} \in \mathscr{T}_h$ be the element such that $\Gamma \subset \partial K_\Gamma^{(L)} \cap \partial\Omega$. Then for $v \in H^1(\Omega, \mathscr{T}_h)$ we introduce the following notation:

$$\begin{aligned} v_\Gamma^{(L)} &= \text{the trace of } v|_{K_\Gamma^{(L)}} \text{ on } \Gamma, \qquad (2.33)\\ \langle v\rangle_\Gamma &= [v]_\Gamma = v_\Gamma^{(L)}. \end{aligned}$$

If $\Gamma \in \mathscr{F}_h^B$, then by $v_\Gamma^{(R)}$ we formally denote the exterior trace of v on Γ given either by a boundary condition or by an extrapolation from the interior of Ω.

In case that $\Gamma \in \mathscr{F}_h$ and $[\cdot]_\Gamma$, $\langle\cdot\rangle_\Gamma$ and $\boldsymbol{n}_\Gamma$ appear in integrals $\int_\Gamma \ldots \,\mathrm{d}S$, then we usually omit the subscript Γ and simply write $[\cdot]$, $\langle\cdot\rangle$ and $\boldsymbol{n}$, respectively.

The discontinuous Galerkin method can be characterized as a finite element technique using piecewise polynomial approximations, in general discontinuous on interfaces between neighbouring elements. Therefore, we introduce a finite-dimensional subspace of $H^k(\Omega, \mathscr{T}_h)$, where the approximate solution will be sought.

Let $\mathscr{T}_h$ be a triangulation of Ω introduced in Sect. 2.3.1 and let $p \geq 0$ be an integer. We define the space of discontinuous piecewise polynomial functions

$$S_{hp} = \{v \in L^2(\Omega);\, v|_K \in P_p(K)\ \forall\, K \in \mathscr{T}_h\}, \tag{2.34}$$

where $P_p(K)$ denotes the space of all polynomials of degree $\le p$ on K. We call the number p the *degree of polynomial approximation*. Obviously, $S_{hp} \subset H^k(\Omega, \mathscr{T}_h)$ for any $k \ge 1$ and its dimension $\dim S_{hp} < \infty$.

2.4 DGM Based on a Primal Formulation

In this section we introduce the so-called discontinuous Galerkin method (DGM) based on a *primal formulation* for the solution of problem (2.1). The approximate solution will be sought in the space $S_{hp} \subset H^1(\Omega, \mathscr{T}_h)$. In contrast to the standard (conforming) finite element method, the weak formulation (2.2) given in Sect. 2.1 is not suitable for the derivation of the DGM, because (2.2) does not make sense for $u \in H^1(\Omega, \mathscr{T}_h) \not\subset H^1(\Omega)$. Therefore, we introduce a "weak form of (2.1) in the sense of broken Sobolev spaces".

Let us assume that u is a sufficiently regular solution of (2.1), namely, let $u \in H^2(\Omega)$. Then we speak of a *strong solution*. In deriving the DGM we proceed in the following way. We multiply (2.1a) by a function $v \in H^1(\Omega, \mathscr{T}_h)$, integrate over $K \in \mathscr{T}_h$ and use Green's theorem. Summing over all $K \subset \mathscr{T}_h$, we obtain the identity

$$\sum_{K \in \mathscr{T}_h} \int_K \nabla u \cdot \nabla v \, \mathrm{d}x - \sum_{K \in \mathscr{T}_h} \int_{\partial K} (\boldsymbol{n}_K \cdot \nabla u)\, v \, \mathrm{d}S = \int_\Omega f\, v \, \mathrm{d}x, \tag{2.35}$$

where $\boldsymbol{n}_K$ denotes the outer unit normal to ∂K. The surface integrals over ∂K make sense due to the regularity of u. (Since $u \in H^2(K)$, the derivatives $\partial u/\partial x_i$ have the trace on ∂K and $\partial u/\partial x_i|_{\partial K} \in L^2(\partial K)$ for $i = 1, \dots, d$; see Theorem 1.1 on traces.) We rewrite the surface integrals over ∂K according to the type of faces $\Gamma \in \mathscr{F}_h$ that form the boundary of the element $K \in \mathscr{T}_h$:

$$\begin{aligned} \sum_{K \in \mathscr{T}_h} \int_{\partial K} (\boldsymbol{n}_K \cdot \nabla u)\, v \, \mathrm{d}S = & \sum_{\Gamma \in \mathscr{F}_h^D} \int_\Gamma (\boldsymbol{n}_\Gamma \cdot \nabla u)\, v \, \mathrm{d}S + \sum_{\Gamma \in \mathscr{F}_h^N} \int_\Gamma (\boldsymbol{n}_\Gamma \cdot \nabla u)\, v \, \mathrm{d}S \\ & + \sum_{\Gamma \in \mathscr{F}_h^I} \int_\Gamma \boldsymbol{n}_\Gamma \cdot \left((\nabla u_\Gamma^{(L)})\, v_\Gamma^{(L)} - (\nabla u_\Gamma^{(R)})\, v_\Gamma^{(R)} \right) \mathrm{d}S. \end{aligned} \tag{2.36}$$

(There is the sign "−" in the last integral, since $\boldsymbol{n}_\Gamma$ is the outer unit normal to $\partial K_\Gamma^{(L)}$ but the inner unit normal to $\partial K_\Gamma^{(R)}$, see Sect. 2.3.1 or Fig. 2.2.)

Due to the assumption that $u \in H^2(\Omega)$, we have

$$[u]_\Gamma = [\nabla u]_\Gamma = 0, \quad \nabla u_\Gamma^{(L)} = \nabla u_\Gamma^{(R)} = \langle \nabla u \rangle_\Gamma, \quad \Gamma \in \mathscr{F}_h^I. \tag{2.37}$$

Thus, the integrand of the last integral in (2.36) can be written in the form

$$\boldsymbol{n}_\Gamma \cdot (\nabla u)_\Gamma^{(L)} v_\Gamma^{(L)} - \boldsymbol{n}_\Gamma \cdot (\nabla u)_\Gamma^{(R)} v_\Gamma^{(R)} = \boldsymbol{n}_\Gamma \cdot \langle \nabla u \rangle_\Gamma [v]_\Gamma. \tag{2.38}$$

By virtue of the Neumann boundary condition (2.1c),

$$\sum_{\Gamma \in \mathscr{F}_h^N} \int_\Gamma (\boldsymbol{n}_\Gamma \cdot \nabla u) v \,\mathrm{d}S = \int_{\partial\Omega_N} g_N v \,\mathrm{d}S. \tag{2.39}$$

Now, (2.33) and (2.35)–(2.39) imply that

$$\begin{aligned}
&\sum_{K \in \mathscr{T}_h} \int_K \nabla u \cdot \nabla v \,\mathrm{d}x - \sum_{\Gamma \in \mathscr{F}_h^I} \int_\Gamma \boldsymbol{n} \cdot \langle \nabla u \rangle [v] \,\mathrm{d}S - \sum_{\Gamma \in \mathscr{F}_h^D} \int_\Gamma \boldsymbol{n} \cdot \nabla u \, v \,\mathrm{d}S \\
&\quad = \sum_{K \in \mathscr{T}_h} \int_K \nabla u \cdot \nabla v \,\mathrm{d}x - \sum_{\Gamma \in \mathscr{F}_h^{ID}} \int_\Gamma \boldsymbol{n} \cdot \langle \nabla u \rangle [v] \,\mathrm{d}S \qquad (2.40)\\
&\quad = \int_\Omega f v \,\mathrm{d}x + \int_{\partial\Omega_N} g_N v \,\mathrm{d}S, \qquad v \in H^1(\Omega, \mathscr{T}_h).
\end{aligned}$$

Here and in what follows, in integrals over Γ the symbol $\boldsymbol{n}$ means $\boldsymbol{n}_\Gamma$.

Relation (2.40) is the basis of the DG discretization of problem (2.1). However, in order to guarantee the existence of the approximate solution and its convergence to the exact one, some additional terms have to be included in the DG formulation.

In order to mimic the continuity of the approximate solution in a weaker sense, we define the *interior and boundary penalty bilinear form*

$$\begin{aligned}
J_h^\sigma(u, v) &= \sum_{\Gamma \in \mathscr{F}_h^I} \int_\Gamma \sigma [u]\,[v] \,\mathrm{d}S + \sum_{\Gamma \in \mathscr{F}_h^D} \int_\Gamma \sigma u \, v \,\mathrm{d}S \qquad (2.41)\\
&= \sum_{\Gamma \in \mathscr{F}_h^{ID}} \int_\Gamma \sigma [u]\,[v] \,\mathrm{d}S, \qquad u, v \in H^1(\Omega, \mathscr{T}_h).
\end{aligned}$$

The boundary penalty is associated with the boundary linear form

$$J_D^\sigma(v) = \sum_{\Gamma \in \mathscr{F}_h^D} \int_\Gamma \sigma u_D \, v \,\mathrm{d}S. \tag{2.42}$$

Here $\sigma > 0$ is a penalty weight. Its choice will be discussed in Sect. 2.6. Obviously, for the exact strong solution $u \in H^2(\Omega)$,

$$J_h^\sigma(u, v) = J_D^\sigma(v) \qquad \forall\, v \in H^1(\Omega, \mathscr{T}_h), \tag{2.43}$$

since $[u]_\Gamma = 0$ for $\Gamma \in \mathscr{F}_h^I$ and $[u]_\Gamma = u_\Gamma = u_D$ for $\Gamma \in \mathscr{F}_h^D$.

The interior penalty replaces the continuity of the approximate solution on interior faces, which is required in the standard conforming finite element method. The boundary penalty introduces the Dirichlet boundary condition in the discrete problem.

Moreover, the left-hand side of (2.40) is not symmetric with respect to u and v. In the theoretical analysis, it is advantageous to have some type of symmetry. Hence, it is desirable to include some additional term, which "symmetrizes" the left-hand side of (2.40) and which vanishes for the exact solution. Therefore, let $u \in H^1(\Omega) \cap H^2(\Omega, \mathscr{T}_h)$ be a function which satisfies the Dirichlet boundary condition (2.1b). Then we use the identity

$$\sum_{\Gamma \in \mathscr{F}_h^{ID}} \int_\Gamma \boldsymbol{n} \cdot \langle \nabla v \rangle [u] \,\mathrm{d}S = \sum_{\Gamma \in \mathscr{F}_h^{D}} \int_\Gamma \boldsymbol{n} \cdot \nabla v \, u_D \,\mathrm{d}S \quad \forall v \in H^2(\Omega, \mathscr{T}_h), \tag{2.44}$$

which is valid since $[u]_\Gamma = 0$ for $\Gamma \in \mathscr{F}_h^I$, $[u]_\Gamma = u_\Gamma = u_D$ for $\Gamma \in \mathscr{F}_h^D$ and $\langle \nabla v \rangle_\Gamma = \nabla v_\Gamma$ for $\Gamma \in \mathscr{F}_h^D$ by definition.

Now, without a deeper motivation, we introduce five variants of the *discontinuous Galerkin weak formulation*. Each particular method is commented on in Remark 2.10. Hence, we sum identity (2.40) with -1, 1 or 0-multiple of (2.44) and possibly add equality (2.43). This leads us to the following notation. For $u, v \in H^2(\Omega, \mathscr{T}_h)$ we introduce the bilinear *diffusion forms*

$$a_h^{\mathrm{s}}(u, v) = \sum_{K \in \mathscr{T}_h} \int_K \nabla u \cdot \nabla v \,\mathrm{d}x - \sum_{\Gamma \in \mathscr{F}_h^{ID}} \int_\Gamma (\boldsymbol{n} \cdot \langle \nabla u \rangle [v] + \boldsymbol{n} \cdot \langle \nabla v \rangle [u]) \,\mathrm{d}S, \tag{2.45a}$$

$$a_h^{\mathrm{n}}(u, v) = \sum_{K \in \mathscr{T}_h} \int_K \nabla u \cdot \nabla v \,\mathrm{d}x - \sum_{\Gamma \in \mathscr{F}_h^{ID}} \int_\Gamma (\boldsymbol{n} \cdot \langle \nabla u \rangle [v] - \boldsymbol{n} \cdot \langle \nabla v \rangle [u]) \,\mathrm{d}S, \tag{2.45b}$$

$$a_h^{\mathrm{i}}(u, v) = \sum_{K \in \mathscr{T}_h} \int_K \nabla u \cdot \nabla v \,\mathrm{d}x - \sum_{\Gamma \in \mathscr{F}_h^{ID}} \int_\Gamma \boldsymbol{n} \cdot \langle \nabla u \rangle [v] \,\mathrm{d}S, \tag{2.45c}$$

and the right-hand side linear forms

$$F_h^{\mathrm{s}}(v) = \int_\Omega f v \,\mathrm{d}x + \sum_{\Gamma \in \mathscr{F}_h^N} \int_\Gamma g_N v \,\mathrm{d}S - \sum_{\Gamma \in \mathscr{F}_h^D} \int_\Gamma \boldsymbol{n} \cdot \nabla v \, u_D \,\mathrm{d}S, \tag{2.46a}$$

$$F_h^{\mathrm{n}}(v) = \int_\Omega f v \,\mathrm{d}x + \sum_{\Gamma \in \mathscr{F}_h^N} \int_\Gamma g_N v \,\mathrm{d}S + \sum_{\Gamma \in \mathscr{F}_h^D} \int_\Gamma \boldsymbol{n} \cdot \nabla v \, u_D \,\mathrm{d}S, \tag{2.46b}$$

$$F_h^{\mathrm{i}}(v) = \int_\Omega f v \,\mathrm{d}x + \sum_{\Gamma \in \mathscr{F}_h^N} \int_\Gamma g_N v \,\mathrm{d}S. \tag{2.46c}$$

Moreover, for $u, v \in H^2(\Omega, \mathscr{T}_h)$ let us define the bilinear forms

$$A_h^{\mathrm{s}}(u, v) = a_h^{\mathrm{s}}(u, v), \tag{2.47a}$$
$$A_h^{\mathrm{n}}(u, v) = a_h^{\mathrm{n}}(u, v), \tag{2.47b}$$
$$A_h^{\mathrm{s},\sigma}(u, v) = a_h^{\mathrm{s}}(u, v) + J_h^{\sigma}(u, v), \tag{2.47c}$$
$$A_h^{\mathrm{n},\sigma}(u, v) = a_h^{\mathrm{n}}(u, v) + J_h^{\sigma}(u, v), \tag{2.47d}$$
$$A_h^{\mathrm{i},\sigma}(u, v) = a_h^{\mathrm{i}}(u, v) + J_h^{\sigma}(u, v), \tag{2.47e}$$

and the linear forms

$$\ell_h^{\mathrm{s}}(v) = F_h^{\mathrm{s}}(v), \tag{2.48a}$$
$$\ell_h^{\mathrm{n}}(v) = F_h^{\mathrm{n}}(v), \tag{2.48b}$$
$$\ell_h^{\mathrm{s},\sigma}(v) = F_h^{\mathrm{s}}(v) + J_D^{\sigma}(v), \tag{2.48c}$$
$$\ell_h^{\mathrm{n},\sigma}(v) = F_h^{\mathrm{n}}(v) + J_D^{\sigma}(v), \tag{2.48d}$$
$$\ell_h^{\mathrm{i},\sigma}(v) = F_h^{\mathrm{i}}(v) + J_D^{\sigma}(v). \tag{2.48e}$$

Since $S_{hp} \subset H^2(\Omega, \mathscr{T}_h)$, the forms (2.47) make sense for $u_h, v_h \in S_{hp}$. Consequently, we define five numerical schemes.

Definition 2.7 A function $u_h \in S_{hp}$ is called a *DG approximate solution* of problem (2.1), if it satisfies one of the following identities:

$$\text{(i)} \quad A_h^{\mathrm{s}}(u_h, v_h) = \ell_h^{\mathrm{s}}(v_h) \quad \forall v_h \in S_{hp}, \tag{2.49a}$$
$$\text{(ii)} \quad A_h^{\mathrm{n}}(u_h, v_h) = \ell_h^{\mathrm{n}}(v_h) \quad \forall v_h \in S_{hp}, \tag{2.49b}$$
$$\text{(iii)} \quad A_h^{\mathrm{s},\sigma}(u_h, v_h) = \ell_h^{\mathrm{s},\sigma}(v_h) \quad \forall v_h \in S_{hp}, \tag{2.49c}$$
$$\text{(iv)} \quad A_h^{\mathrm{n},\sigma}(u_h, v_h) = \ell_h^{\mathrm{n},\sigma}(v_h) \quad \forall v_h \in S_{hp}, \tag{2.49d}$$
$$\text{(v)} \quad A_h^{\mathrm{i},\sigma}(u_h, v_h) = \ell_h^{\mathrm{i},\sigma}(v_h) \quad \forall v_h \in S_{hp}, \tag{2.49e}$$

where the forms $A_h^{\mathrm{s}}, A_h^{\mathrm{n}}, \ldots,$ and $\ell_h^{\mathrm{s}}, \ell_h^{\mathrm{n}}, \ldots,$ are defined by (2.47) and (2.48), respectively.

The diffusion forms $a_h^{\mathrm{s}}, a_h^{\mathrm{n}}, a_h^{\mathrm{i}}$ defined by (2.45) can be simply written in the form

$$a_h(u, v) = \sum_{K \in \mathscr{T}_h} \int_K \nabla u \cdot \nabla v \, \mathrm{d}x - \sum_{\Gamma \in \mathscr{F}_h^{ID}} \int_\Gamma (\boldsymbol{n} \cdot \langle \nabla u \rangle [v] + \Theta \boldsymbol{n} \cdot \langle \nabla v \rangle [u]) \, \mathrm{d}S, \tag{2.50}$$

where $\Theta = 1$ in the case of the form a_h^{s}, $\Theta = -1$ for a_h^{n} and $\Theta = 0$ for a_h^{i} and the bilinear forms $A_h^{\mathrm{s}}, A_h^{\mathrm{n}}, A_h^{\mathrm{s},\sigma}, A_h^{\mathrm{n},\sigma}$ and $A_h^{\mathrm{i},\sigma}$ defined by (2.47) can be written in the form

$$A_h(u, v) = a_h(u, v) + \vartheta J_h^{\sigma}(u, v), \tag{2.51}$$

where $\vartheta = 0$ for A_h^{s} and A_h^{n} and $\vartheta = 1$ for $A_h^{\mathrm{s},\sigma}$, $A_h^{\mathrm{n},\sigma}$ and $A_h^{\mathrm{i},\sigma}$.

Similarly we can write

$$F_h(v) = \int_\Omega f\, v\, \mathrm{d}x + \sum_{\Gamma\in\mathscr{F}_h^N} \int_\Gamma g_N\, v\, \mathrm{d}S - \Theta \sum_{\Gamma\in\mathscr{F}_h^D} \int_\Gamma \boldsymbol{n}\cdot\nabla v\, u_D\, \mathrm{d}S, \tag{2.52}$$

with $\Theta = 1$ for F_h^{s}, $\Theta = -1$ for F_h^{n} and $\Theta = 0$ for F_h^{i}, and then the right-hand side form reads

$$\ell_h(v) = F_h(v) + \vartheta J_D^\sigma(v), \tag{2.53}$$

where $\vartheta = 0$ for ℓ_h^{s} and ℓ_h^{n} and $\vartheta = 1$ for $\ell_h^{\mathrm{s},\sigma}$, $\ell_h^{\mathrm{n},\sigma}$ and $\ell_h^{\mathrm{i},\sigma}$.

The form a_h^{n} ($\Theta = -1$), a_h^{i} ($\Theta = 0$) and a_h^{s} ($\Theta = 1$) represents the so-called *nonsymmetric*, *incomplete* and *symmetric* variant of the diffusion discretization, respectively.

If we denote by A_h any form defined by (2.47) and by ℓ_h, we denote the form defined by (2.53), i.e., any form given by (2.48), the *discrete problem* (2.49) can be formulated to find $u_h \in S_{hp}$ satisfying the identity

$$A_h(u_h, v_h) = \ell_h(v_h) \quad \forall v_h \in S_{hp}. \tag{2.54}$$

The discrete problem (2.54) is equivalent to a system of linear algebraic equations, which can be solved by a suitable direct or iterative method. Namely, let $\{\varphi_i,\ i = 1,\dots,N_h\}$ be a basis of the space S_{hp}, where $N_h = \dim S_{hp}$ (= dimension of S_{hp}). The approximate solution u_h is sought in the form $u_h(x) = \sum_{j=1}^{N_h} u^j \varphi_j(x)$, where $u^j,\ j = 1,\dots,N_h$, are unknown real coefficients. Then, due to the linearity of the form A_h, the discrete problem (2.54) is equivalent to the system

$$\sum_{j=1}^{N_h} A_h(\varphi_j,\varphi_i) u^j = \ell_h(\varphi_j), \quad j = 1,\dots,N_h. \tag{2.55}$$

It can be written in the matrix form

$$\mathbb{A}U = L,$$

where $\mathbb{A} = (a_{ij})_{i,j=1}^{N_h} = (\mathbb{A}_h(\varphi_j,\varphi_i))_{i,j=1}^{N_h}$, $U = (u^j)_{j=1}^{N_h}$ and $L = (\ell_h(\varphi_j))_{j=1}^{N_h}$.

From the construction of the forms A_h and ℓ_h, one can see that the strong solution $u \in H^2(\Omega)$ of problem (2.1) satisfies the identity

$$A_h(u, v) = \ell_h(v) \quad \forall v \in H^2(\Omega, \mathscr{T}_h), \tag{2.56}$$

which represents the *consistency* of the method. Relations (2.54) and (2.56) imply the so-called *Galerkin orthogonality* of the error $e_h = u_h - u$ of the method:

$$A_h(e_h, v_h) = 0 \quad \forall v_h \in S_{hp}, \tag{2.57}$$

which will be used in analysing error estimates.

Remark 2.8 Comparing the above process of the derivation of the DG schemes with the abstract numerical method in Sect. 2.2, we see that we can define the function spaces

$$V = H^2(\Omega), \quad W_h = H^2(\Omega, \mathscr{T}_h), \quad V_h = S_{hp}. \tag{2.58}$$

However, as we will see later, the space W_h will not be equipped with the norm $\|\cdot\|_{H^2(\Omega,\mathscr{T}_h)}$ defined by (2.30), but by another norm introduced later in (2.103) will be used.

Remark 2.9 The interior and boundary penalty form J_h^σ together with the form J_D^σ replace the continuity of conforming finite element approximate solutions and represent Dirichlet boundary conditions. Thus, in contrast to standard conforming finite element techniques, both Dirichlet and Neumann boundary conditions are included automatically in the formulation (2.54) of the discrete problem. This is an advantage particularly in the case of nonhomogeneous Dirichlet boundary conditions, because it is not necessary to construct subsets of finite element spaces formed by functions approximating the Dirichlet boundary condition in a suitable way.

Remark 2.10 Method (2.49a) was introduced by Delves et al. ([76, 77, 172, 173]), who called it a *global element method*. Its advantage is the symmetry of the discrete problem due to the third term on the right-hand side of (2.45a). On the other hand, a significant disadvantage is that the bilinear form A_h^{s} is indefinite. This causes difficulties when dealing with time-dependent problems, because some eigenvalues of the operator associated with the form A_h can have negative real parts and then the resulting space-time discrete schemes become unconditionally unstable. Therefore, we prove in Lemma 2.36 the continuity of the bilinear form A_h^{s}, but further on we are not concerned with this method any more.

Scheme (2.49b) was introduced by Baumann and Oden in [12, 230] and is usually called the *Baumann–Oden method*. It is straightforward to show that the corresponding bilinear form A_h^{n} is positive semidefinite due to the third term on the right-hand side of (2.45b). An interesting property of this method is that it is unstable for piecewise linear approximations, i.e., for $p = 1$.

Scheme (2.49c) is called the *symmetric interior penalty Galerkin* (SIPG) method. It was derived by Arnold ([7]) and Wheeler ([283]) by adding penalty terms to the form A_h^{s}. (In this case a_h and F_h are defined by (2.50) and (2.52) with $\Theta = 1$.) This formulation leads to a symmetric bilinear form, which is coercive, if the penalty parameter σ is sufficiently large. Moreover, the Aubin–Nitsche duality technique (also called Aubin–Nitsche trick) can be used to obtain an optimal error estimate in the $L^2(\Omega)$-norm.

Method (2.49d), called the *nonsymmetric interior penalty Galerkin* (NIPG) method, was proposed by Girault, Rivière and Wheeler in [239]. (Here $\Theta = -1$.) In this case the bilinear form $A_h^{\mathrm{n},\sigma}$ is nonsymmetric and does not allow one to obtain an optimal error estimate in the $L^2(\Omega)$-norm with the aid of the Aubin-Nitsche trick. However, numerical experiments show that in some situations (for example, if uniform grids are used) the odd degrees of the polynomial approximation give the optimal order of convergence. On the other hand, a favorable property of the NIPG method is the coercivity of $A_h^{\mathrm{n},\sigma}(\cdot,\cdot)$ for any penalty parameter $\sigma > 0$.

Finally, method (2.49e), called the *incomplete interior penalty Galerkin* (IIPG) method ($\Theta = 0$), was studied in [74, 263, 265]. In this case the bilinear form $A_h^{\mathrm{i},\sigma}$ is nonsymmetric and does not allow one to obtain an optimal error estimate in the $L^2(\Omega)$-norm. The penalty parameter σ has to be chosen sufficiently large in order to guarantee the coercivity of $A_h^{\mathrm{i},\sigma}$. The advantage of the IIPG method is the simplicity of the discrete diffusion operator, because the expressions from (2.44) do not appear in (2.45c). This is particularly advantageous in the case when the diffusion operator is nonlinear with respect to ∇u. (See, e.g., [87] or Chap. 9 of this book.)

It would also be possible to define the scheme $A_h^{\mathrm{i}}(u,v) = \ell_h^{\mathrm{i}}(v)\ \forall v \in S_{hp}$, where $A_h^{\mathrm{i}}(u,v) = a_h^{\mathrm{i}}(u,v)$ and $\ell_h^{\mathrm{i}}(v) = F_h^{\mathrm{i}}(v)$, but this method does not make sense, because it does not contain the Dirichlet boundary data u_D from condition (2.1b).

In the following, we deal with the theoretical analysis of the DGM applied to the numerical solution of the model problem (2.1). Namely, we pay attention to the existence and uniqueness of the approximate solution defined by (2.54) and derive error estimates.

2.5 Basic Tools of the Theoretical Analysis of DGM

Theoretical analysis of the DG method presented in this book is based on three fundamental tools: the *multiplicative trace inequality*, the *inverse inequality*, and the *approximation properties* of the spaces of piecewise polynomial functions. In this section we introduce and prove these important tools under the assumptions about the meshes in Sect. 2.3.2.

Our first objective will be to summarize some important concepts and results from finite element theory, treated, e.g., in [52].

Definition 2.11 Let $n > 0$ be an integer. We say that sets $\omega, \widehat{\omega} \subset \mathbb{R}^n$ are *affine equivalent*, if there exists an invertible affine mapping $F_\omega : \widehat{\omega} \to \omega$ such that $F_\omega(\widehat{\omega}) = \omega$ and

$$x = F_\omega(\hat{x}) = \mathbb{B}_\omega \hat{x} + b_\omega \in \omega, \quad \hat{x} \in \widehat{\omega}, \tag{2.59}$$

where $\mathbb{B}_\omega$ is an $n \times n$ nonsingular matrix and $b_\omega \in \mathbb{R}^n$.

If $\hat{v} : \widehat{\omega} \to \mathbb{R}$, then the inverse mapping F_{ω}^{-1} allows us to transform the function $\hat{v}$ to $v : \omega \to \mathbb{R}$ by the relation

$$v(x) = \hat{v}(F_{\omega}^{-1}(x)), \quad x \in \omega. \tag{2.60}$$

Hence,

$$v = \hat{v} \circ F_{\omega}^{-1}, \quad \hat{v} = v \circ F_{\omega} \tag{2.61}$$

and

$$\hat{v}(\hat{x}) = v(x) \text{ for all } \hat{x},\ x \text{ in the correspondence (2.59).}$$

If $\mathbb{B}$ is an $n \times n$ matrix, then its norm associated with the Euclidean norm $|\cdot|$ in $\mathbb{R}^n$ is defined as $\|\mathbb{B}\| = \sup_{0 \neq x \in \mathbb{R}^n} |\mathbb{B}x|/|x|$.

The following lemmas give us bounds for the norms of matrices $\mathbb{B}_{\omega}$ and $\mathbb{B}_{\omega}^{-1}$ and the relations between Sobolev seminorms of functions v and $\hat{v}$ satisfying (2.61). First, we introduce the following notation for bounded domains $\omega,\ \widehat{\omega}$:

$$h_{\omega} = \mathrm{diam}(\omega),\ h_{\widehat{\omega}} = \mathrm{diam}(\widehat{\omega}), \tag{2.62}$$

$$\rho_{\omega} = \text{radius of the largest ball inscribed into } \overline{\omega}, \tag{2.63}$$

$$\rho_{\widehat{\omega}} = \text{radius of the largest ball inscribed into } \overline{\widehat{\omega}}.$$

Lemma 2.12 *Let $\omega,\ \widehat{\omega} \subset \mathbb{R}^n$ be affine-equivalent bounded domains with the invertible mapping $F_{\omega}(\hat{x}) = \mathbb{B}_{\omega}\hat{x} + b_{\omega} \in \omega$ for $\hat{x} \in \widehat{\omega}$. Then*

$$\|\mathbb{B}_{\omega}\| \leq \frac{h_{\omega}}{2\rho_{\widehat{\omega}}}, \quad \|\mathbb{B}_{\omega}^{-1}\| \leq \frac{h_{\widehat{\omega}}}{2\rho_{\omega}}. \tag{2.64}$$

Further, the substitution theorem implies that

$$|\det(\mathbb{B}_{\omega})| = |\omega|/|\widehat{\omega}|, \tag{2.65}$$

where $|\omega|$ and $|\widehat{\omega}|$ denote the n-dimensional Lebesgue measure of ω and $\widehat{\omega}$, respectively.

For the proof of (2.64) see [52, Theorem 3.1.3]. The proof of (2.65) is a consequence of the substitution theorem. Further, we cite here Theorem 3.1.2 from [52].

Lemma 2.13 *Let $\omega,\ \widehat{\omega} \subset \mathbb{R}^n$ be affine-equivalent bounded domains with the invertible mapping $F_{\omega}(\hat{x}) = \mathbb{B}_{\omega}\hat{x} + b_{\omega} \in \omega$ for $\hat{x} \in \widehat{\omega}$. If $v \in W^{m,\alpha}(\omega)$ for some integer $m \geq 0$ and some $\alpha \in [1, \infty]$, then the function $\hat{v} = v \circ F_{\omega} \in W^{m,\alpha}(\widehat{\omega})$. Moreover, there exists a constant C depending on m and d only such that*

$$|\hat{v}|_{W^{m,\alpha}(\widehat{\omega})} \leq C\, \|\mathbb{B}_{\omega}\|^m\, |\det(\mathbb{B}_{\omega})|^{-1/\alpha}\, |v|_{W^{m,\alpha}(\omega)}, \tag{2.66}$$

$$|v|_{W^{m,\alpha}(\omega)} \leq C\, \|\mathbb{B}_{\omega}^{-1}\|^m\, |\det(\mathbb{B}_{\omega})|^{1/\alpha}\, |\hat{v}|_{W^{m,\alpha}(\widehat{\omega})}. \tag{2.67}$$

In our finite element analysis, we have $n = d$ and the set ω represents an element $K \in \mathscr{T}_h$ and $\widehat{\omega}$ is chosen as a reference element $\widehat{K}$, i. e., the simplex with vertices

$$\begin{aligned} \hat{a}_1 =&(0,0,\dots,0), \quad \hat{a}_2 = (1,0,\dots,0), \quad \hat{a}_3 = (0,1,0,\dots,0),\dots \\ &\dots, \quad \hat{a}_{d+1} = (0,0,\dots,1) \in \mathbb{R}^d. \end{aligned} \tag{2.68}$$

The elements K and $\widehat{K}$ are considered as closed sets. The Sobolev spaces over K and $\widehat{K}$ are defined as the spaces over the interiors of these sets. (In Sect. 7.3, we will also apply the above results to the case with $n = 1$, $\omega = \Gamma \in \mathscr{F}_h$ and $\widehat{\omega} = (0, 1)$.)

As a consequence of the above results we can formulate the following assertions.

Corollary 2.14 *If $K \in \mathscr{T}_h$ and $v \in H^m(K)$, where $m \geq 0$ is an integer, then the function $\hat{v}(\hat{x}) = v(F_K(\hat{x})) \in H^m(\widehat{K})$ and*

$$|v|_{H^m(K)} \leq c_c h_K^{\frac{d}{2}-m} |\hat{v}|_{H^m(\widehat{K})}, \tag{2.69}$$

$$|\hat{v}|_{H^m(\widehat{K})} \leq c_c h_K^{m-\frac{d}{2}} |v|_{H^m(K)}, \tag{2.70}$$

where $c_c > 0$ depends on the shape regularity constant C_R but not on K and v.

Exercises 2.15 Prove (2.69) and (2.70) using the shape-regularity assumption (2.19) and the results of Lemmas 2.12 and 2.13.

In deriving error estimates we apply the following important result from [52, Theorem 3.1.4].

Theorem 2.16 *Let $\widehat{\omega} \subset \mathbb{R}^n$ be a bounded domain and for some integers $p \geq 0$ and $m \geq 0$ and some numbers $\alpha, \beta \in [1, \infty]$, let the spaces $W^{p+1,\alpha}(\widehat{\omega})$ and $W^{m,\beta}(\widehat{\omega})$ satisfy the continuous embedding*

$$W^{p+1,\alpha}(\widehat{\omega}) \hookrightarrow W^{m,\beta}(\widehat{\omega}). \tag{2.71}$$

Let $\widehat{\Pi}$ be a continuous linear mapping of $W^{p+1,\alpha}(\widehat{\omega})$ into $W^{m,\beta}(\widehat{\omega})$ such that

$$\widehat{\Pi}\hat{\phi} = \hat{\phi} \quad \forall\, \hat{\phi} \in P_p(\widehat{\omega}). \tag{2.72}$$

Let a set ω be affine-equivalent to the set $\widehat{\omega}$. This means that there exists an affine mapping $x = F_\omega(\hat{x}) = \mathbb{B}_\omega \hat{x} + b_\omega \in \omega$ for $\hat{x} \in \widehat{\omega}$, where $\mathbb{B}_\omega$ is a nonsingular $n \times n$ matrix and $b_\omega \in \mathbb{R}^n$. Let the mapping Π_ω be defined by

$$\Pi_\omega v(x) = (\widehat{\Pi}\hat{v})(F_\omega^{-1}(x)), \tag{2.73}$$

for all functions $\hat{v} \in W^{p+1,\alpha}(\widehat{\omega})$ *and* $v \in W^{p+1,\alpha}(\omega)$ *such that* $\hat{v}(\hat{x}) = v(F_\omega(\hat{x})) = v(x)$. *Then there exists a constant* $C(\widehat{\Pi}, \widehat{\omega})$ *such that*

$$|\widehat{\Pi}\hat{v} - \hat{v}|_{W^{m,\beta}(\widehat{\omega})} \leq C(\widehat{\Pi}, \widehat{\omega})|\hat{v}|_{W^{p+1,\alpha}(\widehat{\omega})}, \tag{2.74}$$

and

$$|v - \Pi_\omega v|_{W^{m,\beta}(\omega)} \leq C(\widehat{\Pi}, \widehat{\omega})\, |\omega|^{(1/\beta)-(1/\alpha)}\, \frac{h_\omega^{p+1}}{\rho_\omega^m} |v|_{W^{p+1,\alpha}(\omega)} \qquad \forall\, v \in W^{p+1,\alpha}(\omega), \tag{2.75}$$

with $h_\omega = \mathrm{diam}(\omega)$, ρ_ω *defined as the radius of the largest ball inscribed into* $\overline{\omega}$ *and* $|\omega|$ *defined as the n-dimensional Lebesgue measure of the set* ω. *We set* $1/\infty := 0$.

Exercises 2.17 Prove (2.75) using (2.74), (2.66), (2.67), (2.64) and (2.65).

Another important result used often in finite element theory is the Bramble–Hilbert lemma (see [52, Theorem 4.1.3] or [287, Theorem 9.3]).

Theorem 2.18 (Bramble–Hilbert lemma) *Let us assume that* $\omega \subset \mathbb{R}^n$ *is a bounded domain with Lipschitz boundary. Let* $p \geq 0$ *be an integer and* $\alpha \in [1, \infty]$ *and let* f *be a continuous linear functional on the space* $W^{p+1,\alpha}(\Omega)$ *(i.e.,* $f \in (W^{p+1,\alpha}(\omega))^*$*) satisfying the condition*

$$f(v) = 0 \quad \forall\, v \in P_p(\omega). \tag{2.76}$$

Then there exists a constant $C_{BH} > 0$ *depending only on* ω *such that*

$$|f(v)| \leq C_{BH} \|f\|_{(W^{p+1,\alpha}(\omega))^*} |v|_{W^{p+1,\alpha}(\omega)} \quad \forall\, v \in W^{p+1,\alpha}(\omega). \tag{2.77}$$

2.5.1 Multiplicative Trace Inequality

The forms a_h and J_h^σ given by (2.45) and (2.41), respectively, contain several integrals over faces. Therefore, in the theoretical analysis we need to estimate norms over faces by norms over elements. These estimates are usually obtained using the *multiplicative trace inequality*. In the literature, it is possible to find several variants of the multiplicative trace inequality. Here, we present the variant, which suits our considerations.

Lemma 2.19 (Multiplicative trace inequality) *Let the shape-regularity assumption (2.19) be satisfied. Then there exists a constant* $C_M > 0$ *independent of* v, h *and* K *such that*

$$\|v\|^2_{L^2(\partial K)} \le C_M \left(\|v\|_{L^2(K)} \, |v|_{H^1(K)} + h_K^{-1} \|v\|^2_{L^2(K)} \right), \qquad (2.78)$$
$$K \in \mathscr{T}_h, \; v \in H^1(K), \; h \in (0, \bar{h}).$$

Proof Let $K \in \mathscr{T}_h$ be arbitrary but fixed. We denote by $\boldsymbol{x}_K$ the center of the largest d-dimensional ball inscribed into the simplex K. Without loss of generality we suppose that $\boldsymbol{x}_K$ is the origin of the coordinate system.

Since the space $C^\infty(K)$ is dense in $H^1(K)$, it is sufficient to prove (2.78) for $v \in C^\infty(K)$. We start from the following relation obtained from Green's identity (1.23):

$$\int_{\partial K} v^2 \boldsymbol{x} \cdot \boldsymbol{n} \, \mathrm{d}S = \int_K \nabla \cdot (v^2 \boldsymbol{x}) \, \mathrm{d}x, \qquad v \in C^\infty(K), \qquad (2.79)$$

where $\boldsymbol{n}$ denotes here the outer unit normal to ∂K. Let $\boldsymbol{n}_\Gamma$ be the outer unit normal to K on a side Γ of K. Then

$$\boldsymbol{x} \cdot \boldsymbol{n}_\Gamma = |\boldsymbol{x}| |\boldsymbol{n}_\Gamma| \cos\alpha = |\boldsymbol{x}| \cos\alpha = \rho_K, \quad \boldsymbol{x} \in \Gamma, \qquad (2.80)$$

see Fig. 2.3. From (2.80) we have

$$\int_{\partial K} v^2 \boldsymbol{x} \cdot \boldsymbol{n} \, \mathrm{d}S = \sum_{\Gamma \subset \partial K} \int_\Gamma v^2 \boldsymbol{x} \cdot \boldsymbol{n}_\Gamma \, \mathrm{d}S = \rho_K \sum_{\Gamma \subset \partial K} \int_\Gamma v^2 \, \mathrm{d}S = \rho_K \|v\|^2_{L^2(\partial K)}. \qquad (2.81)$$

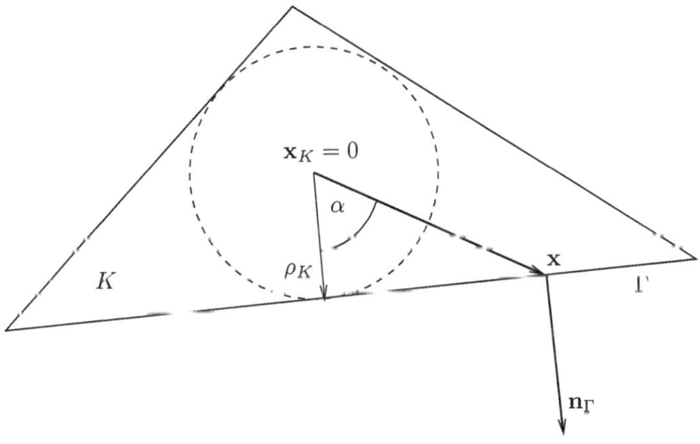

Fig. 2.3 Simplex K with its face Γ

Moreover,

$$\int_K \nabla \cdot (v^2 \boldsymbol{x})\,\mathrm{d}x = \int_K \left(v^2 \nabla \cdot \boldsymbol{x} + \boldsymbol{x} \cdot \nabla v^2\right)\,\mathrm{d}x \tag{2.82}$$
$$= d \int_K v^2\,\mathrm{d}x + 2\int_K v\boldsymbol{x} \cdot \nabla v\,\mathrm{d}x \le d\|v\|^2_{L^2(K)} + 2\int_K |v\boldsymbol{x} \cdot \nabla v|\,\mathrm{d}x.$$

With the aid of the Cauchy inequality, the second term of (2.82) is estimated as

$$2\int_K |v\boldsymbol{x} \cdot \nabla v|\,\mathrm{d}x \le 2 \sup_{x \in K} |\boldsymbol{x}| \int_K |v||\nabla v|\,\mathrm{d}x \le 2h_K \|v\|_{L^2(K)} |v|_{H^1(K)}. \tag{2.83}$$

Then (2.19), (2.79), (2.81)–(2.83) give

$$\|v\|^2_{L^2(\partial K)} \le \frac{1}{\rho_K}\left[2h_K \|v\|_{L^2(K)} |v|_{H^1(K)} + d\|v\|^2_{L^2(K)}\right] \tag{2.84}$$
$$\le C_R\left[2\|v\|_{L^2(K)} |v|_{H^1(K)} + \frac{d}{h_K}\|v\|^2_{L^2(K)}\right],$$

which proves (2.78) with $C_M = C_R \max\{2, d\}$. □

Exercises 2.20 Prove that the multiplicative trace inequality is valid also for vector-valued functions $\boldsymbol{v} : \Omega \to \mathbb{R}^n$, i.e.,

$$\|\boldsymbol{v}\|^2_{L^2(\partial K)} \le C_M \left(\|\boldsymbol{v}\|_{L^2(K)} |\boldsymbol{v}|_{H^1(K)} + h_K^{-1} \|\boldsymbol{v}\|^2_{L^2(K)}\right), \quad \boldsymbol{v} \in (H^1(K))^n,\ K \in \mathscr{T}_h. \tag{2.85}$$

Hint: Use (2.78) for each component of $\boldsymbol{v} = (v_1, \dots, v_n)$, sum these inequalities and apply the discrete Cauchy inequality (1.52).

2.5.2 *Inverse Inequality*

In deriving error estimates, we need to estimate the H^1-seminorm of a polynomial function by its L^2-norm, i.e., we apply the so-called *inverse inequality*.

Lemma 2.21 (Inverse inequality) *Let the shape-regularity assumption (2.19) be satisfied. Then there exists a constant $C_I > 0$ independent of v, h and K such that*

$$|v|_{H^1(K)} \le C_I h_K^{-1} \|v\|_{L^2(K)} \quad \forall v \in P_p(K),\ \forall K \in \mathscr{T}_h,\ \forall h \in (0, \bar{h}). \tag{2.86}$$

Proof Let $\widehat{K}$ be a reference triangle and $F_K : \widehat{K} \to K$, $K \in \mathscr{T}_h$ be an affine mapping such that $F_K(\widehat{K}) = K$. By (2.69) (for $m = 1$) and (2.70) (for $m = 0$) we have

$$|v|_{H^1(K)} \leq c_c h_K^{\frac{d}{2}-1} |\hat{v}|_{H^1(\widehat{K})}, \qquad \|\hat{v}\|_{L^2(\widehat{K})} \leq c_c h_K^{-\frac{d}{2}} \|v\|_{L^2(K)}. \tag{2.87}$$

From [253, Theorem 4.76], we have

$$|\hat{v}|_{H^1(\widehat{K})} \leq c_s p^2 \|\hat{v}\|_{L^2(\widehat{K})}, \quad \hat{v} \in P_p(\widehat{K}), \tag{2.88}$$

where $c_s > 0$ depends on d but not on $\hat{v}$ and p. A simple combination of (2.87) and (2.88) proves (2.86) with $C_I = c_s c_c^2 p^2$. Let us note that (2.88) is a consequence of the norm equivalence on finite-dimensional spaces. □

Other inverse inequalities will appear in Sect. 7.3, Lemma 7.35.

2.5.3 Approximation Properties

With respect to the error analysis of the abstract numerical method treated in Sect. 2.2, a suitable S_{hp}-interpolation has to be introduced. Let $\mathscr{T}_h$ be a given triangulation of the domain Ω. Then for each $K \in \mathscr{T}_h$, we define the mapping $\pi_{K,p} : L^2(K) \to P_p(K)$ such that for every $\varphi \in L^2(K)$

$$\pi_{K,p}\varphi \in P_p(K), \quad \int_K (\pi_{K,p}\varphi)v \, dx = \int_K \varphi v \, dx \quad \forall v \in P_p(K). \tag{2.89}$$

On the basis of the mappings $\pi_{K,p}$ we introduce the S_{hp}-interpolation Π_{hp}, defined for all $\varphi \in L^2(\Omega)$ by

$$(\Pi_{hp}\varphi)|_K = \pi_{K,p}(\varphi|_K) \quad \forall K \in \mathscr{T}_h. \tag{2.90}$$

It can be easily shown that if $\varphi \in L^2(\Omega)$, then

$$\Pi_{hp}\varphi \in S_{hp}, \quad \int_\Omega (\Pi_{hp}\varphi)v \, dx = \int_\Omega \varphi v \, dx \quad \forall v \in S_{hp}. \tag{2.91}$$

Hence, Π_{hp} is the $L^2(\Omega)$-projection on the space S_{hp}.

The approximation properties of the interpolation operators $\pi_{K,p}$ and Π_{hp} are the consequence of Theorem 2.16.

Lemma 2.22 *Let the shape-regularity assumption (2.19) be valid and let p, q, s be integers, $p \geq 0$, $0 \leq q \leq \mu$, where $\mu = \min(p+1, s)$. Then there exists a constant $C_A > 0$ such that*

$$|\pi_{K,p}v - v|_{H^q(K)} \leq C_A h_K^{\mu-q} |v|_{H^\mu(K)} \quad \forall v \in H^s(K)\ \forall K \in \mathscr{T}_h\ \forall h \in (0, \bar{h}). \tag{2.92}$$

Hence, if $p \geq 1$ and $s \geq 2$, then

$$\|\pi_{K,p}v - v\|_{L^2(K)} \leq C_A h_K^{\mu} |v|_{H^{\mu}(K)} \quad \forall v \in H^s(K)\ \forall K \in \mathscr{T}_h\ \forall h \in (0, \bar{h}), \tag{2.93}$$

$$|\pi_{K,p}v - v|_{H^1(K)} \leq C_A h_K^{\mu-1} |v|_{H^{\mu}(K)} \quad \forall v \in H^s(K)\ \forall K \in \mathscr{T}_h\ \forall h \in (0, \bar{h}), \tag{2.94}$$

$$|\pi_{K,p}v - v|_{H^2(K)} \leq C_A h_K^{\mu-2} |v|_{H^{\mu}(K)} \quad \forall v \in H^s(K)\ \forall K \in \mathscr{T}_h\ \forall \hbar \in (0, \bar{h}). \tag{2.95}$$

Moreover, we have

$$\|\pi_{K,1}v - v\|_{L^\infty(K)} \leq C_A h_K |v|_{W^{1,\infty}(K)}\ \forall v \in W^{1,\infty}(K)\ \forall K \in \mathscr{T}_h\ \forall h \in (0, \bar{h}). \tag{2.96}$$

Exercises 2.23 Prove Lemma 2.22 using Theorem 2.16 and assumption (2.19).

The above results immediately imply the approximation properties of the operator Π_{hp}.

Lemma 2.24 *Let the shape-regularity assumption (2.19) be satisfied and let p, q, s be integers, $p \geq 0$, $0 \leq q \leq \mu$, where $\mu = \min(p + 1, s)$. Then*

$$\left|\Pi_{hp}v - v\right|_{H^q(\Omega,\mathscr{T}_h)} \leq C_A h^{\mu-q} |v|_{H^{\mu}(\Omega,\mathscr{T}_h)}, \quad v \in H^s(\Omega, \mathscr{T}_h),\ h \in (0, \bar{h}), \tag{2.97}$$

where C_A is the constant from (2.92). Hence, if $p \geq 1$ and $s \geq 2$, then

$$\|\Pi_{hp}v - v\|_{L^2(\Omega)} \leq C_A h^{\mu} |v|_{H^{\mu}(\Omega,\mathscr{T}_h)}, \quad v \in H^s(\Omega, \mathscr{T}_h),\ h \in (0, \bar{h}), \tag{2.98}$$

$$\left|\Pi_{hp}v - v\right|_{H^1(\Omega,\mathscr{T}_h)} \leq C_A h^{\mu-1} |v|_{H^{\mu}(\Omega,\mathscr{T}_h)}, \quad v \in H^s(\Omega, \mathscr{T}_h),\ h \in (0, \bar{h}), \tag{2.99}$$

$$\left|\Pi_{hp}v - v\right|_{H^2(\Omega,\mathscr{T}_h)} \leq C_A h^{\mu-2} |v|_{H^{\mu}(\Omega,\mathscr{T}_h)}, \quad v \in H^s(\Omega, \mathscr{T}_h),\ h \in (0, \bar{h}). \tag{2.100}$$

Proof Using (2.90), definition of the seminorm in a broken Sobolev space (2.31) and the approximation properties (2.92), we obtain (2.97). This immediately implies (2.98)–(2.100). □

Moreover, using the combination of the multiplicative trace inequality (2.78) and Lemma 2.22, we can prove the approximation properties of the operator Π_{hp} in the norms defined over the boundaries of elements.

Lemma 2.25 *Let the shape-regularity assumption (2.19) be satisfied and let $p \geq 1$, $s \geq 2$ be integers and $\alpha \geq -1$. Then*

$$\sum_{K\in\mathscr{T}_h} h_K^\alpha \|\Pi_{hp}v - v\|^2_{L^2(\partial K)} \le 2C_M C_A^2 h^{2\mu-1+\alpha}|v|^2_{H^\mu(\Omega,\mathscr{T}_h)}, \tag{2.101}$$

$$\sum_{K\in\mathscr{T}_h} h_K^\alpha \|\nabla(\Pi_{hp}v - v)\|^2_{L^2(\partial K)} \le 2C_M C_A^2 h^{2\mu-3+\alpha}|v|^2_{H^\mu(\Omega,\mathscr{T}_h)}, \tag{2.102}$$

$$v \in H^s(\Omega, \mathscr{T}_h),\ h \in (0, \bar{h}),$$

where $\mu = \min(p+1, s)$, C_M *is the constant from (2.78) and* C_A *is the constant from (2.92).*

Proof (i) Let $v \in H^s(\Omega, \mathscr{T}_h)$. For simplicity we put $\eta = \Pi_{hp}v - v$. Then relation (2.90) implies that $\eta|_K = \pi_{K,p}v|_K - v|_K$ for $K \in \mathscr{T}_h$. Using the multiplicative trace inequality (2.78), the approximation property (2.92), and the seminorm definition (2.31), we have

$$\begin{aligned}
\sum_{K\in\mathscr{T}_h} h_K^\alpha \|\eta\|^2_{L^2(\partial K)} &\le C_M \sum_{K\in\mathscr{T}_h} h_K^\alpha \left(\|\eta\|_{L^2(K)}\,|\eta|_{H^1(K)} + h_K^{-1}\|\eta\|^2_{L^2(K)}\right)\\
&\le C_M \sum_{K\in\mathscr{T}_h} h_K^\alpha C_A^2 \left(h_K^\mu h_K^{\mu-1} + h_K^{-1}h_K^{2\mu}\right)|v|^2_{H^\mu(K)}\\
&\le 2C_M C_A^2 h^{2\mu-1+\alpha}|v|^2_{H^\mu(\Omega,\mathscr{T}_h)}.
\end{aligned}$$

(ii) Similarly as above, using the vector-valued variant of the multiplicative trace inequality (2.85), identities (1.21) and the approximation property (2.92), we get

$$\begin{aligned}
\sum_{K\in\mathscr{T}_h} h_K^\alpha \|\nabla\eta\|^2_{L^2(\partial K)} &\le C_M \sum_{K\in\mathscr{T}_h} h_K^\alpha \left(\|\nabla\eta\|_{L^2(K)}\,|\nabla\eta|_{H^1(K)} + h_K^{-1}\|\nabla\eta\|^2_{L^2(K)}\right)\\
&= C_M \sum_{K\in\mathscr{T}_h} h_K^\alpha \left(|\eta|_{H^1(K)}\,|\eta|_{H^2(K)} + h_K^{-1}|\eta|^2_{H^1(K)}\right)\\
&\le C_M \sum_{K\in\mathscr{T}_h} h_K^\alpha C_A^2 \left(h_K^{\mu-1} h_K^{\mu-2} + h_K^{-1}h_K^{2(\mu-1)}\right)|v|^2_{H^\mu(K)}\\
&\le 2C_M C_A^2 h^{2\mu-3+\alpha}|v|^2_{H^\mu(\Omega,\mathscr{T}_h)}.
\end{aligned}$$

□

2.6 Existence and Uniqueness of the Approximate Solution

We start with the theoretical analysis of the DGM, namely we prove the existence of a numerical solution defined by (2.54). Then, in Sect. 2.7, we derive error estimates. We follow the formal analysis of the abstract numerical methods in Sect. 2.2. Therefore, we show the *continuity* and the *coercivity* of the form A_h given by (2.47) in a suitable norm. This norm should reflect the discontinuity of functions from the broken Sobolev spaces $H^1(\Omega, \mathscr{T}_h)$. To this end, we define the following mesh-dependent norm

$$|||u|||_{\mathscr{T}_h} = \left(|u|^2_{H^1(\Omega,\mathscr{T}_h)} + J_h^\sigma(u,u) \right)^{1/2}, \tag{2.103}$$

where $|\cdot|_{H^1(\Omega,\mathscr{T}_h)}$ and J_h^σ are given by (2.31) and (2.41), respectively.

In what follows, because there is no danger of misunderstanding, we omit the subscript $\mathscr{T}_h$. This means that we simply write $|||\cdot||| = |||\cdot|||_{\mathscr{T}_h}$. We call $|||\cdot|||$ the *DG-norm.*

Exercises 2.26 Prove that $|||\cdot|||$ is a norm in the spaces $H^1(\Omega, \mathscr{T}_h)$ and S_{hp}.

2.6.1 The Choice of Penalty Weight σ

In the following considerations we assume that the system $\{\mathscr{T}_h\}_{h\in(0,\bar{h})}$ of triangulations satisfies the shape-regularity assumption (2.19) and the equivalence condition (2.20).

We consider the penalty weight $\sigma : \cup_{\Gamma\in\mathscr{F}_h^{ID}} \to \mathbb{R}$ in the form

$$\sigma|_\Gamma = \sigma_\Gamma = \frac{C_W}{h_\Gamma}, \quad \Gamma \in \mathscr{F}_h^{ID}, \tag{2.104}$$

where $C_W > 0$ is the *penalization constant* and $h_\Gamma (\sim h)$ is the quantity given by one of the possibilities from (2.24)–(2.27) with respect to the considered mesh assumptions (MA1)–(MA4), see Lemma 2.5. Let us note that in some cases it is possible to consider a different form of the penalty parameter σ, as mentioned in Remark 2.51.

Under the introduced notation, in view of (2.41), (2.42) and (2.104), the interior and boundary penalty form and the associated boundary linear form read as

$$J_h^\sigma(u,v) = \sum_{\Gamma\in\mathscr{F}_h^{ID}} \int_\Gamma \frac{C_W}{h_\Gamma} [u]\,[v]\,\mathrm{d}S, \qquad J_D^\sigma(v) = \sum_{\Gamma\in\mathscr{F}_h^{D}} \int_\Gamma \frac{C_W}{h_\Gamma} u_D\, v\,\mathrm{d}S. \tag{2.105}$$

In what follows, we introduce technical lemmas, which will be useful in the theoretical analysis.

Lemma 2.27 *Let (2.20) be valid. Then for each $v \in H^1(\Omega, \mathscr{T}_h)$ we have*

$$\sum_{\Gamma\in\mathscr{F}_h^{ID}} h_\Gamma^{-1} \int_\Gamma [v]^2\,\mathrm{d}S \le \frac{2}{C_T} \sum_{K\in\mathscr{T}_h} h_K^{-1} \int_{\partial K} |v|^2\,\mathrm{d}S, \tag{2.106}$$

$$\sum_{\Gamma \in \mathscr{F}_h^{ID}} h_\Gamma \int_\Gamma \langle v \rangle^2 \, \mathrm{d}S \le C_G \sum_{K \in \mathscr{T}_h} h_K \int_{\partial K} |v|^2 \, \mathrm{d}S. \tag{2.107}$$

Hence,

$$\sum_{\Gamma \in \mathscr{F}_h^{ID}} \sigma_\Gamma \, \|[v]\|^2_{L^2(\Gamma)} \le \frac{2C_W}{C_T} \sum_{K \in \mathscr{T}_h} h_K^{-1} \|v\|^2_{L^2(\partial K)}, \tag{2.108}$$

$$\sum_{\Gamma \in \mathscr{F}_h^{ID}} \frac{1}{\sigma_\Gamma} \|\langle v \rangle\|^2_{L^2(\Gamma)} \le \frac{C_G}{C_W} \sum_{K \in \mathscr{T}_h} h_K \|v\|^2_{L^2(\partial K)}. \tag{2.109}$$

Proof (i) By definition (2.32), the inequality

$$(\gamma + \delta)^2 \le 2(\gamma^2 + \delta^2), \quad \gamma, \delta \in \mathbb{R}, \tag{2.110}$$

and (2.20) we have

$$\begin{aligned}
&\sum_{\Gamma \in \mathscr{F}_h^{ID}} h_\Gamma^{-1} \int_\Gamma [v]^2 \, \mathrm{d}S \\
&\quad = \sum_{\Gamma \in \mathscr{F}_h^{I}} h_\Gamma^{-1} \int_\Gamma \left| v_\Gamma^{(L)} - v_\Gamma^{(R)} \right|^2 \mathrm{d}S + \sum_{\Gamma \in \mathscr{F}_h^{D}} h_\Gamma^{-1} \int_\Gamma \left| v_\Gamma^{(L)} \right|^2 \mathrm{d}S \\
&\quad \le 2 \sum_{\Gamma \in \mathscr{F}_h^{I}} h_\Gamma^{-1} \int_\Gamma \left(\left| v_\Gamma^{(L)} \right|^2 + \left| v_\Gamma^{(R)} \right|^2 \right) \mathrm{d}S + \sum_{\Gamma \in \mathscr{F}_h^{D}} h_\Gamma^{-1} \int_\Gamma \left| v_\Gamma^{(L)} \right|^2 \mathrm{d}S \\
&\quad \le 2C_T^{-1} \sum_{\Gamma \in \mathscr{F}_h^{ID}} h^{-1}_{K_\Gamma^{(L)}} \int_\Gamma \left| v_\Gamma^{(L)} \right|^2 \mathrm{d}S + 2C_T^{-1} \sum_{\Gamma \in \mathscr{F}_h^{I}} h^{-1}_{K_\Gamma^{(R)}} \int_\Gamma \left| v_\Gamma^{(R)} \right|^2 \mathrm{d}S \\
&\quad \le 2C_T^{-1} \sum_{K \in \mathscr{T}_h} h_K^{-1} \int_{\partial K} |v|^2 \, \mathrm{d}S.
\end{aligned}$$

This and (2.104) immediately imply (2.108).

(ii) In the proof of (2.107) we proceed similarly, using (2.32), (2.20) and (2.110). Inequalities (2.108) and (2.109) are obtained from (2.106), (2.107) and (2.104). □

2.6.2 Continuity of Diffusion Bilinear Forms

First, we prove several auxiliary assertions.

Lemma 2.28 *Any form a_h defined by (2.45) satisfies the estimate*

$$|a_h(u, v)| \le \|u\|_{1,\sigma} \, \|v\|_{1,\sigma} \quad \forall u, v \in H^2(\Omega, \mathscr{T}_h), \tag{2.111}$$

where

$$\|v\|_{1,\sigma}^2 = |||v|||^2 + \sum_{\Gamma \in \mathscr{F}_h^{ID}} \int_\Gamma \sigma^{-1} (\boldsymbol{n} \cdot \langle \nabla v \rangle)^2 \, \mathrm{d}S \tag{2.112}$$
$$= |v|_{H^1(\Omega, \mathscr{T}_h)}^2 + J_h^\sigma(v, v) + \sum_{\Gamma \in \mathscr{F}_h^{ID}} \int_\Gamma \sigma^{-1} (\boldsymbol{n} \cdot \langle \nabla v \rangle)^2 \mathrm{d}S.$$

Proof It follows from (2.45) that

$$|a_h(u, v)| \le \underbrace{\sum_{K \in \mathscr{T}_h} \int_K |\nabla u \cdot \nabla v| \, \mathrm{d}x}_{\chi_1} \tag{2.113}$$
$$+ \underbrace{\sum_{\Gamma \in \mathscr{F}_h^{ID}} \int_\Gamma |\boldsymbol{n} \cdot \langle \nabla u \rangle \, [v]| \, \mathrm{d}S}_{\chi_2} + \underbrace{\sum_{\Gamma \in \mathscr{F}_h^{ID}} \int_\Gamma |\boldsymbol{n} \cdot \langle \nabla v \rangle \, [u]| \, \mathrm{d}S}_{\chi_3}.$$

(For the form a_h^{i} the term χ_3 vanishes, of course.) Obviously, the Cauchy inequality, the discrete Cauchy inequality, and (2.31) imply that

$$\chi_1 \le \sum_{K \in \mathscr{T}_h} |u|_{H^1(K)} |v|_{H^1(K)} \le |u|_{H^1(\Omega, \mathscr{T}_h)} |v|_{H^1(\Omega, \mathscr{T}_h)}. \tag{2.114}$$

Further, by the Cauchy inequality,

$$\chi_2 \le \sum_{\Gamma \in \mathscr{F}_h^{ID}} \left(\int_\Gamma \sigma^{-1} (\boldsymbol{n} \cdot \langle \nabla u \rangle)^2 \, \mathrm{d}S \right)^{1/2} \left(\int_\Gamma \sigma [v]^2 \, \mathrm{d}S \right)^{1/2} \tag{2.115}$$
$$\le \left(\sum_{\Gamma \in \mathscr{F}_h^{ID}} \int_\Gamma \sigma^{-1} (\boldsymbol{n} \cdot \langle \nabla u \rangle)^2 \, \mathrm{d}S \right)^{1/2} \left(\sum_{\Gamma \in \mathscr{F}_h^{ID}} \int_\Gamma \sigma [v]^2 \, \mathrm{d}S \right)^{1/2},$$

and

$$\chi_3 \le \left(\sum_{\Gamma \in \mathscr{F}_h^{ID}} \int_\Gamma \sigma^{-1} (\boldsymbol{n} \cdot \langle \nabla v \rangle)^2 \, \mathrm{d}S \right)^{1/2} \left(\sum_{\Gamma \in \mathscr{F}_h^{ID}} \int_\Gamma \sigma [u]^2 \, \mathrm{d}S \right)^{1/2}. \tag{2.116}$$

Using the discrete Cauchy inequality, from (2.114)–(2.116) we derive the bound

$$|a_h(u, v)| \le |u|_{H^1(\Omega, \mathscr{T}_h)} |v|_{H^1(\Omega, \mathscr{T}_h)} \tag{2.117}$$

$$+\left(\sum_{\Gamma\in\mathscr{F}_h^{ID}}\int_\Gamma \sigma^{-1}(\boldsymbol{n}\cdot\langle\nabla u\rangle)^2\,\mathrm{d}S\right)^{1/2}\left(\sum_{\Gamma\in\mathscr{F}_h^{ID}}\int_\Gamma \sigma[v]^2\,\mathrm{d}S\right)^{1/2}$$

$$+\left(\sum_{\Gamma\in\mathscr{F}_h^{ID}}\int_\Gamma \sigma^{-1}(\boldsymbol{n}\cdot\langle\nabla v\rangle)^2\,\mathrm{d}S\right)^{1/2}\left(\sum_{\Gamma\in\mathscr{F}_h^{ID}}\int_\Gamma \sigma[u]^2\,\mathrm{d}S\right)^{1/2}$$

$$\leq\left(|u|^2_{H^1(\Omega,\mathscr{T}_h)}+\sum_{\Gamma\in\mathscr{F}_h^{ID}}\int_\Gamma \sigma^{-1}(\boldsymbol{n}\cdot\langle\nabla u\rangle)^2\mathrm{d}S+J_h^\sigma(u,u)\right)^{1/2}$$

$$\times\left(|v|^2_{H^1(\Omega,\mathscr{T}_h)}+\sum_{\Gamma\in\mathscr{F}_h^{ID}}\int_\Gamma \sigma^{-1}(\boldsymbol{n}\cdot\langle\nabla v\rangle)^2\mathrm{d}S+J_h^\sigma(v.v)\right)^{1/2}$$

$$=\|u\|_{1,\sigma}\,\|v\|_{1,\sigma}.$$

□

Exercises 2.29 Prove that $\|\cdot\|_{1,\sigma}$ introduced by (2.112) defines a norm in the broken Sobolev space $H^2(\Omega,\mathscr{T}_h)$.

Corollary 2.30 *By virtue of (2.47a) and (2.47b), Lemma 2.28 and Exercise 2.29, the bilinear forms* $\mathrm{A}_h^{\mathrm{s}}$ *and* $\mathrm{A}_h^{\mathrm{n}}$ *are bounded with respect to the norm* $\|\cdot\|_{1,\sigma}$ *in the broken Sobolev space* $H^2(\Omega,\mathscr{T}_h)$.

Exercises 2.31 Prove Corollary 2.30.

Further, we pay attention on the expression $J_h^\sigma(u,v)$ for $u,v\in H^1(\Omega,\mathscr{T}_h)$.

Lemma 2.32 *Let assumptions (2.104), (2.19) and (2.20) be satisfied. Then*

$$|J_h^\sigma(u,v)|\leq J_h^\sigma(u,u)^{1/2}J_h^\sigma(v,v)^{1/2}\quad\forall\,u,v\in H^1(\Omega,\mathscr{T}_h),\tag{2.118}$$

and

$$J_h^\sigma(v,v)\leq\frac{2C_WC_M}{C_T}\sum_{K\in\mathscr{T}_h}\left(h_K^{-2}\|v\|^2_{L^2(K)}+h_K^{-1}\|v\|_{L^2(K)}|v|_{H^1(K)}\right)\tag{2.119}$$

$$\leq\frac{C_WC_M}{C_T}\sum_{K\in\mathscr{T}_h}\left(3h_K^{-2}\|v\|^2_{L^2(K)}+|v|^2_{H^1(K)}\right)\qquad\forall\,v\in H^1(\Omega,\mathscr{T}_h).$$

Proof Let $u,v\in H^1(\Omega,\mathscr{T}_h)$. By the definition (2.41) of the form J_h^σ and the Cauchy inequality,

$$|J_h^\sigma(u,v)| \le \sum_{\Gamma\in\mathscr{F}_h^{ID}} \int_\Gamma \sigma|[u]\,[v]|\mathrm{d}S \tag{2.120}$$
$$\le \left(\sum_{\Gamma\in\mathscr{F}_h^{ID}} \int_\Gamma \sigma[u]^2\,\mathrm{d}S\right)^{1/2} \left(\sum_{\Gamma\in\mathscr{F}_h^{ID}} \int_\Gamma \sigma[v]^2\,\mathrm{d}S\right)^{1/2}$$
$$= J_h^\sigma(u,u)^{1/2} J_h^\sigma(v,v)^{1/2}.$$

Further, the definition of the form J_h^σ, (2.104), (2.20) and (2.108) imply that

$$J_h^\sigma(v,v) = \sum_{\Gamma\in\mathscr{F}_h^{ID}} \int_\Gamma \sigma[v]^2\,\mathrm{d}S = \sum_{\Gamma\in\mathscr{F}_h^{ID}} \frac{C_W}{h_\Gamma}\|[v]^2\|_{L^2(\Gamma)} \le \frac{2C_W}{C_T}\sum_{K\in\mathscr{T}_h} h_K^{-1}\|v\|^2_{L^2(\partial K)}.$$

Now, using the multiplicative trace inequality (2.78), we get

$$J_h^\sigma(v,v) \le \frac{2C_W C_M}{C_T}\sum_{K\in\mathscr{T}_h}\left(h_K^{-2}\|v\|^2_{L^2(K)} + h_K^{-1}\|v\|_{L^2(K)}|v|_{H^1(K)}\right). \tag{2.121}$$

The last relation in (2.119) follows from (2.121) and the Young inequality. □

Lemmas 2.28 and 2.32 immediately imply the boundedness also of the forms $A_h^{\mathrm{s},\sigma}$, $A_h^{\mathrm{n},\sigma}$ and $A_h^{\mathrm{i},\sigma}$ with respect to the norm $\|\cdot\|_{1,\sigma}$.

Corollary 2.33 *Let assumptions (2.104), (2.19) and (2.20) be satisfied. Then the forms A_h defined by (2.47) satisfy the estimate*

$$|A_h(u,v)| \le 2\|u\|_{1,\sigma}\|v\|_{1,\sigma} \quad \forall u,v\in H^2(\Omega,\mathscr{T}_h). \tag{2.122}$$

Proof For the boundedness of $A_h = A_h^{\mathrm{s}}$ and $A_h = A_h^{\mathrm{n}}$, see Corollary (2.30). Let $A_h = A_h^{\mathrm{s},\sigma}$ or $A_h = A_h^{\mathrm{n},\sigma}$ or $A_h = A_h^{\mathrm{i},\sigma}$. Then, by virtue of (2.47c)–(2.47e), Lemmas 2.28 and 2.32 we have

$$|A_h(u,v)| \le |a_h(u,v)| + |J_h^\sigma(u,v)| \le \|u\|_{1,\sigma}\,\|v\|_{1,\sigma} + J_h^\sigma(u,u)^{1/2}J_h^\sigma(v,v)^{1/2}$$
$$\le \|u\|_{1,\sigma}\,\|v\|_{1,\sigma} + \|u\|_{1,\sigma}\,\|v\|_{1,\sigma} = 2\|u\|_{1,\sigma}\,\|v\|_{1,\sigma}.$$

□

The following lemma allows us to estimate the expressions with integrals over $\Gamma\in\mathscr{F}_h$ in terms of norms over elements $K\in\mathscr{T}_h$.

Lemma 2.34 *Let the weight σ be defined by (2.104). Then, under assumptions (2.19) and (2.20), for any $v\in H^2(\Omega,\mathscr{T}_h)$ the following estimate holds:*

$$\sum_{\Gamma\in\mathscr{F}_h^{ID}}\int_\Gamma \sigma^{-1}(\boldsymbol{n}\cdot\langle\nabla v\rangle)^2\,\mathrm{d}S \le \frac{C_G C_M}{C_W}\sum_{K\in\mathscr{T}_h}\left(h_K\|\nabla v\|_{L^2(K)}|\nabla v|_{H^1(K)} + \|\nabla v\|^2_{L^2(K)}\right)$$
$$= \frac{C_G C_M}{C_W}\sum_{K\in\mathscr{T}_h}\left(h_K|v|_{H^1(K)}|v|_{H^2(K)} + |v|^2_{H^1(K)}\right)$$
$$\le \frac{C_G C_M}{2C_W}\sum_{K\in\mathscr{T}_h}\left(h_K^2|v|^2_{H^2(K)} + 3|v|^2_{H^1(K)}\right). \tag{2.123}$$

Moreover, if $v\in S_{hp}$, then

$$\sum_{\Gamma\in\mathscr{F}_h^{ID}}\int_\Gamma \sigma^{-1}(\boldsymbol{n}\cdot\langle\nabla v_h\rangle)^2\,\mathrm{d}S \le \frac{C_G C_M}{C_W}(C_I+1)|v_h|^2_{H^1(\Omega,\mathscr{T}_h)}. \tag{2.124}$$

Proof Using (2.109) and the multiplicative trace inequality (2.78), we find that

$$\sum_{\Gamma\in\mathscr{F}_h^{ID}}\int_\Gamma \sigma^{-1}(\boldsymbol{n}\cdot\langle\nabla v\rangle)^2\,\mathrm{d}S$$
$$\le \frac{C_G}{C_W}\sum_{K\in\mathscr{T}_h} h_K\|\nabla v\|^2_{L^2(\partial K)}$$
$$\le \frac{C_G C_M}{C_W}\sum_{K\in\mathscr{T}_h} h_K\left(\|\nabla v\|_{L^2(K)}|\nabla v|_{H^1(K)} + h_K^{-1}\|\nabla v\|^2_{L^2(K)}\right),$$

which is the first inequality in (2.123). The second one directly follows from the Young inequality.

If $v\in S_{hp}$, then (2.123) and the inverse inequality (2.86) imply that

$$\sum_{\Gamma\in\mathscr{F}_h^{ID}}\int_\Gamma \sigma^{-1}(\boldsymbol{n}\cdot\langle\nabla v_h\rangle)^2\,\mathrm{d}S \le \frac{C_G C_M}{C_W}\sum_{K\in\mathscr{T}_h}\left(C_I\|\nabla v_h\|^2_{L^2(K)} + \|\nabla v_h\|^2_{L^2(K)}\right)$$
$$= \frac{C_G C_M}{C_W}(C_I+1)\sum_{K\in\mathscr{T}_h}\|\nabla v_h\|^2_{L^2(K)} = \frac{C_G C_M}{C_W}(C_I+1)|v_h|^2_{H^1(\Omega,\mathscr{T}_h)},$$

which we wanted to prove. □

We continue in the derivation of various inequalities based on the estimation of the $\|\cdot\|_{1,\sigma}$-norm.

Lemma 2.35 *Under assumptions of Lemma 2.34, there exist constants C_σ, $\tilde{C}_\sigma > 0$ such that*

$$J_h^\sigma(u,u)^{1/2} \le |||u||| \le \|u\|_{1,\sigma} \le C_\sigma\, R_a(u)\quad \forall\, u\in H^2(\Omega,\mathscr{T}_h),\ h\in(0,h), \tag{2.125}$$

$$J_h^\sigma(v_h, v_h)^{1/2} \le |||v_h||| \le \|v_h\|_{1,\sigma} \le \tilde{C}_\sigma |||v_h||| \quad \forall v_h \in S_{hp},\ h \in (0, \bar{h}), \tag{2.126}$$

where

$$R_a(u) = \Big(\sum_{K\in\mathscr{T}_h} \Big(|u|^2_{H^1(K)} + h_K^2 |u|^2_{H^2(K)} + h_K^{-2} \|u\|^2_{L^2(K)} \Big) \Big)^{1/2}, \quad u \in H^2(\Omega, \mathscr{T}_h). \tag{2.127}$$

Proof The first two inequalities in (2.125) as well as in (2.126) follow immediately from the definition of the DG-norm (2.103) and the $\|\cdot\|_{1,\sigma}$-norm (2.112). Moreover, in view of (2.123) and (2.119), for $u \in H^2(\Omega, \mathscr{T}_h)$ we have

$$\begin{aligned} \|u\|^2_{1,\sigma} =& |u|^2_{H^1(\Omega,\mathscr{T}_h)} + J_h^\sigma(u,u) + \sum_{\Gamma\in\mathscr{F}_h^{ID}} \int_\Gamma \sigma^{-1} (\boldsymbol{n}\cdot\langle\nabla u\rangle)^2 \mathrm{d}S \\ \le& \sum_{K\in\mathscr{T}_h} |u|^2_{H^1(K)} + \frac{C_W C_M}{C_T} \sum_{K\in\mathscr{T}_h} \Big(3h_K^{-2} \|u\|^2_{L^2(K)} + |u|^2_{H^1(K)} \Big) \\ &+ \frac{C_G C_M}{2C_W} \sum_{K\in\mathscr{T}_h} \Big(h_K^2 |u|^2_{H^2(K)} + 3|u|^2_{H^1(K)} \Big). \end{aligned}$$

Now, after a simple manipulation, we get

$$\begin{aligned} \|u\|^2_{1,\sigma} \le \sum_{K\in\mathscr{T}_h} \Big(& |u|^2_{H^1(K)} \Big(1 + \frac{3C_G C_M}{2C_W} + \frac{C_W C_M}{C_T} \Big) \\ & + |u|^2_{H^2(K)} h_K^2 \frac{C_G C_M}{2C_W} + \|u\|^2_{L^2(K)} h_K^{-2} \frac{3C_W C_M}{C_T} \Big). \end{aligned}$$

Hence, (2.125) holds with

$$C_\sigma = \Big(\max \Big(1 + \frac{3C_G C_M}{2C_W} + \frac{C_W C_M}{C_T}, \frac{C_G C_M}{2C_W}, \frac{3C_W C_M}{C_T} \Big) \Big)^{1/2}.$$

Further, if $v_h \in S_{hp}$, then (2.112), (2.124) and (2.103) immediately imply (2.126) with $\tilde{C}_\sigma = (1 + C_G C_M (C_I + 1)/C_W)^{1/2}$. □

In what follows, we are concerned with properties of the bilinear forms A_h defined by (2.47). First, we prove the continuity of the bilinear forms A_h defined by (2.47) in the space S_{hp} with respect to the norm $|||\cdot|||$.

Lemma 2.36 *Let assumptions (2.104), (2.19) and (2.20) be satisfied. Then there exists a constant $C_B > 0$ such that the form A_h defined by (2.47) satisfies the estimate*

$$|A_h(u_h, v_h)| \leq C_B \, |||u_h||| \, |||v_h||| \quad \forall u_h, v_h \in S_{hp}. \tag{2.128}$$

Proof Estimates (2.122) and (2.126) give (2.128) with $C_B = 2\tilde{C}_\sigma^2$. □

Further, we prove an inequality similar to (2.128) replacing $u_h \in S_{hp}$ by $u \in H^2(\Omega, \mathscr{T}_h)$.

Lemma 2.37 *Let assumptions (2.19), (2.20) and (2.104) be satisfied. Then there exists a constant $\tilde{C}_B > 0$ such that*

$$|A_h(u, v_h)| \leq \tilde{C}_B \, R_a(u) \, |||v_h||| \quad \forall \, u \in H^2(\Omega, \mathscr{T}_h) \, \forall \, v_h \in S_{hp} \, \forall \, h(0, \bar{h}), \tag{2.129}$$

where R_a is defined by (2.127).

Proof By (2.122) and (2.125),

$$|A_h(u, v_h)| \leq 2\|u\|_{1,\sigma} \|v_h\|_{1,\sigma} \leq 2C_\sigma \tilde{C}_\sigma R_a(u) |||v_h|||,$$

which is (2.129) with $\tilde{C}_B = 2C_\sigma \tilde{C}_\sigma$. □

2.6.3 Coercivity of Diffusion Bilinear Forms

Lemma 2.38 (NIPG coercivity) *For any $C_W > 0$ the bilinear form $A_h^{\mathrm{n},\sigma}$ defined by (2.47d) satisfies the coercivity condition*

$$A_h^{\mathrm{n},\sigma}(v, v) \geq |||v|||^2 \quad \forall v \in H^2(\Omega, \mathscr{T}_h). \tag{2.130}$$

Proof From (2.45b) and (2.47d) it immediately follows that

$$A_h^{\mathrm{n},\sigma}(v, v) = a_h^{\mathrm{n}}(v, v) + J_h^\sigma(v, v) = |v|^2_{H^1(\Omega, \mathscr{T}_h)} + J_h^\sigma(v, v) = |||v|||^2, \tag{2.131}$$

which we wanted to prove. □

The proof of the coercivity of the symmetric bilinear form $A_h^{\mathrm{s},\sigma}$ is more complicated.

Lemma 2.39 (SIPG coercivity) *Let assumptions (2.19) and (2.20) be satisfied, let*

$$C_W \geq 4C_G C_M(1 + C_I), \tag{2.132}$$

where C_M, C_I and C_G are the constants from (2.78), (2.86) and (2.20), respectively, and let the penalty parameter σ be given by (2.101) for all $\Gamma \subset \mathscr{F}_h^{ID}$. Then

$$A_h^{\mathrm{s},\sigma}(v_h, v_h) \geq \frac{1}{2} |||v_h|||^2 \quad \forall v_h \in S_{hp} \, \forall h \in (0, \bar{h}).$$

Proof Let $\delta > 0$. Then from (2.41), (2.104), (2.45a) and the Cauchy and Young inequalities it follows that

$$
\begin{aligned}
a_h^s(v_h, v_h) &= |v_h|^2_{H^1(\Omega,\mathscr{T}_h)} - 2 \sum_{\Gamma\in\mathscr{F}_h^{ID}} \int_\Gamma \boldsymbol{n}\cdot\langle\nabla v_h\rangle[v_h]\,\mathrm{d}S \qquad (2.133)\\
&\geq |v_h|^2_{H^1(\Omega,\mathscr{T}_h)} - 2\left\{\frac{1}{\delta}\sum_{\Gamma\in\mathscr{F}_h^{ID}}\int_\Gamma h_\Gamma(\boldsymbol{n}\cdot\langle\nabla v_h\rangle)^2\,\mathrm{d}S\right\}^{\frac{1}{2}}\left\{\delta\sum_{\Gamma\in\mathscr{F}_h^{ID}}\int_\Gamma \frac{1}{h_\Gamma}[v_h]^2\,\mathrm{d}S\right\}^{\frac{1}{2}}\\
&\geq |v_h|^2_{H^1(\Omega,\mathscr{T}_h)} - \omega - \frac{\delta}{C_W} J_h^\sigma(v_h, v_h),
\end{aligned}
$$

where

$$
\omega = \frac{1}{\delta}\sum_{\Gamma\in\mathscr{F}_h^{ID}}\int_\Gamma h_\Gamma|\langle\nabla v_h\rangle|^2\,\mathrm{d}S. \qquad (2.134)
$$

Further, from assumption (2.20), inequality (2.107), the multiplicative trace inequality (2.78) and the inverse inequality (2.86) we get

$$
\begin{aligned}
\omega &\leq \frac{C_G}{\delta}\sum_{K\in\mathscr{T}_h} h_K\|\nabla v_h\|^2_{L^2(\partial K)} \qquad (2.135)\\
&\leq \frac{C_G C_M}{\delta}\sum_{K\in\mathscr{T}_h} h_K\left(|v_h|_{H^1(K)}|\nabla v_h|_{H^1(K)} + h_K^{-1}|v_h|^2_{H^1(K)}\right)\\
&\leq \frac{C_G C_M(1+C_I)}{\delta}|v_h|^2_{H^1(\Omega,\mathscr{T}_h)}.
\end{aligned}
$$

Now let us choose

$$
\delta = 2C_G C_M(1+C_I). \qquad (2.136)
$$

Then it follows from (2.132) and (2.133)–(2.136) that

$$
\begin{aligned}
a_h^s(v_h, v_h) &\geq \frac{1}{2}\left(|v_h|^2_{H^1(\Omega,\mathscr{T}_h)} - \frac{4C_G C_M(1+C_I)}{C_W} J_h^\sigma(v_h, v_h)\right) \qquad (2.137)\\
&\geq \frac{1}{2}\left(|v_h|^2_{H^1(\Omega,\mathscr{T}_h)} - J_h^\sigma(v_h, v_h)\right).
\end{aligned}
$$

Finally, definition (2.47c) of the form $A_h^{s,\sigma}$ and (2.137) imply that

$$
\begin{aligned}
A_h^{s,\sigma}(v_h, v_h) &= a_h^s(v_h, v_h) + J_h^\sigma(v_h, v_h) \qquad (2.138)\\
&\geq \frac{1}{2}\left(|v_h|^2_{H^1(\Omega,\mathscr{T}_h)} + J_h^\sigma(v_h, v_h)\right) = \frac{1}{2}|||v_h|||^2,
\end{aligned}
$$

which we wanted to prove. □

Lemma 2.40 (IIPG coercivity) *Let assumptions (2.19) and (2.20) be satisfied, let*

$$C_W \geq C_G C_M (1 + C_I), \tag{2.139}$$

where C_M, C_I and C_G are constants from (2.78), (2.86) and (2.20), respectively, and let the penalty parameter σ be given by (2.104) for all $\Gamma \in \mathscr{F}_h^{ID}$. Then

$$A_h^{\mathrm{i},\sigma}(v_h, v_h) \geq \frac{1}{2} |||v_h|||^2 \quad \forall v_h \in S_{hp}.$$

Proof The proof is almost identical with the proof of the previous lemma. □

Corollary 2.41 *We can summarize the above results in the following way. We have*

$$A_h(v_h, v_h) \geq C_C |||v_h|||^2 \quad \forall v_h \in S_{hp}, \tag{2.140}$$

with

$$\begin{array}{lll} C_C = 1 & \textit{for } A_h = A_h^{\mathrm{n},\sigma} & \textit{if } C_W > 0, \\ C_C = 1/2 & \textit{for } A_h = A_h^{\mathrm{s},\sigma} & \textit{if } C_W \geq 4 C_G C_M (1 + C_I), \\ C_C = 1/2 & \textit{for } A_h = A_h^{\mathrm{i},\sigma} & \textit{if } C_W \geq C_G C_M (1 + C_I). \end{array}$$

Corollary 2.42 *By virtue of Corollary 1.7, the coercivity of the forms A_h implies the* existence and uniqueness *of the solution of the discrete problems (2.49c)–(2.49e) (SIPG, NIPG and IIPG method).*

2.7 Error Estimates

In this section, we derive error estimates of the SIPG, NIPG and IIPG variants of the DGM applied to the numerical solution of the Poisson problem (2.1). Namely, the error $u_h - u$ will be estimated in the DG-norm and the $L^2(\Omega)$-norm.

2.7.1 Estimates in the DG-Norm

Let $u \in H^2(\Omega)$ denote the exact strong solution of problem (2.1) and let and $u_h \in S_{hp}$ be the approximate solution obtained by method (2.54), where the forms A_h and ℓ_h are defined by (2.47c)–(2.47e) and (2.48c)–(2.48e), respectively. The error of the method is defined as the function $e_h = u_h - u \in H^2(\Omega, \mathscr{T}_h)$. It can be written in the form

$$e_h = \xi + \eta, \quad \text{with } \xi = u_h - \Pi_{hp} u \in S_{hp}, \ \eta = \Pi_{hp} u - u \in H^2(\Omega, \mathscr{T}_h), \tag{2.141}$$

where Π_{hp} is the S_{hp}-interpolation defined by (2.90). Hence, we split the error into two parts ξ and η. The term η represents the error of the S_{hp}-interpolation of the function u. (It is possible to say that η approximates the *distance* of the exact solution from the space S_{hp}, where the approximate solution is sought.) The term η can be simply estimated on the basis of the approximation properties (2.92) and (2.97). On the other hand, the term ξ represents the *distance* between the approximate solution u_h and the projection of the exact solution on the space S_{hp}. The estimation of ξ is sometimes more complicated.

We suppose that the system of triangulations $\{\mathscr{T}_h\}_{h\in(0,\bar{h})}$ satisfies the shape-regularity assumption (2.19) and that the equivalence condition (2.20) holds.

First, we prove the so-called *abstract error estimate*, representing a bound of the error in terms of the S_{hp}-interpolation error η.

Theorem 2.43 *Let assumptions (2.19) and (2.20) be satisfied and let the exact solution of problem (2.1) satisfy the condition $u \in H^2(\Omega)$. Then there exists a constant $C_{\mathrm{AE}} > 0$ such that*

$$|||e_h||| \le C_{\mathrm{AE}}\, R_a(\eta) = C_{\mathrm{AE}}\, R_a(\Pi_{hp}u - u), \quad h \in (0, \bar{h}), \tag{2.142}$$

where $R_a(\eta)$ is given by (2.127).

Proof We express the error by (2.141), i.e., $e_h = u_h - u = \xi + \eta$. The error e_h satisfies the Galerkin orthogonality condition (2.57), which is equivalent to the relation

$$A_h(\xi, v_h) = -A_h(\eta, v_h) \qquad \forall\, v_h \in S_{hp}. \tag{2.143}$$

If we set $v_h := \xi \in S_{hp}$ in (2.143) and use (2.47c)–(2.47e) and the coercivity (2.140), we find that

$$C_C |||\xi|||^2 \le A_h(\xi, \xi) = -A_h(\eta, \xi). \tag{2.144}$$

Now we apply Lemma 2.37 and get

$$|A_h(\eta, \xi)| \le \tilde{C}_B\, R_a(\eta)\, |||\xi|||.$$

The above and (2.144) already imply that

$$|||\xi||| \le \frac{\tilde{C}_B}{C_C}\, R_a(\eta). \tag{2.145}$$

Obviously,

$$|||e_h||| \le |||\xi||| + |||\eta|||. \tag{2.146}$$

Finally, (2.125) gives

$$|||\eta||| \leq C_\sigma R_a(\eta). \tag{2.147}$$

Hence, (2.146), (2.145) and (2.147) yield the abstract error estimate (2.142) with $C_{\mathrm{AE}} = C_\sigma + \tilde{C}_B/C_C$. □

The abstract error estimate is the basis for estimating the error e_h in terms of the mesh-size h.

Theorem 2.44 (DG-norm error estimate) *Let us assume that $s \geq 2$, $p \geq 1$, are integers, $u \in H^s(\Omega)$ is the solution of problem (2.1), $\{\mathscr{T}_h\}_{h\in(0,\bar{h})}$ is a system of triangulations of the domain Ω satisfying the shape-regularity condition (2.19), and the equivalence condition (2.20) (cf. Lemma 2.5). Moreover, let the penalty constant C_W satisfy the conditions from Corollary 2.41. Let $u_h \in S_{hp}$ be the approximate solution obtained by using of the SIPG, NIPG or IIPG method (2.49c)–(2.49e). Then the error $e_h = u_h - u$ satisfies the estimate*

$$|||e_h||| \leq C_1 h^{\mu-1}|u|_{H^\mu(\Omega)}, \quad h \in (0, \bar{h}), \tag{2.148}$$

where $\mu = \min(p+1, s)$ and C_1 is a constant independent of h and u. Hence, if $s \geq p+1$, we get the error estimate

$$|||e_h||| \leq C_1 h^p |u|_{H^{p+1}(\Omega)}.$$

Proof It is enough to use the abstract error estimate (2.142), where the expressions $|\eta|_{H^1(K)}$, $|\eta|_{H^2(K)}$ and $\|\eta\|_{L^2(K)}$, $K \in \mathscr{T}_h$, are estimated on the basis of the approximation properties (2.93)–(2.95), rewritten for $\eta|_K = (\Pi_{hp}u - u)|_K = \pi_{K,p}(u|_K) - u|_K$ and $K \in \mathscr{T}_h$:

$$\begin{aligned} \|\eta\|_{L^2(K)} &\leq C_A h_K^\mu |u|_{H^\mu(K)}, \\ |\eta|_{H^1(K)} &\leq C_A h_K^{\mu-1} |u|_{H^\mu(K)}, \\ |\eta|_{H^2(K)} &\leq C_A h_K^{\mu-2} |u|_{H^\mu(K)}. \end{aligned} \tag{2.149}$$

Thus, the inequality $h_K \leq h$ and the relation $\sum_{K\in\mathscr{T}_h} |u|^2_{H^\mu(K)} = |u|^2_{H^\mu(\Omega)}$ imply

$$\begin{aligned} R_a(\eta) &= \left(\sum_{K\in\mathscr{T}_h} \left(|\eta|^2_{H^1(K)} + h_K^2 |\eta|^2_{H^2(K)} + h_K^{-2} \|\eta\|_{L^2(K)} \right) \right)^{1/2} \\ &\leq \sqrt{3} C_A h^{\mu-1} |u|_{H^\mu(\Omega)}, \end{aligned} \tag{2.150}$$

which together with (2.142) gives (2.148) with the constant $C_1 = \sqrt{3} C_{\mathrm{AE}} C_A$. □

In order to derive an error estimate in the $L^2(\Omega)$-norm we present the following result.

Lemma 2.45 (Broken Poincaré inequality) *Let the system $\{\mathscr{T}_h\}_{h\in(0,\bar{h})}$ of triangulations satisfy the shape-regularity assumption (2.19). Then there exists a constant $C > 0$ independent of h and v_h such that*

$$\|v_h\|^2_{L^2(\Omega)} \le C\left(\sum_{K\in\mathscr{T}_h} |v_h|^2_{H^1(K)} + \sum_{\Gamma\in\mathscr{F}_h^{ID}} \frac{1}{\mathrm{diam}(\Gamma)}\|[v_h]\|^2_{L^2(\Gamma)}\right) \tag{2.151}$$
$$\forall v_h \in S_{hp}\ \forall h \in (0,\bar{h}).$$

The proof of the broken Poincaré inequality (2.151) was carried out in [7] in the case where Ω is a convex polygonal domain, $\partial\Omega_D = \partial\Omega$ and the assumption (MA2) in Sect. 2.3.2 is satisfied. The proof of inequality (2.151) in a general case with the nonempty Neumann part of the boundary can be found in [36].

From Theorem 2.44 and (2.151) we obtain the following result.

Corollary 2.46 ($L^2(\Omega)$-(suboptimal) error estimate) *Let the assumptions of Theorem 2.44 be satisfied. Then*

$$\|e_h\|_{L^2(\Omega)} \le C_2 h^{\mu-1}|u|_{H^\mu(\Omega)}, \quad h \in (0,\bar{h}), \tag{2.152}$$

where C_2 is a constant independent of h. Hence, if $s \ge p+1$, we get the error estimate

$$\|e_h\|_{L^2(\Omega)} \le C_2 h^p |u|_{H^{p+1}(\Omega)}. \tag{2.153}$$

Remark 2.47 The error estimate (2.153), which is of order $O(h^p)$, is suboptimal with respect to the approximation property (2.97) with $q = 0$, $\mu = p+1 \le s$ of the space S_{hp} giving the order $O(h^{p+1})$. In the next section we prove an optimal error estimate in the $L^2(\Omega)$-norm for SIPG method using the Aubin–Nitsche technique.

2.7.2 Optimal $L^2(\Omega)$-Error Estimate

Our further aim is to derive the optimal error estimate in the $L^2(\Omega)$-norm. It will be based on the *duality technique* sometimes called the *Aubin–Nitsche trick*. Since this approach requires the symmetry of the corresponding bilinear form and the regularity of the exact solution to the dual problem, we consider the SIPG method applied to problem (2.1) with $\partial\Omega_D = \partial\Omega$ and $\partial\Omega_N = \emptyset$. This means that we seek u satisfying

$$-\Delta u = f \quad \text{in } \Omega, \tag{2.154a}$$
$$u = u_D \quad \text{on } \partial\Omega. \tag{2.154b}$$

Moreover, for an arbitrary $z \in L^2(\Omega)$, we consider the *dual problem*: Given $z \in L^2(\Omega)$, find ψ such that

$$-\Delta\psi = z \quad \text{in } \Omega, \quad \psi = 0 \quad \text{on } \partial\Omega. \tag{2.155}$$

Under the notation

$$V = H_0^1(\Omega) = \left\{v \in H^1(\Omega);\ v = 0 \text{ on } \partial\Omega\right\}, \tag{2.156}$$

the weak formulation of (2.155) reads: Find $\psi \in V$ such that

$$\int_\Omega \nabla\psi \cdot \nabla v \,\mathrm{d}x = \int_\Omega zv \,\mathrm{d}x = (z, v)_{L^2(\Omega)} \quad \forall\, v \in V. \tag{2.157}$$

Let us assume that $\psi \in H^2(\Omega)$ and that there exists a constant $C_D > 0$, independent of z, such that

$$\|\psi\|_{H^2(\Omega)} \le C_D \|z\|_{L^2(\Omega)}. \tag{2.158}$$

This is true provided the polygonal (polyhedral) domain Ω is convex, as follows from [153]. (See Remark 2.50.) Let us note that $H^2(\Omega) \subset C(\overline{\Omega})$, if $d \le 3$.

Let A_h be the symmetric bilinear form given by (2.47c), i.e.,

$$A_h(u, v) = a_h^{\mathrm{s}}(u, v) + J_h^\sigma(u, v), \qquad u, v \in H^2(\Omega, \mathscr{T}_h), \tag{2.159}$$

where a_h^{s} and J_h^σ are defined by (2.45a) and (2.105), respectively.

First, we prove the following auxiliary result.

Lemma 2.48 *Let $\psi \in H^2(\Omega)$ be the solution of problem (2.155). Then*

$$A_h(\psi, v) = (v, z)_{L^2(\Omega)} \quad \forall\, v \in H^2(\Omega, \mathscr{T}_h). \tag{2.160}$$

Proof The function $\psi \in H^2(\Omega)$ satisfies the conditions

$$[\psi]_\Gamma = 0 \quad \forall\, \Gamma \in \mathscr{F}_h^I, \qquad \psi|_{\partial\Omega} = 0. \tag{2.161}$$

Let $v \in H^2(\Omega, \mathscr{T}_h)$. Using (2.155), (2.161) and Green's theorem, we obtain

$$\begin{aligned}
(v, z)_{L^2(\Omega)} &= \int_\Omega zv \,\mathrm{d}x = -\int_\Omega \Delta\psi v \,\mathrm{d}x \\
&= \sum_{K \in \mathscr{T}_h} \int_K \nabla\psi \cdot \nabla v \,\mathrm{d}x - \sum_{K \in \mathscr{T}_h} \int_{\partial K} \nabla\psi \cdot \boldsymbol{n}\, v \,\mathrm{d}S \\
&= \sum_{K \in \mathscr{T}_h} \int_K \nabla\psi \cdot \nabla v \,\mathrm{d}x
\end{aligned}$$

$$
\begin{aligned}
&-\left(\sum_{\Gamma\in\mathscr{F}_h^I}\int_\Gamma \langle\nabla\psi\rangle\cdot\boldsymbol{n}\,[v]\,\mathrm{d}S+\sum_{\Gamma\in\mathscr{F}_h^I}\int_\Gamma \langle\nabla v\rangle\cdot\boldsymbol{n}\,[\psi]\,\mathrm{d}S\right)\\
&-\left(\sum_{\Gamma\in\mathscr{F}_h^B}\int_\Gamma \nabla\psi\cdot\boldsymbol{n}\,v\,\mathrm{d}S+\sum_{\Gamma\in\mathscr{F}_h^B}\int_\Gamma \nabla v\cdot\boldsymbol{n}\,\psi\,\mathrm{d}S\right)\\
&+\left(\sum_{\Gamma\in\mathscr{F}_h^I}\int_\Gamma \sigma\,[\psi]\,[v]\,\mathrm{d}S+\sum_{\Gamma\in\mathscr{F}_h^B}\int_\Gamma \sigma\,\psi\,v\,\mathrm{d}S\right).
\end{aligned}
$$

Hence, in view of the definition of the form A_h, we have (2.160). □

Theorem 2.49 ($L^2(\Omega)$-optimal error estimate) *Let us assume that $s \geq 2$, $p \geq 1$, are integers, Ω is a bounded convex polyhedral domain, $u \in H^s(\Omega)$ is the solution of problem (2.1), $\{\mathscr{T}_h\}_{h\in(0,\bar{h})}$ is a system of triangulations of the domain Ω satisfying the shape-regularity condition (2.19), and the equivalence condition (2.20) (cf. Lemma 2.5). Moreover, let the penalty constant C_W satisfy the condition from Corollary 2.41. Let $u_h \in S_{hp}$ be the approximate solution obtained using the SIPG method (2.49c) (i.e., $\Theta = 1$ and the form $A_h = A_h^{\sigma,s}$ is given by (2.45a) and (2.47c). Then*

$$
\|e_h\|_{L^2(\Omega)} \leq C_3 h^{\mu} |u|_{H^{\mu}(\Omega)}, \tag{2.162}
$$

where $e_h = u_h - u$, $\mu = \min\{p+1, s\}$ and C_3 is a constant independent of h and u.

Proof Let $\psi \in H^2(\Omega)$ be the solution of the dual problem (2.157) with $z := e_h = u_h - u \in L^2(\Omega)$ and let $\Pi_{h1}\psi \in S_{h1}$ be the approximation of ψ defined by (2.90) with $p = 1$. By (2.160), we have

$$
A_h(\psi, v) = (e_h, v)_{L^2(\Omega)} \qquad \forall v \in H^2(\Omega, \mathscr{T}_h). \tag{2.163}
$$

The symmetry of the form A_h, the Galerkin orthogonality (2.57) of the error and (2.163) with $v := e_h$ yield

$$
\begin{aligned}
\|e_h\|^2_{L^2(\Omega)} &= A_h(\psi, e_h) = A_h(e_h, \psi) \\
&= A_h(e_h, \psi - \Pi_{h1}\psi).
\end{aligned} \tag{2.164}
$$

Moreover, from (2.122), it follows that

$$
A_h(e_h, \psi - \Pi_{h1}\psi) \leq 2\|e_h\|_{1,\sigma}\, \|\psi - \Pi_{h1}\psi\|_{1,\sigma}, \tag{2.165}
$$

where, by (2.112),

$$\|v\|_{1,\sigma}^2 = |||v|||^2 + \sum_{\Gamma \in \mathscr{F}_h^{ID}} \int_\Gamma \sigma^{-1} (\boldsymbol{n} \cdot \langle \nabla v \rangle)^2 \, \mathrm{d}S. \tag{2.166}$$

By (2.125) and (2.150) (with $\mu = 2$), we have

$$\|\psi - \Pi_{h1}\psi\|_{1,\sigma} \le C_\sigma \, R_a(\psi - \Pi_{h1}\psi) \le \sqrt{3} C_\sigma C_A h |\psi|_{H^2(\Omega)}. \tag{2.167}$$

Now, the inverse inequality (2.86) and estimates (2.100), (2.99) imply that

$$\begin{aligned} |\nabla e_h|_{H^1(K)} &= |\nabla(u - u_h)|_{H^1(K)} \\ &\le |\nabla(u - \Pi_{hp}u)|_{H^1(K)} + |\nabla(\Pi_{hp}u - u_h)|_{H^1(K)} \\ &\le |u - \Pi_{hp}u|_{H^2(K)} + C_I h_K^{-1} \|\nabla(\Pi_{hp}u - u_h)\|_{L^2(K)} \\ &\le C_A h_K^{\mu-2} |u|_{H^\mu(K)} + C_I h_K^{-1} \left(\|\nabla(\Pi_{hp}u - u)\|_{L^2(K)} + \|\nabla(u - u_h)\|_{L^2(K)} \right) \\ &\le C_A(1 + C_I) h_K^{\mu-2} |u|_{H^\mu(K)} + C_I h_K^{-1} \|\nabla e_h\|_{L^2(K)}. \end{aligned} \tag{2.168}$$

By (2.123), (2.168) and the discrete Cauchy inequality,

$$\begin{aligned} &\sum_{\Gamma \in \mathscr{F}_h^{ID}} \int_\Gamma \sigma^{-1} (\boldsymbol{n} \cdot \langle \nabla e_h \rangle)^2 \, \mathrm{d}S \\ &\quad \le \frac{C_G C_M}{C_W} \sum_{K \in \mathscr{T}_h} \left(h_K \|\nabla e_h\|_{L^2(K)} |\nabla e_h|_{H^1(K)} + \|\nabla e_h\|_{L^2(K)}^2 \right) \\ &\quad \le \frac{C_G C_M}{C_W} \left\{ C_A(1 + C_I) h^{\mu-1} |e_h|_{H^1(\Omega, \mathscr{T}_h)} |u|_{H^\mu(\Omega)} + (1 + C_I) |e_h|_{H^1(\Omega, \mathscr{T}_h)}^2 \right\}. \end{aligned} \tag{2.169}$$

Since $|e_h|_{H^1(\Omega, \mathscr{T}_h)} \le |||e_h|||$, using (2.148) and (2.169), we have

$$\sum_{\Gamma \in \mathscr{F}_h^{ID}} \int_\Gamma \sigma^{-1} (\boldsymbol{n} \cdot \langle \nabla e_h \rangle)^2 \, \mathrm{d}S \le \frac{C_G C_M}{C_W} C_1 (1 + C_I)(C_1 + C_A) h^{2(\mu-1)} |u|_{H^\mu(\Omega)}^2.$$

Thus, (2.148) and (2.166) yield the estimate

$$\|e_h\|_{1,\sigma}^2 \le C_5 h^{2(\mu-1)} |u|_{H^\mu(\Omega)}^2 \tag{2.170}$$

with $C_5 = C_1 \left\{ 1 + C_G C_M C_W^{-1} (1 + C_I)(C_1 + C_A) \right\}$. It follows from (2.165), (2.167), and (2.170) that

$$A_h(e_h, \psi - \Pi_{h1}\psi) \le C_6 h^\mu |\psi|_{H^2(\Omega)} \, |u|_{H^\mu(\Omega)}, \tag{2.171}$$

where $C_6 = 2\sqrt{3}C_\sigma C_A \sqrt{C_5}$.

Finally, by (2.164), (2.171), and (2.158) with $z = e_h$,

$$\|e_h\|_{L^2(\Omega)}^2 \le C_D C_6 h^\mu |u|_{H^\mu(\Omega)} \|e_h\|_{L^2(\Omega)}, \tag{2.172}$$

which already implies estimate (2.162) with $C_3 = C_D C_6$. □

Remark 2.50 As we see from the above results, if the exact solution $u \in H^{p+1}(\Omega)$ and the finite elements of degree p are used, the error is of the optimal order $O(h^{p+1})$ in the $L^2(\Omega)$-norm. In the case, when the polygonal domain is not convex and/or the Neumann and Dirichlet parts of the boundary $\Omega_N \neq \emptyset$ and $\Omega_D \neq \emptyset$, the exact solution ψ of the dual problem (2.155) is not an element of the space $H^2(\Omega)$. Then it is necessary to work in the Sobolev–Slobodetskii spaces of functions with *noninteger derivatives* and the error in the $L^2(\Omega)$-norm is not of the optimal order $O(h^{p+1})$. The analysis of error estimates for the DG discretization of boundary value problems with boundary singularities is the subject of works [137, 284], where optimal error estimates were obtained with the aid of a suitable graded mesh refinement. The main tools are here the Sobolev–Slobodetskii spaces and weighted Sobolev spaces. For the definitions and properties of these spaces, see [37, 209].

Remark 2.51 In [240] the Neumann problem (i.e., $\partial\Omega = \partial\Omega_N$) was solved by the NIPG approach, where the penalty coefficient σ was chosen in the form

$$\sigma|_\Gamma = \frac{C_W}{h_\Gamma^\beta}, \qquad \Gamma \in \mathscr{F}_h, \tag{2.173}$$

instead of (2.104), where $\beta \ge 1/2$. If triangular grids do not contain any hanging nodes (i.e., the triangulations $\mathscr{T}_h$ are conforming), then an optimal error estimate in the $L^2(\Omega)$-norm of this analogue of the NIPG method was proven provided that $\beta \ge 3$ for $d = 2$ and $\beta \ge 3/2$ for $d = 3$. In this case the interior penalty is so strong that the DG methods behave like the standard conforming (i.e., continuous) finite element schemes. On the other hand, the stronger penalty causes worse computational properties of the corresponding algebraic system, see [41].

2.8 Baumann–Oden Method

In this section we analyze the Baumann–Oden scheme (2.49b). Hence, we seek $u_h \in S_{hp}$ such that

$$A_h(u_h, v_h) = \ell_h(v_h) \qquad \forall v_h \in S_{hp}, \tag{2.174}$$

where $A_h(\cdot,\cdot)$ and ℓ_h are given by (2.47b) and (2.48b), respectively:

$$A_h(u, v) = \sum_{K \in \mathscr{T}_h} \int_K \nabla u \cdot \nabla v \, \mathrm{d}x - \sum_{\Gamma \in \mathscr{F}_h^{ID}} \int_\Gamma (\boldsymbol{n} \cdot \langle \nabla u \rangle [v] - \boldsymbol{n} \cdot \langle \nabla v \rangle [u]) \, \mathrm{d}S, \tag{2.175}$$

$$\ell_h(v) = \int_\Omega f v \, \mathrm{d}x + \sum_{\Gamma \in \mathscr{F}_h^N} \int_\Gamma g_N v \, \mathrm{d}S + \sum_{\Gamma \in \mathscr{F}_h^D} \int_\Gamma (\boldsymbol{n} \cdot \nabla v) u_D \, \mathrm{d}S.$$

Obviously, (2.175) gives

$$A_h(v, v) \geq |v|^2_{H^1(\Omega, \mathscr{T}_h)} \quad \forall v \in H^2(\Omega, \mathscr{T}_h), \tag{2.176}$$

where only a seminorm stands on the right-hand side. We speak about a *weak coercivity*. (The above inequality is valid with the sign = of course.) Therefore, it is possible to derive error estimates in a seminorm only.

This method was presented and analyzed for one-dimensional diffusion problem in [12]. In [239], Rivière, Wheeler, and Girault showed how to obtain error estimates under the assumption that the polynomial degree $p > 2$ and the mesh is conforming. The analysis carried out in [239, Lemma 5.1] is based on the existence of an interpolation operator $I_{hp} : H^2(\Omega, \mathscr{T}_h) \to S_{hp}$ for $p \geq 2$ such that

$$\int_\Gamma \langle \nabla(v - I_{hp}v) \rangle \cdot \boldsymbol{n} \, \mathrm{d}S = 0 \qquad \forall \Gamma \in \mathscr{F}_h, \ v \in H^2(\Omega, \mathscr{T}_h), \tag{2.177}$$

$$\left| I_{hp}v - v \right|_{H^q(\Omega, \mathscr{T}_h)} \leq \bar{C}_A h^{\mu - q} |v|_{H^\mu(\Omega, \mathscr{T}_h)}, \quad v \in H^s(\Omega, \mathscr{T}_h), \ h \in (0, \bar{h}), \tag{2.178}$$

where $\mu = \min(p + 1, s)$, $s \geq 2$, $q = 0, 1, 2$ and $\bar{C}_A$ is a constant.

In the following, we present the error estimate for the Baumann–Oden method. The proof differs from the technique in [239].

Theorem 2.52 *Let $u \in H^s(\Omega)$ with $s \geq 2$ be the exact solution of problem (2.1). Let the system of triangulations $\{\mathscr{T}_h\}_{h \in (0, \bar{h})}$ satisfy the shape-regularity assumption (2.19) and the conformity assumption (MA4) from Sect. 2.3.2, and let $u_h \in S_{hp}$, $p \geq 2$, be the approximate solution given by (2.174). Then there exists a constant $C_{\mathrm{BO}} > 0$ independent of $h \in (0, \bar{h})$ and u, such that*

$$|u - u_h|_{H^1(\Omega, \mathscr{T}_h)} \leq C_{\mathrm{BO}} h^{\mu - 1} |u|_{H^\mu(\Omega)}. \tag{2.179}$$

Proof Let I_{hp} be the interpolation operator satisfying (2.177) and (2.178). We put $\eta = I_{hp}u - u \in H^1(\Omega, \mathscr{T}_h)$ and $\xi = u_h - I_{hp}u \in S_{hp}$. Then $e_h = u_h - u = \eta + \xi$. From the definition (2.175) of the form A_h and the Galerkin orthogonality (2.57), we have

$$|\xi|^2_{H^1(\Omega,\mathscr{T}_h)} = |I_{hp}u - u_h|^2_{H^1(\Omega,\mathscr{T}_h)} = A_h(I_{hp}u - u_h, I_{hp}u - u_h) \tag{2.180}$$
$$= A_h(I_{hp}u - u, I_{hp}u - u_h) = A_h(\eta, \xi).$$

Moreover, in view of (2.175) and (2.177),

$$A_h(I_{hp}u - u, v_h) = A_h(\eta, v_h) = 0 \quad \forall v_h \in S_{h0}, \tag{2.181}$$

where S_{h0} denotes the space of piecewise constant functions on $\mathscr{T}_h$. Hence, if Π_{h0} is the orthogonal projection of $L^2(\Omega)$ onto S_{h0}, then (2.47b), (2.111) and (2.181) imply that

$$|A_h(\eta, \xi)| \le |A_h(\eta, \xi - \Pi_{h0}\xi)| + |A_h(\eta, \Pi_{h0}\xi)|$$
$$\le \|\eta\|_{1,\sigma} \, \|\xi - \Pi_{h0}\xi\|_{1,\sigma}, \tag{2.182}$$

where, by (2.112),

$$\|v\|^2_{1,\sigma} = |||v|||^2 + \sum_{\Gamma \in \mathscr{F}_h^{ID}} \int_\Gamma \sigma^{-1} (\boldsymbol{n} \cdot \langle \nabla v \rangle)^2 \, \mathrm{d}S. \tag{2.183}$$

Since $\Pi_{h0}|_K$ is constant on each $K \in \mathscr{T}_h$, obviously

$$|\xi - \Pi_{h0}\xi|_{H^1(K)} = |\xi|_{H^1(K)}, \quad K \in \mathscr{T}_h. \tag{2.184}$$

Moreover, it follows from the approximation properties (2.90) and (2.93) (with $\mu = 1,\ p = 0$) that

$$\|\xi - \Pi_{h0}\xi\|_{L^2(K)} \le C_A h_K |\xi|_{H^1(K)}, \quad K \in \mathscr{T}_h. \tag{2.185}$$

Let $\psi \in H^1(\Omega, \mathscr{T}_h)$. Then, using the definition (2.105) of the form J_h^σ, the definition (2.104) of the weight σ, inequality (2.108), and the multiplicative trace inequality (2.78), we find that

$$|||\psi|||^2 = |\psi|^2_{H^1(\Omega,\mathscr{T}_h)} + J_h^\sigma(\psi, \psi) \tag{2.186}$$
$$\le |\psi|^2_{H^1(\Omega,\mathscr{T}_h)} + \frac{2C_W}{C_T} \sum_{K \in \mathscr{T}_h} h_K^{-1} \|\psi\|^2_{L^2(\partial K)}$$
$$\le |\psi|^2_{H^1(\Omega,\mathscr{T}_h)} + \frac{2C_W C_M}{C_T} \sum_{K \in \mathscr{T}_h} \left(h_K^{-2} \|\psi\|^2_{L^2(K)} + h_K^{-1} \|\psi\|_{L^2(K)} |\psi|_{H^1(K)} \right).$$

Let us set $\psi = \xi - \Pi_{h0}\xi$ in (2.186). Then, in view of (2.184) and (2.185), we get

$$|||\xi - \Pi_{h0}\xi|||^2 \le (1 + 2(1 + C_A)C_A C_W C_M / C_T) \sum_{K \in \mathscr{T}_h} |\xi|^2_{H^1(K)} \tag{2.187}$$
$$= (1 + 2(1 + C_A)C_A C_W C_M / C_T)|\xi|^2_{H^1(\Omega, \mathscr{T}_h)}.$$

Moreover, using the relation $\nabla \Pi_{h0}\xi = 0$ and (2.124), we have

$$\sum_{\Gamma \in \mathscr{F}_h^{ID}} \int_\Gamma \sigma^{-1}(\boldsymbol{n} \cdot \langle \nabla(\xi - \Pi_{h0}\xi)\rangle)^2 \, \mathrm{d}S \tag{2.188}$$
$$= \sum_{\Gamma \in \mathscr{F}_h^{ID}} \int_\Gamma \sigma^{-1}(\boldsymbol{n} \cdot \langle \nabla\xi\rangle)^2 \, \mathrm{d}S \le (C_G C_M (C_I + 1)/C_W)|\xi|^2_{H^1(\Omega, \mathscr{T}_h)}.$$

Therefore, (2.183), (2.187) and (2.188) imply that

$$\|\xi - \Pi_{h0}\xi\|_{1,\sigma} \le C_7 |\xi|_{H^1(\Omega, \mathscr{T}_h)}, \tag{2.189}$$

where $C_7 = (1 + 2(1 + C_A)C_A C_W C_M / C_T + C_G C_M (C_I + 1)/C_W)^{1/2}$.

On the other hand, if we set $\psi := \eta$ in (2.186), then by (2.178) we obtain

$$|||\eta|||^2 \le \bar{C}_A^2 h^{2(\mu-1)} |u|^2_{H^\mu(\Omega)} + 4\bar{C}_A^2 C_W C_M / C_T h^{2(\mu-1)} |u|^2_{H^\mu(\Omega)} \tag{2.190}$$
$$= \bar{C}_A^2 (1 + 4 C_W C_M / C_T) h^{2(\mu-1)} |u|^2_{H^\mu(\Omega)}.$$

Similarly, inequalities (2.123) and (2.178) give

$$\sum_{\Gamma \in \mathscr{F}_h^{ID}} \int_\Gamma \sigma^{-1}(\boldsymbol{n} \cdot \langle \nabla\eta\rangle)^2 \, \mathrm{d}S \tag{2.191}$$
$$\le \frac{C_G C_M}{C_W} \sum_{K \in \mathscr{T}_h} \left(\|\nabla\eta\|^2_{L^2(K)} + h_K \|\nabla\eta\|_{L^2(K)} |\nabla\eta|_{H^1(K)} \right)$$
$$\le \frac{2 C_G C_M \bar{C}_A^2}{C_W} h^{2(\mu-1)} |u|^2_{H^\mu(\Omega)}.$$

Then (2.183), (2.190) and (2.191) yield

$$\|\eta\|_{1,\sigma} \le C_8 h^{\mu-1} |u|_{H^\mu(\Omega)}, \tag{2.192}$$

where $C_8 = \bar{C}_A((1 + 4C_W C_M / C_T) + 2 C_G C_M / C_W)^{1/2}$.

Further, from (2.180), (2.182), (2.189) and (2.192), we have

$$|\xi|^2_{H^1(\Omega, \mathscr{T}_h)} = |A_h(\eta, \xi)| \le C_7 C_8 h^{\mu-1} |\xi|_{H^1(\Omega, \mathscr{T}_h)} |u|_{H^\mu(\Omega)} \tag{2.193}$$

and thus,

$$|\xi|_{H^1(\Omega,\mathscr{T}_h)} \leq C_7 C_8 h^{\mu-1} |u|_{H^\mu(\Omega)}. \tag{2.194}$$

Finally, the triangle inequality, the definition of η and ξ, (2.178), and (2.194) imply that

$$\begin{aligned} |u - u_h|_{H^1(\Omega,\mathscr{T}_h)} &= |u - I_{hp}u|_{H^1(\Omega,\mathscr{T}_h)} + |I_{hp}u - u_h|_{H^1(\Omega,\mathscr{T}_h)} \\ &\leq (\bar{C}_A + C_7 C_8) h^{\mu-1} |u|_{H^\mu(\Omega)}, \end{aligned} \tag{2.195}$$

which proves the theorem with $C_{\mathrm{BO}} := \bar{C}_A + C_7 C_8$. □

2.9 Numerical Examples

In this section, we demonstrate by numerical experiments the error estimates (2.148), (2.152) and (2.162). In the first example, we assume that the exact solution is sufficiently regular. We show that the use of a higher degree of polynomial approximation increases the rate of convergence of the method. In the second example, the exact solution has a singularity. Then the order of convergence does not increase with the increasing degree of the polynomial approximation used. The computational results are in agreement with theory and show that the accuracy of the method is determined by the degree of the polynomial approximation as well as the regularity of the solution.

2.9.1 Regular Solution

Let us consider the problem of finding a function $u : \Omega = (0, 1) \times (0, 1) \to \mathbb{R}$ such that

$$\begin{aligned} -\Delta u &= 8\pi^2 \sin(2\pi x_1) \sin(2\pi x_2) \quad \text{in } \Omega, \\ u &= 0 \quad \text{on } \partial\Omega. \end{aligned} \tag{2.196}$$

It is easy to verify that the exact solution of (2.196) has the form

$$u = \sin(2\pi x_1) \sin(2\pi x_2), \quad (x_1, x_2) \in \Omega. \tag{2.197}$$

Obviously, $u \in C^\infty(\overline{\Omega})$.

We investigate the *experimental order of convergence* (EOC) of the SIPG, NIPG and IIPG methods defined by (2.49c)–(2.49e). We assume that a (semi)norm $\|e_h\|$ of the computational error behaves according to the formula

$$\|e_h\| = C h^{\mathrm{EOC}}, \tag{2.198}$$

where $C > 0$ is a constant, $h = \max_{K \in \mathscr{T}_h} h_K$, and EOC $\in \mathbb{R}$ is the experimental order of convergence. Since the exact solution is known and therefore $\|e_h\|$ can be exactly evaluated, it is possible to evaluate EOC in the following way. Let $\|e_{h_1}\|$ and $\|e_{h_2}\|$ be computational errors of the numerical solutions obtained on two different meshes $\mathscr{T}_{h_1}$ and $\mathscr{T}_{h_2}$, respectively. Then from (2.198), eliminating the constant C, we obtain

$$\mathrm{EOC} = \frac{\log(\|e_{h_1}\|/\|e_{h_2}\|)}{\log(h_1/h_2)}. \tag{2.199}$$

Moreover, we evaluate the *global experimental order of convergence* (GEOC) from the approximation of (2.198) with the aid of the least squares method, where all computed pairs $[h, e_h]$ are taken into account simultaneously.

We used a set of four uniform triangular grids having 128, 512, 2048, and 8192 elements, shown in Fig. 2.4. The meshes consist of right-angled triangles with the diameter $h = \sqrt{2}/\sqrt{\#\mathscr{T}_h/2}$, where $\#\mathscr{T}_h$ is the number of elements of $\mathscr{T}_h$. EOC is evaluated according to (2.199) for all pairs of "neighbouring" grids. Tables 2.1 and 2.2 show the computational errors in the $L^2(\Omega)$-norm and the $H^1(\Omega, \mathscr{T}_h)$-seminorm and EOC obtained by the SIPG, NIPG and IIPG methods using the P_p, $p = 1, \ldots, 6$, polynomial approximations. These results are also visualized in Fig. 2.5.

Fig. 2.4 Computational grids used for the numerical solution of problems (2.196) and (2.201)

Table 2.1 Computational errors and EOC in the $L^2(\Omega)$-norm for the regular solution of problem (2.196)

		SIPG		NIPG		IIPG	
p	$h/\sqrt{2}$	$\|e_h\|_{L^2(\Omega)}$	EOC	$\|e_h\|_{L^2(\Omega)}$	EOC	$\|e_h\|_{L^2(\Omega)}$	EOC
1	1/8	6.7452E–02	–	2.9602E–02	–	6.3939E–02	–
1	1/16	1.8745E–02	1.85	7.6200E–03	1.96	1.7383E–02	1.88
1	1/32	4.8463E–03	1.95	1.9292E–03	1.98	4.4579E–03	1.96
1	1/64	1.2252E–03	1.98	4.8536E–04	1.99	1.1239E–03	1.99
	GEOC		1.93		1.98		1.95
2	1/8	3.9160E–03	–	1.0200E–02	–	4.7447E–03	–
2	1/16	4.9164E–04	2.99	2.5723E–03	1.99	8.4877E–04	2.48
2	1/32	6.1644E–05	3.00	6.4259E–04	2.00	1.8081E–04	2.23
2	1/64	7.7184E–06	3.00	1.6032E–04	2.00	4.2670E–05	2.08
	GEOC		3.00		2.00		2.26
3	1/8	3.1751E–04	–	5.5550E–04	–	3.2684E–04	–
3	1/16	1.9150E–05	4.05	3.4481E–05	4.01	2.0077E–05	4.02
3	1/32	1.1775E–06	4.02	2.1333E–06	4.01	1.2414E–06	4.02
3	1/64	7.3124E–08	4.01	1.3250E–07	4.01	7.7176E–08	4.01
	GEOC		4.03		4.01		4.02
4	1/8	2.3496E–05	–	3.7990E–05	–	2.7046E–05	–
4	1/16	7.5584E–07	4.96	2.4304E–06	3.97	1.2929E–06	4.39
4	1/32	2.3824E–08	4.99	1.5512E–07	3.97	7.2190E–08	4.16
4	1/64	7.4627E–10	5.00	9.7626E–09	3.99	4.3310E–09	4.06
	GEOC		4.98		3.97		4.20
5	1/8	1.4133E–06	–	2.3017E–06	–	1.6501E–06	–
5	1/16	2.2193E–08	5.99	3.6590E–08	5.98	2.6160E–08	5.98
5	1/32	3.4686E–10	6.00	5.7147E–10	6.00	4.0753E–10	6.00
5	1/64	5.4139E–12	6.00	8.8468E–12	6.01	6.3670E–12	6.00
	GEOC		6.00		6.00		6.00
6	1/8	7.3313E–08	–	1.1239E–07	–	9.5990E–08	–
6	1/16	5.8381E–10	6.97	1.5138E–09	6.21	1.1620E–09	6.37
6	1/32	4.5855E–12	6.99	2.2864E–11	6.05	1.6380E–11	6.15
6	1/64	3.8771E–14	6.89	3.5354E–13	6.02	2.4417E–13	6.07
	GEOC		6.95		6.09		6.19

We observe that EOC of the SIPG technique are in a good agreement with the theoretical ones, i.e., $O(h^{p+1})$ in the $L^2(\Omega)$-norm (estimate (2.162)) and $O(h^p)$ in the $H^1(\Omega, \mathscr{T}_h)$-seminorm (estimate (2.148)). On the other hand, the experimental order of convergence of the NIPG and IIPG techniques measured in the $L^2(\Omega)$-norm is better than the theoretical estimate (2.152). We deduce that

Table 2.2 Computational errors and EOC in the $H^1(\Omega, \mathscr{T}_h)$-seminorm for the regular solution of problem (2.196)

		SIPG		NIPG		IIPG	
p	$h/\sqrt{2}$	$\|e_h\|_{H^1(\Omega,\mathscr{T}_h)}$	EOC	$\|e_h\|_{H^1(\Omega,\mathscr{T}_h)}$	EOC	$\|e_h\|_{H^1(\Omega,\mathscr{T}_h)}$	EOC
1	1/8	1.5018E+00	–	1.2423E+00	–	1.4946E+00	–
1	1/16	7.7679E–01	0.95	6.4615E–01	0.94	7.7519E–01	0.95
1	1/32	3.9214E–01	0.99	3.2741E–01	0.98	3.9181E–01	0.98
1	1/64	1.9666E–01	1.00	1.6450E–01	0.99	1.9658E–01	1.00
	GEOC		0.98		0.97		0.98
2	1/8	2.4259E–01	–	1.9985E–01	–	2.1634E–01	–
2	1/16	6.2760E–02	1.95	5.0217E–02	1.99	5.5693E–02	1.96
2	1/32	1.5849E–02	1.99	1.2536E–02	2.00	1.4053E–02	1.99
2	1/64	3.9743E–03	2.00	3.1305E–03	2.00	3.5244E–03	2.00
	GEOC		1.98		2.00		1.98
3	1/8	2.5610E–02	–	2.4029E–02	–	2.3425E–02	–
3	1/16	3.2202E–03	2.99	3.0531E–03	2.98	2.9699E–03	2.98
3	1/32	4.0238E–04	3.00	3.8298E–04	2.99	3.7253E–04	3.00
3	1/64	5.0260E–05	3.00	4.7890E–05	3.00	4.6607E–05	3.00
	GEOC		3.00		2.99		2.99
4	1/8	2.2049E–03	–	2.2096E–03	–	2.0645E–03	–
4	1/16	1.4023E–04	3.97	1.3801E–04	4.00	1.3039E–04	3.98
4	1/32	8.8035E–06	3.99	8.5962E–06	4.00	8.1650E–06	4.00
4	1/64	5.5077E–07	4.00	5.3601E–07	4.00	5.1038E–07	4.00
	GEOC		3.99		4.00		3.99
5	1/8	1.5680E–04	–	1.6457E–04	–	1.5090E–04	–
5	1/16	4.9305E–06	4.99	5.1666E–06	4.99	4.7527E–06	4.99
5	1/32	1.5413E–07	5.00	1.6126E–07	5.00	1.4865E–07	5.00
5	1/64	4.8146E–09	5.00	5.0316E–09	5.00	4.6439E–09	5.00
	GEOC		5.00		5.00		5.00
6	1/8	9.5245E–06	–	1.0198E–05	–	9.3719E–06	–
6	1/16	1.5092E–07	5.98	1.5951E–07	6.00	1.4762E–07	5.99
6	1/32	2.3666E–09	5.99	2.4862E–09	6.00	2.3083E–09	6.00
6	1/64	3.7008E–11	6.00	3.8770E–11	6.00	3.6051E–11	6.00
	GEOC		5.99		6.00		6.00

$$\|e_h\|_{L^2(\Omega)} = O(h^{\bar{p}}), \quad \bar{p} = \begin{cases} p+1 & \text{for } p \text{ odd}, \\ p & \text{for } p \text{ even}. \end{cases} \tag{2.200}$$

This interesting property of the NIPG and IIPG techniques was observed by many authors (cf. [183, 230]), but up to now a theoretical justification has been missing, see Sect. 2.9.3 for some comments. The EOC in the $H^1(\Omega, \mathscr{T}_h)$-seminorm of NIPG and IIPG methods is in agreement with (2.148).

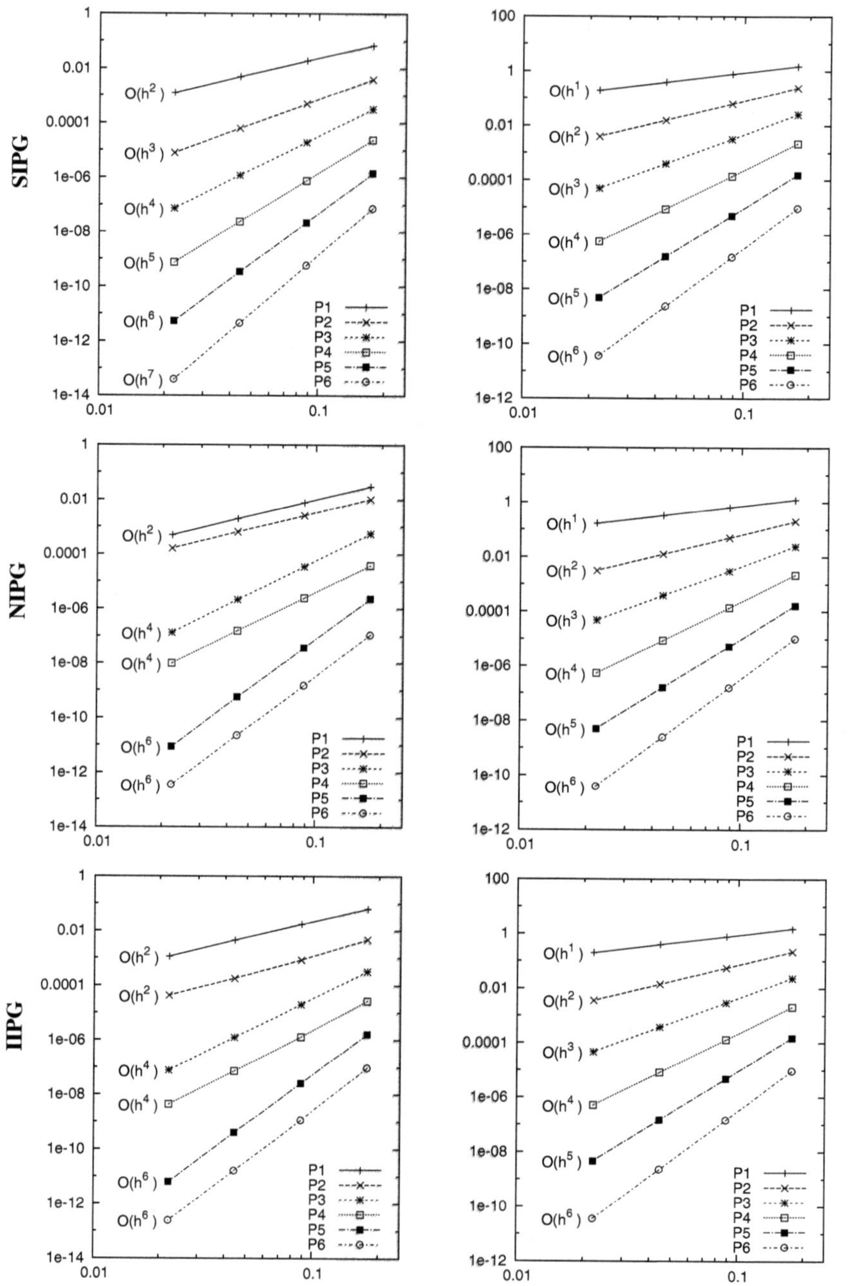

Fig. 2.5 Computational errors and EOC in the $L^2(\Omega)$-norm (*left*) and in the $H^1(\Omega, \mathscr{T}_h)$-seminorm (*right*) for the regular solution of problem (2.196)

2.9.2 Singular Case

In the domain $\Omega = (0, 1) \times (0, 1)$ we consider the Poisson problem

$$
\begin{aligned}
-\Delta u &= g \quad \text{in } \Omega, \\
u &= 0 \quad \text{on } \partial\Omega,
\end{aligned}
\tag{2.201}
$$

with the right-hand side g chosen in such a way that the exact solution has the form

$$
u(x_1, x_2) = 2r^{\alpha} x_1 x_2 (1 - x_1)(1 - x_2) = r^{\alpha+2} \sin(2\varphi)(1 - x_1)(1 - x_2), \tag{2.202}
$$

where r, φ are the polar coordinates ($r = (x_1^2 + x_2^2)^{1/2}$) and $\alpha \in \mathbb{R}$ is a constant. The function u is equal to zero on $\partial\Omega$ and its regularity depends on the value of α. Namely, by [15],

$$
u \in H^{\beta}(\Omega) \quad \forall \beta \in (0, \alpha + 3), \tag{2.203}
$$

where $H^{\beta}(\Omega)$ denotes the Sobolev–Slobodetskii space of functions with *noninteger derivatives*.

We present numerical results obtained for $\alpha = -3/2$ and $\alpha = 1/2$. If $\alpha = -3/2$, then $u \in H^{\beta}(\Omega)$ for all $\beta \in (0, 3/2)$, whereas for the value $\alpha = 1/2$, we have $u \in H^{\beta}(\Omega)$ for all $\beta \in (0, 7/2)$. Figure 2.6 shows the function u for both values of α.

We carried out computations on 4 triangular grids introduced in Sect. 2.9.1 by the SIPG, NIPG and IIPG technique with the aid of P_p, $p = 1, \ldots, 6$, polynomial approximations. Tables 2.3, 2.4 and Tables 2.5, 2.6 show the computational errors in the $L^2(\Omega)$-norm as well as the $H^1(\Omega, \mathscr{T}_h)$-seminorm, and the corresponding experimental orders of convergence for $\alpha = 1/2$ and $\alpha = -3/2$, respectively. These values are visualized in Figs. 2.7 and 2.8 in which the achieved experimental order of convergence is easy to observe.

These results lead us to the proposition that for the SIPG method the error behaves like

$$
\begin{aligned}
\|u - u_h\|_{L^2(\Omega)} &= O(h^{\mu}), \qquad u \in H^{\beta}(\Omega) \\
|u - u_h|_{H^1(\Omega)} &= O(h^{\mu-1}), \qquad u \in H^{\beta}(\Omega),
\end{aligned}
\tag{2.204}
$$

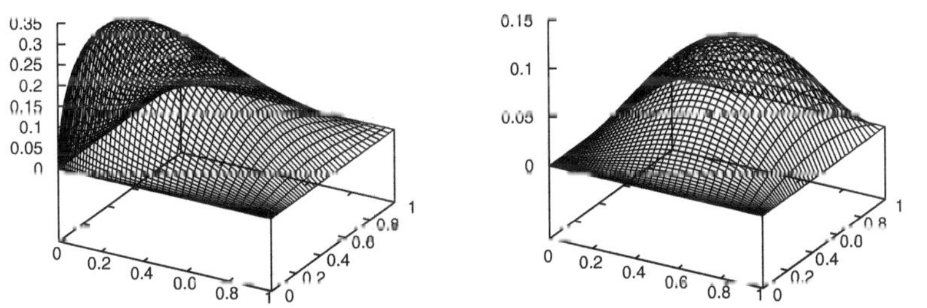

Fig. 2.6 Exact solution (2.202) for $\alpha = -3/2$ (*left*) and $\alpha = 1/2$ (*right*)

where $\mu = \min(p+1, \beta)$, and for the IIPG and NIPG methods like

$$\|u - u_h\|_{L^2(\Omega)} = O(h^{\bar{\mu}}), \qquad u \in H^\beta(\Omega) \tag{2.205}$$
$$|u - u_h|_{H^1(\Omega)} = O(h^{\mu-1}), \qquad u \in H^\beta(\Omega),$$

where $\mu = \min(p+1, \beta)$, $\bar{\mu} = \min(\bar{p}, \beta)$, and $\bar{p}$ is given by (2.200). The statements (2.204) and (2.205) are in agreement with numerical experiments (not presented here) carried out by other authors for additional values of α.

Moreover, the experimental order of convergence of the SIPG technique given by (2.204) corresponds to the result in [121], where for any $\beta \in (1, 3/2)$ we get

$$\|v - I_h v\|_{L^2(\Omega)} \le C(\beta) h^\mu \|v\|_{H^\beta(\Omega)}, \quad v \in H^\beta(\Omega), \tag{2.206}$$
$$|v - I_h v|_{H^1(\Omega)} \le C(\beta) h^{\mu-1} \|v\|_{H^\beta(\Omega)}, \quad v \in H^\beta(\Omega),$$

where $I_h v$ is a piecewise polynomial Lagrange interpolation to v of degree $\le p$, $\mu = \min(p+1, \beta)$ and $C(\beta)$ is a constant independent of h and v. By [13, Sect. 3.3] and the references therein, where the interpolation in the so-called Besov spaces is used, the precise error estimate of order $O(h^{3/2})$ in the $L^2(\Omega)$-norm and $O(h^{1/2})$ in the $H^1(\Omega, \mathscr{T}_h)$-seminorm can be established, which corresponds to our numerical experiments.

Finally, the experimental order of convergence of the NIPG and IIPG techniques given by (2.205) corresponds to (2.206) and results (2.200).

2.9.3 A Note on the $L^2(\Omega)$-Optimality of NIPG and IIPG

Numerical experiments from Sect. 2.9.1 lead us to the observation (2.200), which was presented, e.g., in [12, 238] and the references cited therein. The optimal order of convergence for the odd degrees of approximation was theoretically justified in [211], where NIPG and IIPG methods were analyzed for uniform partitions of the one-dimensional domain. See also [50], where similar results were obtained.

On the other hand, several examples of 1D special non-uniform (but quasi-uniform) meshes were presented in [157], where the NIPG method gives the error in the $L^2(\Omega)$-norm of order $O(h^p)$ even for odd p. A suboptimal EOC can also be obtained for the IIPG method using these meshes, see [238, Sect. 1.5, Table 1.2].

In [101], it was shown that the use of odd degrees of polynomial approximation of IIPG method leads to the optimal order of convergence in the $L^2(\Omega)$-norm on 1D quasi-uniform grids if and only if the penalty parameter (of order $O(h^{-1})$) is chosen in a special way. These results lead us to the hypothesis that the observation (2.200) is not valid in general.

However, extending theoretical results either to NIPG method or to higher dimensions is problematic. Some attempt was presented in [82], where the optimal order of convergence in the $L^2(\Omega)$-norm on equilateral triangular grids was proved for the IIPG method with reduced interior and boundary penalties.

Table 2.3 Computational errors and EOC in the $L^2(\Omega)$-norm for the solution of problem (2.201) with $\alpha = 1/2$

		SIPG		NIPG		IIPG	
p	$h/\sqrt{2}$	$\|e_h\|_{L^2(\Omega)}$	EOC	$\|e_h\|_{L^2(\Omega)}$	EOC	$\|e_h\|_{L^2(\Omega)}$	EOC
1	1/8	2.1789E–03	–	8.1338E–04	–	1.8698E–03	–
1	1/16	5.7581E–04	1.92	2.1069E–04	1.95	4.8403E–04	1.95
1	1/32	1.4740E–04	1.97	5.3806E–05	1.97	1.2267E–04	1.98
1	1/64	3.7248E–05	1.98	1.3609E–05	1.98	3.0848E–05	1.99
	GEOC		1.96		1.97		1.97
2	1/8	5.7796E–05	–	1.0098E–04	–	5.9762E–05	–
2	1/16	7.2545E–06	2.99	2.6758E–05	1.92	1.1004E–05	2.44
2	1/32	9.1150E–07	2.99	6.9525E–06	1.94	2.4341E–06	2.18
2	1/64	1.1434E–07	2.99	1.7734E–06	1.97	5.8760E–07	2.05
	GEOC		2.99		1.94		2.22
3	1/8	2.6233E–06	–	4.0597E–06	–	2.7474E–06	–
3	1/16	1.9366E–07	3.76	3.3583E–07	3.60	2.1985E–07	3.64
3	1/32	1.4898E–08	3.70	2.8012E–08	3.58	1.7889E–08	3.62
3	1/64	1.1930E–09	3.64	2.3717E–09	3.56	1.4838E–09	3.59
	GEOC		3.70		3.58		3.62
4	1/8	2.6498E–07	–	4.1937E–07	–	3.0663E–07	–
4	1/16	2.1097E–08	3.65	3.4292E–08	3.61	2.4522E–08	3.64
4	1/32	1.7819E–09	3.57	2.8705E–09	3.58	2.0460E–09	3.58
4	1/64	1.5429E–10	3.53	2.4482E–10	3.55	1.7516E–10	3.55
	GEOC		3.58		3.58		3.59
5	1/8	5.8491E–08	–	9.3494E–08	–	7.2011E–08	–
5	1/16	4.9611E–09	3.56	8.1022E–09	3.53	6.1832E–09	3.54
5	1/32	4.2999E–10	3.53	7.0989E–10	3.51	5.3944E–10	3.52
5	1/64	3.7656E–11	3.51	6.2465E–11	3.51	4.7387E–11	3.51
	GEOC		3.53		3.52		3.52
6	1/8	1.9318E–08	–	2.9767E–08	–	2.6495E–08	–
6	1/16	1.6677E–09	3.53	2.6000E–09	3.52	2.3079E–09	3.52
6	1/32	1.4570E–10	3.52	2.2856E–10	3.51	2.0259E–10	3.51
6	1/64	1.2809E–11	3.51	2.0149E–11	3.50	1.7847E–11	3.50
	GEOC		3.52		3.51		3.51

Table 2.4 Computational errors and EOC in the $H^1(\Omega, \mathscr{T}_h)$-seminorm for the solution of problem (2.201) with $\alpha = 1/2$

		SIPG		NIPG		IIPG	
p	$h/\sqrt{2}$	$\|e_h\|_{H^1(\Omega,\mathscr{T}_h)}$	EOC	$\|e_h\|_{H^1(\Omega,\mathscr{T}_h)}$	EOC	$\|e_h\|_{H^1(\Omega,\mathscr{T}_h)}$	EOC
1	1/8	5.0805E–02	–	4.2283E–02	–	5.0531E–02	–
1	1/16	2.5722E–02	0.98	2.1564E–02	0.97	2.5653E–02	0.98
1	1/32	1.2919E–02	0.99	1.0877E–02	0.99	1.2902E–02	0.99
1	1/64	6.4715E–03	1.00	5.4607E–03	0.99	6.4674E–03	1.00
	GEOC		0.99		0.98		0.99
2	1/8	4.0313E–03	–	3.2281E–03	–	3.5738E–03	–
2	1/16	1.0230E–03	1.98	8.0878E–04	2.00	9.0960E–04	1.97
2	1/32	2.5750E–04	1.99	2.0223E–04	2.00	2.2938E–04	1.99
2	1/64	6.4585E–05	2.00	5.0547E–05	2.00	5.7592E–05	1.99
	GEOC		1.99		2.00		1.99
3	1/8	2.2371E–04	–	2.2267E–04	–	2.0664E–04	–
3	1/16	3.2897E–05	2.77	3.2455E–05	2.78	3.0237E–05	2.77
3	1/32	5.0341E–06	2.71	4.9281E–06	2.72	4.5992E–06	2.72
3	1/64	8.0276E–07	2.65	7.8150E–07	2.66	7.2933E–07	2.66
	GEOC		2.71		2.72		2.72
4	1/8	2.8019E–05	–	2.6863E–05	–	2.3759E–05	–
4	1/16	4.5630E–06	2.62	4.3388E–06	2.63	3.8426E–06	2.63
4	1/32	7.7950E–07	2.55	7.3892E–07	2.55	6.5504E–07	2.55
4	1/64	1.3572E–07	2.52	1.2850E–07	2.52	1.1398E–07	2.52
	GEOC		2.56		2.57		2.57
5	1/8	8.0765E–06	–	8.3686E–06	–	7.0904E–06	–
5	1/16	1.3891E–06	2.54	1.4415E–06	2.54	1.2239E–06	2.53
5	1/32	2.4249E–07	2.52	2.5191E–07	2.52	2.1413E–07	2.51
5	1/64	4.2611E–08	2.51	4.4293E–08	2.51	3.7673E–08	2.51
	GEOC		2.52		2.52		2.52
6	1/8	3.2423E–06	–	3.4916E–06	–	2.9734E–06	–
6	1/16	5.6456E–07	2.52	6.0843E–07	2.52	5.1885E–07	2.52
6	1/32	9.9090E–08	2.51	1.0684E–07	2.51	9.1177E–08	2.51
6	1/64	1.7456E–08	2.50	1.8826E–08	2.50	1.6072E–08	2.50
	GEOC		2.51		2.51		2.51

Table 2.5 Computational errors and EOC in the $L^2(\Omega)$-norm for the solution of problem (2.201) with $\alpha = -3/2$

		SIPG		NIPG		IIPG	
p	$h/\sqrt{2}$	$\|e_h\|_{L^2(\Omega)}$	EOC	$\|e_h\|_{L^2(\Omega)}$	EOC	$\|e_h\|_{L^2(\Omega)}$	EOC
1	1/8	9.2233E–03	–	1.4850E–02	–	7.9896E–03	–
1	1/16	3.2898E–03	1.49	5.3458E–03	1.47	2.8145E–03	1.51
1	1/32	1.1569E–03	1.51	1.8699E–03	1.52	9.8230E–04	1.52
1	1/64	4.0594E–04	1.51	6.5039E–04	1.52	3.4327E–04	1.52
	GEOC		1.50		1.51		1.51
2	1/8	2.3410E–03	–	4.6812E–03	–	1.7779E–03	–
2	1/16	8.1979E–04	1.51	1.6138E–03	1.54	6.0110E–04	1.56
2	1/32	2.8885E–04	1.50	5.6696E–04	1.51	2.0820E–04	1.53
2	1/64	1.0199E–04	1.50	2.0059E–04	1.50	7.2989E–05	1.51
	GEOC		1.51		1.51		1.53
3	1/8	9.7871E–04	–	3.1394E–03	–	1.0279E–03	–
3	1/16	3.4597E–04	1.50	1.1136E–03	1.50	3.6119E–04	1.51
3	1/32	1.2235E–04	1.50	3.9426E–04	1.50	1.2736E–04	1.50
3	1/64	4.3269E–05	1.50	1.3948E–04	1.50	4.4971E–05	1.50
	GEOC		1.50		1.50		1.50
4	1/8	6.4002E–04	–	1.6788E–03	–	7.8547E–04	–
4	1/16	2.2608E–04	1.50	5.9262E–04	1.50	2.7649E–04	1.51
4	1/32	7.9902E–05	1.50	2.0934E–04	1.50	9.7529E–05	1.50
4	1/64	2.8245E–05	1.50	7.3980E–05	1.50	3.4442E–05	1.50
	GEOC		1.50		1.50		1.50
5	1/8	3.8770E–04	–	1.1048E–03	–	6.0190E–04	–
5	1/16	1.3695E–04	1.50	3.9046E–04	1.50	2.1214E–04	1.50
5	1/32	4.8400E–05	1.50	1.3801E–04	1.50	7.4886E–05	1.50
5	1/64	1.7109E–05	1.50	4.8784E–05	1.50	2.6455E–05	1.50
	GEOC		1.50		1.50		1.50
6	1/8	2.7881E–04	–	7.5211E–04	–	5.2298E–04	–
6	1/16	9.8519E–05	1.50	2.6580E–04	1.50	1.8457E–04	1.50
6	1/32	3.4822E–05	1.50	9.3954E–05	1.50	6.5195E–05	1.50
6	1/64	1.2310E–05	1.50	3.3215E–05	1.50	2.3039E–05	1.50
	GEOC		1.50		1.50		1.50

Table 2.6 Computational errors and EOC in the $H^1(\Omega, \mathscr{T}_h)$-seminorm for the solution of problem (2.201) with $\alpha = -3/2$

		SIPG		NIPG		IIPG	
p	$h/\sqrt{2}$	$\|e_h\|_{H^1(\Omega,\mathscr{T}_h)}$	EOC	$\|e_h\|_{H^1(\Omega,\mathscr{T}_h)}$	EOC	$\|e_h\|_{H^1(\Omega,\mathscr{T}_h)}$	EOC
1	1/8	4.0604E–01	–	3.9606E–01	–	4.0035E–01	–
1	1/16	2.8999E–01	0.49	2.8508E–01	0.47	2.8631E–01	0.48
1	1/32	2.0555E–01	0.50	2.0312E–01	0.49	2.0309E–01	0.50
1	1/64	1.4539E–01	0.50	1.4413E–01	0.50	1.4370E–01	0.50
	GEOC		0.49		0.49		0.49
2	1/8	1.9294E–01	–	2.3736E–01	–	1.8460E–01	–
2	1/16	1.3627E–01	0.50	1.6750E–01	0.50	1.3052E–01	0.50
2	1/32	9.6419E–02	0.50	1.1842E–01	0.50	9.2389E–02	0.50
2	1/64	6.8224E–02	0.50	8.3741E–02	0.50	6.5385E–02	0.50
	GEOC		0.50		0.50		0.50
3	1/8	1.4304E–01	–	2.3656E–01	–	1.5217E–01	–
3	1/16	1.0145E–01	0.50	1.6731E–01	0.50	1.0794E–01	0.50
3	1/32	7.1853E–02	0.50	1.1833E–01	0.50	7.6459E–02	0.50
3	1/64	5.0852E–02	0.50	8.3679E–02	0.50	5.4113E–02	0.50
	GEOC		0.50		0.50		0.50
4	1/8	9.4937E–02	–	1.7438E–01	–	1.0791E–01	–
4	1/16	6.7297E–02	0.50	1.2334E–01	0.50	7.6474E–02	0.50
4	1/32	4.7649E–02	0.50	8.7229E–02	0.50	5.4139E–02	0.50
4	1/64	3.3715E–02	0.50	6.1686E–02	0.50	3.8306E–02	0.50
	GEOC		0.50		0.50		0.50
5	1/8	7.8490E–02	–	1.4046E–01	–	9.6583E–02	–
5	1/16	5.5605E–02	0.50	9.9348E–02	0.50	6.8396E–02	0.50
5	1/32	3.9357E–02	0.50	7.0261E–02	0.50	4.8400E–02	0.50
5	1/64	2.7843E–02	0.50	4.9686E–02	0.50	3.4238E–02	0.50
	GEOC		0.50		0.50		0.50
6	1/8	6.4288E–02	–	1.2563E–01	–	9.3368E–02	–
6	1/16	4.5518E–02	0.50	8.8855E–02	0.50	6.6077E–02	0.50
6	1/32	3.2208E–02	0.50	6.2836E–02	0.50	4.6744E–02	0.50
6	1/64	2.2782E–02	0.50	4.4434E–02	0.50	3.3060E–02	0.50
	GEOC		0.50		0.50		0.50

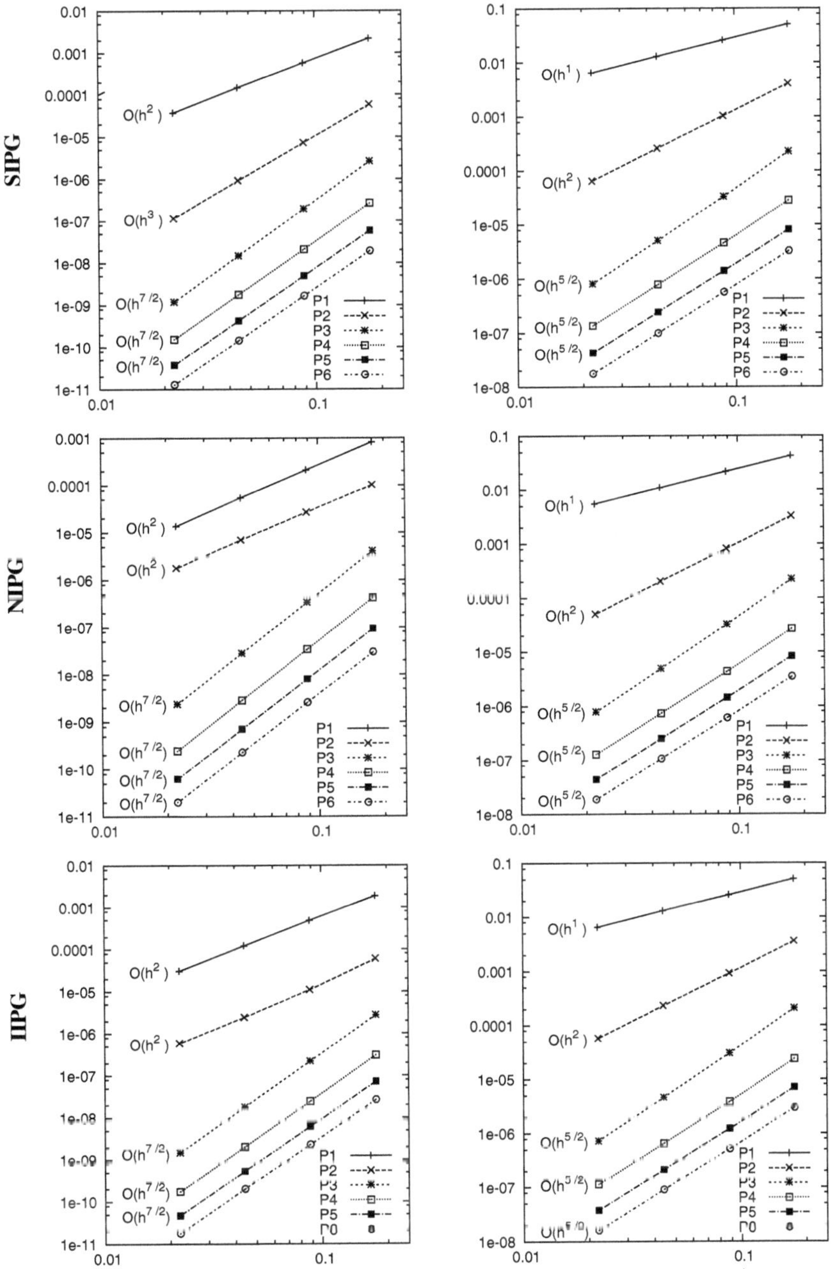

Fig. 2.7 Computational errors and EOC in the $L^2(\Omega)$-norm (*left*) and the $H^1(\Omega, \mathscr{T}_h)$-seminorm (*right*) for the the solution of problem (2.201) with $\alpha = 1/2$

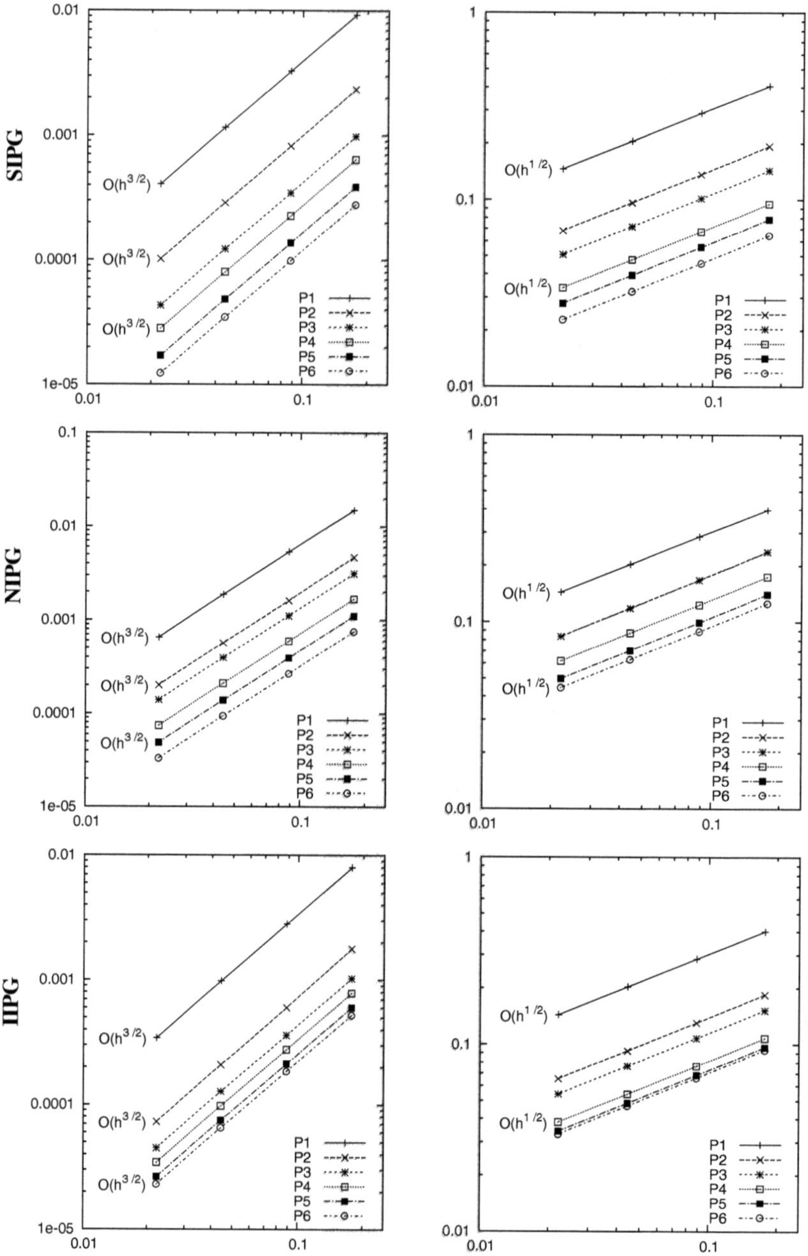

Fig. 2.8 Computational errors and EOC in the $L^2(\Omega)$-norm (*left*) and the $H^1(\Omega, \mathscr{T}_h)$-seminorm (*right*) for the the solution of problem (2.201) with $\alpha = -3/2$

Chapter 3
Methods Based on a Mixed Formulation

In this chapter we introduce two types of the DG discretization that were derived with the aid of a mixed formulation. Numerical methods based on mixed formulations are very often used for solving problems in hydrology (e.g., porous medial flows), where not only the sought solution, but also fluxes defined on the basis of first-order derivatives have to be evaluated. In this chapter we describe basic results of two techniques based on a mixed formulation: the *Bassi–Rebay* (BR) method and the *local discontinuous Galerkin* (LDG) methods.

Let us note that all DG methods can be reformulated either as variational or as mixed problems, see [8], where a unified analysis of DG methods was presented. For simplicity, in this whole Chapter , we confine ourselves to conforming triangular grids satisfying assumption (MA4) from Sect. 2.3.2.

3.1 A General Mixed DG Method

Let us consider problem (2.1). Introducing the auxiliary vector variable $\boldsymbol{q} := \nabla u$, this problem can be rewritten in the form

$$\nabla u = \boldsymbol{q} \quad \text{in } \Omega, \tag{3.1a}$$

$$\nabla \cdot \boldsymbol{q} = f \quad \text{in } \Omega, \tag{3.1b}$$

$$u = u_D \quad \text{on } \partial\Omega_D, \tag{3.1c}$$

$$\boldsymbol{q} \cdot \boldsymbol{n} = g_N \quad \text{on } \partial\Omega_N, \tag{3.1d}$$

which represents a system of first-order differential equations. Problem (3.1) is called the *mixed formulation* of the Poisson problem (2.1), see e.g., [38, 232]. For some DG variants based on the mixed formulation, the auxiliary variable $\boldsymbol{q}$ can be eliminated from the equations, and thus the implementation becomes simpler. This is usually not the case for classical mixed methods.

V. Dolejší and M. Feistauer, *Discontinuous Galerkin Method*,
Springer Series in Computational Mathematics 48,
DOI 10.1007/978-3-319-19267-3_3

Let $\{\mathscr{T}_h\}_{h\in(0,\bar{h})}$, $\bar{h} > 0$, be a system of triangulations of the domain $\Omega \subset \mathbb{R}^d$ satisfying the shape-regularity assumption (2.19) and the assumption of the conformity (MA4) from Sect. 2.3.2. Let S_{hp} be the space of piecewise polynomial functions defined by (2.34), where we seek an approximation of degree $\leq p$ of the primal function u.

Moreover, we define the space $\boldsymbol{\Sigma}_{hp}$ of vector-valued piecewise polynomial functions, where we seek an approximation of the auxiliary function $\boldsymbol{q}$:

$$\boldsymbol{\Sigma}_{hp} = \{\boldsymbol{r} : \Omega \to \mathbb{R}^d;\ \boldsymbol{r}|_K \in (P_{p^\star}(K))^d\ \forall\, K \in \mathscr{T}_h\}, \tag{3.2}$$

where $P_{p^\star}(K)$ denotes the space of all polynomials on K of degree $\leq p^\star$. In order to have a well-posed problem, the value $p^\star$ has to be chosen according to p (the polynomial degree of the primal space S_{hp}). A natural requirement is that $\nabla v \in \boldsymbol{\Sigma}_{hp}$ for any $v \in S_{hp}$.

In this section, we consider the case $p^\star = p$, i.e.,

$$\boldsymbol{\Sigma}_{hp} = (S_{hp})^d. \tag{3.3}$$

Then, both the approximation of u as well as the approximations to each of the components of $\boldsymbol{q}$ belong to the same space. Therefore, implementing the corresponding methods is much simpler than that in standard mixed methods, especially for high-degree polynomial approximations. For the case $p^\star \neq p$ we refer, e.g., to [73, 270, 271].

In order to derive a *mixed discrete formulation*, we multiply relations (3.1a) and (3.1b) by $\boldsymbol{r} \in H^1(\Omega, \mathscr{T}_h)^d$ and by $v \in H^1(\Omega, \mathscr{T}_h)$, respectively, integrate over $K \in \mathscr{T}_h$, and use Green's theorem. Then we obtain

$$\begin{aligned} \int_K \boldsymbol{q}\cdot\boldsymbol{r}\,\mathrm{d}x &= -\int_K u\,\nabla\cdot\boldsymbol{r}\,\mathrm{d}x + \int_{\partial K} u\,\boldsymbol{r}\cdot\boldsymbol{n}\,\mathrm{d}S, \\ \int_K \boldsymbol{q}\cdot\nabla v\,\mathrm{d}x &= \int_K f\,v\,\mathrm{d}x + \int_{\partial K} v\,\boldsymbol{q}\cdot\boldsymbol{n}\,\mathrm{d}S, \end{aligned} \tag{3.4}$$

where $\boldsymbol{n}$ denotes the outer unit normal to ∂K.

Now, according to [8], we consider the following *abstract mixed discrete formulation*: find $u_h \in S_{hp}$ and $\boldsymbol{q}_h \in \boldsymbol{\Sigma}_{hp}$ such that

$$\sum_{K\in\mathscr{T}_h}\int_K \boldsymbol{q}_h\cdot\boldsymbol{r}_h\,\mathrm{d}x = -\sum_{K\in\mathscr{T}_h}\int_K u_h\,\nabla\cdot\boldsymbol{r}_h\,\mathrm{d}x + \sum_{K\in\mathscr{T}_h}\int_{\partial K} \hat{u}\,\boldsymbol{r}_h\cdot\boldsymbol{n}\,\mathrm{d}S\quad \forall\boldsymbol{r}_h\in\boldsymbol{\Sigma}_{hp}, \tag{3.5}$$

$$\sum_{K\in\mathscr{T}_h}\int_K \boldsymbol{q}_h\cdot\nabla v_h\,\mathrm{d}x = \int_\Omega f\,v_h\,\mathrm{d}x + \sum_{K\in\mathscr{T}_h}\int_{\partial K} v_h\,\hat{\boldsymbol{q}}\cdot\boldsymbol{n}\,\mathrm{d}S\quad \forall v_h\in S_{hp}, \tag{3.6}$$

where $\hat{\boldsymbol{q}} = \hat{\boldsymbol{q}}(u_h, \boldsymbol{q}_h) : \cup_{K\in\mathscr{T}_h} \partial K \to \mathbb{R}^d$ and $\hat{u} = \hat{u}(u_h, \boldsymbol{q}_h) : \cup_{K\in\mathscr{T}_h} \partial K \to \mathbb{R}$ are *numerical fluxes* approximating $\boldsymbol{q}_h$ and u_h, respectively, on the boundary of $K \in \mathscr{T}_h$. It means that $\hat{\boldsymbol{q}}$ and $\hat{u}$ are double-valued functions on $\Gamma \in \mathscr{F}_h^I$ and single-valued functions on $\Gamma \in \mathscr{F}_h^B$. To complete the specification of a DG method we have to express the numerical fluxes $\hat{\boldsymbol{q}}$ and $\hat{u}$ in terms of $\boldsymbol{q}_h$ and u_h and in terms of the boundary conditions. We present and analyze two approaches: the BR2 and LDG methods in Sects. 3.2 and 3.3, respectively.

3.1.1 Equivalent Formulations

For further analysis, we reformulate the abstract problem (3.5) and (3.6) in a more appropriate form. We use the average and jump operators defined by (2.32) and (2.33) which also make sense for functions from $\boldsymbol{\Sigma}_{hp}$. Let $\varphi \in H^1(\Omega, \mathscr{T}_h)$ and $\boldsymbol{\theta} \in H^1(\Omega, \mathscr{T}_h)^d$. Then a straightforward computation gives

$$\sum_{K\in\mathscr{T}_h} \int_{\partial K} \varphi \boldsymbol{\theta} \cdot \boldsymbol{n} \,\mathrm{d}S = \sum_{\Gamma\in\mathscr{F}_h} \int_{\Gamma} [\varphi] \langle\boldsymbol{\theta}\rangle \cdot \boldsymbol{n} \,\mathrm{d}S + \sum_{\Gamma\in\mathscr{F}_h^I} \int_{\Gamma} \langle\varphi\rangle [\boldsymbol{\theta}] \cdot \boldsymbol{n} \,\mathrm{d}S. \tag{3.7}$$

After a simple application of this identity, from (3.5) and (3.6) we get

$$\begin{aligned} &\sum_{K\in\mathscr{T}_h} \int_K \boldsymbol{q}_h \cdot \boldsymbol{r}_h \,\mathrm{d}x + \sum_{K\in\mathscr{T}_h} \int_K u_h \nabla \cdot \boldsymbol{r}_h \,\mathrm{d}x \\ &\quad = \sum_{\Gamma\in\mathscr{F}_h} \int_\Gamma [\hat{u}] \langle\boldsymbol{r}_h\rangle \cdot \boldsymbol{n} \,\mathrm{d}S + \sum_{\Gamma\in\mathscr{F}_h^I} \int_\Gamma \langle\hat{u}\rangle [\boldsymbol{r}_h] \cdot \boldsymbol{n} \,\mathrm{d}S \quad \forall \boldsymbol{r}_h \in \boldsymbol{\Sigma}_{hp}, \end{aligned} \tag{3.8}$$

$$\begin{aligned} &\sum_{K\in\mathscr{T}_h} \int_K \boldsymbol{q}_h \cdot \nabla v_h \,\mathrm{d}x \\ &\quad = \int_\Omega f v_h \,\mathrm{d}x + \sum_{\Gamma\in\mathscr{F}_h} \int_\Gamma \langle\hat{\boldsymbol{q}}\rangle \cdot \boldsymbol{n} [v_h] \,\mathrm{d}S + \sum_{\Gamma\in\mathscr{F}_h^I} \int_\Gamma [\hat{\boldsymbol{q}}] \cdot \boldsymbol{n} \langle v_h\rangle \,\mathrm{d}S \quad \forall v_h \in S_{hp}. \end{aligned} \tag{3.9}$$

Now we express $\boldsymbol{q}_h$ solely in terms of u_h. If we take $\varphi := v_h$ and $\boldsymbol{\theta} := \boldsymbol{r}_h$ in (3.7) and use Green's theorem, then for all $\boldsymbol{r}_h \in H^1(\Omega, \mathscr{T}_h)^d$ and $v_h \in H^1(\Omega, \mathscr{T}_h)$ we obtain

$$\begin{aligned} &-\sum_{K\in\mathscr{T}_h} \int_K v_h \nabla \cdot \boldsymbol{r}_h \,\mathrm{d}x \\ &\quad = \sum_{K\in\mathscr{T}_h} \int_K \boldsymbol{r}_h \cdot \nabla v_h \,\mathrm{d}x - \sum_{\Gamma\in\mathscr{F}_h} \int_\Gamma \langle\boldsymbol{r}_h\rangle \cdot \boldsymbol{n} [v_h] \,\mathrm{d}S - \sum_{\Gamma\in\mathscr{F}_h^I} \int_\Gamma [\boldsymbol{r}_h] \cdot \boldsymbol{n} \langle v_h\rangle \,\mathrm{d}S. \end{aligned} \tag{3.10}$$

Using (3.10) with $v_h := u_h$ and identity (3.8), we have

$$\sum_{K\in\mathscr{T}_h}\int_K (\boldsymbol{q}_h - \nabla u_h)\cdot \boldsymbol{r}_h\,\mathrm{d}x = \sum_{\Gamma\in\mathscr{F}_h}\int_\Gamma [\hat{u}-u_h]\,\langle \boldsymbol{r}_h\rangle\cdot\boldsymbol{n}\,\mathrm{d}S + \sum_{\Gamma\in\mathscr{F}_h^I}\int_\Gamma \langle \hat{u}-u_h\rangle\,[\boldsymbol{r}_h]\cdot\boldsymbol{n}\,\mathrm{d}S \quad \forall \boldsymbol{r}_h\in\boldsymbol{\Sigma}_{hp}. \tag{3.11}$$

Therefore, the original abstract problem (3.5) and (3.6) is equivalent to (3.8) and (3.9) as well as to (3.9) and (3.11).

Finally, in order to express $\boldsymbol{q}_h$ in terms of u_h, we put $\boldsymbol{r}_h := \nabla v_h$ in (3.11), subtract from (3.9), and get

$$\begin{aligned}\sum_{K\in\mathscr{T}_h}\int_K \nabla u_h\cdot\nabla v_h\,\mathrm{d}x &+ \sum_{\Gamma\in\mathscr{F}_h}\int_\Gamma \big([\hat{u}-u_h]\,\langle\nabla v_h\rangle\cdot\boldsymbol{n} - \langle\hat{\boldsymbol{q}}\rangle\cdot\boldsymbol{n}\,[v_h]\big)\,\mathrm{d}S\\ &+ \sum_{\Gamma\in\mathscr{F}_h^I}\int_\Gamma \big(\langle\hat{u}-u_h\rangle\,[\nabla v_h]\cdot\boldsymbol{n} - [\hat{\boldsymbol{q}}]\cdot\boldsymbol{n}\langle v_h\rangle\big)\,\mathrm{d}S\\ &= \int_\Omega f\,v_h\,\mathrm{d}x \qquad \forall v_h\in S_{hp}.\end{aligned} \tag{3.12}$$

This relation is already independent of $\boldsymbol{q}_h$ and it will be used in the following analysis.

3.1.2 Lifting Operators

In order to rewrite the previous formulation in a more compact form, we define the *lifting operators* $\boldsymbol{L}_{u_D} : H^1(\Omega,\mathscr{T}_h)\to\boldsymbol{\Sigma}_{hp}$ and $\boldsymbol{L} : H^1(\Omega,\mathscr{T}_h)\to\boldsymbol{\Sigma}_{hp}$ by

$$\sum_{K\in\mathscr{T}_h}\int_K \boldsymbol{L}_{u_D}(\varphi)\cdot\boldsymbol{r}_h\,\mathrm{d}x = \sum_{\Gamma\in\mathscr{F}_h^D}\int_\Gamma u_D\boldsymbol{r}_h\cdot\boldsymbol{n}\,\mathrm{d}S - \sum_{\Gamma\in\mathscr{F}_h^{ID}}\int_\Gamma [\varphi]\boldsymbol{n}\cdot\langle\boldsymbol{r}_h\rangle\,\mathrm{d}S \quad \forall\boldsymbol{r}_h\in\boldsymbol{\Sigma}_{hp}, \tag{3.13}$$

and

$$\sum_{K\in\mathscr{T}_h}\int_K \boldsymbol{L}(\varphi)\cdot\boldsymbol{r}_h\,\mathrm{d}x = -\sum_{\Gamma\in\mathscr{F}_h^{ID}}\int_\Gamma [\varphi]\boldsymbol{n}\cdot\langle\boldsymbol{r}_h\rangle\,\mathrm{d}S \quad \forall\boldsymbol{r}_h\in\boldsymbol{\Sigma}_{hp}, \tag{3.14}$$

for $\varphi\in H^1(\Omega,\mathscr{T}_h)$. From (3.13) and (3.14), we see that

$$\sum_{K\in\mathscr{T}_h}\int_K \boldsymbol{L}_{u_D}(\varphi)\cdot\boldsymbol{r}_h\,\mathrm{d}x = \sum_{K\in\mathscr{T}_h}\int_K \boldsymbol{L}(\varphi)\cdot\boldsymbol{r}_h\,\mathrm{d}x + \sum_{\Gamma\in\mathscr{F}_h^D}\int_\Gamma u_D\boldsymbol{r}_h\cdot\boldsymbol{n}\,\mathrm{d}S \quad \forall\boldsymbol{r}_h\in\boldsymbol{\Sigma}_{hp}. \tag{3.15}$$

Further, for each $\Gamma \in \mathscr{F}_h$ we define the operator $\boldsymbol{l}_{\Gamma,u_D} : H^1(\Omega, \mathscr{T}_h) \to \boldsymbol{\Sigma}_{hp}$ by

$$\sum_{K\in\mathscr{T}_h} \int_K \boldsymbol{l}_{\Gamma,u_D}(\varphi)\cdot \boldsymbol{r}_h \, \mathrm{d}x = \begin{cases} -\int_\Gamma [\varphi]\,\boldsymbol{n}\cdot\langle \boldsymbol{r}_h\rangle \, \mathrm{d}S & \text{for } \Gamma \in \mathscr{F}_h^I \\ \int_\Gamma (u_D-\varphi)\,\boldsymbol{n}\cdot \boldsymbol{r}_h \, \mathrm{d}S & \text{for } \Gamma\in\mathscr{F}_h^D \\ 0 & \text{for } \Gamma\in\mathscr{F}_h^N \end{cases} \quad \forall \boldsymbol{r}_h \in \boldsymbol{\Sigma}_{hp}. \tag{3.16}$$

Moreover, we set $\boldsymbol{l}_\Gamma := \boldsymbol{l}_{\Gamma,0}$ ($= \boldsymbol{l}_{\Gamma,u_D}$ with $u_D = 0$) and for $K \in \mathscr{T}_h$ we use the notation $\mathscr{F}_h(K) = \{\Gamma \in \mathscr{F}_h;\, \Gamma \subset \partial K\}$. Obviously, if $u_D = 0$, then $\boldsymbol{l}_\Gamma = \boldsymbol{l}_{\Gamma,u_D}$ for any $\Gamma \in \mathscr{F}_h$. If $u_D \neq 0$, then $\boldsymbol{l}_\Gamma = \boldsymbol{l}_{\Gamma,u_D}$ only for any $\Gamma \in \mathscr{F}_h^I \cup \mathscr{F}_h^N$.

If $\varphi \in H^1(\Omega, \mathscr{T}_h)$, then the support of $\boldsymbol{l}_{\Gamma,u_D}(\varphi) \in \boldsymbol{\Sigma}_{hp}$ is the union of (one or two) elements $K \in \mathscr{T}_h$ sharing the face Γ. This follows from the following reasoning. Let $\Gamma \in \mathscr{F}_h$ be arbitrary but fixed, let $K' \in \mathscr{T}_h$ be an element not having Γ as its face, and let $\boldsymbol{r}_h' \in \boldsymbol{\Sigma}_{hp}$ be such that its support is K'. Then the right-hand side of (3.16) vanishes for $\boldsymbol{r}_h'$ and, therefore, $\boldsymbol{l}_{\Gamma,u_D}(\varphi)|_{K'}$ appearing on the left-hand side of (3.16) has to be equal to zero.

This property leads to the identity

$$\sum_{K\in\mathscr{T}_h}\sum_{\Gamma\in\mathscr{F}_h(K)} \int_K \boldsymbol{l}_{\Gamma,u_D}(u_h)\cdot\boldsymbol{l}_\Gamma(v_h)\,\mathrm{d}x = \sum_{\Gamma\in\mathscr{F}_h}\int_\Omega \boldsymbol{l}_{\Gamma,u_D}(u_h)\cdot\boldsymbol{l}_\Gamma(v_h)\,\mathrm{d}x, \tag{3.17}$$

valid for all $u_h, v_h \in S_{hp}$. Finally, for each $K \in \mathscr{T}_h$ and $\varphi \in H^1(\Omega, \mathscr{T}_h)$ we have

$$\sum_{\Gamma\in\mathscr{F}_h(K)} \boldsymbol{l}_{\Gamma,u_D}(\varphi) = \boldsymbol{L}_{u_D}(\varphi) \quad \text{in } K \tag{3.18}$$

and

$$\sum_{\Gamma\in\mathscr{F}_h(K)} \boldsymbol{l}_\Gamma(\varphi) = \boldsymbol{L}(\varphi) \quad \text{in } K. \tag{3.19}$$

Moreover, putting $\boldsymbol{r}_h = \boldsymbol{l}_\Gamma(v_h)$ in (3.16) with $v_h \in S_{hp}$ and summing over all $\Gamma \subset \mathscr{F}_h$, we obtain another useful identity

$$\sum_{\Gamma\in\mathscr{F}_h}\int_\Omega \boldsymbol{l}_{\Gamma,u_D}(\varphi)\cdot\boldsymbol{l}_\Gamma(v_h)\,\mathrm{d}x = \sum_{\Gamma\in\mathscr{F}_h}\int_\Omega \boldsymbol{l}_\Gamma(\varphi)\cdot\boldsymbol{l}_\Gamma(v_h)\,\mathrm{d}x + \sum_{\Gamma\in\mathscr{F}_h^D}\int_\Gamma u_D\boldsymbol{n}\cdot\boldsymbol{l}_\Gamma(v_h)\,\mathrm{d}S. \tag{3.20}$$

In what follows we present several possible specifications of the numerical fluxes $\hat{\boldsymbol{q}}$ and $\hat{u}$.

3.2 Bassi–Rebay Methods

In 1997, Bassi and Rebay proposed in [23] the following simple and natural choice of the numerical fluxes $\hat{u}$ and $\hat{\boldsymbol{q}}$:

$$\hat{u}|_\Gamma = \begin{cases} \langle u_h \rangle & \text{if } \Gamma \in \mathscr{F}_h^I \\ u_D & \text{if } \Gamma \in \mathscr{F}_h^D \\ u_h & \text{if } \Gamma \in \mathscr{F}_h^N \end{cases}, \quad \hat{\boldsymbol{q}} \cdot \boldsymbol{n}|_\Gamma = \begin{cases} \langle \boldsymbol{q}_h \rangle \cdot \boldsymbol{n} & \text{if } \Gamma \in \mathscr{F}_h^I \\ \boldsymbol{q}_h \cdot \boldsymbol{n} & \text{if } \Gamma \in \mathscr{F}_h^D \\ g_N & \text{if } \Gamma \in \mathscr{F}_h^N \end{cases}, \tag{3.21}$$

for each $\Gamma \subset \partial K$ and each $K \in \mathscr{T}_h$. This means that both values of the double-valued fluxes $\hat{u}$ and $\hat{\boldsymbol{q}}$ on interior faces are identical and thus $[\hat{u}] = [\hat{\boldsymbol{q}}] = 0$.

In the following, we introduce the Bassi–Rebay method both as mixed and variational formulations.

3.2.1 Mixed Formulation

We use (3.8) and (3.9), which are equivalent to the abstract problem. Inserting the definition (3.21) of the numerical fluxes into (3.8) and (3.9), we obtain

$$\begin{aligned} &\sum_{K \in \mathscr{T}_h} \int_K \boldsymbol{q}_h \cdot \boldsymbol{r}_h \,\mathrm{d}x + \sum_{K \in \mathscr{T}_h} \int_K u_h \nabla \cdot \boldsymbol{r}_h \,\mathrm{d}x \\ &\quad = \sum_{\Gamma \in \mathscr{F}_h^D} \int_\Gamma u_D \boldsymbol{r}_h \cdot \boldsymbol{n} \,\mathrm{d}S + \sum_{\Gamma \in \mathscr{F}_h^N} \int_\Gamma u_h \boldsymbol{r}_h \cdot \boldsymbol{n} \,\mathrm{d}S + \sum_{\Gamma \in \mathscr{F}_h^I} \int_\Gamma \langle u_h \rangle [\boldsymbol{r}_h] \cdot \boldsymbol{n} \,\mathrm{d}S, \end{aligned} \tag{3.22}$$

and

$$\begin{aligned} \sum_{K \in \mathscr{T}_h} \int_K \boldsymbol{q}_h \cdot \nabla v_h \,\mathrm{d}x &= \sum_{\Gamma \in \mathscr{F}_h^I} \int_\Gamma \langle \boldsymbol{q}_h \rangle \cdot \boldsymbol{n} \, [v_h] \,\mathrm{d}S + \sum_{\Gamma \in \mathscr{F}_h^D} \int_\Gamma \boldsymbol{q}_h \cdot \boldsymbol{n} \, v_h \,\mathrm{d}S \\ &\quad + \int_\Omega f \, v_h \,\mathrm{d}x + \sum_{\Gamma \in \mathscr{F}_h^N} \int_\Gamma g_N v_h \,\mathrm{d}S, \end{aligned} \tag{3.23}$$

respectively.

In order to introduce the mixed formulation, we define the forms

$$\begin{aligned} a(\boldsymbol{q}_h, \boldsymbol{r}_h) &= \sum_{K \in \mathscr{T}_h} \int_K \boldsymbol{q}_h \cdot \boldsymbol{r}_h \,\mathrm{d}x, \\ b(u_h, \boldsymbol{r}_h) &= \sum_{K \in \mathscr{T}_h} \int_K u_h \nabla \cdot \boldsymbol{r}_h \,\mathrm{d}x - \sum_{\Gamma \in \mathscr{F}_h^N} \int_\Gamma u_h \boldsymbol{r}_h \cdot \boldsymbol{n} \,\mathrm{d}S - \sum_{\Gamma \in \mathscr{F}_h^I} \int_\Gamma \langle u_h \rangle [\boldsymbol{r}_h] \cdot \boldsymbol{n} \,\mathrm{d}S. \end{aligned} \tag{3.24}$$

Moreover, using Green's theorem and (3.7), we obtain the identity

$$\sum_{K\in\mathscr{T}_h}\int_K u_h \nabla\cdot \boldsymbol{r}_h \,\mathrm{d}x - \sum_{\Gamma\in\mathscr{F}_h^N}\int_\Gamma u_h \boldsymbol{r}_h\cdot\boldsymbol{n}\,\mathrm{d}S \tag{3.25}$$
$$= -\sum_{K\in\mathscr{T}_h}\int_K \nabla u_h\cdot\boldsymbol{r}_h\,\mathrm{d}x + \sum_{\Gamma\in\mathscr{F}_h^{ID}}\int_\Gamma \langle\boldsymbol{r}_h\rangle\cdot\boldsymbol{n}\,[u_h]\,\mathrm{d}S + \sum_{\Gamma\in\mathscr{F}_h^{I}}\int_\Gamma \langle u_h\rangle\,[\boldsymbol{r}_h]\cdot\boldsymbol{n}\,\mathrm{d}S,$$

which implies that the form b from (3.24) becomes the form

$$b(u_h,\boldsymbol{r}_h) = -\sum_{K\in\mathscr{T}_h}\int_K \nabla u_h\cdot\boldsymbol{r}_h\,\mathrm{d}x + \sum_{\Gamma\in\mathscr{F}_h^{ID}}\int_\Gamma \langle\boldsymbol{r}_h\rangle\cdot\boldsymbol{n}\,[u_h]\,\mathrm{d}S. \tag{3.26}$$

Therefore, (3.22) and (3.23) can be reformulated as the *saddle-point problem*:

Definition 3.1 (*BR method*) We say that $(u_h,\boldsymbol{q}_h)\in S_{hp}\times\boldsymbol{\Sigma}_{hp}$ is the *approximate solution* of the mixed formulation of the *Bassi–Rebay* method, if

$$a(\boldsymbol{q}_h,\boldsymbol{r}_h)+b(u_h,\boldsymbol{r}_h) = F(\boldsymbol{r}_h)\qquad \forall\,\boldsymbol{r}_h\in\boldsymbol{\Sigma}_{hp}, \tag{3.27}$$
$$-\,b(v_h,\boldsymbol{q}_h) = G(v_h)\qquad \forall\,v_h\in S_{hp}, \tag{3.28}$$

where the bilinear forms a and b are given by (3.24) and (3.26), respectively, and

$$F(\boldsymbol{r}_h) = \sum_{\Gamma\in\mathscr{F}_h^D}\int_\Gamma u_D\boldsymbol{r}_h\cdot\boldsymbol{n}\,\mathrm{d}S, \tag{3.29}$$

$$G(v_h) = \int_\Omega f\,v_h\,\mathrm{d}x + \sum_{\Gamma\in\mathscr{F}_h^N}\int_\Gamma g_N v_h\,\mathrm{d}S. \tag{3.30}$$

We recall that similarly, as in the IPG methods the Dirichlet as well as Neumann boundary conditions are incorporated in the right-hand sides of (3.27) and (3.28), which is not the case of the classical mixed formulation.

It is clear that the corresponding *inf-sup condition* should be satisfied in order to ensure the existence and uniqueness of the solution of (3.27) and (3.28), see e.g., [38, 232]. Prior to discussing the existence of the solution of (3.27) and (3.28), we introduce an equivalent formulation.

3.2.2 Variational Formulation

With the aid of (3.26) and (3.24), relation (3.27) can be written as

$$\sum_{K\in\mathscr{T}_h}\int_K (\boldsymbol{q}_h-\nabla u_h)\cdot\boldsymbol{r}_h\,\mathrm{d}x = \sum_{\Gamma\in\mathscr{F}_h^D}\int_\Gamma u_D\boldsymbol{r}_h\cdot\boldsymbol{n}\,\mathrm{d}S - \sum_{\Gamma\in\mathscr{F}_h^{ID}}\int_\Gamma \langle\boldsymbol{r}_h\rangle\cdot\boldsymbol{n}\,[u_h]\,\mathrm{d}S \tag{3.31}$$

Putting $\boldsymbol{r}_h = \nabla v_h$ in (3.31) and subtracting from (3.23), we get

$$\sum_{K \in \mathscr{T}_h} \int_K \nabla u_h \cdot \nabla v_h \,\mathrm{d}x - \sum_{\Gamma \in \mathscr{F}_h^{ID}} \int_\Gamma \big(\langle \nabla v_h \rangle \cdot \boldsymbol{n} \, [u_h] + \langle \boldsymbol{q}_h \rangle \cdot \boldsymbol{n} \, [v_h] \big) \,\mathrm{d}S \tag{3.32}$$
$$= \int_\Omega f v_h \,\mathrm{d}x - \sum_{\Gamma \in \mathscr{F}_h^{D}} \int_\Gamma u_D \nabla v_h \cdot \boldsymbol{n} \,\mathrm{d}S + \sum_{\Gamma \in \mathscr{F}_h^{N}} \int_\Gamma g_N v_h \,\mathrm{d}S.$$

Moreover, relation (3.31) together with the definition (3.13) of the lifting operator $\boldsymbol{L}_{u_D}$ gives

$$\boldsymbol{L}_{u_D}(u_h) = \boldsymbol{q}_h - \nabla u_h. \tag{3.33}$$

Using (3.33) and (3.14), we rewrite the last term on the left-hand side of (3.32) in the form

$$- \sum_{\Gamma \in \mathscr{F}_h^{ID}} \int_\Gamma \langle \boldsymbol{q}_h \rangle \cdot \boldsymbol{n} \, [v_h] \,\mathrm{d}S = - \sum_{\Gamma \in \mathscr{F}_h^{ID}} \int_\Gamma \langle \boldsymbol{L}_{u_D}(u_h) + \nabla u_h \rangle \cdot \boldsymbol{n} \, [v_h] \,\mathrm{d}S$$
$$= \sum_{K \in \mathscr{T}_h} \int_K \big(\boldsymbol{L}_{u_D}(u_h) + \nabla u_h \big) \cdot \boldsymbol{L}(v_h) \,\mathrm{d}x. \tag{3.34}$$

Similarly, from (3.13) we obtain the identity

$$\sum_{\Gamma \in \mathscr{F}_h^{D}} \int_\Gamma u_D \nabla v_h \cdot \boldsymbol{n} \,\mathrm{d}S - \sum_{\Gamma \in \mathscr{F}_h^{ID}} \int_\Gamma [u_h] \langle \nabla v_h \rangle \cdot \boldsymbol{n} \,\mathrm{d}S = \sum_{K \in \mathscr{T}_h} \int_K \boldsymbol{L}_{u_D}(u_h) \cdot \nabla v_h \,\mathrm{d}x. \tag{3.35}$$

Finally, inserting (3.34) and (3.35) into (3.32) we find that

$$\sum_{K \in \mathscr{T}_h} \int_K \big(\nabla u_h + \boldsymbol{L}_{u_D}(u_h) \big) \cdot \big(\nabla v_h + \boldsymbol{L}(v_h) \big) \,\mathrm{d}x \tag{3.36}$$
$$= \int_\Omega f v_h \,\mathrm{d}x + \sum_{\Gamma \in \mathscr{F}_h^{N}} \int_\Gamma g_N v_h \,\mathrm{d}S \qquad \forall v_h \in S_{hp},$$

which represents a variational formulation of the BR method. It is equivalent to the saddle-point problem (3.27) and (3.28).

This approach was proposed by Bassi and Rebay in [23] for the solution of the compressible Navier–Stokes equations. Although this method gives good results in some cases, its application to the Poisson problem (2.1) is rather problematic, because the corresponding *inf-sup* condition does not hold, as was shown in [39] through a counter-example. Therefore, it seems quite natural to add a suitable stabilization term to the left-hand side of (3.36).

In [27], Bassi et al. proposed replacing the term

$$\int_K \boldsymbol{L}_{u_D}(u_h) \cdot \boldsymbol{L}(v_h)\,\mathrm{d}x \tag{3.37}$$

in (3.36) by

$$\zeta \sum_{\Gamma \in \partial K} \int_\Omega \boldsymbol{l}_{\Gamma,u_D}(u_h) \cdot \boldsymbol{l}_\Gamma(v_h)\,\mathrm{d}x \tag{3.38}$$

for each $K \in \mathscr{T}_h$, where $\zeta > 0$ is a stabilization parameter. Then, using (3.17), we obtain the following formulation

$$\begin{aligned} &\sum_{K \in \mathscr{T}_h} \int_K \big(\nabla u_h \cdot \nabla v_h + \boldsymbol{L}_{u_D}(u_h) \cdot \nabla v_h + \nabla u_h \cdot \boldsymbol{L}(v_h)\big)\,\mathrm{d}x \\ &+ \zeta \sum_{\Gamma \in \mathscr{F}_h} \int_\Omega \boldsymbol{l}_{\Gamma,u_D}(u_h) \cdot \boldsymbol{l}_\Gamma(v_h)\,\mathrm{d}x \\ =& \int_\Omega f\, v_h\,\mathrm{d}x + \sum_{\Gamma \in \mathscr{F}_h^N} \int_\Gamma g_N v_h\,\mathrm{d}S \qquad \forall v_h \in S_{hp}. \end{aligned} \tag{3.39}$$

In [27], this stabilization was introduced with the value of stabilization parameter $\zeta = 1$. However, Brezzi et al. in [40] proved the error estimates of scheme (3.39) under the assumption that $\zeta > 3$ (for conforming triangular grids). It is not clear whether the variant proposed in [27] (with the value $\zeta = 1$) is convergent or not. However, let us mention an advantage of scheme (3.39). The stiffness matrix corresponding to (3.38) is much more sparse than the matrix corresponding to (3.37). Indeed, if we take v_h having its support inside one element $K \in \mathscr{T}_h$ (far from the boundary), then, in general, 10 elements are involved in (3.37), while only 4 elements are involved in (3.38), see Fig. 3.1. Therefore, the stabilization term (3.38) leads to the same sparsity as the DG methods based on the primal formulation (2.49a)–(2.49e).

Formulation (3.39) is equivalent to problem (3.5) and (3.6) with the numerical flux $\hat{u}$ given by (3.21) and the numerical flux $\hat{\boldsymbol{q}}$ given by

$$\hat{\boldsymbol{q}} \cdot \boldsymbol{n}|_\Gamma = \begin{cases} \big(\langle \boldsymbol{q}_n \rangle - \zeta \langle \boldsymbol{l}_\Gamma(u_h) \rangle\big) \cdot \boldsymbol{n} & \text{if } \Gamma \in \mathscr{F}_h^{ID} \\ g_N & \text{if } \Gamma \in \mathscr{F}_h^N \end{cases}, \tag{3.40}$$

for all $\Gamma \subset \partial K,\ K \in \mathscr{T}_h$.

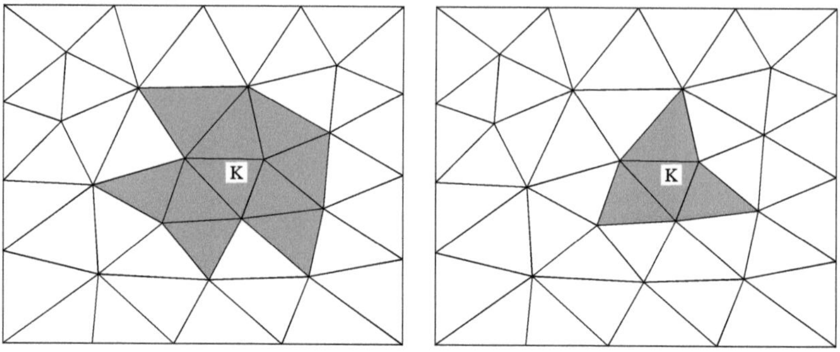

Fig. 3.1 Elements involved in the stabilization terms (3.37) (*left*) and (3.38) (*right*)

Let us define the bilinear form $A_h^{\mathrm{BR}} : H^1(\Omega, \mathscr{T}_h) \times H^1(\Omega, \mathscr{T}_h) \to \mathbb{R}$ by

$$A_h^{\mathrm{BR}}(u, v) = \sum_{K \in \mathscr{T}_h} \int_K \nabla u \cdot \nabla v \, dx - \sum_{\Gamma \in \mathscr{F}_h^{ID}} \int_\Gamma \langle \nabla v \rangle \cdot \boldsymbol{n} \, [u] \, \mathrm{d}S - \sum_{\Gamma \in \mathscr{F}_h^{ID}} \int_\Gamma \langle \nabla u \rangle \cdot \boldsymbol{n} \, [v] \, \mathrm{d}S$$
$$+ \zeta \sum_{\Gamma \in \mathscr{F}_h} \int_\Omega \boldsymbol{l}_\Gamma(u) \cdot \boldsymbol{l}_\Gamma(v) \, \mathrm{d}x, \qquad u, v \in H^1(\Omega, \mathscr{T}_h), \tag{3.41}$$

where $\zeta > 0$ is a given constant. By virtue of (3.14), the form A_h^{BR} restricted to the finite-dimensional space $S_{hp} \times S_{hp}$ can be rewritten in the equivalent form

$$A_h^{\mathrm{BR}}(u_h, v_h) = \sum_{K \in \mathscr{T}_h} \int_K \big(\nabla u_h \cdot \nabla v_h + \boldsymbol{L}(u_h) \cdot \nabla v_h + \nabla u_h \cdot \boldsymbol{L}(v_h)\big) \, \mathrm{d}x$$
$$+ \zeta \sum_{\Gamma \in \mathscr{F}_h} \int_\Omega \boldsymbol{l}_\Gamma(u_h) \cdot \boldsymbol{l}_\Gamma(v_h) \, \mathrm{d}x, \quad u_h, v_h \in S_{hp}. \tag{3.42}$$

Therefore, taking (3.15) and (3.20) into account, we rewrite (3.39) in the form

$$A_h^{\mathrm{BR}}(u_h, v_h) = \ell_h^{\mathrm{BR}}(v_h) \qquad \forall \, v_h \in S_{hp}, \tag{3.43}$$

where A_h^{BR} is given by (3.41) and $\ell_h^{\mathrm{BR}} : S_{hp} \to \mathbb{R}$ is a linear operator defined by

$$\ell_h^{\mathrm{BR}}(v_h) = \int_\Omega f \, v_h \, \mathrm{d}x + \sum_{\Gamma \in \mathscr{F}_h^N} \int_\Gamma g_N v_h \, \mathrm{d}S \tag{3.44}$$
$$- \sum_{\Gamma \in \mathscr{F}_h^D} \int_\Gamma u_D \, \nabla v_h \cdot \boldsymbol{n} \, \mathrm{d}S - \zeta \sum_{\Gamma \in \mathscr{F}_h^D} \int_\Gamma u_D \, \boldsymbol{l}_\Gamma(v_h) \cdot \boldsymbol{n} \, \mathrm{d}S.$$

Hence, we introduce the following definition:

Definition 3.2 (*BR2 method*) Let the forms A_h^{BR} and ℓ_h^{BR} be defined by (3.41) and (3.44), respectively. We say that $u_h \in S_{hp}$ is the *approximate solution* obtained by the BR2 method, if identity (3.43) is satisfied.

Obviously, relation (3.43) represents a system of linear algebraic equations (cf., e.g., (2.55)), which can be solved by a suitable solver.

3.2.3 Theoretical Analysis

In the following, we deal with the analysis of the previous numerical scheme, namely with the existence of an approximate solution and error estimates. We employ some results from [40], where numerical analysis was carried out for homogeneous Dirichlet boundary condition considered on the whole boundary $\partial\Omega$. For simplicity, similarly as in [40], we confine ourselves to conforming triangular meshes ($d = 2$). Hence, we assume in the following that we have a system of triangulations $\{\mathscr{T}_h\}_{h\in(0,\bar{h})}$ satisfying the shape-regularity assumption (2.19) and the assumption (MA4) of the mesh conformity in Sect. 2.3.2.

We start from the consistency results.

Lemma 3.3 (Consistency) *The BR2 method (3.43) is consistent, i.e., if $u \in H^2(\Omega)$ is the weak solution of (2.1), then*

$$A_h^{\mathrm{BR}}(u, v_h) = \ell_h^{\mathrm{BR}}(v_h) \qquad \forall\, v_h \in S_{hp}, \tag{3.45}$$

where A_h^{BR} and ℓ_h^{BR} are given by (3.41) *and* (3.44), *respectively.*

Proof If $u \in H^2(\Omega)$ is the weak solution of (2.1), then u satisfies (2.37), (2.40) and $u = u_D$ on $\partial\Omega_D$. Therefore, due to (3.16) and the identity

$$0 = \int_\Gamma [u]\boldsymbol{n} \cdot \langle \nabla v_h \rangle \,\mathrm{d}S, \quad \Gamma \in \mathscr{F}_h^I, \tag{3.46}$$

we obtain

$$\int_\Omega \boldsymbol{l}_\Gamma(u) \cdot \boldsymbol{l}_\Gamma(v_h)\,\mathrm{d}x = \begin{cases} 0 & \text{for } \Gamma \in \mathscr{F}_h^I \cup \mathscr{F}_h^N, \\ -\int_\Gamma u\, \boldsymbol{l}_\Gamma(v_h) \cdot \boldsymbol{n}\,\mathrm{d}S & \text{for } \Gamma \in \mathscr{F}_h^D. \end{cases} \tag{3.47}$$

Moreover, using (3.41), (3.44), (3.46), (3.47), the identity $u = u_D$ on $\partial\Omega_D$ and (2.40), we get

$$
\begin{aligned}
&A_h^{\mathrm{BR}}(u, v_h) - \ell_h^{\mathrm{BR}}(v_h) \\
&\quad = \sum_{K\in\mathscr{T}_h} \int_K \nabla u \cdot \nabla v_h \,\mathrm{d}x - \sum_{\Gamma\in\mathscr{F}_h^D} \int_\Gamma \nabla v_h \cdot \boldsymbol{n}\, u \,\mathrm{d}S - \sum_{\Gamma\in\mathscr{F}_h^{ID}} \int_\Gamma \langle \nabla u \cdot \boldsymbol{n}\rangle [v_h] \,\mathrm{d}S \\
&\qquad - \zeta \sum_{\Gamma\in\mathscr{F}_h^D} \int_\Gamma u\, \boldsymbol{l}_\Gamma(v_h) \cdot \boldsymbol{n} \,\mathrm{d}S - \int_\Omega f\, v_h \,\mathrm{d}x - \sum_{\Gamma\in\mathscr{F}_h^N} \int_\Gamma g_N v_h \,\mathrm{d}S \\
&\qquad + \sum_{\Gamma\in\mathscr{F}_h^D} \int_\Gamma u_D \nabla v_h \cdot \boldsymbol{n} \,\mathrm{d}S + \zeta \sum_{\Gamma\in\mathscr{F}_h^D} \int_\Gamma u_D\, \boldsymbol{l}_\Gamma(v_h) \cdot \boldsymbol{n} \,\mathrm{d}S \\
&\quad = \sum_{K\in\mathscr{T}_h} \int_K \nabla u \cdot \nabla v_h \,\mathrm{d}x - \sum_{\Gamma\in\mathscr{F}_h^{ID}} \int_\Gamma \langle \nabla u \cdot \boldsymbol{n}\rangle [v_h] \,\mathrm{d}S \\
&\qquad - \int_\Omega f\, v_h \,\mathrm{d}x - \sum_{\Gamma\in\mathscr{F}_h^N} \int_\Gamma g_N v_h \,\mathrm{d}S = 0, \quad v_h \in S_{hp},
\end{aligned}
$$

which proves the lemma. □

Corollary 3.4 *The linearity of $A_h^{\mathrm{BR}}(\cdot,\cdot)$ and Lemma 3.3 gives the Galerkin orthogonality of the numerical method (3.43), i.e.,*

$$
A_h^{\mathrm{BR}}(u - u_h, v_h) = A_h^{\mathrm{BR}}(u, v_h) - A_h^{\mathrm{BR}}(u_h, v_h) = 0 \quad v_h \in S_{hp}, \tag{3.48}
$$

where u is the regular weak solution of (2.1) and u_h is the BR2 approximate solution.

The error estimates will be proven in the norm

$$
|||v|||_{BR}^2 := |v|_{H^1(\Omega,\mathscr{T}_h)}^2 + \|v\|_{\mathrm{li}}^2, \qquad v \in H^1(\Omega, \mathscr{T}_h), \tag{3.49}
$$

where the broken Sobolev seminorm $|\cdot|_{H^1(\Omega,\mathscr{T}_h)}$ is defined in (2.31) and

$$
\|v\|_{\mathrm{li}} := \left(\sum_{\Gamma\in\mathscr{F}_h} \|\boldsymbol{l}_\Gamma(v)\|_{L^2(\Omega)}^2 \right)^{1/2}, \qquad v \in H^1(\Omega, \mathscr{T}_h), \tag{3.50}
$$

with the lifting operator $\boldsymbol{l}_\Gamma(\cdot)$ defined by (3.16).

Exercise 3.5 Prove that $|||\cdot|||_{BR}$ is a norm in $H^1(\Omega, \mathscr{T}_h)$.

Lemma 3.6 (Coercivity and continuity) *Let $\mathscr{T}_h$, $h \in (0, \bar{h})$, be a system of conforming shape-regular triangulations (cf. assumptions (2.19) and (MA4) from Sect. (2.3.2)). Let $A_h^{\mathrm{BR}}(\cdot,\cdot)$ be the bilinear form defined by (3.42) with $\zeta > 3$. Then there exist constants $C_B > 0$ and $C_C > 0$ such that*

$$
\begin{aligned}
\left| A_h^{\mathrm{BR}}(u_h, v_h) \right| &\le C_B |||u_h|||_{BR}\, |||v_h|||_{BR} \qquad \forall\, u_h, v_h \in S_{hp}, &(3.51)\\
A_h^{\mathrm{BR}}(v_h, v_h) &\ge C_C |||v_h|||_{BR}^2 \qquad \forall\, v_h \in S_{hp}. &(3.52)
\end{aligned}
$$

Proof (i) Due to assumption (MA4) from Sect. 2.3.2, each $K \in \mathscr{T}_h$ has 3 faces. Then, as a simple consequence of (3.19) and the inequality $(a_1 + a_2 + a_3)^2 \leq 3(a_1^2 + a_2^2 + a_3^2)$ we get

$$\|\boldsymbol{L}(v_h)\|_{L^2(K)}^2 \leq 3 \sum_{\Gamma \in \mathscr{F}_h(K)} \|\boldsymbol{l}_\Gamma(v_h)\|_{L^2(K)}^2, \quad v_h \in S_{hp}, \tag{3.53}$$

$$\|\boldsymbol{L}(v_h)\|_{L^2(\Omega)}^2 \leq 3 \sum_{\Gamma \in \mathscr{F}_h} \|\boldsymbol{l}_\Gamma(v_h)\|_{L^2(\Omega)}^2 = 3\|v_h\|_{\mathrm{li}}^2, \quad v_h \in S_{hp}, \tag{3.54}$$

where the second inequality follows from summing the first inequality over all $K \in \mathscr{T}_h$ and reordering of addends. Then (3.53) together with (3.42), (3.17) and the Cauchy inequality imply that

$$\begin{aligned}
\left|A_h^{\mathrm{BR}}(u_h, v_h)\right| &\leq \sum_{K \in \mathscr{T}_h} \big(\|\nabla u_h\|_{L^2(K)} \|\nabla v_h\|_{L^2(K)} + \|\boldsymbol{L}(u_h)\|_{L^2(K)} |\nabla v_h|_{L^2(K)} \\
&\quad + \|\nabla u_h\|_{L^2(K)} \|\boldsymbol{L}(v_h)\|_{L^2(K)}\big) + \zeta \sum_{\Gamma \in \mathscr{F}_h} \|\boldsymbol{l}_\Gamma(u_h)\|_{L^2(\Omega)} \|\boldsymbol{l}_\Gamma(v_h)\|_{L^2(\Omega)} \\
&\leq |u_h|_{H^1(\Omega,\mathscr{T}_h)} |v_h|_{H^1(\Omega,\mathscr{T}_h)} + |u_h|_{H^1(\Omega,\mathscr{T}_h)} \left(3 \sum_{\Gamma \in \mathscr{F}_h} \|\boldsymbol{l}_\Gamma(v_h)\|_{L^2(\Omega)}^2\right)^{1/2} \\
&\quad + |v_h|_{H^1(\Omega,\mathscr{T}_h)} \left(3 \sum_{\Gamma \in \mathscr{F}_h} \|\boldsymbol{l}_\Gamma(u_h)\|_{L^2(\Omega)}^2\right)^{1/2} + \zeta \|u_h\|_{\mathrm{li}} \|v_h\|_{\mathrm{li}}
\end{aligned} \tag{3.55}$$

Now, the discrete Cauchy inequality immediately gives (3.51) with $C_B = 3 + \zeta$.
(ii) Let $\delta > 0$. Then from (3.42), (3.17), (3.53) and the Young inequality we derive

$$\begin{aligned}
A_h^{\mathrm{BR}}(v_h, v_h) &= \sum_{K \in \mathscr{T}_h} \left(|\nabla v_h|_{H^1(K)}^2 + 2\int_K \boldsymbol{L}(v_h) \cdot \nabla v_h \,\mathrm{d}x + \zeta \sum_{\Gamma \in \mathscr{F}_h} \|\boldsymbol{l}_\Gamma(v_h)\|_{L^2(K)}^2\right) \\
&\geq \sum_{K \in \mathscr{T}_h} \left((1-\delta)|\nabla v_h|_{H^1(K)}^2 - \frac{1}{\delta}\|\boldsymbol{L}(v_h)\|_{L^2(K)}^2 + \zeta \sum_{\Gamma \in \mathscr{F}_h} \|\boldsymbol{l}_\Gamma(v_h)\|_{L^2(K)}^2\right) \\
&\geq \sum_{K \in \mathscr{T}_h} \left((1-\delta)|\nabla v_h|_{H^1(K)}^2 + \left(\zeta - \frac{3}{\delta}\right) \sum_{\Gamma \in \mathscr{F}_h} \|\boldsymbol{l}_\Gamma(v_h)\|_{L^2(K)}^2\right).
\end{aligned}$$

This implies that (3.52) holds with $C_C = \min(1 - \delta, \zeta - \frac{3}{\delta})$, which is positive for $\delta \in (3/\zeta, 1)$. □

Corollary 3.7 *By virtue of Corollary 1.7, the coercivity of the form A_h^{BR} implies the existence and uniqueness of the solution of the discrete problem (3.43).*

Remark 3.8 Lemma 3.6 can be easily extended to nonconforming triangulations with hanging nodes. In this case, we have to assume that the parameter $\zeta > g_{\max}$, where $g_{\max} = \max_{K\in\mathscr{T}_h} g_K$ and g_K is the number of faces of the element $K \in \mathscr{T}_h$. Then the proof of Lemma 3.6 remains the same with $C_B = g_{\max} + \zeta$ and $C_C = \min(1-\delta, \zeta - \frac{g_{\max}}{\delta})$, which is positive for $\delta \in (g_{\max}/\zeta, 1)$.

In order to derive the error estimates of the BR2 method, we use the following results. Similarly as in Sect. 2.6, for $\Gamma \in \mathscr{F}_h$ we define h_Γ either by (2.25) or by (2.26). Thus,

$$C_T h_K \le h_\Gamma \le C_G h_K, \quad C_G^{-1} h_K^{-1} \le h_\Gamma^{-1} \le C_T^{-1} h_K^{-1} \qquad (3.56)$$
$$\forall \Gamma \in \mathscr{F}_h,\ \Gamma \subset \partial K,\ K \in \mathscr{T}_h,$$

where C_T and C_G are the constants from (2.20).

Lemma 3.9 *Let $\Gamma \in \mathscr{F}_h$, $\varphi \in H^1(\Omega, \mathscr{T}_h)$ and let $\boldsymbol{l}_\Gamma(\varphi) \in \boldsymbol{\Sigma}_{hp}$ be defined by (3.16). Then there exists a constant $C_s > 0$ such that*

$$\|\langle \boldsymbol{l}_\Gamma(\varphi)\rangle\|_{L^2(\Gamma)} \le C_s h_\Gamma^{-1/2} \|\boldsymbol{l}_\Gamma(\varphi)\|_{L^2(\Omega)}. \qquad (3.57)$$

Proof We consider the case $\Gamma \in \mathscr{F}_h^I$ (for $\Gamma \in \mathscr{F}_h^D$ the proof is analogous and for $\Gamma \in \mathscr{F}_h^N$ inequality (3.57) is trivial, since $\boldsymbol{l}_\Gamma = 0$).

Let $K_\Gamma^{(L)}$ and $K_\Gamma^{(R)}$ denote two elements sharing $\Gamma \in \mathscr{F}_h^I$. From (3.16) it follows that $K_\Gamma^{(L)} \cup K_\Gamma^{(R)}$ is the support of $\boldsymbol{l}_\Gamma(\varphi)$. Then

$$\begin{aligned}\|\langle \boldsymbol{l}_\Gamma(\varphi)\rangle\|_{L^2(\Gamma)}^2 &= \int_\Gamma \langle \boldsymbol{l}_\Gamma(\varphi)\rangle^2 \,\mathrm{d}S = \int_\Gamma \frac{1}{4}(\boldsymbol{l}_\Gamma(\varphi)|_{K_\Gamma^{(L)}} + \boldsymbol{l}_\Gamma(\varphi)|_{K_\Gamma^{(R)}})^2 \,\mathrm{d}S \\ &\le \frac{1}{2}\left(\|\boldsymbol{l}_\Gamma(\varphi)\|_{L^2(\partial K_\Gamma^{(L)})}^2 + \|\boldsymbol{l}_\Gamma(\varphi)\|_{L^2(\partial K_\Gamma^{(R)})}^2\right).\end{aligned} \qquad (3.58)$$

Since $\boldsymbol{l}_\Gamma(\varphi)$ is piecewise polynomial, the combination of (2.78) and (2.86) gives

$$\|\boldsymbol{l}_\Gamma(\varphi)\|_{L^2(\partial K_\Gamma^{(L)})}^2 \le C_M(1 + C_I)\,\mathrm{diam}(K_\Gamma^{(L)})^{-1} \,\|\boldsymbol{l}_\Gamma(\varphi)\|_{L^2(K_\Gamma^{(L)})}^2 \qquad (3.59)$$

and an analogous relation for $K_\Gamma^{(R)}$. Hence, (3.58) and (3.59) and (3.56) yield (3.57) with $C_s = (C_G C_M(1 + C_I))^{1/2}$. □

Lemma 3.10 *There exists a constant $C_\ell > 0$ such that*

$$\|[v_h]\|_{L^2(\Gamma)} \le C_\ell h_\Gamma^{1/2} \|\boldsymbol{l}_\Gamma(v_h)\|_{L^2(\Omega)} \quad \forall v_h \in S_{hp}\ \forall \Gamma \in \mathscr{F}_h^{ID}, \qquad (3.60)$$
$$\|\boldsymbol{l}_\Gamma(\varphi)\|_{L^2(\Omega)} \le C_s h_\Gamma^{-1/2} \|[\varphi]\|_{L^2(\Gamma)} \quad \forall \varphi \in H^1(\Omega, \mathscr{T}_h)\ \forall \Gamma \in \mathscr{F}_h^{ID}, \qquad (3.61)$$

where C_s is the constant from Lemma 3.9.

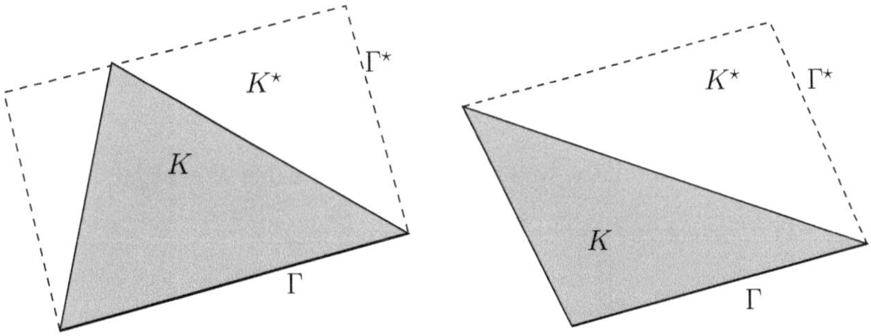

Fig. 3.2 Proof of Lemma 3.10: 2D examples of an element K, a face Γ and the corresponding parallelogram $K^\star$ with the face $\Gamma^\star$

Proof (i) We follow the proof in [40]. Let $\Gamma \in \mathscr{F}_h^{ID}$ be a face of an element $K \in \mathscr{T}_h$ and $\varphi_h \in P_p(\Gamma)$ be a polynomial function of degree $\leq p$ on Γ. Let $K^\star$ be a parallelogram such that $K \subset K^\star$, Γ is one of its faces, and let $\Gamma^\star$ be the face of K^* neighbouring to Γ but not a subset of K (see Fig. 3.2 showing possible 2D situations). Let $\mathscr{P}(\varphi_h) \in P_p(K^\star)$ be the extension of φ_h on $K^\star$, which is constant along each line parallel with $\Gamma^\star$. Then, using (3.56), we have

$$\|\mathscr{P}(\varphi_h)\|_{L^2(K)}^2 \leq \int_{K^\star} \mathscr{P}(\varphi_h)^2 \,\mathrm{d}x \leq h_K \int_\Gamma \varphi_h^2 \,\mathrm{d}S \leq C_T^{-1} h_\Gamma \|\varphi_h\|_{L^2(\Gamma)}^2. \quad (3.62)$$

Into (3.16) we substitute $\varphi := v_h \in S_{hp}$ and $\boldsymbol{r}_h \in \boldsymbol{\Sigma}_{hp}$ such that

$$\boldsymbol{r}_h := \begin{cases} \mathscr{P}([v_h]_\Gamma)\boldsymbol{n}_\Gamma & \text{on } K, \\ 0 & \text{elsewhere,} \end{cases} \quad (3.63)$$

where $\boldsymbol{n}_\Gamma$ is the outer unit normal corresponding to Γ. It means that $\boldsymbol{r}_h$ is parallel with $\boldsymbol{n}_\Gamma$ and its absolute value is equal to $\mathscr{P}([v_h])$. Thus we have

$$\int_\Gamma [v_h] \langle \mathscr{P}([v_h])\boldsymbol{n} \rangle \cdot \boldsymbol{n} \,\mathrm{d}S = -\int_K \mathscr{P}([v_h])\boldsymbol{n} \cdot \boldsymbol{l}_\Gamma(v_h) \,\mathrm{d}x. \quad (3.64)$$

Since $\mathscr{P}([v_h]) = [v_h]$ on Γ, the left-hand side of (3.64) is equal to $\|[v_h]\|_{L^2(\Gamma)}^2$. Therefore, the Cauchy inequality, (3.62) and (3.64) give

$$\begin{aligned} \|[v_h]\|_{L^2(\Gamma)}^2 &\leq \|\mathscr{P}([v_h])\|_{L^2(K)} \|\boldsymbol{l}_\Gamma(v_h)\|_{L^2(K)} \\ &\leq C_T^{-1/2} h_\Gamma^{1/2} \|[v_h]\|_{L^2(\Gamma)} \|\boldsymbol{l}_\Gamma(v_h)\|_{L^2(K)}, \end{aligned} \quad (3.65)$$

which proves (3.60) with $C_\ell = C_T^{-1/2}$.

(ii) In order to prove (3.61), let $\Gamma \in \mathscr{F}_h^I$ and $\varphi \in H^1(\Omega, \mathscr{T}_h)$. Putting $\boldsymbol{r}_h := \boldsymbol{l}_\Gamma(\varphi)$ in (3.16) and using (3.57), we obtain

$$\|\boldsymbol{l}_\Gamma(\varphi)\|^2_{L^2(\Omega)} = -\int_\Gamma [\varphi]\boldsymbol{n}\cdot\langle\boldsymbol{l}_\Gamma(\varphi)\rangle\,\mathrm{d}S \le \|[\varphi]\|_{L^2(\Gamma)}\|\langle\boldsymbol{l}_\Gamma(\varphi)\rangle\|_{L^2(\Gamma)}$$
$$\le C_s h_\Gamma^{-1/2}\|\boldsymbol{l}_\Gamma(\varphi)\|_{L^2(\Omega)}\|[\varphi]\|_{L^2(\Gamma)}, \qquad (3.66)$$

which proves (3.61). Similarly, we prove (3.61) in case that $\Gamma \in \mathscr{F}_h^D$. □

Lemma 3.11 *There exists a constant* $C_L > 0$ *such that*

$$|||v - \Pi_{hp}v|||_{BR} \le C_L h^{\mu-1}|v|_{H^\mu(\Omega)} \quad \forall v \in H^s(\Omega), \qquad (3.67)$$

where $\mu = \min(p+1, s)$ *and* Π_{hp} *is the operator defined by (2.90) and has the approximation properties (2.97).*

Proof By (3.49), (3.50) and (3.16), we have

$$|||v - \Pi_{hp}v|||^2_{BR} = |v - \Pi_{hp}v|^2_{H^1(\Omega,\mathscr{T}_h)} + \sum_{\Gamma\in\mathscr{F}_h^{ID}} \|\boldsymbol{l}_\Gamma(v - \Pi_{hp}v)\|^2_{L^2(\Omega)}. \qquad (3.68)$$

The use of (3.61) with $\varphi := v - \Pi_{hp}v$, (3.56) and (2.101) (with $\alpha = -1$) imply that

$$\sum_{\Gamma\in\mathscr{F}_h^{ID}} \|\boldsymbol{l}_\Gamma(v - \Pi_{hp}v)\|^2_{L^2(\Omega)} \le C_s^2 \sum_{\Gamma\in\mathscr{F}_h^{ID}} h_\Gamma^{-1}\left\|[v - \Pi_{hp}v]\right\|^2_{L^2(\Gamma)} \qquad (3.69)$$
$$\le C_s^2 C_T^{-1}\sum_{K\in\mathscr{T}_h} h_K^{-1}\left\|v - \Pi_{hp}v\right\|^2_{L^2(\partial K)} \le 2C_s^2 C_T^{-1} C_M C_A^2 h^{2\mu-2}|v|^2_{H^\mu(\Omega,\mathscr{T}_h)}.$$

Finally, from (3.68), (3.69) and (2.97) applied to the first term on the right-hand side of (3.68), we obtain (3.67) with $C_L = C_A(1 + 2C_s^2 C_T^{-1} C_M)^{1/2}$. □

Now, we are ready to formulate the following error estimate.

Theorem 3.12 ($|||\cdot|||_{BR}$-norm error estimate) *Let us assume that* $s \ge 2$ *is an integer,* $u \in H^s(\Omega)$ *is the strong weak solution of problem (2.1),* $\{\mathscr{T}_h\}_{h\in(0,\bar{h})}$ *is the system of conforming shape-regular triangulations of* Ω, S_{hp} *is the space of discontinuous piecewise polynomial functions (2.34),* $u_h \in S_{hp}$ *is its BR2 approximate solution defined by (3.42) and (3.44) and* $\mu = \min(p+1, s)$. *Then there exists a constant* $C_1 > 0$ *independent of* h *and* u *such that the estimate*

$$|||u - u_h|||_{BR} \le C_1 h^{\mu-1}|u|_{H^\mu(\Omega)} \qquad (3.70)$$

holds for all $h \in (0, \bar{h})$.

Proof From the triangle inequality, we have

$$|||u - u_h|||_{BR} \le |||u - \Pi_{hp}u\,|||_{BR} + |||\Pi_{hp}u - u_h|||_{BR}. \qquad (3.71)$$

The coercivity property (3.52), the Galerkin orthogonality (3.48) and the continuity property (3.51) imply that

$$
\begin{aligned}
C_C|||\Pi_{hp}u - u_h|||_{BR}^2 &\le A_h^{\mathrm{BR}}(\Pi_{hp}u - u_h, \Pi_{hp}u - u_h) \qquad (3.72)\\
&= A_h^{\mathrm{BR}}(\Pi_{hp}u - u, \Pi_{hp}u - u_h) + \underbrace{A_h^{\mathrm{BR}}(u - u_h, \Pi_{hp}u - u_h)}_{=0}\\
&\le C_B|||\Pi_{hp}u - u|||_{BR}\,|||\Pi_{hp}u - u_h|||_{BR}.
\end{aligned}
$$

Hence,

$$
|||\Pi_{hp}u - u_h|||_{BR} \le \frac{C_B}{C_C}|||\Pi_{hp}u - u|||_{BR}. \qquad (3.73)
$$

Finally, from (3.71), (3.73) and (3.67) we get

$$
|||u - u_h|||_{BR} \le \left(1 + \frac{C_B}{C_C}\right)|||u - \Pi_{hp}u|||_{BR} \le \left(1 + \frac{C_B}{C_C}\right)C_L h^{\mu-1}|u|_{H^\mu(\Omega)},
$$

which proves the theorem with $C_1 = (1 + C_B/C_C)/C_L$. □

In what follows, using the duality technique, we prove the optimal order of convergence in the $L^2(\Omega)$-norm. To this end, similarly as in Sect. 2.7.2, we consider the dual problem: Given $z \in L^2(\Omega)$, find ψ satisfying (2.155). We assume that its weak solution is regular, namely, $\psi \in H^2(\Omega)$ and moreover, there exists a constant $C_D > 0$, independent of z and such that estimate (2.158) is valid. To this end, we assume that the domain Ω is convex. Then, by [153], the mentioned assumptions are satisfied.

Theorem 3.13 ($L^2(\Omega)$-error estimate) *Let us assume that $s \ge 2$ is an integer, $u \in H^s(\Omega)$ is the strong weak solution of problem (2.1), $\{\mathscr{T}_h\}_{h\in(0,\bar{h})}$ is the system of conforming shape-regular triangulations of Ω, S_{hp} is the space of discontinuous piecewise polynomial functions (2.34), $u_h \in S_{hp}$ is its BR2 approximate solution defined by (3.42) and (3.44) and $\mu = \min(p+1, s)$. Let there exist the weak solution of the dual problem (2.155) from $H^2(\Omega)$ satisfying (2.158). Then there exists a constant $C > 0$, independent of h and u, such that the estimate*

$$
\|u - u_h\|_{L^2(\Omega)} \le C_2 h^\mu |u|_{H^s(\Omega)}, \qquad (3.74)
$$

holds for all $h \in (0, \bar{h})$.

Proof The proof is based on the duality technique, which was applied in the proof of Theorem 2.49. Let $\psi \in H^2(\Omega)$ be the solution of the dual problem (2.157) with $z := e_h = u_h - u \in L^2(\Omega)$. Then

$$|\psi|_{H^2(\Omega)} \le C_D \|e_h\|_{L^2(\Omega)}, \tag{3.75}$$

where C_D is the constant from (2.158).

Moreover, let $\Pi_{h1}\psi \in S_{h1}$ be the approximation of ψ defined by (2.90) with $p = 1$. Taking the regularity of the solution ψ of problem (2.155) and the consistency of the BR2 method (3.45) into account, we can show that

$$A_h^{\mathrm{BR}}(\psi, v) = (u_h - u, v)_{L^2(\Omega)} \qquad \forall v \in H^2(\Omega, \mathscr{T}_h), \tag{3.76}$$

where the form A_h^{BR} is given by (3.41). The symmetry of A_h^{BR}, the Galerkin orthogonality of the error (3.48) and (3.76) with $v := e_h$ yield

$$\begin{aligned} \|e_h\|^2_{L^2(\Omega)} &= A_h^{\mathrm{BR}}(\psi, e_h) = A_h^{\mathrm{BR}}(e_h, \psi) \\ &= A_h^{\mathrm{BR}}(e_h, \psi - \Pi_{h1}\psi). \end{aligned} \tag{3.77}$$

By virtue of (3.41) we can write

$$\begin{aligned} A_h^{\mathrm{BR}}(e_h, \psi - \Pi_{h1}\psi) = &\sum_{K\in\mathscr{T}_h} \int_K \nabla e_h \cdot \nabla(\psi - \Pi_{h1}\psi)\, dx \qquad (=: \chi_1) \\ &- \sum_{\Gamma\in\mathscr{F}_h^{ID}} \int_\Gamma \langle\nabla(\psi - \Pi_{h1}\psi)\rangle \cdot \boldsymbol{n}\, [e_h]\, \mathrm{d}S \qquad (=: \chi_2) \\ &- \sum_{\Gamma\in\mathscr{F}_h^{ID}} \int_\Gamma \langle\nabla e_h\rangle \cdot \boldsymbol{n}\, [\psi - \Pi_{h1}\psi]\, \mathrm{d}S \qquad (=: \chi_3) \\ &+ \zeta \sum_{\Gamma\in\mathscr{F}_h} \int_\Omega \boldsymbol{l}_\Gamma(e_h) \cdot \boldsymbol{l}_\Gamma(\psi - \Pi_{h1}\psi)\, \mathrm{d}x. \qquad (=: \chi_4) \end{aligned} \tag{3.78}$$

We estimate the individual terms on the right-hand side of (3.78). The Cauchy inequality, (3.49), (3.67) and (3.70) imply that

$$\begin{aligned} |\chi_1| &\le |e_h|_{H^1(\Omega,\mathscr{T}_h)} |\psi - \Pi_{h1}\psi|_{H^1(\Omega,\mathscr{T}_h)} \\ &\le |||e_h|||_{BR}\, |||\psi - \Pi_{h1}\psi|||_{BR} \le C_L C_1 h^\mu |u|_{H^\mu(\Omega)} |\psi|_{H^2(\Omega)}. \end{aligned} \tag{3.79}$$

Moreover, let h_Γ associated with $\Gamma \in \mathscr{F}_h$ be given either by (2.25) or by (2.26). Then it satisfies (3.56). The Cauchy inequality implies that

$$|\chi_2| \le \left(\sum_{\Gamma\in\mathscr{F}_h^{ID}} \int_\Gamma h_\Gamma (\langle\nabla(\psi - \Pi_{h1}\psi)\rangle \cdot \boldsymbol{n})^2\, \mathrm{d}S \right)^{1/2} \left(\sum_{\Gamma\in\mathscr{F}_h^{ID}} \int_\Gamma h_\Gamma^{-1} [e_h]^2\, \mathrm{d}S \right)^{1/2}. \tag{3.80}$$

Combining (2.107) and (2.102) (with $\alpha = 1$ and $p = 1$), we get

$$\sum_{\Gamma \in \mathscr{F}_h^{ID}} \int_\Gamma h_\Gamma (\langle \nabla(\psi - \Pi_{h1}\psi)\rangle \cdot \boldsymbol{n})^2 \, \mathrm{d}S \tag{3.81}$$
$$\le C_G \sum_{K \in \mathscr{T}_h} h_K \int_{\partial K} |\nabla(\psi - \Pi_{h1}\psi)|^2 \, \mathrm{d}S \le 2C_G C_M C_A^2 h^2 |\psi|^2_{H^2(\Omega)}.$$

Moreover, the identity $e_h = u_h - u = u_h - \Pi_{hp}u + \Pi_{hp}u - u$, the Cauchy inequality, relations (2.106), (2.101) (with $\alpha = -1$) and (3.60) give

$$\sum_{\Gamma \in \mathscr{F}_h^{ID}} \int_\Gamma h_\Gamma^{-1} [e_h]^2 \, \mathrm{d}S \tag{3.82}$$
$$\le 2 \sum_{\Gamma \in \mathscr{F}_h^{ID}} \int_\Gamma h_\Gamma^{-1} [u - \Pi_{hp}u]^2 \, \mathrm{d}S + 2 \sum_{\Gamma \in \mathscr{F}_h^{ID}} \int_\Gamma h_\Gamma^{-1} [\Pi_{hp}u - u_h]^2 \, \mathrm{d}S$$
$$\le 4C_T^{-1} \sum_{K \in \mathscr{T}_h} h_K^{-1} \int_{\partial K} (u - \Pi_{hp}u)^2 \, \mathrm{d}S + 2 \sum_{\Gamma \in \mathscr{F}_h^{ID}} \int_\Gamma h_\Gamma^{-1} [\Pi_{hp}u - u_h]^2 \, \mathrm{d}S$$
$$\le 8C_A^2 C_M C_T^{-1} h^{2(\mu-1)} |u|^2_{H^\mu(\Omega)} + 2C_l^2 \sum_{\Gamma \in \mathscr{F}_h^{ID}} \|\boldsymbol{l}_\Gamma(\Pi_{hp}u - u_h)\|^2_{L^2(\Omega)}.$$

Further, (3.49), (3.50), (3.67) and (3.70) imply that

$$\sum_{\Gamma \in \mathscr{F}_h^{ID}} \|\boldsymbol{l}_\Gamma(\Pi_{hp}u - u_h)\|^2_{L^2(\Omega)} \le |||\Pi_{hp}u - u_h|||^2_{BR} \tag{3.83}$$
$$\le 2|||\Pi_{hp}u - u|||^2_{BR} + 2|||u - u_h|||^2_{BR} \le 2(C_L^2 + C_1^2) h^{2(\mu-1)} |u|^2_{H^\mu(\Omega)}.$$

Finally, (3.80)–(3.83) give

$$|\chi_2| \le C_3 h^\mu |u|_{H^\mu(\Omega)} |\psi|_{H^2(\Omega)} \tag{3.84}$$

with $C_3 := \{2C_G C_M C_A^2 (8C_A^2 C_M C_T^{-1} + 4C_l^2 (C_L^2 + C_1^2))\}^{1/2}$. Similarly, from (3.80) and the Cauchy inequality it follows that

$$|\chi_3| \le \left(\sum_{\Gamma \in \mathscr{F}_h^{ID}} \int_\Gamma h_\Gamma (\langle \nabla e_h \rangle \cdot \boldsymbol{n})^2 \, \mathrm{d}S \right)^{1/2} \left(\sum_{\Gamma \in \mathscr{F}_h^{ID}} \int_\Gamma h_\Gamma^{-1} [\psi - \Pi_{h1}\psi]^2 \, \mathrm{d}S \right)^{1/2}. \tag{3.85}$$

Using (2.168), similarly as in (2.169), we obtain

$$\sum_{\Gamma \in \mathscr{F}_h^{ID}} \int_\Gamma h_\Gamma(\langle \nabla e_h \rangle \cdot \boldsymbol{n})^2 \, dS \tag{3.86}$$

$$\leq C_G C_M \sum_{K \in \mathscr{T}_h} h_K \left(\|\nabla e_h\|_{L^2(K)} |\nabla e_h|_{H^1(K)} + h_K^{-1} \|\nabla e_h\|^2_{L^2(K)} \right)$$

$$\leq C_G C_M \left\{ C_A(1 + C_I) h^{\mu-1} |e_h|_{H^1(\Omega,\mathscr{T}_h)} |u|_{H^\mu(\Omega)} + (1 + C_I)|e_h|^2_{H^1(\Omega,\mathscr{T}_h)} \right\}.$$

Further, the combination of (2.106) and (2.101) (with $\alpha = -1$ and $p = 1$) yields

$$\sum_{\Gamma \in \mathscr{F}_h^{ID}} \int_\Gamma h_\Gamma^{-1} [\psi - \Pi_{h1}\psi]^2 \, dS \tag{3.87}$$

$$\leq \sum_{K \in \mathscr{T}_h} 2h_K^{-1} C_T^{-1} \int_{\partial K} |\psi - \Pi_{h1}\psi|^2 \, dS \leq 4 C_M C_A^2 C_T^{-1} h^2 |\psi|^2_{H^2(\Omega)}.$$

Thus, using (3.49), (3.70), (3.85)–(3.87), we get

$$|\chi_3| \leq C_4 h^\mu |u|_{H^\mu(\Omega)} |\psi|_{H^2(\Omega)} \tag{3.88}$$

with $C_4 = 2C_M C_A \big(C_G C_T^{-1}(C_1 C_A + C_1^2)(1 + C_I)\big)^{1/2}$.

Finally, it follows from the Cauchy inequality, (3.49), (3.67) and (3.70) that

$$|\chi_4| \leq \zeta \left(\sum_{\Gamma \in \mathscr{F}_h} \int_\Omega \|\boldsymbol{l}_\Gamma(e_h)\|^2 \, dx \right)^{1/2} \left(\sum_{\Gamma \in \mathscr{F}_h} \int_\Omega \|\boldsymbol{l}_\Gamma(\psi - \Pi_{h1}\psi)\|^2 \, dx \right)^{1/2}$$

$$\leq \zeta |||e_h|||_{BR} \, |||\psi - \Pi_{h1}\psi|||_{BR} \leq \zeta C_L C_1 h^\mu |u|_{H^\mu(\Omega)} |\psi|_{H^2(\Omega)}. \tag{3.89}$$

In order to conclude the proof, we combine (3.77)–(3.79), (3.84), (3.88), (3.89) and (3.75) to obtain (3.74) with

$$C_2 = ((1 + \zeta) C_L C_1 + C_3 + C_4) C_D.$$

□

3.3 Local Discontinuous Galerkin Method

The subject of this section is the definition and analysis of a popular DG technique called the *local discontinuous Galerkin* (LDG) method. We start again from the abstract formulation (3.5) and (3.6), where the fluxes $\hat{u}$ and $\hat{\boldsymbol{q}}$ are defined in general by the relations

$$\hat{u} = \begin{cases} \langle u_h \rangle - [u_h]\boldsymbol{\beta} \cdot \boldsymbol{n} - \gamma [\boldsymbol{q}_h] \cdot \boldsymbol{n} & \text{on } \Gamma \in \mathscr{F}_h^I, \\ u_D & \text{on } \Gamma \in \mathscr{F}_h^D, \\ u_h - \gamma(\boldsymbol{q}_h \cdot \boldsymbol{n} - g_N) & \text{on } \Gamma \in \mathscr{F}_h^N, \end{cases}$$

$$\hat{\boldsymbol{q}} \cdot \boldsymbol{n} = \begin{cases} \langle \boldsymbol{q}_h \rangle \cdot \boldsymbol{n} + \boldsymbol{\beta}[\boldsymbol{q}_h \cdot \boldsymbol{n}] \cdot \boldsymbol{n} - \alpha [u_h] & \text{on } \Gamma \in \mathscr{F}_h^I, \\ \boldsymbol{q}_h \cdot \boldsymbol{n} - \alpha(u_h - u_D) & \text{on } \Gamma \in \mathscr{F}_h^D, \\ g_N & \text{on } \Gamma \in \mathscr{F}_h^N, \end{cases} \tag{3.90}$$

where $\alpha : \cup_{\Gamma \in \mathscr{F}_h} \Gamma \to \mathbb{R}$, $\boldsymbol{\beta} : \cup_{\Gamma \in \mathscr{F}_h} \Gamma \to \mathbb{R}^d$ and $\gamma : \cup_{\Gamma \in \mathscr{F}_h} \Gamma \to \mathbb{R}$ are suitable functions. Again both values of the double-valued fluxes $\hat{u}$ and $\hat{\boldsymbol{q}}$ on interior faces are identical and thus $[\hat{u}] = [\hat{\boldsymbol{q}}] = 0$.

Remark 3.14 The role of the auxiliary parameters α and γ is to ensure the stability and, hence, the accuracy of the method (see the following analysis). If $\gamma = 0$, then formulation (3.5) and (3.6) with fluxes (3.90) represents the LDG method which was studied in [8]. In this case the numerical flux $\hat{u}$ does not depend on $\boldsymbol{q}_h$ and the auxiliary variable $\boldsymbol{q}_h$ can be eliminated, as we show in the following section. This unusual local solvability property gives its name to the LDG method. Finally, let us mention that also the BR2 method has the same property.

In the following, we introduce the mixed formulation of the LDG method and also its variational formulation for $\gamma = 0$.

3.3.1 Mixed Formulation

Similarly as in Sect. 3.2, we start from relations (3.8) and (3.9), which are equivalent to the abstract problem (3.5) and (3.6). Substituting (3.90) into (3.8) and (3.9), we have

$$\begin{aligned} &\sum_{K \in \mathscr{T}_h} \int_K \boldsymbol{q}_h \cdot \boldsymbol{r}_h \, \mathrm{d}x + \sum_{K \in \mathscr{T}_h} \int_K u_h \nabla \cdot \boldsymbol{r}_h \, \mathrm{d}x \\ &\quad = \sum_{\Gamma \in \mathscr{F}_h^D} \int_\Gamma u_D \boldsymbol{r}_h \cdot \boldsymbol{n} \, \mathrm{d}S + \sum_{\Gamma \in \mathscr{F}_h^N} \int_\Gamma \big(u_h - \gamma(\boldsymbol{q}_h \cdot \boldsymbol{n} - g_N)\big) \boldsymbol{r}_h \cdot \boldsymbol{n} \, \mathrm{d}S \\ &\quad + \sum_{\Gamma \in \mathscr{F}_h^I} \int_\Gamma \big(\langle u_h \rangle - [u_h]\boldsymbol{\beta} \cdot \boldsymbol{n} - \gamma [\boldsymbol{q}_h] \cdot \boldsymbol{n}\big) [\boldsymbol{r}_h] \cdot \boldsymbol{n} \, \mathrm{d}S, \end{aligned} \tag{3.91}$$

and

$$\sum_{K\in\mathscr{T}_h}\int_K \boldsymbol{q}_h\cdot\nabla v_h\,\mathrm{d}x \tag{3.92}$$
$$=\sum_{\Gamma\in\mathscr{F}_h^I}\int_\Gamma \big(\langle\boldsymbol{q}_h\rangle\cdot\boldsymbol{n}+\boldsymbol{\beta}[\boldsymbol{q}_h\cdot\boldsymbol{n}]\cdot\boldsymbol{n}-\alpha[u_h]\big)\,[v_h]\,\mathrm{d}S$$
$$+\sum_{\Gamma\in\mathscr{F}_h^D}\int_\Gamma \big(\boldsymbol{q}_h\cdot\boldsymbol{n}-\alpha(u_h-u_D)\big)\,v_h\,\mathrm{d}S+\sum_{\Gamma\in\mathscr{F}_h^N}\int_\Gamma g_N v_h\,\mathrm{d}S+\int_\Omega f\,v_h\,\mathrm{d}x,$$

respectively. In order to introduce the mixed formulation, we define the forms

$$a(\boldsymbol{q}_h,\boldsymbol{r}_h)=\sum_{K\in\mathscr{T}_h}\int_K \boldsymbol{q}_h\cdot\boldsymbol{r}_h\,\mathrm{d}x+\sum_{\Gamma\in\mathscr{F}_h^I}\int_\Gamma (\gamma[\boldsymbol{q}_h]\cdot\boldsymbol{n})\,([\boldsymbol{r}_h]\cdot\boldsymbol{n})\,\mathrm{d}S \tag{3.93}$$
$$+\sum_{\Gamma\in\mathscr{F}_h^N}\int_\Gamma \gamma(\boldsymbol{q}_h\cdot\boldsymbol{n})(\boldsymbol{r}_h\cdot\boldsymbol{n})\,dS,\quad \boldsymbol{q}_h,\,\boldsymbol{r}_h\in\boldsymbol{\Sigma}_{hp},$$
$$b(u_h,\boldsymbol{r}_h)=\sum_{K\in\mathscr{T}_h}\int_K u_h\,\nabla\cdot\boldsymbol{r}_h\,\mathrm{d}x-\sum_{\Gamma\in\mathscr{F}_h^N}\int_\Gamma u_h\,\boldsymbol{r}_h\cdot\boldsymbol{n}\,\mathrm{d}S$$
$$-\sum_{\Gamma\in\mathscr{F}_h^I}\int_\Gamma (\langle u_h\rangle-[u_h]\boldsymbol{\beta}\cdot\boldsymbol{n})\,[\boldsymbol{r}_h]\cdot\boldsymbol{n}\,\mathrm{d}S,\quad u_h\in S_{hp},\ \boldsymbol{r}_h\in\boldsymbol{\Sigma}_{hp},$$
$$c(u_h,v_h)=\sum_{\Gamma\in\mathscr{F}_h^{ID}}\int_\Gamma \alpha[u_h][v_h]\,\mathrm{d}S,\quad u_h,\,v_h\in S_{hp}.$$

Using identity (3.25), we find that

$$b(u_h,\boldsymbol{r}_h)=-\sum_{K\in\mathscr{T}_h}\int_K \nabla u_h\cdot\boldsymbol{r}_h\,\mathrm{d}x+\sum_{\Gamma\in\mathscr{F}_h^D}\int_\Gamma \boldsymbol{r}_h\cdot\boldsymbol{n}\,u_h\,\mathrm{d}S \tag{3.94}$$
$$+\sum_{\Gamma\in\mathscr{F}_h^I}\int_\Gamma (\langle\boldsymbol{r}_h\rangle\cdot\boldsymbol{n}+\boldsymbol{\beta}\cdot\boldsymbol{n}[\boldsymbol{r}_h]\cdot\boldsymbol{n})\,[u_h]\,\mathrm{d}S,\ u_h\in S_{hp},\ \boldsymbol{r}_h\in\boldsymbol{\Sigma}_{hp}.$$

Therefore, (3.91) and (3.92) can be reformulated as a *saddle-point problem.*

Definition 3.15 (*LDG method*) We say that $(u_h,\boldsymbol{q}_h)\in S_{hp}\times\boldsymbol{\Sigma}_{hp}$ is the *approximate solution* of the mixed formulation of the *local discontinuous Galerkin method*, if

$$a(\boldsymbol{q}_h,\boldsymbol{r}_h)+b(u_h,\boldsymbol{r}_h)=F(\boldsymbol{r}_h)\qquad\forall\,\boldsymbol{r}_h\in\boldsymbol{\Sigma}_{hp}, \tag{3.95a}$$
$$-b(v_h,\boldsymbol{q}_h)+c(u_h,v_h)=G(v_h)\qquad\forall\,v_h\in S_{hp}, \tag{3.95b}$$

where the bilinear forms a, b and c are given by (3.93) and

$$F(\boldsymbol{r}_h) = \sum_{\Gamma\in\mathcal{F}_h^D} \int_\Gamma u_D \boldsymbol{r}_h \cdot \boldsymbol{n}\, \mathrm{d}S + \sum_{\Gamma\in\mathcal{F}_h^N} \int_\Gamma (\gamma g_N) \boldsymbol{r}_h \cdot \boldsymbol{n}\, \mathrm{d}S, \tag{3.96}$$

$$G(v_h) = \sum_{\Gamma\in\mathcal{F}_h^D} \int_\Gamma \alpha u_D v_h\, \mathrm{d}S + \sum_{\Gamma\in\mathcal{F}_h^N} \int_\Gamma g_N v_h\, \mathrm{d}S + \int_\Omega f\, v_h\, \mathrm{d}x.$$

We recall that similarly as in IPG methods the Dirichlet as well as Neumann boundary conditions are incorporated in the right-hand sides of (3.95), which is not the case of the classical mixed formulation.

We are able to establish the existence and uniqueness of the approximate solution.

Lemma 3.16 *Let S_{hp} and $\boldsymbol{\Sigma}_{hp}$ be spaces of piecewise polynomial functions defined by (2.34) and (3.3), respectively. Let a, b and c be bilinear forms given by (3.93), and F and G linear forms given by (3.96). If $\alpha > 0$, $\gamma \geq 0$ and $\boldsymbol{\beta} \in \mathbb{R}^d$ is arbitrary for each $\Gamma \in \mathcal{F}_h$, then problem (3.95) has a unique approximate solution $(u_h, \boldsymbol{q}_h) \in S_{hp} \times \boldsymbol{\Sigma}_{hp}$.*

Proof Since problem (3.95) is linear and finite dimensional, it is sufficient to show that the only solution of (3.95) with $f = 0$, $u_D = 0$ and $g_N = 0$ is $u_h = 0$ and $\boldsymbol{q}_h = 0$. Obviously, this setting implies that $F(\boldsymbol{r}_h) = 0$ for all $\boldsymbol{r}_h \in \boldsymbol{\Sigma}_{hp}$ and $G(v_h) = 0$ for all $v_h \in S_{hp}$. Putting $v_h := u_h$ and $\boldsymbol{r}_h := \boldsymbol{q}_h$ in (3.95) and summing both relations, we obtain

$$a(\boldsymbol{q}_h, \boldsymbol{q}_h) + c(u_h, u_h) = 0. \tag{3.97}$$

This implies that $a(\boldsymbol{q}_h, \boldsymbol{q}_h) = 0$ and $c(u_h, u_h) = 0$ since $\gamma \geq 0$ and $\alpha > 0$, respectively. Obviously, $\boldsymbol{q}_h = 0$. Moreover, (3.97) gives

$$[u_h]_\Gamma = 0 \text{ for } \Gamma \in \mathcal{F}_h^I, \qquad u_h|_\Gamma = 0 \text{ for } \Gamma \in \mathcal{F}_h^D, \tag{3.98}$$

since $\alpha > 0$. Furthermore, (3.95b) and the equivalent definition (3.94) of b give

$$b(u_h, \boldsymbol{r}_h) = -\sum_{K\in\mathcal{T}_h} \int_K \nabla u_h \cdot \boldsymbol{r}_h\, \mathrm{d}x = 0 \quad \forall \boldsymbol{r}_h \in \boldsymbol{\Sigma}_{hp}. \tag{3.99}$$

Hence, $\nabla u_h = 0$ on each $K \in \mathcal{T}_h$ and thus u_h is piecewise constant in Ω, which together with (3.98) implies $u_h = 0$. □

3.3.2 Variational Formulation

If the parameter $\gamma = 0$, then, by virtue of Remark 3.14, the auxiliary variable $\boldsymbol{q}_h$ can be eliminated from the formulation. Let us define the *lifting operator* $\boldsymbol{M} : H^1(\Omega, \mathscr{T}_h)^d \to \boldsymbol{\Sigma}_{hp}$ by

$$\sum_{K \in \mathscr{T}_h} \int_K \boldsymbol{M}(\boldsymbol{\theta}) \cdot \boldsymbol{r}_h \, \mathrm{d}x = - \sum_{\Gamma \in \mathscr{F}_h^I} \int_\Gamma [\boldsymbol{\theta} \cdot \boldsymbol{n}] \, [\boldsymbol{r}_h \cdot \boldsymbol{n}] \, \mathrm{d}S \qquad \forall \boldsymbol{r}_h \in \boldsymbol{\Sigma}_{hp}. \tag{3.100}$$

Equation (3.95a) with the equivalent definition (3.94) of $b(u_h, \boldsymbol{r}_h)$ yield the identity

$$\begin{aligned} &\sum_{K \in \mathscr{T}_h} \int_K (\boldsymbol{q}_h - \nabla u_h) \cdot \boldsymbol{r}_h \, \mathrm{d}x \qquad\qquad (3.101)\\ &= - \sum_{\Gamma \in \mathscr{F}_h^{ID}} \int_\Gamma [u_h] \, \langle \boldsymbol{r}_h \rangle \cdot \boldsymbol{n} \, \mathrm{d}S + \sum_{\Gamma \in \mathscr{F}_h^D} \int_\Gamma u_D \, \boldsymbol{r}_h \cdot \boldsymbol{n} \, \mathrm{d}S - \sum_{\Gamma \in \mathscr{F}_h^I} \int_\Gamma \boldsymbol{\beta} \cdot \boldsymbol{n}[u_h] \, [\boldsymbol{r}_h] \cdot \boldsymbol{n} \, \mathrm{d}S \\ &= \sum_{K \in \mathscr{T}_h} \int_K \boldsymbol{L}_{u_D}(u_h) \cdot \boldsymbol{r}_h \, \mathrm{d}x + \sum_{K \in \mathscr{T}_h} \int_K \boldsymbol{M}(u_h \boldsymbol{\beta}) \cdot \boldsymbol{r}_h \, \mathrm{d}x, \qquad \boldsymbol{r}_h \in \boldsymbol{\Sigma}_{hp}, \end{aligned}$$

where the last equality follows from the definitions of the lifting operators (3.13) and (3.100). Hence, (3.101) gives

$$\boldsymbol{q}_h = \nabla u_h + \boldsymbol{L}_{u_D}(u_h) + \boldsymbol{M}(u_h \boldsymbol{\beta}). \tag{3.102}$$

Moreover, putting $\boldsymbol{r}_h := \nabla v_h$ in (3.101) and subtracting from (3.92), after some manipulation we obtain the identity

$$\begin{aligned} &\sum_{K \in \mathscr{T}_h} \int_K \nabla u_h \cdot \nabla v_h \, \mathrm{d}x - \sum_{\Gamma \in \mathscr{F}_h^{ID}} \int_\Gamma \big([u_h] \, \langle \nabla v_h \rangle \cdot \boldsymbol{n} + \langle \boldsymbol{q}_h \rangle \cdot \boldsymbol{n}[v_h] - \alpha [u_h][v_h]\big) \, \mathrm{d}S \\ &- \sum_{\Gamma \in \mathscr{F}_h^I} \int_\Gamma \big(\boldsymbol{\beta} \cdot \boldsymbol{n}[u_h] \, [\nabla v_h] \cdot \boldsymbol{n} + [\boldsymbol{q}_h \cdot \boldsymbol{n}] \boldsymbol{\beta}[v_h] \cdot \boldsymbol{n}\big) \, \mathrm{d}S \qquad\qquad (3.103) \\ &= \int_\Omega f \, v_h \, \mathrm{d}x + \sum_{\Gamma \in \mathscr{F}_h^N} \int_\Gamma g_N v_h \, \mathrm{d}S - \sum_{\Gamma \in \mathscr{F}_h^D} \int_\Gamma u_D (\nabla v_h \cdot \boldsymbol{n} - \alpha v_h) \, \mathrm{d}S. \end{aligned}$$

Taking into account the definitions (3.13), (3.14) and (3.100) of the lifting operators $\boldsymbol{L}_{u_D}$, $\boldsymbol{L}$ and $\boldsymbol{M}$, respectively, we rewrite the identity (3.103) in the form

$$\sum_{K\in\mathscr{T}_h}\int_K \left(\nabla u_h + \boldsymbol{L}_{u_D}(u_h) + \boldsymbol{M}(u_h\boldsymbol{\beta})\right)\cdot\nabla v_h\,\mathrm{d}x \tag{3.104}$$
$$+\sum_{K\in\mathscr{T}_h}\int_K (\boldsymbol{L}(v_h)+\boldsymbol{M}(v_h\boldsymbol{\beta}))\cdot\boldsymbol{q}_h\,\mathrm{d}x + \sum_{\Gamma\in\mathscr{F}_h^{ID}}\int_\Gamma \alpha[u_h][v_h]\,\mathrm{d}S$$
$$=\int_\Omega f\,v_h\,\mathrm{d}x + \sum_{\Gamma\in\mathscr{F}_h^N}\int_\Gamma g_N\,v_h\,\mathrm{d}S + \sum_{\Gamma\in\mathscr{F}_h^D}\int_\Gamma \alpha u_D v_h\,\mathrm{d}S.$$

Putting $\boldsymbol{r}_h := \boldsymbol{L}(v_h)+\boldsymbol{M}(v_h\boldsymbol{\beta})$ in (3.101), we find that

$$\sum_{K\in\mathscr{T}_h}\int_K \boldsymbol{q}_h\cdot\left(\boldsymbol{L}(v_h)+\boldsymbol{M}(v_h\boldsymbol{\beta})\right)\mathrm{d}x \tag{3.105}$$
$$=\sum_{K\in\mathscr{T}_h}\int_K \left(\nabla u_h + \boldsymbol{L}_{u_D}(u_h)+\boldsymbol{M}(u_h\boldsymbol{\beta})\right)\cdot\left(\boldsymbol{L}(v_h)+\boldsymbol{M}(v_h\boldsymbol{\beta})\right)\mathrm{d}x.$$

Furthermore, from (3.104) and (3.105) we have

$$\sum_{K\in\mathscr{T}_h}\int_K \left(\nabla u_h + \boldsymbol{L}_{u_D}(u_h) + \boldsymbol{M}(u_h\boldsymbol{\beta})\right)\cdot\nabla v_h\,\mathrm{d}x \tag{3.106}$$
$$+\sum_{K\in\mathscr{T}_h}\int_K \left(\nabla u_h + \boldsymbol{L}_{u_D}(u_h)+\boldsymbol{M}(u_h\boldsymbol{\beta})\right)\cdot(\boldsymbol{L}(v_h)+\boldsymbol{M}(v_h\boldsymbol{\beta}))\,\mathrm{d}x$$
$$+\sum_{\Gamma\in\mathscr{F}_h^{ID}}\int_\Gamma \alpha[u_h][v_h]\,\mathrm{d}S$$
$$=\int_\Omega f\,v_h\,\mathrm{d}x + \sum_{\Gamma\in\mathscr{F}_h^N}\int_\Gamma g_N\,v_h\,\mathrm{d}S + \sum_{\Gamma\in\mathscr{F}_h^D}\int_\Gamma \alpha u_D v_h\,\mathrm{d}S.$$

Finally, with the aid of (3.15), we rewrite (3.106) in the form

$$A_h^{\mathrm{LDG}}(u_h, v_h) = \ell_h^{\mathrm{LDG}}(v_h) \qquad \forall\, v_h\in S_{hp}, \tag{3.107}$$

where $A_h^{\mathrm{LDG}} : H^1(\Omega,\mathscr{T}_h)\times H^1(\Omega,\mathscr{T}_h)\to\mathbb{R}$ and $\ell_h^{\mathrm{LDG}} : H^1(\Omega,\mathscr{T}_h)\to\mathbb{R}$ are linear operators defined by

$$A_h^{\mathrm{LDG}}(u_h, v_h) = \sum_{\Gamma\in\mathscr{F}_h^{ID}}\int_\Gamma \alpha[u_h][v_h]\,\mathrm{d}S \tag{3.108}$$
$$+\sum_{K\in\mathscr{T}_h}\int_K (\nabla u_h + \boldsymbol{L}(u_h) + \boldsymbol{M}(u_h\boldsymbol{\beta}))\cdot(\nabla v_h + \boldsymbol{L}(v_h)+\boldsymbol{M}(v_h\boldsymbol{\beta}))\,\mathrm{d}x,$$

$$\ell_h^{\mathrm{LDG}}(v_h) = \int_\Omega f\, v_h \,\mathrm{d}x + \sum_{\Gamma\in\mathscr{F}_h^N} \int_\Gamma g_N\, v_h \,\mathrm{d}S \tag{3.109}$$
$$- \sum_{\Gamma\in\mathscr{F}_h^D} \int_\Gamma u_D\Big(\big(\nabla v_h + \boldsymbol{L}(v_h) + \boldsymbol{M}(v_h\boldsymbol{\beta})\big)\cdot\boldsymbol{n} - \alpha v_h\Big)\,\mathrm{d}S.$$

The above considerations lead us to the following definition.

Definition 3.17 (*LDG method*) Let the forms A_h^{LDG} and $\boldsymbol{l}_h^{\mathrm{LDG}}$ be defined by (3.108) and (3.109), respectively. We say that $u_h \in S_{hp}$ is an *LDG approximate solution* of problem (2.1), if it satisfies condition (3.107).

3.3.3 Theoretical Analysis

In the sequel, we analyze the LDG method. Similarly, as in [8] we confine ourselves to the case $\gamma = 0$ in (3.90). The general case was studied in [43]. We start with several lemmas.

First, we introduce a local variant of the lifting operator $\boldsymbol{M}$. Namely, for each $\Gamma \in \mathscr{F}_h^I$ we define the operator $\boldsymbol{m}_\Gamma : H^1(\Omega, \mathscr{T}_h)^d \to \boldsymbol{\Sigma}_{hp}$ by

$$\sum_{K\in\mathscr{T}_h} \int_K \boldsymbol{m}_\Gamma(\boldsymbol{\theta})\cdot\boldsymbol{r}_h \,\mathrm{d}x = -\int_\Gamma [\boldsymbol{\theta}\cdot\boldsymbol{n}]\,[\boldsymbol{r}_h\cdot\boldsymbol{n}]\,\mathrm{d}S \qquad \forall \boldsymbol{r}_h \in \boldsymbol{\Sigma}_{hp}. \tag{3.110}$$

Obviously, if $\boldsymbol{\theta} \in H^1(\Omega, \mathscr{T}_h)^d$, then the support of $\boldsymbol{m}_\Gamma(\boldsymbol{\theta}) \in \boldsymbol{\Sigma}_{hp}$, where $\Gamma \in \mathscr{F}_h^I$ is the union of two elements $K \in \mathscr{T}_h$ sharing the face Γ. Moreover, due to (3.100), we have

$$\boldsymbol{M}(\boldsymbol{\theta}) = \sum_{\Gamma\in\mathscr{F}_h^I} \boldsymbol{m}_\Gamma(\boldsymbol{\theta}) \qquad \forall \boldsymbol{\theta} \in \boldsymbol{\Sigma}_{hp}. \tag{3.111}$$

Furthermore, let $K \in \mathscr{T}_h$ and $\Gamma \in \mathscr{F}_h(K) = \{\Gamma \in \mathscr{F}_h;\ \Gamma \subset \partial K\}$ be arbitrary but fixed. Let $\boldsymbol{r}_h \in H^1(\Omega, \mathscr{T}_h)^d$ vanish outside K. Then $[\boldsymbol{r}_h]_\Gamma\cdot\boldsymbol{n}_\Gamma = 2\langle\boldsymbol{r}_h\rangle_\Gamma\cdot\boldsymbol{n}_\Gamma$. Moreover, for $\varphi_h \in H^1(\Omega, \mathscr{T}_h)$, by virtue of (3.16) and (3.110), we have

$$\int_K \boldsymbol{m}_\Gamma(\varphi_h\boldsymbol{n}_\Gamma)\cdot\boldsymbol{r}_h \,\mathrm{d}x = -\int_\Gamma [\varphi_h\boldsymbol{n}\cdot\boldsymbol{n}]\,[\boldsymbol{r}_h\cdot\boldsymbol{n}]\,\mathrm{d}S \tag{3.112}$$
$$= -2\int_\Gamma [\varphi_h\boldsymbol{n}\cdot\boldsymbol{n}]\,\langle\boldsymbol{r}_h\cdot\boldsymbol{n}\rangle\,\mathrm{d}S = 2\int_K \boldsymbol{l}_\Gamma(\varphi_h)\cdot\boldsymbol{r}_h\,\mathrm{d}x.$$

Since $\boldsymbol{r}_h$ and K are arbitrary, we conclude that

$$\boldsymbol{m}_\Gamma(\varphi_h\boldsymbol{n}_\Gamma) = 2\boldsymbol{l}_\Gamma(\varphi_h) \ \ \forall \varphi_h \in H^1(\Omega, \mathscr{T}_h),\ \Gamma \in \mathscr{F}_h^I. \tag{3.113}$$

Now, similarly as in Sect. 3.2, we present the consistency, continuity and coercivity of the LDG method.

Lemma 3.18 (Consistency) *The LDG method (3.107)–(3.109) is consistent, i.e., if* $u \in H^2(\Omega)$ *is the weak solution of (2.1), then*

$$A_h^{\mathrm{LDG}}(u, v_h) = \ell_h^{\mathrm{LDG}}(v_h) \qquad \forall v_h \in S_{hp}, \tag{3.114}$$

where $A_h^{\mathrm{LDG}}(\cdot,\cdot)$ *and* $\ell_h^{\mathrm{LDG}}(\cdot)$ *are given by (3.108) and (3.109), respectively.*

Proof Let $u \in H^2(\Omega)$ be the weak solution of (2.1). Obviously, $[u]_\Gamma = 0$ for $\Gamma \in \mathscr{F}_h^I$. Moreover, $\boldsymbol{M}(\boldsymbol{\beta} u) = 0$,

$$\sum_{K\in\mathscr{T}_h} \int_K \boldsymbol{L}(u) \boldsymbol{r}_h \, \mathrm{d}x = - \sum_{\Gamma\in\mathscr{F}_h^D} \int_\Gamma u \boldsymbol{r}_h \cdot \boldsymbol{n} \, \mathrm{d}S \qquad \forall \boldsymbol{r}_h \in \boldsymbol{\Sigma}_{hp}. \tag{3.115}$$

and

$$\sum_{K\in\mathscr{T}_h} \int_K \nabla u \cdot \boldsymbol{M}(\boldsymbol{\beta} v_h) \, \mathrm{d}x = - \sum_{\Gamma\in\mathscr{F}_h^I} \int_\Gamma [\boldsymbol{\beta} v_h \cdot \boldsymbol{n}][\nabla u \cdot \boldsymbol{n}] \, \mathrm{d}S = 0 \qquad \forall v_h \in S_{hp}, \tag{3.116}$$

which follows from the definition of the lifting operators (3.14) and (3.100). Then, (3.108), together with (3.115) and (3.116), give

$$\begin{aligned} A_h^{\mathrm{LDG}}(u, v_h) &= \sum_{K\in\mathscr{T}_h} \int_K \nabla u \cdot (\nabla v_h + \boldsymbol{L}(v_h) + \boldsymbol{M}(v_h \boldsymbol{\beta})) \, \mathrm{d}x \\ &\quad + \sum_{K\in\mathscr{T}_h} \int_K \boldsymbol{L}(u) \cdot (\nabla v_h + \boldsymbol{L}(v_h) + \boldsymbol{M}(v_h \boldsymbol{\beta})) \, \mathrm{d}x + \sum_{\Gamma\in\mathscr{F}_h^D} \int_\Gamma \alpha u v_h \, \mathrm{d}S \\ &= \sum_{K\in\mathscr{T}_h} \int_K \nabla u \cdot \nabla v_h \, \mathrm{d}x - \sum_{\Gamma\in\mathscr{F}_h^{ID}} \int_\Gamma [v_h] \nabla u \cdot \boldsymbol{n} \, \mathrm{d}S \\ &\quad \sum_{\Gamma\in\mathscr{F}_h^D} \int_\Gamma u \Big((\nabla v_h + \boldsymbol{L}(v_h) + \boldsymbol{M}(v_h \boldsymbol{\beta})) \cdot \boldsymbol{n} - \alpha v_h \Big) \, \mathrm{d}S. \end{aligned} \tag{3.117}$$

Finally, since $u = u_D$ on $\partial\Omega_D$, the weak solution u satisfies (2.40) and $\langle \nabla u \rangle_\Gamma = \nabla u|_\Gamma$ for $\Gamma \in \mathscr{F}_h^{ID}$, relations (3.117) and (3.109) imply that

$$A_h^{\mathrm{LDG}}(u, v_h) - \ell_h^{\mathrm{LDG}}(v_h) = \sum_{K\in\mathscr{T}_h}\int_K \nabla u\cdot\nabla v_h\,\mathrm{d}x - \sum_{\Gamma\in\mathscr{F}_h^{ID}}\int_\Gamma [v_h]\nabla u\cdot\boldsymbol{n}\,\mathrm{d}S$$
$$-\int_\Omega f\,v_h\,\mathrm{d}x - \sum_{\Gamma\in\mathscr{F}_h^N}\int_\Gamma g_N\,v_h\,\mathrm{d}S = 0.$$

□

Corollary 3.19 *The linearity of* A_h^{LDG} *and Lemma 3.18 gives the Galerkin orthogonality of the local discontinuous Galerkin method, i.e.,*

$$A^{\mathrm{LDG}}(u-u_h, v_h) = A^{\mathrm{LDG}}(u, v_h) - A^{\mathrm{LDG}}(u_h, v_h) = 0 \quad \forall v_h \in S_{hp}, \tag{3.118}$$

where u *is the strong weak solution of (2.1) and* u_h *is its approximation given by (3.107)–(3.109).*

In what follows, we again employ the norm defined by (3.49), i.e.,

$$|||v|||_{BR}^2 := |v|_{H^1(\Omega,\mathscr{T}_h)}^2 + \|v\|_{\mathrm{li}}^2, \qquad v\in H^1(\Omega,\mathscr{T}_h), \tag{3.119}$$

where the broken Sobolev seminorm $|\cdot|_{H^1(\Omega,\mathscr{T}_h)}$ is defined by (2.31) and $\|\cdot\|_{\mathrm{li}}$ is defined by (3.50).

Lemma 3.20 (Coercivity and continuity) *Let* A_h^{LDG} *be the form defined by (3.108) with* $\alpha|_\Gamma = C_W/h_\Gamma$, $\Gamma\in\mathscr{F}_h$, $C_W>0$ *and* $\|\boldsymbol{\beta}\|_\infty := \max_{\Gamma\in\mathscr{F}_h}\|\boldsymbol{\beta}\|_{L^\infty(\Gamma)} < \infty$. *If*

$$C_W \ge C_S^2\left(\frac{1}{2} + 9\,(1+2\|\boldsymbol{\beta}\|_\infty)^2\right), \tag{3.120}$$

where C_S *is given by (3.57), then there exist constants* $C_B>0$ *and* $C_C>0$ *such that*

$$\left|A_h^{\mathrm{LDG}}(u_h, v_h)\right| \le C_B\,|||u_h|||_{BR}\,|||v_h|||_{BR} \quad \forall u_h, v_h \in S_{hp}, \tag{3.121}$$
$$A_h^{\mathrm{LDG}}(v_h, v_h) \ge C_C\,|||v_h|||_{BR}^2 \quad \forall v_h\in S_{hp}. \tag{3.122}$$

Proof (i) Let $u_h, v_h\in S_{hp}$. Then from (3.108) we have

$$A_h^{\mathrm{LDG}}(u_h, v_h) \tag{3.123}$$
$$= \sum_{K\in\mathscr{T}_h}\int_K \nabla u_h\cdot\nabla v_h\,\mathrm{d}x \qquad (=: \chi_1)$$
$$+ \sum_{K\in\mathscr{T}_h}\int_K (\boldsymbol{L}(u_h) + \boldsymbol{M}(u_h\boldsymbol{\beta}))\cdot\nabla v_h\,\mathrm{d}x \qquad (=: \chi_2)$$
$$+ \sum_{K\in\mathscr{T}_h}\int_K (\boldsymbol{L}(v_h) + \boldsymbol{M}(v_h\boldsymbol{\beta}))\cdot\nabla u_h\,\mathrm{d}x \qquad (=: \chi_3)$$

$$+ \sum_{K \in \mathscr{T}_h} \int_K (\boldsymbol{L}(u_h) + \boldsymbol{M}(u_h \boldsymbol{\beta})) \cdot (\boldsymbol{L}(v_h) + \boldsymbol{M}(v_h \boldsymbol{\beta}))\, \mathrm{d}x \qquad (=: \chi_4)$$

$$+ \sum_{\Gamma \in \mathscr{F}_h^{ID}} \int_\Gamma \alpha [u_h][v_h]\, \mathrm{d}S. \qquad (=: \chi_5)$$

Now we estimate the individual terms on the right-hand side. Obviously, due to the Cauchy inequality, (3.60), (3.50) and the fact that the support of $\boldsymbol{l}_\Gamma$ consists of at most two elements, we have

$$\begin{aligned} |\chi_1| &\leq |u_h|_{H^1(\Omega, \mathscr{T}_h)} |v_h|_{H^1(\Omega, \mathscr{T}_h)}, \\ |\chi_5| &\leq \sum_{\Gamma \in \mathscr{F}_h} \alpha \|[u_h]\|_{L^2(\Gamma)} \|[v_h]\|_{L^2(\Gamma)} \leq 2C_w C_l^2 \|u_h\|_{\mathrm{li}} \|v_h\|_{\mathrm{li}}. \end{aligned} \tag{3.124}$$

Due to assumption (MA4) from Sect. 2.3.2, each $K \in \mathscr{T}_h$ has 3 faces. As a simple consequence of (3.111) and the inequality $(a_1 + a_2 + a_3)^2 \leq 3(a_1^2 + a_2^2 + a_3^2)$ we obtain

$$\begin{aligned} \|\boldsymbol{M}(\boldsymbol{r}_h)\|_{L^2(K)}^2 &\leq 3 \sum_{\Gamma \in \mathscr{F}_h(K)} \|\boldsymbol{m}_\Gamma(\boldsymbol{r}_h)\|_{L^2(K)}^2, \quad \boldsymbol{r}_h \in H^1(\Omega, \mathscr{T}_h)^d, \\ \|\boldsymbol{M}(\boldsymbol{r}_h)\|_{L^2(\Omega)}^2 &\leq 3 \sum_{\Gamma \in \mathscr{F}_h} \|\boldsymbol{m}_\Gamma(\boldsymbol{r}_h)\|_{L^2(\Omega)}^2, \quad \boldsymbol{r}_h \in H^1(\Omega, \mathscr{T}_h)^d, \end{aligned} \tag{3.125}$$

where the second relation follows from the summation of the first inequality over all $K \in \mathscr{T}_h$ and reordering of addends. Now, (3.125) together with (3.50) and (3.113) give

$$\|\boldsymbol{M}(\boldsymbol{\beta} v_h)\|_{L^2(\Omega)}^2 \leq 12 \|\boldsymbol{\beta}\|_\infty^2 \sum_{\Gamma \in \mathscr{F}_h} \|\boldsymbol{l}_\Gamma(v_h)\|_{L^2(\Omega)}^2 = 12 \|\boldsymbol{\beta}\|_\infty^2 \|v_h\|_{\mathrm{li}}^2. \tag{3.126}$$

Hence, from the triangle inequality, (3.54) and (3.126) it follows that

$$\|\boldsymbol{L}(v_h) + \boldsymbol{M}(\boldsymbol{\beta} v_h)\|_{L^2(\Omega)} \leq \sqrt{3}(1 + 2\|\boldsymbol{\beta}\|_\infty) \|v_h\|_{\mathrm{li}} \quad \forall\, v_h \in S_{hp}. \tag{3.127}$$

Then, the Cauchy inequality, (3.123) and (3.127) yield

$$\begin{aligned} |\chi_2| &\leq \|\boldsymbol{L}(u_h) + \boldsymbol{M}(u_h \boldsymbol{\beta})\|_{L^2(\Omega)} |v_h|_{H^1(\Omega, \mathscr{T}_h)} \\ &\leq \sqrt{3}\,(1 + 2\|\boldsymbol{\beta}\|_\infty)\, \|u_h\|_{\mathrm{li}} |v_h|_{H^1(\Omega, \mathscr{T}_h)}, \end{aligned} \tag{3.128}$$

and similarly

$$|\chi_3| \leq \sqrt{3}\,(1 + 2\|\boldsymbol{\beta}\|_\infty)\, \|v_h\|_{\mathrm{li}} |u_h|_{H^1(\Omega, \mathscr{T}_h)}. \tag{3.129}$$

Using the same process, we obtain

$$\begin{aligned}|\chi_4| &\leq \|\boldsymbol{L}(u_h)+\boldsymbol{M}(u_h\boldsymbol{\beta})\|_{L^2(\Omega)}\|\boldsymbol{L}(v_h)+\boldsymbol{M}(v_h\boldsymbol{\beta})\|_{L^2(\Omega)} \\ &\leq 3\,(1+2\|\boldsymbol{\beta}\|_\infty)^2\,\|u_h\|_{\mathrm{li}}\|v_h\|_{\mathrm{li}}.\end{aligned} \tag{3.130}$$

Finally, by (3.123), (3.124), (3.128)–(3.130), we have (3.121) with $C_B = 2C_wC_l^2 + \left(1+\sqrt{3}(1+2\|\boldsymbol{\beta}\|_\infty)\right)^2$.

(ii) Let $\delta > 0$ and let $v_h \in S_{hp}$. We derive from (3.108), the Cauchy inequality, (3.127), (3.61), (3.50) and the Young inequality the following relations:

$$\begin{aligned}A_h^{\mathrm{LDG}}(v_h, v_h) &= \sum_{K\in\mathscr{T}_h}\int_K |\nabla v_h|^2\,\mathrm{d}x + 2\sum_{K\in\mathscr{T}_h}\int_K (\boldsymbol{L}(v_h)+\boldsymbol{M}(v_h\boldsymbol{\beta}))\cdot\nabla v_h\,\mathrm{d}x \\ &\quad + \sum_{K\in\mathscr{T}_h}\int_K |\boldsymbol{L}(v_h)+\boldsymbol{M}(v_h\boldsymbol{\beta})|^2\,\mathrm{d}x + \sum_{\Gamma\in\mathscr{F}_h^{ID}}\int_\Gamma \alpha[v_h]^2\,\mathrm{d}S \\ &\geq |v_h|^2_{H^1(\Omega,\mathscr{T}_h)} - 2|v_h|_{H^1(\Omega,\mathscr{T}_h)}\|\boldsymbol{L}(v_h)+\boldsymbol{M}(v_h\boldsymbol{\beta})\|_{L^2(\Omega)} \\ &\quad + \|\boldsymbol{L}(v_h)+\boldsymbol{M}(v_h\boldsymbol{\beta})\|^2_{L^2(\Omega)} + \sum_{\Gamma\in\mathscr{F}_h^{ID}}\frac{C_w}{C_s^2}\|\boldsymbol{l}_\Gamma(v_h)\|^2_{L^2(\Omega)} \\ &\geq |v_h|^2_{H^1(\Omega,\mathscr{T}_h)} - 2|v_h|_{H^1(\Omega,\mathscr{T}_h)}\sqrt{3}(1+2\|\boldsymbol{\beta}\|_\infty)\|v_h\|_{\mathrm{li}} \\ &\quad - 3(1+2\|\boldsymbol{\beta}\|_\infty)^2\|v_h\|^2_{\mathrm{li}} + \frac{C_w}{C_s^2}\|v_h\|^2_{\mathrm{li}} \\ &\geq |v_h|^2_{H^1(\Omega,\mathscr{T}_h)}(1-\delta) + \|v_h\|^2_{\mathrm{li}}\left\{\left(-1-\frac{1}{\delta}\right)3(1+2\|\boldsymbol{\beta}\|_\infty)^2 + \frac{C_w}{C_s^2}\right\}.\end{aligned} \tag{3.131}$$

Putting $\delta = 1/2$ and using the choice (3.120), we obtain (3.122) with $C_C = 1/2$. □

Corollary 3.21 *The Corollary 1.7 and the coercivity of the form A_h^{LDG} imply the existence and uniqueness of the solution of the discrete problem (3.107).*

Let $\Pi_{hp}u$ be the S_{hp}-interpolation of u given by (2.90). Similarly, as in Sect. 3.2, we can use estimate (3.67).

Theorem 3.22 ($|||\cdot|||_{BR}$-norm error estimate) *Let $u \in H^s(\Omega)$ with $s \geq 2$ be the weak solution of (2.1) and let $u_h \in S_{hp}$ be the LDG approximate solution defined by (3.107)–(3.109). Then there exists a constant $C > 0$ such that*

$$|||u - u_h|||_{BR} \leq Ch^{\mu-1}|u|_{H^s(\Omega)}, \tag{3.132}$$

where $\mu = \min(p+1, s)$.

Exercise 3.23 Prove Theorem 3.22. Hint: Follow the proof of Theorem 3.12. Only the coercivity and continuity of the form A_h^{BR} have to be replaced by the coercivity and continuity of A_h^{LDG}.

Theorem 3.24 ($L^2(\Omega)$-error estimates) *Let $u \in H^s(\Omega)$ with $s \geq 2$ be the weak solution of (2.1) and let $u_h \in S_{hp}$ be its LDG approximation given by (3.107)–(3.109). Then there exists a constant $C > 0$ such that the following estimate holds:*

$$\|u - u_h\|_{L^2(\Omega)} \leq Ch^{\mu}|u|_{H^s(\Omega)}, \tag{3.133}$$

where $\mu = \min(p + 1, s)$.

Exercise 3.25 Prove Theorem 3.24. Hint: Use the duality arguments similarly as in Theorem 3.13.

Chapter 4
DGM for Convection-Diffusion Problems

The next Chaps. 4–6 will be devoted to the DGM for the solution of nonstationary, in general nonlinear, convection-diffusion initial-boundary value problems. Some equations treated here can serve as a simplified model of the Navier–Stokes system describing compressible flow, but the subject of convection-diffusion problems is important for a number of areas in science and technology, as is mentioned in the introduction.

In this chapter we are concerned with the analysis of the DGM applied to the space discretization of nonstationary linear and nonlinear convection-diffusion equations. The time variable will be left as continuous. This means that we deal with the so-called *space semidiscretization*, also called the *method of lines*. The full space-time discretization will be the subject of Chaps. 5 and 6.

The diffusion terms are discretized by interior penalty Galerkin techniques (SIPG, NIPG and IIPG) introduced in Chap. 2. A special attention is paid to the discretization of convective terms, where the concept of the numerical flux (well-known from the finite volume method) is used. We derive error estimates for a nonlinear equation discretized by all three mentioned techniques. These estimates are suboptimal in the $L^\infty(L^2)$-norm and they are not uniform with respect to the diffusion coefficient. However, for the symmetric SIPG variant, the optimal error estimate in the $L^\infty(L^2)$-norm is derived. Finally, for a linear convection-diffusion equation, we derive error estimates uniform with respect to the diffusion coefficient.

4.1 Scalar Nonlinear Nonstationary Convection-Diffusion Equation

Let $\Omega \subset \mathbb{R}^d$, $d = 2, 3$, be a bounded polygonal (if $d = 2$) or polyhedral (if $d = 3$) domain with Lipschitz boundary $\partial\Omega = \partial\Omega_D \cup \partial\Omega_N$, $\partial\Omega_D \cap \partial\Omega_N = \emptyset$, and $T > 0$. We assume that the $(d-1)$-dimensional measure of $\partial\Omega_D$ is positive. Let us denote $Q_T = \Omega \times (0, T)$.

V. Dolejší and M. Feistauer, *Discontinuous Galerkin Method*,
Springer Series in Computational Mathematics 48,
DOI 10.1007/978-3-319-19267-3_4

We are concerned with the following nonstationary nonlinear convection-diffusion problem with initial and mixed Dirichlet–Neumann boundary conditions: Find $u : \overline{Q}_T \to \mathbb{R}$ such that

$$\frac{\partial u}{\partial t} + \sum_{s=1}^{d} \frac{\partial f_s(u)}{\partial x_s} = \varepsilon \Delta u + g \quad \text{in } Q_T, \tag{4.1a}$$

$$u\big|_{\partial\Omega_D \times (0,T)} = u_D, \tag{4.1b}$$

$$\varepsilon \boldsymbol{n} \cdot \nabla u\big|_{\partial\Omega_N \times (0,T)} = g_N, \tag{4.1c}$$

$$u(x, 0) = u^0(x), \quad x \in \Omega. \tag{4.1d}$$

We assume that the data satisfy the following conditions:

$$\boldsymbol{f} = (f_1, \ldots, f_d),\ f_s \in C^1(\mathbb{R}),\ f_s' \text{ are bounded},\ f_s(0) = 0,\ s = 1, \ldots, d, \tag{4.2a}$$

$$\varepsilon > 0, \tag{4.2b}$$

$$g \in C([0, T]; L^2(\Omega)), \tag{4.2c}$$

$$u_D = \text{trace of some } u^* \in C([0, T]; H^1(\Omega)) \cap L^\infty(Q_T) \text{ on } \partial\Omega_D \times (0, T), \tag{4.2d}$$

$$g_N \in C([0, T]; L^2(\partial\Omega_N)), \tag{4.2e}$$

$$u^0 \in L^2(\Omega). \tag{4.2f}$$

The constant ε is a diffusion coefficient, f_s, $s = 1, \ldots, d$, are nonlinear convective fluxes and g is a source term. It can be seen that the assumption that $f_s(0) = 0$ is not limiting. If u satisfies (4.1a), then it also satisfies the equation

$$\frac{\partial u}{\partial t} + \sum_{s=1}^{d} \frac{\partial (f_s(u) - f_s(0))}{\partial x_s} = \varepsilon \Delta u + g,$$

and the new convective fluxes $\tilde{f}_s(u) := f_s(u) - f_s(0)$, $s = 1, \ldots, d$, satisfy (4.2a). Let us note that in Sect. 6.2 we will be concerned with more complicated situation, where both convection and diffusion terms are nonlinear.

It is suitable to introduce the concept of a *weak solution*. To this end, we define the space

$$H^1_{0D}(\Omega) = \{v \in H^1(\Omega);\ v|_{\partial\Omega_D} = 0\},$$

and the following forms:

$$(u, v) = (u, v)_{L^2(\Omega)} = \int_\Omega uv \, dx, \quad u, v \in L^2(\Omega),$$

$$a(u, v) = \varepsilon \int_\Omega \nabla u \cdot \nabla v \, dx, \quad u, v \in H^1(\Omega),$$

$$b(u, v) = \int_\Omega \sum_{s=1}^d \frac{\partial f_s(u)}{\partial x_s} v \, dx, \quad u \in H^1(\Omega) \cap L^\infty(\Omega), \; v \in L^2(\Omega),$$

$$(u, v)_N = \int_{\partial\Omega_N} u v \, dS, \quad u, v \in L^2(\partial\Omega_N).$$

Definition 4.1 A function u is called the *weak solution* of problem (4.1), if it satisfies the conditions

$$u - u^* \in L^2(0, T; H^1_{0D}(\Omega)), \quad u \in L^\infty(Q_T), \tag{4.3a}$$

$$\frac{d}{dt}(u(t), v) + b(u(t), v) + a(u(t), v) = (g(t), v) + (g_N(t), v)_N \; \forall v \in H^1_{0D}(\Omega)$$
$$\text{(in the sense of distributions in } (0, T)), \tag{4.3b}$$

$$u(0) = u_0 \quad \text{in } \Omega. \tag{4.3c}$$

Let us recall that by $u(t)$ we denote the function in Ω such that $u(t)\,(x) = u(x, t)$, $x \in \Omega$.

With the aid of techniques from [235], [217] or [246], it is possible to prove that for a function u satisfying (4.3a) and (4.3b) we have $u \in C([0, T]; L^2(\Omega))$, which means that condition (4.3c) makes sense, and that there exists a unique solution of problem (4.3). Moreover, it satisfies the condition $\partial u/\partial t \in L^2(Q_T)$. Then (4.3b) can be rewritten as

$$\left(\frac{\partial u(t)}{\partial t}, v\right) + b(u(t), v) + a(u(t), v) = (g(t), v) + (g_N(t), v)_N \tag{4.4}$$
$$\forall v \in H^1_{0D}(\Omega) \text{ and almost every } t \in (0, T).$$

We say that u satisfying (4.3) is a *strong solution*, if

$$u \in L^2(0, T; H^2(\Omega)), \quad \frac{\partial u}{\partial t} \in L^2(0, T; H^1(\Omega)). \tag{4.5}$$

It is possible to show that the strong solution u satisfies Eq. (4.1) pointwise (almost everywhere) and $u \in C([0, T], H^1(\Omega))$.

4.2 Discretization

In this section we introduce a DG space semidiscretization of problem (4.1). We use the notation and auxiliary results from Sects. 2.3–2.5.

By $\mathscr{T}_h$ $(h > 0)$ we denote a triangulation of the domain Ω introduced in Sect. 2.3.1. We start from the strong solution u satisfying (4.5), multiply equation (4.1a) by an arbitrary $v \in H^2(\Omega, \mathscr{T}_h)$, integrate over each $K \in \mathscr{T}_h$, and apply Green's theorem. We obtain the identity

$$\int_K \frac{\partial u(t)}{\partial t} v \,\mathrm{d}x + \int_{\partial K} \sum_{s=1}^{d} f_s(u(t)) n_s v \,\mathrm{d}S - \int_K \sum_{s=1}^{d} f_s(u(t)) \frac{\partial v}{\partial x_s} \,\mathrm{d}x \\ + \varepsilon \int_K \nabla u(t) \cdot \nabla v \,\mathrm{d}x - \varepsilon \int_{\partial K} (\nabla u(t) \cdot \boldsymbol{n}) v \,\mathrm{d}S = \int_K g(t) v \,\mathrm{d}x. \tag{4.6}$$

Here $\boldsymbol{n} = (n_1, \ldots, n_d)$ denotes the outer unit normal to ∂K. It is possible to write

$$\sum_{s=1}^{d} f_s(u) n_s = \boldsymbol{f}(u) \cdot \boldsymbol{n}, \quad \sum_{s=1}^{d} f_s(u) \frac{\partial v}{\partial x_s} = \boldsymbol{f}(u) \cdot \nabla v. \tag{4.7}$$

Summing (4.6) over all $K \in \mathscr{T}_h$, using the technique introduced in Sect. 2.4 for the discretization of the diffusion term, we obtain the identity

$$\left(\frac{\partial u(t)}{\partial t}, v \right) + A_h(u(t), v) + \tilde{b}_h(u(t), v) = \ell_h(v)(t), \tag{4.8}$$

where

$$A_h(w, v) = \varepsilon a_h(w, v) + \varepsilon J_h^\sigma(w, v), \tag{4.9}$$

$$a_h(u, v) = \sum_{K \in \mathscr{T}_h} \int_K \nabla u \cdot \nabla v \,\mathrm{d}x - \sum_{\Gamma \in \mathscr{F}_h^{ID}} \int_\Gamma (\langle \nabla u \rangle \cdot \boldsymbol{n}[v] + \Theta \langle \nabla v \rangle \cdot \boldsymbol{n}[u]) \,\mathrm{d}S, \tag{4.10}$$

$$J_h^\sigma(u, v) = \sum_{\Gamma \in \mathscr{F}_h^{ID}} \int_\Gamma \sigma [u]\, [v] \,\mathrm{d}S, \tag{4.11}$$

$$\tilde{b}_h(u, v) = \sum_{K \in \mathscr{T}_h} \left\{ \int_{\partial K} \sum_{s=1}^{d} f_s(u(t)) n_s v \,\mathrm{d}S - \int_K \sum_{s=1}^{d} f_s(u(t)) \frac{\partial v}{\partial x_s} \,\mathrm{d}x \right\}, \tag{4.12}$$

$$\ell_h(v)\,(t) = (g(t), v) + (g_N(t), v)_N + \varepsilon \sum_{\Gamma \in \mathscr{F}_h^D} \int_\Gamma \Big(\sigma v - \Theta (\nabla v \cdot \boldsymbol{n}) \Big) u_D(t) \,\mathrm{d}S. \tag{4.13}$$

(The symbols $\langle \cdot \rangle$, $[\cdot]$ are defined in (2.32) and (2.33).) We call a_h and J_h the diffusion form and the interior and boundary penalty form, respectively. Similarly as in (2.104), the *penalty weight* σ is given by

$$\sigma|_\Gamma = \sigma_\Gamma = \frac{C_W}{h_\Gamma}, \quad \Gamma \in \mathscr{F}_h^{ID}, \tag{4.14}$$

where h_Γ characterizes the "size" of $\Gamma \in \mathscr{F}_h$ defined in Sect. 2.6 and $C_W > 0$ is a suitable constant. The symbol $\tilde{b}_h$ corresponds to the convection terms. It will be further discretized.

Similarly, as in Sect. 2.4, for $\Theta = -1$, $\Theta = 0$ and $\Theta = 1$ the form a_h (together with the form J_h^σ) represents the nonsymmetric variant (NIPG), incomplete variant (IIPG) and symmetric variant (SIPG), respectively, of the diffusion form.

Remark 4.2 Let us note that in contrast to Chap. 2, the form A_h contains the diffusion coefficient ε, compare (2.45a)–(2.45c) with (4.9). Therefore, the estimates from Chap. 2, which will be used here, have to be equipped with the multiplication factor $\varepsilon > 0$. We do not emphasize it in the following.

Now we pay a special attention to the approximation of the convective terms represented by the form $\tilde{b}_h$. The integrals $\int_{\partial K} \sum_{s=1}^d f_s(u(t)) n_s v \, \mathrm{d}S$ can be expressed in terms of the expressions $\int_\Gamma \sum_{s=1}^d f_s(u(t)) n_s v \, \mathrm{d}S$, which will be approximated with the aid of the so-called *numerical flux* $H(u, w, \boldsymbol{n})$:

$$\int_\Gamma \sum_{s=1}^d f_s(u(t)) n_s v \, \mathrm{d}S \approx \int_\Gamma H(u_\Gamma^{(L)}, u_\Gamma^{(R)}, \boldsymbol{n}) \, v_\Gamma^{(L)} \, \mathrm{d}S, \qquad \Gamma \in \mathscr{F}_h. \tag{4.15}$$

Here $H : \mathbb{R} \times \mathbb{R} \times \mathrm{B}_1 \to \mathbb{R}$ is a suitably defined function and $\mathrm{B}_1 = \{\boldsymbol{n} \in \mathbb{R}^d;\ |\boldsymbol{n}| = 1\}$ is the unit sphere in $\mathbb{R}^d$. The simplest are the *central numerical fluxes* given by

$$H(v_1, v_2, \boldsymbol{n}) = \sum_{s=1}^d f_s\left(\frac{v_1 + v_2}{2}\right) n_s, \qquad H(v_1, v_2, \boldsymbol{n}) = \sum_{s=1}^d \frac{f_s(v_1) + f_s(v_2)}{2} n_s.$$

However, in the most of applications it is suitable to use *upwinding*[1] numerical fluxes as, for example,

$$H(u_1, u_2, \boldsymbol{n}) = \begin{cases} \sum_{s=1}^d f_s(u_1) n_s, & \text{if } P > 0 \\ \sum_{s=1}^d f_s(u_2) n_s, & \text{if } P \le 0 \end{cases}, \quad \text{where } P = \sum_{s=1}^d f_s'\left(\frac{u_1 + u_2}{2}\right) n_s. \tag{4.16}$$

[1] The concept of upwinding is based on the idea that the information on properties of a quantity u is propagated in the flow direction. Therefore, discretization of convective terms is carried out with the aid of data located in the upwind direction from the points in consideration.

or the *Lax–Friedrichs numerical flux*

$$H(v_1, v_2, \boldsymbol{n}) = \sum_{s=1}^{d} \frac{f_s(v_1) + f_s(v_2)}{2} n_s - \lambda |v_1 - v_2|,$$

where $\lambda > 0$ has to be chosen in an appropriate way. For more examples and theoretical background of numerical fluxes we refer to [127].

If $\Gamma \in \mathscr{F}_h^B$, then it is necessary to specify the meaning of $u_\Gamma^{(R)}$ in (4.15). It is possible to use the *extrapolation* from the interior of the computational domain

$$u_\Gamma^{(R)} := u_\Gamma^{(L)}, \qquad \Gamma \in \mathscr{F}_h^B. \tag{4.17}$$

In the theoretical analysis, we assume that the numerical flux satisfies the following properties:

1. *continuity*: $H(u, v, \boldsymbol{n})$ is *Lipschitz-continuous* with respect to u, v: there exists a constant $L_H > 0$ such that

$$\begin{aligned} &|H(u, v, \boldsymbol{n}) - H(u^*, v^*, \boldsymbol{n})| \le L_H(|u - u^*| + |v - v^*|), \\ &u,\ v,\ u^*,\ v^* \in \mathbb{R},\ \boldsymbol{n} \in \mathrm{B}_1. \end{aligned} \tag{4.18}$$

2. *consistency*:

$$H(u, u, \boldsymbol{n}) = \sum_{s=1}^{d} f_s(u) n_s, \quad u \in \mathbb{R},\ \boldsymbol{n} = (n_1, \ldots, n_d) \in \mathrm{B}_1. \tag{4.19}$$

3. *conservativity*:

$$H(u, v, \boldsymbol{n}) = -H(v, u, -\boldsymbol{n}), \quad u,\ v \in \mathbb{R},\ \boldsymbol{n} \in \mathrm{B}_1. \tag{4.20}$$

By virtue of (4.18) and (4.19), the functions $f_s, s = 1, \ldots, d$, are Lipschitz-continuous with constant $L_f = 2L_H$. From (4.2a) and (4.19) we see that

$$H(0, 0, \boldsymbol{n}) = 0 \quad \forall \boldsymbol{n} \in \mathrm{B}_1. \tag{4.21}$$

Using the conservativity (4.20) of H and notation (2.32) and (2.33), we find that

$$\begin{aligned} &\sum_{K \in \mathscr{T}_h} \sum_{\Gamma \subset \partial K, \Gamma \in \mathscr{F}_h} \int_\Gamma H(u_\Gamma^{(L)}, u_\Gamma^{(R)}, \boldsymbol{n}) v_\Gamma^{(L)} \, \mathrm{d}S \\ &= \sum_{\Gamma \in \mathscr{F}_h^I} \int_\Gamma H(u_\Gamma^{(L)}, u_\Gamma^{(R)}, \boldsymbol{n}) \left(v_\Gamma^{(L)} - v_\Gamma^{(R)} \right) \mathrm{d}S + \sum_{\Gamma \in \mathscr{F}_h^B} \int_\Gamma H(u_\Gamma^{(L)}, u_\Gamma^{(R)}, \boldsymbol{n}) v_\Gamma^{(L)} \, \mathrm{d}S \\ &= \sum_{\Gamma \in \mathscr{F}_h} \int_\Gamma H(u_\Gamma^{(L)}, u_\Gamma^{(R)}, \boldsymbol{n}) [v] \, \mathrm{d}S \end{aligned} \tag{4.22}$$

Let us recall that in integrals $\int_\Gamma$ the symbol $\boldsymbol{n}$ denotes the normal $\boldsymbol{n}_\Gamma$.

Then, by virtue of (4.15) and (4.22), we define the *convection form* $b_h(u, v)$ approximating $\tilde{b}_h(u, v)$:

$$b_h(u, v) = \sum_{\Gamma \in \mathscr{F}_h} \int_\Gamma H(u_\Gamma^{(L)}, u_\Gamma^{(R)}, \boldsymbol{n})\, [v]\, \mathrm{d}S - \sum_{K \in \mathscr{T}_h} \int_K \boldsymbol{f}(u) \cdot \nabla v\, \mathrm{d}x, \tag{4.23}$$
$$u, v \in H^1(\Omega, \mathscr{T}_h), \quad u \in L^\infty(\Omega).$$

By the definitions (4.12), (4.23) and the consistency (4.19), we have

$$b_h(u, v) = \tilde{b}_h(u, v) \quad \forall u \in H^2(\Omega)\, \forall v \in H^2(\Omega, \mathscr{T}_h). \tag{4.24}$$

Let S_{hp} be the space of discontinuous piecewise polynomial functions (2.34). Since $S_{hp} \subset H^2(\Omega, \mathscr{T}_h) \cap L^\infty(\Omega)$, the forms (4.10), (4.11), (4.13) and (4.23) make sense for $u := u_h$, $v := v_h \in S_{hp}$. Then, we introduce the space DG-discretization of (4.1).

Definition 4.3 We define the *semidiscrete approximate solution* as a function $u_h : Q_T \to \mathbb{R}$ satisfying the conditions

$$u_h \in C^1([0, T]; S_{hp}), \tag{4.25a}$$
$$\left(\frac{\partial u_h(t)}{\partial t}, v_h\right) + A_h(u_h(t), v_h) + b_h(u_h(t), v_h) = \ell_h(v_h)\,(t) \tag{4.25b}$$
$$\forall v_h \in S_{hp},\ \forall t \in [0, T],$$
$$(u_h(0), v_h) = (u^0, v_h) \quad \forall v_h \in S_{hp}. \tag{4.25c}$$

We see that the initial condition (4.25c) can be written as $u_h(0) = \Pi_{hp} u^0$, where Π_{hp} is the operator of the $L^2(\Omega)$-projection on the space S_{hp} (cf. (2.90)).

The discrete problem (4.25) is equivalent to an initial value problem for a system of ordinary differential equations (ODEs). Namely, let $\{\varphi_i,\ i = 1, \ldots, N_h\}$ be a basis of the space S_{hp}, where $N_h = \dim S_{hp}$. The approximate solution u_h is sought in the form

$$u_h(x, t) = \sum_{j-1}^{N_h} u^j(t) \varphi_j(x), \tag{4.26}$$

where $u^j(t) : [0, T] \to \mathbb{R}$, $j = 1, \ldots, N_h$, are unknown functions. For simplicity, wc put

$$B_h(u_h, v_h) = \ell_h(v_h) - A_h(u_h, v_h) - b_h(u_h, v_h), \quad u_h, v_h \in S_{hp}.$$

Now, substituting (4.26) into (4.25b) and putting $v_h := \varphi_i$, we get

$$\sum_{j=1}^{N_h} \frac{\mathrm{d}u^j(t)}{\mathrm{d}t} \left(\varphi_j, \varphi_i\right) = B_h \left(\sum_{j=1}^{N_h} u^j(t)\varphi_j, \varphi_i \right), \qquad i = 1, \ldots, N_h, \tag{4.27}$$

which is the system of the ODEs for the unknown functions u^j, $j = 1, \ldots, N_h$. This approach to the numerical solution of initial boundary value problems via the space semidiscretization is called the *method of lines*.

If we apply some ODE solver to problem (4.27), we obtain a fully discrete problem. In Chap. 5 we will pay attention to some full space-time discretization techniques. In what follows we are concerned with the analysis of the semidiscrete problem (4.25).

Taking into account that the exact solution with property (4.5) satisfies $[u]_\Gamma = 0$ for $\Gamma \in \mathscr{F}_h^I$, $u|_{\partial\Omega_D \times (0,T)} = u_D$ and using (4.8) and (4.24), we find that u satisfies the *consistency* identity

$$\left(\frac{\partial u(t)}{\partial t}, v_h \right) + A_h(u(t), v_h) + b_h(u(t), v_h) = \ell_h(v_h)\,(t) \tag{4.28}$$

for all $v_h \in S_{hp}$ and almost all $t \in (0, T)$. This will be used in the error analysis.

Exercise 4.4 Verify the relation (4.28).

4.3 Abstract Error Estimate

In this section we analyze the behaviour of the error in method (4.25). We use results derived in Sects. 2.6 and 2.7 dealing with the properties of the diffusion form a_h and the penalty form J_h^σ. Similarly as in (2.103), we use the DG-norm

$$|||v||| = \left(|v|^2_{H^1(\Omega, \mathscr{T}_h)} + J_h^\sigma(v, v) \right)^{1/2}, \qquad v \in H^1(\Omega, \mathscr{T}_h). \tag{4.29}$$

In the error analysis we suppose that the following basic assumptions are satisfied.

Assumptions 4.5 Let the following assumptions be satisfied:

- assumptions (4.2) on data of problem (4.1),
- properties (4.18)–(4.20) of the numerical flux H,
- $\{\mathscr{T}_h\}_{h \in (0, \bar{h})}$ is a system of triangulations of the domain Ω satisfying the shape-regularity assumption (2.19) and the equivalence condition (2.20) of h_Γ and h_K (cf. Lemma 2.5),
- the penalization constant C_W satisfies the conditions from Corollary 2.41 for SIPG, NIPG and IIPG versions of the diffusion form a_h.

We apply again the multiplicative trace inequality (2.78), the inverse inequality (2.86) and the approximation properties (2.93)–(2.95) and (2.98)–(2.100).

4.3.1 Consistency of the Convection Form in the Case of the Dirichlet Boundary Condition

We are concerned with Lipschitz-continuity and consistency of the form b_h. The consistency analysis is split in two cases. In this section we consider the case when the Dirichlet boundary condition is considered on the whole boundary $\partial\Omega$, i.e., $\partial\Omega_D = \partial\Omega$ and $\partial\Omega_N = \emptyset$. Analyzing the consistency of the form b_h in the case of mixed boundary conditions is more complicated and is presented in Sect. 4.3.2.

In what follows we assume that $s \geq 2,\ p \geq 1$ are integers.

Lemma 4.6 *Let $\Gamma_N = \emptyset$ (then $\mathscr{F}_h = \mathscr{F}_h^{ID}$). Then there exist constants $C_{b1}, \dots, C_{b4} > 0$ such that*

$$|b_h(u,v) - b_h(\bar{u},v)| \leq C_{b1} |||v||| \left(\|u-\bar{u}\|^2_{L^2(\Omega)} + \sum_{K\in\mathscr{T}_h} h_K \|u-\bar{u}\|^2_{L^2(\partial K)} \right)^{1/2}, \tag{4.30}$$

$$u,\ \bar{u} \in H^1(\Omega,\mathscr{T}_h)\cap L^\infty(\Omega),\ v\in H^1(\Omega,\mathscr{T}_h),\ h\in(0,\bar{h}),$$

$$|b_h(u_h,v_h) - b_h(\bar{u}_h,v_h)| \leq C_{b2} |||v_h|||\ \|u_h-\bar{u}_h\|_{L^2(\Omega)}, \tag{4.31}$$

$$u_h,\ \bar{u}_h,\ v_h \in S_{hp},\ h\in(0,\bar{h}).$$

If $\Pi_{hp}u$ is the S_{hp}-interpolant of $u \in H^s(\Omega)$ defined by (2.90) and we put $\eta = u - \Pi_{hp}u$, then

$$|b_h(u,v_h) - b_h(\Pi_{hp}u,v_h)| \leq C_{b3} R_b(\eta) |||v_h|||,\quad v_h\in S_{hp},\ h\in(0,\bar{h}), \tag{4.32}$$

where

$$R_b(\eta) = \left(\sum_{K\in\mathscr{T}_h} \left(\|\eta\|^2_{L^2(K)} + h_K^2 |\eta|^2_{H^1(K)} \right) \right)^{1/2}. \tag{4.33}$$

Moreover, if $\xi = u_h - \Pi_{hp}u$, then under the above assumptions,

$$|b_h(u,v_h) - b_h(u_h,v_h)| \leq C_{b4} |||v_h||| \left(R_b(\eta) + \|\xi\|_{L^2(\Omega)} \right),\quad v_h\in S_{hp},\ h\in(0,\bar{h}). \tag{4.34}$$

Proof (i) By (4.23), for $u,\ \bar{u},\ v \in H^1(\Omega, \mathscr{T}_h)$,

$$b_h(u,v) - b_h(\bar{u},v) = \underbrace{-\sum_{K\in\mathscr{T}_h}\int_K (\boldsymbol{f}(u) - \boldsymbol{f}(\bar{u}))\cdot\nabla v\,\mathrm{d}x}_{=:\sigma_1} \tag{4.35}$$
$$+ \underbrace{\sum_{\Gamma\in\mathscr{F}_h}\int_\Gamma \left(H(u_\Gamma^{(L)}, u_\Gamma^{(R)}, \boldsymbol{n}) - H(\bar{u}_\Gamma^{(L)}, \bar{u}_\Gamma^{(R)}, \boldsymbol{n})\right)[v]\,\mathrm{d}S}_{=:\sigma_2}.$$

Let us recall that for $\Gamma \in \mathscr{F}_h^B$ we define the functions $u_\Gamma^{(R)}$ and $\bar{u}_\Gamma^{(R)}$ by extrapolation: $u_\Gamma^{(R)} = u_\Gamma^{(L)}$ and $\bar{u}_\Gamma^{(R)} = \bar{u}_\Gamma^{(L)}$.

From the Lipschitz-continuity of the functions $f_s,\ s = 1,\dots,d$, and the discrete Cauchy inequality we have

$$|\sigma_1| \le L_f \sum_{K\in\mathscr{T}_h}\int_K \sum_{s=1}^d |u-\bar{u}|\left|\frac{\partial v}{\partial x_s}\right| \mathrm{d}x \le \sqrt{d}L_f\|u-\bar{u}\|_{L^2(\Omega)}|v|_{H^1(\Omega,\mathscr{T}_h)}. \tag{4.36}$$

Relation (4.35), the Lipschitz-continuity (4.18) of H, the Cauchy inequality, (2.20), (4.11) and (4.14) imply that

$$\begin{aligned}
|\sigma_2| &\le L_H \sum_{\Gamma\in\mathscr{F}_h}\int_\Gamma \left(\left|u_\Gamma^{(L)} - \bar{u}_\Gamma^{(L)}\right| + \left|u_\Gamma^{(R)} - \bar{u}_\Gamma^{(R)}\right|\right)|[v]|\,\mathrm{d}S \qquad (4.37)\\
&\le L_H\left(\sum_{\Gamma\in\mathscr{F}_h}\int_\Gamma \frac{[v]^2}{h_\Gamma}\,\mathrm{d}S\right)^{\frac{1}{2}}\left(\sum_{\Gamma\in\mathscr{F}_h}\int_\Gamma h_\Gamma\left(\left|u_\Gamma^{(L)} - \bar{u}_\Gamma^{(L)}\right| + \left|u_\Gamma^{(R)} - \bar{u}_\Gamma^{(R)}\right|\right)^2\mathrm{d}S\right)^{\frac{1}{2}}\\
&\le L_H\sqrt{\frac{C_G}{C_W}}J_h^\sigma(v,v)^{1/2}\left(\sum_{K\in\mathscr{T}_h}\int_{\partial K} 2h_K|u-\bar{u}|^2\,\mathrm{d}S\right)^{1/2}\\
&= L_H\sqrt{\frac{2C_G}{C_W}}J_h^\sigma(v,v)^{1/2}\left(\sum_{K\in\mathscr{T}_h} h_K\|u-\bar{u}\|^2_{L^2(\partial K)}\right)^{1/2}.
\end{aligned}$$

(Let us note that the third inequality in (4.37) is valid only if $\mathscr{F}_h = \mathscr{F}_h^{ID}$.) Taking into account (4.35)–(4.37) and using the discrete Cauchy inequality, we get

$$\begin{aligned}
&|b_h(u,v) - b_h(\bar{u},v)| \qquad (4.38)\\
&\le\sqrt{d}L_f\|u-\bar{u}\|_{L^2(\Omega)}|v|_{H^1(\Omega,\mathscr{T}_h)} + L_H\sqrt{\frac{2C_G}{C_W}}J_h^\sigma(v,v)^{1/2}\left(\sum_{K\in\mathscr{T}_h} h_K\|u-\bar{u}\|^2_{L^2(\partial K)}\right)^{\frac{1}{2}}
\end{aligned}$$

$$\leq \left(dL_f^2 \|u-\bar{u}\|^2_{L^2(\Omega)} + L_H^2 \frac{2C_G}{C_W} \sum_{K\in\mathscr{T}_h} h_K \|u-\bar{u}\|^2_{L^2(\partial K)} \right)^{\frac{1}{2}} \left(|v|^2_{H^1(\Omega,\mathscr{T}_h)} + J_h^\sigma(v,v) \right)^{\frac{1}{2}}.$$

This immediately implies (4.30) with $C_{b1} = \left(\max(dL_f^2, 2L_H^2 C_G/C_W)\right)^{1/2}$.

(ii) Further, let $u_h, \bar{u}_h, v_h \in S_{hp}$. Using the multiplicative trace inequality (2.78) and the inverse inequality (2.86), for $\varphi \in S_{hp}$ we obtain

$$\sum_{K\in\mathscr{T}_h} h_K \|\varphi\|^2_{L^2(\partial K)} \leq C_M \sum_{K\in\mathscr{T}_h} \left(\|\varphi\|^2_{L^2(K)} + h_K \|\varphi\|_{L^2(K)} |\varphi|_{H^1(K)} \right) \tag{4.39}$$

$$\leq C_M \sum_{K\in\mathscr{T}_h} \left(\|\varphi\|^2_{L^2(K)} + C_I \|\varphi\|^2_{L^2(K)} \right) = C_M(1+C_I)\|\varphi\|^2_{L^2(\Omega)}.$$

Now, if we set $\varphi := u_h - \bar{u}_h$ and use (4.30) with $u := u_h, \bar{u} := \bar{u}_h$ and $v := v_h$, we get (4.31) with $C_{b2} = C_{b1}(1 + C_M(1 + C_I))^{1/2}$.

(iii) In order to prove (4.32), we start from (4.30) with $u \in H^s(\Omega), \bar{u} := \Pi_{hp} u$ and $v := v_h \in S_{hp}$. Using the multiplicative trace inequality (2.78) and the Young inequality, we find that

$$\sum_{K\in\mathscr{T}_h} h_K \|u - \Pi_{hp}u\|^2_{L^2(\partial K)} = \sum_{K\in\mathscr{T}_h} h_K \|\eta\|^2_{L^2(\partial K)} \tag{4.40}$$

$$\leq C_M \sum_{K\in\mathscr{T}_h} \left(\|\eta\|^2_{L^2(K)} + h_K \|\eta\|_{L^2(K)} |\eta|_{H^1(K)} \right)$$

$$\leq C_M \sum_{K\in\mathscr{T}_h} \left(\|\eta\|^2_{L^2(K)} + \frac{1}{2}\|\eta\|^2_{L^2(K)} + \frac{1}{2} h_K^2 |\eta|^2_{H^1(K)} \right) \leq \frac{3}{2} C_M R_b(\eta)^2,$$

where $R_b(\eta)$ is defined in (4.33). Consequently,

$$\|u - \Pi_{hp}\|^2_{L^2(\Omega)} + \sum_{K\in\mathscr{T}_h} h_K \|u - \Pi_{hp}u\|^2_{L^2(\partial K)} \leq (1 + \frac{3}{2} C_M) R_b(\eta)^2,$$

which together with (4.30) immediately yield (4.32) with $C_{b3} = C_{b1}(1+3C_M/2)^{1/2}$.

(iv) The triangle inequality gives

$$|b_h(u, v_h) - b_h(u_h, v_h)| \leq |b_h(u, v_h) - b_h(\Pi_{hp}u, v_h)| + |b_h(\Pi_{hp}u, v_h) - b_h(u_h, v_h)|.$$

From relations (4.32) and (4.31) with $\bar{u}_h = \Pi_{hp}u$ and $\xi = u_h - \Pi_{hp}u$, we get (4.34) with $C_{b4} = \max(C_{b2}, C_{b3})$. □

4.3.2 Consistency of the Convective Form in the Case of Mixed Boundary Conditions

Since Lemma 4.6 is valid only if a Dirichlet boundary condition is prescribed on $\partial\Omega$, we are concerned here with the consistency of the form b_h in the case of a nonempty Neumann part $\partial\Omega_N$ of the boundary $\partial\Omega$. We start from several auxiliary results.

The first lemma shows the existence of a vector-valued function with suitable properties. Its proof is based on the usual definition of a domain with the Lipschitz boundary.

Lemma 4.7 *There exists a vector-valued function $\boldsymbol{\varphi} \in (W^{1,\infty}(\Omega))^d$ such that*

$$\boldsymbol{\varphi} \cdot \boldsymbol{n} \geq 1 \quad \text{on } \partial\Omega, \tag{4.41}$$

where $\boldsymbol{n}$ is the unit outer normal to $\partial\Omega$.

Proof By [208] or [227], it follows from the Lipschitz-continuity of $\partial\Omega$ that there exist numbers $\alpha, \beta > 0$, Cartesian coordinate systems

$$X_r = (x_{r,1}, \ldots, x_{r,d-1}, x_{r,d})^{\mathrm{T}} = (x_r', x_{r,d})^{\mathrm{T}}, \tag{4.42}$$

Lipschitz-continuous functions

$$a_r : \Delta_r = \left\{x_r' = (x_{r,1}, \ldots, x_{r,d-1})^{\mathrm{T}}; |x_{r,i}| < \alpha, \, i = 1, \ldots, d-1\right\} \to \mathbb{R} \tag{4.43}$$

with a Lipschitz constant $L > 0$, and orthogonal transformations $A_r : \mathbb{R}^d \to \mathbb{R}^d$, $r = 1, \ldots, m$, such that

$$\forall x \in \partial\Omega \quad \exists r \in \{1, \ldots, m\} \quad \exists x_r' \in \Delta_r : \, x = A_r^{-1}\left(x_r', a_r(x_r')\right). \tag{4.44}$$

Under the notation

$$\begin{aligned} \widehat{V}_r^+ &= \left\{(x_r', x_{r,d}) \in \mathbb{R}^d; a_r(x_r') < x_{r,d} < a_r(x_r') + \beta, \, x_r' \in \Delta_r\right\}, \\ \widehat{V}_r^- &= \left\{(x_r', x_{r,d}) \in \mathbb{R}^d; a_r(x_r') - \beta < x_{r,d} < a_r(x_r'), \, x_r' \in \Delta_r\right\}, \\ \widehat{\Lambda}_r &= \left\{(x_r', x_{r,d}); x_{r,d} = a_r(x_r') \in \mathbb{R}, \, x_r' \in \Delta_r\right\}, \end{aligned} \tag{4.45}$$

we have

$$\widehat{V}_r^+ \subset A_r(\Omega), \quad \widehat{\Lambda}_r \subset A_r(\partial\Omega), \quad \widehat{V}_r^- \subset A_r(\mathbb{R}^d \setminus \overline{\Omega}), \quad \partial\Omega \subset \bigcup_{r=1}^m U_r, \tag{4.46}$$

where the sets U_r are defined by the relations

$$\widehat{U}_r = \widehat{V}_r^+ \cup \widehat{\Lambda}_r \cup \widehat{V}_r^-, \quad U_r = A_r^{-1}(\widehat{U}_r). \tag{4.47}$$

The mappings A_r can be written in the form

$$A_r(x) = \mathbb{Q}_r x + x_r^0, \quad x \in \mathbb{R}^d, \tag{4.48}$$

where $x_r^0 \in \mathbb{R}^d$ and $\mathbb{Q}_r$ are orthogonal $d \times d$ matrices, i. e., $\mathbb{Q}_r \mathbb{Q}_r^{\mathrm{T}} = \mathbb{I} =$ unit matrix. Then the transformation of a d-dimensional vector $y \in \mathbb{R}^d$ reads as

$$y \in \mathbb{R}^d \to \mathbb{Q}_r y \in \mathbb{R}^d. \tag{4.49}$$

The sets U_r are open. There exists an open set U_0 such that

$$\overline{U}_0 \subset \Omega, \quad \overline{\Omega} \subset \bigcup_{r=0}^{m} U_r. \tag{4.50}$$

By the theorem on partition of unity [208], there exist functions $\varphi_r \in C_0^\infty(U_r)$, $r = 0, \ldots, m$, such that $0 \le \varphi_r \le 1$ and

$$\sum_{r=0}^{m} \varphi_r(x) = 1 \text{ for } x \in \overline{\Omega} \quad \text{and} \quad \sum_{r=1}^{m} \varphi_r(x) = 1 \text{ for } x \in \partial\Omega. \tag{4.51}$$

Since the functions a_r are Lipschitz-continuous in Δ_r, they are differentiable almost everywhere in Δ_r. Hence, there exists the gradient

$$\nabla a_r(x_r') = \left(\frac{\partial a_r}{\partial x_{r,1}}(x_r'), \ldots, \frac{\partial a_r}{\partial x_{r,d-1}}(x_r')\right)^{\mathrm{T}} \quad \text{for a. e. } x_r' \in \Delta_r, \tag{4.52}$$

and

$$|\nabla a_r| \le L \quad \text{a. e. in } \Delta_r, \ r = 1, \ldots, m. \tag{4.53}$$

(Here a. e. is meant with respect to $(d-1)$ dimensional measure.) Then there exists an outer unit normal

$$\boldsymbol{n}_r\left(x_r', a_r(x_r')\right) = \frac{1}{\sqrt{1 + |\nabla a_r(x_r')|^2}} (\nabla a_r(x_r'), -1) \tag{4.54}$$

to $\partial \widehat{V}_r^+$ for a. e. $X_r = (x_r', a_r(x_r')) \in \widehat{\Lambda}_r$ (with respect to $(d-1)$-dimensional measure defined on $\widehat{\Lambda}_r$—cf. [208]) and

$$\boldsymbol{n}(x) = \mathbb{Q}_r^{\mathrm{T}} \boldsymbol{n}_r(A_r(x)), \quad \text{a. e. } x \in \partial\Omega, \ A_r(x) \in \widehat{\Lambda}_r, \tag{4.55}$$

is the outer unit normal to $\partial\Omega$.

If we set $\boldsymbol{e}_d = (0, \ldots, 0, -1)^{\mathrm{T}} \in \mathbb{R}^d$, then by (4.52) and (4.53)

$$\boldsymbol{n}_r(X_r) \cdot \boldsymbol{e}_d = \frac{1}{\sqrt{1 + |\nabla a_r(x_r')|^2}} \geq \frac{1}{\sqrt{1+L^2}}, \quad X_r \in \widehat{\Lambda}_r, \quad r = 1, \ldots, m. \tag{4.56}$$

By virtue of the orthogonality of $\mathbb{Q}_r$, for a. e. $x \in \partial\Omega$, with $A_r(x) \in \widehat{\Lambda}_r$, we have

$$\begin{aligned} \boldsymbol{n}(x) \cdot (\mathbb{Q}_r^{\mathrm{T}} \boldsymbol{e}_d) &= \left(\mathbb{Q}_r^{\mathrm{T}} \boldsymbol{n}_r(A_r(x))\right) \cdot \left(\mathbb{Q}_r^{\mathrm{T}} \boldsymbol{e}_d\right) \\ &= \left(\mathbb{Q}_r^{\mathrm{T}} \boldsymbol{n}_r(A_r(x))\right)^{\mathrm{T}} \cdot \left(\mathbb{Q}_r^{\mathrm{T}} \boldsymbol{e}_d\right) \\ &= \left(\boldsymbol{n}_r(A_r(x))^{\mathrm{T}} \mathbb{Q}_r\right) \cdot \left(\mathbb{Q}_r^{\mathrm{T}} \boldsymbol{e}_d\right) \\ &= \boldsymbol{n}_r(A_r(x)) \cdot \boldsymbol{e}_d \geq \frac{1}{\sqrt{1+L^2}}, \quad r = 1, \ldots, m. \end{aligned} \tag{4.57}$$

Now we define the function $\boldsymbol{\varphi}$ by

$$\boldsymbol{\varphi}(x) = \sqrt{1+L^2} \sum_{r=1}^{m} \varphi_r(x)\, \mathbb{Q}_r^T \boldsymbol{e}_d, \quad x \in \mathbb{R}^d. \tag{4.58}$$

Obviously, $\boldsymbol{\varphi} \in (C_0^\infty(\mathbb{R}^d))^d$ and thus $\boldsymbol{\varphi} \in W^{1,\infty}(\Omega)^d$. Moreover, by (4.51), (4.57) and (4.58),

$$\boldsymbol{\varphi}(x) \cdot \boldsymbol{n}(x) \geq \sum_{r=1}^{m} \varphi_r(x) = 1, \quad x \in \partial\Omega,$$

what we wanted to prove. □

Now we prove a "global version" of the multiplicative trace inequality.

Lemma 4.8 *There exists a constant $C_M' > 0$ such that*

$$\|v\|^2_{L^2(\partial\Omega)} \leq C_M' \left\{ |||v||| \left(\|v\|^2_{L^2(\Omega)} + \sum_{K\in\mathscr{T}_h} h_K \|v\|^2_{L^2(\partial K)} \right)^{1/2} + \|v\|^2_{L^2(\Omega)} \right\},$$
$$v \in H^1(\Omega, \mathscr{T}_h), \ h \in (0, \bar{h}). \tag{4.59}$$

Proof Let $v \in H^1(\Omega, \mathscr{T}_h)$, $h \in (0, \bar{h})$ and $K \in \mathscr{T}_h$. Let $\boldsymbol{\varphi} \in (W^{1,\infty}(\Omega))^d$ be the function from Lemma 4.7. By Green's theorem,

$$\int_{\partial K} v^2 \boldsymbol{\varphi} \cdot \boldsymbol{n} \, \mathrm{d}S = \int_K \nabla \cdot (v^2 \boldsymbol{\varphi}) \, \mathrm{d}x = \int_K (v^2 \nabla \cdot \boldsymbol{\varphi} + 2v\boldsymbol{\varphi} \cdot \nabla v) \, \mathrm{d}x.$$

The summation over all $K \in \mathscr{T}_h$ implies that

$$\int_{\partial\Omega} v^2 \boldsymbol{\varphi} \cdot \boldsymbol{n} \, \mathrm{d}S + \sum_{\Gamma \in \mathscr{F}_h^I} \int_\Gamma [v^2] \boldsymbol{\varphi} \cdot \boldsymbol{n} \, \mathrm{d}S = \sum_{K \in \mathscr{T}_h} \int_K \left(v^2 \, \nabla \cdot \boldsymbol{\varphi} + 2v\boldsymbol{\varphi} \cdot \nabla v \right) \mathrm{d}x. \tag{4.60}$$

In view of (4.41) and (4.60),

$$\int_{\partial\Omega} v^2 \, \mathrm{d}S \le \int_{\partial\Omega} v^2 \boldsymbol{\varphi} \cdot \boldsymbol{n} \, \mathrm{d}S \le \sum_{K \in \mathscr{T}_h} \int_K |v^2 \nabla \cdot \boldsymbol{\varphi} + 2v\boldsymbol{\varphi} \cdot \nabla v| \mathrm{d}x + \sum_{\Gamma \in \mathscr{F}_h^I} \int_\Gamma \left| [v^2] \right| \, |\boldsymbol{\varphi}| \, \mathrm{d}S.$$

Taking into account that $\boldsymbol{\varphi} \in (W^{1,\infty}(\Omega))^d$ and using the Cauchy and Young inequalities, we find that

$$\|v\|^2_{L^2(\partial\Omega)} \le \|\boldsymbol{\varphi}\|_{(W^{1,\infty}(\Omega))^d} \Big(\sum_{\Gamma \in \mathscr{F}_h^I} \int_\Gamma \left| [v^2] \right| \mathrm{d}S + \|v\|^2_{L^2(\Omega)} + 2 \sum_{K \in \mathscr{T}_h} \|v\|_{L^2(K)} \, |v|_{H^1(K)} \Big). \tag{4.61}$$

Further, by the Cauchy inequality, (2.20), (2.107), (4.11) and (4.14), we have

$$\begin{aligned} \sum_{\Gamma \in \mathscr{F}_h^I} \int_\Gamma \left| [v^2] \right| \, \mathrm{d}S &= 2 \sum_{\Gamma \in \mathscr{F}_h^I} \int_\Gamma |[v] \, \langle v \rangle | \, \mathrm{d}S \\ &\le 2 \left(\sum_{\Gamma \in \mathscr{F}_h^I} \int_\Gamma \sigma [v]^2 \mathrm{d}S \right)^{1/2} \left(\sum_{\Gamma \in \mathscr{F}_h^I} \int_\Gamma \sigma^{-1} \langle v \rangle^2 \mathrm{d}S \right)^{1/2} \\ &\le 2 C_W^{-1/2} C_G^{1/2} \, J_h^\sigma(v, v)^{1/2} \left(\sum_{K \in \mathscr{T}_h} h_K \|v\|^2_{L^2(\partial K)} \right)^{1/2} . \end{aligned} \tag{4.62}$$

Now, it follows from (4.61) and (4.62) and the discrete Cauchy inequality that

$$\begin{aligned} \|v\|^2_{L^2(\partial\Omega)} \le \|\boldsymbol{\varphi}\|_{(W^{1,\infty}(\Omega))^d} &\left\{ 2 C_W^{-1/2} C_G^{1/2} \, J_h^\sigma(v, v)^{1/2} \left(\sum_{K \in \mathscr{T}_h} h_K \|v\|^2_{L^2(\partial K)} \right)^{1/2} \right. \\ &\left. + \|v\|^2_{L^2(\Omega)} + 2 \|v\|_{L^2(\Omega)} \, |v|_{H^1(\Omega, \mathscr{T}_h)} \right\}, \end{aligned}$$

which implies (4.59) with $C'_M = \max(2C_W^{-1/2}C_G^{1/2}, 2)\|\boldsymbol{\varphi}\|_{(W^{1,\infty}(\Omega))^d}$. □

Now we apply the above results to the derivation of the consistency estimate of the form b_h. This form can be expressed as

$$b_h(w, v) = b_h^{ID}(w, v) + b_h^N(w, v), \tag{4.63}$$

where

$$b_h^{ID}(w, v) = -\sum_{K\in\mathscr{T}_h}\int_K \sum_{s=1}^d f_s(w)\frac{\partial v}{\partial x_s}\,\mathrm{d}x + \sum_{\Gamma\in\mathscr{F}_h^I}\int_\Gamma H(w|_\Gamma^{(L)}, w|_\Gamma^{(R)}, \boldsymbol{n})[v]_\Gamma\,\mathrm{d}S$$
$$+\sum_{\Gamma\in\mathscr{F}_h^D}\int_\Gamma H(w|_\Gamma^{(L)}, w|_\Gamma^{(L)}, \boldsymbol{n})\,v|_\Gamma^{(L)}\,\mathrm{d}S \tag{4.64}$$

and, due to (4.19),

$$b_h^N(w, v) = \sum_{\Gamma\in\mathscr{F}_h^N}\int_\Gamma H(w|_\Gamma^{(L)}, w|_\Gamma^{(L)}, \boldsymbol{n})v|_\Gamma^{(L)}\,\mathrm{d}S = \sum_{\Gamma\in\mathscr{F}_h^N}\int_\Gamma \sum_{s=1}^d f_s(w|_\Gamma^{(L)})n_s\,v|_\Gamma^{(L)}\,\mathrm{d}S. \tag{4.65}$$

Let us set $\xi = u_h - \Pi_{hp}u \in S_{hp}$. We are interested in estimating the expression

$$b_h(u, \xi) - b_h(u_h, \xi) = \left(b_h^{ID}(u, \xi) - b_h^{ID}(u_h, \xi)\right) + \left(b_h^N(u, \xi) - b_h^N(u_h, \xi)\right). \tag{4.66}$$

Then, by (4.34) with $v_h = \xi$,

$$\left|b_h^{ID}(u, \xi) - b_h^{ID}(u_h, \xi)\right| \le C_{b4}|||\xi|||\left(R_b(\eta) + \|\xi\|_{L^2(\Omega)}\right), \tag{4.67}$$

where $R_b(\eta)$ is defined by (4.33).

It remains to estimate the second term on the right-hand side of (4.66).

Lemma 4.9 *Let $u \in H^s(\Omega)$, $u_h \in S_{hp}$, $\xi = u_h - \Pi_{hp}u$. Then*

$$\left|b_h^N(u, \xi) - b_h^N(u_h, \xi)\right| \le C_N\left(R_c(\eta)^2 + |||\xi|||\,\|\xi\|_{L^2(\Omega)} + \|\xi\|_{L^2(\Omega)}^2\right), \tag{4.68}$$

where

$$R_c(\eta) = \Big(\sum_{K\in\mathscr{T}_h}\left(h_K^{-1}\|\eta\|_{L^2(K)}^2 + h_K|\eta|_{H^1(K)}^2\right)\Big)^{1/2} \tag{4.69}$$

and C_N is a constant independent of u, u_h and h.

Proof By (4.65), Lipschitz-continuity (4.18), Cauchy and Young inequalities, and the relation $u_h - u = \eta + \xi$, where $\eta = \Pi_{hp}u - u$, we get

$$\left|b_h^N(u,\xi) - b_h^N(u_h,\xi)\right| \leq C_L \|u - u_h\|_{L^2(\partial\Omega_N)} \|\xi\|_{L^2(\partial\Omega_N)} \quad (4.70)$$
$$\leq C_L \|u - u_h\|_{L^2(\partial\Omega)} \|\xi\|_{L^2(\partial\Omega)} \leq C_L \left(\frac{1}{2}\|\eta\|^2_{L^2(\partial\Omega)} + \frac{3}{2}\|\xi\|^2_{L^2(\partial\Omega)}\right)$$

with $C_L = 2L_H$. Moreover, using the multiplicative trace inequality (2.78) and the Young inequality, we find that

$$\|\eta\|^2_{L^2(\partial\Omega)} \leq \sum_{K\in\mathscr{T}_h} \|\eta\|^2_{L^2(\partial K)} \leq C_M \sum_{K\in\mathscr{T}_h} \left(h_K^{-1}\|\eta\|^2_{L^2(K)} + \|\eta\|_{L^2(K)}|\eta|_{H^1(K)}\right)$$
$$\leq C_M \sum_{K\in\mathscr{T}_h} \left(h_K^{-1}\|\eta\|^2_{L^2(K)} + \frac{1}{2}h_K^{-1}\|\eta\|^2_{L^2(K)} + \frac{1}{2}h_K|\eta|^2_{H^1(K)}\right)$$
$$\leq \frac{3}{2}C_M R_c(\eta)^2, \quad (4.71)$$

where $R_c(\eta)$ is defined in (4.69).

We estimate $\|\xi\|^2_{L^2(\partial\Omega)}$ according to Lemma 4.8. Taking into account that $\xi \in S_{hp}$ and using the multiplicative trace inequality (2.78) and the inverse inequality (2.86), we find that

$$\sum_{K\in\mathscr{T}_h} h_K\|\xi\|^2_{L^2(\partial K)} \leq C_M \sum_{K\in\mathscr{T}_h} h_K \left(\|\xi\|_{L^2(K)}|\xi|_{H^1(K)} + h_K^{-1}\|\xi\|^2_{L^2(K)}\right) \quad (4.72)$$
$$\leq C_M(1 + C_I)\,\|\xi\|^2_{L^2(\Omega)}.$$

Hence, in view of (4.59) and (4.72), we have

$$\|\xi\|^2_{L^2(\partial\Omega)} \leq C'_M \left\{(C_M(1 + C_I) + 1)^{1/2}\; |||\xi|||\; \|\xi\|_{L^2(\Omega)} + \|\xi\|^2_{L^2(\Omega)}\right\} \quad (4.73)$$
$$\leq C^* \left(|||\xi|||\,\|\xi\|_{L^2(\Omega)} + \|\xi\|^2_{L^2(\Omega)}\right),$$

where $C^* = C'_M(C_M(1+C_I)+1)^{1/2}$. Finally, (4.70), (4.71) and (4.73) yield estimate (4.68) with $C_N = \frac{1}{2}C_L \max(2C_M, 3C^*)$, which we wanted to prove. □

Let us summarize the above results.

Corollary 4.10 *Let $u \in H^s(\Omega)$, $s \geq 2$, $u_h \in S_{hp}$, $\xi = u_h - \Pi_{hp}u$, $\eta = \Pi_{hp}u - u$. Then*

$$|b_h(u,\xi) - b_h(u_h,\xi)| \quad (4.74)$$
$$\leq C_b \left(|||\xi|||\left(R_b(\eta) + \|\xi\|_{L^2(\Omega)}\right) + \delta_N\left(R_c(\eta)^2 + \|\xi\|^2_{L^2(\Omega)}\right)\right),$$

where $\delta_N = 0$, *if* $\partial\Omega_N = \emptyset$, *and* $\delta_N = 1$, *if* $\partial\Omega_N \neq \emptyset$.

Proof Estimate (4.74) is an immediàte consequence of (4.67) and (4.68) with the constant $C_b = C_{b4} + C_N$. □

4.3.3 Error Estimates for the Method of Lines

Now we derive the error estimates of the method of lines (4.25) under the assumption that the exact solution u satisfies the condition

$$\frac{\partial u}{\partial t} \in L^2(0, T; H^s(\Omega)), \tag{4.75}$$

where $s \geq 2$ is an integer. Assumption (4.75) implies that $u \in C([0, T]; H^s(\Omega))$.

Let $\Pi_{hp}u(t)$ be the S_{hp}-interpolation of $u(t)$ $(t \in [0, T])$ from (2.90). We set

$$\xi = u_h - \Pi_{hp}u \in S_{hp}, \quad \eta = \Pi_{hp}u - u \in H^s(\Omega, \mathscr{T}_h). \tag{4.76}$$

Then the error e_h can be expressed as

$$e_h = u_h - u = \xi + \eta. \tag{4.77}$$

Subtracting (4.28) from (4.25b), where we substitute $v_h := \xi$, we get

$$\left(\frac{\partial \xi}{\partial t}, \xi\right) + A_h(\xi, \xi) = b_h(u, \xi) - b_h(u_h, \xi) - \left(\frac{\partial \eta}{\partial t}, \xi\right) - A_h(\eta, \xi). \tag{4.78}$$

(Of course, $\xi = \xi(t)$, $\eta = \eta(t)$ for $t \in [0, T]$, but we do not emphasize the dependence on t by our notation, if it is not necessary.) In what follows we estimate the individual terms on the right-hand side of (4.78).

The Cauchy inequality implies that

$$\left|\left(\frac{\partial \eta}{\partial t}, \xi\right)\right| \leq \left\|\frac{\partial \eta}{\partial t}\right\|_{L^2(\Omega)} \|\xi\|_{L^2(\Omega)}. \tag{4.79}$$

Moreover, using the result of Lemma 2.37, we have

$$|A_h(\eta, \xi)| \leq \varepsilon \widetilde{C}_B R_a(\eta) |||\xi|||, \tag{4.80}$$

where $\widetilde{C}_B$ is the constant from (2.129) and

$$R_a(\eta) = \left(\sum_{K \in \mathscr{T}_h} \left(|\eta|^2_{H^1(K)} + h_K^2 |\eta|^2_{H^2(K)} + h_K^{-2} \|\eta\|^2_{L^2(K)}\right)\right)^{1/2}. \tag{4.81}$$

Finally, we define the term

$$R_Q(\eta) = \frac{2C_1^2}{\varepsilon C_C}\Big(R_b(\eta) + \varepsilon R_a(\eta)\Big)^2 + 2C_1\left(R_c(\eta)^2 + \left\|\frac{\partial \eta}{\partial t}\right\|^2_{L^2(\Omega)}\right), \tag{4.82}$$

where $R_b(\eta)$ is defined by (4.33), $R_c(\eta)$ is defined by (4.69), and the constant C_1 is defined as $C_1 = \max(C_b + 1, \widetilde{C}_B)$. This notation will be useful in the following.

Now we prove the so-called *abstract error estimate*, representing a bound of the error in terms of the S_{hp}-interpolation error η. Let us recall that in order to increase the readability of the derivation of the error estimate, we number constants appearing in the proofs.

Theorem 4.11 *Let Assumptions 4.5 from Sect. 4.3 be satisfied. Let u be the exact strong solution of problem (4.1) satisfying (4.75) and let u_h be the approximate solution obtained by scheme (4.25). Then the error $e_h = u_h - u$ satisfies the estimate*

$$\begin{aligned} &\|e_h(t)\|^2_{L^2(\Omega)} + \varepsilon C_C \varepsilon \int_0^T |||e_h(\vartheta)|||^2 \,\mathrm{d}\vartheta \qquad (4.83)\\ &\quad\le C_2(\varepsilon)\left(\int_0^T R_Q(\eta(t))\,\mathrm{d}t + \|\eta(t)\|^2_{L^2(\Omega)} + C_C\int_0^T |||\eta(\vartheta)|||^2\,\mathrm{d}\vartheta\right),\\ &\qquad\qquad t \in (0,T),\quad h \in (0,\bar{h}), \end{aligned}$$

where C_C is the constant from the coercivity inequality (2.140) of the form $\frac{1}{\varepsilon}A_h = a_h + J_h^\sigma$, $R_Q(\eta)$ is given by (4.82) and $C_2(\varepsilon)$ is a constant independent of h and u, but depending on ε (see (4.93)).

Proof As in (4.76), we set $\xi = u_h - \Pi_{hp}u \in S_{hp}$, $\eta = \Pi_{hp}u - u$. Then (4.77) holds: $e_h = u_h - u = \xi + \eta$. Due to the coercivity (2.140) of the form A_h,

$$\varepsilon C_C |||\xi|||^2 \le A_h(\xi,\xi). \tag{4.84}$$

It follows from (4.78), (4.84) and the relation

$$\left(\frac{\partial \xi}{\partial t}, \xi\right) = \frac{1}{2}\frac{\mathrm{d}}{\mathrm{d}t}\|\xi\|^2_{L^2(\Omega)}, \tag{4.85}$$

that

$$\frac{1}{2}\frac{\mathrm{d}}{\mathrm{d}t}\|\xi\|^2_{L^2(\Omega)} + \varepsilon\, C_C |||\xi|||^2 \le b_h(u,\xi) - b_h(u_h,\xi) - \left(\frac{\partial \eta}{\partial t},\xi\right) - A_h(\eta,\xi). \tag{4.86}$$

Now from (4.74), (4.79) and (4.80), using the inequality $(\gamma+\delta)^2 \le 2(\gamma^2+\delta^2)$ and Cauchy and Young inequalities, we derive the estimates

$$
\begin{aligned}
&\frac{\mathrm{d}}{\mathrm{d}t}\|\xi\|^2_{L^2(\Omega)} + 2\varepsilon C_C |||\xi|||^2 && (4.87)\\
&\quad\le 2C_b\Big(|||\xi|||\left(R_b(\eta) + \|\xi\|_{L^2(\Omega)}\right) + R_c(\eta)^2 + \|\xi\|^2_{L^2(\Omega)}\Big)\\
&\qquad + 2\|\partial_t\eta\|_{L^2(\Omega)}\|\xi\|_{L^2(\Omega)} + 2\varepsilon\widetilde{C}_B R_a(\eta)|||\xi|||\\
&\quad\le 2C_1\Big\{ |||\xi|||\left(\|\xi\|_{L^2(\Omega)} + R_b(\eta) + \varepsilon R_a(\eta)\right)\\
&\qquad + R_c(\eta)^2 + \|\xi\|^2_{L^2(\Omega)} + \|\partial_t\eta\|^2_{L^2(\Omega)}\Big\}\\
&\quad\le \varepsilon C_C|||\xi|||^2 + \frac{C_1^2}{\varepsilon C_C}\left(\|\xi\|_{L^2(\Omega)} + R_b(\eta) + \varepsilon R_a(\eta)\right)^2\\
&\qquad + 2C_1\Big\{ R_c(\eta)^2 + \|\xi\|^2_{L^2(\Omega)} + \|\partial_t\eta\|^2_{L^2(\Omega)}\Big\}\\
&\quad\le \varepsilon\, C_C|||\xi|||^2 + C_3\left(1+\frac{1}{\varepsilon\, C_C}\right)\|\xi\|^2_{L^2(\Omega)} + R_Q(\eta),
\end{aligned}
$$

where $C_1 = \max(C_b+1, \widetilde{C}_B)$, $C_3 = 2\max(C_1, C_1^2)$ and $R_Q(\eta)$ is given by (4.82). Hence,

$$
\frac{\mathrm{d}}{\mathrm{d}t}\|\xi(t)\|^2_{L^2(\Omega)} + \varepsilon C_C|||\xi(t)|||^2 \le C_3\left(1+\frac{1}{\varepsilon\, C_C}\right)\|\xi(t)\|^2_{L^2(\Omega)} + R_Q(\eta(t)). \tag{4.88}
$$

Since $u, \frac{\partial u}{\partial t} \in L^2(0,T;H^\mu(\Omega))$, the right-hand side of (4.88) is integrable over $(0,T)$. From (4.76) and (4.25c) we see that $\xi(0)=0$. The integration of (4.88) from 0 to $t\in[0,T]$ yields

$$
\begin{aligned}
&\|\xi(t)\|^2_{L^2(\Omega)} + \varepsilon C_C\int_0^t |||\xi(\vartheta)|||^2\,\mathrm{d}\vartheta && (4.89)\\
&\quad\le C_3\left(1+\frac{1}{\varepsilon\, C_C}\right)\int_0^t \|\xi(\vartheta)\|^2_{L^2(\Omega)}\,\mathrm{d}\vartheta + \int_0^t R_Q(\eta(\vartheta))\,\mathrm{d}\vartheta.
\end{aligned}
$$

Now we apply the Gronwall Lemma 1.9 with

$$
\begin{aligned}
y(t) &= \|\xi(t)\|^2_{L^2(\Omega)}, & q(t) &= \varepsilon C_C\textstyle\int_0^t |||\xi(\vartheta)|||^2\,\mathrm{d}\vartheta,\\
r(t) &= C_3\frac{\varepsilon C_C+1}{\varepsilon\, C_C}, & z(t) &= \textstyle\int_0^t R_Q(\eta(\vartheta))\,\mathrm{d}\vartheta.
\end{aligned}
$$

Further, let us set

$$R(\eta, \varepsilon) = \int_0^T R_Q(\eta(\vartheta))\, \mathrm{d}\vartheta, \tag{4.90}$$
$$c_1(\varepsilon) = 1 + C_3 \frac{\varepsilon C_C + 1}{\varepsilon C_C} T \exp\left(C_3 \frac{\varepsilon C_C + 1}{\varepsilon C_C} T\right).$$

We easily show that

$$z(t) \le \int_0^T R_Q(\eta(\vartheta))\, \mathrm{d}\vartheta = R(\eta, \varepsilon), \quad \exp\left(\int_\vartheta^t r(s)\mathrm{d}s\right) \le \exp\left(C_3 \frac{\varepsilon C_C + 1}{\varepsilon C_C} T\right),$$
$$z(t) + \int_0^t r(\vartheta) z(\vartheta) \exp\left(\int_\vartheta^t r(s)\mathrm{d}s\right) \mathrm{d}\vartheta \le R(\eta, \varepsilon) c_1(\varepsilon).$$

This, (4.89) and the Gronwall Lemma 1.9 yield the estimate

$$\|\xi(t)\|^2_{L^2(\Omega)} + \varepsilon C_C \int_0^t |||\xi(\vartheta)|||^2 \mathrm{d}\vartheta \le R(\eta, \varepsilon) c_1(\varepsilon), \quad t \in [0, T],\ h \in (0, \bar{h}). \tag{4.91}$$

By virtue of the relation $e_h = \xi + \eta$ and the inequality $(\gamma + \delta)^2 \le 2(\gamma^2 + \delta^2)$, we can write

$$\|e_h\|^2_{L^2(\Omega)} \le 2\left(\|\xi\|^2_{L^2(\Omega)} + \|\eta\|^2_{L^2(\Omega)}\right), \qquad |||e_h|||^2 \le 2\left(|||\xi|||^2 + |||\eta|||^2\right).$$

Using (4.91), we deduce that

$$\|e_h(t)\|^2_{L^2(\Omega)} + \varepsilon C_C \int_0^t |||e_h(\vartheta)|||^2\, \mathrm{d}\vartheta \tag{4.92}$$
$$\le 2\left(R(\eta, \varepsilon) c_1(\varepsilon) + \|\eta(t)\|^2_{L^2(\Omega)} + \varepsilon C_C \int_0^t |||\eta(\vartheta)|||^2\, \mathrm{d}\vartheta\right), \ t \in [0, T],\ h \in (0, \bar{h}),$$

which already implies estimate (4.83) with the constant

$$C_2(\varepsilon) = 2\left(1 + C_3 \frac{\varepsilon C_C + 1}{\varepsilon C_C} T \exp\left(C_3 \frac{\varepsilon C_C + 1}{\varepsilon C_C} T\right)\right). \tag{4.93}$$

□

4.4 Error Estimates in Terms of h

Now we derive the first main result of this chapter on the error estimate of the method of lines for the solution of the nonlinear convection-diffusion problem. It will be obtained by estimating the right-hand side of (4.83) in terms of h.

We assume that $s \geq 2$ and the exact solution u satisfies the regularity assumption

$$\frac{\partial u}{\partial t} \in L^2(0, T; H^s(\Omega)). \tag{4.94}$$

Then $u \in C([0, T], H^s(\Omega))$. As usual, we put $\eta(t) = u(u) - \Pi_{hp}u(t)$, $t \in (0, T)$, and $\mu = \min(p + 1, s)$. Recalling (2.149), we have

$$\begin{aligned} \|\eta(t)\|_{L^2(K)} &\leq C_A\, h_K^{\mu} |u(t)|_{H^\mu(K)}, \quad && K \in \mathscr{T}_h,\ t \in (0, T), \\ |\eta(t)|_{H^1(K)} &\leq C_A\, h_K^{\mu-1} |u(t)|_{H^\mu(K)}, \quad && K \in \mathscr{T}_h,\ t \in (0, T), \\ |\eta(t)|_{H^2(K)} &\leq C_A\, h_K^{\mu-2} |u(t)|_{H^\mu(K)}, \quad && K \in \mathscr{T}_h,\ t \in (0, T), \end{aligned} \tag{4.95}$$

where C_A is the constant from Lemma 2.22. Then, a simple manipulation gives

$$\sum_{K \in \mathscr{T}_h} \left(|\eta(t)|^2_{H^1(K)} + h_K^2 |\eta(t)|^2_{H^2(K)} + h_K^{-2} \|\eta(t)\|_{L^2(K)} \right) \leq 3C_A^2 h^{2(\mu-1)} |u(t)|^2_{H^\mu(\Omega)},$$

for any $t \in (0, T)$. This together with (4.81) implies that

$$R_a(\eta(t)) = R_a\left(u(t) - \Pi_{hp}u(t)\right) \leq \sqrt{3} C_A h^{\mu-1} |u(t)|_{H^\mu(\Omega)}, \quad t \in (0, T). \tag{4.96}$$

Similarly, from (4.33), we obtain

$$R_b(\eta(t)) = R_b\left(u(t) - \Pi_{hp}u(t)\right) \leq \sqrt{2} C_A h^{\mu} |u(t)|_{H^\mu(\Omega)}, \quad t \in (0, T). \tag{4.97}$$

Moreover, (4.69) and (4.95) give

$$R_c(\eta(t)) \leq \sqrt{2} C_A h^{\mu-1/2} |u(t)|_{H^\mu(\Omega)}, \quad t \in (0, T). \tag{4.98}$$

Further, we use the notation $\partial_t u = \partial u/\partial t$ and $\partial_t(\Pi_{hp}u) = \partial(\Pi_{hp}u)/\partial t$. Then definition (2.90) of the interpolation operator Π_{hp} and the relation

$$\partial_t(\Pi_{hp}u(t)) = \Pi_{hp}(\partial_t u(t)) \in S_{hp} \tag{4.99}$$

imply that

$$\|\partial_t \eta\|_{L^2(\Omega)} = \left\|\partial_t(\Pi_{hp}u - u)\right\|_{L^2(\Omega)} = \left\|\Pi_{hp}(\partial_t u) - \partial_t u\right\|_{L^2(\Omega)} \leq C_A h^{\mu} \, |\partial_t u|_{H^\mu(\Omega)} \, . \tag{4.100}$$

Exercise 4.12 Using the theorem on differentiating an integral with respect to a parameter, prove (4.99).

Summarizing (4.82) with (4.96), (4.97), (4.98) and (4.100), we see that for $t \in (0, T)$, we have

$$
\begin{aligned}
R_Q(\eta(t)) &= \frac{2C_1^2}{\varepsilon C_C}\left(R_b(\eta(t)) + \varepsilon R_a(\eta(t))\right)^2 + 2C_1\left(R_c(\eta(t))^2 + \|\partial_t \eta(t)\|^2_{L^2(\Omega)}\right) \\
&\le \frac{2C_1^2}{\varepsilon C_C}\left(\sqrt{2}C_A h^\mu |u(t)|_{H^\mu(\Omega)} + \varepsilon\sqrt{3}C_A h^{\mu-1}|u(t)|_{H^\mu(\Omega)}\right)^2 \\
&\quad + 4C_1C_A^2h^{2\mu-1}|u(t)|^2_{H^\mu(\Omega)} + 2C_1C_A^2h^{2\mu}\,|\partial_t u(t)|^2_{H^\mu(\Omega)} \\
&\le \frac{2C_1^2C_A^2}{\varepsilon C_C}h^{2(\mu-1)}|u(t)|^2_{H^\mu(\Omega)}(2h^2 + 2\sqrt{6}\varepsilon h + 3\varepsilon^2) \\
&\quad + 4C_1C_A^2h^{2(\mu-1)}\left(|u(t)|^2_{H^\mu(\Omega)} + |\partial_t u(t)|^2_{H^\mu(\Omega)}\right)(h + h^2) \\
&\le C_4 h^{2(\mu-1)}\left(\varepsilon^{-1}h^2 + h + \varepsilon + h^2\right)\left(|u(t)|^2_{H^\mu(\Omega)} + |\partial_t u(t)|^2_{H^\mu(\Omega)}\right),
\end{aligned}
\tag{4.101}
$$

where

$$
C_4 = 4C_A^2 \max\left(\frac{\sqrt{6}C_1^2}{C_C}, C_1\right). \tag{4.102}
$$

The integration of (4.101) over $(0, T)$ yields

$$
\begin{aligned}
&\int_0^T R_Q(\eta(t))\,\mathrm{d}t \\
&\quad \le C_4 h^{2\mu-2}\left(\varepsilon^{-1}h^2 + h + \varepsilon + h^2\right)\left(|u|^2_{L^2(0,T;H^\mu(\Omega))} + |\partial_t u|^2_{L^2(0,T;H^\mu(\Omega))}\right).
\end{aligned}
\tag{4.103}
$$

Furthermore, using (4.29), (2.119) and (4.95), we get

$$
\begin{aligned}
|||\eta(t)|||^2 &= |\eta(t)|^2_{H^1(\Omega,\mathscr{T}_h)} + J_h^\sigma(\eta(t), \eta(t)) \\
&\le \sum_{K\in\mathscr{T}_h}\left(|\eta(t)|^2_{H^1(K)} + C_W C_M C_T^{-1}\left(3h_K^{-2}\|\eta(t)\|^2_{L^2(K)} + |\eta(t)|^2_{H^1(K)}\right)\right) \\
&\le C_5 h^{2(\mu-1)}|u(t)|^2_{H^\mu(\Omega)}, \quad t \in (0, T),
\end{aligned}
\tag{4.104}
$$

where $C_5 = C_A^2(4C_W C_M C_T^{-1} + 1)$. Hence,

$$
\varepsilon C_C \int_0^T |||\eta(t)|||^2\,\mathrm{d}t \le \varepsilon C_C C_5 h^{2(\mu-1)}|u|^2_{L^2(0,T;H^\mu(\Omega))}. \tag{4.105}
$$

Remark 4.13 The above estimates illustrate a typical situation in numerical analysis, where a number of constants appear. They are often defined recursively in a complicated way on the basis of constants introduced before. As an example we illustrate this situation by the process leading to the determination of the constant C_4 defined by (4.102). This relation contains the constant C_A appearing in Lemmas 2.22 and 2.24 and the constant C_1, which is defined recursively in the following way:

$$\begin{aligned}
C_1 &= \max(C_b + 1, \widetilde{C}_B),\\
C_b &= C_{b4} + C_N,\\
C_{b4} &= \max(C_{b2}, C_{b3}),\\
C_N &= \frac{1}{2} C_L \max(2C_M, 3C^*),\\
C_L &= 2L_H,\\
C^* &= C'_M(C_M(1 + C_I) + 1)^{1/2},\\
C'_M &= \max(2C_W^{-1/2} C_G^{1/2}, 2)\|\boldsymbol{\varphi}\|_{(W^{1,\infty}(\Omega))^d},\\
C_{b2} &= C_{b1}(1 + C_M(1 + C_I))^{1/2},\\
C_{b3} &= C_{b1}(1 + 3C_M/3)^{1/2},\\
C_{b1} &= (\max(d\, L_f^2, 2L_H^2\, C_G/C_W))^{1/2},
\end{aligned}$$

where $\widetilde{C}_B$ is the constant from Lemma 2.37, C_M is the constant from Lemma 2.37 (multiplicative trace inequality), L_H is the constant from the Lipschitz continuity (4.18) of the numerical flux H, C_I is the constant from the inverse inequality (2.86), C_W is the constant from the definition (2.104) of the weight in the penalty form J_h^σ, C_G is the constant from the equivalence condition (2.20), $\boldsymbol{\varphi}$ is the function from Lemma 4.7 and L_f is the constant from the Lipschitz continuity of the convective fluxes f_s, $s = 1, \ldots, d$.

Now we are ready to present the final error estimates.

Theorem 4.14 *Let Assumptions 4.5 from Sect. 4.3 be satisfied. Let u be the exact strong solution of problem (4.1) satisfying (4.75) and let u_h be the approximate solution obtained by the scheme (4.25). Then the error $e_h = u_h - u$ satisfies the estimate*

$$\begin{aligned}
&\max_{t\in[0,T]} \|e_h(t)\|^2_{L^2(\Omega)} + C_C\varepsilon \int_0^T |||e_h(\vartheta)|||^2 \,\mathrm{d}\vartheta &(4.106)\\
&\le \tilde{C}_2(\varepsilon)h^{2(\mu-1)} \left(|u|^2_{L^2(0,T;H^\mu(\Omega))} + |\partial_t u|^2_{L^2(0,T;H^\mu(\Omega))}\right), \quad h \in (0, \bar{h}),
\end{aligned}$$

where C_C is the constant from the coercivity inequality (2.140) of the form $\frac{1}{\varepsilon}A_h = a_h + J_h^\sigma$ and $\tilde{C}_2(\varepsilon)$ is a constant independent of h and u, specified in the proof.

Proof If $t \in [0, T]$, then the estimation of the right-hand side of (4.83) by (4.103), (4.105) and (4.95) implies that

$$
\begin{aligned}
&\|e_h(t)\|^2_{L^2(\Omega)} + C_C\varepsilon \int_0^T |||e_h(\vartheta)|||^2 \, \mathrm{d}\vartheta \\
&\le C_2(\varepsilon)\left(\int_0^T R_Q(\eta(t))\,\mathrm{d}t + \|\eta(t)\|^2_{L^2(\Omega)} + \varepsilon C_C \int_0^T |||\eta(\vartheta)|||^2 \, \mathrm{d}\vartheta\right), \\
&\le \tilde{C}_2(\varepsilon) h^{2\mu-2}\left(|u|^2_{L^2(0,T;H^\mu(\Omega))} + |\partial_t u|^2_{L^2(0,T;H^\mu(\Omega))}\right),
\end{aligned}
$$

where $C_2(\varepsilon)$ is the constant from Theorem 4.11 given by (4.93) and

$$
\tilde{C}_2(\varepsilon) = C_2(\varepsilon)(C_4 + C_C C_5 + C_A^2)\left(\varepsilon^{-1}\bar{h}^2 + \bar{h} + \varepsilon + \bar{h}^2\right). \tag{4.107}
$$

This proves (4.106). □

Remark 4.15 Estimate (4.106) implies that

$$
\|u - u_h\|_{L^\infty(0,T;L^2(\Omega))} = O(h^{\mu-1}) \quad \text{for } h \to 0+. \tag{4.108}
$$

This is in contrast to the approximation properties (2.98) implying that

$$
\|u - \Pi_{hp} u\|_{L^\infty(0,T;L^2(\Omega))} = O(h^\mu). \tag{4.109}
$$

Numerical experiments presented in the next section demonstrate that the error estimate (4.106) is suboptimal in the $L^\infty(0, T; L^2(\Omega))$-norm. Similarly as in Sect. 2.7.2 we can derive optimal error estimate in this norm. This is the subject of the next section.

Remark 4.16 From (4.107) and (4.93) we can see that the error estimate (4.106) cannot be used for ε very small, because the definition (4.93) of the constant $C_2(\varepsilon)$ contains the term of the form $\exp(C/\varepsilon)$, which blows up exponentially for $\varepsilon \to 0+$. This is caused by the technique used in the theoretical analysis (application of the Young inequality and the Gronwall lemma) in order to overcome the nonlinearity in the convective terms. The nonlinearity of the convective terms represents a serious obstacle for obtaining a uniform error estimate with respect to $\varepsilon \to 0+$. In Sect. 4.6 we are concerned with error estimates of the DGM applied to the numerical solution of a linear convection-diffusion-reaction equation, uniform with respect to the diffusion parameter $\varepsilon \to 0+$.

4.5 Optimal $L^\infty(0, T; L^2(\Omega))$-Error Estimate

With respect to Remark 4.15, in this section we derive an optimal error estimate in the $L^\infty(0, T; L^2(\Omega))$-norm. Similarly as in Sect. 2.7.2, the analysis is based on the *duality technique*. Therefore, we consider only the SIPG variant of the DGM and the Dirichlet boundary condition on the whole boundary $\partial\Omega$.

Let $\Omega \subset \mathbb{R}^d$, $d = 2, 3$, be a *bounded convex polygonal* (if $d = 2$) or *polyhedral* (if $d = 3$) domain with Lipschitz boundary $\partial\Omega$ and $T > 0$. We are concerned with the nonstationary nonlinear convection-diffusion problem to find $u : Q_T = \Omega \times (0, T) \to \mathbb{R}$ such that

$$\frac{\partial u}{\partial t} + \sum_{s=1}^{d} \frac{\partial f_s(u)}{\partial x_s} = \varepsilon \Delta u + g \quad \text{in } Q_T, \tag{4.110a}$$

$$u\big|_{\partial\Omega\times(0,T)} = u_D, \tag{4.110b}$$

$$u(x, 0) = u^0(x), \quad x \in \Omega. \tag{4.110c}$$

The diffusion coefficient $\varepsilon > 0$ is a given constant, $g : Q_T \to \mathbb{R}$, $u_D : \partial\Omega \times (0, T) \to \mathbb{R}$ and $u^0 : \Omega \to \mathbb{R}$ are given functions satisfying (4.2c), (4.2d) with $\partial\Omega_D = \partial\Omega$, (4.2f), and $f_s \in C^1(\mathbb{R})$, $s = 1, \dots, d$, are fluxes satisfying (4.2a).

Let us recall the definitions of the forms introduced in Sect. 4.1 by (4.9), (4.10) (with $\Theta = 1$), (4.13), (4.11) and (4.23). Namely, for functions $u, \varphi \in H^2(\Omega, \mathscr{T}_h)$ we write

$$A_h(w, v) = \varepsilon a_h(w, v) + \varepsilon J_h^\sigma(w, v), \tag{4.111}$$

$$a_h(u, \varphi) = \sum_{K\in\mathscr{T}_h} \int_K \nabla u \cdot \nabla\varphi \,\mathrm{d}x - \sum_{\Gamma\in\mathscr{F}_h} \int_\Gamma (\langle\nabla u\rangle \cdot \boldsymbol{n}[\varphi] + \langle\nabla\varphi\rangle \cdot \boldsymbol{n}[u]) \,\mathrm{d}S, \tag{4.112}$$

$$J_h^\sigma(u, \varphi) = \sum_{\Gamma\in\mathscr{F}_h} \int_\Gamma \sigma[u]\,[\varphi] \,\mathrm{d}S, \tag{4.113}$$

$$\ell_h(\varphi)(t) = \int_\Omega g(t)\,\varphi \,\mathrm{d}x + \varepsilon \sum_{\Gamma\in\mathscr{F}_h^B} \int_\Gamma \Big(\sigma\varphi - (\nabla\varphi\cdot\boldsymbol{n})\Big) u_D(t) \,\mathrm{d}S, \tag{4.114}$$

$$\begin{aligned} b_h(u, \varphi) = &- \sum_{K\in\mathscr{T}_h} \int_K \sum_{s=1}^{d} f_s(u) \frac{\partial\varphi}{\partial x_s} \,\mathrm{d}x + \sum_{\Gamma\in\mathscr{F}_h^I} \int_\Gamma H\left(u|_\Gamma^{(L)}, u|_\Gamma^{(R)}, \boldsymbol{n}\right) [\varphi]_\Gamma \,\mathrm{d}S \\ &+ \sum_{\Gamma\in\mathscr{F}_h^B} \int_\Gamma H(u|_\Gamma^{(L)}, u|_\Gamma^{(L)}, \boldsymbol{n})\varphi_\Gamma^{(L)} \,\mathrm{d}S. \end{aligned}$$

By $(\cdot, \cdot)$ we denote the scalar product in the space $L^2(\Omega)$. The weight σ is again defined by (4.14). We assume that the numerical flux H has properties (4.18)–(4.20) from Sect. 4.2.

Let the exact solution u of problem (4.110) satisfy the regularity condition (4.94). Moreover, let $u_h \in C^1([0, T]; S_{hp})$ denote the approximate solution defined by (4.25) and let Π_{hp} be the operator of the $L^2(\Omega)$-projection on the space S_{hp} (cf. (2.90)).

In Sect. 4.3.3, we derived the (sub-optimal) estimate from identity (4.78). The term $A_h(\Pi_{hp}u - u, \xi)$ appearing on the right-hand side of (4.78) cannot be estimated in "an optimal way" (i.e., of order $O(h^\mu)$), because, by virtue of (4.80) and (4.96),

$$|A_h(\Pi_{hp}u - u, \xi)| = |A_h(\eta, \xi)| \le \varepsilon \widetilde{C}_B R_a(\eta) |||\xi|||,$$

and $R_a(\eta) = O(h^{\mu-1})$. Therefore, instead of the $L^2(\Omega)$-projection Π_{hp}, we introduce a new projection P_{hp}, for which the terms mentioned above vanish.

Hence, for every $h \in (0, \bar{h})$ and $t \in [0, T]$, we define the function $\mathrm{P}_{hp}u(t)$ as the *A_h-projection* of $u(t)$ on S_{hp}, i.e., a function satisfying the conditions

$$\mathrm{P}_{hp}u(t) \in S_{hp}, \qquad A_h(\mathrm{P}_{hp}u(t), \varphi_h) = A_h(u(t), \varphi_h) \quad \forall \varphi_h \in S_{hp}. \tag{4.115}$$

We are interested in estimates of the functions

$$\chi(t) = u(t) - \mathrm{P}_{hp}u(t) \text{ and } \partial_t \chi(t) = \frac{\partial}{\partial t}\chi(t) = \frac{\partial}{\partial t}\left(u(t) - \mathrm{P}_{hp}u(t)\right), \quad t \in [0, T],$$

in the DG-norm $|||\cdot|||$ given by (4.29) and in the $L^2(\Omega)$-norm. First, we derive estimates of these functions in the DG-norm.

Lemma 4.17 *There exists a constant $C_{P,e} > 0$ independent of u, ε and h such that*

$$|||\chi(t)||| \le C_{P,e} h^{\mu-1} |u(t)|_{H^\mu(\Omega)}, \quad t \in [0, T], \tag{4.116}$$

$$|||\partial_t \chi(t)||| \le C_{P,e} h^{\mu-1} |\partial_t u(u)|_{H^\mu(\Omega)}, \quad t \in [0, T], \tag{4.117}$$

for all $h \in (0, \bar{h})$.

Proof In what follows we usually omit the argument t of the functions u, $\mathrm{P}_{hp}u$, $\Pi_{hp}u$, etc. By (2.138) and (4.115), we obtain

$$\begin{aligned} \frac{\varepsilon}{2}|||\Pi_{hp}u - \mathrm{P}_{hp}u|||^2 &\le A_h(\Pi_{hp}u - \mathrm{P}_{hp}u,\, \Pi_{hp}u - \mathrm{P}_{hp}u) \qquad (4.118)\\ &= A_h(\Pi_{hp}u - \mathrm{P}_{hp}u,\, \Pi_{hp}u - \mathrm{P}_{hp}u) + \underbrace{A_h(\mathrm{P}_{hp}u - u,\, \Pi_{hp}u - \mathrm{P}_{hp}u)}_{=0}\\ &= A_h(\Pi_{hp}u - u,\, \Pi_{hp}u - \mathrm{P}_{hp}u). \end{aligned}$$

Using the result of Lemma 2.37, we find that

$$A_h(\Pi_{hp}u - u,\ \Pi_{hp}u - \mathrm{P}_{hp}u) \le \varepsilon \widetilde{C}_B\, R_a(\Pi_{hp}u - u)\, |||\Pi_{hp}u - \mathrm{P}_{hp}u|||,$$

where R_a is given by (4.81). This and (4.118) imply that

$$|||\Pi_{hp}u - \mathrm{P}_{hp}u||| \le 2\widetilde{C}_B\, R_a(\Pi_{hp}u - u). \tag{4.119}$$

Further, recalling (2.125), we have

$$|||u - \Pi_{hp}u||| \le C_\sigma R_a(u - \Pi_{hp}u). \tag{4.120}$$

Now it is sufficient to use the triangle inequality

$$|||\chi||| = |||u - \mathrm{P}_{hp}u||| \le |||u - \Pi_{hp}u||| + |||\Pi_{hp}u - \mathrm{P}_{hp}u|||,$$

which implies that

$$|||\chi||| \le (C_\sigma + 2\widetilde{C}_B)\, R_a(\Pi_{hp}u - u). \tag{4.121}$$

Finally, the combination of (4.96) and (4.121) gives

$$|||\chi(t)||| \le \sqrt{3}C_A(C_\sigma + 2\widetilde{C}_B)h^{\mu-1}|u(t)|_{H^\mu(\Omega)}, \quad t \in (0,T),$$

which proves (4.116) with $C_{P,\mathrm{e}} = \sqrt{3}C_A(C_\sigma + 2\widetilde{C}_B)$.

Let us deal now with the norm $|||\partial_t\chi|||$. As

$$A_h\left(u(t) - \mathrm{P}_{hp}u(t),\ \varphi_h\right) = 0 \qquad \forall \varphi_h \in S_{hp},\ \forall t \in (0,T),$$

from the definitions (4.111) of A_h, for all $\varphi_h \in S_{hp}$, we have

$$0 = \frac{\mathrm{d}}{\mathrm{d}t}\left(A_h\left(u(t) - \mathrm{P}_{hp}u(t),\ \varphi_h\right)\right) = A_h\left(\frac{\partial(u(t) - \mathrm{P}_{hp}u(t))}{\partial t},\ \varphi_h\right), \tag{4.122}$$

i.e.,

$$A_h(\partial_t\chi, \varphi_h) = 0 \qquad \forall \varphi_h \in S_{hp}. \tag{4.123}$$

Similarly as in (4.118), using the coercivity (2.140) of the form A_h and relation (2.129) from Lemma 2.37, we find that

$$\frac{\varepsilon}{2}|||\partial_t(\Pi_{hp}u - \mathrm{P}_{hp}u)|||^2$$
$$\le A_h\left(\partial_t(\Pi_{hp}u - \mathrm{P}_{hp}u),\ \partial_t(\Pi_{hp}u - \mathrm{P}_{hp}u)\right) + \underbrace{A_h\left(\partial_t(\mathrm{P}_{hp}u - u),\ \partial_t(\Pi_{hp}u - \mathrm{P}_{hp}u)\right)}_{=0}$$

$$= A_h\left(\partial_t(\Pi_{hp}u - u), \partial_t(\Pi_{hp}u - \mathrm{P}_{hp}u)\right)$$
$$\leq \varepsilon\widetilde{C}_B\, R_a\left(\partial_t(\Pi_{hp}u - u)\right) |||\partial_t(\Pi_{hp}u - \mathrm{P}_{hp}u)|||.$$

Hence, we have

$$|||\partial_t(\Pi_{hp}u - \mathrm{P}_{hp}u)||| \leq 2\widetilde{C}_B\, R_a\left(\partial_t(\Pi_{hp}u - u)\right).$$

Then, similarly as in (4.120), we get

$$|||\partial_t(u - \Pi_{hp}u)||| \leq C_\sigma R_a\left(\partial_t(u - \Pi_{hp}u)\right),$$

which together with the triangle inequality gives

$$\begin{aligned} |||\partial_t(u - \mathrm{P}_{hp}u)(t)||| &\leq |||\partial_t(u - \Pi_{hp}u)(t)||| + |||\partial_t(\Pi_{hp}u - \mathrm{P}_{hp}u)(t)||| \\ &\leq (2\widetilde{C}_B + C_\sigma)R_a\left(\partial_t(u - \Pi_{hp}u)(t)\right), \quad t \in (0, T). \end{aligned} \tag{4.124}$$

Finally, we use relation (4.99) and estimate (4.96) rewritten for $\partial_t u(t) - \Pi_{hp}(\partial_t u(t))$:

$$R_a\left(\partial_t u(t) - \Pi_{hp}(\partial_t u(t))\right) \leq \sqrt{3}C_A h^{\mu-1}\, |\partial_t u(t)|_{H^\mu(\Omega)}.$$

This and (4.124) already give (4.117). □

In what follows, for an arbitrary $z \in L^2(\Omega)$ we consider the elliptic *dual problem* (2.155): Given $z \in L^2(\Omega)$, find ψ such that

$$-\Delta\psi = z \quad \text{in } \Omega, \quad \psi|_{\partial\Omega} = 0. \tag{4.125}$$

Similarly as in (2.157), the weak formulation of problem (4.125) reads: Find $\psi \in H_0^1(\Omega)$ such that

$$\int_\Omega \nabla\psi \cdot \nabla v \,\mathrm{d}x = \int_\Omega z\, v \,\mathrm{d}x \qquad \forall\, v \in H_0^1(\Omega). \tag{4.126}$$

As the domain Ω is convex, for every $z \in L^2(\Omega)$ the weak solution ψ is regular, i.e., $\psi \subset H^2(\Omega)$, and there exists a constant $C_D > 0$, independent of z such that

$$\|\psi\|_{H^2(\Omega)} \leq C_D\|z\|_{L^2(\Omega)}, \tag{4.127}$$

as follows from [153]. Let us note that $H^2(\Omega) \subset C(\overline{\Omega})$.

Further, let $\Pi_{h1}\psi$ be the piecewise linear $L^2(\Omega)$-projection of the function ψ on S_{h1} (cf. (2.91)). Obviously, using (2.125), and (4.96) with $\mu = 2$, we have

$$\|\psi - \Pi_{h1}\psi\|_{1,\sigma} \leq C_\sigma R_a(\psi - \Pi_{h1}\psi) \leq \sqrt{3}C_A C_\sigma h|\psi|_{H^2(\Omega)}. \tag{4.128}$$

Finally, taking into account that the form A_h is the ε multiple of the form A_h from Chap. 2 and using estimate 2.122, we have

$$|A_h(u, v)| \le 2\,\varepsilon\|u\|_{1,\sigma}\|v\|_{1,\sigma} \quad \forall u, v \in H^2(\Omega, \mathscr{T}_h). \tag{4.129}$$

Now we use the dual problem (4.125) to obtain $L^2(\Omega)$-optimal error estimates for $\chi = u - \mathrm{P}_{hp}u$ and $\partial_t\chi = (u - \mathrm{P}_{hp}u)_t$.

Lemma 4.18 *There exists a constant $C_{P,L} > 0$ independent of ε such that*

$$\|\chi(t)\|_{L^2(\Omega)} \le C_{P,L}h^{\mu}|u(t)|_{H^{\mu}(\Omega)}, \qquad t \in (0, T), \tag{4.130}$$

$$\|\partial_t\chi(t)\|_{L^2(\Omega)} \le C_{P,L}h^{\mu}|\partial_t u(t)|_{H^{\mu}(\Omega)}, \quad t \in (0, T), \tag{4.131}$$

for all $h \in (0, \bar{h})$.

Proof We have

$$\|\chi\|_{L^2(\Omega)} = \sup_{0\neq z\in L^2(\Omega)} \frac{|(\chi, z)|}{\|z\|_{L^2(\Omega)}}. \tag{4.132}$$

Taking into account that the form A_h is the ε multiple of the form A_h from Chap. 2, we see that by Lemma 2.48, for $z \in L^2(\Omega)$ and ψ satisfying (4.125), we have

$$(\chi, z) = \frac{1}{\varepsilon}\, A_h(\psi, \chi). \tag{4.133}$$

Further, the symmetry of A_h and relation (4.115) give

$$A_h(\Pi_{h1}\psi, \chi) = A_h(\chi, \Pi_{h1}\psi) = A_h(u - \mathrm{P}_{hp}u,\ \Pi_{h1}\psi) = 0, \tag{4.134}$$

and therefore,

$$(\chi, z) = \frac{1}{\varepsilon}\, A_h(\psi - \Pi_{h1}\psi, \chi). \tag{4.135}$$

Now, using (4.129), we have

$$|(\chi, z)| = \frac{1}{\varepsilon}\,|A_h(\psi - \Pi_{h1}\psi, \chi)| \le 2\|\psi - \Pi_{h1}\psi\|_{1,\sigma}\,\|\chi\|_{1,\sigma}. \tag{4.136}$$

Moreover, by (4.128) and (4.127), we obtain

$$\|\psi - \Pi_{h1}\psi\|_{1,\sigma} \le \sqrt{3}C_A C_\sigma h|\psi|_{H^2(\Omega)} \le \sqrt{3}C_A C_\sigma C_D h\|z\|_{L^2(\Omega)}. \tag{4.137}$$

Triangle inequality, (2.125), (2.126), (4.96) and (4.116) imply the estimate

$$\begin{aligned}\|\chi(t)\|_{1,\sigma} &= \left\|u - \mathrm{P}_{hp}u\right\|_{1,\sigma} \le \left\|u - \Pi_{hp}u\right\|_{1,\sigma} + \left\|\Pi_{hp}u - \mathrm{P}_{hp}u\right\|_{1,\sigma} \\ &\le C_\sigma R_a(u - \Pi_{hp}u) + \widetilde{C}_\sigma |||\Pi_{hp}u - \mathrm{P}_{hp}u||| \\ &\le C_\sigma R_a(u - \Pi_{hp}u) + \widetilde{C}_\sigma |||\Pi_{hp}u - u||| + \widetilde{C}_\sigma |||u - \mathrm{P}_{hp}u||| \\ &\le C_\sigma R_a(u - \Pi_{hp}u) + \widetilde{C}_\sigma C_\sigma R_a(u - \Pi_{hp}u) + \widetilde{C}_\sigma |||\chi||| \\ &\le C_\sigma(1 + \widetilde{C}_\sigma)\sqrt{3} C_A h^{\mu-1} |u(t)|_{H^\mu(\Omega)} + \widetilde{C}_\sigma C_{P,\mathrm{e}}\, h^{\mu-1} |u(t)|_{H^\mu(\Omega)} \\ &= C_6 h^{\mu-1} |u(t)|_{H^\mu(\Omega)}, \quad t \in (0, T),\end{aligned} \tag{4.138}$$

where $C_6 = C_\sigma(1 + \widetilde{C}_\sigma)\sqrt{3}C_A + \widetilde{C}_\sigma C_{P,\mathrm{e}}$. Summarizing (4.136), (4.137) and (4.138), we find that

$$\begin{aligned}(\chi(t), z) &\le 2\sqrt{3} C_A C_\sigma C_D h \|z\|_{L^2(\Omega)} C_6 h^{\mu-1} |u(t)|_{H^\mu(\Omega)} \\ &= C_{P,\mathrm{L}}\, h^\mu |u(t)|_{H^\mu(\Omega)} \|z\|_{L^2(\Omega)}, \quad t \in (0, T),\end{aligned}$$

where $C_{P,\mathrm{L}} = 2\sqrt{3} C_A C_\sigma C_D C_6$. Hence,

$$\|\chi(t)\|_{L^2(\Omega)} = \sup_{0 \ne z \in L^2(\Omega)} \frac{|(\chi(t), z)|}{\|z\|_{L^2(\Omega)}} \le C_{P,\mathrm{L}}\, h^\mu |u(t)|_{H^\mu(\Omega)}, \quad t \in (0, T),$$

which completes the proof of (4.130).

Finally, let us prove estimate (4.131). Differentiating (4.115) with respect to t yields

$$A_h(\partial_t \chi, \varphi_h) = 0 \quad \forall \varphi_h \in S_{hp}. \tag{4.139}$$

We have

$$\|\partial_t \chi\|_{L^2(\Omega)} = \sup_{0 \ne z \in L^2(\Omega)} \frac{|(\partial_t \chi, z)|}{\|z\|_{L^2(\Omega)}}. \tag{4.140}$$

Similarly as in (4.133), we get

$$(\partial_t \chi, z) = \frac{1}{\varepsilon} A_h(\psi, \partial_t \chi). \tag{4.141}$$

The symmetry of A_h and relation (4.139) imply that

$$A_h(\Pi_{h1}\psi, \partial_t \chi) = A_h(\partial_t \chi, \Pi_{h1}\psi) = A_h\left(\partial_t(u - \mathrm{P}_{hp}u), \Pi_{h1}\psi\right) = 0.$$

These relations, (4.141) and (4.129) yield

$$|(\partial_t \chi, z)| = \frac{1}{\varepsilon}|A_h(\psi - \Pi_{h1}\psi, \partial_t \chi)| \le 2\|\psi - \Pi_{h1}\psi\|_{1,\sigma}\|\partial_t \chi\|_{1,\sigma}. \tag{4.142}$$

The term $\|\psi - \Pi_{h1}\psi\|_{1,\sigma}$ is estimated by (4.137) and similarly as in (4.138), we obtain

$$\|\partial_t \chi(t)\|_{1,\sigma} \le C_6 h^{\mu-1}|\partial_t u(t)|_{H^\mu(\Omega)}, \quad t \in (0, T). \tag{4.143}$$

Finally, from (4.140), (4.142), (4.137) and (4.143), we arrive at estimate (4.131). □

Let us note that assuming the symmetry of the form A_h is crucial in the presented proof. It enables us to exchange arguments in (4.134). This is not possible in the NIPG and IIPG methods, where the analysis of optimal $L^\infty(L^2)$-error estimates still represents an open problem.

Lemma 4.19 *Let us assume that u is the solution of the continuous problem (4.110) satisfying condition (4.75), u_h is the solution of the discrete problem (4.25), $\mathrm{P}_{hp}u$ is defined by (4.115), and $\zeta = \mathrm{P}_{hp}u - u_h \in S_{hp}$. Then there exists a constant $C_b > 0$, independent of $h \in (0, \bar{h})$, such that*

$$|b_h(u, \zeta) - b_h(u_h, \zeta)| \le C_b |||\zeta||| \left(h^\mu |u|_{H^\mu(\Omega)} + \|\zeta\|_{L^2(\Omega)}\right). \tag{4.144}$$

Proof We proceed similarly as in the proof of Lemma 4.6. The triangle inequality gives

$$|b_h(u, \zeta) - b_h(u_h, \zeta)| \le |b_h(u, \zeta) - b_h(\mathrm{P}_{hp}u, \zeta)| + |b_h(\mathrm{P}_{hp}u, \zeta) - b_h(u_h, \zeta)|. \tag{4.145}$$

Applying (4.30) with $\bar{u} := \mathrm{P}_{hp}u$ and $v := \zeta \in S_{hp}$, we get

$$|b_h(u, \zeta) - b_h(\mathrm{P}_{hp}u, \zeta)| \le C_{b1} |||\zeta||| \left(\|\chi\|^2_{L^2(\Omega)} + \sum_{K \in \mathscr{T}_h} h_K \|\chi\|^2_{L^2(\partial K)} \right)^{1/2}. \tag{4.146}$$

(Let us recall that $\chi = u - \mathrm{P}_{hp}u$). The multiplicative trace inequality (2.78) and the Cauchy inequality give

$$\begin{aligned}
\sum_{K \in \mathscr{T}_h} h_K \|\chi\|^2_{L^2(\partial K)} &\le C_M \sum_{K \in \mathscr{T}_h} \left(h_K |\chi|_{H^1(K)} \|\chi\|_{L^2(K)} + \|\chi\|^2_{L^2(K)} \right) \\
&\le C_M \left(h \left(\sum_{K \in \mathscr{T}_h} |\chi|^2_{H^1(K)} \right)^{1/2} \left(\sum_{K \in \mathscr{T}_h} \|\chi\|^2_{L^2(K)} \right)^{1/2} + \sum_{K \in \mathscr{T}_h} \|\chi\|^2_{L^2(K)} \right) \\
&\le C_M \left(h |\chi|_{H^1(\Omega, \mathscr{T}_h)} \|\chi\|_{L^2(\Omega)} + \|\chi\|^2_{L^2(\Omega)} \right).
\end{aligned}$$

The above relations, the inequality $|\chi|_{H^1(\Omega,\mathscr{T}_h)} \le |||\chi|||$ and estimates (4.116) and (4.130) imply that

$$\sum_{K\in\mathscr{T}_h} h_K \|\chi(t)\|^2_{L^2(\partial K)} \le C_M \left(C_{P,\mathrm{e}} C_{P,\mathrm{L}}\, h\, h^{\mu-1}\, h^\mu + C^2_{P,\mathrm{L}} h^{2\mu}\right) |u(t)|^2_{H^\mu(\Omega)}$$

$$= C_7 h^{2\mu} |u(t)|^2_{H^\mu(\Omega)}, \quad t \in (0, T), \tag{4.147}$$

where $C_7 = C_M(C_{P,\mathrm{e}} C_{P,\mathrm{L}} + C^2_{P,\mathrm{L}})$. Furthermore, (4.146), (4.130) and (4.147) give

$$|b_h(u, \zeta) - b_h(\mathrm{P}_{hp}u, \zeta)| \le C_{b1} \left(C^2_{P,\mathrm{L}} + C_7\right)^{1/2} h^\mu |||\zeta||| |u(t)|_{H^\mu(\Omega)}. \tag{4.148}$$

Furthermore, estimate (4.31) with $\bar{u}_h := \mathrm{P}_{hp}u \in S_{hp}$ and $v_h := \zeta \in S_{hp}$ gives

$$|b_h(\mathrm{P}_{hp}u, \zeta) - b_h(u_h, \zeta)| \le C_{b2} |||\zeta|||\, \|u_h - \mathrm{P}_{hp}u\|_{L^2(\Omega)} = C_{b2} |||\zeta|||\, \|\zeta\|_{L^2(\Omega)}. \tag{4.149}$$

Finally, inserting estimates (4.148) and (4.149) into (4.145), we obtain inequality (4.144) with $C_b = \max\left(C_{b2}, C_{b1}(C^2_{P,\mathrm{L}} + C_7)^{1/2}\right)$. □

Now we can proceed to the *main result*, which is the optimal error estimate in the norm of the space $L^\infty(0, T; L^2(\Omega))$ of the DG method (4.25) applied on nonconforming meshes.

Theorem 4.20 *Let $\Omega \subset \mathbb{R}^d$, $d = 2, 3$, be a bounded convex polygonal (if $d = 2$) or* polyhedral *(if $d = 3$) domain with Lipschitz boundary $\partial\Omega$. Let Assumptions 4.5 in Sect. 4.3 be satisfied. Let u be the exact solution of problem (4.1), where $\partial\Omega_D = \partial\Omega$ and $\partial\Omega_N = \emptyset$, satisfying the regularity condition (4.94) and let u_h be the approximate solution obtained by scheme (4.25) with the SIPG version of the diffusion terms and the constant C_W satisfying (2.132). Then the error $e_h = u_h - u$ satisfies the estimate*

$$\|e_h\|_{L^\infty(0,T;L^2(\Omega))} \le C_8 h^\mu, \quad h \in (0, \bar{h}), \tag{4.150}$$

with a constant $C_8 > 0$ independent of h.

Proof Let $\mathrm{P}_{hp}u$ be defined by (4.115) and let χ and ζ be as in Lemmas 4.17, 4.18 and 4.19, i.e., $\chi = u - \mathrm{P}_{hp}u$, $\zeta = \mathrm{P}_{hp}u - u_h$. Then $e_h = u_h - u = -\chi - \zeta$. Let us subtract (4.25b) from (4.28), substitute $\zeta \in S_{hp}$ for v_h, and use the relations

$$\left(\frac{\partial\zeta(t)}{\partial t}, \zeta(t)\right) = \frac{1}{2}\frac{\mathrm{d}}{\mathrm{d}t} \|\zeta(t)\|^2_{L^2(\Omega)}, \qquad A_h(u(t) - \mathrm{P}_{hp}u(t), \zeta(t)) = 0.$$

Then we get

$$\frac{1}{2}\frac{\mathrm{d}}{\mathrm{d}t}\|\zeta(t)\|^2_{L^2(\Omega)} + A_h(\zeta(t),\, \zeta(t)) \\ = (b_h(u_h(t),\, \zeta(t)) - b_h(u(t),\, \zeta(t))) - (\partial_t \chi(t),\, \zeta(t)). \tag{4.151}$$

The first term on the right-hand side can be estimated by Lemma 4.19 and the Young inequality. In estimating the second term on the right-hand side we use the Cauchy and Young inequalities and Lemma 4.18. Finally, the coercivity property (2.140) (where $C_C = 1/2$) of $\frac{1}{\varepsilon}A_h = a_h + J_h^\sigma$ gives the estimate on the left-hand side of (4.151). On the whole, after some manipulation, we get

$$\begin{aligned}
&\frac{\mathrm{d}}{\mathrm{d}t}\|\zeta(t)\|^2_{L^2(\Omega)} + \varepsilon|||\zeta(t)|||^2 \\
&\le 2\,|b_h(u_h(t),\, \zeta(t)) - b_h(u(t),\, \zeta(t))| + 2|(\partial_t\chi(t),\, \zeta(t))| \\
&\le 2C_b|||\zeta|||\left(h^\mu|u|_{H^\mu(\Omega)} + \|\zeta\|_{L^2(\Omega)}\right) + 2\|\partial_t\chi(t)\|_{L^2(\Omega)}\|\zeta(t)\|_{L^2(\Omega)} \\
&\le \varepsilon|||\zeta(t)|||^2 + \frac{2C_b^2}{\varepsilon}h^{2\mu}|u|^2_{H^\mu(\Omega)} + \frac{2C_b^2}{\varepsilon}\|\zeta\|^2_{L^2(\Omega)} + C_{P,\mathrm{L}}^2 h^{2\mu}|\partial_t u|^2_{H^\mu(\Omega)} + \|\zeta(t)\|^2_{L^2(\Omega)} \\
&\le \varepsilon|||\zeta(t)|||^2 + C_9 h^{2\mu}\left(\frac{1}{\varepsilon}|u|^2_{H^\mu(\Omega)} + |\partial_t u|^2_{H^\mu(\Omega)}\right) + C_9\left(1 + \frac{1}{\varepsilon}\right)\|\zeta\|^2_{L^2(\Omega)},
\end{aligned} \tag{4.152}$$

where $C_9 = \max(2C_b^2,\, C_{P,\mathrm{L}}^2,\, 1)$. This implies that

$$\frac{\mathrm{d}}{\mathrm{d}t}\|\zeta(t)\|^2_{L^2(\Omega)} \le C_9 h^{2\mu}\left(\frac{1}{\varepsilon}|u|^2_{H^\mu(\Omega)} + |\partial_t u|^2_{H^\mu(\Omega)}\right) + C_9\left(1 + \frac{1}{\varepsilon}\right)\|\zeta\|^2_{L^2(\Omega)}. \tag{4.153}$$

Using (4.25c), (2.97), (4.130), we have

$$\begin{aligned}
\|\zeta(0)\|^2_{L^2(\Omega)} &= \|\mathrm{P}_{hp}u(0) - u_h(0)\|^2_{L^2(\Omega)} = \|\mathrm{P}_{hp}u(0) - \Pi_{hp}u(0)\|^2_{L^2(\Omega)} \\
&\le 2\|\mathrm{P}_{hp}u(0) - u(0)\|^2_{L^2(\Omega)} + 2\|u(0) - \Pi_{hp}u(0)\|^2_{L^2(\Omega)} \\
&\le 2(C_A^2 + C_{P,\mathrm{L}}^2)h^{2\mu}|u^0|^2_{H^\mu(\Omega)} = C_{10}h^{2\mu}|u^0|^2_{H^\mu(\Omega)},
\end{aligned} \tag{4.154}$$

where $C_{10} = 2(C_A^2 + C_{P,\mathrm{L}}^2)$.

Integrating of (4.153) from 0 to $t \in [0, T]$ and (4.154) yield

$$\begin{aligned}
\|\zeta(t)\|^2_{L^2(\Omega)} &\le C_9\, h^{2\mu}\left(\frac{1}{\varepsilon}\int_0^t |u(\vartheta)|^2_{H^\mu(\Omega)}\,\mathrm{d}\vartheta + \int_0^t |\partial_t u(\vartheta)|^2_{H^\mu(\Omega)}\,\mathrm{d}\vartheta\right) \\
&\quad + C_9\left(1 + \frac{1}{\varepsilon}\right)\int_0^t \|\zeta(\vartheta)\|^2_{L^2(\Omega)}\,\mathrm{d}\vartheta + C_{10}\, h^{2\mu}|u^0|^2_{H^\mu(\Omega)}. \\
&\le C_9\left(1 + \frac{1}{\varepsilon}\right)\int_0^t \|\zeta(\vartheta)\|^2_{L^2(\Omega)}\,\mathrm{d}\vartheta + C_{11}h^{2\mu}N(\varepsilon, u),
\end{aligned} \tag{4.155}$$

where $C_{11} = \max(C_9, C_{10})$ and

$$N(\varepsilon, u) = \frac{1}{\varepsilon}\int_0^t |u(\vartheta)|^2_{H^\mu(\Omega)}\,\mathrm{d}\vartheta + \int_0^t |\partial_t u(\vartheta)|^2_{H^\mu(\Omega)}\,\mathrm{d}\vartheta + |u^0|^2_{H^\mu(\Omega)}.$$

Now we apply the Gronwall Lemma 1.9, where we put

$$\begin{aligned} y(t) &= \|\zeta(t)\|^2_{L^2(\Omega)}, & q(t) &= 0,\\ r(t) &= C_9\left(1 + 1/\varepsilon\right), & z(t) &= C_{11}h^{2\mu}N(\varepsilon, u). \end{aligned}$$

Then, after some manipulation, we obtain the estimate

$$\|\zeta(t)\|^2_{L^2(\Omega)} \le C_{11}h^{2\mu}\, N(\varepsilon, u)\, \exp\left(C_9\left(1 + \frac{1}{\varepsilon}\right)t\right). \tag{4.156}$$

Since $e_h = -\chi - \zeta$, to complete the proof, it is sufficient to combine (4.156) with the estimate (4.130) of $\|\chi(t)\|_{L^2(\Omega)}$ in Lemma 4.18. □

Exercise 4.21 Prove estimates (4.155) and (4.156) in detail.

4.6 Uniform Error Estimates with Respect to the Diffusion Coefficient

In Sects. 4.1–4.5, error estimates for the space DG semidiscretization were derived in the case of nonlinear convection-diffusion problems. From the presented analysis we can see that the constants in these estimates blow up exponentially if the diffusion coefficient $\varepsilon \to 0+$. This means that these estimates are not applicable, if $\varepsilon > 0$ is very small. (See also Remark 4.16.) There is question as to whether it is possible to obtain error estimates that are uniform with respect to the diffusion coefficient $\varepsilon \to 0+$ of convection-diffusion problems.

In this section we are concerned with the error analysis of the DGM of lines applied to a linear convection-diffusion equation, which also contains a reaction term, and its coefficients satisfy some special assumptions used in works analyzing numerical methods for linear convection-diffusion problems (cf. [245, Chap. III], or [183]). As a result, we obtain error estimates, uniform with respect to the diffusion coefficient $\varepsilon \to 0+$, and valid even for $\varepsilon = 0$.

4.6.1 Continuous Problem

Let $\Omega \subset \mathbb{R}^d$ ($d = 2$ or 3) be a bounded polygonal (for $d = 2$) or polyhedral (for $d = 3$) domain with Lipschitz boundary $\partial\Omega$ and $T > 0$. We set $Q_T = \Omega \times (0, T)$.

Let $\boldsymbol{v} : \overline{Q_T} = \overline{\Omega} \times [0, T] \to \mathbb{R}^d$ be a given *transport flow velocity*. We assume that $\partial\Omega = \partial\Omega^- \cup \partial\Omega^+$, and for all $t \in (0, T)$,

$$\begin{aligned} \boldsymbol{v}(x,t) \cdot \boldsymbol{n}(x) &< 0 \text{ on } \partial\Omega^-, \\ \boldsymbol{v}(x,t) \cdot \boldsymbol{n}(x) &\geq 0 \text{ on } \partial\Omega^+, \end{aligned} \tag{4.157}$$

where $\boldsymbol{n}(x)$ denotes the outer unit normal to the boundary of Ω. We assume that the parts $\partial\Omega^-$ and $\partial\Omega^+$ are independent of time. With respect to our former notation, we can write $\partial\Omega_D = \partial\Omega^-$ and $\partial\Omega_N = \partial\Omega^+$. The part $\partial\Omega^-$ of the boundary $\partial\Omega$ represents the inlet through which the fluid enters the domain Ω. The part of $\partial\Omega^+$, where $\boldsymbol{v} \cdot \boldsymbol{n} > 0$, represents the outlet through which the fluid leaves the domain Ω, and the part on which $\boldsymbol{v} \cdot \boldsymbol{n} = 0$ represents impermeable walls.

We consider the following linear initial-boundary value convection-diffusion-reaction problem: Find $u : Q_T \to \mathbb{R}$ such that

$$\frac{\partial u}{\partial t} + \boldsymbol{v} \cdot \nabla u - \varepsilon \Delta u + cu = g \quad \text{in } Q_T, \tag{4.158a}$$

$$u = u_D \quad \text{on } \partial\Omega^- \times (0, T), \tag{4.158b}$$

$$\varepsilon \boldsymbol{n} \cdot \nabla u = g_N \quad \text{on } \partial\Omega^+ \times (0, T), \tag{4.158c}$$

$$u(x, 0) = u^0(x), \quad x \in \Omega. \tag{4.158d}$$

In the case $\varepsilon = 0$, we put $g_N = 0$ and ignore the Neumann condition (4.158c).

Equation (4.158a) describes the transport and diffusion in a fluid of a quantity u as, for example, temperature or concentration of some material. The constant $\varepsilon \geq 0$ is the diffusion coefficient, c represents a reaction coefficient, and g defines the source of the quantity u. Such equations appear, for example, in fluid dynamics or heat and mass transfer.

We assume that the data satisfy the following conditions:

$$g \in C([0, T]; L^2(\Omega)), \tag{4.159a}$$

$$u_0 \in L^2(\Omega), \tag{4.159b}$$

$$u_D \text{ is the trace of some } u^* \in C([0, T]; H^1(\Omega)) \cap L^\infty(Q_T) \text{ on } \partial\Omega^- \times (0, T), \tag{4.159c}$$

$$\boldsymbol{v} \in C([0, T]; W^{1,\infty}(\Omega)),\ |\boldsymbol{v}| \leq C_{\boldsymbol{v}} \text{ in } \overline{\Omega} \times [0, T],\ |\nabla \boldsymbol{v}| \leq C_{\boldsymbol{v}} \text{ a. e. in } Q_T, \tag{4.159d}$$

$$c \in C([0, T]; L^\infty(\Omega)),\ |c(x, t)| \leq C_c \text{ a. e. in } Q_T, \tag{4.159e}$$

$$c - \frac{1}{2}\nabla \cdot \boldsymbol{v} \geq \gamma_0 > 0 \text{ in } Q_T \text{ with a constant } \gamma_0, \tag{4.159f}$$

$$g_N \in C([0, T]; L^2(\partial\Omega^+)), \tag{4.159g}$$

$$\varepsilon \geq 0. \tag{4.159h}$$

Assumption (4.159f) is not restrictive, because using the transformation $u = e^{\alpha t} w$, $\alpha = \text{const}$ substituted into (4.158) leads to the equation for w in the form

$$\frac{\partial w}{\partial t} + \boldsymbol{v} \cdot \nabla w - \varepsilon \Delta w + (c + \alpha) w = g e^{-\alpha t}.$$

Condition (4.159f) now reads $c + \alpha - \frac{1}{2} \nabla \cdot \boldsymbol{v} \geq \gamma_0 > 0$ and is satisfied if we choose $\alpha > 0$ large enough.

The weak formulation is derived in a standard way. Equation (4.158) is multiplied by any $\varphi \in V = \{\varphi \in H^1(\Omega); \varphi|_{\partial\Omega^-} = 0\}$, Green's theorem is applied and condition (4.158c) is used.

Definition 4.22 We say that a function u is the *weak solution* to (4.158) if it satisfies the conditions

$$u - u^* \in L^2(0, T; V), \quad u \in L^\infty(Q_T), \tag{4.160a}$$

$$\begin{aligned}
&\frac{d}{dt} \int_\Omega u \varphi \, \mathrm{d}x + \varepsilon \int_\Omega \nabla u \cdot \nabla \varphi \, \mathrm{d}x + \int_{\partial\Omega^+} (\boldsymbol{v} \cdot \boldsymbol{n}) u \varphi \, \mathrm{d}S - \int_\Omega u \nabla \cdot (\varphi \boldsymbol{v}) \, \mathrm{d}x \\
&\quad + \int_\Omega c u \varphi \, \mathrm{d}x = \int_\Omega g \varphi \, \mathrm{d}x + \int_{\partial\Omega^+} g_N \varphi \, \mathrm{d}S \\
&\quad \text{for all } \varphi \in V \text{ in the sense of distributions on } (0, T),
\end{aligned} \tag{4.160b}$$

$$u(0) = u^0 \text{ in } \Omega. \tag{4.160c}$$

We assume that the weak solution u exists and is sufficiently regular, namely,

$$\frac{\partial u}{\partial t} \in L^2(0, T; H^s(\Omega)), \tag{4.161}$$

where $s \geq 2$ is an integer. Then also $u \in C([0, T); H^s(\Omega))$ and it is possible to show that this solution u satisfies equation (4.158a) pointwise (almost everywhere). (If $\varepsilon > 0$, then with the aid of techniques from [217, 235, 246], it is possible to prove that there exists a unique weak solution. Moreover, it satisfies the condition $\partial u / \partial t \in L^2(Q_T)$.)

4.6.2 Discretization of the Problem

Let $\mathscr{T}_h$ be a standard conforming triangulation of the closure of the domain Ω into a finite number of closed triangles ($d = 2$) or tetrahedra ($d = 3$). Hence, the mesh $\mathscr{T}_h$ satisfies assumption (MA4) in Sect. 2.3.2. This means that we do not consider hanging nodes (or hanging edges) in this case. Otherwise we use the same notation as in Sect. 4.2.

We assume that the conforming triangulations satisfy the shape-regularity assumption (2.19). For $K \in \mathscr{T}_h$ we set

$$\partial K^-(t) = \{x \in \partial K;\ \boldsymbol{v}(x,t) \cdot \boldsymbol{n}(x) < 0\}, \tag{4.162}$$

$$\partial K^+(t) = \{x \in \partial K;\ \boldsymbol{v}(x,t) \cdot \boldsymbol{n}(x) \geq 0\}, \tag{4.163}$$

where $\boldsymbol{n}$ denotes the outer unit normal to ∂K. Hence, $\partial K^-(t)$ and $\partial K^+(t)$ denote the inlet and outlet parts of the boundary of K, respectively. In what follows we do not emphasize the dependence of ∂K^+ and ∂K^- on time by notation.

In order to derive error estimates that are uniform with respect to ε, we discretize the convective terms using the idea of the *upwinding* (see (4.16)). This choice allows us to avoid using the Gronwall lemma, which causes the non-uniformity of the error estimates in Sects. 4.3 and 4.5 (see Remark 4.30). Multiplying the convective term $\boldsymbol{v} \cdot \nabla u$ by any $\varphi \in H^2(\Omega, \mathscr{T}_h)$, integrating over element K and applying Green's theorem, we get

$$\begin{aligned} \int_K (\boldsymbol{v} \cdot \nabla u)\varphi \, \mathrm{d}x &= -\int_K u \, \nabla \cdot (\varphi \boldsymbol{v}) \, \mathrm{d}x + \int_{\partial K} (\boldsymbol{v} \cdot \boldsymbol{n}) u \varphi \, \mathrm{d}S \\ &= -\int_K u \, \nabla \cdot (\varphi \boldsymbol{v}) \, \mathrm{d}x + \int_{\partial K^-} (\boldsymbol{v} \cdot \boldsymbol{n}) u \varphi \, \mathrm{d}S + \int_{\partial K^+} (\boldsymbol{v} \cdot \boldsymbol{n}) u \varphi \, \mathrm{d}S. \end{aligned} \tag{4.164}$$

On the inflow part of the boundary of K we use information from outside of the element K. Therefore, we write there u^- instead of u. If $x \in \partial\Omega^-$, then we set $u^-(x) := u_D(x)$. The integrals over ∂K^+, where the information "flows out" of the element, remain unchanged. We take into account that $[u] = 0$ on $\Gamma \in \mathscr{F}_h^I$ and $u|_{\partial\Omega^-}$ satisfies the Dirichlet condition (4.158b). We further rearrange the terms in (4.164) and find that

$$\begin{aligned} &\int_K (\boldsymbol{v} \cdot \nabla u)\varphi \, \mathrm{d}x \\ &\quad = -\int_K u \, \nabla \cdot (\varphi \boldsymbol{v}) \, \mathrm{d}x + \int_{\partial K^-} (\boldsymbol{v} \cdot \boldsymbol{n}) u^- \varphi \, \mathrm{d}S + \int_{\partial K^+} (\boldsymbol{v} \cdot \boldsymbol{n}) u \varphi \, \mathrm{d}S \\ &\quad = -\int_K u \, \nabla \cdot (\varphi \boldsymbol{v}) \, \mathrm{d}x + \underbrace{\int_{\partial K} (\boldsymbol{v} \cdot \boldsymbol{n}) u \varphi \, \mathrm{d}S - \int_{\partial K^+ \cup \partial K^-} (\boldsymbol{v} \cdot \boldsymbol{n}) u \varphi \, \mathrm{d}S}_{=0} \\ &\qquad + \int_{\partial K^-} (\boldsymbol{v} \cdot \boldsymbol{n}) u^- \varphi \, \mathrm{d}S + \int_{\partial K^+} (\boldsymbol{v} \cdot \boldsymbol{n}) u \varphi \, \mathrm{d}S \\ &\quad = \int_K (\boldsymbol{v} \cdot \nabla u)\varphi \, \mathrm{d}x + \int_{\partial K^-} (\boldsymbol{v} \cdot \boldsymbol{n})(u^- - u)\varphi \, \mathrm{d}S \\ &\quad = \int_K (\boldsymbol{v} \cdot \nabla u)\varphi \, \mathrm{d}S - \int_{\partial K^- \setminus \partial\Omega} (\boldsymbol{v} \cdot \boldsymbol{n})[u]\varphi \, \mathrm{d}S - \int_{\partial K^- \cap \partial\Omega} (\boldsymbol{v} \cdot \boldsymbol{n})(u - u_D)\varphi \, \mathrm{d}S, \end{aligned} \tag{4.165}$$

where we set $[u] = u - u^-$ on $\partial K^- \setminus \partial\Omega$.

Remark 4.23 Let us note that identity (4.165) can be derived from the relation

$$\int_K (\boldsymbol{v} \cdot \nabla u)\varphi \, \mathrm{d}x = -\int_K u \nabla \cdot (\varphi \boldsymbol{v}) \, \mathrm{d}x + \sum_{\Gamma \subset \partial K} \int_\Gamma H(u_\Gamma^{(L)}, u_\Gamma^{(R)}, \boldsymbol{n}_\Gamma)\varphi \, \mathrm{d}S,$$

where H is the numerical flux given (in analogy to (4.16)) by

$$H(u_1, u_2, \boldsymbol{n}) = \begin{cases} \boldsymbol{v} \cdot \boldsymbol{n} \, u_1, & \text{if } \boldsymbol{v} \cdot \boldsymbol{n} > 0 \\ \boldsymbol{v} \cdot \boldsymbol{n} \, u_2, & \text{if } \boldsymbol{v} \cdot \boldsymbol{n} \le 0 \end{cases} \tag{4.166}$$

and $H(u_1, u_2, \boldsymbol{n}) = \boldsymbol{v} \cdot \boldsymbol{n} \, u_D$ on $\partial K^- \cap \partial \Omega$.

Exercise 4.24 Verify Remark 4.23.

Now we proceed to the derivation of the discrete problem. We start from Eq. (4.158a) under assumption (4.161), multiply it by any $\varphi \in H^2(\Omega, \mathscr{T}_h)$, integrate over each element K, apply Green's theorem to the diffusion and convective terms and sum over all elements $K \in \mathscr{T}_h$. Then we use the identity (4.165) for convective terms, add some terms to both sides of the resulting identity or vanishing terms (similarly as in Sect. 2.4 in the discretization of the diffusion term) and use the boundary conditions (we recall that $\partial \Omega_D = \partial \Omega^- = \cup_{K \in \mathscr{T}_h} \partial K^- \cap \partial \Omega$). After some manipulation we find that the exact solution u satisfies the following identity for $\varphi \in H^2(\Omega, \mathscr{T}_h)$:

$$\left(\frac{\partial u(t)}{\partial t}, \varphi \right) + A_h(u(t), \varphi) + b_h(u(t), \varphi) + c_h(u(t), \varphi) = \ell_h(\varphi)(t) \tag{4.167}$$

$$\text{for a. e. } t \in (0, T),$$

where the forms in (4.167) are defined in the following way:

$$(u, \varphi) = \int_\Omega u\varphi \, \mathrm{d}x, \tag{4.168}$$

$$A_h(u, \varphi) = \varepsilon a_h(u, \varphi) + \varepsilon J_h^\sigma(u, \varphi), \tag{4.169}$$

$$\begin{aligned} a_h(u, \varphi) = &\sum_{K \in \mathscr{T}_h} \int_K \nabla u \cdot \nabla \varphi \, \mathrm{d}x - \sum_{\Gamma \in \mathscr{F}_h^I} \int_\Gamma (\langle \nabla u \rangle \cdot \boldsymbol{n} [\varphi] + \Theta \langle \nabla \varphi \rangle \cdot \boldsymbol{n} [u]) \, \mathrm{d}S \\ &- \sum_{K \subset \mathscr{T}_h} \int_{\partial K^- \cap \partial \Omega} ((\nabla u \cdot \boldsymbol{n})\varphi + \Theta (\nabla \varphi \cdot \boldsymbol{n}) u) \, \mathrm{d}S, \end{aligned} \tag{4.170}$$

$$J_h^\sigma(u, \varphi) = \sum_{\Gamma \in \mathscr{F}_h^I} \int_\Gamma \sigma [u][\varphi] \, \mathrm{d}S + \sum_{K \in \mathscr{T}_h} \int_{\partial K^- \cap \partial \Omega} \sigma u \varphi \, \mathrm{d}S, \tag{4.171}$$

$$b_h(u,\varphi) = \sum_{K\in\mathscr{T}_h}\int_K (\boldsymbol{v}\cdot\nabla u)\varphi\,\mathrm{d}x - \sum_{K\in\mathscr{T}_h}\int_{\partial K^-\setminus\partial\Omega}(\boldsymbol{v}\cdot\boldsymbol{n})[u]\varphi\,\mathrm{d}S \tag{4.172}$$
$$- \sum_{K\in\mathscr{T}_h}\int_{\partial K^-\cap\partial\Omega}(\boldsymbol{v}\cdot\boldsymbol{n})u\varphi\,\mathrm{d}S,$$

$$c_h(u,\varphi) = \int_\Omega cu\varphi\,\mathrm{d}x, \tag{4.173}$$

$$\ell_h(\varphi)(t) = \int_\Omega g(t)\varphi\,\mathrm{d}x + \sum_{K\in\mathscr{T}_h}\int_{\partial K^+\cap\partial\Omega} g_N(t)\varphi\,\mathrm{d}S \tag{4.174}$$
$$+\varepsilon\sum_{K\in\mathscr{T}_h}\int_{\partial K^-\cap\partial\Omega}\sigma u_D(t)\varphi\,\mathrm{d}S$$
$$+\varepsilon\Theta\sum_{K\in\mathscr{T}_h}\int_{\partial K^-\cap\partial\Omega} u_D(t)(\nabla\varphi\cdot\boldsymbol{n})\,\mathrm{d}S$$
$$-\sum_{K\in\mathscr{T}_h}\int_{\partial K^-\cap\partial\Omega}(\boldsymbol{v}\cdot\boldsymbol{n})u_D(t)\varphi\,\mathrm{d}S.$$

In the form representing the discretization of the diffusion term we use the nonsymmetric (NIPG) formulation for $\Theta = -1$, and the incomplete (IIPG) formulation for $\Theta = 0$ or symmetric formulation (SIPG) for $\Theta = 1$.

The weight $\sigma|_\Gamma$ is defined by (2.104), where h_Γ is given by (2.24) or (2.25) or (2.26) and satisfies (2.20). The constant $C_W > 0$ from (2.104) is arbitrary for the NIPG version, and it satisfies condition (2.132) or (2.139) for the SIPG or IIPG version, respectively.

The approximate solution will be sought for each $t \in (0,T)$ in the finite dimensional space

$$S_{hp} = \left\{\varphi \in L^2(\Omega);\, \varphi|_K \in P_p(K)\ \ \forall K \in \mathscr{T}_h\right\}, \tag{4.175}$$

where $p \geq 1$ is an integer and $P_p(K)$ is the space of polynomials on K of degree at most p.

Definition 4.25 The *DG approximate solution* of problem (4.158) is defined as a function u_h such that

$$u_h \in C^1([0,T); S_{hp}), \tag{4.176a}$$
$$\left(\frac{\partial u_h(t)}{\partial t}, \varphi_h\right) + A_h(u_h(t),\varphi_h) + b_h(u_h(t),\varphi_h) + c_h(u_h(t),\varphi_h) = \ell_h(\varphi_h)(t)$$
$$\forall \varphi_h \in S_{hp}\ \forall t \in (0,T), \tag{4.176b}$$
$$(u_h(0),\varphi_h) = (u^0,\varphi_h) \qquad \forall \varphi_h \in S_{hp}. \tag{4.176c}$$

If $\varepsilon = 0$, we can also choose $p = 0$. In this case we get the finite volume method using piecewise constant approximations. Thus, the finite volume method is a special case of the DGM.

4.6.3 Error Estimates

Let us consider a system $\{\mathscr{T}_h\}_{h\in(0,\bar{h})}$, $\bar{h} > 0$, of *conforming* triangulations of Ω satisfying the shape-regularity assumption (2.19). By Π_{hp} we again denote the S_{hp}-interpolation defined by (2.90) with approximation properties formulated in Lemma 2.24. Thus, if $\mu = \min(p+1, s)$, $s \geq 2$ and $v \in H^s(K)$, then (2.93)–(2.95) hold.

If we denote

$$\xi = u_h - \Pi_{hp}u, \qquad \eta = \Pi_{hp}u - u, \tag{4.177}$$

where u is the exact solution satisfying the regularity conditions (4.161) and u_h is the approximate solution, then the error $e_h = u_h - u = \xi + \eta$. By (2.93)–(2.95) and (4.100), for all $K \in \mathscr{T}_h$ and $h \in (0, \bar{h})$ we have

$$\|\eta\|_{L^2(K)} \leq C_A h^{\mu} |u|_{H^{\mu}(K)}, \tag{4.178}$$

$$|\eta|_{H^1(K)} \leq C_A h^{\mu-1} |u|_{H^{\mu}(K)}, \tag{4.179}$$

$$|\eta|_{H^2(K)} \leq C_A h^{\mu-2} |u|_{H^{\mu}(K)}, \tag{4.180}$$

$$\|\eta\|_{L^2(\Omega)} \leq C_A h^{\mu} |u|_{H^{\mu}(\Omega)}, \tag{4.181}$$

$$\|\partial_t \eta\|_{L^2(\Omega)} \leq C_A h^{\mu} \, |\partial_t u|_{H^{\mu}(\Omega)}, \tag{4.182}$$

almost everywhere in $(0, T)$, where $\partial_t \eta = \partial\eta/\partial t$ and $\partial_t u = \partial u/\partial t$. If $p \geq 0$ and $s \geq 1$, then (4.178), (4.179), (4.181) and (4.182) hold as well, as follows from (2.92).

In the error analysis we use the multiplicative trace inequality (2.78), the inverse inequality (2.86) and the modified variant of the Gronwall Lemma 1.10. For simplicity of notation we introduce the following norm over a subset ω of either $\partial\Omega$ or ∂K:

$$\|\varphi\|_{v,\omega} = \|\sqrt{|v \cdot n|}\, \varphi\|_{L^2(\omega)}, \tag{4.183}$$

where n is the corresponding outer unit normal.

Now we prove the following property of the form b_h given by (4.172).

Lemma 4.26 *There exist positive constants $\overline{C}_b'$ and C_b independent of u, h, ε such that*

$$|b_h(\eta, \xi)| \leq \frac{1}{4} \sum_{K\in\mathscr{T}_h} \left(\|\xi\|^2_{v,\partial K^+\cap\partial\Omega} + \|[\xi]\|^2_{v,\partial K^-\setminus\partial\Omega} \right) \tag{4.184}$$
$$+ \overline{C}_b \sum_{K\in\mathscr{T}_h} \|\eta\|_{L^2(K)} \|\xi\|_{L^2(K)} + R_2(\eta),$$

where

$$R_2(\eta) = \overline{C}_b' \sum_{K\in\mathscr{T}_h} \left(\|\eta\|_{L^2(K)} |\eta|_{H^1(K)} + h_K^{-1} \|\eta\|^2_{L^2(K)} \right), \tag{4.185}$$

$$\overline{C}_b = C_v(1 + C_A C_I), \quad \overline{C}_b' = C_v C_M \tag{4.186}$$

and C_v is the constant in assumption (4.159d).

Proof Using (4.172) and Green's theorem, we find that

$$\begin{aligned} b_h(\eta, \xi) = & \sum_{K\in\mathscr{T}_h} \left(\int_K (\boldsymbol{v}\cdot\nabla\eta)\xi \,\mathrm{d}x \right. \\ & \left. - \int_{\partial K^- \cap \partial\Omega} (\boldsymbol{v}\cdot\boldsymbol{n})\xi\eta \,\mathrm{d}S - \int_{\partial K^- \setminus \partial\Omega} (\boldsymbol{v}\cdot\boldsymbol{n})\xi(\eta - \eta^-) \,\mathrm{d}S \right) \\ = & \sum_{K\in\mathscr{T}_h} \left(\int_{\partial K} (\boldsymbol{v}\cdot\boldsymbol{n})\xi\eta \,\mathrm{d}S - \int_K \eta(\boldsymbol{v}\cdot\nabla\xi)\,\mathrm{d}x - \int_K \eta\xi\,\nabla\cdot\boldsymbol{v}\,\mathrm{d}x \right. \\ & \left. - \int_{\partial K^- \cap \partial\Omega} (\boldsymbol{v}\cdot\boldsymbol{n})\xi\eta \,\mathrm{d}S - \int_{\partial K^- \setminus \partial\Omega} (\boldsymbol{v}\cdot\boldsymbol{n})\xi(\eta - \eta^-) \,\mathrm{d}S \right), \end{aligned} \tag{4.187}$$

where the superscript $^-$ denotes the values on ∂K from outside the element K. Hence,

$$\begin{aligned} |b_h(\eta, \xi)| \le & \left| \sum_{K\in\mathscr{T}_h} \int_K \eta(\boldsymbol{v}\cdot\nabla\xi)\,\mathrm{d}x \right| + \left| \sum_{K\in\mathscr{T}_h} \int_K \eta\xi\,\nabla\cdot\boldsymbol{v}\,\mathrm{d}x \right| \\ & + \left| \sum_{K\in\mathscr{T}_h} \left(\int_{\partial K} (\boldsymbol{v}\cdot\boldsymbol{n})\xi\eta \,\mathrm{d}S - \int_{\partial K^- \cap \partial\Omega} (\boldsymbol{v}\cdot\boldsymbol{n})\xi\eta \,\mathrm{d}S \right. \right. \\ & \left. \left. - \int_{\partial K^- \setminus \partial\Omega} (\boldsymbol{v}\cdot\boldsymbol{n})\xi(\eta - \eta^-) \,\mathrm{d}S \right) \right|. \end{aligned} \tag{4.188}$$

The second term on the right-hand side of (4.188) is estimated easily with the aid of the Cauchy inequality and assumption (4.159d):

$$\left| \sum_{K\in\mathscr{T}_h} \int_K \eta\xi\,\nabla\cdot\boldsymbol{v}\,\mathrm{d}x \right| \le C_v \sum_{K\in\mathscr{T}_h} \|\eta\|_{L^2(K)} \|\xi\|_{L^2(K)}. \tag{4.189}$$

Since

$$\sum_{K\in\mathscr{T}_h} \int_{\partial K^+ \setminus \partial\Omega} (\boldsymbol{v}\cdot\boldsymbol{n})\xi\eta \,\mathrm{d}S = - \sum_{K\in\mathscr{T}_h} \int_{\partial K^- \setminus \partial\Omega} (\boldsymbol{v}\cdot\boldsymbol{n})\xi^-\eta^- \,\mathrm{d}S \tag{4.190}$$

and $\boldsymbol{v} \cdot \boldsymbol{n} \geq 0$ on ∂K^+, with the aid of the Young inequality, the set decomposition

$$\partial K = \partial K^+ \cup (\partial K^- \cap \partial\Omega) \cup (\partial K^- \setminus \partial\Omega)$$

and notation (4.183), the third term on the right-hand side of (4.188) can be rewritten and then estimated in the following way:

$$\begin{aligned}
&\left| \sum_{K\in\mathscr{T}_h} \left(\int_{\partial K^+} (\boldsymbol{v}\cdot\boldsymbol{n})\xi\eta \,\mathrm{d}S + \int_{\partial K^-\setminus\partial\Omega} \left\{ (\boldsymbol{v}\cdot\boldsymbol{n})\xi\eta - (\boldsymbol{v}\cdot\boldsymbol{n})\xi(\eta-\eta^-) \right\} \mathrm{d}S \right) \right| \\
&= \left| \sum_{K\in\mathscr{T}_h} \left(\int_{\partial K^+\cap\partial\Omega} (\boldsymbol{v}\cdot\boldsymbol{n})\xi\eta \,\mathrm{d}S + \int_{\partial K^+\setminus\partial\Omega} (\boldsymbol{v}\cdot\boldsymbol{n})\xi\eta \,\mathrm{d}S + \int_{\partial K^-\setminus\partial\Omega} (\boldsymbol{v}\cdot\boldsymbol{n})\eta^-\xi \,\mathrm{d}S \right) \right| \\
&= \left| \sum_{K\in\mathscr{T}_h} \left(\int_{\partial K^+\cap\partial\Omega} (\boldsymbol{v}\cdot\boldsymbol{n})\xi\eta \,\mathrm{d}S + \int_{\partial K^-\setminus\partial\Omega} (\boldsymbol{v}\cdot\boldsymbol{n})\eta^-(\xi-\xi^-) \,\mathrm{d}S \right) \right| \\
&\leq \frac{1}{4} \sum_{K\in\mathscr{T}_h} \left(\int_{\partial K^+\cap\partial\Omega} (\boldsymbol{v}\cdot\boldsymbol{n})\xi^2 \,\mathrm{d}S + \int_{\partial K^-\setminus\partial\Omega} |\boldsymbol{v}\cdot\boldsymbol{n}|[\xi]^2 \,\mathrm{d}S \right) \qquad (4.191)\\
&\quad + \sum_{K\in\mathscr{T}_h} \left(\int_{\partial K^+\cap\partial\Omega} (\boldsymbol{v}\cdot\boldsymbol{n})\eta^2 \,\mathrm{d}S + \int_{\partial K^-\setminus\partial\Omega} |\boldsymbol{v}\cdot\boldsymbol{n}|(\eta^-)^2 \,\mathrm{d}S \right) \\
&\leq \frac{1}{4} \sum_{K\in\mathscr{T}_h} \left(\|\xi\|^2_{\boldsymbol{v},\partial K^+\cap\partial\Omega} + \|[\xi]\|^2_{\boldsymbol{v},\partial K^-\setminus\partial\Omega} \right) \\
&\quad + \sum_{K\in\mathscr{T}_h} \left(\|\eta\|^2_{\boldsymbol{v},\partial K^+\cap\partial\Omega} + \|\eta^-\|^2_{\boldsymbol{v},\partial K^-\setminus\partial\Omega} \right).
\end{aligned}$$

Using the multiplicative trace inequality, the boundedness of $\boldsymbol{v}$ and estimates (4.178) and (4.179), we get

$$\begin{aligned}
&\sum_{K\in\mathscr{T}_h} \left(\|\eta\|^2_{\boldsymbol{v},\partial K^+\cap\partial\Omega} + \|\eta^-\|^2_{\boldsymbol{v},\partial K^-\setminus\partial\Omega} \right) \qquad (4.192)\\
&\quad \leq C_{\boldsymbol{v}} \sum_{K\in\mathscr{T}_h} \|\eta\|^2_{L^2(\partial K)} \leq C_{\boldsymbol{v}} C_M \sum_{K\in\mathscr{T}_h} \left(\|\eta\|_{L^2(K)} |\eta|_{H^1(K)} + h_K^{-1} \|\eta\|^2_{L^2(K)} \right).
\end{aligned}$$

By virtue of the definition (4.177) of η and (2.89) and (2.90), the first term on the right-hand side of (4.188) vanishes if the vector $\boldsymbol{v}$ is piecewise linear, because $\boldsymbol{v} \cdot \nabla\xi|_K \in P_p(K)$ in this case. If this is not the case, we have to proceed in a more sophisticated way. For every $t \in [0, T)$ we introduce a function $\Pi_{h1}\boldsymbol{v}(t)$ which is a piecewise linear $L^2(\Omega)$-projection of $\boldsymbol{v}(t)$ on the space S_{hp}. Under assumption (4.159d), by (2.96),

$$\|\boldsymbol{v} - \Pi_{h1}\boldsymbol{v}\|_{L^\infty(K)} \leq C_A h_K |\boldsymbol{v}|_{W^{1,\infty}(K)}, \quad K \in \mathscr{T}_h, \ h \in (0, \bar{h}). \qquad (4.193)$$

The first term in (4.188) is then estimated with the aid of (2.89), (2.86), (4.193), the Cauchy inequality and assumption (4.159d) in the following way:

$$\left| \sum_{K \in \mathscr{T}_h} \int_K \eta(\boldsymbol{v} \cdot \nabla \xi) \, \mathrm{d}x \right| \tag{4.194}$$

$$\leq \sum_{K \in \mathscr{T}_h} \left| \int_K \eta(\Pi_{h1}\boldsymbol{v} \cdot \nabla \xi) \, \mathrm{d}x \right| + \sum_{K \in \mathscr{T}_h} \left| \int_K \eta((\boldsymbol{v} - \Pi_{h1}\boldsymbol{v}) \cdot \nabla \xi) \, \mathrm{d}x \right|$$

$$= \sum_{K \in \mathscr{T}_h} \left| \int_K \eta((\boldsymbol{v} - \Pi_{h1}\boldsymbol{v}) \cdot \nabla \xi) \, \mathrm{d}x \right| \leq \sum_{K \in \mathscr{T}_h} \|\boldsymbol{v} - \Pi_{h1}\boldsymbol{v}\|_{L^\infty(K)} \|\eta\|_{L^2(K)} |\xi|_{H^1(K)}$$

$$\leq \sum_{K \in \mathscr{T}_h} C_A h_K |\boldsymbol{v}|_{W^{1,\infty}(K)} \, \|\eta\|_{L^2(K)} \, C_I h_K^{-1} \|\xi\|_{L^2(K)}$$

$$\leq C_{\boldsymbol{v}} C_A C_I \sum_{K \in \mathscr{T}_h} \|\eta\|_{L^2(K)} \|\xi\|_{L^2(K)}.$$

Using (4.189), (4.191) and (4.194) in (4.188), we obtain (4.184) with constants defined in (4.186). This finishes the proof of Lemma 4.26. □

Further, by (4.80) and the Young inequality,

$$|A_h(\eta, \xi)| \leq \varepsilon \widetilde{C}_B R_a(\eta) |||\xi||| \leq \frac{\varepsilon}{4} |||\xi|||^2 + \varepsilon \widetilde{C}_B^2 R_a(\eta)^2 = \frac{\varepsilon}{4} |||\xi|||^2 + \varepsilon R_1(\eta), \tag{4.195}$$

where

$$R_1(\eta) = \widetilde{C}_B^2 \sum_{K \in \mathscr{T}_h} \left(|\eta|_{H^1(K)}^2 + h_K^2 |\eta|_{H^2(K)}^2 + h_K^{-2} \|\eta\|_{L^2(K)}^2 \right). \tag{4.196}$$

Finally, the Cauchy inequality gives

$$|c_h(\eta, \xi)| \leq C_c \|\eta\|_{L^2(\Omega)} \|\xi\|_{L^2(\Omega)}, \tag{4.197}$$

$$|(\partial_t \eta, \xi)| \leq \|\partial_t \eta\|_{L^2(\Omega)} \, \|\xi\|_{L^2(\Omega)}. \tag{4.198}$$

Now we can formulate the *abstract error estimate*.

Theorem 4.27 *Let us assume that $\{\mathscr{T}_h\}_{h \in (0,\bar{h})}$ is a system of conforming shape-regular triangulations (cf. (2.19)) of the domain Ω and let assumptions (4.159) be satisfied. Let us assume that the constant $C_W > 0$ satisfies the conditions in Corollary 2.41 for NIPG, SIPG and IIPG versions of the diffusion form. Let the exact solution u of problem (4.158) be regular in the sense of (4.161) and let u_h be the approximate solution obtained by the method of lines (4.176). Then the error $e_h = u_h - u$ satisfies the estimate*

$$\begin{aligned}
&\Big(\|e_h(t)\|^2_{L^2(\Omega)} + \frac{\varepsilon}{2}\int_0^t |||e_h(\vartheta)|||^2 \mathrm{d}\vartheta + 2\gamma_0 \int_0^t \|e_h(\vartheta)\|^2_{L^2(\Omega)}\,\mathrm{d}\vartheta \\
&\quad + \frac{1}{2}\int_0^t \sum_{K\in\mathscr{T}_h} \Big(\|e_h(\vartheta)\|^2_{\boldsymbol{v}(\vartheta),\partial K\cap\partial\Omega} + \|[e_h(\vartheta)]\|^2_{\boldsymbol{v}(\vartheta),\partial K^-(\vartheta)\setminus\partial\Omega}\Big)\,\mathrm{d}\vartheta\Big)^{1/2} \\
&\le \sqrt{2}\Big(\int_0^t (\varepsilon R_1(\eta(\vartheta)) + R_2(\eta(\vartheta))\mathrm{d}\vartheta\Big)^{1/2} \\
&\quad + 2\sqrt{2}\int_0^t \Big(\|\eta(\vartheta)\|_{L^2(\Omega)}(C_c + \overline{C}_b) + \Big\|\partial_t \eta(\vartheta)\Big\|_{L^2(\Omega)}\Big)\,\mathrm{d}\vartheta \\
&\quad + \sqrt{2}\Big(\|\eta(\vartheta)\|^2_{L^2(\Omega)} + \int_0^t (\frac{\varepsilon}{2}|||\eta(\vartheta)|||^2 + 2\gamma_0\|\eta(\vartheta)\|^2_{L^2(\Omega)} + R_2(\eta(\vartheta)))\,\mathrm{d}\vartheta\Big)^{1/2}, \\
&\quad t\in[0,T],\quad h\in(0,\bar{h}),
\end{aligned} \tag{4.199}$$

where R_1 and R_2 are given by (4.196) *and* (4.185), *respectively.*

Proof The proof will be carried out in several steps.
We subtract Eq. (4.167) from (4.176b) and for arbitrary but fixed $t \in [0, T]$, we put $\varphi := \xi(t)$ to get

$$\begin{aligned}
&(\partial_t \xi, \xi) + A_h(\xi, \xi) + b_h(\xi, \xi) + c_h(\xi, \xi) \\
&\quad = -(\partial_t \eta, \xi) - A_h(\eta, \xi) - b_h(\eta, \xi) - c_h(\eta, \xi).
\end{aligned} \tag{4.200}$$

Obviously,

$$(\partial_t \xi, \xi) = \frac{1}{2}\frac{\mathrm{d}}{\mathrm{d}t}\|\xi\|^2_{L^2(\Omega)}, \tag{4.201}$$

and, in view of Corollary 2.41,

$$A_h(\xi, \xi) \ge \frac{\varepsilon}{2}|||\xi|||^2. \tag{4.202}$$

Further, let us rearrange the terms in the form b_h. We have

$$\begin{aligned}
b_h(\xi, \xi) = &\sum_{K\in\mathscr{T}_h} \Big(\int_K (\boldsymbol{v}\cdot\nabla\xi)\xi\,\mathrm{d}x - \int_{\partial K^-\cap\partial\Omega} (\boldsymbol{v}\cdot\boldsymbol{n})\xi^2\,\mathrm{d}S - \int_{\partial K^-\setminus\partial\Omega} (\boldsymbol{v}\cdot\boldsymbol{n})[\xi]\xi\,\mathrm{d}S\Big) \\
&- \sum_{K\in\mathscr{T}_h} \Big(-\frac{1}{2}\int_K (\nabla\cdot\boldsymbol{v})\,\xi^2\,\mathrm{d}x + \frac{1}{2}\int_{\partial K} (\boldsymbol{v}\cdot\boldsymbol{n})\xi^2\,\mathrm{d}S - \int_{\partial K^-\cap\partial\Omega} (\boldsymbol{v}\cdot\boldsymbol{n})\xi^2\,\mathrm{d}S \\
&- \int_{\partial K^-\setminus\partial\Omega} (\boldsymbol{v}\cdot\boldsymbol{n})\xi(\xi - \xi^-)\,\mathrm{d}S\Big).
\end{aligned}$$

Using the decomposition $\partial K = \partial K^- \cup \partial K^+$, we get

$$b_h(\xi,\xi) = \sum_{K\in\mathscr{T}_h} \frac{1}{2}\Big(-\int_K \xi^2 \nabla\cdot \boldsymbol{v}\,\mathrm{d}x - \int_{\partial K^-\cap\partial\Omega} (\boldsymbol{v}\cdot\boldsymbol{n})\xi^2\,\mathrm{d}S \\ - \int_{\partial K^-\setminus\partial\Omega} (\boldsymbol{v}\cdot\boldsymbol{n})(\xi^2 - 2\xi\xi^-)\,\mathrm{d}S \\ + \int_{\partial K^+\cap\partial\Omega} (\boldsymbol{v}\cdot\boldsymbol{n})\xi^2\,\mathrm{d}S + \int_{\partial K^+\setminus\partial\Omega} (\boldsymbol{v}\cdot\boldsymbol{n})\xi^2\,\mathrm{d}S \Big).$$

Now, by virtue of the relation

$$\sum_{K\in\mathscr{T}_h} \int_{\partial K^+\setminus\partial\Omega} (\boldsymbol{v}\cdot\boldsymbol{n})\xi^2\,\mathrm{d}S = -\sum_{K\in\mathscr{T}_h} \int_{\partial K^-\setminus\partial\Omega} (\boldsymbol{v}\cdot\boldsymbol{n})(\xi^-)^2\,\mathrm{d}S, \tag{4.203}$$

definition (4.162) and (4.183), we find that

$$\begin{aligned} b_h(\xi,\xi) &= \frac{1}{2}\sum_{K\in\mathscr{T}_h} \Big(-\int_K \xi^2 \nabla\cdot\boldsymbol{v}\,\mathrm{d}x - \int_{\partial K^-\cap\partial\Omega} (\boldsymbol{v}\cdot\boldsymbol{n})\xi^2\,\mathrm{d}S \\ &\quad - \int_{\partial K^-\setminus\partial\Omega} (\boldsymbol{v}\cdot\boldsymbol{n})(\xi^2 - 2\xi\xi^- + (\xi^-)^2)\,\mathrm{d}S + \int_{\partial K^+\cap\partial\Omega} (\boldsymbol{v}\cdot\boldsymbol{n})\xi^2\,\mathrm{d}S \Big) \\ &= \frac{1}{2}\sum_{K\in\mathscr{T}_h} \Big(-\int_K \xi^2 \nabla\cdot\boldsymbol{v}\,\mathrm{d}x + \int_{\partial K^-\cap\partial\Omega} |(\boldsymbol{v}\cdot\boldsymbol{n})|\xi^2\,\mathrm{d}S \\ &\quad + \int_{\partial K^-\setminus\partial\Omega} |(\boldsymbol{v}\cdot\boldsymbol{n})|(\xi^2 - 2\xi\xi^- + (\xi^-)^2)\,\mathrm{d}S + \int_{\partial K^+\cap\partial\Omega} |(\boldsymbol{v}\cdot\boldsymbol{n})|\xi^2\,\mathrm{d}S \Big) \\ &= \frac{1}{2}\sum_{K\in\mathscr{T}_h} \Big(\|\xi\|^2_{\boldsymbol{v},\partial K^-\cap\partial\Omega} + \|[\xi]\|^2_{\boldsymbol{v},\partial K^-\setminus\partial\Omega} + \|\xi\|^2_{\boldsymbol{v},\partial K^+\cap\partial\Omega} \Big) \\ &\quad - \frac{1}{2}\int_\Omega \xi^2 \nabla\cdot\boldsymbol{v}\,\mathrm{d}x. \end{aligned} \tag{4.204}$$

Finally,

$$c_h(\xi,\xi) = \int_\Omega c\xi^2\,\mathrm{d}x. \tag{4.205}$$

On the basis of (4.200)–(4.202), (4.204) and (4.205) we obtain the inequality

$$\begin{aligned} \frac{1}{2}\frac{\mathrm{d}}{\mathrm{d}t}\|\xi\|^2_{L^2(\Omega)} &+ \frac{\varepsilon}{2}|||\xi|||^2 + \int_\Omega (c - \frac{1}{2}\nabla\cdot\boldsymbol{v})\xi^2\,\mathrm{d}x \\ &+ \frac{1}{2}\sum_{K\in\mathscr{T}_h} \Big(\|\xi\|^2_{\boldsymbol{v},\partial K\cap\partial\Omega} + \|[\xi]\|^2_{\boldsymbol{v},\partial K^-\setminus\partial\Omega} \Big) \\ &\le |(\partial_t\eta,\xi)| + |A_h(\eta,\xi)| + |b_h(\eta,\xi)| + |c_h(\eta,\xi)|. \end{aligned} \tag{4.206}$$

Now, assumptions (4.159e), (4.159f) and inequalities (4.184), (4.195), (4.197) and (4.198) imply that

$$\frac{\mathrm{d}}{\mathrm{d}t}\|\xi\|^2_{L^2(\Omega)} + \frac{\varepsilon}{2}|||\xi|||^2 + 2\gamma_0\|\xi\|^2_{L^2(\Omega)} + \frac{1}{2}\sum_{K\in\mathscr{T}_h}\left(\|\xi\|^2_{\nu,\partial K\cap\partial\Omega} + \|[\xi]\|^2_{\nu,\partial K^-\setminus\partial\Omega}\right)$$
$$\leq 2\|\xi\|_{L^2(\Omega)}\left(\|\eta\|_{L^2(\Omega)}(C_c+\overline{C}_b) + \|\partial_t\eta\|_{L^2(\Omega)}\right) + 2\varepsilon R_1(\eta) + 2R_2(\eta). \tag{4.207}$$

Integrating (4.207) over $(0,t)$ and using the relation $\xi(0)=0$, we get

$$\|\xi(t)\|^2_{L^2(\Omega)} + \int_0^t \frac{\varepsilon}{2}|||\xi(\vartheta)|||^2\,\mathrm{d}\vartheta + 2\gamma_0\|\xi\|^2_{L^2(Q_t)} \tag{4.208}$$
$$+\frac{1}{2}\int_0^t\sum_{K\in\mathscr{T}_h}\left(\|\xi(\vartheta)\|^2_{\nu(\vartheta),\partial K\cap\partial\Omega} + \|[\xi(\vartheta)]\|^2_{\nu(\vartheta),\partial K^-(\vartheta)\setminus\partial\Omega}\right)\mathrm{d}\vartheta$$
$$\leq 2\int_0^t \|\xi(\vartheta)\|_{L^2(\Omega)}\left(\|\eta(\vartheta)\|_{L^2(\Omega)}(C_c+\overline{C}_b) + \|\partial_t\eta(\vartheta)\|_{L^2(\Omega)}\right)\mathrm{d}\vartheta$$
$$+2\int_0^t(\varepsilon R_1(\eta(\vartheta)) + R_2(\eta(\vartheta)))\mathrm{d}\vartheta.$$

As the last step we make use of the modified the Gronwall Lemma 1.10 with

$$\chi(t) = \|\xi(t)\|_{L^2(\Omega)}, \tag{4.209}$$
$$R(t) = \frac{\varepsilon}{2}\int_0^t |||\xi(\vartheta)|||^2\,\mathrm{d}\vartheta + 2\gamma_0\|\xi\|^2_{L^2(Q_t)}$$
$$+\frac{1}{2}\int_0^t\sum_{K\in\mathscr{T}_h}\left(\|\xi(\vartheta)\|^2_{\nu(\vartheta),\partial K\cap\partial\Omega} + \|[\xi(\vartheta)]\|^2_{\nu(\vartheta),\partial K^-(\vartheta)\setminus\partial\Omega}\right)\mathrm{d}\vartheta,$$
$$A(t) = 2\int_0^t(\varepsilon R_1(\eta(\vartheta)) + R_2(\eta(\vartheta)))\,\mathrm{d}\vartheta,$$
$$B(t) = \|\eta(t)\|_{L^2(\Omega)}(C_c+\overline{C}_b) + \|\partial_t\eta(t)\|_{L^2(\Omega)}.$$

For simplicity, we denote the left-hand side of inequality (4.208) as $L(\xi,t)$. Then for $t\in[0,T]$ we get

$$\sqrt{L(\xi,t)} \leq \left(2\int_0^t(\varepsilon R_1(\eta(t)) + R_2(\eta(t)))\,\mathrm{d}\vartheta\right)^{1/2} \tag{4.210}$$
$$+\int_0^t\left(\|\eta(t)\|_{L^2(\Omega)}(C_c+\overline{C}_b) + \|\partial_t\eta(t)\|_{L^2(\Omega)}\right)\mathrm{d}\vartheta.$$

To obtain the estimate of $e_h = u_h - u = \xi + \eta$, we note that

$$\begin{aligned}
\|e_h\|^2_{L^2(\Omega)} &\le 2\left(\|\xi\|^2_{L^2(\Omega)} + \|\eta\|^2_{L^2(\Omega)}\right),\\
|||e_h|||^2 &\le 2\left(|||\xi|||^2 + |||\eta|||^2\right),\\
\|e_h\|^2_{\nu,\partial K\cap\partial\Omega} &\le 2\left(\|\xi\|^2_{\nu,\partial K\cap\partial\Omega} + \|\eta\|^2_{\nu,\partial K\cap\partial\Omega}\right),\\
\|[e_h]\|^2_{\nu,\partial K^-\backslash\partial\Omega} &\le 2\left(\|[\xi]\|^2_{\nu,\partial K^-\backslash\partial\Omega} + \|[\eta]\|^2_{\nu,\partial K^-\backslash\partial\Omega}\right).
\end{aligned}$$

We can find that

$$\sqrt{L(e_h,t)} \le \sqrt{2}\sqrt{L(\xi,t)+L(\eta,t)} \le \sqrt{2}\left(\sqrt{L(\xi,t)} + \sqrt{L(\eta,t)}\right). \tag{4.211}$$

Similarly as in the proof of (4.192), under the notation (4.185) and (4.186), we find that

$$\sum_{K\in\mathscr{T}_h}\left(\|\eta\|^2_{\nu,\partial K\cap\partial\Omega} + \|[\eta]\|^2_{\nu,\partial K^-\backslash\partial\Omega}\right) \le 2R_2(\eta). \tag{4.212}$$

Now, from (4.210), (4.211) and (4.212) it follows that

$$\begin{aligned}
\sqrt{L(e_h,t)} \le\ & 2\Big(\int_0^t (\varepsilon R_1(\eta(t)) + R_2(\eta(t)))\,\mathrm{d}\vartheta\Big)^{1/2} \\
&+ \sqrt{2}\int_0^t \left(\|\eta(t)\|_{L^2(\Omega)}(C_c + \overline{C}_b) + \left\|\partial_t\eta(t)\right\|_{L^2(\Omega)}\right)\mathrm{d}\vartheta \\
&+ \sqrt{2}\left(\|\eta(t)\|^2_{L^2(\Omega)} + \int_0^t\left(\frac{\varepsilon}{2}|||\eta(\vartheta)|||^2 + 2\gamma_0\|\eta\|^2_{L^2(\Omega)} + R_2(\eta(\vartheta))\right)\mathrm{d}\vartheta\right)^{1/2},
\end{aligned} \tag{4.213}$$

which is the desired result (4.199). □

Now, we formulate the main result of this section, representing the error estimate in terms of the mesh-size h.

Theorem 4.28 *Let us assume that $\{\mathscr{T}_h\}_{h\in(0,\bar{h})}$ is a system of conforming shape-regular triangulations (cf. (2.19)) of the domain Ω and let assumption (4.159) be satisfied. Let us assume that the constant $C_W > 0$ satisfies the conditions from Corollary 2.41 for NIPG, SIPG and IIPG versions of the diffusion form. Let the exact solution u of problem (4.158) be regular in the sense of (4.161) and let u_h be the approximate solution obtained by the method of lines (4.176). Then the error $e_h = u_h - u$ satisfies the estimate*

$$\max_{t\in(0,T)}\|e_h(t)\|_{L^2(\Omega)} + \left(\frac{\varepsilon}{2}\int_0^T |||e_h(\vartheta)|||^2\,\mathrm{d}\vartheta + 2\gamma_0\|e_h\|^2_{L^2(Q_T)}\right)^{1/2} \tag{4.214}$$

$$+\left(\frac{1}{2}\sum_{K\in\mathscr{T}_h}\int_0^T\left(\|e_h(t)\|^2_{\nu(t),\partial K\cap\partial\Omega}+\|[e_h(t)]\|^2_{\nu(t),\partial K^-(t)\setminus\partial\Omega}\right)\mathrm{d}t\right)^{1/2}$$
$$\leq \widetilde{C}h^{\mu-1}(\sqrt{\varepsilon}+\sqrt{h}),$$

where $\widetilde{C} > 0$ is a constant independent of ε and h.

Proof Estimate (4.214) will be derived from the abstract error estimate (4.199) and estimates (4.178)–(4.182) of the term η.

By (4.196), (4.185), (4.178)–(4.180), the inequality $h_K \leq h$ and the relation

$$\sum_{K\in\mathscr{T}_h}|u|^2_{H^\mu(K)}=|u|^2_{H^\mu(\Omega)}, \tag{4.215}$$

we have

$$R_1(\eta)\leq 3\widetilde{C}_B^2C_A^2h^{2(\mu-1)}|u|^2_{H^\mu(\Omega)}, \tag{4.216}$$
$$R_2(\eta)\leq 2\overline{C}_b'C_A^2h^{2\mu-1}|u|^2_{H^\mu(\Omega)}. \tag{4.217}$$

From (4.104), we have

$$|||\eta|||^2\leq C_A^2\left(1+\frac{4C_WC_M}{C_T}\right)h^{2(\mu-1)}|u|^2_{H^\mu(\Omega)}. \tag{4.218}$$

Now, estimates (4.178), (4.182), (4.199), (4.216)–(4.218) and the inequality $\sqrt{a}+\sqrt{b}+\sqrt{c}\leq\sqrt{3}(a+b+c)^{1/2}$ valid for $a, b, c \geq 0$, imply that

$$\max_{t\in(0,T)}\|e_h(t)\|_{L^2(\Omega)}+\left(\frac{\varepsilon}{2}\int_0^T|||e_h(\vartheta)|||^2\,\mathrm{d}\vartheta+2\gamma_0\|e_h\|^2_{L^2(Q_T)}\right)^{1/2} \tag{4.219}$$
$$+\left(\frac{1}{2}\sum_{K\in\mathscr{T}_h}\int_0^T\left(\|e_h(t)\|^2_{\nu(t),\partial K\cap\partial\Omega}+\|[e_h(t)]\|^2_{\nu(t),\partial K^-(t)\setminus\partial\Omega}\right)\mathrm{d}t\right)^{1/2}$$
$$\leq\Bigg\{\sqrt{6}\left(\left(3\varepsilon\widetilde{C}_B^2C_A^2h^{2(\mu-1)}+2\overline{C}_b'C_A^2h^{2\mu-1}\right)\int_0^T|u(\vartheta)|^2_{H^\mu(\Omega)}\,\mathrm{d}\vartheta\right)^{1/2}$$
$$+2\sqrt{6}\left(C_A(C_c+\overline{C}_h)h^\mu\int_0^T|u(\vartheta)|_{H^\mu(\Omega)}\,\mathrm{d}\vartheta+C_Ah^\mu\int_0^T|\partial_tu(\vartheta)|_{H^\mu(\Omega)}\,\mathrm{d}\vartheta\right)$$

$$+\sqrt{6}\Big(C_A^2 h^{2\mu} \max_{t\in[0,T]} |u(t)|^2_{H^\mu(\Omega)}$$
$$+ C_A^2\Big(2\gamma_0 h^{2\mu} + \frac{\varepsilon}{2}\Big(1 + \frac{4C_W C_M}{C_T}\Big)h^{2(\mu-1)} + 2\overline{C}_b' h^{2\mu-1}\Big)\int_0^T |u(\vartheta)|^2_{H^\mu(\Omega)}\,\mathrm{d}\vartheta\Big)^{\frac{1}{2}}\Big\}.$$

The above inequality and the inequality $h < \bar{h}$ already imply estimate (4.214) with a constant $\widetilde{C}$ depending on the constants $\widetilde{C}_B, C_A, \overline{C}_b', C_c, \overline{C}_b, \bar{h}, \gamma_0, C_W, C_M, C_T$ and the seminorms

$$|u|_{L^2(0,T;H^\mu(\Omega))},\ |u|_{L^1(0,T;H^\mu(\Omega))},\ |u|_{C([0,T];H^\mu(\Omega))},\ |\partial_t u|_{L^1(0,T;H^\mu(\Omega))}. \qquad \square$$

Exercise 4.29 (i) Prove estimate (4.212) in detail.
(ii) Verify relations (4.211).
(iii) Express the constant $\widetilde{C}$ from the error estimate (4.214) in terms of the constants $\widetilde{C}_B$, $C_A, \ldots$, and the norms of u and $\partial_t u$.
(iv) Prove relations (4.190) and (4.203) in detail.

Remark 4.30 Let us omit the integrals over $\partial K^- \cap \partial\Omega$ and $\partial K^- \setminus \partial\Omega$ in the form b_h and the corresponding terms on the right-hand side in the definition of the approximate solution u_h (which means that we cancel upwinding). Proceeding in the same way as before, we obtain the estimate of the type

$$\frac{\mathrm{d}}{\mathrm{d}t}\|\xi\|^2_{L^2(\Omega)} + \varepsilon|||\xi|||^2 + 2\int_\Omega\Big(c - \frac{1}{2}\nabla\cdot\boldsymbol{v}\Big)\xi^2\,\mathrm{d}x + \sum_{K\in\mathscr{T}_h}\int_{\partial K}(\boldsymbol{v}\cdot\boldsymbol{n})\xi^2\,\mathrm{d}S$$
$$\le C\varepsilon h^{2(\mu-1)} + Ch^{2\mu} + \|\xi\|^2_{L^2(\Omega)}. \tag{4.220}$$

We can see that it is difficult to handle the terms $\int_\Gamma(\boldsymbol{v}\cdot\boldsymbol{n})\xi^2\,\mathrm{d}S$ on the left-hand side, as $\boldsymbol{v}\cdot\boldsymbol{n}$ may be both positive and negative. We can make some rearrangements, but then it is necessary to use the standard Gronwall Lemma 1.9 and we obtain a term like $\exp(cT/\varepsilon)$ on the right-hand side of the final estimate, which is not desirable, especially for small ε. The use of upwinding is therefore important for obtaining the error estimate uniform with respect to the diffusion coefficient ε. Similar result is valid even on an infinite time interval $[0, +\infty)$ as was shown in [139].

Exercise 4.31 Prove estimate (4.220) and the error estimate following from (4.220).

4.7 Numerical Examples

In Chap. 2 we presented numerical experiments which demonstrate the high order of convergence of the discontinuous Galerkin method (DGM). However, similar results can be obtained for the standard *conforming finite element method* (FEM) (e.g., [52]).

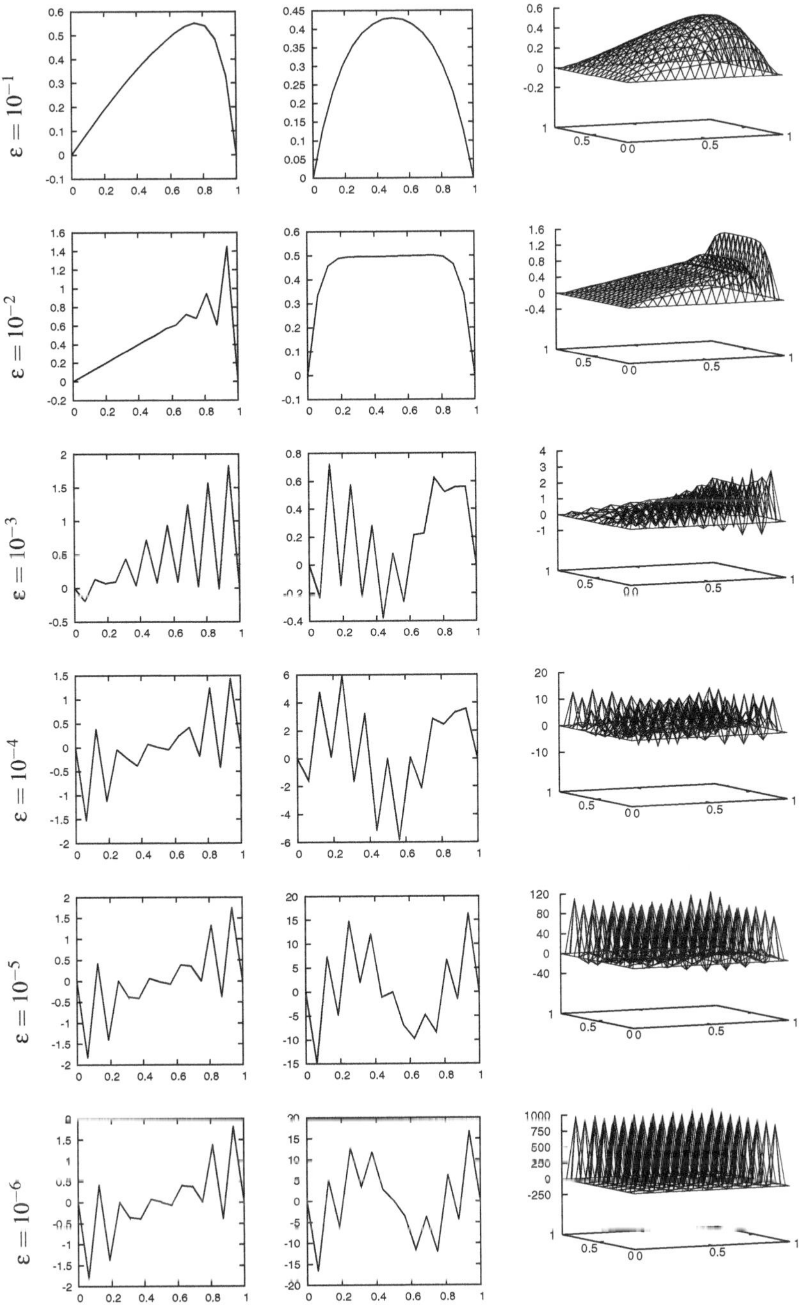

Fig. 4.1 Linear convection-diffusion equation, P_1 **conforming finite element method**, horizontal cut at $x_2 = 0.5$ (*left*), vertical cut at $x_1 = 0.5$ (*center*), 3D view (*right*), for $\varepsilon = 10^{-1}$, 10^{-2}, 10^{-3}, 10^{-4}, 10^{-5} and 10^{-6}

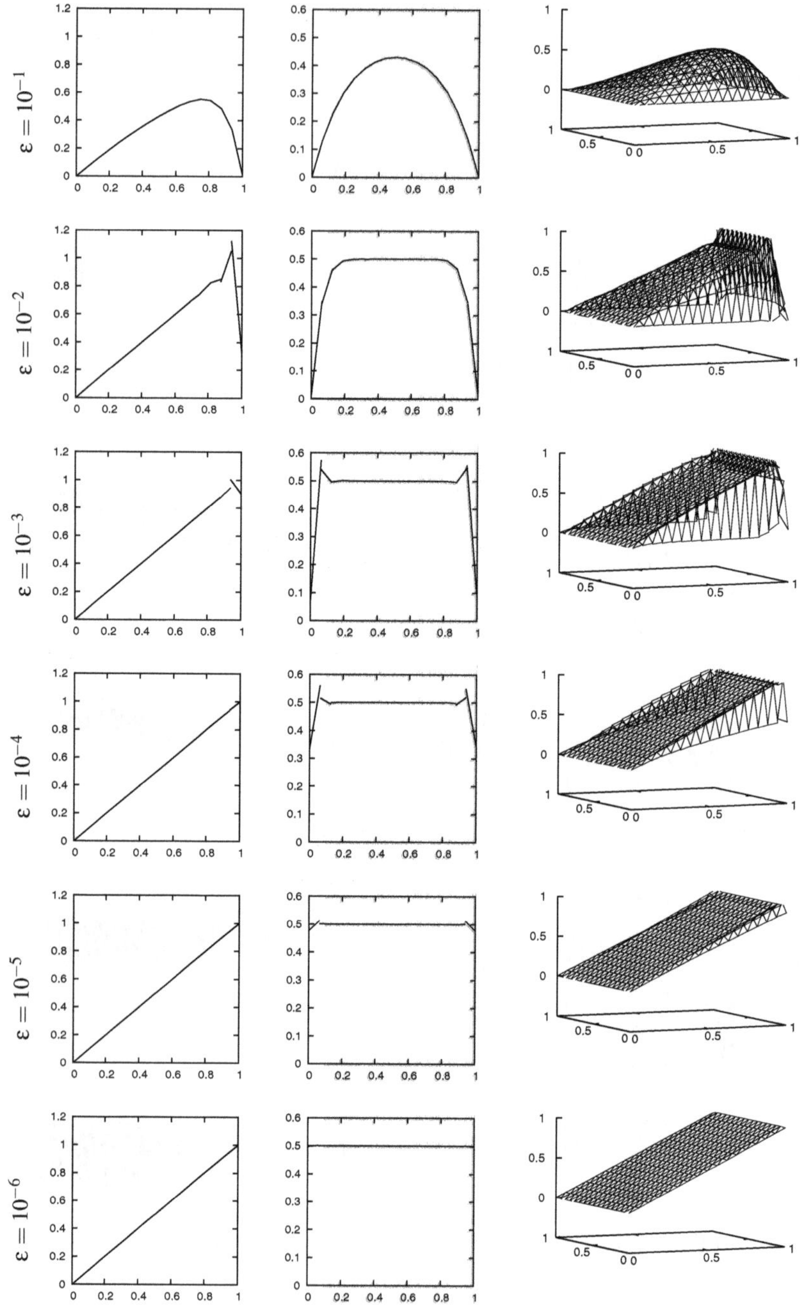

Fig. 4.2 Linear convection-diffusion equation, P_1 **discontinuous Galerkin method**, horizontal cut at $x_2 = 0.5$ (*left*), vertical cut at $x_1 = 0.5$ (*center*), 3D view (*right*), for $\varepsilon = 10^{-1}$, 10^{-2}, 10^{-3}, 10^{-4}, 10^{-5} and 10^{-6}

Moreover, in comparison with conforming FEM, DGM requires more degrees of freedom for obtaining the same level of computational error. On the other hand, the numerical solutions obtained by the conforming FEM and DGM are completely different in the case of convection-diffusion problems, particularly for dominating convection.

Let us consider a simple stationary linear convection-diffusion boundary value problem to find such a function u that

$$\begin{aligned} \frac{\partial u}{\partial x_1} - \varepsilon \Delta u &= 1 \quad \text{in} \quad \Omega = (0,1) \times (0,1), \\ u &= 0 \quad \text{on} \quad \partial\Omega, \end{aligned} \tag{4.221}$$

where $\varepsilon > 0$ is a diffusion coefficient. The exact solution possesses an exponential boundary layer along $x_1 = 1$ and two parabolic boundary layers along $x_2 = 0$ and $x_2 = 1$ (cf. [244]). In the interior grid points the solution $u(x_1, x_2) \approx x_1$.

We solved this problem with the aid of the conforming FEM and the IIPG variant of DGM on a uniform triangular grid with spacing $h = 1/16$ with the aid of piecewise linear approximation. Figures 4.1 and 4.2 show the approximate solutions for $\varepsilon = 10^{-1}, 10^{-2}, 10^{-3}, 10^{-4}, 10^{-5}$ and 10^{-6} obtained by FEM and DGM, respectively.

We can see that the conforming finite element solutions suffer from spurious oscillations whose amplitude increases with decreasing diffusion coefficient. On the other hand, for $\varepsilon = 10^{-1}, 10^{-2}$ and 10^{-3} the discontinuous Galerkin solution contains spurious overshoots and undershoots only in the vicinity of the boundary layers, but inside the domain there are no spurious oscillations. These overshoots and undershoots completely disappear for $\varepsilon \ll 1$. It is caused by the fact that the Dirichlet boundary condition is imposed in a weak sense with the aid of the boundary penalty. From this point of view, the DGM does not require such sophisticated stabilization techniques as the conforming FEM (see [191] for an overview).

Chapter 5
Space-Time Discretization by Multistep Methods

In practical computations, it is necessary to discretize nonstationary initial-boundary value problems both in space and time. In the previous chapter only the DG space semidiscretization, called the *method of lines*, was applied, leading to a large system of ordinary differential equations (ODEs). These systems can be solved by suitable schemes for solving ordinary differential equations. We can mention, for example, explicit Runge–Kutta methods, which are particularly popular in computational fluid dynamics, see e.g., [282]. The drawback of these methods is the *conditional stability* which requires strong limitation of the time step. Therefore, in some cases also other (implicit) methods are used. In this chapter we introduce and analyze full space-time discretizations based on the DGM in space, in combination with multistep methods applied to the resulting system of ODEs.

5.1 Semi-implicit Backward Euler Time Discretization

This section is concerned with the analysis of the full space-time semi-implicit backward Euler time discretization combined with the discontinuous Galerkin space discretization of a scalar nonstationary nonlinear convection-diffusion equation with linear diffusion and nonlinear convection. In this case the diffusion and additional stabilization and penalty terms are treated implicitly, whereas the nonlinear convective terms are treated explicitly. We derive a priori error estimates in the discrete $L^\infty(0, T; L^2(\Omega))$-norm, $L^2(0, T; H^1(\Omega, \mathscr{T}_h))$-norm and $L^\infty(0, T; H^1(\Omega, \mathscr{T}_h))$-norm with respect to the mesh-size h and time step τ. Although this approach is a special case of the higher-order backward difference formula—discontinuous Galerkin method (BDF-DGM) analyzed in Sect. 5.2, we present it separately here in order to explain the theoretical analysis in a clearer way.

Let us consider the *initial-boundary value problem* (4.1) to find $u : Q_T = \Omega \times (0, T) \to \mathbb{R}$ such that

V. Dolejší and M. Feistauer, *Discontinuous Galerkin Method*,
Springer Series in Computational Mathematics 48,
DOI 10.1007/978-3-319-19267-3_5

$$\frac{\partial u}{\partial t} + \sum_{s=1}^{d} \frac{\partial f_s(u)}{\partial x_s} = \varepsilon \Delta u + g \quad \text{in } Q_T, \tag{5.1a}$$

$$u\big|_{\partial\Omega_D \times (0,T)} = u_D, \tag{5.1b}$$

$$\varepsilon \boldsymbol{n} \cdot \nabla u|_{\partial\Omega_N \times (0,T)} = g_N, \tag{5.1c}$$

$$u(x,0) = u^0(x), \quad x \in \Omega. \tag{5.1d}$$

We assume that the data satisfy conditions (4.2), i.e.,

$$\begin{aligned}
&f = (f_1, \dots, f_d),\ f_s \in C^1(\mathbb{R}),\ f_s' \text{ are bounded},\ f_s(0) = 0,\ s = 1, \dots, d,\\
&u_D = \text{trace of some } u^* \in C([0,T]; H^1(\Omega)) \cap L^\infty(Q_T) \text{ on } \partial\Omega_D \times (0,T),\\
&\varepsilon > 0,\ \ g \in C([0,T]; L^2(\Omega)),\ \ g_N \in C([0,T]; L^2(\partial\Omega_N)),\ \ u^0 \in L^2(\Omega).
\end{aligned}$$

We assume that problem (5.1) has a weak solution satisfying the regularity conditions

$$u \in L^\infty(0,T; H^s(\Omega)), \quad \frac{\partial u}{\partial t} \in L^\infty(0,T; H^s(\Omega)), \quad \frac{\partial^2 u}{\partial t^2} \in L^\infty(0,T; L^2(\Omega)), \tag{5.2}$$

where $s \geq 2$ is an integer. Such a solution satisfies problem (5.1) pointwise. Under (5.2),

$$u \in C([0,T]; H^s(\Omega)), \quad \frac{\partial u}{\partial t} \in C([0,T]; L^2(\Omega)). \tag{5.3}$$

Let us recall that again $(\cdot,\cdot)$ denotes the scalar product in the space $L^2(\Omega)$.

5.1.1 Discretization of the Problem

We use the same notation and assumptions as in Sects. 2.4 and 4.2. This means that we suppose that the domain Ω is polygonal if $d = 2$, or polyhedral if $d = 3$, with Lipschitz boundary. By $\mathscr{T}_h$ we denote a partition of the domain Ω and by S_{hp} we denote the space of discontinuous piecewise polynomial functions defined in (2.34).

Further, we use the diffusion, convection, penalty and right-hand side forms a_h, b_h, J_h^σ and ℓ_h defined in Sect. 4.2 by (4.10), (4.13), (4.11) and (4.23), respectively. As in previous sections, by $(\cdot,\cdot)$ we denote the $L^2(\Omega)$-scalar product. Moreover, by (4.9), $A_h = \varepsilon a_h + \varepsilon J_h^\sigma$. Let us recall that the functions f_s, $s = 1, \dots, d$, are Lipschitz-continuous with constant $L_f = 2L_H$, where the constant L_H is introduced in (4.18). As was already shown, the exact solution u with property (5.2) satisfies the *consistency* identity (4.28):

$$\left(\frac{\partial u}{\partial t}(t), v_h\right) + A_h(u(t), v_h) + b_h(u(t), v_h) = \ell_h(v_h)\,(t) \tag{5.4}$$

for all $v_h \in S_{hp}$ and all $t \in (0, T)$.

In order to introduce the fully discretized problem, we consider a partition $0 = t_0 < t_1 < \ldots, t_r = T$ of the time interval $[0, T]$ and set $\tau_k = t_{k+1} - t_k$ for $k = 0, 1, \ldots, r-1$. The exact solution $u(t_k)$ will be approximated by an element $u_h^k \in S_{hp}$, and the time derivative in (5.4) will be approximated by the backward difference. In order to obtain a stable, efficient and simple scheme, the forms A_h and ℓ_h will be treated implicitly, whereas the nonlinear terms represented by the form b_h will be treated explicitly. In this way we arrive at the following *semi-implicit backward Euler-discontinuous Galerkin method* (BE-DGM).

Definition 5.1 We define the *approximate solution* of problem (5.1) by the semi-implicit BE-DGM as functions $u_h^k \in S_{hp}$, $t_k \in [0, T]$, satisfying the conditions

$$u_h^0 = \Pi_{hp} u^0, \tag{5.5a}$$

$$\left(\frac{u_h^{k+1} - u_h^k}{\tau_k}, v_h\right) + A_h(u_h^{k+1}, v_h) + b_h(u_h^k, v_h) = \ell_h(v_h)\,(t_{k+1}) \tag{5.5b}$$

$$\forall v_h \in S_{hp}\ \forall t_{k+1} \in (0, T].$$

(The S_{hp}-interpolation operator Π_{hp} is defined in (2.90)). The function u_h^k is called the approximate solution at time t_k.

For each $t_{k+1} \in (0, T]$, problem (5.5b) is equivalent to a system of linear algebraic equations with a matrix, which is, in general, nonsymmetric but positive definite. This implies the following result.

Lemma 5.2 *Under the assumptions on C_W from Corollary 2.41, the discrete problem (5.5) has a unique solution.*

Exercise 5.3 Prove Lemma 5.2. Hint: Use Corollary 1.7.

5.1.2 Error Estimates

In what follows, we analyze the error estimates of the approximate solution u_h^k, $k = 0, 1, \ldots$, obtained by method (5.5) under the assumption that the exact solution u satisfies (5.2). For simplicity, we assume that the Dirichlet condition is prescribed on the whole boundary $\partial\Omega$. Hence, $\partial\Omega_D = \partial\Omega$ and $\partial\Omega_N = \emptyset$. This means that we want to find $u : Q_T \to \mathbb{R}$ such that

$$\frac{\partial u}{\partial t} + \sum_{s=1}^{d} \frac{\partial f_s(u)}{\partial x_s} = \varepsilon\, \Delta u + g \quad \text{in } Q_T, \tag{5.6a}$$

$$u\,\big|_{\partial\Omega\times(0,T)} = u_D, \tag{5.6b}$$

$$u(x,0) = u^0(x), \quad x \in \Omega. \tag{5.6c}$$

(In this case, the form ℓ_h is defined by (4.114)). Moreover, we consider a uniform partition $t_k = k\tau$, $k = 0, 1, \ldots, r$, of the time interval $[0, T]$ with time step $\tau = T/r$, where $r > 1$ is an integer.

In the error analysis we use similar techniques as in Sect. 4.3. This means that we consider Assumptions 4.5 from Sect. 4.3 and use the results formulated in Lemma 4.6, the approximation properties (2.93)–(2.95), (2.98)–(2.100), the multiplicative trace inequality (2.78), and the inverse inequality (2.86). Moreover, we suppose that the regularity conditions (5.2) are satisfied.

Let $\Pi_{hp} u^k$ be the S_{hp}-interpolation of $u^k = u(t_k)$ $(k = 0, \ldots, r)$, where Π_{hp} is the operator of the $L^2(\Omega)$-projection satisfying estimates (2.98)–(2.100). We set

$$\xi^k = u_h^k - \Pi_{hp} u^k \in S_{hp}, \quad \eta^k = \Pi_{hp} u^k - u^k \in H^s(\Omega, \mathscr{T}_h). \tag{5.7}$$

Then the error $e_h^k = u_h^k - u^k$ can be expressed as

$$e_h^k = \xi^k + \eta^k, \quad k = 0, \ldots, r. \tag{5.8}$$

Setting $v_h := \xi^{k+1}$ in (5.5b), we get

$$\left(u_h^{k+1} - u_h^k, \xi^{k+1}\right) + \tau \left(A_h(u_h^{k+1}, \xi^{k+1}) + b_h(u_h^k, \xi^{k+1}) - \ell_h(\xi^{k+1})\,(t_{k+1})\right) = 0, \quad t_k, t_{k+1} \in [0, T]. \tag{5.9}$$

Moreover, setting $t := t_{k+1}$ and $v_h := \xi^{k+1}$ in (5.4), we obtain

$$\left(u'(t_{k+1}), \xi^{k+1}\right) + A_h(u^{k+1}, \xi^{k+1}) + b_h(u^{k+1}, \xi^{k+1}) - \ell_h(\xi^{k+1})\,(t_{k+1}) = 0, \quad t_{k+1} \in [0, T], \tag{5.10}$$

where $u' = \partial u/\partial t$. Multiplying (5.10) by τ and subtracting from (5.9), we get

$$\left(u_h^{k+1} - u_h^k, \xi^{k+1}\right) - \tau \left(u'(t_{k+1}), \xi^{k+1}\right) \tag{5.11}$$
$$+ \tau \left(A_h(u_h^{k+1} - u^{k+1}, \xi^{k+1}) + b_h(u_h^k, \xi^{k+1}) - b_h(u^{k+1}, \xi^{k+1})\right) = 0,$$

By (5.7) and (5.8), from (5.11) we have

$$\begin{aligned}
&\left(\xi^{k+1} - \xi^{k}, \xi^{k+1}\right) + \tau A_h(\xi^{k+1}, \xi^{k+1}) \\
&\quad = \tau(u'(t_{k+1}), \xi^{k+1}) - (u^{k+1} - u^{k}, \xi^{k+1}) - (\eta^{k+1} - \eta^{k}, \xi^{k+1}) \\
&\quad + \tau\left(b_h(u^{k+1}, \xi^{k+1}) - b_h(u_h^{k}, \xi^{k+1}) - A_h(\eta^{k+1}, \xi^{k+1})\right).
\end{aligned} \tag{5.12}$$

In what follows, we estimate the individual terms on the right-hand side of (5.12). First, for $v \in L^{\infty}(0, T; H^{s}(\Omega))$, $s \geq 2$, we put

$$|v|_R = \max\left(\|v\|_{L^{\infty}(0,T;L^2(\Omega))},\ |v|_{L^{\infty}(0,T;H^1(\Omega))},\ |v|_{L^{\infty}(0,T;H^{\mu}(\Omega))}\right), \tag{5.13}$$

where $\mu = \min(p+1, s)$. For the definitions of seminorms, see (1.34). Obviously, if u satisfies assumptions (5.2), then we can put

$$\|u\|_d := \max\left(|u|_R,\ \left|u'\right|_R,\ \left|u''\right|_R\right) < \infty, \tag{5.14}$$

where $u' = \partial u/\partial t$, $u'' = \partial^2 u/\partial t^2$.

Now we estimate the terms arising from the approximation of the time derivative.

Lemma 5.4 *Under assumptions (5.2), for t_k, $t_{k+1} \in [0, T]$ we have*

$$\left|(u^{k+1} - u^{k}, \xi^{k+1}) - \tau(u'(t_{k+1}), \xi^{k+1})\right| \leq \tau^2 \left|u''\right|_R \|\xi^{k+1}\|_{L^2(\Omega)}, \tag{5.15}$$

$$\|u^{k+1} - u^{k}\|_{L^2(\Omega)} \leq \tau \left|u'\right|_R, \tag{5.16}$$

$$|u^{k+1} - u^{k}|_{H^1(\Omega)} \leq \tau \left|u'\right|_R, \tag{5.17}$$

$$|u^{k+1} - u^{k}|_{H^{\mu}(\Omega)} \leq \tau \left|u'\right|_R. \tag{5.18}$$

Proof (i) The proof of (5.15) is based on the following result (see [122], Par. 8,2, or [144]): If $w : (0, T) \to L^2(\Omega)$ is such that $w,\ w' \in L^1(0, T; L^2(\Omega))$, $v \in L^2(\Omega)$, then $(w', v) \in L^1(0, T)$ and for $t_1, t_2 \in [0, T]$,

$$\int_{t_1}^{t_2} (w'(t), v)\, \mathrm{d}t = (w(t_2) - w(t_1), v). \tag{5.19}$$

The above and (5.2) imply that

$$(u(t_{k+1}) - u(t_k), \xi) = \int_{t_k}^{t_{k+1}} (u'(t), \xi)\, \mathrm{d}t, \tag{5.20}$$

and hence,

$$(u^{k+1} - u^{k}, \xi) - \tau(u'(t_{k+1}), \xi) = \int_{t_k}^{t_{k+1}} (u'(t) - u'(t_{k+1}), \xi)\, \mathrm{d}t. \tag{5.21}$$

Since $u'' \in L^\infty(0,T;L^2(\Omega))$,

$$(u'(t) - u'(t_{k+1}), \xi) = \int_{t_{k+1}}^{t} (u''(\vartheta), \xi)\,\mathrm{d}\vartheta, \tag{5.22}$$

and

$$\int_{t_k}^{t_{k+1}} (u'(t) - u'(t_{k+1}), \xi)\,\mathrm{d}t = \int_{t_k}^{t_{k+1}} \left(\int_{t_{k+1}}^{t} (u''(\vartheta), \xi)\mathrm{d}\vartheta \right) \mathrm{d}t. \tag{5.23}$$

Then, (5.21)–(5.23), the Cauchy inequality, and assumption (5.2) imply that

$$\left|(u^{k+1} - u^k, \xi) - \tau(u'(t_{k+1}), \xi)\right| \le \tau^2 \|u''\|_{L^\infty(0,T;L^2(\Omega))} \|\xi\|_{L^2(\Omega)}, \tag{5.24}$$

which is (5.15).

(ii) Since $u' \in L^\infty(0,T;H^\mu(\Omega)) \subset L^\infty(0,T;L^2(\Omega))$, we can write

$$\|u^{k+1} - u^k\|_{L^2(\Omega)} = \left\| \int_{t_k}^{t_{k+1}} u'(t)\,\mathrm{d}t \right\|_{L^2(\Omega)} \le \tau \|u'\|_{L^\infty(0,T;L^2(\Omega))}, \tag{5.25}$$

which proves (5.16).

(iii) Since $u' \in L^\infty(0,T;H^\mu(\Omega)) \subset L^\infty(0,T;H^1(\Omega))$ and

$$\frac{\partial}{\partial t}\left(\frac{\partial u}{\partial x_l}\right) = \frac{\partial}{\partial x_l}\left(\frac{\partial u}{\partial t}\right), \qquad l = 1, \ldots, d, \tag{5.26}$$

in the sense of distributions, we obtain

$$\begin{aligned} |u^{k+1} - u^k|_{H^1(\Omega)} &= \|\nabla u^{k+1} - \nabla u^k\|_{L^2(\Omega)} = \left\| \int_{t_k}^{t_{k+1}} \frac{\partial}{\partial t} \nabla u(t)\,\mathrm{d}t \right\|_{L^2(\Omega)} \\ &\le \int_{t_k}^{t_{k+1}} \left\| \frac{\partial}{\partial t} \nabla u(t) \right\|_{L^2(\Omega)} \mathrm{d}t = \int_{t_k}^{t_{k+1}} \|\nabla u'(t)\|_{L^2(\Omega)}\,\mathrm{d}t \\ &= \int_{t_k}^{t_{k+1}} |u'(t)|_{H^1(\Omega)}\,\mathrm{d}t \le \tau \|u'\|_{L^\infty(0,T;H^1(\Omega))}, \end{aligned} \tag{5.27}$$

which is (5.17).

(iv) Using a similar argumentation as in (5.27), we derive (5.18). □

Lemma 5.5 *Under assumptions (5.2), for t_k, $t_{k+1} \in [0,T]$ we have*

$$|(\eta^{k+1} - \eta^k, \xi^{k+1})| \le C_A \tau h^\mu \|\xi^{k+1}\|_{L^2(\Omega)} |u'|_R, \tag{5.28}$$

where C_A is the constant from the approximation properties (2.93)–(2.95).

Proof The Cauchy inequality, relations (5.7), (2.93)–(2.95) and (5.18) imply that

$$
\begin{aligned}
|(\eta^{k+1} - \eta^k, \xi^{k+1})| &\le \|\eta^{k+1} - \eta^k\|_{L^2(\Omega)} \|\xi^{k+1}\|_{L^2(\Omega)} \\
&= \|\Pi_{hp}(u^{k+1} - u^k) - (u^{k+1} - u^k)\|_{L^2(\Omega)} \|\xi^{k+1}\|_{L^2(\Omega)} \\
&\le C_A h^\mu |u^{k+1} - u^k|_{H^\mu(\Omega)} \|\xi^{k+1}\|_{L^2(\Omega)} \\
&\le C_A \tau h^\mu |u'|_R \|\xi^{k+1}\|_{L^2(\Omega)},
\end{aligned}
\tag{5.29}
$$

which proves the lemma. □

Exercise 5.6 Prove that (5.26) holds in the sense of distributions.

Further, using the results from the previous chapter, we simply estimate the term $A_h(\eta^k, \xi^k)$ in (5.12). By (4.80) and (4.96),

$$
\begin{aligned}
|A_h(\eta^k, \xi^k)| &\le \varepsilon \tilde{C}_B R_a(\eta^k) |||\xi^k||| \le \sqrt{3}\varepsilon C_A \tilde{C}_B h^{\mu-1} |u^k|_{H^\mu(\Omega)} |||\xi^k||| \\
&\le \varepsilon C_1 h^{\mu-1} |u|_R |||\xi^k|||, \quad h \in (0, \bar{h}), \ t_k \in [0, T],
\end{aligned}
\tag{5.30}
$$

where $C_1 = \sqrt{3} C_A \tilde{C}_B$.

Finally, we derive the estimates of the convective form b_h.

Lemma 5.7 *For* $h \in (0, \bar{h})$, t_k, $t_{k+1} \in [0, T]$ *we have*

$$
\left| b_h(u^{k+1}, \xi^{k+1}) - b_h(u_h^k, \xi^{k+1}) \right| \le C_2 |||\xi^{k+1}||| \left(\|\xi^k\|_{L^2(\Omega)} + (h^\mu + \tau)|u|_R \right),
\tag{5.31}
$$

where $C_2 > 0$ *is independent of* h, τ, k, ξ.

Proof We can write

$$
\begin{aligned}
b_h(u^{k+1}, \xi^{k+1}) - b_h(u_h^k, \xi^{k+1}) &= b_h(u^{k+1}, \xi^{k+1}) - b_h(u^k, \xi^{k+1}) && (=: \Psi_1) \\
&\quad + b_h(u^k, \xi^{k+1}) - b_h(\Pi_{hp} u^k, \xi^{k+1}) && (=: \Psi_2) \\
&\quad + b_h(\Pi_{hp} u^k, \xi^{k+1}) - b_h(u_h^k, \xi^{k+1}) && (=: \Psi_3).
\end{aligned}
\tag{5.32}
$$

We estimate the individual terms in (5.32). By virtue of (4.30),

$$
|\Psi_1| \le C_{b1} |||\xi^{k+1}||| \Big(\|u^{k+1} - u^k\|^2_{L^2(\Omega)} + \sum_{K \in \mathscr{T}_h} h_K \|u^{k+1} - u^k\|^2_{L^2(\partial K)} \Big)^{1/2}.
\tag{5.33}
$$

Using the multiplicative trace inequality (2.78), (5.16) and (5.17), we find that

$$\sum_{K\in\mathscr{T}_h} h_K \|u^{k+1} - u^k\|^2_{L^2(\partial K)} \tag{5.34}$$

$$\le C_M \sum_{K\in\mathscr{T}_h} \left(h_K \|u^{k+1} - u^k\|_{L^2(K)} |u^{k+1} - u^k|_{H^1(K)} + \|u^{k+1} - u^k\|^2_{L^2(K)}\right)$$

$$\le C_M(\bar{h}+1)\tau^2 |u'|^2_R .$$

Then (5.33), (5.34) and (5.16) give

$$|\Psi_1| \le C_{b1} |||\xi^{k+1}||| \left(\tau^2 |u'|^2_R + C_M(\bar{h}+1)\tau^2 |u'|^2_R\right)^{1/2} \le C_3 \tau |||\xi^{k+1}||| \, |u'|_R, \tag{5.35}$$

where $C_3 = C_{b1}(1 + C_M(\bar{h}+1))^{1/2}$.

Moreover, from Lemma 4.6 and (4.97), (5.7) and (5.13), we deduce that

$$|\Psi_2| \le C_{b3} |||\xi^{k+1}||| \; R_b(\Pi_{hp} u^k - u^k) \le \sqrt{2} C_{b3} C_A h^\mu |||\xi^{k+1}||| \; |u|_R, \tag{5.36}$$

$$|\Psi_3| \le C_{b2} |||\xi^{k+1}||| \; \|\Pi_{hp} u^k - u^k_h\|_{L^2(\Omega)} = C_{b2} |||\xi^{k+1}||| \; \|\xi^k\|_{L^2(\Omega)}.$$

By (5.32), (5.35) and (5.36), we obtain (5.31) with $C_2 := \max(C_3, C_{b2}, \sqrt{2} C_{b3} C_A)$.

□

Now we formulate the *main result* regarding the error estimate of the BE-DGM.

Theorem 5.8 *Let Assumptions 4.5 from Sect. 4.3 be satisfied. Let u be the exact solution of problem (5.1) satisfying (5.2). Let $t_k = k\tau$, $k = 0, 1, \ldots, r$, $\tau = T/r$, be a partition of $[0, T]$ and let u^k_h, $k = 0, \ldots, r$, be the approximate solution defined by (5.5). Let us set*

$$e = \{e^k_h\}_{k=0}^r = \{u^k_h - u^k\}_{k=0}^r, \quad \|e\|^2_{h,\tau,L^\infty(L^2)} = \max_{k=0,\ldots,r} \|e^k_h\|^2_{L^2(\Omega)}, \tag{5.37}$$

$$\|e\|^2_{h,\tau,L^2(H^1)} = \varepsilon\tau \sum_{k=0}^{r} |||e^k_h|||^2,$$

$$\|e\|^2_{h,\tau,L^\infty(H^1)} = \varepsilon \max_{k=0,\ldots,r} |||e^k_h|||^2.$$

Then there exist constants $\tilde{C}$, $\hat{C} > 0$ such that

$$\|e\|^2_{h,\tau,L^\infty(L^2)} \le \tilde{C}\left(h^{2(\mu-1)}\left(\varepsilon + h^2 + h^2/\varepsilon\right) + \tau^2 (1 + 1/\varepsilon)\right), \tag{5.38}$$

$$\|e\|^2_{h,\tau,L^2(H^1)} \le \hat{C}\left(h^{2(\mu-1)}\left(\varepsilon + h^2 + h^2/\varepsilon\right) + \tau^2 (1 + 1/\varepsilon)\right), \tag{5.39}$$

$$\forall h \in (0, \bar{h}) \; \forall \tau \in (0, T), \; \tau \le \frac{1}{2}.$$

Moreover, provided

$$h \le C_{IS}\,\tau \tag{5.40}$$

with a constant C_{IS} independent of h and τ, there exists a constant $\bar{C} > 0$ such that

$$\|e\|^2_{h,\tau,L^\infty(H^1)} \le \bar{C}\left(h^{2(\mu-1)}\left(1+\varepsilon+h+h/\varepsilon+h^2+h^2/\varepsilon+h^2/\varepsilon^2\right)\right. \tag{5.41}$$
$$\left.+\tau\left(1+1/\varepsilon+1/\varepsilon^2\right)\right), \quad \forall h \in (0,\bar{h})\ \forall \tau \in (0,T),\ \tau \le \frac{1}{2}.$$

Proof Let $\xi^k = u_h^k - \Pi_{hp}u^k \in S_{hp}$, $\eta^k = \Pi_{hp}u^k - u^k$, $k = 0,\ldots,r$ (cf. (5.7)). Then (5.8) holds: $e_h^k = u_h^k - u^k = \xi^k + \eta^k$. From (5.12) and the relations

$$A_h(\xi^{k+1},\xi^{k+1}) \ge \varepsilon C_C |||\xi^{k+1}|||^2 \tag{5.42}$$

(cf. (4.84)) and

$$2(\xi^{k+1}-\xi^k,\xi^{k+1}) = \left(\|\xi^{k+1}\|^2_{L^2(\Omega)} - \|\xi^k\|^2_{L^2(\Omega)} + \|\xi^{k+1}-\xi^k\|^2_{L^2(\Omega)}\right), \tag{5.43}$$

for $k = 0,\ldots,r-1$, we get

$$\|\xi^{k+1}\|^2_{L^2(\Omega)} - \|\xi^k\|^2_{L^2(\Omega)} + \|\xi^{k+1}-\xi^k\|^2_{L^2(\Omega)} + 2\tau\varepsilon C_C|||\xi^{k+1}|||^2 \tag{5.44}$$
$$\le 2\left(\tau(u'(t_{k+1}),\xi^{k+1}) - (u^{k+1}-u^k,\xi^{k+1}) - \left(\eta^{k+1}-\eta^k,\xi^{k+1}\right)\right)$$
$$+2\tau\left(b_h(u^{k+1},\xi^{k+1}) - b_h(u_h^k,\xi^{k+1}) - A_h(\eta^{k+1},\xi^{k+1})\right) =: \mathrm{RHS}.$$

With the aid of Lemmas 5.4 and 5.5, estimates (5.30) and (5.31) and using notation (5.14), we estimate the right-hand side of (5.44):

$$|\mathrm{RHS}| \le 2\left(\tau^2|u''|_R\|\xi^{k+1}\|_{L^2(\Omega)} + C_A\tau h^\mu|u'|_R\|\xi^{k+1}\|_{L^2(\Omega)}\right)$$
$$+2\tau\left(C_2|||\xi^{k+1}|||\left(\|\xi^k\|_{L^2(\Omega)} + (h^\mu+\tau)|u|_R\right) + C_1\varepsilon h^{\mu-1}|u|_R|||\xi^{k+1}|||\right)$$
$$\le 2(1+C_A)\tau(\tau+h^\mu)\|\xi^{k+1}\|_{L^2(\Omega)}\|u\|_d$$
$$+2\tau(C_2+C_1)|||\xi^{k+1}|||\left(\|\xi^k\|_{L^2(\Omega)} + (h^\mu+\varepsilon h^{\mu-1}+\tau)\|u\|_d\right).$$

Application of the Young inequality gives

$$|\mathrm{RHS}| \le \tau\|\xi^{k+1}\|^2_{L^2(\Omega)} + \tau(1+C_A)^2(\tau+h^\mu)^2\|u\|_d^2 \tag{5.45}$$
$$+\tau\frac{\varepsilon}{2}|||\xi^{k+1}|||^2 + \frac{4\tau}{\varepsilon}(C_2+C_1)^2\left(\|\xi^k\|^2_{L^2(\Omega)} + (h^\mu+\varepsilon h^{\mu-1}+\tau)^2\|u\|_d^2\right).$$

We put $C_4 = 4(C_2 + C_1)^2$ and

$$q(\varepsilon, h, \tau) = 2h^{2(\mu-1)}\left((1 + C_A)^2 h^2 + 2C_4(\varepsilon + h^2\varepsilon^{-1})\right)\|u\|_d^2 + 2\tau^2\left((1 + C_A)^2 + \frac{C_4}{\varepsilon}\right)\|u\|_d^2. \tag{5.46}$$

Using (5.44), (5.45) and the relation $C_C \geq 1/2$, we obtain

$$\begin{aligned} &\|\xi^{k+1}\|^2_{L^2(\Omega)} - \|\xi^k\|^2_{L^2(\Omega)} + \frac{\tau\varepsilon}{2}|||\xi^{k+1}|||^2 \\ &\quad\leq \tau\|\xi^{k+1}\|^2_{L^2(\Omega)} + \frac{\tau C_4}{\varepsilon}\|\xi^k\|^2_{L^2(\Omega)} + \tau q(\varepsilon, h, \tau). \end{aligned} \tag{5.47}$$

Hence,

$$(1-\tau)\|\xi^{k+1}\|^2_{L^2(\Omega)} + \frac{\tau\varepsilon}{2}|||\xi^{k+1}|||^2 \leq (1 + \tau C_4/\varepsilon)\|\xi^k\|^2_{L^2(\Omega)} + \tau q(\varepsilon, h, \tau). \tag{5.48}$$

Moreover, the following inequalities are valid:

$$\begin{aligned} \|\xi^k + \eta^k\|^2_{L^2(\Omega)} &\leq 2\left(\|\xi^k\|^2_{L^2(\Omega)} + \|\eta^k\|^2_{L^2(\Omega)}\right), \\ |||\xi^k + \eta^k|||^2 &\leq 2\left(|||\xi^k|||^2 + |||\eta^k|||^2\right). \end{aligned} \tag{5.49}$$

Now we prove the error estimates (5.38)–(5.41).
(i) By (5.48) (using the assumption that $0 < \tau \leq 1/2$),

$$\|\xi^{k+1}\|^2_{L^2(\Omega)} \leq \frac{1 + \tau C_4/\varepsilon}{1-\tau}\|\xi^k\|^2_{L^2(\Omega)} + \frac{\tau}{1-\tau}q(\varepsilon, h, \tau), \quad k = 0, \ldots, r-1. \tag{5.50}$$

If we set

$$B = \frac{1 + \tau C_4/\varepsilon}{1-\tau}, \tag{5.51}$$

then by induction we derive from (5.50) that

$$\|\xi^k\|^2_{L^2(\Omega)} \leq B^k\|\xi^0\|^2_{L^2(\Omega)} + \frac{B^k - 1}{B-1}\frac{\tau\, q(\varepsilon, h, \tau)}{1-\tau}, \quad k = 0, \ldots, r. \tag{5.52}$$

By (5.51),

$$\frac{\tau}{(B-1)(1-\tau)} = \frac{1}{1 + C_4/\varepsilon} \leq 1. \tag{5.53}$$

As $\tau \leq 1/2$, then $1-\tau \geq 1/2$ and

$$B \leq 1 + 2\tau(1 + C_4/\varepsilon) \leq \exp(2\tau(1 + C_4/\varepsilon)). \tag{5.54}$$

From (5.52)–(5.54) we have

$$\|\xi^k\|^2_{L^2(\Omega)} \leq \exp(2\tau k(1 + C_4/\varepsilon)) \left(\|\xi^0\|^2_{L^2(\Omega)} + q(\varepsilon, h, \tau)\right). \tag{5.55}$$

Further, $k\tau \leq T$ for $k = 0, \dots, r$, and (5.5a) implies that $\xi^0 = 0$. The above and (5.55) yield

$$\|\xi^k\|^2_{L^2(\Omega)} \leq \exp(2T(1 + C_4/\varepsilon))q(\varepsilon, h, \tau), \; k = 0, \dots, r. \tag{5.56}$$

By (5.7), (2.98) and (5.14),

$$\|\eta^k\|^2_{L^2(\Omega)} \leq C_A^2 h^{2\mu} |u^k|^2_{H^\mu(\Omega)} \leq C_A^2 h^{2\mu} \|u\|^2_d. \tag{5.57}$$

Using (5.37), (5.49), (5.56) and (5.57), we find that

$$\begin{aligned} \|e\|^2_{h,\tau,L^\infty(L^2)} &\leq 2 \max_{k=0,\dots,r} \left(\|\xi^k\|^2_{L^2(\Omega)} + \|\eta^k\|^2_{L^2(\Omega)}\right) \\ &\leq 2\left(\exp(2T(1 + C_4/\varepsilon))q(\varepsilon, h, \tau) + C_A^2 h^{2\mu}\|u\|^2_d\right), \end{aligned} \tag{5.58}$$

which implies estimate (5.38).

(ii) Now let as prove (5.39). Summing (5.47) over $k = 0, \dots, r-1$ and taking into account that $\xi^0 = 0$, we get

$$\begin{aligned} &\|\xi^r\|^2_{L^2(\Omega)} + \frac{\tau\varepsilon}{2} \sum_{k=0}^{r-1} |||\xi^{k+1}|||^2 \\ &\quad \leq \tau(1 + C_4/\varepsilon) \sum_{k=0}^{r-1} \left(\|\xi^{k+1}\|^2_{L^2(\Omega)} + \|\xi^k\|^2_{L^2(\Omega)}\right) + Tq(\varepsilon, h, \tau). \end{aligned} \tag{5.59}$$

The above and (5.56) imply that

$$\tau\varepsilon \sum_{k=1}^{r} |||\xi^k|||^2 \leq (4(1 + C_4/\varepsilon)\exp(2T(1 + C_4/\varepsilon)) + 1)\, Tq(\varepsilon, h, \tau). \tag{5.60}$$

In view of (2.125), (4.96) and (5.14), we have

$$\varepsilon|||\eta^k|||^2 \leq \varepsilon C_\sigma^2 R_a(\eta^k)^2 \leq \varepsilon 3 C_A^2 C_\sigma^2 h^{2(\mu-1)}\|u\|^2_d = \varepsilon C_5 h^{2(\mu-1)}\|u\|^2_d, \tag{5.61}$$

where $C_5 = 3C_A^2 C_\sigma^2$. Hence, since $(k+1)\tau = T + \tau$,

$$\tau\varepsilon \sum_{k=0}^{r} |||\eta^k|||^2 \le \varepsilon C_5 h^{2(\mu-1)}(T+\tau)\|u\|_d^2. \tag{5.62}$$

Using (5.37), (5.7), (5.49), (5.60) and (5.62), we arrive at the estimate

$$\begin{aligned}\|e\|^2_{h,\tau,L^2(H^1)} &\le 2\tau\varepsilon \sum_{k=0}^{r} |||\xi^k|||^2 + 2\tau\varepsilon \sum_{k=0}^{r} |||\eta^k|||^2 \\ &\le 2\left(4(1+C_4/\varepsilon)\exp(2T(1+C_4/\varepsilon)) + 1\right) Tq(\varepsilon,h,\tau) \\ &\quad + 2\varepsilon C_5(T+\tau)h^{2(\mu-1)}\|u\|_d^2.\end{aligned} \tag{5.63}$$

Now, assertion (5.39) of the theorem follows from (5.63) and (5.46).

(iii) In what follows, let assumption (5.40) be valid. Then $C_{IS}^{-1}h \le \tau \le 1/2$. Therefore, (5.48) implies that

$$\varepsilon|||\xi^{k+1}|||^2 \le 2\left(\frac{1}{\tau} + \frac{C_4}{\varepsilon}\right)\|\xi^k\|^2_{L^2(\Omega)} + 2q(\varepsilon,h,\tau). \tag{5.64}$$

Using (5.56), we obtain

$$\varepsilon|||\xi^{k+1}|||^2 \le \left(\frac{1}{\tau} + \frac{C_4}{\varepsilon}\right)\exp(2T(1+C_4/\varepsilon))q(\varepsilon,h,\tau) + 2q(\varepsilon,h,\tau). \tag{5.65}$$

Now, according to (5.37), (5.7), (5.49), (5.65) and (5.61), we arrive at the inequalities

$$\|e\|^2_{h,\tau,L^\infty(H^1)} \le \max_{k=0,\dots,r} 2\varepsilon\left(\sum_{k=0}^{r} |||\xi^k|||^2 + \sum_{k=0}^{r} |||\eta^k|||^2\right)$$
$$\le 2(\varepsilon + \tau C_4)\exp(2T(1+C_4/\varepsilon))\frac{q(\varepsilon,h,\tau)}{\varepsilon\tau} + 2q(\varepsilon,h,\tau) + 2C_5\varepsilon h^{2(\mu-1)}\|u\|_d^2.$$

Finally, this, (5.46) and assumption (5.40) yield (5.41). □

Remark 5.9 For simplicity we analyzed here the problem with the Dirichlet condition prescribed on the whole boundary $\partial\Omega$. Using the results from Sect. 4.3.2, it is possible to obtain error estimates in the case of the problem (5.1) with mixed Dirichlet–Neumann boundary conditions.

Remark 5.10 Estimate (5.38) implies that

$$\|u - u_h\|_{L^\infty(0,T;L^2(\Omega))} = O(h^{\mu-1} + \tau) \quad \text{for } h, \tau \to 0+. \tag{5.66}$$

Comparing this result with the approximation properties (2.93)–(2.95) implying that

$$\|u - \Pi_{hp} u\|_{L^\infty(0,T;L^2(\Omega))} = O(h^\mu), \tag{5.67}$$

we see that the error estimate (5.38) is suboptimal with respect to h. There is a question as to whether this estimate can be improved. Numerical experiments carried out in Sect. 5.2 (see also [94]) indicate that the actual order of convergence in the $L^\infty(0, T; L^2(\Omega))$-norm in the case of an odd degree of approximation is better than the theoretically derived estimate.

Remark 5.11 It is clear that estimates (5.38)–(5.41) cannot be used for $\varepsilon \to 0+$, because they blow up exponentially with respect to $1/\varepsilon$. This is a consequence of the application of the Young inequality and the Gronwall lemma, necessary for overcoming the nonlinearity of the convective terms. This nonlinearity represents a serious obstacle for obtaining a uniform error estimate with respect to $\varepsilon \to 0+$. Recently, in [207], error estimates uniform with respect to $\varepsilon \to 0+$ were derived also in the case of nonlinear convection terms.

Estimate (5.41) in $L^\infty(0, T; H^1(\Omega))$-norm has been obtained under assumption (5.40), i.e., $h \le C_{IS}\,\tau$. This nonstandard "inverse stability condition" also appears in [244] and [126].

Remark 5.12 Estimates (5.38)–(5.41) were derived without any additional restriction on the time step $\tau \le 1/2$. It is possible to show that for a fixed $\varepsilon > 0$, the semi-implicit scheme is unconditionally stable. However, there is a natural question as to what happens when $\varepsilon \to 0+$ and in the limit we obtain an explicit scheme for a nonlinear conservation law. Its stability requires using a CFL condition limiting the length of the time step. Our results are not in contradiction with this fact, because, due to Remark 5.11, the error estimate blows up and the scheme may lose the unconditional stability for $\varepsilon \to 0+$.

5.2 Backward Difference Formula for the Time Discretization

In Sect. 5.1, we presented the full space-time discretization of the nonstationary initial-boundary value problem (5.1) by the semi-implicit backward Euler time scheme (5.5). This scheme has a high-order of convergence (depending on the degree of polynomial approximation) with respect to the mesh-size h, but only the first order of convergence with respect to the time step τ.

In many applications, computations with a scheme having the first order of convergence with respect to τ are very inefficient. In this section we introduce a method for solving the nonstationary initial-boundary value problem (4.1) which is based on a combination of the discontinuous Galerkin method for the space semidiscretization and the *k-step backward difference formula* (BDF) for the time discretization. We call this technique as BDF-DGM. The BDF methods are widely used for solving stiff ODEs, see [161, 162].

Similarly as in Sect. 5.1, the diffusion, penalty and stabilization terms are treated implicitly, whereas the nonlinear convective terms are treated by a higher-order explicit extrapolation method. This leads to the necessity to solve only a linear algebraic problem at each time step. We analyze this scheme and derive error estimates in the discrete $L^\infty(0, T; L^2(\Omega))$-norm and in the $L^2(0, T; H^1(\Omega, \mathscr{T}_h))$-norm with respect to the mesh-size h and time step τ for $k = 2, 3$. Mostly, we follow the strategy from [104]. In this section we analyze only the SIPG technique which allows us to obtain h-optimal error estimates in the $L^2(\Omega)$-norm. Concerning NIPG and IIPG approaches, see Remark 5.37.

We consider again the *initial-boundary value problem* (4.1) to find $u : Q_T \to \mathbb{R}$ such that

$$\frac{\partial u}{\partial t} + \sum_{s=1}^{d} \frac{\partial f_s(u)}{\partial x_s} = \varepsilon\, \Delta u + g \quad \text{in } Q_T, \tag{5.68a}$$

$$u\,\big|_{\partial\Omega_D \times (0,T)} = u_D, \tag{5.68b}$$

$$\varepsilon \boldsymbol{n} \cdot \nabla u\,\big|_{\partial\Omega_N \times (0,T)} = g_N, \tag{5.68c}$$

$$u(x, 0) = u^0(x), \quad x \in \Omega. \tag{5.68d}$$

We assume that the data satisfy conditions (4.2), i.e.,

$$\begin{aligned}
&f = (f_1, \ldots, f_d),\ f_s \in C^1(\mathbb{R}),\ f_s' \text{ are bounded},\ f_s(0) = 0,\ s = 1, \ldots, d,\\
&u_D = \text{trace of some } u^* \in C([0, T]; H^1(\Omega)) \cap L^\infty(Q_T) \text{ on } \partial\Omega + D \times (0, T),\\
&\varepsilon > 0,\ \ g \in C([0, T]; L^2(\Omega)),\ \ g_N \in C([0, T]; L^2(\partial\Omega_N)),\ \ u^0 \in L^2(\Omega).
\end{aligned}$$

We suppose that there exists a weak solution u of (5.68) which is sufficiently regular, namely,

$$u \in W^{1,\infty}(0, T; H^s(\Omega)) \cap W^{k,\infty}(0, T; H^1(\Omega)) \cap W^{k+1,\infty}(0, T; L^2(\Omega)), \tag{5.69}$$

where $s \geq 2$ is an integer. Such a solution satisfies problem (5.68) pointwise. Under (5.69), we have $u \in C([0, T]; H^s(\Omega))$, $u' \in C([0, T]; L^2(\Omega))$, where u' means the derivative $\partial u(t)/\partial t$. (For the definitions of the above function spaces, see Sect. 1.3.5).

The symbol $(\cdot, \cdot)$ denotes the scalar product in the space $L^2(\Omega)$.

5.2.1 Discretization of the Problem

We use the same notation and assumptions as in Sects. 2.4 and 4.2. This means that we suppose that the domain Ω is polygonal if $d = 2$, or polyhedral if $d = 3$, with Lipschitz boundary. By $\mathscr{T}_h$ we denote a partition of the domain Ω and use the diffusion, penalty, right-hand side and convection forms $A_h, a_h, \ell_h, J_h^\sigma, b_h$, defined in Sect. 4.2 by relations (4.9)–(4.13) and (4.23). Let $p \geq 1$ be an integer and let S_{hp} be the space of discontinuous piecewise polynomial functions (2.34). Moreover, we assume that Assumptions 4.5 in Sect. 4.3 are satisfied. Let us recall that the functions $f_s,\ s = 1, \ldots, d$, are Lipschitz-continuous with constant $L_f = 2L_H$, where the constant L_H is introduced in (4.18).

Furthermore, as was already shown (cf. (4.28)), the exact solution u with property (5.69) satisfies the *consistency* identity

$$\left(\frac{\partial u}{\partial t}(t), v_h\right) + A_h(u(t), v_h) + b_h(u(t), v_h) = \ell_h(v_h)\,(t) \quad \forall v_h \in S_{hp}\ \forall t \in (0, T). \tag{5.70}$$

Now, because of time discretization, we consider a uniform partition of the time interval $[0, T]$ formed by the time instants $t_j = j\tau,\ j = 0, 1, \ldots, r$, with a time step $\tau = T/r$, where $r > k$ is an integer. The value $u(t_j)$ of the exact solution will be approximated by an element $u_h^j \in S_{hp},\ j = 0, \ldots, r$.

Let $k \geq 1$ be an integer. The time derivative in (5.70) will be approximated by a high-order k-step *backward difference formula*

$$\frac{\partial u}{\partial t}(t_{j+k}) \approx \frac{1}{\tau}\left(\alpha_k u_h^{j+k} + \alpha_{k-1} u_h^{j+k-1} + \cdots + \alpha_0 u_h^j\right) = \frac{1}{\tau}\sum_{i=0}^{k} \alpha_i u_h^{j+i}, \tag{5.71}$$

where $u_h^{j+l} \approx u(t_{j+l})$ and $\alpha_i,\ i = 0, \ldots, k$, are the so-called *BDF coefficients* given by

$$\alpha_k = \sum_{i=1}^{k} \frac{1}{i}, \quad \alpha_i = (-1)^{k-i}\binom{k}{i}\frac{1}{k-i}, \quad i = 0, \ldots, k-1. \tag{5.72}$$

In order to obtain an accurate, stable, efficient and simple scheme, the forms A_h and ℓ_h will be treated implicitly, whereas the nonlinear terms represented by the form b_h will be treated explicitly. In order to keep the high order of the scheme with respect to the time step, in b_h we employ a *high-order explicit extrapolation*

$$u(t_{j+k}) \approx \left(\beta_1 u_h^{j+k-1} + \beta_2 u_h^{j+k-2} + \cdots + \beta_k u_h^j\right) = \sum_{i=1}^{k} \beta_i u_h^{j+k-i}, \tag{5.73}$$

where $\beta_i,\ i = 1, \ldots, k$, are the coefficients given by

$$\beta_i = (-1)^{i+1}\binom{k}{i} = -\alpha_{k-i} i, \quad i = 1, \ldots, k. \tag{5.74}$$

Table 5.1 Values of the coefficients $\alpha_i,\ i = 0, \ldots, k$, and $\beta_i,\ i = 1, \ldots, k$, for $k = 1, 2, 3$

k	$\alpha_i,\ i{=}k, k{-}1, \ldots, 0$	$\beta_i,\ i = 1, \ldots, k$
1	$1\ \ -1$	1
2	$\frac{3}{2}\ \ -2\ \ \frac{1}{2}$	$2\ \ -1$
3	$\frac{11}{6}\ \ -3\ \ \frac{3}{2}\ \ -\frac{1}{3}$	$3\ \ -3\ \ 1$

Table 5.1 shows the values of $\alpha_i,\ i = 0, \ldots, k$, and $\beta_i,\ i = 1, \ldots, k$, for $k = 1, 2, 3$.

Now we are ready to introduce the full space-time BDF-DG discretization of problem (5.68).

Definition 5.13 Let $k \geq 1$ be an integer and let $u_h^1, \ldots, u_h^{k-1} \in S_{hp}$ be given. We define the *approximate solution* of problem (5.68) obtained by the *semi-implicit k-step BDF-DG* method as functions u_h^{l+k}, $t_{l+k} \in [0, T]$, satisfying the conditions

$$u_h^{l+k} \in S_{hp}, \tag{5.75a}$$

$$\frac{1}{\tau}\left(\sum_{i=0}^{k} \alpha_i u_h^{l+i}, v_h\right) + A_h(u_h^{l+k}, v_h) + b_h(E^{l+k}(u_h), v_h) = \ell_h(v_h)(t_{l+k}) \tag{5.75b}$$

$$\forall v_h \in S_{hp},\ l = 0, 1, 2, \ldots, r-k,$$

where E^m denotes the high-order explicit extrapolation operator at the time level t_m given by

$$E^m(u_h) = \sum_{i=1}^{k} \beta_i u_h^{m-i}, \tag{5.76}$$

and $\alpha_i,\ i = 0, \ldots, k$, and $\beta_i,\ i = 1, \ldots, k$, are given by (5.72) and (5.74), respectively. The function u_h^l is called the *approximate solution* at time $t_l,\ l = 0, \ldots, r$.

Remark 5.14 (i) We see that the high-order explicit extrapolation $E^{l+k}(u_h)$ depends on $u_h^l, \ldots, u_h^{l+k-1}$ and is independent of u_h^{l+k}.
(ii) Since scheme (5.75) represents a k-step formula, we have to define the approximate solution $u_h^0, u_h^1, \ldots, u_h^{k-1}$ at times $t_0 = 0, t_1, \ldots, t_{k-1}$. The initial value u_h^0 is defined as the $L^2(\Omega)$ projection of the initial data u^0 on the space S_{hp}. This means that $u_h^0 \in S_{hp}$ and

$$(u_h^0 - u^0, v_h) = 0 \quad \forall v_h \in S_{hp}.$$

The values $u_h^1, \ldots, u_h^{k-1}$ have to be determined, e.g., by a one-step method as, for example, a kth-order Runge–Kutta scheme, see Sect. 5.2.1.1.

(iii) The discrete problem (5.75) is equivalent to a system of linear algebraic equations for each $t_{l+k} \in [0, T]$. The existence and uniqueness of the solution of this linear algebraic problem is proved in Sect. 5.2.2.

(iv) The explicit extrapolation can also be applied to $u \in C([0,T]; L^2(\Omega))$ by

$$E^{l+k}(u) = \sum_{i=1}^{k} \beta_i u^{l+k-i}, \qquad t_l,\ t_{l+k} \in [0,T]. \tag{5.77}$$

5.2.1.1 Runge–Kutta Schemes

There are several Runge–Kutta (RK) schemes,which can be used for obtaining initial values $u_h^1, \dots, u_h^{k-1}$ for the BDF-DGM method (5.75). Here we mention some versions of the RK method.

Let $k \in \mathbb{N}$. For $v_h, \varphi_h \in S_{hp}$, we set

$$B_h(v_h, \varphi_h) = \ell_h(\varphi_h) - A_h(v_h, \varphi_h) - b_h(v_h, , \varphi_h). \tag{5.78}$$

Then we consider the problem of finding functions $u_h^n \in S_{hp}$, $n = 0, 1, \dots$, representing the approximations at the time instants t_n of the function $u_h(t)$, $t \geq 0$, such that

$$\left(\frac{\partial u_h}{\partial t}, \varphi_h\right) = B_h(u_h, \varphi_h) \quad \forall \varphi_h \in S_{hp}, \tag{5.79}$$
$$u_h(0) = u_h^0.$$

Let us suppose that the approximate solution $u_h^n \in S_{hp}$, at time instant t_n was already computed. A general k-stage RK scheme for solving problem (5.79) at the time instant t_{n+1} can be written in the form

$$u_h^{n,0} = u_n^n, \tag{5.80}$$
$$(u_h^{n,i}, \varphi_h) = \sum_{j=0}^{i-1} \left((\alpha_{ij} u_h^{n,j}, \varphi_h) + \tau \beta_{ij} B_h(u_h^{n,j}, \varphi_h) \right) \quad \forall \varphi_h \in S_{hp},\ i = 1, \dots, k,$$
$$u_h^{n+1} = u_h^{n,k}.$$

We specify here the following RK schemes.

(1) 2-stage RK scheme

$$(u_h^{n,1}, \varphi_h) - (u_h^n, \varphi_h) + \tau B_h(u_h^n, \varphi_h), \tag{5.81}$$
$$(u_h^{n+1}, \varphi_h) = \frac{1}{2}(u_h^n + u_h^{n,1}) + \frac{1}{2}\tau B_h(u_h^{n,1}, \varphi_h), \quad \forall \varphi_h \in S_{hp}.$$

(2) 3-stage RK scheme (i)

$$(u_h^{n,1}, \varphi_h) = (u_h^n, \varphi_h) + \tau B_h(u_h^n, \varphi_h), \tag{5.82}$$

$$(u_h^{n,2}, \varphi_h) = \frac{1}{2}(u_h^n + u_h^{n,1}, \varphi_h) + \frac{1}{2}\tau B_h(u_h^{n,1}, \varphi_h),$$
$$(u_h^{n+1}, \varphi_h) = \frac{1}{3}(u_h^n + u_h^{n,1} + u_h^{n,2}, \varphi_h) + \frac{1}{3}\tau B_h(u_h^{n,2}, \varphi_h), \quad \forall \varphi_h \in S_{hp}.$$

(3) 3-stage RK scheme (ii)

$$(u_h^{n,1}, \varphi_h) = (u_h^n, \varphi_h) + \tau B_h(u_h^n, \varphi_h), \tag{5.83}$$
$$(u_h^{n,2}, \varphi_h) = (\frac{3}{4}u_h^n + \frac{1}{4}u_h^{n,1}, \varphi_h) + \frac{1}{4}\tau B_h(u_h^{n,1}, \varphi_h),$$
$$(u_h^{n+1}, \varphi_h) = (\frac{1}{3}u_h^n + \frac{2}{3}u_h^{n,2}, \varphi_h) + \frac{2}{3}\tau B_h(u_h^{n,2}, \varphi_h), \quad \forall \varphi_h \in S_{hp}.$$

5.2.2 Theoretical Analysis

In what follows we are concerned with the analysis of method (5.75) for the SIPG variant of the DGM. Hence, we set $\Theta = 1$ in the definitions (4.10) and (4.13) of the forms A_h and ℓ_h. Moreover, we confine our considerations to the case when $\partial\Omega_N = \emptyset$. This means that we analyze problem (5.6) from Sect. 5.1. Other possibilities will be mentioned in Remark 5.37.

Similarly, as in the analysis of schemes for the numerical solution of ordinary differential equations, we introduce the concept of stability of the BDF method.

Definition 5.15 The BDF method (5.75) is stable (by Dahlquist), if all roots of the polynomial $\rho(\xi) = \sum_{j=0}^{k} \alpha_j \xi^j$ lie in the unit closed circle $\{z \in \mathbb{C}; |z| \le 1\}$ and the roots satisfying the condition $|\xi| = 1$ are simple (the symbol $\mathbb{C}$ denotes the set of complex numbers).

First, we present several results which will be used in Sect. 5.2.3 for deriving error estimates of method (5.75).

In what follows, as usual, we set $\mu = \min(s, p+1)$. We recall that due to (2.128) and (2.140), the form A_h satisfies the inequalities

$$A_h(\varphi_h, \varphi_h) \ge \frac{\varepsilon}{2} |||\varphi_h|||^2 \quad \forall \varphi_h \in S_{hp}, \quad \forall h \in (0, \bar{h}), \tag{5.84}$$
$$|A_h(\varphi_h, \psi_h)| \le \varepsilon C_B |||\varphi_h||| \, |||\psi_h||| \quad \forall \varphi_h, \psi_h \in S_{hp}, \tag{5.85}$$

where $C_B > 0$ is a constant independent of h, φ_h and ψ_h. Since $\alpha_k > 0$, these properties imply the *existence and uniqueness* of the approximate solution of (5.75).

Moreover, similarly as in (4.115), for each $h \in (0, \bar{h})$, we denote by $\mathrm{P}_{hp}w$ the "A_h-projection" of $w \in H^2(\Omega)$ on S_{hp}, i.e.,

$$\mathrm{P}_{hp}w \in S_{hp}, \quad A_h(\mathrm{P}_{hp}w, \varphi_h) = A_h(w, \varphi_h) \quad \forall \varphi_h \in S_{hp}. \tag{5.86}$$

Due to (4.116) and (4.130), we have

$$|||\mathrm{P}_{hp}w - w||| \le C_{P,\mathrm{e}} h^{\mu-1}|w|_{H^\mu(\Omega)} \ \forall w \in H^\mu(\Omega), \tag{5.87}$$

$$\|\mathrm{P}_{hp}w - w\|_{L^2(\Omega)} \le C_{P,\mathrm{L}} h^{\mu}|w|_{H^\mu(\Omega)} \ \forall w \in H^\mu(\Omega), \tag{5.88}$$

where $C_{P,\mathrm{e}}$ and $C_{P,\mathrm{L}}$ are constants independent of h and w.

Let us recall that Assumptions 4.5 from Sect. 4.3 are considered. Then, by (4.30)–(4.31),

$$|b_h(u,w) - b_h(\overline{u},w)| \le C_{b1}|||w||| \left(\|u-\overline{u}\|_{L^2(\Omega)} + \left(\sum_{K\in\mathscr{T}_h} h_K \|u-\overline{u}\|^2_{L(\partial K)} \right)^{1/2} \right) \tag{5.89}$$

and

$$|b_h(u_h, v_h) - b_h(\overline{u}_h, v_h)| \le C_{b2}|||v_h||| \, \|u_h - \overline{u}_h\|_{L^2(\Omega)}, \tag{5.90}$$

where C_{b1} and C_{b2} are constants independent of $u,\ u,\ w \in H^1(\Omega, \mathscr{T}_h), u_h,\ \overline{u}_h,\ v_h \subset S_{hp}$, and $h \in (0, \bar{h})$. Finally, from (4.148), we obtain

$$|b_h(u, v_h) - b_h(\mathrm{P}_{hp}u, v_h)| \le C_1 h^\mu |u|_{H^\mu(\Omega)} |||v_h|||, \tag{5.91}$$

where $C_1 = C_{b1}(C^2_{P,\mathrm{L}} + C_M(C_{P,\mathrm{e}}C_{P,\mathrm{L}} + C^2_{P,\mathrm{L}}))^{1/2}$ is a constant independent of h, $v_h \in S_{hp}$ and $u \in H^\mu(\Omega)$.

Exercise 5.16 Prove estimate (5.91) in detail.

Finally, let us note that due to the linearity of E and P_{hp} and time-independence of P_{hp}, the extrapolation E^{l+k} and the projection P_{hp} commute:

$$\begin{aligned} E^{l+k}(\mathrm{P}_{hp}v) &= \sum_{i=1}^{k} \beta_i (\mathrm{P}_{hp}v)(t_{l+k-i}) = \sum_{i=1}^{k} \beta_i (\mathrm{P}_{hp}\, v(t_{l+k-i})) \\ &= \sum_{i=1}^{k} \mathrm{P}_{hp}(\beta_i v(t_{l+k-i})) = \mathrm{P}_{hp} \sum_{i=1}^{k} \beta_i v(t_{l+k-i}) = \mathrm{P}_{hp} E^{l+k}(v), \end{aligned} \tag{5.92}$$

for any $v \in C([0,T], L^2(\Omega))$.

5.2.2.1 Properties of the k-step BDF Method

In this section we prove some properties of the k-step BDF method which will be used in Sect. 5.2.3. Although the final error estimates in Sect. 5.2.3 are obtained for $k = 2, 3$, some assertions from this section are valid in general for all $k \ge 1$.

First, we present some properties of the coefficients $\alpha_0, \ldots, \alpha_k$ of the k-step BDF method.

Lemma 5.17 *Let $k \geq 1$ and $\alpha_0, \ldots, \alpha_k$ be the coefficients of the k-step BDF given by (5.72). Then*

$$\sum_{i=0}^{k} \alpha_i = 0, \tag{5.93}$$

$$\sum_{i=0}^{k} \alpha_i (k-i) = \sum_{i=0}^{k-1} \alpha_i (k-i) = -1, \tag{5.94}$$

$$\sum_{i=0}^{k} \alpha_i (k-i)^l = \sum_{i=0}^{k-1} \alpha_i (k-i)^l = 0, \quad \text{for } l = 2, \ldots, k. \tag{5.95}$$

Proof Based on the values of coefficients $\alpha_j,\ j = 0, \ldots, k$, from Table 5.1 it is easy to see that (5.93)–(5.95) are valid for $k = 1, 2, 3$. For $k > 3$ these relations will be proved by mathematical induction using the binomial theorem.

(i) Let us assume that (5.93) is valid for $k - 1 \geq 1$, i.e., using (5.72) we have an induction assumption

$$\sum_{i=1}^{k-1} \frac{1}{i} + \sum_{i=0}^{k-2} (-1)^{k-1-i} \binom{k-1}{i} \frac{1}{k-1-i} = 0. \tag{5.96}$$

Using (5.72), after some manipulation we get

$$\begin{aligned}
\sum_{i=0}^{k-1} \alpha_i &= \sum_{i=0}^{k-1} (-1)^{k-i} \binom{k}{i} \frac{1}{k-i} \\
&= \frac{(-1)^k}{k} + \sum_{i=1}^{k-1} \frac{(-1)^{k-1-(i-1)} k(k-1)!}{i(i-1)!(k-1-(i-1))!} \frac{1}{k-1-(i-1)} \\
&= \frac{(-1)^k}{k} + \sum_{i=0}^{k-2} (-1)^{k-1-i} \binom{k-1}{i} \frac{1}{k-1-i} \frac{k}{i+1},
\end{aligned}$$

which together with (5.72) gives

$$\sum_{i=0}^{k} \alpha_i = \sum_{i=1}^{k} \frac{1}{i} + \frac{(-1)^k}{k} + \sum_{i=0}^{k-2} (-1)^{k-1-i} \binom{k-1}{i} \frac{k}{i+1} \frac{1}{k-1-i}. \tag{5.97}$$

Subtracting the vanishing left-hand side of the induction assumption (5.96) from the right-hand side of (5.97), we have

$$
\begin{aligned}
\sum_{i=0}^{k}\alpha_i &= \frac{1}{k}+\frac{(-1)^k}{k}+\sum_{i=0}^{k-2}\binom{k-1}{i}\frac{(-1)^{k-1-i}}{k-1-i}\left(\frac{k}{i+1}-1\right)\\
&= \frac{1}{k}+\frac{(-1)^k}{k}+\sum_{i=0}^{k-2}(-1)^{k-1-i}\binom{k-1}{i}\frac{1}{i+1}\\
&= \frac{1}{k}+\frac{(-1)^k}{k}+\sum_{i=0}^{k-2}(-1)^{k-1-i}\frac{(k-1)!}{(i+1)!(k-1-i)!}\\
&= \frac{1}{k}\left(1+(-1)^k+\sum_{i=1}^{k-1}(-1)^{k-i}\frac{k!}{i!(k-i)!}\right)\\
&= \frac{1}{k}\sum_{i=0}^{k}(-1)^{k-i}\binom{k}{i}=\frac{1}{k}(-1+1)^k=0,
\end{aligned}
\tag{5.98}
$$

which proves (5.93).

(ii) Using (5.72), we evaluate the left hand-side of (5.94) in the form

$$
\begin{aligned}
\sum_{i=0}^{k-1}\alpha_i(k-i) &= \sum_{i=0}^{k-1}(-1)^{k-i}\binom{k}{i}=\sum_{i=0}^{k}(-1)^{k-i}\binom{k}{i}-1\\
&= (-1+1)^k-1=-1.
\end{aligned}
\tag{5.99}
$$

(iii) We denote by $\alpha_i^j,\ i=0,\dots,j,\ 2\le j\le k$, the coefficients $\alpha_i,\ i=0,\dots,j$, of the j-step BDF. Then (5.95) can be rewritten as

$$
\sum_{i=0}^{j}\alpha_i^j(j-i)^l=0,\qquad l=2,\dots,j,\quad 2\le j\le k.
\tag{5.100}
$$

First, we prove that (5.100) is valid for $l=2$. Putting (5.72) into (5.100) (with $l=2$) we have

$$
\begin{aligned}
\sum_{i=0}^{j-1}\alpha_i^j(j-i)^2 &= \sum_{i=0}^{j-1}(-1)^{j-i}\binom{j}{i}(j-i)\\
&= -\sum_{i=0}^{j-1}(-1)^{j-1-i}\frac{j(j-1)!}{i!(j-1-i)!}=-j\sum_{i=0}^{j-1}(-1)^{j-1-i}\binom{j-1}{i}\\
&= -j(1-1)^{j-1}=0.
\end{aligned}
$$

Let us assume that (5.100) is valid for $l-1$ and $2\le j\le k$. Taking into account this induction assumption and (5.72), we successively find that

$$
\begin{aligned}
\sum_{i=0}^{k} \alpha_i^k (k-i)^l &= \sum_{i=0}^{k} \alpha_i^k (k-i)(k-i)^{l-1} \\
&= k \sum_{i=0}^{k} \alpha_i^k (k-i)^{l-1} - \sum_{i=0}^{k} \alpha_i^k (k-i)^{l-1} i \\
&= 0 - \sum_{i=1}^{k-1} \alpha_i^k (k-i)^{l-1} i = - \sum_{i=1}^{k-1} (-1)^{k-i} \binom{k}{i} (k-i)^{l-2} i \\
&= - \sum_{i=1}^{k-1} (-1)^{k-1-(i-1)} \frac{k(k-1)!\,(k-1-(i-1))^{l-2}}{(i-1)!(k-1-(i-1))!} \\
&= -k \sum_{i=0}^{k-2} (-1)^{k-1-i} \frac{(k-1)!}{i!(k-1-i)!} (k-1-i)^{l-2} \\
&= -k \sum_{i=0}^{k-2} \alpha_i^{k-1} (k-1-i)^{l-1} = 0,
\end{aligned}
$$

which proves (5.100) for $l = 2, \ldots, j, 2 \le j \le k$, and therefore (5.95) is valid. □

Lemma 5.18 *Let $k \ge 1$, $\alpha_0, \ldots, \alpha_k$, be coefficients of the k-step BDF method given by (5.72) and let $\beta_1, \ldots, \beta_k$ be given by (5.74). Then*

$$
\sum_{i=0}^{k} |\alpha_i| \le A := 2(2^k - 1), \quad \sum_{i=1}^{k} |\beta_i| \le k(2^k - 1) = \frac{k}{2} A. \tag{5.101}
$$

Proof Taking into account (5.72) and (5.93) and using the binomial theorem, we get

$$
\begin{aligned}
\sum_{i=0}^{k} |\alpha_i| &= |\alpha_k| + \sum_{i=0}^{k-1} |\alpha_i| = \left| - \sum_{i=0}^{k-1} \alpha_i \right| + \sum_{i=0}^{k-1} |\alpha_i| \le 2 \sum_{i=0}^{k-1} |\alpha_i| \\
&\le 2 \sum_{i=0}^{k-1} \binom{k}{i} = 2 \sum_{i=0}^{k} \binom{k}{i} - 2 = 2(2^k - 1),
\end{aligned} \tag{5.102}
$$

which proves the first assertion. Moreover, from (5.74) we have

$$
\sum_{i=1}^{k} |\beta_i| = \sum_{i=1}^{k} i |\alpha_{k-i}| = \sum_{i=0}^{k-1} (k-i) |\alpha_i| \le k \sum_{i=0}^{k-1} |\alpha_i|. \tag{5.103}
$$

In view of (5.102), the last term in (5.103) can be estimated by

$$
k \sum_{i=0}^{k-1} |\alpha_i| \le k2(2^{k-1} - 1) = k(2^k - 2) < k(2^k - 1). \tag{5.104}
$$

Hence, (5.103)–(5.104) give the second assertion of (5.101). □

We recall that a k-step BDF scheme is *stable* if and only if $k \leq 6$, see [161, Chap. III, Theorem 3.4]. The following two lemmas will be used in Sect. 5.2.3 for proving local error estimates of the k-step BDF methods.

Lemma 5.19 *Let $k \geq 1$ and $\alpha_0, \ldots, \alpha_k$ be the coefficients of the k-step BDF given by (5.72). Moreover, let $u \in W^{k,2}(0, T; L^2(\Omega))$, $w \in L^2(\Omega)$, $t \in (0, T)$ and $\tau > 0$. By $\partial_t^j u$, $j = 0, \ldots, k$, we denote the jth-order distributional derivative of u with respect to t. Then the following identity is valid:*

$$\tau \left(u'(t), w\right) = \alpha_k (u(t), w) + \sum_{i=0}^{k-1} \alpha_i \sum_{j=0}^{k} (-1)^j \frac{\tau^j (k-i)^j}{j!} (\partial_t^j u(t), w), \quad (5.105)$$

provided $\partial_t^k u(t)$ makes sense.

Proof From (5.95) we have

$$0 = \sum_{i=0}^{k-1} \alpha_i (k-i)^j = (-1)^j \frac{\tau^j}{j!} (\partial_t^j u(t), w) \sum_{i=0}^{k-1} \alpha_i (k-i)^j, \quad (5.106)$$

for $j = 2, \ldots, k$. Summing (5.106) over $j = 2, \ldots, k$, we get

$$\begin{aligned} 0 &= \sum_{j=2}^{k} \sum_{i=0}^{k-1} \alpha_i (-1)^j \frac{\tau^j (k-i)^j}{j!} (\partial_t^j u(t), w) \\ &= \sum_{i=0}^{k-1} \alpha_i \sum_{j=2}^{k} (-1)^j \frac{\tau^j (k-i)^j}{j!} (\partial_t^j u(t), w). \end{aligned} \quad (5.107)$$

Further, taking into account (5.94), we have

$$\begin{aligned} \sum_{i=0}^{k-1} \alpha_i \sum_{j=1}^{1} (-1)^j \frac{\tau^j (k-i)^j}{j!} (\partial_t^j u(t), w) &= -\tau (u'(t), w) \sum_{i=0}^{k-1} \alpha_i (k-i) \\ &= \tau (u'(t), w). \end{aligned} \quad (5.108)$$

Moreover, using (5.93), we derive

$$\sum_{i=0}^{k-1} \alpha_i \sum_{j=0}^{0} (-1)^j \frac{\tau^j (k-i)^j}{j!} (\partial_t^j u(t), w) = (u(t), w) \sum_{i=0}^{k-1} \alpha_i = -\alpha_k (u(t), w). \quad (5.109)$$

Finally, from (5.107)–(5.109) we obtain

$$\sum_{i=0}^{k-1} \alpha_i \sum_{j=0}^{k} (-1)^j \frac{\tau^j (k-i)^j}{j!} (\partial_t^j u(t), w) = \tau(u'(t), w) - \alpha_k (u(t), w),$$

which proves the lemma. □

Lemma 5.20 *Let $n \geq 0$ be an integer and $f \in W^{n+1,1}(0,T)$. Then for all $t, \vartheta \in [0,T]$*

$$-f(t) + \sum_{j=0}^{n} (-1)^j f^{(j)}(\vartheta) \frac{(\vartheta - t)^j}{j!} \tag{5.110}$$
$$= (-1)^n \int_t^{\vartheta} \int_{z_1}^{\vartheta} \cdots \int_{z_n}^{\vartheta} f^{(n+1)}(z_{n+1})\, dz_{n+1} \ldots dz_1.$$

Proof If $f \in W^{n+1,1}(0,T)$, then the identity

$$f^{(m)}(\vartheta) - f^{(m)}(t) = \int_t^{\vartheta} f^{(m+1)}(z_{m+1})\, \mathrm{d}z_{m+1}, \quad m = 0, \ldots, n \tag{5.111}$$

is valid for all $t, \vartheta \in [0,T]$.

We prove (5.110) by the mathematical induction. The case $n = 0$ is a consequence of (5.111) for $m = 0$. Let us suppose that (5.110) is valid for $n - 1$. Then we have

$$-f(t) + \sum_{j=0}^{n} (-1)^j f^{(j)}(\vartheta) \frac{(\vartheta - t)^j}{j!} \tag{5.112}$$
$$= -f(t) + \sum_{j=0}^{n-1} (-1)^j f^{(j)}(\vartheta) \frac{(\vartheta - t)^j}{j!} + (-1)^n f^{(n)}(\vartheta) \frac{(\vartheta - t)^n}{n!}$$
$$= (-1)^{n-1} \int_t^{\vartheta} \int_{z_1}^{\vartheta} \cdots \int_{z_{n-1}}^{\vartheta} f^{(n)}(z_n)\, \mathrm{d}z_n \ldots \mathrm{d}z_1 - (-1)^{n-1} f^{(n)}(\vartheta) \frac{(\vartheta - t)^n}{n!} =: \omega.$$

Applying the identity $\int_t^{\vartheta} \int_{z_1}^{\vartheta} \cdots \int_{z_{n-1}}^{\vartheta} 1\, \mathrm{d}z_n \ldots \mathrm{d}z_1 = (\vartheta - t)^n / n!$ to the last term of equality (5.112), we get

$$\omega = (-1)^{n-1} \int_t^{\vartheta} \int_{z_1}^{\vartheta} \cdots \int_{z_{n-1}}^{\vartheta} \left(f^{(n)}(z_n) - f^{(n)}(\vartheta) \right) \mathrm{d}z_n \ldots \mathrm{d}z_1. \tag{5.113}$$

Finally, applying (5.111) with $m = n$ to the integrand of (5.113) we obtain (5.110). □

Furthermore, for the k-step BDF scheme ($k \geq 1$), we introduce a sequence $\{\gamma_j\}_{j=0}^{\infty}$.

Lemma 5.21 *Let $k \geq 1$ and $\alpha_0, \ldots, \alpha_k$ be the coefficients of a stable k-step BDF scheme (cf. Definition 5.15). Moreover, let γ_j, $j = 0, 1, \ldots$, be real coefficients defined by*

$$\frac{1}{\alpha_k + \alpha_{k-1}z + \ldots + \alpha_0 z^k} = \sum_{j=0}^{\infty} \gamma_j z^j. \tag{5.114}$$

Then there exists $G > 0$ such that

$$|\gamma_j| < G, \quad j = 0, 1 \ldots . \tag{5.115}$$

Proof Let us denote

$$\hat{\rho}(\xi) = \alpha_k + \alpha_{k-1}\xi + \cdots + \alpha_0 \xi^k = \xi^k \rho(\xi^{-1}). \tag{5.116}$$

The roots of the polynomial $\hat{\rho}$ are the reciprocal values of the nonzero roots of the polynomial ρ. Since ρ has all roots in the circle $\{\xi \in \mathbb{C}; |\xi| \leq 1\}$ (by $\mathbb{C}$ we denote the set of all complex numbers), then all roots of $\hat{\rho}$ lie in the set $\{\xi \in \mathbb{C}; |\xi| \geq 1\}$. Those roots $\xi_1, \ldots, \xi_m$ ($m \geq 0$) of the polynomial ρ, for which $|\xi_i| = 1$, are simple. Hence, the function $1/\hat{\rho}(\xi)$ is holomorphic in the set $\{\xi \in \mathbb{C}; |\xi| < 1\}$ and has a finite number of simple poles ξ_i^{-1}, $i = 0, \ldots, m$, on the unit circle line $|\xi| = 1$. Therefore, there exist constants $\delta_i \in \mathbb{C}$, $i = 0, \ldots, m$, such that the function

$$f(\xi) = \frac{1}{\hat{\rho}(\xi)} - \sum_{i=1}^{m} \frac{\delta_i}{\xi - \xi_i^{-1}} \tag{5.117}$$

is holomorphic in the set $\{\xi \in \mathbb{C}; |\xi| < 1 + \varepsilon\}$, where $\varepsilon > 0$ is sufficiently small. Then the coefficients a_n in the Taylor expansion of the function f at the point $\xi = 0$ can be expressed with the aid of the Cauchy formula in the form

$$a_n = \frac{1}{2\pi i} \int_{\varphi} \frac{f(z)}{z^{n+1}} \, \mathrm{d}z, \tag{5.118}$$

where the curve φ is the unit positively oriented circle line. This implies that the sequence of the coefficients a_n is bounded. Further, it is possible to show that for $i = 1, \ldots, m$ and $|\xi| < 1$,

$$\frac{1}{\xi - \xi_i^{-1}} = -\xi_i \frac{1}{1 - (\xi/\xi_i^{-1})} = -\xi_i \sum_{j=0}^{\infty} \beta_j^i \xi^j, \tag{5.119}$$

where $\beta_j^i = (\xi_i)^j$. Hence, since $|\xi_i| = 1$, also $|\beta_j^i| = 1$. The above considerations immediately imply the assertion of the lemma. □

Remark 5.22 Let $\tilde{\rho}(z)$ be the denominator of the left-hand side of (5.114) and let us set $\gamma_j = 0$ for $j = -k, \ldots, -1$. Then from (5.114) we get

$$1 = \tilde{\rho}(z) \sum_{j=0}^{\infty} \gamma_j z^j = \tilde{\rho}(z) \sum_{j=-k}^{\infty} \gamma_j z^j.$$

Comparing the coefficients with the same powers m of the variable z, we get the relations

$$m = 0 : \sum_{i=0}^{k} \alpha_i \gamma_{m-k+i} = \alpha_k \gamma_m + \ldots + \alpha_0 \gamma_{m-k} = \alpha_k \gamma_0 = 1,$$

$$m \geq 1 : \sum_{i=0}^{k} \alpha_i \gamma_{m-k+i} = \alpha_k \gamma_m + \ldots + \alpha_0 \gamma_{m-k} = 0.$$

From this we see that γ_j can be defined as the solution of the system of linear difference equations

$$\sum_{i=0}^{k} \alpha_i \gamma_{j+i} = 0, \qquad j = -(k-1), \ldots, -1, 0, 1, \ldots, \tag{5.120}$$

with initial conditions $\gamma_{-(k-1)} = \cdots = \gamma_{-1} = 0,\ \gamma_0 = \alpha_k^{-1}$.

We proved in Lemma 5.21 that the sequence $\{\gamma_j\}_{j=0}^{\infty}$ given by (5.114) is bounded for any stable k-step BDF method, i.e., for $k \leq 6$. Finally, in order to derive the error estimates, in Sect. 5.2.3, we use the fact that $\gamma_j \geq 0$ for $j = 0, 1 \ldots$. Since the proof is rather technical, we present this property only for $k = 1, 2, 3$, which is sufficient for further considerations in Sect. 5.2.3.

Lemma 5.23 *Let* $1 \leq k \leq 3$ *and* α_i, $i = 0, \ldots, k$, *be the coefficients of* k*-step BDF and let* $\{\gamma_j\}_{j=0}^{\infty}$ *be the sequence defined by (5.114). Then*

$$\gamma_j > 0, \quad j = 0, 1, \ldots . \tag{5.121}$$

Proof On the basis of Remark 5.22 we put $\gamma_j = 0$ for $j = -(k-2), \ldots, -1$. Then the sequence $\{\gamma_j\}_{j=-k+2}^{\infty}$ satisfies (5.120).

Let $k = 1$. Then $\gamma_j,\ j = 0, 1, \ldots$, satisfy the difference equations (5.120) in the form

$$\gamma_{j+1} - \gamma_j = 0, \quad j = 0, 1, 2, \ldots \tag{5.122}$$

with the initial condition $\gamma_0 = 1$.

Let $k = 2$. Then $\gamma_j,\ j = -1, 0, 1, \ldots$, satisfy the difference equations (5.120) in the form

$$\frac{3}{2}\gamma_{j+2} - 2\gamma_{j+1} + \frac{1}{2}\gamma_j = 0, \quad j = -1, 0, 1, 2, \ldots \tag{5.123}$$

with the initial conditions $\gamma_{-1} = 0$ and $\gamma_0 = 2/3$ (See Table 5.1). The corresponding characteristic polynom $3\zeta^2 - 4\zeta + 1 = 0$ has roots $\zeta_1 = 1$ and $\zeta_2 = 1/3$. Fulfilling the initial conditions, we find that

$$\gamma_j = 1 - \frac{1}{3^{j+1}} \quad j = -1, 0, 1, \ldots . \tag{5.124}$$

Obviously, $\gamma_j > 0$ for $j = 0, 1, \ldots$.

Let $k = 3$. Then $\gamma_j,\ j = -2, -1, \ldots$, satisfy the difference equations (5.120) in the form

$$\frac{11}{6}\gamma_{j+3} - 3\gamma_{j+2} + \frac{3}{2}\gamma_{j+1} - \frac{1}{3}\gamma_j = 0, \quad j = -2, -1, 0, 1, \ldots \tag{5.125}$$

with the initial conditions $\gamma_{-2} = \gamma_{-1} = 0$ and $\gamma_0 = \frac{6}{11}$. By directly using (5.125) we find that

$$\gamma_j > 0, \ j = 1, 2, 3 \ \text{ and } \ \gamma_4 > 1. \tag{5.126}$$

We still have to prove that $\gamma_j > 0$ for $j = 5, 6, \ldots$. To this end, we define a sequence $\chi_j = \gamma_{j-1} - \gamma_{j-2},\ j = 0, 1, \ldots$. Then, we can find that

$$\gamma_j = \gamma_4 + \sum_{i=6}^{j+1} \chi_i, \quad j = 5, 6, \ldots . \tag{5.127}$$

From (5.125) we have

$$\begin{aligned} 0 &= \frac{11}{6}\gamma_{j+3} - 3\gamma_{j+2} + \frac{3}{2}\gamma_{j+1} - \frac{1}{3}\gamma_j \\ &= \frac{11}{6}(\gamma_{j+3} - \gamma_{j+2}) - \frac{7}{6}(\gamma_{j+2} - \gamma_{j+1}) + \frac{1}{3}(\gamma_{j+1} - \gamma_j) \\ &= \frac{11}{6}\chi_{j+4} - \frac{7}{6}\chi_{j+3} + \frac{1}{3}\chi_{j+2}, \quad j = -2, -1, \ldots , \end{aligned} \tag{5.128}$$

and $\chi_0 = 0,\ \chi_1 = \frac{6}{11}$. The solution of system (5.128) has the form

$$\chi_j = \frac{-6i}{\sqrt{39}}\left(\frac{7 + i\sqrt{39}}{22}\right)^j + \frac{6i}{\sqrt{39}}\left(\frac{7 - i\sqrt{39}}{22}\right)^j, \tag{5.129}$$

which gives

$$
\begin{aligned}
|\chi_j| &\le 2\frac{6}{\sqrt{39}}\left|\frac{7+i\sqrt{39}}{22}\right|^j \le 2\frac{6}{\sqrt{36}}\left(\frac{(49+39)^{1/2}}{22}\right)^j = 2\left(\frac{\sqrt{88}}{22}\right)^j \\
&= 2\left(\frac{2\sqrt{2}\sqrt{11}}{22}\right)^j = 2\left(\sqrt{\frac{2}{11}}\right)^j.
\end{aligned}
\tag{5.130}
$$

Therefore, with the aid of (5.126), (5.127) and (5.130) we estimate $\gamma_j,\ j \ge 5$, by

$$
\begin{aligned}
\gamma_j &= \gamma_4 + \sum_{i=6}^{j+1}\chi_i \ge \gamma_4 - \sum_{i=6}^{j+1}|\chi_i| > 1 - 2\sum_{i=6}^{j+1}\left(\sqrt{\frac{2}{11}}\right)^i \\
&> 1 - 2\frac{8}{1331}\sum_{i=0}^{\infty}\left(\sqrt{\frac{2}{11}}\right)^i = 1 - \frac{16}{1331}\frac{1}{1-\sqrt{\frac{2}{11}}} > 1 - \frac{16}{1331}2 > 0,
\end{aligned}
$$

which proves the lemma. □

Remark 5.24 It follows from the real recursive definitions (5.125) and (5.128) that the sequences $\{\gamma_j\}_{j=0}^{\infty}$ and $\{\chi_j\}_{j=0}^{\infty}$ are real although it may seem from (5.129) that they are complex.

5.2.3 Error Estimates

Our goal is now to derive the error estimates for the approximate solution u_h^l, $l = 0, 1, \dots, r$, obtained by the k-step BDF-DG scheme (5.75) under the assumption that the exact solution u satisfies (5.69). We define the discrete analogues of the $L^\infty(0, T; L^2(\Omega))$-norm and $L^2(0, T; H^1(\Omega, \mathscr{T}_h))$-norm.

Definition 5.25 Let $t_l = l\tau$, $l = 0, 1, \dots, r$, with $\tau = T/r$ be a partition of the time interval $[0, T]$ and let u_h^l, $l = 0, \dots, r$, be the approximate solution defined by (5.75). We set

$$
e_h^l = u_h^l - u^l, \quad l = 0, 1, \dots, r,
\tag{5.131}
$$

$$
\|e\|^2_{h,\tau,L^\infty(L^2)} = \max_{l=0,\dots,r}\|e_h^l\|^2_{L^2(\Omega)}, \quad \|e\|^2_{h,\tau,L^2(H^1)} = \tau\varepsilon\sum_{l=0}^{r}|||e_h^l|||^2.
$$

Let $\mathrm{P}_{hp}u^l$ be the "A_h-projection" of $u^l = u(t_l)$ $(l = 0, \dots, r)$ given by (5.86). We put

$$
\zeta^l = u_h^l - \mathrm{P}_{hp}u^l \in S_{hp}, \quad \chi^l = \mathrm{P}_{hp}u^l - u^l \in H^\mu(\Omega, \mathscr{T}_h).
\tag{5.132}
$$

(We recall that $\mu = \min(s, p+1)$.) Then the error e_h^l can be expressed as

$$e_h^l = \zeta^l + \chi^l, \quad l = 0, \ldots, r. \tag{5.133}$$

Moreover, relations (5.86) and (5.132) imply that

$$A_h(\chi^l, v_h) = 0 \quad \forall v_h \in S_{hp}, \; l = 0, \ldots, r. \tag{5.134}$$

In view of (5.75b),

$$\left(\sum_{i=0}^{k} \alpha_i u_h^{l+i}, v_h\right) + \tau \left(A_h(u_h^{l+k}, v_h) + b_h(E^{l+k}(u_h), v_h) - \ell_h(v_h)\,(t_{l+k})\right) = 0,$$
$$\text{if } t_l, \ldots, t_{l+k} \in [0, T], \tag{5.135}$$

where $E^{l+k}(u_h)$ is defined by (5.76). Moreover, setting $t = t_{l+k}$ in (5.70), we get

$$\left(u'(t_{l+k}), v_h\right) + A_h(u^{l+k}, v_h) + b_h(u^{l+k}, v_h) - \ell_h(v_h)\,(t_{l+k}) = 0. \tag{5.136}$$

Multiplying (5.136) by τ, subtracting from (5.135) and using the linearity of the form $A_h(\cdot, \cdot)$, we get

$$\left(\sum_{i=0}^{k} \alpha_i u_h^{l+i}, v_h\right) - \tau\left(u'(t_{l+k}), v_h\right) + \tau\left(A_h(u_h^{l+k} - u^{l+k}, v_h)\right.$$
$$\left. + b_h(E^{l+k}(u_h), v_h) - b_h(u^{l+k}, v_h)\right) = 0, \quad l = 0, \ldots, r-k. \tag{5.137}$$

Taking into account (5.132)–(5.134), from (5.137) we obtain

$$\left(\sum_{i=0}^{k} \alpha_i \zeta^{l+i}, v_h\right) + \tau A_h(\zeta^{l+k}, v_h) \tag{5.138}$$
$$= \tau(u'(t_{l+k}), v_h) - \left(\sum_{i=0}^{k} \alpha_i u^{l+i}, v_h\right) - \left(\sum_{i=0}^{k} \alpha_i \chi^{l+i}, v_h\right)$$
$$+ \tau\left(b_h(u^{l+k}, v_h) - b_h(E^{l+k}(u_h), v_h)\right).$$

In what follows, we estimate the individual terms on the right-hand side of (5.138).

5.2.3.1 Estimating of Individual Terms

Let u satisfy assumptions (5.69). We set $\mu = \min(p+1, s)$ and use the notation $\partial_t^m u := \partial^m u / \partial t^m$. Similarly as in (5.14), we put

$$\|u\|_d := \max\Big(\|\partial_t^{k+1} u\|_{L^\infty(0,T;L^2(\Omega))},\ \|\partial_t^k u\|_{L^\infty(0,T;L^2(\Omega))}, \tag{5.139}$$
$$\|\partial_t^k u\|_{L^\infty(0,T;H^1(\Omega))},\ \|u'\|_{L^\infty(0,T;H^\mu(\Omega))}\Big) < \infty.$$

In the proofs of the following lemmas and theorems we frequently use the Young inequality and $(a_1 + \cdots + a_n)^2 \le n(a_1^2 + \cdots + a_n^2)$ for $a_1, \ldots, a_n \in \mathbb{R}$. Since these inequalities are standard, we do not emphasize their use. In what follows, we use the simplified notation $u^m := u(t_m)$.

Lemma 5.26 *Let us assume that $u : Q_T \to \mathbb{R}$ is a function satisfying conditions (5.69), $l \in \{0, \ldots, r-k\}$ and α_i, $i = 0, \ldots, k$, and β_i, $i = 1, \ldots, k$, are the coefficients given by (5.72) and (5.74), respectively. Let $E^{l+k}(u)$ be given by (5.77). Then there exist positive constants C_2, C_3 depending on k, but independent of l and τ, such that*

$$\left|\left(\sum_{i=0}^{k} \alpha_i u^{l+i}, v_h\right) - \tau\left(u'(t_{l+k}), v_h\right)\right| \le C_2 \tau^{k+1} \|u\|_d \|v_h\|_{L^2(\Omega)}, \tag{5.140}$$

$$\left\|u^{l+k} - E^{l+k}(u)\right\|_{L^2(\Omega)} \le C_3 \tau^k \|u\|_d, \tag{5.141}$$

$$\left|u^{l+k} - E^{l+k}(u)\right|_{H^1(\Omega)} \le C_3 \tau^k \|u\|_d, \tag{5.142}$$

$$|u^{l+i+1} - u^{l+i}|_{H^\mu(\Omega)} \le C_3 \tau \|u\|_d, \tag{5.143}$$

where $\|u\|_d$ is given by (5.139).

Proof (i) From (5.105), where we put $t = t_{l+k}$, we obtain the relation

$$\left(\sum_{i=0}^{k} \alpha_i u^{l+i}, v_h\right) - \tau\left(u'(t_{l+k}), v_h\right)$$
$$= \sum_{i=0}^{k-1} \alpha_i \left((u^{l+i}, v_h) - \sum_{j=0}^{k} (-1)^j \frac{\tau^j (k-i)^j}{j!} (\partial_t^j u(t_{l+k}), v_h) \right).$$

With the aid of Lemmas 5.18 and 5.20, the identity

$$\int_t^\vartheta \int_{z_1}^\vartheta \cdots \int_{z_{n-1}}^\vartheta 1\, \mathrm{d}z_n \ldots \mathrm{d}z_1 = (\vartheta - t)^n / n!$$

and the assumption $u \in W^{k+1,\infty}(0, T; L^2(\Omega))$, we find that

$$\left|\sum_{i=0}^{k-1}\alpha_i\left((u^{l+i}, v_h) - \sum_{j=0}^{k}(-1)^j\frac{\tau^j(k-i)^j}{j!}(\partial_t^j u(t_{l+k}), v_h)\right)\right|$$
$$\le \sum_{i=0}^{k-1}|\alpha_i|\left|\int_{t_{l+i}}^{t_{l+k}}\int_{z_1}^{t_{l+k}}\dots\int_{z_k}^{t_{l+k}}(\partial_t^{k+1}u(z_{k+1}), v_h)\,\mathrm{d}z_{k+1}\dots\mathrm{d}z_1\right|$$
$$\le A\frac{k^{k+1}}{(k+1)!}\tau^{k+1}\|\partial_t^{k+1}u\|_{L^\infty(0,T;L^2(\Omega))}\|v_h\|_{L^2(\Omega)} \le C_2\tau^{k+1}\|u\|_d\|v_h\|_{L^2(\Omega)},$$

where the constant A is defined in (5.101), $\|u\|_d$ is given by (5.139) and $C_2 = A\frac{k^{k+1}}{(k+1)!}$. This proves (5.140).

(ii) By virtue of (5.74), we have

$$u^{l+k} - E^{l+k}(u) = u^{l+k} - \sum_{i=1}^{k}\beta_i u^{l+k-i} - u^{l+k} + \sum_{i=0}^{k-1}\alpha_i(k-i)u^{l+i}. \tag{5.144}$$

Using (5.94) and (5.95) we derive the identity

$$\begin{aligned}
u^{l+k} + \sum_{i=0}^{k-1}\alpha_i(k-i)u^{l+i} &= \sum_{i=0}^{k-1}\alpha_i(k-i)(u^{l+i} - u^{l+k})\\
&= \sum_{i=0}^{k-1}\alpha_i(k-i)(u^{l+i} - u^{l+k})\\
&\quad - \sum_{j=1}^{k-1}\partial_t^j u(t_{l+k})\frac{\tau^j}{j!}(-1)^j\underbrace{\sum_{i=0}^{k-1}\alpha_i(k-i)^{j+1}}_{=0}\\
&= \sum_{i=0}^{k-1}\alpha_i(k-i)(u^{l+i} - u^{l+k})\\
&\quad - \sum_{i=0}^{k-1}\alpha_i(k-i)\sum_{j=1}^{k-1}\partial_t^j u(t_{l+k})\frac{\tau^j}{j!}(-1)^j(k-i)^j\\
&= \sum_{i=0}^{k-1}\alpha_i(k-i)\left(u^{l+i} - \sum_{j=0}^{k-1}\partial_t^j u(t_{l+k})\frac{\tau^j}{j!}(-1)^j(k-i)^j\right).
\end{aligned} \tag{5.145}$$

Since $u \in W^{k,\infty}(0,T;H^1(\Omega)) \subset W^{k,\infty}(0,T;L^2(\Omega))$, with the aid of Lemmas 5.20 and 5.18 we estimate the norm of the last term in (5.145) by

$$
\begin{aligned}
&\left\| \sum_{i=0}^{k-1} \alpha_i (k-i) \left(u^{l+i} - \sum_{j=0}^{k-1} \partial_t^j u(t_{l+k}) \frac{\tau^j}{j!} (-1)^j (k-i)^j \right) \right\|_{L^2(\Omega)} \\
&= \left\| \sum_{i=0}^{k-1} \alpha_i (k-i) \left(\int_{t_{l+i}}^{t_{l+k}} \int_{z_1}^{t_{l+k}} \dots \int_{z_{k-1}}^{t_{l+k}} \partial_t^k u(z_k)\, \mathrm{d}z_k \dots \mathrm{d}z_1 \right) \right\|_{L^2(\Omega)} \\
&\le A \frac{k^{k+1}}{k!} \tau^k \|\partial_t^k u\|_{L^\infty(0,T;L^2(\Omega))} \le C_3 \tau^k \|u\|_d,
\end{aligned}
$$

where $C_3 = A\frac{k^{k+1}}{k!}$. This proves (5.141).

(iii) Since $u \in W^{k,\infty}(0,T;H^1(\Omega))$, we have

$$
\left| u^{l+k} - E^{l+k}(u) \right|_{H^1(\Omega)} = \left\| \nabla u^{l+k} - \nabla E^{l+k}(u) \right\|_{L^2(\Omega)}.
$$

Now we can use the result (5.141) for ∇u instead of u. Taking into account that

$$
\frac{\partial}{\partial t}\left(\frac{\partial u}{\partial x_l}\right) = \frac{\partial}{\partial x_l}\left(\frac{\partial u}{\partial t}\right), \qquad l = 1, \dots, d, \tag{5.146}
$$

in the sense of distributions (see Exercise 5.6), we get

$$
\begin{aligned}
&\left| u^{l+k} - E^{l+k}(u) \right|_{H^1(\Omega)} \le A \frac{k^{k+1}}{k!} \tau^k \left\| \frac{\partial^k}{\partial t^k} \nabla u \right\|_{L^\infty(0,T;L^2(\Omega))} \qquad (5.147)\\
&= A \frac{k^{k+1}}{k!} \tau^k \|\nabla \partial_t^k u\|_{L^\infty(0,T;L^2(\Omega))} \le A \frac{k^{k+1}}{k!} \tau^k \|\partial_t^k u\|_{L^\infty(0,T;H^1(\Omega))} \le C_3 \tau^k \|u\|_d,
\end{aligned}
$$

where $C_3 = A\frac{k^{k+1}}{k!}$. This proves (5.142).

(iv) Using (5.146), assumption $u \in W^{1,\infty}(0,T;H^\mu(\Omega))$ and a similar process as in (5.147) with $k = 1$, and taking into account that $1 \le \frac{k^{k+1}}{k!}$ for $k = 1, 2, \dots$, we derive (5.143). □

Now, we estimate the "A_h-projection terms". First, we mention an auxiliary lemma.

Lemma 5.27 *Under assumptions (5.69), for $t_l, t_{l+1} \in [0,T]$ and $v_h \in S_{hp}$ we have*

$$
|(\chi^{l+1} - \chi^l, v_h)| \le C_4 \tau h^\mu \|u\|_d \|v_h\|_{L^2(\Omega)}, \tag{5.148}
$$

with $C_4 > 0$.

Proof The Cauchy inequality and relations (5.132), (5.88) and (5.143) imply that

$$\begin{aligned}
|(\chi^{l+1} - \chi^l, v_h)| &\le \|\chi^{l+1} - \chi^l\|_{L^2(\Omega)} \|v_h\|_{L^2(\Omega)} \\
&= \|(u^{l+1} - u^l) - (\mathrm{P}_{hp} u^{l+1} - \mathrm{P}_{hp} u^l)\|_{L^2(\Omega)} \|v_h\|_{L^2(\Omega)} \\
&\le C_{P,\mathrm{L}} h^{\mu} |u^{l+1} - u^l|_{H^{\mu}(\Omega)} \|v_h\|_{L^2(\Omega)} \\
&\le C_{P,\mathrm{L}} C_3 \|u\|_d \tau h^{\mu} \|v_h\|_{L^2(\Omega)},
\end{aligned}$$

which proves the lemma with $C_4 = C_{P,\mathrm{L}} C_3 \|u\|_d$. □

Lemma 5.28 *Let $k \ge 1$ and $\alpha_0, \dots, \alpha_k$ be the coefficients of the k-step BDF given by (5.72). Then under assumptions (5.69), for $t_l, \dots, t_{l+k} \in [0, T]$, $v_h \in S_{hp}$, we have*

$$\left| \left(\sum_{i=0}^{k} \alpha_i \chi^{l+i}, v_h \right) \right| \le C_5 \tau h^{\mu} \|u\|_d \|v_h\|_{L^2(\Omega)}, \tag{5.149}$$

with C_5 depending on C_4 and k.

Proof Let $\rho(y)$ be the characteristic polynomial defined by

$$\rho(y) = \sum_{i=0}^{k} \alpha_i y^i, \qquad y \in \mathbb{R}. \tag{5.150}$$

From (5.93) it follows that $y = 1$ is its root. We define the polynomial π by

$$\pi(y) = \rho(y)(y-1)^{-1}. \tag{5.151}$$

Hence,

$$\pi(y)(y-1) = \rho(y). \tag{5.152}$$

We write $\pi(y) = \sum_{i=0}^{k-1} \delta_i y^i$ and determine successively the coefficients δ_i, $i = 0, \dots, k-1$. Using (5.93), (5.150) and (5.151), it is easy to see that

$$\delta_0 = \pi(0) = -\rho(0) = -\alpha_0 = \sum_{i=1}^{k} \alpha_i \tag{5.153}$$

From (5.152) we express the ith ($i \ge 1$) derivative of $\rho(y)$:

$$\rho^{(i)}(y) = (y-1)\pi^{(i)}(y) + i\pi^{(i-1)}(y). \tag{5.154}$$

Evaluating (5.154) at $y = 0$, we get

$$i!\alpha_i = \rho^{(i)}(0) = -\pi^{(i)}(0) + i\pi^{(i-1)}(0) = -i!\,\delta_i + i!\,\delta_{i-1}.$$

From this and (5.153), we obtain by induction

$$\delta_i = -\sum_{j=0}^{i} \alpha_j = \sum_{j=i+1}^{k} \alpha_j, \quad i = 0, \ldots, k-1, \tag{5.155}$$

where the second equality follows from (5.93). Moreover, from (5.150) and (5.151) it follows that

$$\sum_{i=0}^{k} \alpha_i y^i = \rho(y) = (y-1)\pi(y) = (y-1)\sum_{i=0}^{k-1} \delta_i y^i = \sum_{i=0}^{k-1} \delta_i (y^{i+1} - y^i)$$

and, therefore, we can write

$$\begin{aligned} \left|\left(\sum_{i=0}^{k} \alpha_i \chi^{l+i}, v_h\right)\right| &= \left|\left(\sum_{i=0}^{k-1} \delta_i (\chi^{l+i+1} - \chi^{l+i}), v_h\right)\right| \\ &\le \sum_{i=0}^{k-1} |\delta_i| \left|\left(\chi^{l+i+1} - \chi^{l+i}, v_h\right)\right|. \end{aligned}$$

Using (5.148), (5.155) and (5.101) we get

$$\sum_{i=0}^{k-1} |\delta_i| \left|\left(\chi^{l+i+1} - \chi^{l+i}, v_h\right)\right| \le kAC_4 \tau h^{\mu} \|u\|_d \|v_h\|_{L^2(\Omega)},$$

which proves the lemma with $C_5 = kAC_4$. □

Further, we estimate the convection form.

Lemma 5.29 *Let us assume that for $t_l, t_{l+k} \in [0, T]$ the expression $E^{l+k}(u_h)$ is given by (5.76). Then*

$$\left|b_h(u^{l+k}, v_h) - b_h(E^{l+k}(u_h), v_h)\right| \le C_6 |||v_h||| \left(\sum_{i=0}^{k-1} \|\zeta^{l+i}\|_{L^2(\Omega)} + (h^{\mu} + \tau^k)\|u\|_d\right), \tag{5.156}$$

where $C_6 > 0$ is the constant independent of h, τ, l specified in the proof, ζ^{l+i}, $i = 0, \ldots, k-1$, are defined by (5.132) and $\|u\|_d$ is given by (5.139).

Proof We can write

$$\begin{aligned} & b_h(u^{l+k}, v_h) - b_h(E^{l+k}(u_h), v_h) && (5.157) \\ &= b_h(u^{l+k}, v_h) - b_h(E^{l+k}(u), v_h) && (=: \Psi_1) \end{aligned}$$

$$+ b_h(E^{l+k}(u), v_h) - b_h(E^{l+k}(\mathrm{P}_{hp}u), v_h) \qquad (=: \Psi_2)$$
$$+ b_h(E^{l+k}(\mathrm{P}_{hp}u), v_h) - b_h(E^{l+k}(u_h), v_h) \qquad (=: \Psi_3),$$

where $E^{l+k}(u)$ is given by (5.77). We estimate the individual terms in (5.157). From (5.89) we have

$$|\Psi_1| \le C_{b1} |||v_h||| \left\{ \left\| u^{l+k} - E^{l+k}(u) \right\|_{L^2(\Omega)} + \left(\sum_{K \in \mathscr{T}_k} h_K \left\| u^{l+k} - E^{l+k}(u) \right\|^2_{L^2(\partial K)} \right)^{\frac{1}{2}} \right\}. \tag{5.158}$$

Using (2.78), (5.141) and (5.142), we find that

$$\begin{aligned} &\sum_{K \in \mathscr{T}_h} h_K \left\| u^{l+k} - E^{l+k}(u) \right\|^2_{L^2(\partial K)} \\ &\le C_M \sum_{K \in \mathscr{T}_h} h_K \left\| u^{l+k} - E^{l+k}(u) \right\|_{L^2(K)} \left| u^{l+k} - E^{l+k}(u) \right|_{H^1(K)} \\ &+ C_M \sum_{K \in \mathscr{T}_h} \left\| u^{l+k} - E^{l+k}(u) \right\|^2_{L^2(K)} \le C_M \left(\bar{h} C_3^2 + C_3^2 \right) \tau^{2k} \|u\|_d^2. \end{aligned} \tag{5.159}$$

Then (5.158), (5.159) and (5.141) give

$$|\Psi_1| \le C_{b1} \tau^k (C_3 (C_M (\bar{h} + 1))^{1/2} + C_3) \|u\|_d |||v_h||| = C_7 \tau^k \|u\|_d |||v_h|||, \tag{5.160}$$

where $C_7 = C_{b1}(C_3(C_M(\bar{h}+1))^{1/2} + C_3)$. Moreover, from (5.91), (5.90) and (5.92) we find that

$$\begin{aligned} |\Psi_2| &\le C_1 h^\mu \, |||v_h||| \, |E^{l+k}(u)|_{H^\mu(\Omega)}, \\ |\Psi_3| &\le C_{b2} \, |||v_h||| \, \|E^{l+k}(\mathrm{P}_{hp}u) - E^{l+k}(u_h)\|_{L^2(\Omega)} \end{aligned} \tag{5.161}$$

By (5.77), (5.76), (5.101), (5.139) and (5.132), we have

$$|E^{l+k}(u)|_{H^\mu(\Omega)} \le \frac{k}{2} \Lambda \|u\|_d, \tag{5.162}$$

$$\begin{aligned} \|E^{l+k}(\mathrm{P}_{hp}u) - E^{l+k}(u_h)\|_{L^2(\Omega)} &\le \frac{k}{2} \Lambda \sum_{i=1}^{k} \|\mathrm{P}_{hp}u^{l+k-i} - u_h^{l+k-i}\|_{L^2(\Omega)} \\ &= \frac{k}{2} \Lambda \sum_{i=1}^{k} \|\zeta^{l+k-i}\|_{L^2(\Omega)} = \frac{k}{2} \Lambda \sum_{i=0}^{k-1} \|\zeta^{l+i}\|_{L^2(\Omega)}. \end{aligned}$$

Finally, from (5.157), (5.160), (5.161) and (5.162) we obtain

$$\left|b_h(u^{l+k}, v_h) - b_h(E^{l+k}(u_h), v_h)\right| \le C_6 |||v_h||| \left(\sum_{i=0}^{k-1} \|\zeta^{l+i}\|_{L^2(\Omega)} + (h^\mu + \tau^k)\|u\|_d \right),$$

with $C_6 = \max\left(\frac{Ak}{2} C_{b2},\ \frac{Ak}{2} C_1,\ C_7\right)$, which proves (5.156). □

Further, we present an estimate which will be the basis for the error estimation.

Lemma 5.30 *Let Assumptions 4.5 in Sect. 4.3 be satisfied and let $\partial\Omega_N = \emptyset$. Let u be the exact solution of problem (5.68) satisfying (5.69). Let $t_l = l\tau$, $l = 0, 1, \ldots, r$, $\tau = T/r$, be a time partition of $[0, T]$ and let u_h^l, $l = 0, \ldots, r$, be the approximate solution defined by the k-step BDF (5.75), where $k \ge 1$. Then there exists a constant $K > 0$ independent of h, τ, l and ε such that for $l = 0, \ldots, r-k$,*

$$2\left(\sum_{i=0}^{k} \alpha_i \zeta^{l+i}, v_h \right) + 2\tau A_h(\zeta^{l+k}, v_h) \tag{5.163}$$
$$\le \tau \|v_h\|^2_{L^2(\Omega)} + \tau \frac{K}{\varepsilon} \sum_{i=0}^{k-1} \|\zeta^{l+i}\|^2_{L^2(\Omega)} + \frac{\tau\varepsilon}{2} |||v_h|||^2 + \tau K q(u, \varepsilon, h, \tau, k),$$

where

$$q(u, \varepsilon, h, \tau, k) = \left(h^{2\mu} + \tau^{2k}\right)(1 + 1/\varepsilon)\, \|u\|_d^2, \tag{5.164}$$

$v_h \in S_{hp}$ and ζ^l, $l = 0, \ldots, r$, are given by (5.132).

Proof Multiplying (5.138) by 2, we get

$$2\left(\sum_{i=0}^{k} \alpha_i \zeta^{l+i}, v_h \right) + 2\tau A_h(\zeta^{l+k}, v_h) \tag{5.165}$$
$$= 2\tau(u'(t_{l+k}), v_h) - 2\left(\sum_{i=0}^{k} \alpha_i u^{l+i}, v_h \right) - 2\left(\sum_{i=0}^{k} \alpha_i \chi^{l+i}, v_h \right)$$
$$+ 2\tau \left(b_h(u^{l+k}, v_h) - b_h(E^{l+k}(u_h), v_h) \right) =: \text{RHS}.$$

With the aid of Lemmas 5.26, 5.28 and 5.29, we estimate the right-hand side of (5.165) by

$$|\text{RHS}| \le 2\left(C_2 \tau^{k+1} + C_5 \tau h^\mu \right) \|u\|_d \|v_h\|_{L^2(\Omega)} \tag{5.166}$$
$$+ 2\tau C_6 |||v_h||| \left(\sum_{i=0}^{k-1} \|\zeta^{l+i}\|_{L^2(\Omega)} + (h^\mu + \tau^k)\|u\|_d \right).$$

We estimate both terms from (5.166) by

$$2\left(C_2\tau^{k+1} + C_5\tau h^{\mu}\right)\|u\|_d\|v_h\|_{L^2(\Omega)} \le \tau\|v_h\|^2_{L^2(\Omega)} + 2\tau(C_2^2\tau^{2k} + C_5^2h^{2\mu})\|u\|_d^2, \tag{5.167}$$

and

$$\begin{aligned}
&2\tau|||v_h|||C_6\left((h^{\mu}+\tau^k)\|u\|_d + \sum_{i=0}^{k-1}\|\zeta^{l+i}\|_{L^2(\Omega)}\right) \qquad (5.168)\\
&\le \frac{\tau\varepsilon}{2}|||v_h|||^2 + \frac{2\tau}{\varepsilon}C_6^2\left((h^{\mu}+\tau^k)\|u\|_d + \sum_{i=0}^{k-1}\|\zeta^{l+i}\|_{L^2(\Omega)}\right)^2\\
&\le \frac{\tau\varepsilon}{2}|||v_h|||^2 + 2\tau C_6^2\frac{k+2}{\varepsilon}\left((h^{2\mu}+\tau^{2k})\|u\|_d^2 + \sum_{i=0}^{k-1}\|\zeta^{l+i}\|^2_{L^2(\Omega)}\right).
\end{aligned}$$

Taking (5.165), (5.166), (5.167) and (5.168) together, we get

$$\begin{aligned}
&2\left(\sum_{i=0}^{k}\alpha_i\zeta^{l+i}, v_h\right) + 2\tau A_h(\zeta^{l+k}, v_h)\\
&\le \tau\|v_h\|^2_{L^2(\Omega)} + 2\tau(C_2^2\tau^{2k} + C_5^2h^{2\mu})\|u\|_d^2\\
&\quad + \tau\frac{2(k+2)C_6^2}{\varepsilon}\left(\sum_{i=0}^{k-1}\|\zeta^{l+i}\|^2_{L^2(\Omega)} + (h^{2\mu}+\tau^{2k})\|u\|_d^2\right) + \frac{\tau\varepsilon}{2}|||v_h|||^2.
\end{aligned}$$

The above and (5.164) imply that

$$\begin{aligned}
&2\left(\sum_{i=0}^{k}\alpha_i\zeta^{l+i}, v_h\right) + 2\tau A_h(\zeta^{l+k}, v_h)\\
&\le \tau\|v_h\|^2_{L^2(\Omega)} + \tau\frac{K}{\varepsilon}\sum_{i=0}^{k-1}\|\zeta^{l+i}\|^2_{L^2(\Omega)} + \frac{\tau\varepsilon}{2}|||v_h|||^2 + \tau Kq(u,\varepsilon,h,\tau,k),
\end{aligned}$$

with

$$K = 2\max\left(C_2^2, C_5^2, (k+2)C_6^2\right), \tag{5.169}$$

which proves the lemma. □

Now we are ready to formulate and prove the main results of this section. We already mentioned that a disadvantage of the k-step ($k \ge 2$) BDF schemes is the fact that the first $k-1$ steps have to be computed by a one-step method, e.g., a Runge–Kutta scheme. Therefore, we take into account this fact by including the errors of the first $k-1$ steps in the final error estimate.

5.2.3.2 Error Estimates for the 2-Step BDF-DGM in the $L^\infty(0, T; L^2(\Omega))$-norm

Theorem 5.31 *Let Assumptions 4.5 from Sect. 4.3 be satisfied and let $\partial\Omega_N = \emptyset$. Let u be the exact solution of problem (5.68) satisfying (5.69). Let $t_l = l\tau$, $l = 0, 1, \ldots, r$, $\tau = T/r$, be a time partition of $[0, T]$, let u_h^l, $l = 0, \ldots, r$, be the approximate solution defined by the k-step BDF-DG scheme (5.75) with $k = 2$ and let $\tau \leq 1$. Then there exists a constant $\widetilde{C}_2 = O\left(\exp(3GT(1 + 2K/\varepsilon))\right)$ independent of h and τ such that*

$$\|e\|^2_{h,\tau,L^\infty(L^2)} \leq \widetilde{C}_2\left((h^{2\mu} + \tau^4)(1 + 1/\varepsilon) + \sum_{j=0}^{1} \|e_h^j\|^2_{L^2(\Omega)}\right), \tag{5.170}$$

where K is defined by (5.169) and G by (5.115).

Proof If we choose $\widetilde{C}_2 \geq 1$, then the inequality

$$\|e_h^l\|^2_{L^2(\Omega)} \leq \widetilde{C}_2\left((h^{2\mu} + \tau^4)(1 + 1/\varepsilon) + \sum_{j=0}^{1} \|e_h^j\|^2_{L^2(\Omega)}\right) \tag{5.171}$$

holds for $l = 0, 1$. Hence, we have to prove (5.171) for $l = 2, \ldots, r$. Putting $v_h = \zeta^{l+2}$ in (5.163) and using (5.84) and the relations

$$\begin{aligned}
&2\left(\frac{3}{2}\zeta^{l+2} - 2\zeta^{l+1} + \frac{1}{2}\zeta^l, \zeta^{l+2}\right)\\
&= \frac{3}{2}\|\zeta^{l+2}\|^2_{L^2(\Omega)} - 2\|\zeta^{l+1}\|^2_{L^2(\Omega)} + \frac{1}{2}\|\zeta^l\|^2_{L^2(\Omega)}\\
&\quad + 2\|\zeta^{l+2} - \zeta^{l+1}\|^2_{L^2(\Omega)} - \frac{1}{2}\|\zeta^{l+2} - \zeta^l\|^2_{L^2(\Omega)}\\
&\geq \frac{3}{2}\|\zeta^{l+2}\|^2_{L^2(\Omega)} - 2\|\zeta^{l+1}\|^2_{L^2(\Omega)} + \frac{1}{2}\|\zeta^l\|^2_{L^2(\Omega)}\\
&\quad + 2\|\zeta^{l+2} - \zeta^{l+1}\|^2_{L^2(\Omega)} - \|\zeta^{l+2} - \zeta^{l+1}\|^2_{L^2(\Omega)} - \|\zeta^{l+1} - \zeta^l\|^2_{L^2(\Omega)}\\
&= \frac{3}{2}\|\zeta^{l+2}\|^2_{L^2(\Omega)} - 2\|\zeta^{l+1}\|^2_{L^2(\Omega)} + \frac{1}{2}\|\zeta^l\|^2_{L^2(\Omega)}\\
&\quad + \|\zeta^{l+2} - \zeta^{l+1}\|^2_{L^2(\Omega)} - \|\zeta^{l+1} - \zeta^l\|^2_{L^2(\Omega)}, \quad l = 0, \ldots, r-2,
\end{aligned}$$

we get

$$\frac{3}{2}\|\zeta^{l+2}\|^2_{L^2(\Omega)} - 2\|\zeta^{l+1}\|^2_{L^2(\Omega)} + \frac{1}{2}\|\zeta^{l}\|^2_{L^2(\Omega)} \tag{5.172}$$
$$+ \|\zeta^{l+2} - \zeta^{l+1}\|^2_{L^2(\Omega)} - \|\zeta^{l+1} - \zeta^{l}\|^2_{L^2(\Omega)} + \tau\varepsilon |||\zeta^{l+2}|||^2$$
$$\leq \tau\|\zeta^{l+2}\|^2_{L^2(\Omega)} + \frac{\tau K}{\varepsilon}\sum_{i=0}^{1}\|\zeta^{l+i}\|^2_{L^2(\Omega)} + \frac{\tau\varepsilon}{2}|||\zeta^{l+2}|||^2 + \tau K q(u, \varepsilon, h, \tau, 2).$$

Let $m \in \{2, \ldots, r\}$ be arbitrary, but fixed. By (5.172),

$$\frac{3}{2}\|\zeta^{l+2}\|^2_{L^2(\Omega)} - 2\|\zeta^{l+1}\|^2_{L^2(\Omega)} + \frac{1}{2}\|\zeta^{l}\|^2_{L^2(\Omega)} \tag{5.173}$$
$$+ \|\zeta^{l+2} - \zeta^{l+1}\|^2_{L^2(\Omega)} - \|\zeta^{l+1} - \zeta^{l}\|^2_{L^2(\Omega)}$$
$$\leq \tau\|\zeta^{l+2}\|^2_{L^2(\Omega)} + \tau\frac{K}{\varepsilon}\sum_{i=0}^{1}\|\zeta^{l+i}\|^2_{L^2(\Omega)} + \tau K q(u, \varepsilon, h, \tau, 2),$$

for $l = 0, \ldots, m-2$. Let $\gamma_j,\ j = 0, 1, \ldots,$ be defined by (5.114) for $k = 2$. It follows from (5.124) that

$$\gamma_j = 1 - (1/3)^{j+1} > 0. \tag{5.174}$$

Moreover, by virtue of Remark 5.22, the coefficients $\gamma_j,\ j = 0, 1, \ldots,$ satisfy relations (5.120), which for $k = 2$ have the following form:

$$\frac{3}{2}\gamma_1 - 2\gamma_0 = 0, \quad \frac{3}{2}\gamma_{j+2} - 2\gamma_{j+1} + \frac{1}{2}\gamma_j = 0, \ j = 0, 1, \ldots . \tag{5.175}$$

Multiplying (5.173) by γ_{m-l-2} and summing over $l = 0, \ldots, m-2$, we get

$$\|\zeta^{m}\|^2_{L^2(\Omega)} + (\frac{1}{2}\gamma_{m-3} - 2\gamma_{m-2})\|\zeta^{1}\|^2_{L^2(\Omega)} + \frac{1}{2}\gamma_{m-2}\|\zeta^{0}\|^2_{L^2(\Omega)} \tag{5.176}$$
$$+ \frac{2}{3}\|\zeta^{m} - \zeta^{m-1}\|^2_{L^2(\Omega)} + 2\sum_{l=0}^{m-3}\left(\frac{1}{3}\right)^{l+2}\|\zeta^{m-l-1} - \zeta^{m-l-2}\|^2_{L^2(\Omega)}$$
$$- \gamma_{m-2}\|\zeta^{1} - \zeta^{0}\|^2_{L^2(\Omega)}$$
$$\leq \tau\gamma_0\|\zeta^{m}\|^2_{L^2(\Omega)} + \tau\sum_{l=0}^{m-3}\gamma_{m-2-l}\|\zeta^{l+2}\|^2_{L^2(\Omega)} + \tau K\sum_{l=0}^{m-2}\gamma_{m-2-l}q(u, \varepsilon, h, \tau, 2)$$
$$+ \tau\frac{K}{\varepsilon}\sum_{l=0}^{m-2}\gamma_{m-2-l}\left(\|\zeta^{l+1}\|^2_{L^2(\Omega)} + \|\zeta^{l}\|^2_{L^2(\Omega)}\right).$$

By (5.174), $\gamma_0 = 2/3$. Since $\tau \le 1$, we have $0 < 1-\tau\gamma_0 < 1$. Using (5.115), putting some terms together and omitting some non-negative terms from the left-hand side of (5.176), we obtain

$$(1-\tau\gamma_0)\|\zeta^m\|^2_{L^2(\Omega)} \le \tau G\left(1+\frac{2K}{\varepsilon}\right)\sum_{l=0}^{m-1}\|\zeta^l\|^2_{L^2(\Omega)} + 4G\|\zeta^1\|^2_{L^2(\Omega)} + \frac{3}{2}G\|\zeta^0\|^2_{L^2(\Omega)} + TGKq(u,\varepsilon,h,\tau,2). \tag{5.177}$$

If we set

$$X = G\frac{2K/\varepsilon+1}{1-\tau\gamma_0},\quad Y = G\frac{\frac{3}{2}\|\zeta^0\|^2_{L^2(\Omega)} + 4\|\zeta^1\|^2_{L^2(\Omega)}}{1-\tau\gamma_0},\quad Z = \frac{TGK}{1-\tau\gamma_0}, \tag{5.178}$$

we can write (5.177) in the form

$$\|\zeta^m\|^2_{L^2(\Omega)} \le \tau X\sum_{l=0}^{m-1}\|\zeta^l\|^2_{L^2(\Omega)} + Y + Zq(u,\varepsilon,h,\tau,2),\ m = 2,\dots,r. \tag{5.179}$$

It follows from (5.178) and the inequalities $0 < 1-\tau\gamma_0 < 1$ that (5.179) is valid also for $m = 0, 1$.

Now we apply the discrete Gronwall Lemma 1.11, where we set

$$\begin{aligned} x_m &= \|\zeta^m\|^2_{L^2(\Omega)}, \quad c_m = 0, \\ a_m &= Y + Zq(\varepsilon,h,\tau,2), \quad b_j = \tau X, \end{aligned} \tag{5.180}$$

and use the inequality $(1+\tau X)^m \le \exp(m\tau X) \le \exp(TX)$. Then we get

$$\|\zeta^m\|^2_{L^2(\Omega)} \le (Y + Zq(\varepsilon,h,\tau,2))\mathrm{e}^{TX}. \tag{5.181}$$

Further, from this inequality, the definitions (5.178) of X, Y, and Z and the inequality $(1-\tau\gamma_0)^{-1} \le 3$, following from $\tau \le 1$ and $\gamma_0 = 2/3$ for $k = 2$, we obtain the estimate

$$\begin{aligned} \|\zeta^m\|^2_{L^2(\Omega)} &\le (Y + Zq(u,\varepsilon,h,\tau,2))e^{TX} \\ &\le \left(\frac{9}{2}G\|\zeta^0\|^2_{L^2(\Omega)} + 12G\|\zeta^1\|^2_{L^2(\Omega)} + 3TGKq(u,\varepsilon,h,\tau,2)\right) \\ &\quad e^{3TG(\frac{2K}{\varepsilon}+1)}, \end{aligned} \tag{5.182}$$

valid for $m = 0,\dots,r$. From (5.133), (5.88) and (5.139), for $l = 0,\dots,r$, we have

$$\|\zeta^l\|^2_{L^2(\Omega)} \le 2\|e_h^l\|^2_{L^2(\Omega)} + 2\|\chi^l\|^2_{L^2(\Omega)}, \tag{5.183}$$

$$\|\chi^l\|^2_{L^2(\Omega)} \le C^2_{P,L} h^{2\mu} |u^l|^2_{H^\mu(\Omega)} \le C^2_{P,L} \|u\|^2_d h^{2\mu}, \tag{5.184}$$

and, therefore

$$\|\zeta^l\|^2_{L^2(\Omega)} \le 2\|e^l_h\|^2_{L^2(\Omega)} + 2C^2_{P,L} h^{2\mu} \|u\|^2_d, \quad l = 0, 1. \tag{5.185}$$

Now, using (5.131), (5.182), (5.184) and (5.185), we find that

$$\begin{aligned} \|e\|^2_{h,\tau,L^\infty(L^2)} &\le 2 \max_{l=0,\dots,r} \left(\|\zeta^l\|^2_{L^2(\Omega)} + \|\chi^l\|^2_{L^2(\Omega)} \right) \\ &\le 2\Big(9G\|e^0_h\|^2_{L^2(\Omega)} + 24G\|e^1_h\|^2_{L^2(\Omega)} + C_8 h^{2\mu}\|u\|^2_d \\ &\quad + 3TGKq(u, \varepsilon, h, \tau, 2)\Big) e^{3GT(\frac{2K}{\varepsilon}+1)} \end{aligned} \tag{5.186}$$

with $C_8 = C^2_{P,L}(1 + 33G)$, which implies estimate (5.170). □

5.2.3.3 Error Estimates of the 3-Step BDF-DGM in the $L^\infty(0, T; L^2(\Omega))$-norm

First, we derive the following estimate.

Lemma 5.32 *Let Assumptions 4.5 in Sect. 4.3 be satisfied and let $\partial\Omega_N = \emptyset$. Let u be the exact solution of problem (5.68) satisfying (5.69). Let $t_l = l\tau$, $l = 0, 1, \dots, r$, $\tau = T/r$, be a time partition of $[0, T]$, let u^l_h, $l = 0, \dots, r$, be the approximate solution defined by the k-step BDF-DG scheme (5.75) with $k = 3$ and let $\tau \le 1$. Then for $m = 3, \dots, r$ we have*

$$\begin{aligned} &\sum_{l=2}^{m-1} \|\zeta^l - \zeta^{l-1}\|^2_{L^2(\Omega)} \\ &\le \tau\left(3 + \frac{9K}{4\varepsilon}\right) \sum_{l=0}^{m-1} \|\zeta^l\|^2_{L^2(\Omega)} + \frac{3}{4} TKq(u, \varepsilon, h, \tau, 3) \\ &\quad + \frac{23}{4}\|\zeta^2\|^2_{L^2(\Omega)} + \frac{29}{4}\|\zeta^1\|^2_{L^2(\Omega)} + \frac{3}{2}\|\zeta^0\|^2_{L^2(\Omega)} + \frac{3}{4}\tau c C_B |||\zeta^2|||^2, \end{aligned} \tag{5.187}$$

where K and q are defined by (5.169) and (5.164), respectively.

Proof If we set $v_h = \zeta^{l+3} - \zeta^{l+2}$ in (5.163) and use the notation $y^l = \zeta^l - \zeta^{l-1}$ for $l = 1, \dots, r$, we get

$$2\left(\frac{11}{6}\zeta^{l+3} - \frac{18}{6}\zeta^{l+2} + \frac{9}{6}\zeta^{l+1} - \frac{2}{6}\zeta^l, y^{l+3}\right) + 2\tau A_h(\zeta^{l+3}, y^{l+3}) \tag{5.188}$$

$$\le \tau\|y^{l+3}\|^2_{L^2(\Omega)} + \frac{\tau K}{\varepsilon}\sum_{i=0}^{2}\|\zeta^{l+i}\|^2_{L^2(\Omega)} + \frac{\tau\varepsilon}{2}|||y^{l+3}|||^2 + \tau K q(u,\varepsilon,h,\tau,3).$$

Using the following relations

$$\begin{aligned}
&2\left(\frac{11}{6}\zeta^{l+3} - \frac{18}{6}\zeta^{l+2} + \frac{9}{6}\zeta^{l+1} - \frac{2}{6}\zeta^{l}, y^{l+3}\right)\\
&= 2\left(\frac{11}{6}y^{l+3} - \frac{7}{6}y^{l+2} + \frac{2}{6}y^{l+1}, y^{l+3}\right)\\
&= \frac{17}{6}\|y^{l+3}\|^2_{L^2(\Omega)} - \frac{7}{6}\|y^{l+2}\|^2_{L^2(\Omega)} + \frac{2}{6}\|y^{l+1}\|^2_{L^2(\Omega)}\\
&\quad + \frac{7}{6}\|y^{l+3} - y^{l+2}\|^2_{L^2(\Omega)} - \frac{2}{6}\|y^{l+3} - y^{l+1}\|^2_{L^2(\Omega)}\\
&\ge \frac{17}{6}\|y^{l+3}\|^2_{L^2(\Omega)} - \frac{9}{6}\|y^{l+2}\|^2_{L^2(\Omega)} + \frac{3}{6}\|y^{l+3} - y^{l+2}\|^2_{L^2(\Omega)}\\
&\quad - \frac{3}{6}\|y^{l+2} - y^{l+1}\|^2_{L^2(\Omega)},
\end{aligned}$$

together with the identity

$$2A_h(\zeta^{l+3}, y^{l+3}) = A_h(\zeta^{l+3}, \zeta^{l+3}) - A_h(\zeta^{l+2}, \zeta^{l+2}) + A_h(y^{l+3}, y^{l+3}), \quad (5.189)$$

obtained with the aid of the symmetry and linearity of the form A_h, and the coercivity property (5.84) of the term $A_h(y^{l+3}, y^{l+3})$, from (5.188) we get

$$\begin{aligned}
&\frac{17}{6}\|y^{l+3}\|^2_{L^2(\Omega)} - \frac{9}{6}\|y^{l+2}\|^2_{L^2(\Omega)} + \frac{3}{6}\|y^{l+3} - y^{l+2}\|^2_{L^2(\Omega)} \qquad (5.190)\\
&\quad - \frac{3}{6}\|y^{l+2} - y^{l+1}\|^2_{L^2(\Omega)} + \tau\left(A_h(\zeta^{l+3}, \zeta^{l+3}) - A_h(\zeta^{l+2}, \zeta^{l+2})\right)\\
&\le 2\tau\left(\|\zeta^{l+3}\|^2_{L^2(\Omega)} + \|\zeta^{l+2}\|^2_{L^2(\Omega)}\right) + \tau\frac{K}{\varepsilon}\sum_{i=0}^{2}\|\zeta^{l+i}\|^2_{L^2(\Omega)} + \tau K q(u,\varepsilon,h,\tau,3).
\end{aligned}$$

Let $m \in \{2, \ldots, r\}$. After the summation of (5.190) over $l = 0, \ldots, m-4$ we find that

$$\begin{aligned}
&\frac{8}{6}\sum_{l=0}^{m-4}\|y^{l+3}\|^2_{L^2(\Omega)} + \frac{9}{6}\|y^{m-1}\|^2_{L^2(\Omega)} - \frac{9}{6}\|y^{2}\|^2_{L^2(\Omega)} + \frac{3}{6}\|y^{m-1} - y^{m-2}\|^2_{L^2(\Omega)}\\
&\quad - \frac{3}{6}\|y^{2} - y^{1}\|^2_{L^2(\Omega)} + \tau\left(A_h(\zeta^{m-1}, \zeta^{m-1}) - A_h(\zeta^{2}, \zeta^{2})\right) \qquad (5.191)\\
&\le 2\tau\sum_{l=0}^{m-4}\left(\|\zeta^{l+3}\|^2_{L^2(\Omega)} + \|\zeta^{l+2}\|^2_{L^2(\Omega)}\right) + 3\tau\frac{K}{\varepsilon}\sum_{l=0}^{m-2}\|\zeta^{l}\|^2_{L^2(\Omega)}
\end{aligned}$$

$$+ TKq(u, \varepsilon, h, \tau, 3) \le \tau \left(4 + 3\frac{K}{\varepsilon}\right) \sum_{l=0}^{m-1} \|\zeta^l\|^2_{L^2(\Omega)} + TKq(u, \varepsilon, h, \tau, 3).$$

Due to the coercivity property (5.84), we remove the term $A_h(\zeta^{m-1}, \zeta^{m-1}) \ge 0$ from the left-hand side of (5.191). Moreover, using inequality (5.85) with $\varphi = \psi := \zeta^2$, omitting some non-negative terms on the left-hand side of (5.191), and transferring the initial terms to the right-hand side, we get

$$\begin{aligned}
\frac{8}{6}\sum_{l=2}^{m-1} \|\zeta^l - \zeta^{l-1}\|^2_{L^2(\Omega)} &= \frac{8}{6}\sum_{l=2}^{m-1} \|y^l\|^2_{L^2(\Omega)} \\
&\le \frac{8}{6}\|y^2\|^2_{L^2(\Omega)} + \tau\left(4 + 3\frac{K}{\varepsilon}\right)\sum_{l=0}^{m-1} \|\zeta^l\|^2_{L^2(\Omega)} + TKq(u, \varepsilon, h, \tau, 3) \\
&\quad + \frac{9}{6}\|y^2\|^2_{L^2(\Omega)} + \frac{3}{6}\|y^2 - y^1\|^2_{L^2(\Omega)} + \tau\varepsilon C_B |||\zeta^2|||^2 \\
&\le \tau\left(4 + 3\frac{K}{\varepsilon}\right)\sum_{l=0}^{m-1} \|\zeta^l\|^2_{L^2(\Omega)} + TKq(u, \varepsilon, h, \tau, 3) + \frac{23}{3}\|\zeta^2\|^2_{L^2(\Omega)} \\
&\quad + \frac{29}{3}\|\zeta^1\|^2_{L^2(\Omega)} + 2\|\zeta^0\|^2_{L^2(\Omega)} + \tau\varepsilon C_B |||\zeta^2|||^2,
\end{aligned}$$

which implies inequality (5.187). □

Now, we formulate the $L^\infty(L^2)$-error estimate of the three step method.

Theorem 5.33 *Let Assumptions 4.5 in Sect. 4.3 be satisfied and let $\partial\Omega_N = \emptyset$. Let u be the exact solution of problem (5.68) satisfying (5.69). Let $t_l = l\tau$, $l = 0, 1, \ldots, r$, $\tau = T/r$, be a partition of the time interval $[0, T]$, let u_h^l, $l = 0, \ldots, r$, be defined by the k-step BDF-DG scheme (5.75) with $k = 3$ and let $\tau \le 1$. Then there exists a constant $\widetilde{C}_3 = O(\exp(GT(30 + 117K/4\varepsilon)))$ such that*

$$\|e\|^2_{h,\tau,L^\infty(L^2)} \le \widetilde{C}_3 \left((h^{2\mu} + \tau^6)(1 + 1/\varepsilon) + \sum_{l=0}^{2} \|e_h^l\|^2_{L^2(\Omega)} + \tau\varepsilon |||\zeta^2|||^2 \right), \tag{5.192}$$

where K is defined by (5.169) and $\zeta^2 = u_h^2 - \Pi_{hp}u^2$ is given by (5.132).

Proof By virtue of (5.131), the aim is to prove the inequality

$$\|e_h^m\|^2_{L^2(\Omega)} \le \widetilde{C}_3 \left((h^{2\mu} + \tau^6)\left(1 + \frac{1}{\varepsilon}\right) + \sum_{l=0}^{2} \|e_h^l\|^2_{L^2(\Omega)} + \tau\varepsilon |||\zeta^2|||^2 \right) \tag{5.193}$$

for $m = 0, 1, \ldots, r$. Putting $v_h = \zeta^{l+3}$ and $k = 3$ in (5.163), we get

$$2\left(\frac{11}{6}\zeta^{l+3}-\frac{18}{6}\zeta^{l+2}+\frac{9}{6}\zeta^{l+1}-\frac{2}{6}\zeta^{l},\zeta^{l+3}\right)+2\tau A_h(\zeta^{l+3},\zeta^{l+3}) \tag{5.194}$$
$$\leq \tau\|\zeta^{l+3}\|^2_{L^2(\Omega)}+\frac{\tau K}{\varepsilon}\sum_{i=0}^{2}\|\zeta^{l+i}\|^2_{L^2(\Omega)}+\frac{\tau\varepsilon}{2}|||\zeta^{l+3}|||^2+\tau Kq(u,\varepsilon,h,\tau,3),$$

where by (5.164), $q(u,\varepsilon,h,\tau,3)=(h^{2\mu}+\tau^6)(1+1/\varepsilon)$. With the aid of (5.84) applied to the term $A_h(\zeta^{l+3},\zeta^{l+3})$ and the relations

$$\begin{aligned}
&2\left(\frac{11}{6}\zeta^{l+3}-\frac{18}{6}\zeta^{l+2}+\frac{9}{6}\zeta^{l+1}-\frac{2}{6}\zeta^{l},\zeta^{l+3}\right) \qquad (5.195)\\
&=\frac{11}{6}\|\zeta^{l+3}\|^2_{L^2(\Omega)}-\frac{18}{6}\|\zeta^{l+2}\|^2_{L^2(\Omega)}+\frac{9}{6}\|\zeta^{l+1}\|^2_{L^2(\Omega)}-\frac{2}{6}\|\zeta^{l}\|^2_{L^2(\Omega)}\\
&+\frac{18}{6}\|\zeta^{l+3}-\zeta^{l+2}\|^2_{L^2(\Omega)}-\frac{9}{6}\|\zeta^{l+3}-\zeta^{l+1}\|^2_{L^2(\Omega)}\\
&+\frac{2}{6}\|\zeta^{l+3}-\zeta^{l}\|^2_{L^2(\Omega)}\\
&\geq\frac{11}{6}\|\zeta^{l+3}\|^2_{L^2(\Omega)}-\frac{18}{6}\|\zeta^{l+2}\|^2_{L^2(\Omega)}+\frac{9}{6}\|\zeta^{l+1}\|^2_{L^2(\Omega)}-\frac{2}{6}\|\zeta^{l}\|^2_{L^2(\Omega)}\\
&-\frac{18}{6}\|\zeta^{l+2}-\zeta^{l+1}\|^2_{L^2(\Omega)},
\end{aligned}$$

we arrive at

$$\begin{aligned}
&\frac{11}{6}\|\zeta^{l+3}\|^2_{L^2(\Omega)}-\frac{18}{6}\|\zeta^{l+2}\|^2_{L^2(\Omega)}+\frac{9}{6}\|\zeta^{l+1}\|^2_{L^2(\Omega)}-\frac{2}{6}\|\zeta^{l}\|^2_{L^2(\Omega)}\\
&\leq\tau\|\zeta^{l+3}\|^2_{L^2(\Omega)}+\tau\frac{K}{\varepsilon}\sum_{i=0}^{2}\|\zeta^{l+i}\|^2_{L^2(\Omega)}+3\|\zeta^{l+2}-\zeta^{l+1}\|^2_{L^2(\Omega)}\\
&+\tau Kq(u,\varepsilon,h,\tau,3). \qquad (5.196)
\end{aligned}$$

Let $\gamma_j,\ j=0,1,\ldots,$ be the sequence defined by (5.114) for $k=3$. Due to Lemmas 5.21 and 5.23, the bounds (5.115) and (5.121) are valid, respectively. Moreover, by Remark 5.22, the coefficients $\gamma_j,\ j=0,1,\ldots,$ satisfy relations (5.120) (cf. also Table 5.1). For $k=3$ they have the following form:

$$\begin{aligned}
\gamma_0&=\frac{6}{11}, \qquad (5.197)\\
\frac{11}{6}\gamma_1-\frac{18}{6}\gamma_0&=0,\\
\frac{11}{6}\gamma_2-\frac{18}{6}\gamma_1+\frac{9}{6}\gamma_0&=0,\\
\frac{11}{6}\gamma_{j+3}-\frac{18}{6}\gamma_{j+2}+\frac{9}{6}\gamma_{j+1}-\frac{2}{6}\gamma_j&=0,\quad j=0,1,\ldots.
\end{aligned}$$

Multiplying (5.196) by γ_{m-3-l}, summing over $l = 0, \ldots, m-3$ and using (5.197), we find that

$$\begin{aligned}
&\|\zeta^m\|^2_{L^2(\Omega)} + \left(-\frac{18}{6}\gamma_{m-3} + \frac{9}{6}\gamma_{m-4} - \frac{2}{6}\gamma_{m-5}\right)\|\zeta^2\|^2_{L^2(\Omega)} \\
&\quad + \left(\frac{9}{6}\gamma_{m-3} - \frac{2}{6}\gamma_{m-4}\right)\|\zeta^1\|^2_{L^2(\Omega)} - \frac{2}{6}\gamma_{m-3}\|\zeta^0\|^2_{L^2(\Omega)} \\
&\le \gamma_0\tau\|\zeta^m\|^2_{L^2(\Omega)} + \tau\sum_{l=0}^{m-4}\gamma_{m-3-l}\|\zeta^{l+3}\|^2_{L^2(\Omega)} + \tau\frac{K}{\varepsilon}\sum_{l=0}^{m-3}\gamma_{m-3-l}\sum_{i=0}^{2}\|\zeta^{l+i}\|^2_{L^2(\Omega)} \\
&\quad + 3\sum_{l=0}^{m-3}\gamma_{m-3-l}\|\zeta^{l+2} - \zeta^{l+1}\|^2_{L^2(\Omega)} + \tau K\sum_{l=0}^{m-3}\gamma_{m-3-l}q(u,\varepsilon,h,\tau,3).
\end{aligned} \tag{5.198}$$

Omitting some non-negative terms of the left-hand side of (5.198) and using estimate (5.115), we obtain

$$\begin{aligned}
\|\zeta^m\|^2_{L^2(\Omega)} &\le \tau\gamma_0\|\zeta^m\|^2_{L^2(\Omega)} + \tau G\left(1 + \frac{3K}{\varepsilon}\right)\sum_{l-0}^{m-1}\|\zeta^l\|^2_{L^2(\Omega)} \\
&\quad + 3G\sum_{l=2}^{m-1}\|\zeta^l - \zeta^{l-1}\|^2_{L^2(\Omega)} + TGKq(u,\varepsilon,h,\tau,3) \\
&\quad + \frac{20}{6}G\|\zeta^2\|^2_{L^2(\Omega)} + \frac{2}{6}G\|\zeta^1\|^2_{L^2(\Omega)} + \frac{2}{6}G\|\zeta^0\|^2_{L^2(\Omega)}.
\end{aligned} \tag{5.199}$$

Using estimate (5.187) for the term $\sum_{l=2}^{m-1}\|\zeta^l - \zeta^{l-1}\|^2_{L^2(\Omega)}$, from (5.199) we get

$$\begin{aligned}
\|\zeta^m\|^2_{L^2(\Omega)} &\le \tau\gamma_0\|\zeta^m\|^2_{L^2(\Omega)} + \tau G\left(10 + \frac{39K}{4\varepsilon}\right)\sum_{l=0}^{m-1}\|\zeta^l\|^2_{L^2(\Omega)} \\
&\quad + \frac{13}{4}TGKq(u,\varepsilon,h,\tau,3) + \frac{9}{4}\tau\varepsilon C_B G|||\zeta^2|||^2 \\
&\quad + \frac{247}{12}G\|\zeta^2\|^2_{L^2(\Omega)} + \frac{265}{12}G\|\zeta^1\|^2_{L^2(\Omega)} + \frac{29}{6}G\|\zeta^0\|^2_{L^2(\Omega)}.
\end{aligned} \tag{5.200}$$

If we use the notation

$$\begin{aligned}
X &= G\frac{\frac{39K}{4\varepsilon} + 10}{1 - \tau\gamma_0}, \quad Z = \frac{13}{4}\frac{TGK}{1 - \tau\gamma_0}, \\
Y &= \frac{G}{1 - \tau\gamma_0}\left(\frac{9}{4}\tau\varepsilon C_B|||\zeta^2|||^2 + \frac{247}{12}\|\zeta^2\|^2_{L^2(\Omega)} + \frac{265}{12}\|\zeta^1\|^2_{L^2(\Omega)} + \frac{29}{6}\|\zeta^0\|^2_{L^2(\Omega)}\right),
\end{aligned}$$

we can write (5.200) in the form

$$\|\zeta^m\|^2_{L^2(\Omega)} \le \tau X \sum_{l=0}^{m-1} \|\zeta^l\|^2_{L^2(\Omega)} + Y + Zq(u, \varepsilon, h, \tau, 3), \quad m = 3, \dots, r. \tag{5.201}$$

Now, in the same way as in the proof of Theorem 5.31, with the aid of the discrete Gronwall Lemma 1.11 we derive the estimate

$$\|\zeta^m\|^2_{L^2(\Omega)} \le (Y + Zq(u, \varepsilon, h, \tau, 3))\mathrm{e}^{TX}, \ m = 3, \dots, r. \tag{5.202}$$

This inequality and the inequality $(1 - \tau\gamma_0)^{-1} \le 3$ (which is also valid for $k = 3$ since $\gamma_0 = \frac{6}{11}$ and $\tau \le 1$), imply that

$$\begin{aligned} \|\zeta^m\|^2_{L^2(\Omega)} &\le (Y + Zq(u, \varepsilon, h, \tau, 3))e^{TX} \\ &\le \Big(\frac{29}{2} G\|\zeta^0\|^2_{L^2(\Omega)} + \frac{265}{4} G\|\zeta^1\|^2_{L^2(\Omega)} + \frac{247}{4} G\|\zeta^2\|^2_{L^2(\Omega)} \\ &\quad + \frac{27}{4}\tau\varepsilon G C_B |||\zeta^2|||^2 + \frac{39}{4} TGKq(u, \varepsilon, h, \tau, 3)\Big) e^{TG(117K/4\varepsilon+30)}, \ m = 3, \dots, r. \end{aligned} \tag{5.203}$$

By (5.133), we have $\|e_h^m\|^2_{L^2(\Omega)} \le 2\|\zeta^m\|^2_{L^2(\Omega)} + 2\|\chi^m\|^2_{L^2(\Omega)}$. Now, (5.184), (5.185), and (5.203) imply

$$\begin{aligned} \|e_h^m\|^2_{L^2(\Omega)} &\le 2\left(\|\zeta^m\|^2_{L^2(\Omega)} + \|\chi^m\|^2_{L^2(\Omega)}\right) \\ &\le 2\Big(29G\|e_h^0\|^2_{L^2(\Omega)} + \frac{265}{2} G\|e_h^1\|^2_{L^2(\Omega)} + \frac{247}{2} G\|e_h^2\|^2_{L^2(\Omega)} + \frac{27}{4}\tau\varepsilon G C_B |||\zeta^2|||^2 \\ &\quad + \frac{39}{4} TGKq(u, \varepsilon, h, \tau, 3) + C^2_{P,L} h^{2\mu} \|u\|^2_d\Big) e^{TG(117K/4\varepsilon+30)}, \quad m = 3, \dots, r, \end{aligned}$$

which gives (5.192) with

$$\widetilde{C}_3 := \max\left(265\, G, \left(\frac{39}{2} TGK + C_{P,L}\right) \|u\|^2_d + \frac{27}{2}\right) e^{TG(117K/4\varepsilon+30)}$$

for $m = 3, \dots, r$. Obviously, $\widetilde{C}_3 \ge 1$ and, hance (5.193) is satisfied also for $m = 0, 1, 2$, which proves the theorem. □

5.2.3.4 Error Estimates for the 2- and 3-Step BDF-DGM in the $L^2(0, T; H^1(\Omega, \mathscr{T}_h))$-norm

Theorem 5.34 *Let Assumptions 4.5 in Sect. 4.3 be satisfied and let* $\partial\Omega_N = \emptyset$. *Let* u *be the exact solution of problem (5.68) satisfying (5.69). Let* $t_l = l\tau$, $l = 0, 1, \dots, r$, $\tau = T/r$, *be a partition of* $[0, T]$ *and let* u_h^l, $l = 0, \dots, r$, *be the* *approximate solution defined by the k-step BDF-DG scheme (5.75) with $k = 2, 3$. Then there exists a constant $\widehat{C}$ such that*

$$
\begin{aligned}
&\|e\|^2_{h,\tau,L^2(H^1)} \\
&\le \widehat{C}\left(\varepsilon h^{2(\mu-1)} + (1+1/\varepsilon)^2(h^{2\mu} + \tau^{2k}) + (1+1/\varepsilon)\sum_{j=0}^{k-1}\left(\|e_h^j\|_{L^2(\Omega)} + \tau\varepsilon |||e_h^j|||^2\right)\right).
\end{aligned}
\tag{5.204}
$$

Proof (i) For $k = 2$, we can use relation (5.172), i.e.,

$$
\begin{aligned}
&\frac{3}{2}\|\zeta^{l+2}\|^2_{L^2(\Omega)} - 2\|\zeta^{l+1}\|^2_{L^2(\Omega)} + \frac{1}{2}\|\zeta^l\|^2_{L^2(\Omega)} + \|\zeta^{l+2} - \zeta^{l+1}\|^2_{L^2(\Omega)} \\
&\quad - \|\zeta^{l+1} - \zeta^l\|^2_{L^2(\Omega)} + \frac{\tau\varepsilon}{2}|||\zeta^{l+2}|||^2 \\
&\le \tau\|\zeta^{l+2}\|^2_{L^2(\Omega)} + \tau\frac{K}{\varepsilon}\sum_{i=0}^{1}\|\zeta^{l+i}\|^2_{L^2(\Omega)} + \tau K q(u, \varepsilon, h, \tau, 2), \quad l = 0, \ldots, r-2.
\end{aligned}
\tag{5.205}
$$

Now, after summing (5.205) over $l = 0, \ldots, r-2$, we obtain

$$
\begin{aligned}
&\frac{3}{2}\|\zeta^r\|^2_{L^2(\Omega)} - \frac{1}{2}\|\zeta^{r-1}\|^2_{L^2(\Omega)} - \frac{3}{2}\|\zeta^1\|^2_{L^2(\Omega)} + \frac{1}{2}\|\zeta^0\|^2_{L^2(\Omega)} \\
&\quad + \|\zeta^r - \zeta^{r-1}\|^2_{L^2(\Omega)} - \|\zeta^1 - \zeta^0\|^2_{L^2(\Omega)} + \frac{\tau\varepsilon}{2}\sum_{l=2}^{r}|||\zeta^l|||^2 \\
&\le \tau\sum_{l=2}^{r}\|\zeta^l\|^2_{L^2(\Omega)} + \tau\frac{K}{\varepsilon}\sum_{l=1}^{r-1}\left(\|\zeta^l\|^2_{L^2(\Omega)} + \|\zeta^{l-1}\|^2_{L^2(\Omega)}\right) + TKq(u, \varepsilon, h, \tau, 2),
\end{aligned}
$$

which implies that

$$
\begin{aligned}
\frac{\tau\varepsilon}{2}\sum_{l=2}^{r}|||\zeta^l|||^2 &\le \frac{1}{2}\|\zeta^{r-1}\|^2_{L^2(\Omega)} + \frac{7}{2}\|\zeta^1\|^2_{L^2(\Omega)} + \frac{3}{2}\|\zeta^0\|^2_{L^2(\Omega)} \\
&\quad + \tau\sum_{l=2}^{r}\|\zeta^l\|^2_{L^2(\Omega)} + \tau\frac{2K}{\varepsilon}\sum_{l=0}^{r-1}\|\zeta^l\|^2_{L^2(\Omega)} + TKq(u, \varepsilon, h, \tau, 2).
\end{aligned}
\tag{5.206}
$$

Using (5.185), from (5.206) we get

$$
\begin{aligned}
\frac{\tau\varepsilon}{2}\sum_{l=2}^{r}|||\zeta^l|||^2 &\le \left(11 + 2T + 4T\frac{K}{\varepsilon}\right)\left(\|e\|^2_{h,\tau,L^\infty(L^2)} + C^2_{P,L}h^{2\mu}\|u\|^2_d\right) \\
&\quad + TKq(u, \varepsilon, h, \tau, 2)
\end{aligned}
$$

and therefore,

$$\frac{\tau\varepsilon}{2}\sum_{l=0}^{r}|||\zeta^l|||^2 \le \left(11+2T+4T\frac{K}{\varepsilon}\right)\left(\|e\|^2_{h,\tau,L^\infty(L^2)} + C^2_{P,\mathrm{L}}h^{2\mu}\|u\|^2_d\right)$$
$$+\,TKq(u,\varepsilon,h,\tau,2) + \frac{\tau\varepsilon}{2}\sum_{j=0}^{1}|||\zeta^j|||^2. \quad (5.207)$$

By (5.133), we have $|||\zeta^l|||^2 \le 2|||e_h^l|||^2 + 2|||\chi^l|||^2$. This, (5.87) and (5.139) yield

$$|||\zeta^l|||^2 \le 2|||e_h^l|||^2 + 2C^2_{P,\mathrm{e}}h^{2(\mu-1)}\|u\|^2_d, \quad l=0,1,\dots,r, \quad (5.208)$$

Inequalities (5.207) and (5.208) imply that

$$\frac{\tau\varepsilon}{2}\sum_{l=0}^{r}|||\zeta^l|||^2 \le \left(11+2T+4T\frac{K}{\varepsilon}\right)\left(\|e\|^2_{h,\tau,L^\infty(L^2)} + C^2_{P,\mathrm{L}}h^{2\mu}\|u\|^2_d\right) \quad (5.209)$$
$$+\,TKq(u,\varepsilon,h,\tau,2) + 2\tau\varepsilon C^2_{P,\mathrm{e}}h^{2(\mu-1)}\|u\|^2_d + \tau\varepsilon\sum_{j=0}^{1}|||e_h^j|||^2.$$

Further, summing (5.87) over $l=0,\dots,r$, we get

$$\tau\varepsilon\sum_{l=0}^{r}|||\chi^l|||^2 \le \varepsilon C^2_{P,\mathrm{e}}h^{2(\mu-1)}\|u\|^2_d(T+\tau) \le 2\varepsilon C^2_{P,\mathrm{e}}Th^{2(\mu-1)}\|u\|^2_d. \quad (5.210)$$

Finally, let us set $C_9 = \max(11+2T, 4TK)$. Then, from (5.131), (5.133), (5.209) and (5.170) it follows that

$$\|e\|^2_{h,\tau,L^2(H^1)} \le 2\tau\varepsilon\sum_{l=0}^{r}\left(|||\zeta^l|||^2 + |||\chi^l|||^2\right) \quad (5.211)$$
$$\le 4C_9(1+1/\varepsilon)\left(\widetilde{C}_2(h^{2\mu}+\tau^4)(1+1/\varepsilon) + C^2_{P,\mathrm{L}}h^{2\mu}\|u\|^2_d\right)$$
$$+\,4TKq(u,\varepsilon,h,\tau,2) + 8\tau\varepsilon C^2_{P,\mathrm{e}}h^{2(\mu-1)}\|u\|^2_d + 4\varepsilon C^2_{P,\mathrm{e}}Th^{2(\mu-1)}\|u\|^2_d$$
$$+\,4C_9(1+1/\varepsilon)\widetilde{C}_2\sum_{j=0}^{1}\left(\|e_h^j\|^2_{L^2(\Omega)} + \tau\varepsilon|||e_h^j|||^2\right).$$

Now, for $k=2$, assertion (5.204) of the theorem follows from (5.164) and (5.211) with $\widehat{C} = O\left(\exp(3GT(1+2K/\varepsilon))\right)$ since $\widetilde{C}_2 = O(\exp(3GT(1+2K/\varepsilon)))$.

(ii) For $k=3$, we start from inequality (5.194). Using (5.195), the coercivity (5.84) applied to the term $A_h(\zeta^{l+3},\zeta^{l+3})$ and summing over $l=0,\dots,r-3$, we find that

$$
\begin{aligned}
&\frac{11}{6}\|\zeta^r\|^2_{L^2(\Omega)} - \frac{7}{6}\|\zeta^{r-1}\|^2_{L^2(\Omega)} + \frac{2}{6}\|\zeta^{r-2}\|^2_{L^2(\Omega)} - \frac{11}{6}\|\zeta^2\|^2_{L^2(\Omega)} \\
&\quad + \frac{7}{6}\|\zeta^1\|^2_{L^2(\Omega)} - \frac{2}{6}\|\zeta^0\|^2_{L^2(\Omega)} + \frac{\tau\varepsilon}{2}\sum_{l=3}^{r}|||\zeta^l|||^2 \\
&\leq \tau\sum_{l=3}^{r}\|\zeta^l\|^2_{L^2(\Omega)} + \tau\frac{K}{\varepsilon}\sum_{l=0}^{r-3}\sum_{i=0}^{2}\|\zeta^{l+i}\|^2_{L^2(\Omega)} \qquad (5.212)\\
&\quad + 3\sum_{l=2}^{r-1}\|\zeta^l - \zeta^{l-1}\|^2_{L^2(\Omega)} + TKq(u,\varepsilon,h,\tau,3).
\end{aligned}
$$

By (5.187) and (5.212) we have

$$
\begin{aligned}
&\frac{\tau\varepsilon}{2}\sum_{l=3}^{r}|||\zeta^l|||^2 \leq \frac{7}{6}\|\zeta^{r-1}\|^2_{L^2(\Omega)} + \frac{229}{12}\|\zeta^2\|^2_{L^2(\Omega)} + \frac{247}{12}\|\zeta^1\|^2_{L^2(\Omega)} \qquad (5.213)\\
&\quad + \frac{29}{6}\|\zeta^0\|^2_{L^2(\Omega)} + \tau\sum_{l=3}^{r}\|\zeta^l\|^2_{L^2(\Omega)} + \tau\frac{3K}{\varepsilon}\sum_{l=0}^{r-1}\|\zeta^l\|^2_{L^2(\Omega)} \\
&\quad + \frac{13}{4}TKq(u,\varepsilon,h,\tau,3) + \tau\left(9+\frac{27K}{4\varepsilon}\right)\sum_{l=0}^{r-1}\|\zeta^l\|^2_{L^2(\Omega)} + \frac{9}{4}\tau\varepsilon C_B|||\zeta^2|||^2.
\end{aligned}
$$

This inequality and (5.185) imply that

$$
\begin{aligned}
&\frac{\tau\varepsilon}{2}\sum_{l=3}^{r}|||\zeta^l|||^2 \leq \frac{9}{4}\tau\varepsilon C_B|||\zeta^2|||^2 + \frac{13}{4}TKq(u,\varepsilon,h,\tau,3) \\
&\quad + 2\left(\frac{137}{3} + 10T + T\frac{39K}{4\varepsilon}\right)\left(\|e\|^2_{h,\tau,L^\infty(L^2)} + C^2_{P,\mathrm{L}}h^{2\mu}\|u\|^2_d\right),
\end{aligned}
$$

which together with (5.208) yield

$$
\begin{aligned}
&\frac{\tau\varepsilon}{2}\sum_{l=0}^{r}|||\zeta^l|||^2 \leq \frac{13}{4}TKq(u,\varepsilon,h,\tau,3) + \frac{9}{2}\tau\varepsilon C_B(C^2_{P,\mathrm{e}}h^{2(\mu-1)}\|u\|^2_d + |||e^2_h|||^2) \\
&\qquad + 2\left(\frac{137}{3} + 10T + T\frac{39K}{4\varepsilon}\right)\left(\|e\|^2_{h,\tau,L^\infty(L^2)} + C^2_{P,\mathrm{L}}h^{2\mu}\|u\|^2_d\right) \\
&\qquad + \tau\varepsilon\left(\sum_{l=0}^{2}|||e^l_h|||^2 + 3C^2_{P,\mathrm{e}}h^{2(\mu-1)}\|u\|^2_d\right). \qquad (5.214)
\end{aligned}
$$

Using the notation $C_{10} = 8\max(\frac{137}{3} + 10T, \frac{39TK}{4})$, from (5.214), (5.192) and (5.210), we obtain the estimate

$$\|e\|^2_{h,\tau,L^2(H^1)} \le 2\tau\varepsilon \sum_{l=0}^{r} (|||\zeta^l|||^2 + |||\chi^l|||^2) \tag{5.215}$$
$$\le C_{10}(1+1/\varepsilon)\left(\widetilde{C}_3(h^{2\mu}+\tau^6)(1+1/\varepsilon) + C^2_{P,\mathrm{L}} h^{2\mu}\|u\|^2_d\right)$$
$$+ 13TKq(u,\varepsilon,h,\tau,3) + 18\tau\varepsilon C_B C^2_{P,\mathrm{e}} h^{2(\mu-1)}\|u\|^2_d + 4\varepsilon C^2_{P,\mathrm{e}} T h^{2(\mu-1)}\|u\|^2_d$$
$$+ C_{10}(1+1/\varepsilon)\widetilde{C}_3\left(\sum_{l=0}^{2} \|e^l_h\|^2_{L^2(\Omega)} + \tau\varepsilon|||e^2_h|||^2\right) + 18\tau\varepsilon C_B|||e^2_h|||^2$$
$$+ 4\tau\varepsilon\left(\sum_{l=0}^{2} |||e^l_h|||^2 + 3C^2_{P,\mathrm{e}} h^{2(\mu-1)}\|u\|^2_d\right).$$

Now, for $k = 3$, the assertion (5.204) of the theorem follows from (5.164) and (5.215) with $\widehat{C} = O\,(\exp(GT(30 + 117K/4\varepsilon)))$, since $\widetilde{C}_3 = O(\exp(GT(30+117K/4\varepsilon)))$. □

Remark 5.35 We observe that estimates (5.170), (5.192) and (5.204) are optimal with respect to h as well as τ in the discrete $L^\infty(0,T;L^2(\Omega))$-norm and $L^2(0,T;H^1(\Omega,\mathscr{T}_h))$-norm.

It can be seen that these estimates are not of practical use for $\varepsilon \to 0+$, because they blow up exponentially with respect to $1/\varepsilon$. This is caused by the treatment of nonlinear terms in the error analysis. The nonlinearity of the convective terms represents a serious obstacle for obtaining a uniform error estimate with respect to $\varepsilon \to 0+$.

Remark 5.36 The proven unconditional stability may seem to be in contradiction with the Dahlquist barrier (see [162, Theorem 1.4]) which implies that the 3-step BDF method cannot be unconditionally A-stable. However, in our case, the k-step BDF scheme with $k = 2, 3$ was not applied to a general system of ODEs, but to system (5.68) arising from the space semi-discretization of (5.68) under the assumptions of the symmetry of the form A_h and some favourable properties of the form b_h, which cause that all eigenvalues of the Jacobi matrix of the corresponding ODE system lie in the stability region of the k-step BDF method with $k = 2, 3$ for any $\tau \le 1$ and $h \in (0, \bar{h})$.

Remark 5.37 The presented numerical analysis can be partly extended also to NIPG and IIPG variants of the DG method. However, the determination of error estimates for the 3-step BDF-DG method employs equality (5.189), which is not valid for NIPG and IIPG variants due to their non-symmetry. It is not clear to us whether it is possible to avoid this obstacle.

On the other hand, for the 2-step BDF-DG method, a weaker result than (5.170) can be derived for NIPG and IIPG variants, for example,

$$\|e\|^2_{h,\tau,L^\infty(L^2)} \le \tilde{C}\left((h^{2(\mu-1)} + \tau^4)(1 + 1/\varepsilon) + \sum_{j=0}^{1} \|e_h^j\|^2_{L^2(\Omega)}\right), \qquad (5.216)$$

where $\tilde{C}$ is independent of h and τ. Estimate (5.216) can also be proved in the case of mixed Dirichlet–Neumann boundary conditions, i.e., for nonempty $\partial\Omega_N$.

5.2.4 Numerical Examples

In this section we demonstrate the theoretical error estimates (5.170), (5.192) and (5.204) derived in the previous section. We try to investigate the dependence of the computational error on h and τ independently. Based on (5.170), (5.192) and (5.204) we expect that the computational error $e_{h,\tau}$ in the $L^2(\Omega)$-norm as well as the $H^1(\Omega, \mathscr{T}_h)$-seminorm depends on h and τ according to the formula

$$e_{h,\tau} \approx c_h h^{p+1} + c_\tau \tau^k, \qquad (5.217)$$

where c_h and c_τ are constants independent of h and τ.

In our numerical experiments we solve Eq. (5.68a) in $\Omega = (0, 1)^2$, $\partial\Omega = \partial\Omega_D$, $f_i(u) = u^2/2$, $i = 1, 2$, equipped with the boundary condition (5.68b) and the initial condition (5.68d).

5.2.4.1 Convergence with Respect to τ

In this case we put $\varepsilon = 0.01$, $T = 1$ and the functions u_D, u_0 and g are chosen in such a way that the exact solution has the form $u(x_1, x_2, t) = 16\,(e^{10t} - 1)/(e^{10} - 1)\, x_1 (1 - x_1)x_2(1 - x_2)$.

The computations were carried out on a fine triangular mesh having 4219 elements with a piecewise cubic approximation in space and using 6 different time steps: 1/20, 1/40, 1/80, 1/160, 1/320, 1/640. For such data setting we expect that $c_h h^{p+1} \ll c_\tau \tau^k$ and, therefore the space discretization errors are negligible. Figure 5.1 shows the computational errors at $t = T$ and the corresponding experimental order of convergence with respect to τ in the $L^2(\Omega)$-norm and the $H^1(\Omega, \mathscr{T}_h)$-seminorm for the k-step BDF scheme (5.75) with $k = 1$, $k = 2$ and $k = 3$. The expected order of convergence $O(\tau^k)$ is observed in each case. A smaller decrease of the order of convergence in the $H^1(\Omega, \mathscr{T}_h)$-seminorm for $k = 3$ and $\tau = 1/640$ is caused by the influence of the spatial discretization since in this case the statement $c_h h^{p+1} \ll c_\tau \tau^k$ is no longer valid.

5.2.4.2 Convergence with Respect to h

In this case we put $\varepsilon = 0.1$, $T = 10$ and the functions u_D, u_0 and g are chosen in such a way that the exact solution has the form $u(x_1, x_2, t) = (1 - e^{-10t})(x_1^2 + x_2^2)x_1x_2(1 - x_1)(1 - x_2)$. As we see, we have $\mu = p + 1$.

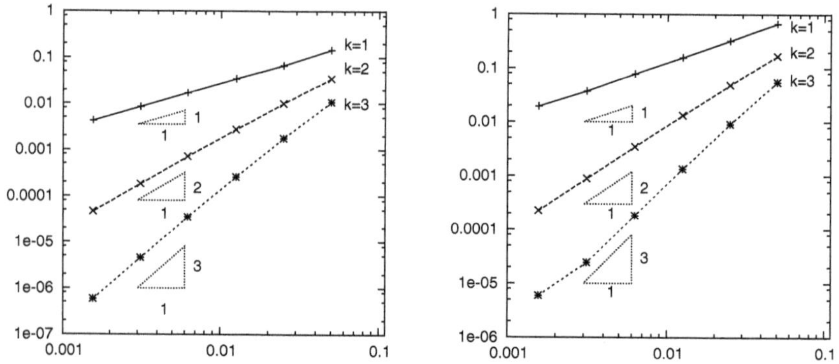

Fig. 5.1 Computational errors and orders of convergence with respect to the time step τ in the $L^2(\Omega)$-norm (*left*) and the $H^1(\Omega, \mathscr{T}_h)$-seminorm (*right*) for scheme (5.75) with $k = 1$ (*full line*), $k = 2$ (*dashed line*) and $k = 3$ (*dotted line*)

The computations were carried out with the 3-step BDF scheme (5.75) on 7 triangular meshes having 128, 288, 512, 1152, 2048, 4608 and 8192 elements, using the time step $\tau = 0.01$. For such data setting we expect that $c_h h^{p+1} \gg c_\tau \tau^k$ and the time discretization errors can be neglected. Figure 5.2 shows the computational errors at $t = T$ and the corresponding experimental order of convergence with respect to h in the $L^2(\Omega)$-norm and the $H^1(\Omega, \mathscr{T}_h)$-seminorm for piecewise linear P_1, quadratic P_2 and cubic P_3 approximations. We observe the order of convergence $O(h^{p+1})$ for $p = 1, 2, 3$ in the $L^2(\Omega)$-norm and $O(h^p)$ in the $H^1(\Omega, \mathscr{T}_h)$-seminorm, which perfectly corresponds to the theoretical results (5.204).

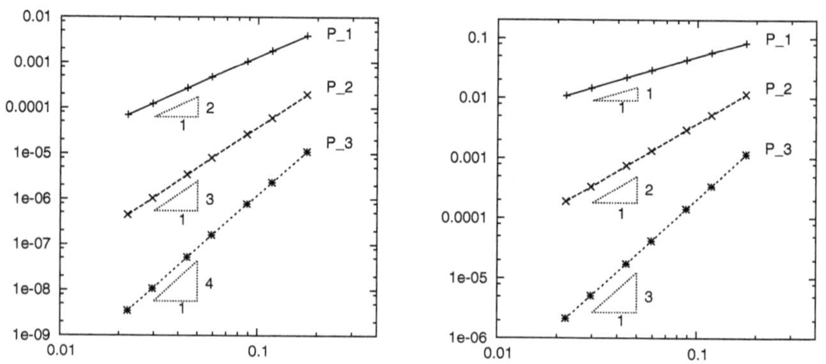

Fig. 5.2 Computational errors and orders of convergence with respect to the mesh-size h in the $L^2(\Omega)$-norm (*left*) and the $H^1(\Omega, \mathscr{T}_h)$-seminorm (*right*) for scheme (5.75) with P_1 (*full line*), P_2 (*dashed line*) and P_3 (*dotted line*) approximations

Chapter 6
Space-Time Discontinuous Galerkin Method

In Chap. 5 we introduced and analyzed methods based on the combination of the DGM space discretization with the backward difference formula in time. Although this approach gives satisfactory results in a number of applications (see Chap. 9), its drawback is a complicated adaptation of the space computational mesh and the time step. From this point of view, a more suitable approach is the *space-time discontinuous Galerkin* method (ST-DGM), where the DGM is applied separately in space and in time.

The ST-DGM can use different triangulations arising on different time levels due to a mesh adaptation and, thus it perfectly suits the numerical solution of nonstationary problems. Moreover, the ST-DGM can (locally) employ different polynomial degrees p and q in space and time discretization, respectively.

Section 6.1 will be concerned with basic ideas and techniques of the ST-DGM applied to a model of a linear heat equation. In Sect. 6.2, we extend the analysis to a more general convection-diffusion problem with nonlinear convection and nonlinear diffusion. Sections 6.3 and 6.4 will be devoted to some special ST-DG techniques.

6.1 Space-Time DGM for a Heat Equation

In this section we present and analyze the *space time discontinuous Galerkin* method applied to a simple model problem represented by the linear heat equation. We explain the main aspects of the ST-DG discretization for this problem and derive the error estimates in the $L^\infty(0, T; L^2(\Omega))$-norm and the DG-norm formed by the $L^2(0, T; H^1(\Omega, \mathscr{T}_h))$-norm and penalty terms.

Let $\Omega \subset \mathbb{R}^d$, $d = 2$ or 3, be a bounded polygonal or polyhedral domain, $T > 0$ and $Q_T := \Omega \times (0, T)$. We consider the problem to find $u : Q_T \to \mathbb{R}$ such that

V. Dolejší and M. Feistauer, *Discontinuous Galerkin Method*,
Springer Series in Computational Mathematics 48,
DOI 10.1007/978-3-319-19267-3_6

$$\frac{\partial u}{\partial t} = \varepsilon\,\Delta u + g \quad \text{in } Q_T, \tag{6.1a}$$

$$u\,\big|_{\partial\Omega_D\times(0,T)} = u_D, \tag{6.1b}$$

$$\nabla u\cdot \boldsymbol{n}\,\big|_{\partial\Omega_N\times(0,T)} = g_N, \tag{6.1c}$$

$$u(x,0) = u^0(x), \quad x\in\Omega. \tag{6.1d}$$

Similarly, as in Sect. 4.2 we assume that the boundary $\partial\Omega$ is formed by two disjoint parts $\partial\Omega_D$ and $\partial\Omega_N$ with $\text{meas}_{d-1}(\partial\Omega_D) > 0$, and that the data satisfy the usual conditions (cf. (4.2)): $u_D =$ trace of some $u^* \in C([0,T]; H^1(\Omega))$ on $\partial\Omega_D\times(0,T)$, $\varepsilon > 0,\ g\in C([0,T]; L^2(\Omega)),\ g_N\in C([0,T]; L^2(\partial\Omega_N))$ and $u^0\in L^2(\Omega)$.

6.1.1 Discretization of the Problem

6.1.1.1 Space-Time Partition and Function Spaces

In order to derive the space-time discontinuous Galerkin discretization, we introduce some notation.

Let $r > 1$ be an integer. In the time interval $[0,T]$ we construct a partition $0 = t_0 < \cdots < t_r = T$ and denote

$$I_m = (t_{m-1}, t_m),\quad \overline{I}_m = [t_{m-1}, t_m],\quad \tau_m = t_m - t_{m-1},\quad \tau = \max_{m=1,\ldots,r} \tau_m.$$

Then

$$[0,T] = \cup_{m=1}^{r}\overline{I}_m,\quad I_m\cap I_n = \emptyset \ \text{for } m\neq n,\ m,n = 1,\ldots,r.$$

If φ is a function defined in $\bigcup_{m=1}^{r} I_m$, we introduce the notation

$$\varphi_m^{\pm} = \varphi(t_m\pm) = \lim_{t\to t_m\pm}\varphi(t),\quad \{\varphi\}_m = \varphi_m^{+} - \varphi_m^{-}, \tag{6.2}$$

provided the one-sided limits $\lim_{t\to t_m\pm}\varphi(t)$ exist.

For each time instant t_m, $m = 0,\ldots,r$, and interval I_m, $m = 1,\ldots,r$, we consider a partition $\mathscr{T}_{h,m}$ (called triangulation) of the closure $\overline{\Omega}$ of the domain Ω into a finite number of closed simplexes (triangles for $d = 2$ and tetrahedra for $d = 3$) with mutually disjoint interiors. The partitions $\mathscr{T}_{h,m}$ may be in general different for different m. Figure 6.1 shows an illustrative example of the space-time partition for $d = 1$.

In what follows we use a similar notation as in Sect. 2.1, only a subscript $_m$ has to be added to the notation because of different grids $\mathscr{T}_{h,m}$. By $\mathscr{F}_{h,m}$ we denote the system of all faces of all elements $K\in\mathscr{T}_{h,m}$. Further, we denote the set of all inner faces by $\mathscr{F}_{h,m}^{I}$ and the set of all boundary faces by $\mathscr{F}_{h,m}^{B}$. Each $\Gamma\in\mathscr{F}_{h,m}$

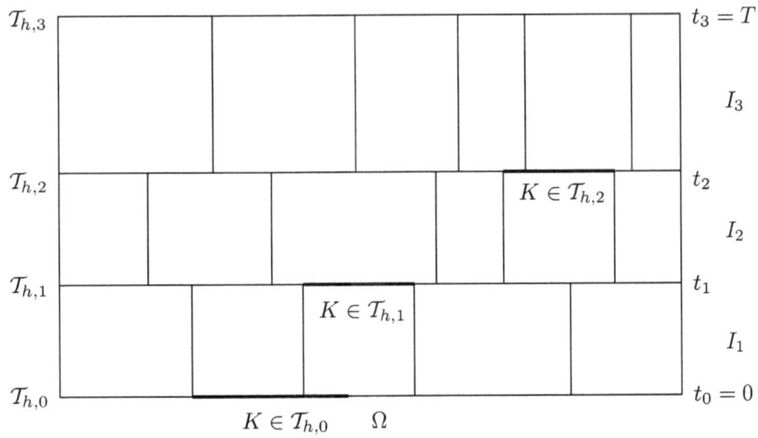

Fig. 6.1 Space-time discretization for space dimension $d = 1$

will be associated with a unit normal vector $\boldsymbol{n}_\Gamma$, which has the same orientation as the outer normal to $\partial\Omega$ for $\Gamma \in \mathscr{F}_{h,m}^B$. In $\Gamma \in \mathscr{F}_{h,m}^B$ we distinguish the subsets the of all "Dirichlet" boundary faces $\mathscr{F}_{h,m}^D = \left\{\Gamma \in \mathscr{F}_{h,m};\ \Gamma \subset \partial\Omega_D\right\}$ and of all "Neumann" boundary faces $\mathscr{F}_{h,m}^N = \left\{\Gamma \in \mathscr{F}_{h,m},\ \Gamma \subset \partial\Omega_N\right\}$. We set

$$h_K = \mathrm{diam}(K) \text{ for } K \in \mathscr{T}_{h,m}, \quad h_m = \max_{K\in\mathscr{T}_{h,m}} h_K, \quad h = \max_{m=1,\dots,r} h_m.$$

By ρ_K we denote the radius of the largest ball inscribed into K.

For any integer $k \geq 1$, over a triangulation $\mathscr{T}_{h,m}$, we define the *broken Sobolev space*

$$H^k(\Omega, \mathscr{T}_{h,m}) = \{v \in L^2(\Omega);\ v|_K \in H^k(K)\ \forall\, K \in \mathscr{T}_{h,m}\}, \tag{6.3}$$

with seminorm

$$|v|_{H^k(\Omega,\mathscr{T}_{h,m})} = \left(\sum_{K\in\mathscr{T}_{h,m}} |v|^2_{H^k(K)}\right)^{1/2}. \tag{6.4}$$

In the same way as in Chap. 2, we use the symbols $\langle v\rangle_\Gamma$ and $[v]_\Gamma$ for the mean value and the jump of $v \in H^k(\Omega, \mathscr{T}_{h,m})$ on the face $\Gamma \in \mathscr{F}_{h,m}$, see (2.32).

Let $p, q \geq 1$ be integers. For every $m = 1, \dots, r$ we define the finite-dimensional space

$$S_{h,m}^p = \left\{\varphi \in L^2(\Omega);\ \varphi|_K \in P_p(K)\ \forall\, K \in \mathscr{T}_{h,m}\right\}. \tag{6.5}$$

Over each mesh $\mathscr{T}_{h,m}$ we use the L^2-projections analogous to $\pi_{K,p}$ and Π_{hp} defined in (2.89) and (2.90). For simplicity we denote these projections by $\Pi_{h,m}$. Hence, if $K \in \mathscr{T}_{h,m}$, $m = 1, \ldots, r$, and $v \in L^2(K)$, then

$$(\Pi_{h,m}v)|_K \in P_p(K), \quad (\Pi_{h,m}v - v, \varphi)_{L^2(K)} = 0 \quad \forall \varphi \in P_p(K), \tag{6.6}$$

and, if $v \in L^2(\Omega)$, then

$$\Pi_{h,m}v \in S^p_{h,m}, \quad (\Pi_{h,m}v - v, \varphi) = 0 \quad \forall \varphi \in S^p_{h,m}. \tag{6.7}$$

As in previous sections, $(\cdot, \cdot)_{L^2(K)}$ and $(\cdot, \cdot)$ denote the $L^2(K)$-scalar product and the $L^2(\Omega)$-scalar product, respectively, and $P_p(K)$ denotes the space of all polynomials on K of degree $\leq p$. Properties of these projections follow from Lemmas 2.22 and 2.24 and they are summarized in (6.28) and (6.29).

The approximate solution will be sought in the space of functions that are piecewise polynomial in space and time:

$$S^{p,q}_{h,\tau} = \Big\{ \varphi \in L^2(Q_T);\ \varphi(x,t)\big|_{I_m} = \sum_{i=0}^{q} t^i\, \varphi_{m,i}(x) \\ \text{with } \varphi_{m,i} \in S^p_{h,m},\ i = 0, \ldots, q,\ \ m = 1, \ldots, r \Big\}. \tag{6.8}$$

6.1.2 Space-Time DG Discretization

We derive the full space-time discontinuous Galerkin discretization in a similar way as the space discretization introduced in detail in Chap. 2. We consider an exact regular solution satisfying the conditions

$$u \in L^2(0, T; H^2(\Omega)), \quad \frac{\partial u}{\partial t} \in L^2(0, T; H^1(\Omega)). \tag{6.9}$$

Then $u \in C([0, T]; H^1(\Omega))$. Such solution satisfies (6.1) pointwise. Moreover, let $m \in \{1, \ldots, r\}$ be arbitrary but fixed. We multiply (6.1a) by $\varphi \in S^{p,q}_{h,\tau}$, integrate over $K \times I_m$, and sum over all elements $K \in \mathscr{T}_{h,m}$. Then

$$\int_{I_m} (u', \varphi)\, \mathrm{d}t + \varepsilon \int_{I_m} \Big(\sum_{K \in \mathscr{T}_{h,m}} \int_K \nabla u \cdot \nabla \varphi\, \mathrm{d}x - \sum_{K \in \mathscr{T}_{h,m}} \int_{\partial K} \nabla u \cdot \boldsymbol{n} \varphi\, \mathrm{d}S \Big)\, \mathrm{d}t \\ = \int_{I_m} (g, \varphi)\, \mathrm{d}t, \tag{6.10}$$

where we use the notation $u' = \partial u / \partial t$.

First, we deal with the time derivative term. With the aid of integration by parts, we have

$$\int_{I_m} (u', \varphi)\, \mathrm{d}t = -\int_{I_m} (u, \varphi')\, \mathrm{d}t + (u_m^-, \varphi_m^-) - (u_{m-1}^+, \varphi_{m-1}^+). \tag{6.11}$$

Since the exact solution u is continuous with respect to t, we have $u_{m-1}^+ = u_{m-1}^-$ (cf. (6.2)) and, thus

$$(u_{m-1}^+, \varphi_{m-1}^+) = (u_{m-1}^-, \varphi_{m-1}^+). \tag{6.12}$$

Substitution of (6.12) into (6.11) and the integration by parts (in the reverse manner) yield

$$\begin{aligned} \int_{I_m} (u', \varphi)\, \mathrm{d}t &= -\int_{I_m} (u, \varphi')\, \mathrm{d}t + (u_m^-, \varphi_m^-) - (u_{m-1}^-, \varphi_{m-1}^+) \\ &= \int_{I_m} (u', \varphi)\, \mathrm{d}t + (u_{m-1}^+, \varphi_{m-1}^+) - (u_{m-1}^-, \varphi_{m-1}^+) \\ &= \int_{I_m} (u', \varphi)\, \mathrm{d}t + \left(\{u\}_{m-1}, \varphi_{m-1}^+\right). \end{aligned} \tag{6.13}$$

Remark 6.1 Identity (6.13) also makes sense for a function u, which is piecewise polynomial with respect to t on I_m, $m = 1, \ldots, r$. Then the equality (6.12) can be interpreted in such a way that the value of the function u at t_{m-1} from the right (on the new time interval) is approximated by the $L^2(\Omega)$-projection of the value of u at t_{m-1} from the left (on the previous time interval). Therefore, we can speak about the "upwinding" with respect to time—compare with the "space upwinding" in (4.16).

The discretization of the diffusion term and the right-hand side in (6.10) is the same as in Chap. 2. Hence, by virtue of (2.41)–(2.42) and (2.50)–(2.53), we define the diffusion, penalty and right-hand side forms as

$$\begin{aligned} a_{h,m}(w, \varphi) = &\sum_{K \in \mathscr{T}_{h,m}} \int_K \nabla w \cdot \nabla \varphi\, \mathrm{d}x - \sum_{\Gamma \in \mathscr{F}_{h,m}^I} \int_\Gamma (\langle \nabla w \rangle \cdot \boldsymbol{n}[\varphi] + \Theta \langle \nabla \varphi \rangle \cdot \boldsymbol{n}\,[w])\, \mathrm{d}S \\ &- \sum_{\Gamma \in \mathscr{F}_{h,m}^D} \int_\Gamma (\nabla w \cdot \boldsymbol{n} \varphi + \Theta \nabla \varphi \cdot \boldsymbol{n} w)\, \mathrm{d}S, \end{aligned} \tag{6.14}$$

$$J_{h,m}^\sigma(w, \varphi) = \sum_{\Gamma \in \mathscr{F}_{h,m}^I} \frac{C_W}{h_\Gamma} \int_\Gamma [w]\,[\varphi]\, \mathrm{d}S + \sum_{\Gamma \in \mathscr{F}_{h,m}^D} \frac{C_W}{h_\Gamma} \int_\Gamma w \varphi\, \mathrm{d}S, \tag{6.15}$$

$$A_{h,m}(w, \varphi) = \varepsilon a_{h,m}(w, \varphi) + \varepsilon J_{h,m}^\sigma(w, \varphi), \tag{6.16}$$

$$\ell_{h,m}(\varphi) = \int_\Omega g \varphi\, \mathrm{d}x + \int_{\partial\Omega_N} g_N \varphi\, \mathrm{d}S \tag{6.17}$$

$$- \varepsilon\Theta \sum_{\Gamma \in \mathscr{F}_{h,m}^{D}} \int_{\Gamma} \nabla\varphi \cdot \boldsymbol{n} u_D \, \mathrm{d}S + \varepsilon \sum_{\Gamma \in \mathscr{F}_{h,m}^{D}} \frac{C_W}{h_\Gamma} \int_{\Gamma} u_D \, \varphi \, \mathrm{d}S,$$

where $C_W > 0$ is a suitable constant and h_Γ characterizes the face Γ (cf. Lemma 2.5). Moreover, in (6.14) and (6.17), we take $\Theta = -1$, $\Theta = 0$ and $\Theta = 1$ and obtain the nonsymmetric (NIPG), incomplete (IIPG) and symmetric (SIPG) variants of the approximation of the diffusion terms, respectively. Obviously, forms (6.14)–(6.17) make sense for $v, w, \varphi \in H^2(\Omega, \mathscr{T}_{h,m})$.

By virtue of (6.10), (6.13) and (6.14)–(6.17), the exact regular solution u satisfies the identity

$$\int_{I_m} \big((u', \varphi) + A_{h,m}(u, \varphi)\big) \, \mathrm{d}t + \big(\{u\}_{m-1}, \varphi_{m-1}^{+}\big) = \int_{I_m} \ell_{h,m}(\varphi) \, \mathrm{d}t \tag{6.18}$$
$$\forall \varphi \in S_{h,\tau}^{p,q}, \ m = 1, \ldots, r, \ \text{with } u(0-) = u^0.$$

Based on (6.18), we introduce the approximate solution.

Definition 6.2 We say that a function U is a ST-DG approximate solution of problem (6.1), if $U \in S_{h,\tau}^{p,q}$ and

$$\int_{I_m} \big((U', \varphi) + A_{h,m}(U, \varphi)\big) \, \mathrm{d}t + \big(\{U\}_{m-1}, \varphi_{m-1}^{+}\big) = \int_{I_m} \ell_{h,m}(\varphi) \, \mathrm{d}t \tag{6.19}$$
$$\forall \varphi \in S_{h,\tau}^{p,q}, \ m = 1, \ldots, r, \ \text{ with } \ U_0^- := \Pi_{h,0} u^0,$$

where $U' = \partial U / \partial t$. We call (6.19) the space-time discontinuous Galerkin discrete problem.

Remark 6.3 The expression $\big(\{U\}_{m-1}, \varphi_{m-1}^{+}\big)$ in (6.19) patches together the approximate solution on neighbouring intervals I_{m-1} and I_m. At time $t = t_0 = 0$ we have $\{U\}_0 = U_0^+ - \Pi_{h,m} u^0$. It is also possible to consider $q = 0$. In this case, scheme (6.19) represents a variant of the backward Euler method analyzed in Sect. 5.1. Therefore, we assume that $q \geq 1$.

Remark 6.4 With respect to notation in previous chapters, we should denote the approximate solution by $u_{h\tau}$, which would express that the approximate solution depends on the space and time discretization parameters h and τ. However, for the sake of simplicity we use the symbol U.

Theorem 6.5 *Let the constant C_W satisfy the conditions from Corollary 2.41. Then there exists a unique approximate solution of (6.19).*

Proof Let $m \in \{1, \ldots, r\}$ be fixed and let U_{m-1}^- be given either by the initial condition or from the previous interval I_{m-1}. Identity (6.19) can be written in the form

$$\mathscr{R}(U, \varphi) = \int_{I_m} \ell_{h,m}(\varphi) \, \mathrm{d}t + \big(U_{m-1}^-, \varphi_{m-1}^{+}\big), \quad \varphi \in S_{h,\tau,m}^{p,q}, \tag{6.20}$$

where

$$\mathscr{R}(U,\varphi) := \int_{I_m} \left((U',\varphi) + A_{h,m}(U,\varphi) \right) \mathrm{d}t + \left(U_{m-1}^{+}, \varphi_{m-1}^{+} \right) \tag{6.21}$$

and

$$S_{h,\tau,m}^{p,q} := \left\{ \varphi \in L^2(\Omega \times I_m);\ \varphi(x,t) = \sum_{i=0}^{q} t^i \varphi_i(x) \text{ with } \varphi_{m,i} \in S_{h,m}^{p},\ i = 0,\ldots,q \right\}. \tag{6.22}$$

Obviously, the form $\mathscr{R}$ is a bilinear form on the finite dimension space $S_{h,\tau,m}^{p,q}$ and the right-hand side of (6.20) is a linear functional depending on $\varphi \in S_{h,\tau,m}^{p,q}$. Then, by virtue of Corollary 1.7, it is sufficient to prove the coercivity of the form $\mathscr{R}$ on $S_{h,\tau,m}^{p,q}$ with respect to a suitable norm. Hence, using (4.85), the coercivity of $A_{h,m}$ following from (2.140) and integration over I_m, we obtain

$$\begin{aligned}
\mathscr{R}(\varphi,\varphi) &= \int_{I_m} \left((\varphi',\varphi) + A_{h,m}(\varphi,\varphi) \right) \mathrm{d}t + \left(\varphi_{m-1}^{+}, \varphi_{m-1}^{+} \right) \qquad (6.23)\\
&= \int_{I_m} \left(\frac{1}{2}\frac{\mathrm{d}}{\mathrm{d}t} \|\varphi\|_{L^2(\Omega)}^2 + A_{h,m}(\varphi,\varphi) \right) \mathrm{d}t + \left\| \varphi_{m-1}^{+} \right\|_{L^2(\Omega)}^2 \\
&= \frac{1}{2}\left(\left\|\varphi_m^{-}\right\|_{L^2(\Omega)}^2 - \left\|\varphi_{m-1}^{+}\right\|_{L^2(\Omega)}^2 \right) + \int_{I_m} A_{h,m}(\varphi,\varphi)\,\mathrm{d}t + \left\|\varphi_{m-1}^{+}\right\|_{L^2(\Omega)}^2 \\
&\geq \frac{1}{2}\left(\left\|\varphi_m^{-}\right\|_{L^2(\Omega)}^2 + \left\|\varphi_{m-1}^{+}\right\|_{L^2(\Omega)}^2 \right) + \varepsilon C_C \int_{I_m} |||\varphi|||^2\,\mathrm{d}t =: \|\varphi\|_{\star}^2.
\end{aligned}$$

It is possible to show that $\|\cdot\|_{\star}$ is a norm on the space $S_{h,\tau,m}^{p,q}$ and, thus the form $\mathscr{R}$ is coercive. Then Corollary 1.7 implies the existence and uniqueness of the approximate solution. □

Exercise 6.6 Show that $\|\cdot\|_{\star}$ defined in (6.23) is a norm on the space $S_{h,\tau,m}^{p,q}$.

Our main goal will be to investigate the qualitative properties of the ST-DG scheme (6.19). In particular, we are concerned with the analysis of error estimates.

First, we recall some results from previous chapters.

6.1.3 Auxiliary Results

In the theoretical analysis, we consider a system of triangulations

$$\{\mathscr{T}_{h\tau}\}_{h\in(0,\bar{h}),\tau\in(0,\bar{\tau})},\quad \bar{h}>0,\ \bar{\tau}>0, \qquad \mathscr{T}_{h\tau} = \{\mathscr{T}_{h,m}\}_{m=0}^{r}, \tag{6.24}$$

satisfying the *shape-regularity assumption* (2.19) and the *equivalence condition* (2.20):

$$\frac{h_K}{\rho_K} \le C_R, \quad K \in \mathscr{T}_{h,m}, \ m = 0, \dots, r, \ h \in (0, \bar{h}), \tag{6.25}$$

$$C_T h_K \le h_\Gamma \le C_G h_K, \quad K \in \mathscr{T}_{h,m}, \ \Gamma \in \mathscr{F}_{h,m}, \ \Gamma \subset K, m = 0, \dots, r, \ h \in (0, \bar{h}), \tag{6.26}$$

with constants $C_R, C_T, C_G > 0$ independent of h, K, Γ, m and r (cf. Lemma 2.5). We also assume that the constant C_W from definition (6.15) of the penalty forms $J^\sigma_{h,m}$, $m = 0, \dots, r$, satisfies conditions from Corollary 2.41.

We again use the DG-norm in the space $H^1(\Omega, \mathscr{T}_{h,m})$, $m = 0, \dots, r$, given by

$$|||\varphi|||_m = \Big(\sum_{K \in \mathscr{T}_{h,m}} |\varphi|^2_{H^1(K)} + J^\sigma_{h,m}(\varphi, \varphi) \Big)^{1/2}. \tag{6.27}$$

If p, $s \ge 1$ are integers and $\mu = \min(s, p+1)$, then it follows from Lemmas 2.22 and 2.24 that for $m = 1, \dots, r$ and any $v \in H^s(\Omega)$, we have the standard error estimates for the space interpolation:

if $s \ge 1$, then

$$\|\Pi_{h,m} v - v\|_{L^2(K)} \le C_A \, h_K^\mu |v|_{H^\mu(K)}, \tag{6.28a}$$

$$|\Pi_{h,m} v - v|_{H^1(K)} \le C_A \, h_K^{\mu-1} |v|_{H^\mu(K)}, \tag{6.28b}$$

if $s \ge 2$, then

$$|\Pi_{h,m} v - v|_{H^2(K)} \le C_A \, h_K^{\mu-2} |v|_{H^\mu(K)}, \tag{6.28c}$$

for $K \in \mathscr{T}_{h,m}$ and $h \in (0, \bar{h})$.

Lemma 6.7 *Let $K \in \mathscr{T}_{h,m}$, $h \in (0, \bar{h})$. Then*

$$\|\Pi_{h,m} v\|_{L^2(K)} \le \|v\|_{L^2(K)} \ \textit{for } v \in L^2(K), \tag{6.29a}$$

$$|\Pi_{h,m} v|_{H^1(K)} \le C_1 |v|_{H^1(K)} \ \textit{for } v \in H^1(K), \tag{6.29b}$$

where C_1 is a constant independent of h, K, v, m and r.

Proof Inequality (6.29a) is a consequence of the definition (6.6) of the operator $\Pi_{h,m}$. Namely, setting $\varphi = \Pi_{h,m} v$ in (6.6) and using the Cauchy inequality, we find that $\|\Pi_{h,m} v\|^2_{L^2(K)} \le \|v\|_{L^2(K)} \|\Pi_{h,m} v\|_{L^2(K)}$, which implies (6.29a). Inequality (6.29b) immediately follows from (2.97) in Sect. 2.5, where we set $s = q = 1$ and, hence $\mu = 1$. Then we get

$$|\Pi_{h,m} v|_{H^1(K)} \le |\Pi_{h,m} v - v|_{H^1(K)} + |v|_{H^1(K)} \le (C_A + 1) |v|_{H^1(K)},$$

which is (6.29b) with $C_1 = C_A + 1$. □

Similarly as in previous chapters, important tools in analyzing the DGM will be the multiplicative trace inequality and the inverse inequality: There exist constants $C_M,\ C_I > 0$ independent of $h \in (0, \bar{h})$, m, r, $K \in \mathscr{T}_{h,m}$ and v such that

$$\|v\|^2_{L^2(\partial K)} \le C_M \left(\|v\|_{L^2(K)} \, |v|_{H^1(K)} + h_K^{-1} \|v\|^2_{L^2(K)} \right), \quad v \in H^1(K), \tag{6.30}$$

and

$$|v|_{H^1(K)} \le C_I h_K^{-1} \|v\|_{L^2(K)}, \quad v \in P_p(K). \tag{6.31}$$

We use also the inequalities

$$|J^\sigma_{h,m}(v, w)| \le J^\sigma_{h,m}(v, v)^{1/2} J^\sigma_{h,m}(w, w)^{1/2} \le \frac{1}{2}(\delta J^\sigma_{h,m}(v, v) + \delta^{-1} J^\sigma_{h,m}(w, w)), \tag{6.32}$$

with an arbitrary $\delta > 0$, which are consequences of the definition of the form $J^\sigma_{h,m}$ and the Cauchy and Young inequalities. (See (2.118).)

6.1.4 Space-Time Projection Operator

In this section we introduce the $S^{p,q}_{h,\tau}$-interpolation defined as the space-time projection operator $\pi : C([0, T]; L^2(\Omega)) \to S^{p,q}_{h,\tau}$ in the following way.

Let $v \in C([0, T]; L^2(\Omega))$. Then

$$\pi\, v \in S^{p,q}_{h,\tau}, \tag{6.33a}$$

$$(\pi\, v)\,(x, t_m-) = \Pi_{h,m}\, v(x, t_m-) \text{ for almost all } x \in \Omega \text{ and all } m = 1, \ldots, r, \tag{6.33b}$$

$$\int_{I_m} (\pi v - v,\ \varphi)\, \mathrm{d}t = 0 \text{ for all } \varphi \in S^{p,q-1}_{h,\tau} \text{ and all } m = 1, \ldots, r, \tag{6.33c}$$

where the operator $\Pi_{h,m}$ is given by (6.7). As we see, condition (6.33c) means that the interpolation error $\pi v - v$ is orthogonal to the polynomials of degree $\le q - 1$ on I_m. The lower degree of test functions φ is compensated by the additional condition (6.33b), which means that the space-time projection πv is equal to the space projection $\Pi_{h,m}\, v$ at the time instants t_m for $t \to t_m-$. Because of our further considerations it is suitable to set

$$(\pi v)(x, 0-) := \Pi_{h,0} v(x, 0). \tag{6.34}$$

In what follows, we are concerned with the existence and uniqueness of the interpolation operator π and its properties.

6.1.4.1 The Existence and Uniqueness of the Space-Time Projection π

We start from the definition of the Legendre polynomials $\mathscr{L}_i(\vartheta)$, $\vartheta \in [-1, +1]$. They are defined by the conditions

$$\begin{aligned} \mathscr{L}_0(\vartheta) &= 1, \\ \mathscr{L}_1(\vartheta) &= \vartheta, \\ \mathscr{L}_{i+1}(\vartheta) &= \frac{2i+1}{i+1}\vartheta \mathscr{L}_i(\vartheta) - \frac{i}{i+1}\mathscr{L}_{i-1}(\vartheta), \quad i = 1, 2, \ldots . \end{aligned} \tag{6.35}$$

By induction it is possible to prove that the polynomials $\mathscr{L}_i$ are L^2-orthogonal in the interval $[-1, 1]$ and satisfy the relation $\mathscr{L}_i(1) = 1$ (see, e.g., [210]). Moreover, the sequence $\{\mathscr{L}_i\}_{i=0}^{\infty}$ is dense in the space $C([-1, 1])$. Hence, any $\varphi \in C([-1, 1]; S^p_{h,m})$, $m = 1, \ldots, r$, can be expressed as

$$\varphi(x, t) = \sum_{i=0}^{\infty} \varphi_i(x) \mathscr{L}_i(t),$$

where $\varphi_i \in S^p_{h,m}$, $i = 0, 1, \ldots$.

Let $q \geq 1$ and $m \in \{1, \ldots, r\}$ be given. Let $P_q(-1, 1; S^p_{h,m})$ be the Bochner space of polynomial functions $v : \Omega \times [-1, 1] \to \mathbb{R}$ of degree $\leq q$ with respect to $t \in [-1, 1]$ given by (1.44), where $v(\cdot, t)$ is a piecewise polynomial function from $S^p_{h,m}$ for all $t \in [-1, 1]$. Further, let us define the operator $\hat{\pi} : C([-1, 1]; L^2(\Omega)) \to P_q(-1, 1; S^p_{h,m})$ by

$$\hat{\pi}\hat{v} \in P_q(-1, 1; S^p_{h,m}), \tag{6.36a}$$

$$(\hat{\pi}\hat{v})(x, 1) = (\Pi_{h,m}\hat{v})(x, 1), \quad x \in \Omega, \tag{6.36b}$$

$$\int_{-1}^{1} (\hat{\pi}\hat{v} - \hat{v}, \varphi)\mathrm{d}\vartheta = 0 \quad \forall \varphi \in P_{q-1}(-1, 1; S^p_{h,m}). \tag{6.36c}$$

Lemma 6.8 *Let $q \geq 1$ and $m \in \{1, \ldots, r\}$ be given. Let $\Pi_{h,m}$ be the space projection operator defined by (6.7). Then the operator $\hat{\pi}$ from (6.36) can be uniquely expressed with the aid of the Legendre polynomials in the following way. If $\hat{v} \in C([-1, 1]; L^2(\Omega))$, $x \in \Omega$ and $t \in [-1, 1]$, then*

$$(\hat{\pi}\hat{v})(x, t) = \sum_{i=0}^{q-1} \hat{v}_i(x) \mathscr{L}_i(t) + \left((\Pi_{h,m}\hat{v})(x, 1) - \sum_{i=0}^{q-1} \hat{v}_i(x) \right) \mathscr{L}_q(t), \tag{6.37}$$

where $\hat{v}_i \in S^p_{h,m}$ are the coefficients in the expansion of the function $\Pi_{h,m}\hat{v}$ in the basis formed by the Legendre polynomials in the form

$$(\Pi_{h,m}\hat{v})(x,t) = \sum_{i=0}^{\infty} \hat{v}_i(x)\,\mathscr{L}_i(t), \quad \hat{v}_i \in S^p_{h,m}. \tag{6.38}$$

Proof First, we show that $\hat{\pi}\hat{v}$ defined by (6.37) and (6.38) satisfies (6.36). It is obvious that $\hat{\pi}\hat{v} \in P_q(-1,1;S^p_{h,m})$ and (omitting the variable x) we have

$$\begin{aligned}\hat{\pi}\hat{v}(1) &= \sum_{i=0}^{q-1} \hat{v}_i\,\mathscr{L}_i(1) + \left(\Pi_{h,m}\hat{v}(1) - \sum_{i=0}^{q-1}\hat{v}_i\right)\mathscr{L}_q(1)\\ &= \sum_{i=0}^{q-1}\hat{v}_i + \Pi_{h,m}\hat{v}(1) - \sum_{i=0}^{q-1}\hat{v}_i = \Pi_{h,m}\hat{v}(1).\end{aligned}$$

Let us choose an arbitrary function $w \in S^p_{h,m}$ and nonnegative integer $k < q$. Using (6.7), (6.37) and (6.38) we successively get

$$\begin{aligned}&\int_{-1}^{1}(\hat{\pi}\hat{v} - \hat{v}, w\mathscr{L}_k)\,\mathrm{d}\vartheta\\ &= \int_{-1}^{1}\left(\sum_{i=0}^{q-1}\hat{v}_i\,\mathscr{L}_i + \left(\Pi_{h,m}\hat{v}(1) - \sum_{i=0}^{q-1}\hat{v}_i\right)\mathscr{L}_q - \Pi_{h,m}\hat{v}, w\mathscr{L}_k\right)\mathrm{d}\vartheta\\ &= \int_{-1}^{1}\left(-\sum_{i=q}^{\infty}\hat{v}_i\,\mathscr{L}_i + \left(\Pi_{h,m}\hat{v}(1) - \sum_{i=0}^{q-1}\hat{v}_i\right)\mathscr{L}_q, w\mathscr{L}_k\right)\mathrm{d}\vartheta = 0,\end{aligned}$$

thanks to the orthogonality of the polynomials $\mathscr{L}_i,\ i = 0, 1, \ldots$.

Now we prove uniqueness. Let us assume that there exist two functions $\tilde{\phi}_1, \tilde{\phi}_2 \in P_q(-1,1;S^p_{h,m})$ satisfying (6.36b) and (6.36c) (i.e., relations (6.36b) and (6.36c) are valid with $\hat{\pi}\hat{v} := \tilde{\phi}_i,\ i = 1,2$). It follows from (6.36b) that $\tilde{\phi}_1(1) = \tilde{\phi}_2(1) \in S^p_{h,m}$. Then there exists $\tilde{w} \in P_{q-1}(-1,1;S^p_{h,m})$ such that $\tilde{\phi}_1(\vartheta) - \tilde{\phi}_2(\vartheta) = (\vartheta - 1)\tilde{w}(\vartheta)$. Further, by (6.36c),

$$0 = \int_{-1}^{1}(\tilde{\phi}_1(\vartheta) - \tilde{\phi}_2(\vartheta), \tilde{w}(\vartheta))\,\mathrm{d}\vartheta = \int_{-1}^{1}(\vartheta - 1)\|\tilde{w}(\vartheta)\|^2\,\mathrm{d}\vartheta.$$

Since the function $(\vartheta - 1)\|\tilde{w}(\vartheta)\|^2$ is continuous and non-positive in the interval $(-1,1)$, necessarily $\|\tilde{w}(\vartheta)\|^2 = 0$ for all $\vartheta \in (-1,1)$ and, hence $\tilde{w} = 0$. □

Theorem 6.9 *The projection π defined by (6.33) exists and is unique. Moreover,*

$$(\pi v)|_{I_m} = \pi(\Pi_{h,m}v)\big|_{I_m} = \Pi_{h,m}(\pi v|_{I_m}), \quad m = 1, \ldots, r. \tag{6.39}$$

Proof First, we express the operator π with the aid of the operator $\hat{\pi}$ from (6.36). Let $m \in \{1, \ldots, r\}$ be fixed. If we introduce the mapping $Q_m : (-1, 1) \to I_m = (t_{m-1}, t_m)$ such that

$$Q_m(\vartheta) = \frac{t_m + t_{m-1} + \vartheta \tau_m}{2}, \quad \vartheta \in (-1, 1),$$

we can put

$$(\pi v)(t) = (\hat{\pi}\hat{v})(Q_m^{-1}(t)) \ \text{ with } \hat{v}(\vartheta) = v(Q_m(\vartheta)), \quad t \in I_m, \ m = 1, \ldots, r,$$

where $\hat{\pi}$ is defined by (6.36). It is obvious that the mapping π defined in this way satisfies (6.33). The uniqueness of π can be proven in an analogous way as the uniqueness of $\hat{\pi}$ in the proof of Lemma 6.8.

Further, we prove (6.39). Obviously, by (6.33b),

$$\pi(\Pi_{h,m}v)\,(t_m-) = \Pi_{h,m}(\Pi_{h,m}v(t_m-)) = (\Pi_{h,m}v)\,(t_m-) = (\pi v)\,(t_m-). \quad (6.40)$$

By the definition of $\Pi_{h,m}(v|_{I_m})$ we have

$$\int_\Omega (\Pi_{h,m}v - v)\,\varphi_h \,\mathrm{d}x = 0 \quad \forall \varphi_h \in S_{h,m}^p. \quad (6.41)$$

Moreover, by virtue of (6.33c),

$$\int_{I_m} \left(\int_\Omega (\pi v(t) - v(t))\,\varphi_h \,\mathrm{d}x \right) t^j \,\mathrm{d}t = 0 \quad \forall \varphi_h \in S_{h,m}^p \ \forall j = 0, \ldots, q-1. \quad (6.42)$$

Similarly, we can write

$$\begin{aligned} 0 &= \int_{I_m} \left(\int_\Omega \left(\pi(\Pi_{h,m}v) - \Pi_{h,m}v\right) \varphi_h \,\mathrm{d}x \right) t^j \,\mathrm{d}t \\ &= \int_{I_m} \left(\int_\Omega \left((\pi(\Pi_{h,m}v)(t) - v(t)) + (v(t) - \Pi_{h,m}v(t))\right) \varphi_h \,\mathrm{d}x \right) t^j \,\mathrm{d}t \\ &= \int_{I_m} \left(\int_\Omega (\pi(\Pi_{h,m}v)\,(t) - v(t))\,\varphi_h \,\mathrm{d}x \right) t^j \,\mathrm{d}t, \end{aligned}$$

as follows from (6.33c) and (6.41). These relations, (6.42) and (6.40) imply that $(\pi v)|_{I_m} = \pi(\Pi_{h,m}v)|_{I_m}$.

The proof of the second relation in (6.39) is even simpler. It is possible to write

$$\pi v|_{I_m} = \sum_{i=0}^{q} v_i t^i,$$

where $v_i \in S^p_{h,m}$. Since $\Pi_{h,m} v_i = v_i$, we have

$$\Pi_{h,m}(\pi v|_{I_m}) = \sum_{i=0}^{q} \Pi_{h,m} v_i t^i = \sum_{i=0}^{q} v_i t^i = \pi v|_{I_m}. \qquad \square$$

6.1.4.2 Approximation Properties of the Space-Time Projection π

Now we derive the approximation properties of the projection π defined by (6.33). First, we present one technical result.

Lemma 6.10 *Let $u \in H^{q+1}(0, T; L^2(\Omega))$. Then, under the notation $\partial_t^{q+1} := \partial^{q+1}/\partial t^{q+1}$, $q = 0, 1, \ldots,$ we have*

$$\partial_t^{q+1}(\Pi_{h,m} u) = \Pi_{h,m}(\partial_t^{q+1} u), \tag{6.43}$$

$$\partial_t^{q+1}(\nabla \Pi_{h,m} u) = \nabla \Pi_{h,m}(\partial_t^{q+1} u). \tag{6.44}$$

Proof Actually, $\Pi_{h,m} u(\cdot, t) \in S^p_{h,m}$ and for all $t \in I_m$,

$$\int_\Omega \big(\Pi_{h,m} u(x,t) - u(x,t)\big)\, \varphi(x)\, \mathrm{d}x = 0 \quad \forall \varphi \in S^p_{h,m}.$$

The differentiation with respect to t yields

$$\int_\Omega \big(\partial_t^{q+1}(\Pi_{h,m} u(x,t)) - \partial_t^{q+1} u(x,t)\big)\, \varphi(x)\, \mathrm{d}x = 0 \quad \forall \varphi \in S^p_{h,m}.$$

Moreover, obviously $\partial_t^{q+1}(\Pi_{h,m} u(t)) \in S^p_{h,m}$ and thus (6.43) holds. Similarly we can prove (6.44). $\square$

Now, we introduce mappings P_m, $m = 1, \ldots, r$, which will serve for a representation of the operator π given by (6.33).

Lemma 6.11 *Let $m \in \{1, \ldots, r\}$ be arbitrary but fixed and $\varphi \in C([t_{m-1}, t_m]; S^p_{h,m})$. Then*

$$\pi\, \varphi(x,t) = \mathrm{P}_m \varphi(x,t) \quad \forall x \in K \quad \forall K \in \mathscr{T}_{h,m} \quad \forall t \in I_m, \tag{6.45}$$

where P_m *is defined in the following way: For* $\zeta \in C([t_{m-1}, t_m])$,

$$\mathrm{P}_m\zeta \in P_q(I_m), \tag{6.46a}$$

$$\int_{I_m} (\mathrm{P}_m\zeta(t) - \zeta(t))t^j \,\mathrm{d}t = 0 \quad \forall j = 0, \ldots, q-1, \tag{6.46b}$$

$$\mathrm{P}_m\zeta(t_m-) = \zeta(t_m-). \tag{6.46c}$$

Proof Let $m \in \{1, \ldots, r\}$. Similarly as above, we can prove that conditions (6.46) uniquely define the operator P_m. From the definition of π and P_m, it follows that on $K \times I_m$ for every $K \in \mathscr{T}_{h,m}$ the functions $\pi\varphi$ and $\mathrm{P}_m\varphi$ are polynomials of degree $\leq q$ in $t \in I_m$ and degree $\leq p$ in $x \in K$. Moreover,

$$\pi\varphi(x, t_m-) = \varphi(x, t_m-) = \mathrm{P}_m\varphi(x, t_m-) \quad \forall x \in K.$$

Obviously, condition (6.33c) is equivalent to

$$\int_{I_m} \left(\int_K (\pi\varphi(x,t) - \varphi(x,t))\, \psi(x)\, \mathrm{d}x \right) t^j \,\mathrm{d}t = 0 \tag{6.47}$$
$$\forall j = 0, \ldots, q-1 \quad \forall \psi \in P_p(K) \quad \forall K \in \mathscr{T}_{h,m}.$$

Further, by (6.46), for any $K \in \mathscr{T}_{h,m}$,

$$\int_{I_m} (\mathrm{P}_m\varphi(x,t) - \varphi(x,t))\, t^j \,\mathrm{d}t = 0, \quad \forall j = 0, \ldots, q-1 \quad \forall x \in K. \tag{6.48}$$

Let $\psi \in P_p(K)$. Then (6.48) and Fubini's theorem imply that

$$0 = \int_K \left(\int_{I_m} (\mathrm{P}_m\varphi(x,t) - \varphi(x,t))\, t^j \,\mathrm{d}t \right) \psi(x)\, \mathrm{d}x \tag{6.49}$$
$$= \int_{I_m} \left(\int_K (\mathrm{P}_m\varphi(x,t) - \varphi(x,t))\psi(x)\, \mathrm{d}x \right) t^j \,\mathrm{d}t$$
$$\forall j = 0, \ldots, q-1 \quad \forall \psi \in P_p(K) \quad \forall K \in \mathscr{T}_{h,m}.$$

Comparing (6.49) with (6.47) and taking into account the fact that the operator π is uniquely determined by conditions (6.33), we immediately get (6.45). □

In what follows we prove error estimate for the time interpolation and the space-time interpolation operator π. First, we prove the estimate of the operator P_m, $m = 1, \ldots, r$.

Lemma 6.12 *Let* $m \in \{1, \ldots, r\}$ *be given and let* P_m *be given by (6.46). If* $\zeta \in H^{q+1}(I_m)$, *then*

$$\|\mathrm{P}_m\zeta - \zeta\|^2_{L^2(I_m)} \leq C_2\, \tau_m^{2(q+1)} \|\partial_t^{q+1}\zeta\|^2_{L^2(I_m)}, \tag{6.50}$$

where $C_2 > 0$ is a constant independent of ζ, m and τ_m.

Proof We proceed in several steps.
(i) We transform the reference interval $[0, 1]$ onto the interval $[t_{m-1}, t_m]$ by the mapping

$$t = t_m - \tau_m \vartheta, \quad \vartheta \in [0, 1]. \tag{6.51}$$

If $\zeta \in H^{q+1}(I_m)$ and $\hat{\zeta}(\vartheta) := \zeta(t_m - \tau_m \vartheta)$, then $\hat{\zeta} \in H^{q+1}(0, 1)$ and

$$\mathrm{P}_m \zeta(t_m - \tau_m \vartheta) = \hat{\mathrm{P}} \hat{\zeta}(\vartheta),$$

where the operator $\hat{\mathrm{P}}$ is defined by

$$\hat{\mathrm{P}} \hat{\zeta} \in P_q(0, 1), \tag{6.52a}$$

$$\int_0^1 \left(\hat{\mathrm{P}} \hat{\zeta}(\vartheta) - \hat{\zeta}(\vartheta) \right) \vartheta^j \, \mathrm{d}\vartheta = 0 \quad \forall\, j = 0, \ldots, q - 1, \tag{6.52b}$$

$$\hat{\mathrm{P}} \hat{\zeta}(0+) = \hat{\zeta}(0+). \tag{6.52c}$$

Moreover, if we set

$$\begin{aligned} Z_m(t) &= \mathrm{P}_m \zeta(t) - \zeta(t), \quad t \in (t_{m-1}, t_m), \\ z(\vartheta) &= \hat{\mathrm{P}} \hat{\zeta}(\vartheta) - \hat{\zeta}(\vartheta), \quad \vartheta \in (0, 1), \end{aligned} \tag{6.53}$$

we have

$$\begin{aligned} z(\vartheta) &= Z_m(t_m - \tau_m \vartheta), \\ \partial_\vartheta^{q+1} z(\vartheta) &= (-1)^{q+1} \tau_m^{q+1} \, \partial_t^{q+1} Z_m(t_m - \tau_m \vartheta), \quad \vartheta \in (0, 1). \end{aligned} \tag{6.54}$$

By the substitution theorem,

$$\begin{aligned} \|z\|_{L^2(0,1)}^2 &= \frac{1}{\tau_m} \|Z_m\|_{L^2(I_m)}^2, \\ \|\partial_\vartheta^{q+1} z\|_{L^2(0,1)}^2 &= \tau_m^{2q+1} \|\partial_t^{q+1} Z_m\|_{L^2(I_m)}^2. \end{aligned} \tag{6.55}$$

(ii) Since conditions (6.52) determine the values of the operator $\hat{\mathrm{P}}$ uniquely, it is clear that

$$\hat{\mathrm{P}} \psi = \psi \quad \text{for} \quad \psi \in P_q(0, 1). \tag{6.56}$$

Now we prove that the operator $\hat{\mathrm{P}}$ is a continuous mapping of the space $H^{q+1}(0,1)$ into $L^2(0,1)$. Let $u_n \in H^{q+1}(0,1)$, $n = 1, 2, \ldots$ and $u_n \to 0$ in $H^{q+1}(0,1)$ as $n \to \infty$. The continuous embedding $H^{q+1}(0,1) \hookrightarrow C([0,1])$ implies that

$$u_n \to 0 \quad \text{uniformly in } [0,1] \tag{6.57}$$

and hence by (6.52c),

$$\hat{\mathrm{P}} u_n(0) \to 0. \tag{6.58}$$

For $j = 0, \ldots, q-1$ we have

$$\int_0^1 \left(\hat{\mathrm{P}} u_n - u_n\right)(\vartheta)\, \vartheta^j \,\mathrm{d}\vartheta = 0.$$

This fact and (6.57) imply that

$$\int_0^1 \hat{\mathrm{P}} u_n(\vartheta)\, \vartheta^j \,\mathrm{d}\vartheta = \int_0^1 u_n(\vartheta)\, \vartheta^j \,\mathrm{d}\vartheta \to 0, \quad j = 0, \ldots, q-1. \tag{6.59}$$

Since $\hat{\mathrm{P}} u_n \in P_q(0,1)$, we can write

$$\hat{\mathrm{P}} u_n(\vartheta) = \sum_{i=1}^{q} c_i^{(n)}\, \vartheta^i + (\hat{\mathrm{P}} u_n)(0), \quad \vartheta \in [0,1], \tag{6.60}$$

where $c_i^{(n)} \in \mathbb{R}$, $i = 1, \ldots, q$. Integration yields

$$\int_0^1 \hat{\mathrm{P}} u_n(\vartheta)\, \vartheta^j \,\mathrm{d}\vartheta = \int_0^1 \sum_{i=1}^{q} c_i^{(n)}\, \vartheta^{i+j} \,\mathrm{d}\vartheta + \hat{\mathrm{P}} u_n(0) \int_0^1 \vartheta^j \,\mathrm{d}\vartheta \tag{6.61}$$

$$= \sum_{i=1}^{q} c_i^{(n)} \frac{1}{i+j+1} + \hat{\mathrm{P}} u_n(0) \frac{1}{j+1}, \qquad j = 0, \ldots, q-1.$$

The matrix $\left(\frac{1}{i+j+1}\right)_{\substack{i=1,\ldots,q,\\ j=0,\ldots,q-1}} = \left(\frac{1}{i+j}\right)_{i,j=1}^{q}$ is the Gram matrix of the linearly independent functions ϑ^i, $i = 0, \ldots, q-1$, defined by the scalar product $((\phi, s)) = \int_0^1 \vartheta \phi(\vartheta) s(\vartheta)\,\mathrm{d}\vartheta$. Hence, this matrix is nonsingular. Using this fact and (6.58), (6.59) and (6.61), we find that

$$c_i^{(n)} \to 0 \quad \text{for} \quad i = 1, \ldots, q \quad \text{as} \quad n \to \infty. \tag{6.62}$$

Now (6.62), (6.58) and (6.60) imply that $\hat{P}u_n \to 0$ uniformly in $[0, 1]$ and, thus $\hat{P}u_n \to 0$ in $L^2(0, 1)$.

(iii) The above results allow us to apply Theorem 2.16 and get the estimate

$$\|z\|_{L^2(0,1)} \leq C\|\partial_\vartheta^{q+1} z\|_{L^2(0,1)} \tag{6.63}$$

with a constant $C > 0$ independent of $z \in H^{q+1}(0, 1)$. This fact and (6.55) imply that

$$\|Z_m\|^2_{L^2(I_m)} \leq C\, \tau_m^{2(q+1)} \|\partial_t^{q+1} Z_m\|^2_{L^2(I_m)}. \tag{6.64}$$

Taking into account that $\partial_t^{q+1} \mathrm{P}_m \zeta = 0$, we immediately get (6.50). □

In Sect. 6.1.11 we give a direct proof of estimate (6.50) without the use of Theorem 2.16. As a consequence of Lemmas 6.11 and 6.12 we get the following result.

Lemma 6.13 *There exists a constant $C_3 > 0$ such that*

$$\|\pi\varphi(x, \cdot) - \varphi(x, \cdot)\|^2_{L^2(I_m)} \leq C_3\, \tau_m^{2(q+1)} \left\|\partial_t^{q+1} \varphi(x, \cdot)\right\|^2_{L^2(I_m)}, \tag{6.65}$$

for all $\varphi \in H^{q+1}(I_m; S^p_{h,m})$, $x \in K$, $K \in \mathscr{T}_{h,m}$, or $x \in \Gamma$, $\Gamma \in \mathscr{F}_{h,m}$ and $m = 1, \ldots, r$.

The error analysis will require the following results, analogous to Lemma 6.13. Their derivation will be based on the continuous embeddings

$$H^1(0, 1; L^2(\Omega)) \hookrightarrow C([0, 1]; L^2(\Omega)) \hookrightarrow L^\infty(0, 1; L^2(\Omega)). \tag{6.66}$$

(See, for example, [195].) Hence, there exists a constant $C > 0$ such that

$$\|v\|_{L^\infty(0,1;L^2(\Omega))} \leq C\|v\|_{H^1(0,1;L^2(\Omega))} \quad \forall v \in H^1(0, 1; L^2(\Omega)), \tag{6.67}$$

where

$$\|v\|^2_{H^1(0,1;L^2(\Omega))} = \|v\|^2_{L^2(0,1,L^2(\Omega))} + \|\partial_\vartheta v\|^2_{L^2(0,1;L^2(\Omega))}. \tag{6.68}$$

Lemma 6.14 *If $\varphi \in W^{q+1,\infty}(I_m; S^p_{h,m})$, then*

$$\|\pi\varphi - \varphi\|_{L^\infty(I_m;L^2(\Omega))} \leq C_4\, \tau_m^{q+1} |\varphi|_{W^{q+1,\infty}(I_m;L^2(\Omega))}, \quad m = 1, \ldots, r, \tag{6.69}$$

where $C_4 > 0$ is a constant independent of φ, m, τ_m.

Proof We proceed in a similar way as in the proof of Lemma 6.12. Let us use the transformation (6.51), i.e., $\vartheta \in (0, 1) \to t = t_m - \tau_m\vartheta \in I_m$. For (almost all) $x \in \Omega, t \in I_m$ we set

$$Z_m(x,t) = \pi\, \varphi(x,t) - \varphi(x,t) \tag{6.70}$$

and

$$z(x,\vartheta) = Z_m(x,\, t_m - \tau_m \vartheta), \quad \vartheta \in (0,1). \tag{6.71}$$

By (6.45), $Z_m(x,t) = \mathrm{P}_m\, \varphi(x,t) - \varphi(x,t)$. Moreover,

$$z(x,\vartheta) = \hat{\mathrm{P}}\, \varphi(x,\, t_m - \tau_m\vartheta) - \varphi(x,\, t_m - \tau_m\vartheta),$$

where the operator $\hat{\mathrm{P}}$ is defined by (6.52). Obviously,

$$\begin{aligned}
\partial_\vartheta z(x,\vartheta) &= -\,\tau_m \partial_t Z_m(x,\, t_m - \tau_m\vartheta),\\
\partial_\vartheta^{q+1} z(x,\vartheta) &= (-1)^{q+1}\tau_m^{q+1}\, \partial_t^{q+1} Z_m(x,\, t_m - \tau_m\vartheta), \quad \vartheta \in (0,1).
\end{aligned}$$

It follows from the above relations that

$$\|z(x,\cdot)\|^2_{L^2(0,1)} = \frac{1}{\tau_m} \|Z_m(x,\cdot)\|^2_{L^2(I_m)}, \tag{6.72a}$$

$$\|\partial_\vartheta z(x,\cdot)\|^2_{L^2(0,1)} = \tau_m \|\partial_t Z_m(x,\cdot)\|^2_{L^2(I_m)}, \tag{6.72b}$$

$$\|\partial_\vartheta^{q+1} z(x,\cdot)\|^2_{L^2(0,1)} = \tau_m^{2q+1} \|\partial_t^{q+1} Z_m(x,\cdot)\|^2_{L^2(I_m)}. \tag{6.72c}$$

From part (ii) of the proof of Lemma 6.12 we deduce that the operator $\hat{\mathrm{P}}$ is a continuous mapping of the space $H^{q+1}(0,1)$ into $H^1(0,1)$. Actually, assuming that $u_n \to 0$ in $H^{q+1}(0,1)$ for $n \to +\infty$, taking into account that $\hat{\mathrm{P}}u_n(0) \to 0$, (6.60) and (6.62), we find that $\hat{\mathrm{P}}u_n \to 0$ and $\partial_\vartheta \hat{P} u_n \to 0$ uniformly in $[0,1]$ as $n \to +\infty$. Hence, $\hat{\mathrm{P}}u_n \to 0$ for $n \to +\infty$ in $H^1(0,1)$.

These results, (6.56) and Theorem 2.16 imply that there exists a constant $C > 0$ independent of z such that

$$\begin{aligned}
\|z(x,\cdot)\|^2_{L^2(0,1)} &\le C \|\partial_\vartheta^{q+1} z(x,\cdot)\|^2_{L^2(0,1)}\\
\|\partial_\vartheta z(x,\cdot)\|^2_{L^2(0,1)} &\le C \|\partial_\vartheta^{q+1} z(x,\cdot)\|^2_{L^2(0,1)}.
\end{aligned} \tag{6.73}$$

Now, by (6.71),

$$\begin{aligned}
\|z\|_{L^\infty(0,1;L^2(\Omega))} &= \operatorname{ess\,sup}_{\vartheta\in(0,1)} \left(\int_\Omega |z(x,\vartheta)|^2\, \mathrm{d}x \right)^{1/2} \\
&= \operatorname{ess\,sup}_{t\in I_m} \left(\int_\Omega |Z_m(x,t)|^2 \mathrm{d}x \right)^{1/2} = \|Z_m\|_{L^\infty(I_m;L^2(\Omega))}.
\end{aligned} \tag{6.74}$$

Further, by (6.72c) and Fubini's theorem we have

$$\begin{aligned}\|\partial_\vartheta^{q+1} z\|^2_{L^2(0,1;L^2(\Omega))} &= \int_0^1 \left(\int_\Omega |\partial_\vartheta^{q+1} z|^2 \,\mathrm{d}x \right) \mathrm{d}\vartheta = \int_\Omega \left(\int_0^1 |\partial_\vartheta^{q+1} z|^2 \,\mathrm{d}\vartheta \right) \mathrm{d}x \\ &= \int_\Omega \|\partial_\vartheta^{q+1} z(x,\cdot)\|^2_{L^2(0,1)} \mathrm{d}x = \tau_m^{2q+1} \int_\Omega \|\partial_t^{q+1} Z_m(x,\cdot)\|^2_{L^2(I_m)} \,\mathrm{d}x \qquad (6.75) \\ &= \tau_m^{2q+1} \|\partial_t^{q+1} Z_m\|^2_{L^2(I_m;L^2(\Omega))}.\end{aligned}$$

Now, taking into account (6.67), (6.68), (6.73)–(6.75), we find that

$$\begin{aligned}\|Z_m\|^2_{L^\infty(I_m;L^2(\Omega))} &= \|z\|^2_{L^\infty(0,1;L^2(\Omega))} \le C \left(\|z\|^2_{L^2(0,1;L^2(\Omega))} + \|\partial_\vartheta z\|^2_{L^2(0,1;L^2(\Omega))} \right) \\ &\le C \|\partial_\vartheta^{q+1} z\|^2_{L^2(0,1;L^2(\Omega))} = C\, \tau_m^{2q+1} \|\partial_t^{q+1} Z_m\|^2_{L^2(I_m;L^2(\Omega))}.\end{aligned}$$

Finally, this estimate, the inequality

$$\|\partial_t^{q+1} Z_m\|^2_{L^2(I_m;L^2(\Omega))} \le \tau_m \|\partial_t^{q+1} Z_m\|^2_{L^\infty(I_m;L^2(\Omega))}, \tag{6.76}$$

and the definition (6.70) of Z_m yield (6.69). □

6.1.5 Abstract Error Estimate

The following sections will be devoted to the estimation of the error $e = U - u$, where u is the exact solution of problem (6.1) and U is the approximate solution obtained by the ST-DGM (6.19).

Subtracting (6.18) from (6.19) and using linearity of the forms, we obtain

$$\int_{I_m} \left((U' - u', \varphi) + A_{h,m}(U - u, \varphi) \right) \mathrm{d}t + \left(\{U\}_{m-1} - \{u\}_{m-1}, \varphi^+_{m-1} \right) = 0, \tag{6.77}$$

$$\varphi \in S^{p,q}_{h,\tau},\ m = 1, \ldots, r.$$

Similarly, as in previous chapters, we decompose the error into two parts:

$$e = U - u = \xi + \eta, \tag{6.78}$$

where

$$\xi = U - \pi u \in S^{p,q}_{h,\tau}, \quad \eta = \pi u - u. \tag{6.79}$$

Then (6.77) and (6.78) yield

$$\int_{I_m} \left((\xi', \varphi) + A_{h,m}(\xi, \varphi)\right) \mathrm{d}t + (\{\xi\}_{m-1}, \varphi_{m-1}^+) \tag{6.80}$$
$$= -\int_{I_m} (\eta', \varphi)\, \mathrm{d}t - (\{\eta\}_{m-1}, \varphi_{m-1}^+) - \int_{I_m} A_{h,m}(\eta, \varphi)\, \mathrm{d}t, \quad \varphi \in S_{h,\tau}^{p,q}, \ m = 1, \ldots, r.$$

By (6.34), we have $(\pi u)_0^- = \Pi_{h,0} u^0$. This relation, (6.19) and (6.33b) imply that

$$\begin{aligned} \xi_0^- &= U_0^- - (\pi u)_0^- = 0, \\ \xi_0^+ &= U_0^+ - (\pi u)_0^+ = U_0^+ - \Pi_{h,0} u^0, \\ \{\xi\}_0 &= \xi_0^+, \end{aligned} \tag{6.81}$$

and

$$\begin{aligned} \eta_0^- &= (\pi u)_0^- - u(0) = \Pi_{h,0} u^0 - u^0, \\ \eta_0^+ &= (\pi u)_0^+ - u(0) = \Pi_{h,0} u^0 - u^0 = \eta_0^-, \\ \{\eta\}_0 &= \eta_0^+ - \eta_0^- = 0. \end{aligned} \tag{6.82}$$

In order to derive the error estimates, it is necessary to estimate the individual terms in (6.80), where a suitable test function φ is used. Let us set $\varphi = \xi$. Then

$$\int_{I_m} \left((\xi', \xi) + A_{h,m}(\xi, \xi)\right) \mathrm{d}t + (\{\xi\}_{m-1}, \xi_{m-1}^+) \tag{6.83}$$
$$= -\int_{I_m} (\eta', \xi)\, \mathrm{d}t - (\{\eta\}_{m-1}, \xi_{m-1}^+) - \int_{I_m} A_{h,m}(\eta, \xi)\, \mathrm{d}t, \quad m = 1, \ldots, r.$$

A simple calculation yields

$$2\int_{I_m} (\xi', \xi)\, \mathrm{d}t + 2\left(\{\xi\}_{m-1}, \xi_{m-1}^+\right) = \int_{I_m} \frac{\mathrm{d}}{\mathrm{d}t} \|\xi\|_{L^2(\Omega)}^2 \mathrm{d}t + 2\left(\{\xi\}_{m-1}, \xi_{m-1}^+\right)$$
$$= \|\xi_m^-\|_{L^2(\Omega)}^2 - \|\xi_{m-1}^+\|_{L^2(\Omega)}^2 + 2\left(\xi_{m-1}^+ - \xi_{m-1}^-, \xi_{m-1}^+\right) \tag{6.84}$$

and

$$2\left(\xi_{m-1}^+ - \xi_{m-1}^-, \xi_{m-1}^+\right) \tag{6.85}$$
$$= \|\xi_{m-1}^+\|_{L^2(\Omega)}^2 - (\xi_{m-1}^-, \xi_{m-1}^+) + (\xi_{m-1}^+ - \xi_{m-1}^-, \xi_{m-1}^+ - \xi_{m-1}^-)$$
$$+ (\xi_{m-1}^+ - \xi_{m-1}^-, \xi_{m-1}^-)$$
$$= \|\xi_{m-1}^+\|_{L^2(\Omega)}^2 + \|\{\xi\}_{m-1}\|_{L^2(\Omega)}^2 - \|\xi_{m-1}^-\|_{L^2(\Omega)}^2 - (\xi_{m-1}^-, \xi_{m-1}^+) + (\xi_{m-1}^+, \xi_{m-1}^-)$$
$$= \|\xi_{m-1}^+\|_{L^2(\Omega)}^2 + \|\{\xi\}_{m-1}\|_{L^2(\Omega)}^2 - \|\xi_{m-1}^-\|_{L^2(\Omega)}^2.$$

Hence, from (6.84) and (6.85), we obtain

$$2\int_{I_m} (\xi', \xi)\,\mathrm{d}t + 2(\{\xi\}_{m-1}, \xi^+_{m-1}) = \|\xi^-_m\|^2_{L^2(\Omega)} - \|\xi^-_{m-1}\|^2_{L^2(\Omega)} + \|\{\xi\}_{m-1}\|^2_{L^2(\Omega)}. \tag{6.86}$$

Now we present a technical lemma.

Lemma 6.15 *If* $\delta > 0$, *then*

$$\left|\int_{I_m} (\eta', \varphi)\,\mathrm{d}t + \left(\{\eta\}_{m-1}, \varphi^+_{m-1}\right)\right| \le \delta \|\eta^-_{m-1}\|^2_{L^2(\Omega)} + \frac{1}{4\delta}\|\{\varphi\}_{m-1}\|^2_{L^2(\Omega)} \tag{6.87}$$
$$\forall\, \varphi \in S^{p,q}_{h,m}.$$

Proof Let $\varphi \in S^{p,q}_{h,\tau}$. Integration by parts yields

$$\int_{I_m} (\eta', \varphi)\,\mathrm{d}t = \left(\eta^-_m, \varphi^-_m\right) - \left(\eta^+_{m-1}, \varphi^+_{m-1}\right) - \int_{I_m} (\eta, \varphi')\,\mathrm{d}t. \tag{6.88}$$

Since $\eta = \pi u - u$ and $\varphi' \in S^{p,q-1}_{h,\tau}$, by the definition (6.33) of the operator π, we have

$$\int_{I_m} (\eta, \varphi')\,\mathrm{d}t = 0.$$

Thus,

$$\begin{aligned}
&\int_{I_m} (\eta', \varphi)\,\mathrm{d}t + \left(\{\eta\}_{m-1}, \varphi^+_{m-1}\right) \qquad (6.89)\\
&= \left(\eta^-_m, \varphi^-_m\right) - \left(\eta^+_{m-1}, \varphi^+_{m-1}\right) + \left(\eta^+_{m-1}, \varphi^+_{m-1}\right) - \left(\eta^-_{m-1}, \varphi^+_{m-1}\right)\\
&= \left(\eta^-_m, \varphi^-_m\right) - \left(\eta^-_{m-1}, \varphi^+_{m-1}\right).
\end{aligned}$$

Further, (6.33b) implies that

$$\eta_m = (\pi u)(t^-_m) - u(t_m) = \Pi_{h,m}\, u(t_m) - u(t_m).$$

Taking into account that $\varphi^-_m \in S^p_{h,m}$ and $\varphi^-_{m-1} \in S^p_{h,m-1}$, from the definition of $\Pi_{h,m}$ and $\Pi_{h,m-1}$ (cf. (6.7)) we get

$$\left(\eta^-_m, \varphi^-_m\right) = 0, \quad \left(\eta_{m-1}, \varphi^-_{m-1}\right) = 0. \tag{6.90}$$

Relations (6.88)–(6.90) imply that

$$\int_{I_m} (\eta', \varphi)\mathrm{d}t + \left(\{\eta\}_{m-1}, \varphi_{m-1}^{+}\right) = -\left(\eta_{m-1}^{-}, \varphi_{m-1}^{+}\right) = -\left(\eta_{m-1}^{-}, \{\varphi\}_{m-1}\right). \quad (6.91)$$

Finally, using the Young inequality on the right-hand side of (6.91) gives (6.87). □

Theorem 6.16 *Let the exact solution u of problem (6.1) satisfy the regularity assumption (6.9) and let U be the approximate solution defined by (6.19). Then the error $e = U - u$ satisfies the estimate*

$$\|e_n^-\|_{L^2(\Omega)}^2 + \frac{\varepsilon}{2}\sum_{m=1}^{n}\int_{I_m} |||e|||_m^2 \,\mathrm{d}t \quad (6.92)$$
$$\leq 2\sum_{m=0}^{n}\|\eta_m^-\|_{L^2(\Omega)}^2 + C\varepsilon\sum_{m=1}^{n}\int_{I_m} R_m(\eta)\,\mathrm{d}t + \varepsilon\sum_{m=1}^{n}\int_{I_m}|||\eta|||_m^2\,\mathrm{d}t,$$
$$n = 1, \dots, r, \; h \in (0, \bar{h}),$$

where

$$R_m(\eta) = \sum_{K\in\mathscr{T}_{h,m}}\left(|\eta|_{H^1(K)}^2 + h_K^2|\eta|_{H^2(K)}^2 + h_K^{-2}\|\eta\|_{L^2(K)}^2\right) \quad (6.93)$$

and $C > 0$ is a constant independent of $h, \tau, r, u, \varepsilon$.

Proof Using relations (6.83), (6.86) and (6.87) with $\delta = 1/2$ and $\varphi = \xi$, we get

$$\|\xi_m^-\|_{L^2(\Omega)}^2 - \|\xi_{m-1}^-\|_{L^2(\Omega)}^2 + \|\{\xi\}_{m-1}\|_{L^2(\Omega)}^2 + 2\int_{I_m} A_{h,m}(\xi,\xi)\,\mathrm{d}t \quad (6.94)$$
$$\leq \|\eta_{m-1}^-\|_{L^2(\Omega)}^2 + \|\{\xi\}_{m-1}\|_{L^2(\Omega)}^2 + 2\left|\int_{I_m} A_{h,m}(\eta,\xi)\,\mathrm{d}t\right|.$$

Further, we use the coercivity of the form $A_{h,m}$, which can be expressed in the same way as in (2.140) (with $C_C = 1/2$) under assumptions on C_W from Corollary 2.41:

$$2A_{h,m}(\xi,\xi) \geq \varepsilon|||\xi|||_m^2. \quad (6.95)$$

Moreover, by Lemma 2.37 and the Young inequality,

$$2|A_{h,m}(\eta,\xi)| \quad (6.96)$$
$$\leq 2\varepsilon\tilde{C}_B^2\sum_{K\in\mathscr{T}_{h,m}}\left(|\eta|_{H^1(K)}^2 + h_K^2|\eta|_{H^2(K)}^2 + h_K^{-2}\|\eta\|_{L^2(K)}^2\right) + \frac{1}{2}\varepsilon|||\xi|||_m^2$$
$$= 2\varepsilon\tilde{C}_B^2 R_m(\eta) + \frac{1}{2}\varepsilon|||\xi|||_m^2.$$

Now (6.94)–(6.96) imply that

$$\|\xi_m^-\|_{L^2(\Omega)}^2 - \|\xi_{m-1}^-\|_{L^2(\Omega)}^2 + \frac{\varepsilon}{2}\int_{I_m} |||\xi|||_m^2 \, dt \le \|\eta_{m-1}^-\|_{L^2(\Omega)}^2 + 2\varepsilon\tilde{C}_B^2 \int_{I_m} R_m(\eta)\, dt. \tag{6.97}$$

Let us note that by (6.81), $\xi_0^- = 0$. The summation of (6.97) over $m = 1, \ldots, n$ $(\le r)$ yields the inequality

$$\|\xi_n^-\|_{L^2(\Omega)}^2 + \frac{\varepsilon}{2}\sum_{m=1}^{n}\int_{I_m} |||\xi|||_m^2 \, dt \le \sum_{m=1}^{n} \|\eta_{m-1}^-\|_{L^2(\Omega)}^2 + 2\varepsilon\tilde{C}_B \sum_{m=1}^{n}\int_{I_m} R_m(\eta)\, dt. \tag{6.98}$$

Finally, since $e = \xi + \eta$, we have

$$\begin{aligned} \|e_j^-\|_{L^2(\Omega)}^2 &\le 2\left(\|\xi_j^-\|_{L^2(\Omega)}^2 + \|\eta_j^-\|_{L^2(\Omega)}^2\right), \\ |||e|||_j^2 &\le 2\left(|||\xi|||_j^2 + |||\eta|||_j^2\right). \end{aligned} \tag{6.99}$$

By (6.99) and (6.98) we get

$$\begin{aligned} &\|e_n^-\|_{L^2(\Omega)}^2 + \frac{\varepsilon}{2}\sum_{m=1}^{n}\int_{I_m} |||e|||_m^2 \, dt \\ &\le 2(\|\xi_n^-\|_{L^2(\Omega)}^2 + \|\eta_n^-\|_{L^2(\Omega)}^2) + \varepsilon \sum_{m=1}^{n}\int_{I_m} (|||\xi|||_m^2 + |||\eta|||_m^2)\, dt \\ &\le 2\sum_{m=0}^{n} \|\eta_m^-\|_{L^2(\Omega)}^2 + 4\varepsilon\tilde{C}_B \sum_{m=1}^{n}\int_{I_m} R_m(\eta)\, dt + \varepsilon \sum_{m=1}^{n}\int_{I_m} |||\eta|||_m^2 \, dt, \end{aligned}$$

which is (6.92), with $C = 4\tilde{C}_B^2$. □

6.1.6 *Estimation of Projection Error in Terms of h and τ*

The abstract error estimate formulated in Theorem 6.16, together with approximation properties of the projection operator π treated in Sect. 6.1.4.2, will allow us to derive error estimates in terms of the sizes of the space and time meshes.

As above, we assume that u and U denote the exact and the approximate solutions satisfying (6.1) and (6.19), respectively. According to (6.78) and (6.79), the error is written as

$$e = U - u = (U - \pi u) + (\pi u - u) = \xi + \eta. \tag{6.100}$$

Moreover, we express the term η in the form

$$\begin{aligned} \eta\big|_{I_m} &= (\pi u - u)\big|_{I_m} = \eta^{(1)} + \eta^{(2)}, \quad m = 1, \ldots, r, \\ \text{with } \eta^{(1)} &= (\Pi_{h,m} u - u)\big|_{I_m}, \ \eta^{(2)} = \left(\pi(\Pi_{h,m} u) - \Pi_{h,m} u\right)\big|_{I_m}, \end{aligned} \tag{6.101}$$

where we used the fact that $\pi\, u\big|_{I_m} = \pi\left(\Pi_{h,m} u\right)\big|_{I_m}$, $m = 1, \ldots, r$. We recall that integers $p, q \geq 1$.

We assume that the weak solution u of (6.1) satisfies the regularity condition

$$u \in H^{q+1}(0, T; H^1(\Omega)) \cap C([0, T]; H^s(\Omega)), \tag{6.102}$$

where $s \geq 2$ is an integer. As usual, we set $\mu = \min(p + 1, s)$. Then u satisfies relations (6.18). Obviously, $C([0, T]; H^s(\Omega)) \subset L^2(0, T; H^s(\Omega))$.

Our further goal is to estimate the expressions

$$\|\eta_m^-\|^2_{L^2(\Omega)}, \ \int_{I_m} \|\eta\|^2_{L^2(K)}\, \mathrm{d}t, \ \int_{I_m} |\eta|^2_{H^1(K)}\, \mathrm{d}t, \ \int_{I_m} |\eta|^2_{H^2(K)}\, \mathrm{d}t, \ J^\sigma_{h,m}(\eta, \eta),$$

which will appear in the error estimation. In what follows, we use the notation

$$|u\|_{H^{q+1}(I_m; H^1(K))} = \left(|u|^2_{H^{q+1}(I_m; L^2(K))} + |u|^2_{H^{q+1}(I_m; H^1(K))}\right)^{1/2}, \tag{6.103}$$

$$|u\|_{H^{q+1}(I_m; H^1(\Omega))} = \left(|u|^2_{H^{q+1}(I_m; L^2(\Omega))} + |u|^2_{H^{q+1}(I_m; H^1(\Omega))}\right)^{1/2}, \tag{6.104}$$

$$|u\|_{H^{q+1}(0,T; H^1(\Omega))} = \left(|u|^2_{H^{q+1}(0,T; L^2(\Omega))} + |u|^2_{H^{q+1}(0,T; H^1(\Omega))}\right)^{1/2}. \tag{6.105}$$

(For the definitions of seminorms in Bochner spaces, see Sect. 1.3.5.)

By (6.101),

$$\begin{aligned} \|\eta\|^2_{L^2(K)} &\leq 2\|\eta^{(1)}\|^2_{L^2(K)} + 2\|\eta^{(2)}\|^2_{L^2(K)}, \\ |\eta|^2_{H^k(K)} &\leq 2|\eta^{(1)}|^2_{H^k(K)} + 2|\eta^{(2)}|^2_{H^k(K)}, \ k = 1, 2. \end{aligned} \tag{6.106}$$

Lemma 6.17 *The following estimates hold:*

$$\|\eta_0^-\|^2_{L^2(\Omega)} \leq C_A^2 h^{2\mu} |u(0)|^2_{H^\mu(\Omega)}, \quad \|\eta_m^-\|^2_{L^2(\Omega)} \leq C_A^2 h^{2\mu} |u(t_m)|^2_{H^\mu(\Omega)}, \tag{6.107}$$

$$\int_{I_m} \|\eta^{(1)}\|^2_{L^2(K)}\,\mathrm{d}t \le C_A^2\, h_K^{2\mu}\,|u|^2_{L^2(I_m;H^\mu(K))}, \tag{6.108}$$

$$\int_{I_m} |\eta^{(1)}|^2_{H^1(K)}\,\mathrm{d}t \le C_A^2\, h_K^{2(\mu-1)}\,|u|^2_{L^2(I_m;H^\mu(K))}, \tag{6.109}$$

$$h_K^2 \int_{I_m} |\eta^{(1)}|^2_{H^2(K)}\,\mathrm{d}t \le C_A^2\, h_K^{2(\mu-1)}\,|u|^2_{L^2(I_m;H^\mu(K))}, \tag{6.110}$$

for $K \in \mathscr{T}_{h,m}$, $m = 1, \dots, r$, *with constant* C_A *from (6.28).*

Proof It is enough to use (6.28) and (6.40). □

The derivation of estimates of terms with $\eta^{(2)}$ is more complicated.

Lemma 6.18 *For* $K \in \mathscr{T}_{h,m}$, $m = 1, \dots, r$, *we have*

$$\int_{I_m} \|\eta^{(2)}\|^2_{L^2(K)}\,\mathrm{d}t \le C_3\, \tau_m^{2(q+1)}\,|u|^2_{H^{q+1}(I_m;L^2(K))}, \tag{6.111}$$

$$\int_{I_m} |\eta^{(2)}|^2_{H^1(K)}\,\mathrm{d}t \le C_5\, \tau_m^{2(q+1)}\,|u|^2_{H^{q+1}(I_m;H^1(K))}, \tag{6.112}$$

$$h_K^2 \int_{I_m} |\eta^{(2)}|^2_{H^2(K)}\,\mathrm{d}t \le C_6\, \tau_m^{2(q+1)}\,|u|^2_{H^{q+1}(I_m;H^1(K))}, \tag{6.113}$$

where $C_3 > 0$ *is the constant from* (6.65) *and* $C_5 > 0$ *and* $C_6 > 0$ *are constants specified in the proof.*

Proof (i) Using Fubini's theorem and relations (6.43), (6.65) and (6.29a) yield the relations

$$\begin{aligned}
&\int_{I_m} \|\eta^{(2)}\|^2_{L^2(K)}\,\mathrm{d}t = \int_{I_m}\Big(\int_K |\eta^{(2)}|^2 \mathrm{d}x\Big)\,\mathrm{d}t \\
&= \int_K\Big(\int_{I_m} |\eta^{(2)}|^2\,\mathrm{d}t\Big)\mathrm{d}x = \int_K \|\pi(\Pi_{h,m}u) - \Pi_{h,m}u\|^2_{L^2(I_m)}\mathrm{d}x \\
&\le C_3\, \tau_m^{2(q+1)} \int_K \|\partial_t^{q+1}(\Pi_{h,m}u)\|^2_{L^2(I_m)}\mathrm{d}x \\
&- C_3\, \tau_m^{2(q+1)} \int_{I_m}\Big(\int_K |\partial_t^{q+1}(\Pi_{h,m}u)|^2 \mathrm{d}x\Big)\,\mathrm{d}t \\
&= C_3\, \tau_m^{2(q+1)} \int_{I_m}\Big(\int_K |\Pi_{h,m}(\partial_t^{q+1}u)|^2 \mathrm{d}x\Big)\,\mathrm{d}t \\
&\le C_3\, \tau_m^{2(q+1)} \int_{I_m}\Big(\int_K |\partial_t^{q+1}u|^2 \mathrm{d}x\Big)\,\mathrm{d}t = C_3\, \tau_m^{2(q+1)}\,|u|^2_{H^{q+1}(I_m;L^2(K))}.
\end{aligned}$$

(ii) Further, due to Fubini's theorem, (6.65), (6.44) and (6.29b), we find that

$$
\begin{aligned}
\int_{I_m} |\eta^{(2)}|^2_{H^1(K)}\,\mathrm{d}t &= \int_{I_m}\left(\int_K \left|\nabla\left(\Pi_{h,m}u - \pi(\Pi_{h,m}u)\right)\right|^2 \mathrm{d}x\right)\mathrm{d}t \\
&= \int_K \left(\int_{I_m} \sum_{j=1}^{d}\left(\frac{\partial}{\partial x_j}(\Pi_{h,m}u) - \pi\left(\frac{\partial}{\partial x_j}(\Pi_{h,m}u)\right)\right)^2 \mathrm{d}t\right)\mathrm{d}x \\
&\le C_3\,\tau_m^{2(q+1)} \int_K |\nabla(\Pi_{h,m}u)|^2_{H^{q+1}(I_m)}\mathrm{d}x \\
&= C_3\,\tau_m^{2(q+1)} \int_K \left(\int_{I_m} \left|\partial_t^{q+1}\nabla(\Pi_{h,m}u)\right|^2 \mathrm{d}t\right)\mathrm{d}x \\
&= C_3\,\tau_m^{2(q+1)} \int_{I_m} \left(\int_K \left|\nabla(\Pi_{h,m}\partial_t^{q+1}u)\right|^2 \mathrm{d}x\right)\mathrm{d}t \\
&= C_3\,\tau_m^{2(q+1)} \int_{I_m} \left|\Pi_{h,m}\left(\partial_t^{q+1}u\right)\right|^2_{H^1(K)} \mathrm{d}t \\
&\le C_3C_1\,\tau_m^{2(q+1)} \int_{I_m} \left|\partial_t^{q+1}u\right|^2_{H^1(K)} \mathrm{d}t = C_3C_1\,\tau_m^{2(q+1)} |u|^2_{H^{q+1}(I_m;H^1(K))},
\end{aligned}
$$

which gives (6.112) with $C_5 = C_3C_1$.

(iii) Using a similar process as in (ii) together with the inverse inequality (6.30) and (6.29), we find that

$$
\begin{aligned}
\int_{I_m} |\eta^{(2)}|_{H^2(K)}\,\mathrm{d}t &\le C_3\,\tau_m^{2(q+1)} \int_{I_m} \left|\Pi_{h,m}\left(\partial_t^{q+1}u\right)\right|^2_{H^2(K)} \mathrm{d}t \\
&\le C_3C_1C_I^2\,\tau_m^{2(q+1)} h_K^{-2} \int_{I_m} \left|\partial_t^{q+1}u\right|^2_{H^1(K)} \mathrm{d}t \\
&= C_3C_1C_I^2\,\tau_m^{2(q+1)} h_K^{-2} |u|^2_{H^{q+1}(I_m;H^1(K))}.
\end{aligned}
$$

This yields (6.113) with $C_6 = C_3C_1C_I^2$. □

Finally, we are concerned with the estimation of $\int_{I_m} J^{\sigma}_{h,m}(\eta,\eta)\,\mathrm{d}t$ under assumption (6.102). It holds that

$$
J^{\sigma}_{h,m}(\eta,\eta) \le 2\big(J^{\sigma}_{h,m}(\eta^{(1)},\eta^{(1)}) + J^{\sigma}_{h,m}(\eta^{(2)},\eta^{(2)})\big). \tag{6.114}
$$

From (2.119) in Sect. 2.6 and estimates (6.108), (6.109) we get

$$
\int_{I_m} J^{\sigma}_{h,m}(\eta^{(1)},\eta^{(1)})\,\mathrm{d}t \le C_J^2\, h^{2(\mu-1)} |u|^2_{L^2(I_m;H^{\mu}(\Omega))}. \tag{6.115}
$$

with the constant $C_J = 2C_A(C_W C_M / C_T)^{1/2}$.

Further, we estimate the expression

$$\int_{I_m} J^\sigma_{h,m}(\eta^{(2)}, \eta^{(2)})\, \mathrm{d}t = \int_{I_m} J^\sigma_{h,m}\big(\pi(\Pi_{h,m}u) - \Pi_{h,m}u,\ \pi(\Pi_{h,m}u) - \Pi_{h,m}u\big)\, \mathrm{d}t.$$

Lemma 6.19 *Let the Dirichlet data $u_D = u_D(x,t)$ have the behaviour in t as a polynomial of degree $\le q$:*

$$u_D(x,t) = \sum_{j=0}^{q} \psi_j(x)\, t^j, \tag{6.116}$$

where $\psi_j \in H^{s-1/2}(\partial\Omega)$ for $j = 0, \ldots, q$ (cf. (1.26)). (Let us note that $u = u_D$ on $\partial\Omega \times (0,T)$.) Then there exists $C_7 > 0$ such that

$$\int_{I_m} J^\sigma_{h,m}(\pi(\Pi_{h,m}u) - \Pi_{h,m}u, \pi(\Pi_{h,m}u) - \Pi_{h,m}u)\, \mathrm{d}t \le C_7 \tau_m^{2(q+1)} |u|^2_{H^{q+1}(I_m;H^1(\Omega))},$$
$$m = 1, \ldots, r \tag{6.117}$$

For general data u_D, if there exists a constant $\overline{C} > 0$ such that

$$\tau_m \le \overline{C} h_{K_\Gamma^{(L)}} \tag{6.118}$$

for all $\Gamma \in \mathscr{F}^D_{h,m}$, $m = 1, \ldots, r$, $h \in (0, \bar{h})$, then there exists $C_8 > 0$ such that

$$\int_{I_m} J^\sigma_{h,m}(\pi(\Pi_{h,m}u) - \Pi_{h,m}u, \pi(\Pi_{h,m}u) - \Pi_{h,m}u)\, \mathrm{d}t \le C_8 \tau_m^{2q} |u\|^2_{H^{q+1}(I_m;H^1(\Omega))},$$
$$m = 1, \ldots, r. \tag{6.119}$$

(The symbol $|u\|_{H^{q+1}(I_m;H^1(\Omega))}$ is defined by (6.104).)

Proof Let us consider an interval I_m and set $\varphi = \Pi_{h,m}u$. Then $\varphi \in H^{q+1}(I_m; S^p_{h,m})$. By (6.15),

$$J^\sigma_{h,m}\big(\pi(\Pi_{h,m}u) - \Pi_{h,m}u,\ \pi(\Pi_{h,m}u) - \Pi_{h,m}u\big) \tag{6.120}$$
$$= C_W \sum_{\Gamma\in\mathscr{F}^I_{h,m}} h_\Gamma^{-1} \int_\Gamma [\pi\varphi - \varphi]^2\, \mathrm{d}S + C_W \sum_{\Gamma\in\mathscr{F}^D_{h,m}} h_\Gamma^{-1} \int_\Gamma |\pi\varphi - \varphi|^2\, \mathrm{d}S.$$

Now we proceed in two steps.

(i) Let $\Gamma \subset \mathscr{F}^I_{h,m}$. If we use the relation $[\pi\varphi - \varphi] = \pi[\varphi] - [\varphi]$, Fubini's theorem and estimate (6.65), we find that

$$\int_{I_m}\left(\int_\Gamma [\pi\varphi-\varphi]^2\,\mathrm{d}S\right)\mathrm{d}t \tag{6.121}$$
$$=\int_\Gamma\left(\int_{I_m}|\pi[\varphi]-[\varphi]|^2\,\mathrm{d}t\right)\mathrm{d}S=\int_\Gamma \|\pi[\varphi(x,\cdot)]-[\varphi(x,\cdot)]\|_{L^2(I_m)}\,\mathrm{d}S$$
$$\le C_3\,\tau_m^{2(q+1)}\int_\Gamma \left\|\partial_t^{q+1}[\varphi(x,\cdot)]\right\|^2_{L^2(I_m)}\,\mathrm{d}S$$
$$=C_3\,\tau_m^{2(q+1)}\int_\Gamma\left(\int_{I_m}|\partial_t^{q+1}[\varphi(x,t)]|^2\,\mathrm{d}t\right)\mathrm{d}S.$$

If we take into account that

$$\partial_t^{q+1}[\varphi(x,\cdot)]=[\partial_t^{q+1}\varphi(x,\cdot)],\quad [\partial_t^{q+1}u]=0, \tag{6.122}$$

and use (6.121) and Fubini's theorem, we obtain

$$\int_{I_m}\left(\int_\Gamma[\pi(\Pi_{h,m}u)-\Pi_{h,m}u]^2\,\mathrm{d}S\right)\mathrm{d}t\le C_3\,\tau_m^{2(q+1)}\int_{I_m}\left(\int_\Gamma[\partial_t^{q+1}(\Pi_{h,m}u-u)]^2\,\mathrm{d}S\right)\mathrm{d}t. \tag{6.123}$$

The application of the multiplicative trace inequality (6.30) implies that

$$\sum_{\Gamma\in\mathscr{F}_{h,m}^I}\int_\Gamma\left[\partial_t^{q+1}(\Pi_{h,m}u-u)\right]^2\mathrm{d}S\le\sum_{K\in\mathscr{T}_{h,m}}\left\|\partial_t^{q+1}(\Pi_{h,m}u-u)\right\|^2_{L^2(\partial K)} \tag{6.124}$$
$$\le C_M\sum_{K\in\mathscr{T}_{h,m}}\Big(\left\|\partial_t^{q+1}(\Pi_{h,m}u-u)\right\|_{L^2(K)}\left|\partial_t^{q+1}(\Pi_{h,m}u-u)\right|_{H^1(K)}$$
$$+h_K^{-1}\left\|\partial_t^{q+1}(\Pi_{h,m}u-u)\right\|^2_{L^2(K)}\Big).$$

By (6.43),

$$\partial_t^{q+1}(\Pi_{h,m}u-u)=\Pi_{h,m}(\partial_t^{q+1}u)-\partial_t^{q+1}u. \tag{6.125}$$

By virtue of (6.102), $\partial_t^{q+1}u\in L^2(I_m;H^1(\Omega))$. This fact and the approximation properties (6.28a)–(6.28b) of $\Pi_{h,m}$, where we consider $s=1$ and thus $\mu=1$, imply that

$$\|\Pi_{h,m}(\partial_t^{q+1}u)-\partial_t^{q+1}u\|_{L^2(K)}\le C_A h_K|\partial_t^{q+1}u|_{H^1(K)}, \tag{6.126}$$
$$|\Pi_{h,m}(\partial_t^{q+1}u)-\partial_t^{q+1}u|_{H^1(K)}\le C_A|\partial_t^{q+1}u|_{H^1(K)}.$$

Summarizing (6.123)–(6.126) and using (6.26), we get

$$\int_{I_m}\Big(\sum_{\Gamma\in\mathscr{F}^I_{h,m}} h_\Gamma^{-1}\int_\Gamma\left[\pi(\Pi_{h,m}u)-\Pi_{h,m}u\right]^2\mathrm{d}S\Big)\,\mathrm{d}t \tag{6.127}$$
$$\le C_{10}\,\tau_m^{2(q+1)}\int_{I_m}\sum_{K\in\mathscr{T}_{h,m}}|\partial_t^{q+1}u|^2_{H^1(K)}\,\mathrm{d}t = C_{10}\,\tau_m^{2(q+1)}|u|^2_{H^{q+1}(I_m;H^1(\Omega))},$$

where $C_{10}=C_T^{-1}C_3C_MC_A^2$.

(ii) In what follows, we assume that $\Gamma\in\mathscr{F}^D_{h,m}$, i.e., $\Gamma\subset\partial\Omega_D\cap\partial K_\Gamma^{(L)}$, and estimate the expression

$$\kappa_{\Gamma,m} := \int_{I_m}\Big(h_\Gamma^{-1}\int_\Gamma|\pi(\Pi_{h,m}u)-\Pi_{h,m}u|^2\,\mathrm{d}S\Big)\,\mathrm{d}t.$$

Proceeding in a similar way as above, using (6.65), we find that

$$\kappa_{\Gamma,m} \le C_3\,\tau_m^{2(q+1)}h_\Gamma^{-1}\int_\Gamma\left\|\partial_t^{q+1}(\Pi_{h,m}u)\right\|^2_{L^2(I_m)}\mathrm{d}S \tag{6.128}$$
$$= C_3\,\tau_m^{2(q+1)}h_\Gamma^{-1}\int_{I_m}\Big(\int_\Gamma\left|\Pi_{h,m}(\partial_t^{q+1}u)\right|^2\mathrm{d}S\Big)\,\mathrm{d}t.$$

If we apply the multiplicative trace inequality (6.30), using assumptions (6.26) and (6.118) and results of Lemma 6.18, we get

$$\kappa_{\Gamma,m} \le C_{11}\,\tau_m^{2q}|u\|^2_{H^{q+1}(I_m;H^1(K_\Gamma^{(L)}))}, \tag{6.129}$$

with $C_{11}=C_3C_T^{-1}C_M\bar{C}$. (The symbol $|\cdot\|$ is defined by (6.103).) Then, the summation over $\Gamma\in\mathscr{F}^D_{h,m}$ yields the estimate

$$\int_{I_m}\Big(\sum_{\Gamma\in\mathscr{F}^D_{h,m}} h_\Gamma^{-1}\int_\Gamma\left|\pi(\Pi_{h,m}u)-\Pi_{h,m}u\right|^2\mathrm{d}s\Big)\,\mathrm{d}t \le C_{11}\,\tau_m^{2q}|u\|^2_{H^{q+1}(I_m;H^1(\Omega))}. \tag{6.130}$$

Now let us assume that the Dirichlet data $u_D=u_D(x,t)$ satisfy (6.116). Then $\partial_t^{q+1}u|_{\partial\Omega}=\partial_t^{q+1}u_D=0$. These relations and (6.128) imply that

$$\kappa_{\Gamma,m} \le C_3\,\tau_m^{2(q+1)}\int_{I_m}\Big(h_\Gamma^{-1}\int_\Gamma\left|\Pi_{h,m}(\partial_t^{q+1}u)-\partial_t^{q+1}u\right|^2\mathrm{d}S\Big)\,\mathrm{d}t. \tag{6.131}$$

Again we use the multiplicative trace inequality and estimates (6.126) and get the estimate

$$\int_{I_m}\Big(\sum_{\Gamma\in\mathscr{F}_{h,m}^D} h_\Gamma^{-1}\int_\Gamma \left|\pi(\Pi_{h,m}u)-\Pi_{h,m}u\right|^2 \mathrm{d}s\Big)\,\mathrm{d}t \le C_{13}\,\tau_m^{2(q+1)}|u|^2_{H^{q+1}(I_m;H^1(\Omega))} \tag{6.132}$$

with $C_{13}=C_3C_T^{-1}C_MC_A^2$. Finally, estimates (6.127) and (6.130) imply (6.117) with $C_7=C_W(C_{10}+C_{11})$ and (6.127) and (6.132) yield (6.119) with $C_8=C_W(C_{10}+C_{13})$. □

Exercise 6.20 Prove in detail estimates (6.128)–(6.132).

From Lemma 6.19 we immediately get the following conclusion.

Corollary 6.21 *If u_D from the boundary condition (6.1b) is defined by (6.116), we put $\gamma=1$. Otherwise, if u_D has a general behaviour, we set $\gamma=0$. Then*

$$\int_{I_m} J_{h,m}^\sigma(\eta^{(2)},\eta^{(2)})\,\mathrm{d}t \le C_9\tau_m^{2(q+\gamma)}\|u\|^2_{H^{q+1}(I_m;H^1(\Omega))}, \tag{6.133}$$

where $C_9=\max(C_7,C_8)$.

From Lemmas 6.17, 6.18, 6.19 and relations (6.100) and (6.101), we can derive the following estimates.

Lemma 6.22 *Let the exact solution u satisfy the regularity condition (6.102). Then there exists a constant C_{12} independent of h, τ_m, m, r and u such that*

$$\begin{aligned}&\int_{I_m}|||u-\pi u|||_m^2\,\mathrm{d}t \\ &\le C_{12}\left(h^{2(\mu-1)}|u|_{L^2(I_m;H^\mu(\Omega))}+\tau_m^{2(q+\gamma)}\|u\|^2_{H^{q+1}(I_m;H^1(\Omega))}\right),\quad m=1,\dots,r.\end{aligned} \tag{6.134}$$

If

$$u\in W^{q+1,\infty}(0,T;L^2(\Omega))\cap C([0,T];H^\mu(\Omega)), \tag{6.135}$$

then there exists a constant C_π independent of h, τ_m, m, r and u such that

$$\|u(t)-\pi u(t)\|_{L^2(\Omega)} \le C_\pi\left(h^\mu|u|_{C([0,T];H^\mu(\Omega))}+\tau_m^{q+1}|u|_{W^{q+1,\infty}(0,T;L^2(\Omega))}\right)\quad \forall t\in I_m,\quad m=1,\dots,r. \tag{6.136}$$

The choice of the parameter $\gamma=0$ or 1 is specified in Corollary 6.21.

Proof (i) From (6.100) and (6.101), we have

$$\pi u - u = \eta = \eta^{(1)} + \eta^{(2)} \tag{6.137}$$

and (6.27) with (6.4) give $|||\eta|||_m^2 = |\eta|^2_{H^1(\Omega,\mathscr{T}_{h,m})} + J^\sigma_{h,m}(\eta,\eta)$. Moreover, (6.106), (6.4), (6.109) and (6.112) yield

$$\begin{aligned}\int_{I_m} |\eta|^2_{H^1(\Omega,\mathscr{T}_{h,m})}\,\mathrm{d}t &\le 2\int_{I_m}\left(|\eta^{(1)}|^2_{H^1(\Omega,\mathscr{T}_{h,m})} + |\eta^{(2)}|^2_{H^1(\Omega,\mathscr{T}_{h,m})}\right)\mathrm{d}t \\ &\le 2C_A^2\, h^{2(\mu-1)}|u|^2_{L^2(I_m;H^\mu(\Omega))} + 2C_5\,\tau_m^{2(q+1)}|u|^2_{H^{q+1}(I_m;H^1(\Omega))}.\end{aligned} \tag{6.138}$$

Furthermore, (6.114), (6.115) and (6.133) give

$$\begin{aligned}\int_{I_m} J^\sigma_{h,m}(\eta,\eta)\,\mathrm{d}t &\le 2\int_{I_m}\left(J^\sigma_{h,m}(\eta^{(1)},\eta^{(1)}) + J^\sigma_{h,m}(\eta^{(2)},\eta^{(2)})\right)\mathrm{d}t \\ &\le 2C_J^2\, h^{2(\mu-1)}|u|^2_{L^2(I_m;H^\mu(\Omega))} + 2C_9\tau_m^{2(q+\gamma)}\|u\|^2_{H^{q+1}(I_m;H^1(\Omega))}.\end{aligned} \tag{6.139}$$

Hence, (6.134) holds with $C_{12} = 2\max(C_A^2 + C_J^2, C_5 + C_9)$.

(ii) Using the triangle inequality, (6.39), (6.28a), (6.29a) and (6.69), we have

$$\begin{aligned}&\|u(t) - \pi u(t)\|_{L^2(\Omega)} \le \|u(t) - \Pi_{h,m}u(t)\|_{L^2(\Omega)} + \|\Pi_{h,m}u(t) - \pi u(t)\|_{L^2(\Omega)} \\ &\le \|u(t) - \Pi_{h,m}u(t)\|_{L^2(\Omega)} + \|\Pi_{h,m}u(t) - \pi(\Pi_{h,m}u(t))\|_{L^2(\Omega)} \\ &\le C_A h^\mu |u|_{C(\overline{I}_m;H^\mu(\Omega))} + C_4\tau_m^{q+1}|u|_{W^{q+1,\infty}(I_m;L^2(\Omega))},\end{aligned}$$

which proves (6.136) with $C_\pi = \max(C_A, C_4)$. □

6.1.7 Error Estimate in the DG-norm

Now we are ready to formulate error estimates of the ST-DGM. First we prove the error estimate in the DG-norm on the basis of the abstract error estimate (6.92). We recall that $e_m^- = (U-u)(t_m-)$

Theorem 6.23 *Let u be the exact solution of problem (6.1) satisfying the regularity condition (6.102) and let $U \in S^{p,q}_{h,\tau}$ be its approximation given by (6.19). Let the inequality*

$$\tau_m \ge C_S h_m^2 \tag{6.140}$$

hold for all $m = 1,\ldots,r$ and let the shape-regularity assumption (6.25) and the equivalence condition (6.26) be satisfied. Then there exists a constant $C_{17} > 0$ independent of h, τ and u such that

$$\|e_m^-\|^2_{L^2(\Omega)} + \frac{\varepsilon}{2}\sum_{j=1}^{m}\int_{I_j} |||e|||_j^2\,\mathrm{d}t \tag{6.141}$$
$$\le C_{17}\varepsilon\left(h^{2(\mu-1)}|u|^2_{C([0,T];H^\mu(\Omega))} + \tau^{2(q+\gamma)}|u\|^2_{H^{q+1}(0,T;H^1(\Omega))}\right),$$
$$h \in (0,\bar{h}), \quad m = 1,\dots,r.$$

Here $\gamma = 0$, if (6.118) holds and the function u_D from the boundary condition (6.1b) has a general behaviour. If u_D is defined by (6.116), then $\gamma = 1$ and condition (6.118) is not required. The symbol $|\cdot\|$ is defined by (6.105).

Proof We start from Theorem 6.16, where we estimate the expressions depending on $\eta = \pi u - u$.

By (6.107), assumption (6.140) and the relation $\sum_{m=1}^r \tau_m = T$, we have

$$\sum_{j=0}^{m}\|\eta_j^-\|^2_{L^2(\Omega)} \le C\sum_{j=0}^{r} h_j^{2\mu}|u(t_j)|^2_{H^\mu(\Omega)} \tag{6.142}$$
$$\le Ch^{2\mu}|u^0|^2_{H^\mu(\Omega)} + C\sum_{j=1}^{r}\tau_j h_j^{2(\mu-1)}|u(t_j)|^2_{H^\mu(\Omega)}$$
$$\le C(T+\bar{h}^2)h^{2(\mu-1)}|u|^2_{C([0,T];H^\mu(\Omega))}.$$

As for the terms R_j, using (6.93) and (6.106)–(6.113), we get

$$\int_{I_j} R_j(\eta)\,\mathrm{d}t \le C\left(h_j^{2(\mu-1)}|u|^2_{L^2(I_j;H^\mu(\Omega))} + \tau_j^{2(q+1)}|u\|^2_{H^{q+1}(I_j;H^1(\Omega))}\right). \tag{6.143}$$

Further, by virtue of Lemma 6.22,

$$\int_{I_j} |||\eta|||_j^2\,\mathrm{d}t \le C\left(h_j^{2(\mu-1)}|u|^2_{L^2(I_j;H^\mu(\Omega))} + \tau_j^{2(q+\gamma)}|u\|^2_{H^{q+1}(I_j;H^1(\Omega))}\right). \tag{6.144}$$

Now we use these estimates in (6.92). Taking into account that for $m = 1,\dots,r$,

$$\sum_{j=1}^{m}|u|^2_{L^2(I_j;H^\mu(\Omega))} \le |u|^2_{L^2(0,T;H^\mu(\Omega))} \le T|u|^2_{C([0,T];H^\mu(\Omega))},$$
$$\sum_{j=1}^{m}|u\|^2_{H^{q+1}(I_j;H^1(\Omega))} \le |u\|^2_{H^{q+1}(0,T;H^1(\Omega))},$$

we immediately arrive at (6.141). □

Remark 6.24 As will be shown in Sect. 6.1.10, assumption (6.140) is not necessary, if the meshes are identical on all time levels, i.e., if $\mathscr{T}_{h,m} = \mathscr{T}_h$ for all $m = 0,\dots,r$.

Estimate (6.141) gives the error bound in the $L^2(\Omega)$-norm only at the nodes t_m, $m = 1, \ldots, r$. Our aim is to derive an error estimate in the $L^\infty(0, T; L^2(\Omega))$-norm, i.e., for all $t \in (0, T)$. This proof is based on results contained in the following section.

6.1.8 Discrete Characteristic Function

In some further considerations, the so-called *discrete characteristic function* will play an important role. This concept was introduced by Chrysafinos and Walkington in [51].

Definition 6.25 Let $y \in [t_{m-1}, t_m]$. We say that $\zeta_y \in S_{h,\tau}^{p,q}$ is a *discrete characteristic function* of the function $\xi \in S_{h,\tau}^{p,q}$ at the point y if

$$\zeta_y(t) = \xi(t) \text{ for } t \in \cup_{i=1}^{m-1} I_i, \quad \zeta = 0 \text{ on the time interval } (t_m, T], \tag{6.145}$$

and in the time interval I_m, it is defined by the conditions

$$\int_{I_m} (\zeta_y, \varphi)\mathrm{d}t = \int_{t_{m-1}}^{y} (\xi, \varphi)\mathrm{d}t \quad \forall \varphi \in S_{h,\tau}^{p,q-1}, \tag{6.146}$$

$$\zeta_y(t_{m-1}^+) = \xi(t_{m-1}^+). \tag{6.147}$$

(The function ζ_y depends also on m, but it is not emphasized by the notation.)

In order to establish the existence and uniqueness of the function ζ_y, we prove the following two lemmas. The first lemma introduces the discrete characteristic function for functions $\phi \in P_q(0, \tau)$.

Lemma 6.26 *Let $\tau > 0$ and $t \in (0, \tau)$. Then for any $\phi \in P_q(0, \tau)$ there exists exactly one $\tilde{\phi} \in P_q(0, \tau)$ satisfying the conditions*

$$\int_0^\tau \tilde{\phi}\, z \mathrm{d}\vartheta = \int_0^t \phi\, z \mathrm{d}\vartheta \quad \forall z \in P_{q-1}(0, \tau), \tag{6.148}$$

$$\tilde{\phi}(0) = \phi(0).$$

The mapping $\tilde{\psi}_t : \phi \mapsto \tilde{\phi}$ is linear and continuous in $P_q(0, \tau)$ and there exists a constant $\tilde{C}_q > 0$ depending on q only such that

$$\left\| \tilde{\phi} - \phi \right\|_{L^2(0,\tau)} \le \tilde{C}_q \|\phi\|_{L^2(t,\tau)}. \tag{6.149}$$

Moreover,

$$\left\|\tilde{\phi}\right\|_{L^2(0,\tau)} \le (1+\tilde{C}_q)\|\phi\|_{L^2(0,\tau)}. \tag{6.150}$$

Proof (i) First, we prove the existence and uniqueness of the function $\tilde{\phi}$. Since $\tilde{\phi} \in P_q(0,\tau)$, for a given ϕ the function $\tilde{\phi}$ can be sought in the form $\tilde{\phi}(\vartheta) = \sum_{i=0}^{q} c_i \vartheta^i$. By virtue of (6.148), $\tilde{\phi}(0) = \phi(0) = c_0$. Taking into account that the functions ϑ^j, $j = 0, ..., q-1$, form a basis in the space $P_{q-1}(0,\tau)$, from (6.148) for all $j = 0, \dots, q-1$ we have

$$\int_0^\tau \sum_{i=0}^{q} c_i \vartheta^i \vartheta^j \mathrm{d}\vartheta = \int_0^t \phi\, \vartheta^j \mathrm{d}\vartheta \tag{6.151}$$
$$\Longleftrightarrow \sum_{i=1}^{q} c_i \int_0^\tau \vartheta^{i+j} \mathrm{d}\vartheta = \int_0^t \phi\, \vartheta^j\, \mathrm{d}\vartheta - c_0 \int_0^\tau \vartheta^j \mathrm{d}\vartheta.$$

Let us set $a_{ij} = \int_0^\tau \vartheta^{i+j}\,\mathrm{d}\vartheta$ and $b_j = \int_0^t \phi\,\vartheta^j\,\mathrm{d}\vartheta - c_0 \int_0^\tau \vartheta^j\,\mathrm{d}\vartheta$, $i = 1, \dots q$, $j = 0, \dots, q-1$. The matrix $\mathbb{A} = (a_{ij})_{i=1,\dots,q}^{j=0,\dots,q-1}$ is nonsingular, because it is the Gram matrix of the elements ϑ^j, $j = 0, \dots, q-1$, in the scalar product

$$((\varphi, \psi)) = \int_0^\tau \vartheta \varphi(\vartheta)\psi(\vartheta)\,\mathrm{d}\vartheta, \qquad \varphi, \psi \in L^2(0,\tau).$$

Hence, the system $\mathbb{A}c = b$ has a unique solution. This proves the existence and uniqueness of the function $\tilde{\phi}$ satisfying conditions (6.148).

(ii) The linearity of the mapping $\tilde{\psi}_t : \phi \mapsto \tilde{\phi}$ is obvious. Let us prove its continuity. Since $\tilde{\phi}(0) = \phi(0)$, we can write $\tilde{\phi} - \phi = \vartheta\overline{\phi}$, where $\overline{\phi} \in P_{q-1}(0,\tau)$. By (6.148), for arbitrary $z \in P_{q-1}(0,\tau)$ we have

$$\int_0^\tau \vartheta\overline{\phi} z\, \mathrm{d}\vartheta = \int_0^\tau (\tilde{\phi} - \phi)\, z\, \mathrm{d}\vartheta \tag{6.152}$$
$$= \int_0^\tau \tilde{\phi}\, z\, \mathrm{d}\vartheta - \int_0^t \phi\, z\, \mathrm{d}\vartheta - \int_t^\tau \phi\, z\, \mathrm{d}\vartheta = -\int_t^\tau \phi\, z\, \mathrm{d}\vartheta.$$

The space $P_{q-1}(0,1)$ is finite-dimensional and the expressions

$$\left(\int_0^1 \hat{\phi}^2\,\mathrm{d}\zeta\right)^{1/2}, \quad \left(\int_0^1 \zeta\,\hat{\phi}^2\,\mathrm{d}\zeta\right)^{1/2} \quad \text{and} \quad \left(\int_0^1 \zeta^2\,\hat{\phi}^2\,\mathrm{d}\zeta\right)^{1/2} \tag{6.153}$$

are norms in $P_{q-1}(0,1)$. It follows from their equivalence that there exist positive constants C_{q1} a C_{q2} depending on q only such that

$$C_{q1}\int_0^1 \hat{\phi}^2\,\mathrm{d}\zeta \le \int_0^1 \zeta\,\hat{\phi}^2\,\mathrm{d}\zeta \quad \forall\hat{\phi}\in P_{q-1}(0,1), \tag{6.154}$$

$$C_{q2}\int_0^1 \zeta^2\hat{\phi}^2\,\mathrm{d}\zeta \le \int_0^1 \hat{\phi}^2\,\mathrm{d}\zeta \quad \forall\hat{\phi}\in P_{q-1}(0,1).$$

As $\zeta^2 \le 1$ for $\zeta \in [0,1]$, we can choose $C_{q2} = 1$. Putting $\zeta = \vartheta/\tau$ and using the substitution theorem, we get

$$C_{q1}\,\tau\int_0^\tau \overline{\phi}^2\,\mathrm{d}\vartheta \le \int_0^\tau \vartheta\,\overline{\phi}^2\,\mathrm{d}\vartheta \quad \forall\overline{\phi}\in P_{q-1}(0,\tau), \tag{6.155}$$

$$\int_0^\tau \vartheta^2\overline{\phi}^2\,\mathrm{d}\vartheta \le \tau^2\int_0^\tau \overline{\phi}^2\,\mathrm{d}\vartheta \quad \forall\overline{\phi}\in P_{q-1}(0,\tau).$$

Now let us set $z := \overline{\phi}$ in (6.152). By (6.155), (6.152) and the Cauchy inequality,

$$C_{q1}\,\tau\int_0^\tau \overline{\phi}^2\,\mathrm{d}\vartheta \le \int_0^\tau \vartheta\,\overline{\phi}^2\,\mathrm{d}\vartheta = -\int_t^\tau \phi\,\overline{\phi}\,\mathrm{d}\vartheta \le \left(\int_t^\tau \phi^2\,\mathrm{d}\vartheta\right)^{1/2}\left(\int_t^\tau \overline{\phi}^2\,\mathrm{d}\vartheta\right)^{1/2}$$
$$\le \left(\int_t^\tau \phi^2\,\mathrm{d}\vartheta\right)^{1/2}\left(\int_0^\tau \overline{\phi}^2\,\mathrm{d}\vartheta\right)^{1/2}.$$

Hence,

$$\left(C_{q1}\,\tau\right)^2\int_0^\tau \overline{\phi}^2\,\mathrm{d}\vartheta \le \int_t^\tau \phi^2\,\mathrm{d}\vartheta.$$

This inequality and (6.155) imply that

$$C_{q1}^2\int_0^\tau (\tilde{\phi}-\phi)^2\,\mathrm{d}\vartheta = C_{q1}^2\int_0^\tau \vartheta^2\overline{\phi}^2\,\mathrm{d}\vartheta \le \left(C_{q1}\,\tau\right)^2\int_0^\tau \overline{\phi}^2\,\mathrm{d}\vartheta \le \int_t^\tau \phi^2\,\mathrm{d}\vartheta,$$

which is (6.149) with $\tilde{C}_q = 1/C_{q1}$. Finally, summarizing the above results, we obtain

$$\left\|\tilde{\phi}\right\|_{L^2(0,\tau)} \le \left\|\tilde{\phi}-\phi\right\|_{L^2(0,\tau)} + \|\phi\|_{L^2(0,\tau)} \le \frac{1}{C_{q1}}\|\phi\|_{L^2(t,\tau)} + \|\phi\|_{L^2(0,\tau)}$$
$$\le (1+\tilde{C}_q)\|\phi\|_{L^2(0,\tau)},$$

which gives (6.150). □

In the previous lemma, we presented the concept of the discrete characteristic function for the function space $P_q(0,\tau)$. Now, we extend this concept to functions from $P_q(0,\tau;S_{h,m}^p)$. In the following considerations we use an orthonormal basis $\{\phi_k(\vartheta),\ k=0,\dots,q\}$ in the space $P_q(0,\tau)$, constructed by the orthonormalization process applied to the sequence $\{\vartheta^k\}_{k=0}^q$. Then, any $v \in P_q(0,\tau;S_{h,m}^p)$ can be written in the form

$$v(x,t) = \sum_{k=0}^{q} \phi_k(t) v_k(x), \quad t \in (0,\tau),\ x \in \Omega, \tag{6.156}$$

where $v_k \in S^p_{h,m}$.

Lemma 6.27 *Let $t \in [0,\tau]$ be arbitrary and fixed. Then for each $v \in P_q(0,\tau; S^p_{h,m})$ there exists exactly one function $\tilde{v} \in P_q(0,\tau; S^p_{h,m})$ satisfying the conditions*

$$\begin{aligned} \int_0^\tau (\tilde{v}, w)\,\mathrm{d}\vartheta &= \int_0^t (v, w)\,\mathrm{d}\vartheta \quad \forall w \in P_{q-1}(0,\tau; S^p_{h,m}), \\ \tilde{v}(0) &= v(0). \end{aligned} \tag{6.157}$$

If we write the function v in the form of (6.156), then

$$\tilde{v} = \sum_{k=0}^{q} \tilde{\phi}_k v_k, \tag{6.158}$$

where $\tilde{\phi}_k = \tilde{\psi}_t(\phi_k)$ and $\tilde{\psi}_t$ is the mapping from Lemma 6.26. Moreover, there exists a constant $C_{CH} > 0$ dependent on q only such that

$$\int_0^\tau |||\tilde{v}|||_m^2\,\mathrm{d}\vartheta \le C_{CH} \int_0^\tau |||v|||_m^2\,\mathrm{d}\vartheta. \tag{6.159}$$

Proof (i) First we show that the function $\tilde{v}$ defined by (6.158) satisfies (6.157). We can see that

$$\tilde{v}(0) = \sum_{k=0}^{q} \tilde{\phi}_k(0) v_k = \sum_{k=0}^{q} \phi_k(0) v_k = v(0).$$

Further, let $w \in P_{q-1}(0,\tau; S^p_{h,m})$. This function can be expressed in the form

$$w(x,\vartheta) = \sum_{n=0}^{q-1} \phi_n(\vartheta) w_n(x), \quad w_n \in S^p_{h,m}.$$

Then, by virtue of Lemma 6.26,

$$\begin{aligned} \int_0^\tau (\tilde{v}, w)\,\mathrm{d}\vartheta &= \int_0^\tau \left(\sum_{k=0}^{q} \tilde{\phi}_k(\vartheta) v_k, \sum_{n=0}^{q-1} \phi_n(\vartheta) w_n \right) \mathrm{d}\vartheta \\ &= \sum_{k=0}^{q} \sum_{n=0}^{q-1} (v_k, w_n) \int_0^\tau \tilde{\phi}_k(\vartheta)\, \phi_n(\vartheta)\mathrm{d}\vartheta = \sum_{k=0}^{q} \sum_{n=0}^{q-1} (v_k, w_n) \int_0^t \phi_k(\vartheta)\, \phi_n(\vartheta)\mathrm{d}\vartheta \end{aligned} \tag{6.160}$$

$$= \int_0^t \left(\sum_{k=0}^{q} \phi_k(\vartheta) v_k, \sum_{n=0}^{q-1} \phi_n(\vartheta) w_n \right) \mathrm{d}\vartheta = \int_0^t (v, w)\, \mathrm{d}\vartheta.$$

(ii) Further, we prove uniqueness. Let us assume that there exist functions $\tilde{v}_1, \tilde{v}_2 \in P_q(0, \tau; S^p_{h,m})$ satisfying (6.157). Since $\tilde{v}_1(0) = \tilde{v}_2(0) = v(0)$, there exists $w_1 \in P_{q-1}(0, \tau; S^p_{h,m})$ such that $\tilde{v}_1(\vartheta) - \tilde{v}_2(\vartheta) = \vartheta w_1(\vartheta)$. Then

$$\int_0^\tau (\tilde{v}_1, w)\mathrm{d}\vartheta = \int_0^\tau (\tilde{v}_2, w)\mathrm{d}\vartheta = \int_0^t (v, w)\mathrm{d}\vartheta \quad \forall w \in P_{q-1}(0, \tau; S^p_{h,m}).$$

It is possible to set $w = w_1$ and get

$$\int_0^\tau (\tilde{v}_1 - \tilde{v}_2, w_1)\mathrm{d}\vartheta = \int_0^\tau (\vartheta w_1, w_1)\mathrm{d}\vartheta = \int_0^\tau \vartheta (w_1, w_1)\mathrm{d}\vartheta.$$

The function $\vartheta \|w_1(\vartheta)\|^2$ is continuous and positive in the interval $(0, \tau)$. This implies that $\|w_1(\vartheta)\|^2 = 0$ for all $\vartheta \in (0, \tau)$. Hence, $w_1(\vartheta) = 0$ and $\tilde{v}_1 = \tilde{v}_2$.

(iii) Using Fubini's theorem, the Cauchy inequality, (6.150), (6.158), the inequality $\sum_{i,j=0}^q a_i a_j \leq (q+1) \sum_{i=0}^q a_i^2$ and the orthonormality of the functions ϕ_k, we find that

$$\begin{aligned}
\sum_{K \in \mathscr{T}_{h,m}} \int_0^\tau |\tilde{v}|^2_{H^1(K)} \mathrm{d}\vartheta &= \sum_{K \in \mathscr{T}_{h,m}} \int_0^\tau \left(\int_K (\nabla \tilde{v} \cdot \nabla \tilde{v}) \mathrm{d}x \right) \mathrm{d}\vartheta \\
&= \sum_{K \in \mathscr{T}_{h,m}} \sum_{k,n=0}^{q} \int_0^\tau \tilde{\phi}_k(\vartheta) \tilde{\phi}_n(\vartheta) \mathrm{d}\vartheta \int_K (\nabla v_k \cdot \nabla v_n) \mathrm{d}x \\
&\leq \sum_{K \in \mathscr{T}_{h,m}} \sum_{k,n=0}^{q} \left\| \tilde{\phi}_k \right\|_{L^2(0,\tau)} \left\| \tilde{\phi}_n \right\|_{L^2(0,\tau)} |v_k|_{H^1(K)} |v_n|_{H^1(K)} \\
&\leq (1 + \tilde{C}_q)^2 \sum_{K \in \mathscr{T}_{h,m}} \sum_{k,n=0}^{q} \|\phi_k\|_{L^2(0,\tau)} \|\phi_n\|_{L^2(0,\tau)} |v_k|_{H^1(K)} |v_n|_{H^1(K)} \\
&\leq (1 + \tilde{C}_q)^2 (q+1) \sum_{K \in \mathscr{T}_{h,m}} \sum_{k=0}^{q} \|\phi_k\|^2_{L^2(0,\tau)} |v_k|^2_{H^1(K)} \\
&= C_{CH} \sum_{K \in \mathscr{T}_{h,m}} \sum_{k=0}^{q} \int_0^\tau \phi_k^2(\vartheta) \mathrm{d}\vartheta \int_K |\nabla v_k|^2 \mathrm{d}x \\
&= C_{CH} \sum_{K \in \mathscr{T}_{h,m}} \sum_{k,n=0}^{q} \int_0^\tau \phi_k(\vartheta) \phi_n(\vartheta) \mathrm{d}\vartheta \int_K (\nabla v_k \cdot \nabla v_n)\, \mathrm{d}x
\end{aligned}$$

$$= C_{CH} \sum_{K \in \mathscr{T}_{h,m}} \int_0^{\tau} \left(\int_K (\nabla v \cdot \nabla v) \mathrm{d}x \right) \mathrm{d}\vartheta.$$

where we set $C_{CH} = (1 + \tilde{C}_q)^2(q+1)$. Hence,

$$\sum_{K \in \mathscr{T}_{h,m}} \int_0^{\tau} |\tilde{v}|^2_{H^1(K)} \mathrm{d}\vartheta \leq C_{CH} \sum_{K \in \mathscr{T}_{h,m}} \int_0^{\tau} |v|^2_{H^1(K)} \mathrm{d}\vartheta. \tag{6.161}$$

Similarly, we can show that

$$\sum_{\Gamma \in \mathscr{F}^I_{h,m}} \int_0^{\tau} \left(\int_{\Gamma} \frac{C_W}{h_{\Gamma}} [\tilde{v}]^2 \mathrm{d}S \right) \mathrm{d}\vartheta$$

$$= \sum_{\Gamma \in \mathscr{F}^I_{h,m}} \int_0^{\tau} \left(\int_{\Gamma} \frac{C_W}{h_{\Gamma}} [\sum_{k=0}^{q} \tilde{\phi}_k(\vartheta) v_k][\sum_{n=0}^{q} \tilde{\phi}_n(\vartheta) v_n] \mathrm{d}S \right) \mathrm{d}\vartheta$$

$$= \sum_{\Gamma \in \mathscr{F}^I_{h,m}} \sum_{k,n=0}^{q} \int_0^{\tau} \tilde{\phi}_k(\vartheta) \tilde{\phi}_n(\vartheta) \mathrm{d}\vartheta \int_{\Gamma} \frac{C_W}{h_{\Gamma}} [v_k][v_n] \mathrm{d}S$$

$$\leq \sum_{\Gamma \in \mathscr{F}^I_{h,m}} \sum_{k,n=0}^{q} \frac{C_W}{h_{\Gamma}} \left\| \tilde{\phi}_k \right\|_{L^2(0,\tau)} \left\| \tilde{\phi}_n \right\|_{L^2(0,\tau)} \|[v_k]\|_{L^2(\Gamma)} \|[v_n]\|_{L^2(\Gamma)}$$

$$\leq (1 + \tilde{C}_q)^2 \sum_{\Gamma \in \mathscr{F}^I_{h,m}} \sum_{k,n=0}^{q} \frac{C_W}{h_{\Gamma}} \|\phi_k\|_{L^2(0,\tau)} \|\phi_n\|_{L^2(0,\tau)} \|[v_k]\|_{L^2(\Gamma)} \|[v_n]\|_{L^2(\Gamma)}$$

$$\leq (1 + \tilde{C}_q)^2 (q+1) \sum_{\Gamma \in \mathscr{F}^I_{h,m}} \sum_{k=0}^{q} \frac{C_W}{h_{\Gamma}} \|\phi_k\|^2_{L^2(0,\tau)} \|[v_k]\|^2_{L^2(\Gamma)}$$

$$= C_{CH} \sum_{\Gamma \in \mathscr{F}^I_{h,m}} \sum_{k=0}^{q} \int_0^{\tau} \phi_k^2(\vartheta) \mathrm{d}\vartheta \int_{\Gamma} \frac{C_W}{h_{\Gamma}} [v_k]^2 \mathrm{d}S$$

$$= C_{CH} \sum_{\Gamma \in \mathscr{F}^I_{h,m}} \sum_{k,n=0}^{q} \int_0^{\tau} \phi_k(\vartheta) \phi_n(\vartheta) \mathrm{d}\vartheta \int_{\Gamma} \frac{C_W}{h_{\Gamma}} [v_k][v_n] \mathrm{d}S$$

$$= C_{CH} \sum_{\Gamma \in \mathscr{F}^I_{h,m}} \int_0^{\tau} \left(\int_{\Gamma} \frac{C_W}{h_{\Gamma}} [\sum_{k=0}^{q} \phi_k(\vartheta) v_k][\sum_{n=0}^{q} \phi_n(\vartheta) v_n] \mathrm{d}S \right) \mathrm{d}\vartheta$$

$$= C_{CH} \sum_{\Gamma \in \mathscr{F}^I_{h,m}} \int_0^{\tau} \left(\int_{\Gamma} \frac{C_W}{h_{\Gamma}} [v]^2 \mathrm{d}S \right) \mathrm{d}\vartheta.$$

As we see, it holds that

$$\sum_{\Gamma\in\mathscr{F}^I_{h,m}} \int_0^\tau \left(\int_\Gamma \frac{C_W}{h_\Gamma}[\tilde{v}]^2 \mathrm{d}S\right)\mathrm{d}\vartheta \le C_{CH} \sum_{\Gamma\in\mathscr{F}^I_{h,m}} \int_0^\tau \left(\int_\Gamma \frac{C_W}{h_\Gamma}[v]^2 \mathrm{d}S\right)\mathrm{d}\vartheta. \quad (6.162)$$

In the same way we find that

$$\sum_{\Gamma\in\mathscr{F}^D_{h,m}} \int_0^\tau \left(\int_\Gamma \frac{C_W}{h_\Gamma}\tilde{v}^2 \mathrm{d}S\right)\mathrm{d}\vartheta \le C_{CH} \sum_{\Gamma\in\mathscr{F}^D_{h,m}} \int_0^\tau \left(\int_\Gamma \frac{C_W}{h_\Gamma}v^2 \mathrm{d}S\right)\mathrm{d}\vartheta. \quad (6.163)$$

Now, (6.15), (6.27), (6.161)–(6.163) already yield (6.159). □

Corollary 6.28 *It is possible to show that the discrete characteristic function is translationally invariant. This means that if we set $\tau = \tau_m$ and apply the linear transformation $t = \vartheta + t_{m-1}$, $\vartheta \in (0, \tau_m)$, and set $v(\vartheta) = \xi(\vartheta + t_{m-1})$, $w(\vartheta) = \varphi(\vartheta + t_{m-1})$, then it follows from Lemma 6.27 that for each $y \in I_m$ there exists a unique function ζ_y satisfying (6.146), (6.147) and*

$$\int_{I_m} |||\zeta_y|||_m^2 \mathrm{d}t \le C_{CH} \int_{I_m} |||\xi|||_m^2 \mathrm{d}t, \quad (6.164)$$

where the constant C_{CH} depends only on q.

Exercise 6.29 Prove relation (6.164) in detail.

6.1.9 Error Estimate in the $L^\infty(0, T; L^2(\Omega))$-norm

Theorem 6.30 *Let u be the exact solution of problem (6.1) satisfying the regularity condition*

$$u \in W^{q+1,\infty}(0, T; L^2(\Omega)) \cap C([0, T]; H^s(\Omega)), \quad (6.165)$$

where $s \ge 2$ is an integer and $\mu = \min(p+1, s)$. Let $U \in S^{p,q}_{h,\tau}$ be its approximation given by (6.19). Let (6.140) hold for all $m = 1, \ldots, r$, and let the shape-regularity assumption (6.25) and the equivalence condition (6.26) be satisfied. Then there exists a constant $C_{18} > 0$ independent of h, τ and u such that

$$\begin{aligned} &\sup_{t\in I_m} \|u(t) - U(t)\|^2_{L^2(\Omega)} \quad (6.166)\\ &\le C_{18}\left(h^{2(\mu-1)}|u|^2_{C([0,T];H^\mu(\Omega))} + \tau_m^{2(q+1)}|u|^2_{W^{q+1,\infty}(0,T;L^2(\Omega))}\right),\\ &\qquad h\in(0,\bar{h}),\quad m = 1,\ldots,r. \end{aligned}$$

Proof We use the notation $\xi = U - \pi u$. Let us recall that $U_0^- = \Pi_{h,0}u^0$, $\xi_0^- = 0$ and $\eta_0^- = u^0 - \Pi_{h,0}u^0$. Moreover, we set $I_0 = \{t_0\} = \{0\}$ (the set formed by $t_0 = 0$) and, hence

$$\sup_{t\in I_0}\|\xi(t)\|_{L^2(\Omega)} = \|\xi_0^-\|_{L^2(\Omega)} = 0. \tag{6.167}$$

Let $m \in \{1,\ldots,r\}$. We define the values $\xi(t_{m-1})$ and $\xi(t_m)$ as the limits $\xi_{m-1}^+ = \lim_{t\to t_{m-1}+}\xi(t)$ and $\xi_m^- = \lim_{t\to t_m-}\xi(t)$, respectively. Then, by the regularity assumption (6.165), the function $\|\xi(t)\|_{L^2(\Omega)}$ is continuous in the closed interval $\overline{I}_m = [t_{m-1}, t_m]$. Now by $y \in \overline{I}_m$ we denote the maximum point of the function $\|\xi(t)\|_{L^2(\Omega)}$, $t \in \overline{I}_m$. We simply write $y = \arg\sup_{t\in\overline{I}_m}\|\xi(t)\|_{L^2(\Omega)}$. Hence

$$\sup_{t\in I_m}\|\xi(t)\|_{L^2(\Omega)} = \|\xi(y)\|_{L^2(\Omega)}.$$

By $\tilde{\xi}$ we denote the discrete characteristic function ζ_y to ξ at the point y, introduced in Definition 6.25. Then, by (6.146) and (6.147),

$$\int_{I_m}(\xi',\tilde{\xi})\,\mathrm{d}t = \int_{t_{m-1}}^{y}(\xi',\xi)\,\mathrm{d}t,\quad \tilde{\xi}_{m-1}^+ = \xi_{m-1}^+,\quad \{\tilde{\xi}\}_{m-1} = \{\xi\}_{m-1}, \tag{6.168}$$

since ξ' is a polynomial of degree $q-1$ with respect to time. In analogy to (6.86) we get

$$\begin{aligned}&2\int_{t_{m-1}}^{y}(\xi',\xi)\,\mathrm{d}t + 2(\{\xi\}_{m-1},\xi_{m-1}^+)\\ &= \|\xi(y)\|_{L^2(\Omega)}^2 - \|\xi_{m-1}^-\|_{L^2(\Omega)}^2 + \|\{\xi\}_{m-1}\|_{L^2(\Omega)}^2.\end{aligned} \tag{6.169}$$

Moreover, the use of identities (6.168) in (6.169) implies the relation

$$\begin{aligned}&\int_{I_m}2(\xi',\tilde{\xi})\,\mathrm{d}t + 2(\{\xi\}_{m-1},\tilde{\xi}_{m-1}^+)\\ &= \|\xi(y)\|_{L^2(\Omega)}^2 - \|\xi_{m-1}^-\|_{L^2(\Omega)}^2 + \|\{\xi\}_{m-1}\|_{L^2(\Omega)}^2.\end{aligned} \tag{6.170}$$

Now identity (6.80), where we set $\varphi = \tilde{\xi}$, gives

$$\begin{aligned}&\int_{I_m}\Big((\xi',\tilde{\xi}) + A_{h,m}(\xi,\tilde{\xi})\Big)\,\mathrm{d}t + (\{\xi\}_{m-1},\tilde{\xi}_{m-1}^+)\\ &= -\int_{I_m}(\eta',\tilde{\xi})\,\mathrm{d}t + (\{\eta\}_{m-1},\tilde{\xi}_{m-1}^+) - \int_{I_m}A_{h,m}(\eta,\tilde{\xi})\,\mathrm{d}t.\end{aligned} \tag{6.171}$$

Then, if we use (6.171), (6.170), (6.87) with $\delta = 1$, (6.96) and notation (6.93), and omit the term with $\|\{\xi\}_{m-1}\|_{L^2(\Omega)}^2$ on the left-hand side of the resulting relation, we get

$$\sup_{t\in I_m} \|\xi(t)\|^2_{L^2(\Omega)} - \|\xi^-_{m-1}\|^2_{L^2(\Omega)} \tag{6.172}$$
$$\le \|\eta^-_{m-1}\|^2_{L^2(\Omega)} - 2\int_{I_m} A_{h,m}(\xi,\tilde{\xi})\,\mathrm{d}t + 2\varepsilon\tilde{C}_B^2 \int_{I_m} R_m(\eta)\,\mathrm{d}t + \frac{\varepsilon}{2}\int_{I_m} |||\tilde{\xi}|||^2_m\,\mathrm{d}t.$$

By (2.128),

$$A_{h,m}(\xi,\tilde{\xi}) \le \varepsilon C_B |||\xi|||_m\ |||\tilde{\xi}|||_m. \tag{6.173}$$

Since $-\sup_{t\in I_{m-1}} \|\xi(t)\|_{L^2(\Omega)} \le -\|\xi^-_{m-1}\|_{L^2(\Omega)}$, by relations (6.164), (6.173) and the Young inequality, from (6.172) we get

$$\sup_{t\in I_m} \|\xi(t)\|^2_{L^2(\Omega)} - \sup_{t\in I_{m-1}} \|\xi(t)\|^2_{L^2(\Omega)} \tag{6.174}$$
$$\le \|\eta^-_{m-1}\|^2_{L^2(\Omega)} + 2\int_{I_m} |A_{h,m}(\xi,\tilde{\xi})|\,\mathrm{d}t + 2\varepsilon\tilde{C}_B^2 \int_{I_m} R_m(\eta)\,\mathrm{d}t + \frac{\varepsilon}{2}\int_{I_m} |||\tilde{\xi}|||^2_m\,\mathrm{d}t$$
$$\le \|\eta^-_{m-1}\|^2_{L^2(\Omega)} + C_B\varepsilon\int_{I_m} |||\xi|||^2_m\,\mathrm{d}t + (\frac{1}{2} + C_B)\varepsilon\int_{I_m} |||\tilde{\xi}|||^2_m\,\mathrm{d}t + 2\varepsilon\tilde{C}_B^2 \int_{I_m} R_m(\eta)\,\mathrm{d}t$$
$$\le \|\eta^-_{m-1}\|^2_{L^2(\Omega)} + K\frac{\varepsilon}{2}\int_{I_m} |||\xi|||^2_m\,\mathrm{d}t + 2\varepsilon\tilde{C}_B^2 \int_{I_m} R_m(\eta)\,\mathrm{d}t,$$

where $K = 2C_B + (1 + 2C_B)C_{CH}$.

Multiplying inequality (6.97) by K and summing with (6.174), we get

$$K\|\xi^-_m\|^2_{L^2(\Omega)} + \sup_{t\in I_m} \|\xi(t)\|^2_{L^2(\Omega)} - K\|\xi^-_{m-1}\|^2_{L^2(\Omega)} - \sup_{t\in I_{m-1}} \|\xi(t)\|^2_{L^2(\Omega)} \tag{6.175}$$
$$\le 2(K+1)\varepsilon\tilde{C}_B^2 \int_{I_m} R_m(\eta)\,\mathrm{d}t + (K+1)\|\eta^-_{m-1}\|^2_{L^2(\Omega)}.$$

The summation of (6.175) over $m = 1,\ldots,n \le r$ and using (6.167) yield

$$K\|\xi^-_n\|^2_{L^2(\Omega)} + \sup_{t\in I_n} \|\xi(t)\|^2_{L^2(\Omega)} \tag{6.176}$$
$$\le 2(K+1)\varepsilon\tilde{C}_B^2 \sum_{m=1}^{n}\int_{I_m} R_m(\eta)\,\mathrm{d}t + (K+1)\sum_{m-1}^{r} \|\eta^-_{m-1}\|^2_{L^2(\Omega)}.$$

Now, if we use estimate (6.143), the inequality

$$\sup_{t\in I_m} \|u(t) - U(t)\|^2_{L^2(\Omega)} \tag{6.177}$$
$$\le 2\left(\sup_{t\in I_m} \|\xi(t)\|^2_{L^2(\Omega)} + \sup_{t\in I_m} \|\eta(t)\|^2_{L^2(\Omega)}\right),$$

and estimates (6.136) and (6.142), we immediately get the error estimate (6.166), which we wanted to prove. □

6.1.10 The Case of Identical Meshes on All Time Levels

In this section we show that if all meshes $\mathscr{T}_{h,m}$, $m = 1 \ldots, r$, are identical, then the assumption (6.140) can be avoided. In order to prove it, we are concerned with the expression

$$\int_{I_m} (\eta', \xi)\, \mathrm{d}t + (\{\eta\}_{m-1}, \xi_{m-1}^{+}), \tag{6.178}$$

which was rewritten in (6.91) in a more general form with the function φ written instead of ξ.

If $\mathscr{T}_{h,m} = \mathscr{T}_h$ for all $m = 1, \ldots, r$, then all spaces $S_{h,m}^p$ and forms $a_{h,m}$, $b_{h,m}, \ldots$ are also identical: $S_{h,m}^p = S_h^p$, $a_{h,m} = a_h$, $J_{h,m}^\sigma = J_h^\sigma, \ldots$ for all $m = 1, \ldots, r$. This implies that $\{\xi\}_{m-1} \in S_h^p$ and, by virtue of (6.100), (6.33b) and (6.7), we have

$$(\eta_{m-1}^{-}, \{\xi\}_{m-1}) = 0. \tag{6.179}$$

Now from (6.179) and (6.91) with $\varphi := \xi$ we immediately see that

$$\int_{I_m} (\eta', \xi)\, \mathrm{d}t + \left(\{\eta\}_{m-1}, \xi_{m-1}^{+}\right) = 0. \tag{6.180}$$

Further, if we follow the proof of Theorem 6.16 on the abstract error estimate, we find out that Lemma 6.15 will not be applied, and regarding the right-hand side of estimate (6.94), the term $\|\eta_{m-1}^{-}\|_{L^2(\Omega)}$ will not appear. This implies that in the abstract error estimate (6.92), the expression $\sum_{m=1}^{n} \|\eta_{m-1}^{-}\|_{L^2(\Omega)}$ is missing, which means that we get the estimate

$$\|e_n^{-}\|_{L^2(\Omega)}^2 + \varepsilon \sum_{m=1}^{n} \int_{I_m} |||e|||_m^2\, \mathrm{d}t \le C\, \varepsilon \sum_{m=1}^{n} \int_{I_m} R_m(\eta)\, \mathrm{d}t + \varepsilon \sum_{m=1}^{n} \int_{I_m} |||\eta|||_m^2\, \mathrm{d}t,$$
$$n = 1, \ldots, r,\ h \in (0, \bar{h}). \tag{6.181}$$

From this estimate, following the proofs of Theorems 6.23 and 6.30, we obtain the error estimates in terms of h and τ without assumption (6.140).

6.1.11 Alternative Proof of Lemma 6.12

Here we prove Lemma 6.12 without using Theorem 3.1.4 from [52].

Lemma 6.31 *Let $s \in C^\infty([0, 1])$, $s(0) = 0$ and*

$$\int_0^1 \vartheta^i s(x)\,\mathrm{d}\vartheta = 0, \quad i = 0, \dots, q-1. \tag{6.182}$$

Then

$$\|s\|_{L^2(0,1)} \leq C\, \|s^{(q+1)}\|_{L^2(0,1)}, \tag{6.183}$$

where C is a constant independent of the function s, and the symbol $s^{(k)}$ denotes the derivative $\mathrm{d}^k s/\mathrm{d}\vartheta^k$.

Proof Let us expand the function with the aid of the Taylor formula with integral remainder

$$s(\vartheta) = s(0) + \dots + \frac{s^{(q)}(0)}{q!}\,\vartheta^q + \int_0^\vartheta \frac{(\vartheta-\tau)^q}{q!} s^{(q+1)}(\tau)\,\mathrm{d}\tau, \quad \vartheta \in [0,1]. \tag{6.184}$$

In the space $L^2(0,1)$ we choose an orthonormal system of polynomials φ_i, $i = 0, 1, \dots$, such that φ_i is a polynomial of degree i and $\varphi_i(0) \neq 0$. (At the end of this section we show how this system can be constructed.) Obviously,

$$\int_0^1 \vartheta^i s(\vartheta)\,\mathrm{d}\vartheta = 0, \quad i = 0, \dots, q-1 \tag{6.185}$$
$$\Longleftrightarrow \quad \int_0^1 \varphi_i(\vartheta) s(\vartheta)\,\mathrm{d}\vartheta = 0, \quad i = 0, \dots, q-1.$$

By virtue of the properties of the system φ_i, $i = 0, 1, \dots$, the expansion (6.184) can be written in the form

$$s(\vartheta) = \sum_{i=0}^{q} c_i \varphi_i(\vartheta) + \int_0^\vartheta \frac{(\vartheta-\tau)^q}{q!} s^{(q+1)}(\tau)\,\mathrm{d}\tau, \quad \vartheta \in [0,1], \tag{6.186}$$

where c_i are constants depending on the values $s(0), s'(0), \dots, s^{(q)}(0)$. From assumption (6.182) and equivalence (6.185) for $j = 0, \dots, q-1$, we get

$$0 = \int_0^1 \varphi_j(\vartheta) s(\vartheta)\,\mathrm{d}\vartheta = c_j + \int_0^1 \varphi_j(\vartheta) \int_0^\vartheta \frac{(\vartheta-\tau)^q}{q!} s^{(q+1)}(\tau)\,\mathrm{d}\tau\,\mathrm{d}\vartheta.$$

The use of Fubini's theorem yields

$$c_j = -\int_0^1 \varphi_j(\vartheta) \int_0^\vartheta \frac{(\vartheta-\tau)^q}{q!} s^{(q+1)}(\tau)\,\mathrm{d}\tau\,\mathrm{d}\vartheta - -\int_0^1 \psi_j(\tau) s^{(q+1)}(\tau)\,\mathrm{d}\tau, \tag{6.187}$$

where

$$\psi_j(\tau) = \frac{1}{q!} \int_\tau^1 \varphi_j(\vartheta)(\vartheta - \tau)^q \,\mathrm{d}\vartheta, \quad j = 0, \ldots, q-1.$$

Since $\varphi_q(0) \neq 0$, from the assumption that $s(0) = 0$ and expansion (6.186), we get

$$c_q = -\frac{1}{\varphi_q(0)} \sum_{i=0}^{q-1} c_i \varphi_i(0) = \int_0^1 \psi_q(\tau) s^{(q+1)}(\tau)\,\mathrm{d}\tau$$

with

$$\psi_q(\tau) = \frac{1}{\varphi_q(0)} \sum_{j=0}^{q-1} \varphi_j(0)\psi_j(\tau).$$

Substituting in expansion (6.186) for $c_i,\ i = 0, \ldots, q$, we find that

$$s(\vartheta) = \int_0^1 k(\vartheta, \tau) s^{(q+1)}(\tau)\,\mathrm{d}\tau, \tag{6.188}$$

where

$$k(\vartheta, \tau) = \begin{cases} \frac{(\vartheta-\tau)^q}{q!} + \varphi_q(\vartheta)\psi_q(\tau) - \sum_{i=0}^{q-1} \varphi_i(\vartheta)\psi_i(\tau) & \text{for} \quad 1 \geq \vartheta > \tau \geq 0, \\ \varphi_q(\vartheta)\psi_q(\tau) - \sum_{i=0}^{q-1} \varphi_i(\vartheta)\psi_i(\tau) & \text{for} \quad 1 \geq \tau > \vartheta \geq 0. \end{cases}$$

The function $k(\vartheta, t)$ is continuous on the set $[0, 1] \times [0, 1]$ and from (6.188) we get (6.183). □

Lemma 6.32 *Let $s \in H^{q+1}(0, 1)$, $s(0) = 0$ and*

$$\int_0^1 \vartheta^i s(\vartheta)\,\mathrm{d}\vartheta = 0, \quad i = 0, \ldots, q-1. \tag{6.189}$$

Then

$$\|s\|_{L^2(0,1)} \leq C\, \|s^{(q+1)}\|_{L^2(0,1)}, \tag{6.190}$$

where $C > 0$ is a constant independent of the function s.

Proof The space $C^\infty([0, 1])$ is dense in $H^{q+1}(0, 1)$. Therefore, there exists a sequence $s_n \in C^\infty([0, 1])$ such that

$$\lim_{n\to\infty} \|s_n - s\|_{H^{q+1}(0,1)} = 0.$$

This fact and Lemma 6.31 imply that

$$\|s_n\|_{L^2(0,1)} \to \|s\|_{L^2(0,1)}, \ \|s_n^{(q+1)}\|_{L^2(0,1)} \to \|s^{(q+1)}\|_{L^2(0,1)} \quad \text{as } n \to \infty,$$
$$\|s_n\|_{L^2(0,1)} \leq C\|s_n^{(q+1)}\|_{L^2(0,1)}, \quad n = 1, 2, \ldots.$$

Hence, (6.190) holds. □

Now we can finish the proof of Lemma 6.12. Using the notation in (6.53), we have $z \in H^{q+1}(0, 1)$, $z(0) = 0$ and

$$\int_0^1 z(\vartheta)\vartheta^j \, d\vartheta = 0 \quad \text{for } j = 0, \ldots, q-1.$$

By Lemma 6.32, the function z satisfies (6.190) (i.e. (6.63)) and, by virtue of (6.55), estimate (6.64) holds, which implies estimate (6.50). This finishes the alternative proof of Lemma 6.12.

Finally, we show how to construct a system of orthonormal polynomials φ_i, $i = 0, 1, \ldots$, in the space $L^2(0, 1)$ such that φ_i is a polynomial of degree i satisfying $\varphi_i(0) \neq 0$. It is possible to put

$$\varphi_i(\vartheta) = \sqrt{2}\,\mathscr{L}_i(1 - 2\vartheta), \quad \vartheta \in [0, 1], \ i = 0, 1, \ldots,$$

where $\mathscr{L}_i$ is the Legendre polynomial defined by (6.35), see [212] or [147]. The system $\mathscr{L}_i$, $i = 0, 1 \ldots$, is a complete orthogonal basis in the space $L^2(-1, 1)$. It is possible to verify that φ_i, $i = 0, 1 \ldots$, form a complete orthonormal basis in $L^2(0, 1)$ and $\varphi_i(0) \neq 0$.

6.2 Space-Time DGM for Nonlinear Convection-Diffusion Problems

In this section we extend the space-time discontinuous Galerkin method (ST-DGM), as explained in the previous section on a simple initial-boundary value problem for the heat equation, to the solution of a more general problem for a convection diffusion equation with *nonlinear convection* and *nonlinear diffusion*. We derive the error estimates in the $L^2(0, T; L^2(\Omega))$-norm and the DG norm formed by the $L^2(0, T; H^1(\Omega))$ norm and penalty terms.

Let $\Omega \subset \mathbb{R}^d$ ($d = 2$ or 3) be a bounded polygonal or polyhedral domain with Lipschitz boundary and $T > 0$. We consider the following initial-boundary value problem: Find $u : Q_T = \Omega \times (0, T) \to \mathbb{R}$ such that

$$\frac{\partial u}{\partial t} + \sum_{s=1}^{d} \frac{\partial f_s(u)}{\partial x_s} - \nabla \cdot (\beta(u)\nabla u) = g \quad \text{in } Q_T, \tag{6.191a}$$

$$u\big|_{\partial\Omega\times(0,T)} = u_D, \tag{6.191b}$$

$$u(x,0) = u^0(x), \quad x \in \Omega. \tag{6.191c}$$

We assume that g, u_D, u^0, f_s are given functions and $f_s \in C^1(\mathbb{R})$, $|f_s'| \le C$, $s = 1, \ldots, d$. Moreover, let

$$\beta : \mathbb{R} \to [\beta_0, \beta_1], \quad 0 < \beta_0 < \beta_1 < \infty, \tag{6.192a}$$

$$|\beta(u_1) - \beta(u_2)| \le L_\beta |u_1 - u_2| \quad \forall u_1, u_2 \in \mathbb{R}. \tag{6.192b}$$

Remark 6.33 In this section we consider problem (6.191) only with a Dirichlet boundary condition. This means that $\partial\Omega_D = \partial\Omega$, $\partial\Omega_N = \emptyset$, $\mathscr{F}_{h,m}^D = \mathscr{F}_{h,m}^B$ and $\mathscr{F}_{h,m}^N = \emptyset$. The analysis of the problem with mixed Dirichlet–Neumann boundary conditions is more complicated due to the properties of the convection form b_h derived in Sect. 4.3.2 and represents an open challenging subject.

In the derivation and analysis of the discrete problem, we assume that the exact solution is regular in the following sense:

$$u \in L^2(0,T; H^2(\Omega)), \quad \frac{\partial u}{\partial t} \in L^2(0,T; H^1(\Omega)), \tag{6.193}$$

$$\|\nabla u(t)\|_{L^\infty(\Omega)} \le C_B \quad \text{for } t \in (0,T). \tag{6.194}$$

6.2.1 Discretization of the Problem

We employ the same notation as in Sect. 6.1. Hence, we consider a partition $0 = t_0 < t_1 < \cdots < t_r = T$ of the time interval $[0, T]$, time subintervals $I_m = (t_{m-1}, t_m)$, $m = 1, \ldots, r$, and triangulations $\mathscr{T}_{h,m}$, $m = 0, \ldots, r$, of the domain Ω associated with time instants t_m, $m = 0, \ldots, r$, and intervals I_m, $m = 1, \ldots, r$. Further, we consider function spaces $S_{h,m}^p$ defined by (6.5) and $S_{h,\tau}^{p,q}$ defined by (6.8) and the projections $\Pi_{h,m}$ and π—see (6.7) and (6.33), respectively.

For the derivation of the space-time discontinuous Galerkin discretization, we assume that $u \in C^1((0,T); H^2(\Omega))$ is an exact solution of problem (6.191). We multiply (6.191a) by $\varphi \in S_{h,\tau}^{p,q}$, integrate over $K \times I_m$, sum over all $K \in \mathscr{T}_{h,m}$, and perform some manipulation. The time-derivative term is discretized in the same manner as in (6.11)–(6.13). Discretization of the convection term and the source term (6.10) is the same as in Chap. 4.

Discretization of the diffusion term is a little more complicated due to the non-linearity of the function β. Using the technique from Sect. 2.4, the application of Green's theorem to the diffusion term gives

$$-\sum_{K\in\mathscr{T}_{h,m}} \int_K \nabla\cdot(\beta(u)\nabla u)\varphi\,\mathrm{d}x \tag{6.195}$$

$$= \sum_{K\in\mathscr{T}_{h,m}} \int_K \beta(u)\nabla u\cdot\nabla\varphi\,\mathrm{d}x - \sum_{\Gamma\in\mathscr{F}_{h,m}^{IB}} \int_\Gamma \langle\beta(u)\nabla u\rangle\cdot\boldsymbol{n}[\varphi]\,\mathrm{d}S.$$

In Sect. 2.4, we add to the right-hand side of (6.195) face integral terms, where the roles of the exact solution u and the test function φ are mutually exchanged. However, in contrast to the case of a linear diffusion (see, e.g., (6.14)), to the right-hand side we cannot add the expression

$$\Theta \sum_{\Gamma\in\mathscr{F}_{h,m}^{I}} \int_\Gamma \langle\beta(\varphi)\nabla\varphi\rangle\cdot\boldsymbol{n}[u]\,\mathrm{d}S + \Theta \sum_{\Gamma\in\mathscr{F}_{h,m}^{B}} \int_\Gamma \beta(\varphi)\nabla\varphi\cdot\boldsymbol{n}(u-u_D)\,\mathrm{d}S,$$

obtained by the mutual exchange of u and φ, because it is not linear with respect to the test function φ. Therefore, in the argument of β we keep the exact solution u, i.e., we use the expression

$$\Theta \sum_{\Gamma\in\mathscr{F}_{h,m}^{I}} \int_\Gamma \langle\beta(u)\nabla\varphi\rangle\cdot\boldsymbol{n}[u]\,\mathrm{d}S + \Theta \sum_{\Gamma\in\mathscr{F}_{h,m}^{B}} \int_\Gamma \beta(u)\nabla\varphi\cdot\boldsymbol{n}(u-u_D)\,\mathrm{d}S, \tag{6.196}$$

which vanishes for a regular function u satisfying the Dirichlet condition (6.191b).

Finally, we arrive at the definition of the following forms. If $v,\ w,\ \varphi\in H^2(\Omega,\ \mathscr{T}_{h,m})$ and $C_W > 0$ is a fixed constant, we define the diffusion, penalty, convection and righ-hand side forms

$$a_{h,m}(v,w,\varphi) = \sum_{K\in\mathscr{T}_{h,m}} \int_K \beta(v)\nabla w\cdot\nabla\varphi\,\mathrm{d}x \tag{6.197}$$
$$- \sum_{\Gamma\in\mathscr{F}_{h,m}^{I}} \int_\Gamma \left(\langle\beta(v)\nabla w\rangle\cdot\boldsymbol{n}[\varphi] + \Theta\langle\beta(v)\nabla\varphi\rangle\cdot\boldsymbol{n}\,[w]\right)\mathrm{d}S$$
$$- \sum_{\Gamma\in\mathscr{F}_{h,m}^{B}} \int_\Gamma \left(\beta(v)\nabla w\cdot\boldsymbol{n}\,\varphi + \Theta\,\beta(v)\nabla\varphi\cdot\boldsymbol{n}\,(w-u_D)\right)\mathrm{d}S,$$

$$J_{h,m}^{\sigma}(w,\varphi) = \sum_{\Gamma\in\mathscr{F}_{h,m}^{I}} \frac{C_W}{h_\Gamma}\int_\Gamma [w]\,[\varphi]\,\mathrm{d}S + \sum_{\Gamma\in\mathscr{F}_{h,m}^{B}} \frac{C_W}{h_\Gamma}\int_\Gamma w\,\varphi\,\mathrm{d}S, \tag{6.198}$$

$$A_{h,m}(w,v,\varphi) = a_{h,m}(w,v,\varphi) + \beta_0 J_{h,m}^{\sigma}(v,\varphi), \tag{6.199}$$

$$b_{h,m}(w,\varphi) = -\sum_{K\in\mathscr{T}_{h,m}} \int_K \sum_{s=1}^{d} f_s(w)\frac{\partial\varphi}{\partial x_s}\,\mathrm{d}x + \sum_{\Gamma\in\mathscr{F}_{h,m}^{I}} \int_\Gamma H\left(w_\Gamma^{(L)}, w_\Gamma^{(R)}, \boldsymbol{n}\right)[\varphi]\,\mathrm{d}S$$
$$+ \sum_{\Gamma\in\mathscr{F}_{h,m}^{D}} \int_\Gamma H\left(w_\Gamma^{(L)}, w_\Gamma^{(L)}, \boldsymbol{n}\right)\varphi\,\mathrm{d}S. \tag{6.200}$$

$$\ell_{h,m}(\varphi) = (g, \varphi) + \beta_0 \sum_{\Gamma \in \mathscr{F}_{h,m}^B} \frac{C_W}{h_\Gamma} \int_\Gamma u_D\, \varphi \,\mathrm{d}S. \tag{6.201}$$

In (6.197), we take $\Theta = -1$, $\Theta = 0$ and $\Theta = 1$ and obtain the nonsymmetric (NIPG), incomplete (IIPG) and symmetric (SIPG) variants of the approximation of the diffusion terms, respectively. In (6.200), H is a numerical flux with the properties (4.18)–(4.20) introduced in Sect. 4.2.

Similarly as in Sect. 6.1, the exact regular solution u of (6.191) satisfies the identity

$$\int_{I_m} \big((u', \varphi) + A_{h,m}(u, u, \varphi) + b_{h,m}(u, \varphi)\big)\,\mathrm{d}t + \big(\{u\}_{m-1}, \varphi_{m-1}^+\big) \tag{6.202}$$
$$= \int_{I_m} \ell_{h,m}(\varphi)\,\mathrm{d}t \quad \forall \varphi \in S_{h,\tau}^{p,q}, \quad \text{with } u(0-) = u(0) = u^0.$$

Here $u' := \partial u/\partial t$ and $(\cdot,\cdot)$ denotes the $L^2(\Omega)$-scalar product.

Based on (6.202), we proceed to the definition of the approximate solution.

Definition 6.34 We say that a function U is an *ST-DG approximate solution* of problem (6.191), if $U \in S_{h,\tau}^{p,q}$ and

$$\int_{I_m} \big((U', \varphi) + A_{h,m}(U, U, \varphi) + b_{h,m}(U, \varphi)\big)\,\mathrm{d}t + \big(\{U\}_{m-1}, \varphi_{m-1}^+\big) \tag{6.203}$$
$$= \int_{I_m} \ell_{h,m}(\varphi)\,\mathrm{d}t \quad \forall \varphi \in S_{h,\tau}^{p,q}, \quad m = 1, \ldots, r, \; U_0^- := \Pi_{h,0} u^0.$$

where $U' = \partial U/\partial t$. We call (6.203) the space-time discontinuous Galerkin discrete problem.

Exercise 6.35 Formulate the ST-DG discrete problem in the case, when mixed Dirichlet–Neumann boundary conditions are used.

In the sequel, we analyze the ST-DGM, namely we derive an estimate of the error $e = U - u$, where u is the exact solution of (6.191) and U is the approximate solution given by (6.203). We assume that the approximate solution U exists and is unique.

6.2.2 Auxiliary Results

In the analysis of the ST-DGM for the nonlinear problem, we proceed in a similar way as in Sect. 6.1 for the heat equation. We consider a system (6.24) of triangulations $\mathscr{T}_{h,m}$, satisfying the conditions of *shape-regularity* (6.25) and of the *equivalence* (6.26). Let $\pi : C([0,T]; L^2(\Omega)) \to S_{h,\tau}^{p,q}$ be the projection operator given by (6.33). The error of the method is expressed again in the form

$$e = U - u = \xi + \eta, \tag{6.204}$$

where

$$\xi = U - \pi u \in S^{p,q}_{h,\tau}, \quad \eta = \pi u - u. \tag{6.205}$$

Then, subtracting (6.202) from (6.203), and using (6.204), for every $\varphi \in S^{p,q}_{h,\tau}$, we find that

$$\begin{aligned}&\int_{I_m} \left((\xi', \varphi) + A_{h,m}(U, U, \varphi) - A_{h,m}(u, u, \varphi)\right) \mathrm{d}t + \left(\{\xi\}_{m-1}, \varphi^+_{m-1}\right) \\ &= \int_{I_m} \left(b_{h,m}(u, \varphi) - b_{h,m}(U, \varphi)\right) \mathrm{d}t - \int_{I_m} (\eta', \varphi)\, \mathrm{d}t - \left(\{\eta\}_{m-1}, \varphi^+_{m-1}\right).\end{aligned} \tag{6.206}$$

Hence, we need to estimate individual terms appearing in (6.206).

The convection form $b_{h,m}$ has the following property.

Lemma 6.36 *For every $k_b > 0$ there exists a constant $C_b > 0$ independent of U, u, h, τ, r and m such that*

$$\begin{aligned}&|b_{h,m}(U, \varphi) - b_{h,m}(u, \varphi)| \\ &\leq \frac{\beta_0}{k_b} |||\varphi|||^2_m + C_b \left(\|\xi\|^2_{L^2(\Omega)} + \|\eta\|^2_{L^2(\Omega)} + \sum_{K \in \mathscr{T}_{h,m}} h^2_K |\eta|^2_{H^1(K)} \right).\end{aligned} \tag{6.207}$$

Proof By (4.30),

$$\begin{aligned}&|b_{h,m}(U, \varphi) - b_{h,m}(u, \varphi)| \\ &\leq \tilde{C} |||\varphi|||_m \left(\|U - u\|_{L^2(\Omega)} + \left(\sum_{K \in \mathscr{T}_{h,m}} h_K \|U - u\|^2_{L^2(\partial K)} \right)^{1/2} \right), \\ &\qquad \varphi \in S^p_{h,m}, \quad m = 1, \ldots, r.\end{aligned} \tag{6.208}$$

The use of (6.205), (6.30), and the Young inequality implies that

$$\|U - u\|_{L^2(\Omega)} \leq \|\xi\|_{L^2(\Omega)} + \|\eta\|_{L^2(\Omega)}, \tag{6.209}$$

$$\begin{aligned}&\sum_{K \in \mathscr{T}_{h,m}} h_K \|U - u\|^2_{L^2(\partial K)} \\ &\leq C \sum_{K \in \mathscr{T}_{h,m}} \left(\|\xi\|^2_{L^2(\Omega)} + h^2_K |\xi|^2_{H^1(K)} + \|\eta\|^2_{L^2(\Omega)} + h^2_K |\eta|^2_{H^1(K)} \right).\end{aligned} \tag{6.210}$$

The above inequality and (6.31) applied to ξ yield

$$\sum_{K\in\mathscr{T}_{h,m}} h_K \|U-u\|^2_{L^2(\partial K)} \le C\left(\|\xi\|^2_{L^2(\Omega)} + \|\eta\|^2_{L^2(\Omega)} + \sum_{K\in\mathscr{T}_{h,m}} h_K^2 |\eta|^2_{H^1(K)}\right). \tag{6.211}$$

Now, using (6.208)–(6.211), we get

$$\begin{aligned} &|b_{h,m}(U,\varphi) - b_{h,m}(u,\varphi)| \\ &\le C|||\varphi|||_m \left(\|\xi\|_{L^2(\Omega)} + \|\eta\|_{L^2(\Omega)} + \left(\sum_{K\in\mathscr{T}_{h,m}} h_K^2|\eta|^2_{H^1(K)}\right)^{1/2}\right). \end{aligned} \tag{6.212}$$

Finally by the Young inequality we obtain (6.207). We can see that the constant C_b has the form $C_b = Ck_b/\beta_0$, where the constant C is independent of U, u, h, τ, r, m, k_b and β_0. □

Let us note that in the following considerations, in some places the simplified form of the Young inequality $ab \le \frac{1}{\delta}a^2 + \delta b^2$ is used.

As for the coercivity of the forms $A_{h,m}$, we can prove the following result.

Lemma 6.37 *Let*

$$C_W > 0, \qquad \text{for } \Theta = -1\ (NIPG), \tag{6.213}$$

$$C_W \ge \left(\frac{4\beta_1}{\beta_0}\right)^2 C_{MI} \quad \text{for } \Theta = 1\ (SIPG), \tag{6.214}$$

$$C_W \ge 2\left(\frac{2\beta_1}{\beta_0}\right)^2 C_{MI} \quad \text{for } \Theta = 0\ (IIPG), \tag{6.215}$$

where $C_{MI} = C_M(C_I+1)C_G$. *Then, for* $m = 1, \ldots, r$,

$$a_{h,m}(U,U,\xi) - a_{h,m}(U,\pi u,\xi) + \beta_0 J^\sigma_{h,m}(\xi,\xi) \ge \frac{\beta_0}{2}|||\xi|||^2_m. \tag{6.216}$$

Proof From (6.197)–(6.198) we immediately see that for $\Theta = -1$ inequality (6.216) holds. Let us assume that $\Theta = 1$. Then, by (6.197) and (6.192a),

$$\begin{aligned} &a_{h,m}(U,U,\xi) - a_{h,m}(U,\pi u,\xi) + \beta_0 J^\sigma_{h,m}(\xi,\xi) \\ &= \sum_{K\in\mathscr{T}_{h,m}} \int_K \beta(U)\nabla\xi\cdot\nabla\xi\,\mathrm{d}x + \beta_0 J^\sigma_{h,m}(\xi,\xi) \\ &\quad - 2\sum_{\Gamma\in\mathscr{F}^I_{h,m}} \int_\Gamma \langle\beta(U)\nabla\xi\rangle\cdot\boldsymbol{n}[\xi]\,\mathrm{d}S - 2\sum_{\Gamma\in\mathscr{F}^B_{h,m}} \int_\Gamma \beta(U)\nabla\xi\cdot\boldsymbol{n}\xi\,\mathrm{d}S \end{aligned} \tag{6.217}$$

$$\geq \beta_0 \sum_{K\in\mathscr{T}_{h,m}} \int_K |\nabla\xi|^2 \mathrm{d}x + \beta_0 J^\sigma_{h,m}(\xi,\xi) - 2\beta_1 \sum_{\Gamma\in\mathscr{F}^I_{h,m}} \int_\Gamma \frac{|\nabla\xi^{(L)}_\Gamma| + |\nabla\xi^{(R)}_\Gamma|}{2} |[\xi]| \mathrm{d}S$$
$$- 2\beta_1 \sum_{\Gamma\in\mathscr{F}^B_{h,m}} \int_\Gamma |\nabla\xi||\xi| \mathrm{d}S.$$

If $\delta > 0$, then the Young inequality implies that

$$\sum_{\Gamma\in\mathscr{F}^I_{h,m}} \int_\Gamma \frac{|\nabla\xi^{(L)}_\Gamma| + |\nabla\xi^{(R)}_\Gamma|}{2} |[\xi]|\,\mathrm{d}S + \sum_{\Gamma\in\mathscr{F}^B_{h,m}} \int_\Gamma |\nabla\xi|\,|\xi|\,\mathrm{d}S \tag{6.218}$$
$$\leq \sum_{\Gamma\in\mathscr{F}^I_{h,m}} \int_\Gamma \frac{h_\Gamma}{\delta C_W} \frac{\left(|\nabla\xi^{(L)}_\Gamma| + |\nabla\xi^{(R)}_\Gamma|\right)^2}{4} \mathrm{d}S + \sum_{\Gamma\in\mathscr{F}^I_{h,m}} \int_\Gamma \frac{\delta C_W}{h_\Gamma} |[\xi]|^2\,\mathrm{d}S$$
$$+ \sum_{\Gamma\in\mathscr{F}^B_{h,m}} \int_\Gamma \frac{h_\Gamma}{\delta C_W} \left|\nabla\xi^{(L)}_\Gamma\right|^2 \mathrm{d}S + \sum_{\Gamma\in\mathscr{F}^B_{h,m}} \int_\Gamma \frac{\delta C_W}{h_\Gamma} |\xi|^2\,\mathrm{d}S$$
$$\leq \sum_{\Gamma\in\mathscr{F}^I_{h,m}} \int_\Gamma \frac{h_\Gamma}{\delta C_W} \frac{|\nabla\xi^{(L)}_\Gamma|^2 + |\nabla\xi^{(R)}_\Gamma|^2}{2} \mathrm{d}S$$
$$+ \sum_{\Gamma\in\mathscr{F}^B_{h,m}} \int_\Gamma \frac{h_\Gamma}{\delta C_W} \left|\nabla\xi^{(L)}_\Gamma\right|^2 \mathrm{d}S + \delta\, J^\sigma_{h,m}(\xi,\xi).$$

Now, by (6.26),

$$\frac{1}{2\delta C_W} \sum_{\Gamma\in\mathscr{F}^I_{h,m}} \int_\Gamma h_\Gamma \left(|\nabla\xi^{(L)}_\Gamma|^2 + |\nabla\xi^{(R)}_\Gamma|^2\right) \mathrm{d}S \tag{6.219}$$
$$+ \frac{1}{\delta C_W} \sum_{\Gamma\in\mathscr{F}^B_{h,m}} \int_\Gamma h_\Gamma \left|\nabla\xi^{(L)}_\Gamma\right|^2 \mathrm{d}S$$
$$\leq \frac{C_G}{2\delta C_W} \sum_{\Gamma\in\mathscr{F}^I_{h,m}} \int_\Gamma \left(h_{K^{(L)}_\Gamma} |\nabla\xi^{(L)}_\Gamma|^2 + h_{K^{(R)}_\Gamma} |\nabla\xi^{(R)}_\Gamma|^2\right) \mathrm{d}S$$
$$+ \frac{C_G}{\delta C_W} \sum_{\Gamma\in\mathscr{F}^B_{h,m}} \int_\Gamma h_{K^{(L)}_\Gamma} \left|\nabla\xi^{(L)}_\Gamma\right|^2 \mathrm{d}S$$
$$\leq \frac{C_G}{\delta C_W} \sum_{K\in\mathscr{T}_{h,m}} \int_{\partial K} h_K |\nabla\xi|^2\,\mathrm{d}S.$$

It follows from (6.217)–(6.219) that

$$
\begin{aligned}
& a_{h,m}(U, U, \xi) - a_{h,m}(U, \pi u, \xi) + \beta_0 J^{\sigma}_{h,m}(\xi, \xi) \\
& \geq \beta_0 |||\xi|||_m^2 - \frac{2\beta_1 C_G}{\delta C_W} \sum_{K \in \mathscr{T}_{h,m}} \int_{\partial K} h_K |\nabla \xi|^2 \mathrm{d}S - 2\beta_1 \delta J^{\sigma}_{h,m}(\xi, \xi).
\end{aligned} \tag{6.220}
$$

If we set $\delta = \frac{\beta_0}{4\beta_1}$ and use inequalities (6.30), (6.31) and assumption (6.214), we get

$$
\begin{aligned}
& a_{h,m}(U, U, \xi) - a_{h,m}(U, \pi u, \xi) + \beta_0 J^{\sigma}_{h,m}(\xi, \xi) \\
& \geq \beta_0 |||\xi|||_m^2 - \frac{8\beta_1^2 C_{MI}}{\beta_0 C_W} \sum_{K \in \mathscr{T}_{h,m}} \int_K |\nabla \xi|^2 \mathrm{d}x - \frac{\beta_0}{2} J^{\sigma}_{h,m}(\xi, \xi) \geq \frac{\beta_0}{2} |||\xi|||_m^2.
\end{aligned} \tag{6.221}
$$

Similarly, we prove (6.216) for $\Theta = 0$, provided (6.215) is satisfied. □

Exercise 6.38 Prove that inequality (6.216) holds in the case $\Theta = 0$ under condition (6.215).

Lemma 6.39 *There exists a constant $C > 0$ independent of U, ξ, φ, h such that*

$$
a_{h,m}(U, U, \varphi) - a_{h,m}(U, \pi u, \varphi) + \beta_0 J^{\sigma}_{h,m}(\xi, \varphi) \leq C(|||\xi|||_m^2 + |||\varphi|||_m^2) \tag{6.222}
$$

for any $\varphi \in S^p_{h,m}$ and $m = 1, \ldots, r$.

Proof By (6.197),

$$
\begin{aligned}
& a_{h,m}(U, U, \varphi) - a_{h,m}(U, \pi u, \varphi) \\
& = \sum_{K \in \mathscr{T}_{h,m}} \int_K \beta(U) \nabla \xi \cdot \nabla \varphi \, \mathrm{d}x \\
& \quad - \sum_{\Gamma \in \mathscr{F}^I_{h,m}} \int_\Gamma (\langle \beta(U) \nabla \xi \rangle \cdot n_\Gamma [\varphi] + \Theta \langle \beta(U) \nabla \varphi \rangle \cdot n_\Gamma [\xi]) \, \mathrm{d}S \\
& \quad - \sum_{\Gamma \in \mathscr{F}^B_{h,m}} \int_\Gamma (\beta(U) \nabla \xi \cdot n_\Gamma \varphi + \Theta \beta(U) \nabla \varphi \cdot n_\Gamma \xi) \, \mathrm{d}S.
\end{aligned} \tag{6.223}
$$

Then, using (6.223), (6.198), (6.192a), (6.26), the Cauchy inequality and Young inequality, we find that

$$
\begin{aligned}
& a_{h,m}(U, U, \varphi) - a_{h,m}(U, \pi u, \varphi) + \beta_0 J^{\sigma}_{h,m}(\xi, \varphi) \\
& \leq \beta_1 \sum_{K \in \mathscr{T}_{h,m}} \int_K \left(|\nabla \xi|^2 + |\nabla \varphi|^2 \right) \mathrm{d}x
\end{aligned}
$$

$$
\begin{aligned}
&+ \beta_1 \sum_{\Gamma \in \mathscr{F}^I_{h,m}} \int_\Gamma \left(\frac{h_\Gamma}{C_W} \left(|\nabla \xi^{(L)}_\Gamma|^2 + |\nabla \xi^{(R)}_\Gamma|^2 \right) + \frac{C_W}{h_\Gamma} |[\varphi]|^2 \right) \mathrm{d}S \\
&+ \beta_1 \sum_{\Gamma \in \mathscr{F}^I_{h,m}} \int_\Gamma \left(\frac{h_\Gamma}{C_W} \left(|\nabla \varphi^{(L)}_\Gamma|^2 + |\nabla \varphi^{(R)}_\Gamma|^2 \right) + \frac{C_W}{h_\Gamma} |[\xi]|^2 \right) \mathrm{d}S \\
&+ \beta_1 \sum_{\Gamma \in \mathscr{F}^B_{h,m}} \int_\Gamma \left(\frac{h_\Gamma}{C_W} |\nabla \xi|^2 + \frac{C_W}{h_\Gamma} |\varphi|^2 \right) \mathrm{d}S \\
&+ \beta_1 \sum_{\Gamma \in \mathscr{F}^B_{h,m}} \int_\Gamma \left(\frac{h_\Gamma}{C_W} |\nabla \varphi|^2 + \frac{C_W}{h_\Gamma} |\xi|^2 \right) \mathrm{d}S \\
&+ \beta_0 J^\sigma_{h,m}(\xi, \varphi) \\
&\le \beta_1 \sum_{K \in \mathscr{T}_{h,m}} \int_K \left(|\nabla \xi|^2 + |\nabla \varphi|^2 \right) \mathrm{d}x + \frac{\beta_1 C_G}{C_W} \sum_{K \in \mathscr{T}_{h,m}} \int_{\partial K} h_K \left(|\nabla \xi|^2 + |\nabla \varphi|^2 \right) \mathrm{d}S \\
&+ \beta_1 J^\sigma_{h,m}(\xi, \xi) + \beta_1 J^\sigma_{h,m}(\varphi, \varphi) + \beta_0 J^\sigma_{h,m}(\xi, \varphi).
\end{aligned}
$$

Now, the above inequalities, (6.30)–(6.32) yield the estimate

$$
\begin{aligned}
&a_{h,m}(U, U, \varphi) - a_{h,m}(U, \pi u, \varphi) + \beta_0 J^\sigma_{h,m}(\xi, \varphi) \\
&\le \beta_1 \sum_{K \in \mathscr{T}_{h,m}} \int_K \left(|\nabla \xi|^2 + |\nabla \varphi|^2 \right) \mathrm{d}x + \frac{\beta_1 C_{MI}}{C_W} \sum_{K \in \mathscr{T}_{h,m}} \int_K \left(|\nabla \xi|^2 + |\nabla \varphi|^2 \right) \mathrm{d}x \\
&+ \beta_1 (J^\sigma_{h,m}(\xi, \xi) + J^\sigma_{h,m}(\varphi, \varphi)) + \beta_0 (J^\sigma_{h,m}(\xi, \xi) + J^\sigma_{h,m}(\varphi, \varphi)) \\
&\le C(|||\xi|||^2_m + |||\varphi|||^2_m),
\end{aligned}
$$

which we wanted to prove. □

Lemma 6.40 *For arbitrary $k_a, k_c > 0$ there exist constants $C_a = C_a(k_a)$, $C_c = C_c(k_c) > 0$ independent of U, ξ, φ and h, such that for each $\varphi \in S^p_{h,m}$ the following estimates hold:*

$$
|a_{h,m}(U, \pi u, \varphi) - a_{h,m}(u, \pi u, \varphi)| \le \frac{\beta_0}{k_a} |||\varphi|||^2_m + C_a(\|\xi\|^2_{L^2(\Omega)} + R_m(\eta)), \tag{6.224}
$$

$$
|a_{h,m}(u, \pi u, \varphi) - a_{h,m}(u, u, \varphi)| \le \frac{\beta_0}{k_c} |||\varphi|||^2_m + C_c \tilde{R}_m(\eta), \tag{6.225}
$$

where

$$
R_m(\eta) = |||\eta|||^2_m + \|\eta\|^2_{L^2(\Omega)} + \sum_{K \in \mathscr{T}_{h,m}} \left(|\eta|^2_{H^1(K)} + h^2_K |\eta|^2_{H^2(K)} \right), \tag{6.226}
$$

$$
\tilde{R}_m(\eta) = |||\eta|||^2_m + \sum_{K \in \mathscr{T}_{h,m}} \left(h^2_K |\eta|^2_{H^2(K)} \right). \tag{6.227}
$$

Proof Since $\nabla\pi u = \nabla\eta + \nabla u,\ [u] = 0, [\pi u] = [\eta]$, we can write

$$
\begin{aligned}
&a_{h,m}(U, \pi u, \varphi) - a_{h,m}(u, \pi u, \varphi) \qquad (6.228)\\
&= \sum_{K\in\mathscr{T}_{h,m}} \int_K ((\beta(U) - \beta(u))\nabla\eta\cdot\nabla\varphi + (\beta(U) - \beta(u))\nabla u\cdot\nabla\varphi)\,\mathrm{d}x\\
&- \sum_{\Gamma\in\mathscr{F}^I_{h,m}} \int_\Gamma (\langle(\beta(U) - \beta(u))\nabla\eta\rangle\cdot\boldsymbol{n}[\varphi] + \langle(\beta(U) - \beta(u))\nabla u\rangle\cdot\boldsymbol{n}[\varphi])\,\mathrm{d}S\\
&- \Theta \sum_{\Gamma\in\mathscr{F}^I_{h,m}} \int_\Gamma \langle(\beta(U) - \beta(u))\nabla\varphi\rangle\cdot\boldsymbol{n}[\eta]\mathrm{d}S\\
&- \sum_{\Gamma\in\mathscr{F}^B_{h,m}} \int_\Gamma ((\beta(U) - \beta(u))\nabla\eta\cdot\boldsymbol{n}\varphi + (\beta(U) - \beta(u))\nabla u\cdot\boldsymbol{n}\varphi)\,\mathrm{d}S\\
&- \Theta \sum_{\Gamma\in\mathscr{F}^B_{h,m}} \int_\Gamma (\beta(U) - \beta(u))\nabla\varphi\cdot\boldsymbol{n}(\pi u - u_D)\mathrm{d}S.
\end{aligned}
$$

This relation, conditions (6.191b) and (6.192) give the inequality

$$
\begin{aligned}
&a_{h,m}(U, \pi u, \varphi) - a_{h,m}(u, \pi u, \varphi)\\
&\le (\beta_1 - \beta_0) \sum_{K\in\mathscr{T}_{h,m}} \int_K |\nabla\eta|\,|\nabla\varphi|\,\mathrm{d}x + L_\beta \sum_{K\in\mathscr{T}_{h,m}} \int_K |U - u|\,|\nabla u|\,|\nabla\varphi|\,\mathrm{d}x\\
&+ \frac{1}{2}(\beta_1 - \beta_0) \sum_{\Gamma\in\mathscr{F}^I_{h,m}} \int_\Gamma \left(|\nabla\eta_\Gamma^{(L)}| + |\nabla\eta_\Gamma^{(R)}|\right) |[\varphi]|\ \mathrm{d}S\\
&+ \frac{1}{2}L_\beta \sum_{\Gamma\in\mathscr{F}^I_{h,m}} \int_\Gamma \left(|(U - u)_\Gamma^{(L)}|\,|\nabla u_\Gamma^{(L)}| + |(U - u)_\Gamma^{(R)}|\,|\nabla u_\Gamma^{(R)}|\right) |[\varphi]|\,\mathrm{d}S\\
&+ \frac{1}{2}(\beta_1 - \beta_0) \sum_{\Gamma\in\mathscr{F}^I_{h,m}} \int_\Gamma \left(|\nabla\varphi_L^{(L)}| + |\nabla\varphi_\Gamma^{(R)}|\right) |[\eta]|\mathrm{d}S\\
&+ (\beta_1 - \beta_0) \sum_{\Gamma\in\mathscr{F}^B_{h,m}} \int_\Gamma |\nabla\eta|\,|\varphi|\,\mathrm{d}S + L_\beta \sum_{\Gamma\in\mathscr{F}^B_{h,m}} \int_\Gamma |U - u|\,|\nabla u|\,|\varphi|\,\mathrm{d}S\\
&+ (\beta_1 - \beta_0) \sum_{\Gamma\in\mathscr{F}^B_{h,m}} \int_\Gamma |\nabla\varphi|\,|\eta|\,\mathrm{d}S.
\end{aligned}
$$

Further, this inequality, (6.194) and the Cauchy and Young inequalities imply that for any $\delta_1, \delta_2, \delta_3 > 0$ we have

$$a_{h,m}(U,\pi u,\varphi) - a_{h,m}(u,\pi u,\varphi) \le (\beta_1-\beta_0)\sum_{K\in\mathscr{T}_{h,m}}\int_K\left(\frac{|\nabla\eta|^2}{\delta_1}+\delta_1|\nabla\varphi|^2\right)\mathrm{d}x$$

$$+ L_\beta C_B\sum_{K\in\mathscr{T}_{h,m}}\int_K\left(\frac{|U-u|^2}{\delta_2}+\delta_2|\nabla\varphi|^2\right)\mathrm{d}x$$

$$+ (\beta_1-\beta_0)\sum_{\Gamma\in\mathscr{F}^I_{h,m}}\int_\Gamma\left(\frac{h_\Gamma}{C_W\delta_1}\left(|\nabla\eta_\Gamma^{(L)}|^2+|\nabla\eta_\Gamma^{(R)}|^2\right)+\frac{C_W\delta_1}{h_\Gamma}|[\varphi]|^2\right)\mathrm{d}S$$

$$+ L_\beta C_B\sum_{\Gamma\in\mathscr{F}^I_{h,m}}\int_\Gamma\left(\frac{h_\Gamma}{C_W\delta_2}\left(|(U-u)_\Gamma^{(L)}|^2+|(U-u)_\Gamma^{(R)}|^2\right)+\frac{C_W\delta_2}{h_\Gamma}|[\varphi]|^2\right)\mathrm{d}S$$

$$+ (\beta_1-\beta_0)\sum_{\Gamma\in\mathscr{F}^I_{h,m}}\int_\Gamma\left(\frac{h_\Gamma\delta_3}{C_W}\left(|\nabla\varphi_\Gamma^{(L)}|^2+|\nabla\varphi_\Gamma^{(R)}|^2\right)+\frac{C_W}{h_\Gamma\delta_3}|[\eta]|^2\right)\mathrm{d}S$$

$$+ (\beta_1-\beta_0)\sum_{\Gamma\in\mathscr{F}^B_{h,m}}\int_\Gamma\left(\frac{h_\Gamma}{C_W\delta_1}|\nabla\eta|^2+\frac{C_W\delta_1}{h_\Gamma}|\varphi|^2\right)\mathrm{d}S$$

$$+ L_\beta C_B\sum_{\Gamma\in\mathscr{F}^B_{h,m}}\int_\Gamma\left(\frac{h_\Gamma}{C_W\delta_2}|U-u|^2+\frac{C_W\delta_2}{h_\Gamma}|\varphi|^2\right)\mathrm{d}S$$

$$+ (\beta_1-\beta_0)\sum_{\Gamma\in\mathscr{F}^B_{h,m}}\int_\Gamma\left(\frac{h_\Gamma\delta_3}{C_W}|\nabla\varphi|^2+\frac{C_W}{h_\Gamma\delta_3}|\eta|^2\right)\mathrm{d}S.$$

Now, using the relations $|U-u|^2 = |\xi+\eta|^2 \le 2\left(|\xi|^2+|\eta|^2\right)$, (6.26), (6.30) and (6.31), we find that

$$\begin{aligned}
&a_{h,m}(U,\pi u,\varphi) - a_{h,m}(u,\pi u,\varphi)\\
&\le \left((\beta_1-\beta_0)\delta_1 + L_\beta C_B\delta_2\right)|||\varphi|||_m^2 + \frac{(\beta_1-\beta_0)C_{MI}\delta_3}{C_W}\sum_{K\in\mathscr{T}_{h,m}}\int_K|\nabla\varphi|^2\mathrm{d}x\\
&\quad+\left(\frac{2L_\beta C_B C_{MI}}{C_W\delta_2}+\frac{2L_\beta C_B}{\delta_2}\right)\sum_{K\in\mathscr{T}_{h,m}}\int_K|\xi|^2\mathrm{d}x + \frac{2L_\beta C_B}{\delta_2}\sum_{K\in\mathscr{T}_{h,m}}\int_K|\eta|^2\mathrm{d}x\\
&\quad+\frac{4L_\beta C_B C_M C_G}{C_W\delta_2}\sum_{K\in\mathscr{T}_{h,m}}\left(\|\eta\|^2_{L^2(K)}+h_K^2|\eta|^2_{H^1(K)}\right)+\frac{\beta_1-\beta_0}{\delta_3}J^\sigma_{h,m}(\eta,\eta)\\
&\quad+\frac{2(\beta_1-\beta_0)C_M C_G}{C_W\delta_1}\sum_{K\in\mathscr{T}_{h,m}}\left(|\eta|^2_{H^1(K)}+h_K^2|\eta|^2_{H^2(K)}\right)+\frac{\beta_1-\beta_0}{\delta_1}\sum_{K\in\mathscr{T}_{h,m}}|\eta|^2_{H^1(K)}.
\end{aligned}$$

Finally, taking into account that $h_K \le h < \bar{h}$ and choosing

$$\delta_1 = \frac{\beta_0}{3k_a(\beta_1-\beta_0)},\quad \delta_2 = \frac{\beta_0}{3k_a L_\beta C_B},\quad \delta_3 = \frac{\beta_0 C_W}{3k_a(\beta_1-\beta_0)C_{MI}},$$

we obtain estimate (6.224) with the constant

$$C_a = \max\left\{\frac{2L_\beta C_B C_{MI}}{C_W\delta_2} + \frac{2L_\beta C_B}{\delta_2}, \frac{2L_\beta C_B}{\delta_2} + \frac{4L_\beta C_B C_M C_G}{C_W\delta_2}, \right. \tag{6.229}$$
$$\left. \frac{\beta_1-\beta_0}{\delta_3}, \frac{2(\beta_1-\beta_0)C_M C_G}{C_W\delta_1} + \frac{\beta_1-\beta_0}{\delta_1} + \frac{4L_\beta C_B C_M C_G \bar{h}^2}{C_W\delta_2}\right\}.$$

We can see that $C_a = \tilde{C}_a k_a/\beta_0$, where the constant $\tilde{C}_a$ depends on C_M, C_I, C_G, C_B, C_W, L_β and $\beta_1 - \beta_0$.

Similarly, we proceed in the proof of (6.225). From the definition of the form $a_{h,m}$ and the properties of the function β it follows that

$$a_{h,m}(u, \pi u, \varphi) - a_{h,m}(u, u, \varphi) = \sum_{K\in\mathscr{T}_{h,m}} \int_K \beta(u)\nabla\eta\cdot\nabla\varphi \mathrm{d}x \tag{6.230}$$
$$- \sum_{\Gamma\in\mathscr{F}_{h,m}^I} \int_\Gamma \langle\beta(u)\nabla\eta\rangle\cdot\boldsymbol{n}[\varphi]\mathrm{d}S - \Theta \sum_{\Gamma\in\mathscr{F}_{h,m}^I} \int_\Gamma \langle\beta(u)\nabla\varphi\rangle\cdot\boldsymbol{n}[\eta]\mathrm{d}S$$
$$- \sum_{\Gamma\in\mathscr{F}_{h,m}^B} \int_\Gamma \beta(u)\nabla\eta\cdot\boldsymbol{n}\varphi\mathrm{d}S - \Theta \sum_{\Gamma\in\mathscr{F}_{h,m}^B} \int_\Gamma \beta(u)\nabla\varphi\cdot\boldsymbol{n}\eta\mathrm{d}S$$
$$\le \beta_1 \sum_{K\in\mathscr{T}_{h,m}} \int_K |\nabla\eta||\nabla\varphi|\mathrm{d}x + \beta_1 \sum_{\Gamma\in\mathscr{F}_{h,m}^I} \int_\Gamma \frac{|\nabla\eta_\Gamma^{(L)}| + |\nabla\eta_\Gamma^{(R)}|}{2}|[\varphi]|\mathrm{d}S$$
$$+ \beta_1 \sum_{\Gamma\in\mathscr{F}_{h,m}^B} \int_\Gamma \frac{|\nabla\varphi_\Gamma^{(L)}| + |\nabla\varphi_\Gamma^{(L)}|}{2}|[\eta]|\mathrm{d}S$$
$$+ \beta_1 \sum_{\Gamma\in\mathscr{F}_{h,m}^B} \int_\Gamma |\nabla\eta||\varphi|\mathrm{d}S + \beta_1 \sum_{\Gamma\in\mathscr{F}_{h,m}^B} \int_\Gamma |\nabla\varphi||\eta|\mathrm{d}S.$$

The use of the Young inequality, the multiplicative trace inequality (6.30) and the inverse inequality (6.31) imply that for arbitrary $\delta_1, \delta_2 > 0$ we get

$$a_{h,m}(u, \pi u, \varphi) - a_{h,m}(u, u, \varphi)$$
$$\le \beta_1 \sum_{K\in\mathscr{T}_{h,m}} \int_K \left(\frac{|\nabla\eta|^2}{\delta_1} + \delta_1|\nabla\varphi|^2\right)\mathrm{d}x$$
$$+ \beta_1 \sum_{\Gamma\in\mathscr{F}_{h,m}^I} \int_\Gamma \left(\frac{h_\Gamma}{C_W\delta_1}\left(|\nabla\eta_\Gamma^{(L)}|^2 + |\nabla\eta_\Gamma^{(R)}|^2\right) + \frac{C_W\delta_1}{h_\Gamma}|[\varphi]|^2\right)\mathrm{d}S$$

$$+ \beta_1 \sum_{\Gamma \in \mathscr{F}_{h,m}^I} \int_\Gamma \left(\frac{h_\Gamma \delta_2}{C_W} \left(|\nabla \varphi_\Gamma^{(L)}|^2 + |\nabla \varphi_\Gamma^{(R)}|^2 \right) + \frac{C_W}{h_\Gamma \delta_2} |[\eta]|^2 \right) \mathrm{d}S$$

$$+ \beta_1 \sum_{\Gamma \in \mathscr{F}_{h,m}^B} \int_\Gamma \left(\frac{h_\Gamma}{C_W \delta_1} |\nabla \eta|^2 + \frac{C_W \delta_1}{h_\Gamma} |\varphi|^2 \right) \mathrm{d}S$$

$$+ \beta_1 \sum_{\Gamma \in \mathscr{F}_{h,m}^B} \int_\Gamma \left(\frac{h_\Gamma \delta_2}{C_W} |\nabla \varphi|^2 + \frac{C_W}{h_\Gamma \delta_2} |\eta|^2 \right) \mathrm{d}S$$

$$\leq \beta_1 \delta_1 |||\varphi|||_m^2 + \frac{\beta_1 C_{MI} \delta_2}{C_W} \sum_{K \in \mathscr{T}_{h,m}} |\varphi|_{H^1(K)}^2 + \frac{\beta_1}{\delta_2} J_h(\eta, \eta)$$

$$+ \frac{\beta_1}{\delta_1} \sum_{K \in \mathscr{T}_{h,m}} |\eta|_{H^1(K)}^2 + \frac{2\beta_1 C_M C_G}{C_W \delta_1} \sum_{K \in \mathscr{T}_{h,m}} \left(|\eta|_{H^1(K)}^2 + h_K^2 |\eta|_{H^2(K)}^2 \right).$$

If we put

$$\delta_1 = \frac{\beta_0}{2k_c \beta_1}, \quad \delta_2 = \frac{C_W \beta_0}{2k_c \beta_1 C_{MI}}, \quad C_c = \max \left\{ \frac{\beta_1}{\delta_2}, \frac{\beta_1}{\delta_1} + \frac{2\beta_1 C_M C_G}{C_W \delta_1} \right\},$$

we get (6.225). We can see that $C_c = \tilde{C}_c k_c / \beta_0$, where $\tilde{C}_c$ depends on C_M, C_I, C_G, C_W, β_0 and β_1. □

Remark 6.41 In view of (6.226), estimate (6.207) can be written as

$$|b_{h,m}(U, \varphi) - b_{h,m}(u, \varphi)| \leq \frac{\beta_0}{k_b} |||\varphi|||_m^2 + C_b \left(\|\xi\|_{L^2(\Omega)}^2 + R_m(\eta) \right). \tag{6.231}$$

6.2.3 Abstract Error Estimate

6.2.3.1 Estimate of ξ

In what follows, we use the conditions (6.25) of the *shape-regularity*, (6.26) of the *equivalence*, and assumptions from Lemma 6.37.

Let us substitute $\varphi := \xi$ in (6.206). From the definition (6.199) of the form $A_{h,m}$ it follows that

$$\int_{I_m} \left((\xi', \xi) + a_{h,m}(U, U, \xi) - a_{h,m}(U, \pi u, \xi) + \beta_0 J_{h,m}^\sigma(\xi, \xi) \right) \mathrm{d}t \tag{6.232}$$
$$+ \left(\{\xi\}_{m-1}, \xi_{m-1}^+ \right)$$
$$= \int_{I_m} \left(-a_{h,m}(U, \pi u, \xi) + a_{h,m}(u, \pi u, \xi) - a_{h,m}(u, \pi u, \xi) + a_{h,m}(u, u, \xi) \right) \mathrm{d}t$$

$$+ \int_{I_m} \left(b_{h,m}(u, \xi) - b_{h,m}(U, \xi) - \beta_0 J^{\sigma}_{h,m}(\eta, \xi) - (\eta', \xi) \right) \mathrm{d}t - \left(\{\eta\}_{m-1}, \xi^+_{m-1} \right).$$

By (6.86), we have

$$\int_{I_m} (\xi', \xi)\mathrm{d}t + \left(\{\xi\}_{m-1}, \xi^+_{m-1} \right) \tag{6.233}$$
$$= \frac{1}{2} \left(\|\xi^-_m\|^2_{L^2(\Omega)} - \|\xi^-_{m-1}\|^2_{L^2(\Omega)} + \|\{\xi\}_{m-1}\|^2_{L^2(\Omega)} \right).$$

Moreover, (6.87) with $\delta := 1$ gives

$$\int_{I_m} (\eta', \varphi)\mathrm{d}t + \left(\{\eta\}_{m-1}, \varphi^+_{m-1} \right) \le \|\eta^-_{m-1}\|^2_{L^2(\Omega)} + \frac{1}{4} \|\{\varphi\}_{m-1}\|^2_{L^2(\Omega)}, \quad \varphi \in S^{p,q}_{h,m}. \tag{6.234}$$

The use of (6.32), (6.231)–(6.234), the Young inequality, and Lemmas 6.37 and 6.40 imply that for arbitrary $\delta, k_a, k_b, k_c > 0$ we have

$$\|\xi^-_m\|^2_{L^2(\Omega)} - \|\xi^-_{m-1}\|^2_{L^2(\Omega)} + \frac{1}{2} \|\{\xi\}_{m-1}\|^2_{L^2(\Omega)}$$
$$+ \beta_0 \left(1 - \frac{2}{k_a} - \frac{2}{k_b} - \frac{2}{k_c} - 2\delta \right) \int_{I_m} |||\xi|||^2_m \mathrm{d}t$$
$$\le C \left(\int_{I_m} \|\xi\|^2_{L^2(\Omega)} \mathrm{d}t + \|\eta^-_{m-1}\|^2_{L^2(\Omega)} + \int_{I_m} R_m(\eta)\mathrm{d}t \right).$$

This and the choice $k_a = k_b = k_c = 16$ and $\delta = \frac{1}{16}$ imply that

$$\|\xi^-_m\|^2_{L^2(\Omega)} - \|\xi^-_{m-1}\|^2_{L^2(\Omega)} + \frac{1}{2} \|\{\xi\}_{m-1}\|^2_{L^2(\Omega)} + \frac{\beta_0}{2} \int_{I_m} |||\xi|||^2_m \, \mathrm{d}t \tag{6.235}$$
$$\le C \left(\int_{I_m} \|\xi\|^2_{L^2(\Omega)} \, \mathrm{d}t + \|\eta^-_{m-1}\|^2_{L^2(\Omega)} + \int_{I_m} R_m(\eta) \, \mathrm{d}t \right), \quad m = 1, \ldots, r.$$

6.2.3.2 Estimate of $\int_{I_m} \|\xi\|^2_{L^2(\Omega)} \mathrm{d}t$

An important task is estimating the term $\int_{I_m} \|\xi\|^2_{L^2(\Omega)} \mathrm{d}t$. The case, when $\beta(u) = \text{const} > 0$, was analyzed in [134] using the approach from [4] based on the application of the so-called Gauss–Radau quadrature and interpolation. However, in the case of nonlinear diffusion, this technique is not applicable. It appears suitable to apply here the approach in [51] based on the concept of discrete characteristic functions constructed to ξ in Sect. 6.1.8.

We proceed in several steps. Let us set

$$t_{m-1+l/q} = t_{m-1} + \frac{l}{q}(t_m - t_{m-1}) \quad \text{for} \quad l = 0, ..., q,$$

and use the notation $\xi_{m-1+l/q} = \xi(t_{m-1+l/q})$, $\xi_{m-1} = \xi_{m-1}^+$, $\xi_m = \xi_m^-$.

Lemma 6.42 *There exist constants $L_q, M_q > 0$ dependent only on q such that*

$$\sum_{l=0}^{q} \|\xi_{m-1+l/q}\|_{L^2(\Omega)}^2 \geq \frac{L_q}{\tau_m} \int_{I_m} \|\xi\|_{L^2(\Omega)}^2 \mathrm{d}t, \tag{6.236}$$

$$\|\xi_{m-1}^+\|_{L^2(\Omega)}^2 \leq \frac{M_q}{\tau_m} \int_{I_m} \|\xi\|_{L^2(\Omega)}^2 \mathrm{d}t. \tag{6.237}$$

Proof Let $\hat{\phi} \in P_q(0, 1)$ be a polynomial depending on $\vartheta \in (0, 1)$ of degree $\leq q$. Since the expressions

$$\left(\sum_{l=0}^{q} \left(\hat{\phi}(l/q)\right)^2\right)^{\frac{1}{2}}, \quad \left(\int_0^1 \hat{\phi}^2 \, \mathrm{d}\vartheta\right)^{\frac{1}{2}}$$

are equivalent norms in the finite-dimensional space $P_q(0, 1)$, there exist constants $L_q, M_q > 0$ dependent only on q such that

$$L_q \int_0^1 \hat{\phi}^2 \, \mathrm{d}\vartheta \leq \sum_{l=0}^{q} \left(\hat{\phi}(l/q)\right)^2 \leq M_q \int_0^1 \hat{\phi}^2 \, \mathrm{d}\vartheta.$$

Putting $\vartheta = \frac{t - t_{m-1}}{\tau_m}$ for $t \in I_m$ and using the substitution theorem, we find that

$$\sum_{l=0}^{q} p^2(t_{m-1+l/q}) \geq \frac{L_q}{\tau_m} \int_{I_m} p^2 \mathrm{d}t \tag{6.238}$$

$$p^2(t_{m-1}) \leq \frac{M_q}{\tau_m} \int_{I_m} p^2 \mathrm{d}t. \tag{6.239}$$

for all $p \in P_q(I_m)$. The application of these inequalities to the function $p(t) = \xi(x, t)$, $t \in [t_{m-1}, t_m]$ considered for each $x \in \Omega$ yields the inequalities

$$\sum_{\ell=0}^{q} \xi^2(x, t_{m-1+\ell/q}) \geq \frac{L_q}{\tau_m} \int_{I_m} \xi^2(x, t) \, \mathrm{d}t,$$

$$\xi^2(x, t_{m-1}) \leq \frac{M_q}{\tau_m} \int_{I_m} \xi^2(x, t) \, \mathrm{d}t, \quad x \in \Omega.$$

Now the integration over Ω with respect to x and Fubini's theorem immediately lead to (6.236) and (6.237). □

Further, we return to identity (6.206), where we set $\varphi := \xi$. It can be written in the form

$$\int_{I_m} \left((\xi', \xi) + a_{h,m}(U, U, \xi) - a_{h,m}(U, \pi u, \xi) + \beta_0 J^{\sigma}_{h,m}(\xi, \xi)\right) \mathrm{d}t + \left(\xi^+_{m-1}, \xi^+_{m-1}\right)$$
$$= \int_{I_m} \left(-a_{h,m}(U, \pi u, \xi) + a_{h,m}(u, \pi u, \xi) - a_{h,m}(u, \pi u, \xi) + a_{h,m}(u, u, \xi)\right) \mathrm{d}t$$
$$+ \int_{I_m} \left(-\beta_0 J^{\sigma}_{h,m}(\eta, \xi) + b_{h,m}(u, \xi) - b_{h,m}(U, \xi) - (\eta', \xi)\right) \mathrm{d}t$$
$$- \left(\{\eta\}_{m-1}, \xi^+_{m-1}\right) + \left(\xi^-_{m-1}, \xi^+_{m-1}\right) \quad \forall \varphi \in S^{p,q}_{h,\tau}.$$

Using the relations (6.91) with $\varphi := \xi$ and

$$\int_{I_m} (\xi, \xi')\mathrm{d}t + \left(\xi^+_{m-1}, \xi^+_{m-1}\right) = \frac{1}{2}\left(\left\|\xi^-_m\right\|^2_{L^2(\Omega)} + \left\|\xi^+_{m-1}\right\|^2_{L^2(\Omega)}\right), \tag{6.240}$$

we get

$$\frac{1}{2}\left(\left\|\xi^-_m\right\|^2_{L^2(\Omega)} + \left\|\xi^+_{m-1}\right\|^2_{L^2(\Omega)}\right)$$
$$+ \int_{I_m} \left(a_{h,m}(U, U, \xi) - a_{h,m}(U, \pi u, \xi) + \beta_0 J^{\sigma}_{h,m}(\xi, \xi)\right) \mathrm{d}t$$
$$\le \int_{I_m} \left(\left|a_{h,m}(U, \pi u, \xi) - a_{h,m}(u, \pi u, \xi)\right| + \left|a_{h,m}(u, \pi u, \xi) - a_{h,m}(u, u, \xi)\right|\right) \mathrm{d}t$$
$$+ \int_{I_m} \left(\beta_0 \left|J^{\sigma}_{h,m}(\eta, \xi)\right| + \left|b_{h,m}(U, \xi) - b_{h,m}(u, \xi)\right|\right) \mathrm{d}t$$
$$+ \left|\left(\eta^-_{m-1}, \xi^+_{m-1}\right)\right| + \left|\left(\xi^-_{m-1}, \xi^+_{m-1}\right)\right|.$$

Now, Lemmas 6.37, 6.40, inequalities (6.32), (6.231) and the Young inequality imply that

$$\frac{1}{2}\left(\left\|\xi^-_m\right\|^2_{L^2(\Omega)} + \left\|\xi^+_{m-1}\right\|^2_{L^2(\Omega)}\right) + \frac{\beta_0}{2}\int_{I_m} |||\xi|||^2_m \mathrm{d}t$$
$$\le \int_{I_m} \left(\frac{\beta_0}{k_a}|||\xi|||^2_m + C_a\|\xi\|^2_{L^2(\Omega)} + C_a R_m(\eta) + \frac{\beta_0}{k_c}|||\xi|||^2_m + C_c \tilde{R}_m(\eta)\right) \mathrm{d}t$$
$$+ \int_{I_m} \left(\frac{\beta_0}{\delta} J^{\sigma}_{h,m}(\eta, \eta) + \delta\beta_0 J^{\sigma}_{h,m}(\xi, \xi) + \frac{\beta_0}{k_b}|||\xi|||^2_m + C_b\|\xi\|^2_{L^2(\Omega)} + C_b R_m(\eta)\right) \mathrm{d}t$$
$$+ \frac{\left\|\eta^-_{m-1}\right\|^2_{L^2(\Omega)}}{\delta_1} + \delta_1\left\|\xi^+_{m-1}\right\|^2_{L^2(\Omega)} + \frac{\left\|\xi^-_{m-1}\right\|^2_{L^2(\Omega)}}{\delta_1} + \delta_1\left\|\xi^+_{m-1}\right\|^2_{L^2(\Omega)}.$$

After some manipulation, taking into account that $\tilde{R}_m(\eta) \le R_m(\eta)$, we get

$$\begin{aligned}
&\|\xi_m^-\|^2_{L^2(\Omega)} + \|\xi^+_{m-1}\|^2_{L^2(\Omega)} + \beta_0\left(1 - \frac{2}{k_a} - \frac{2}{k_b} - \frac{2}{k_c} - 2\delta\right)\int_{I_m} |||\xi|||^2_m \mathrm{d}t \\
&\le 2(C_a + C_b)\int_{I_m} \|\xi\|^2_{L^2(\Omega)}\mathrm{d}t + \left(2(C_a + C_b + C_c) + \frac{\beta_0}{\delta}\right)\int_{I_m} R_m(\eta)\, dt \\
&+ 2\frac{\|\eta^-_{m-1}\|^2_{L^2(\Omega)}}{\delta_1} + 2\frac{\|\xi^-_{m-1}\|^2_{L^2(\Omega)}}{\delta_1} + 4\delta_1\|\xi^+_{m-1}\|^2_{L^2(\Omega)}.
\end{aligned}$$

Finally, the choice $k_a = k_b = k_c = 16$ and $\delta = 1/16$ yields

$$\begin{aligned}
&\|\xi_m^-\|^2_{L^2(\Omega)} + \|\xi^+_{m-1}\|^2_{L^2(\Omega)} + \frac{\beta_0}{2}\int_{I_m} |||\xi|||^2_m \mathrm{d}t \\
&\le C_1\int_{I_m} \|\xi\|^2_{L^2(\Omega)}\mathrm{d}t + C_2\int_{I_m} R_m(\eta)\mathrm{d}t \\
&+ 2\frac{\|\eta^-_{m-1}\|^2_{L^2(\Omega)}}{\delta_1} + 2\frac{\|\xi^-_{m-1}\|^2_{L^2(\Omega)}}{\delta_1} + 4\delta_1\|\xi^+_{m-1}\|^2_{L^2(\Omega)},
\end{aligned} \tag{6.241}$$

with constants $C_1 = 2(C_a + C_b)$, $C_2 = 2(C_a + C_b + C_c) + 16\beta_0$.

Now we prove the following important result.

Lemma 6.43 *There exist constants $C,\ C^* > 0$ such that*

$$\int_{I_m} \|\xi\|^2_{L^2(\Omega)}\, \mathrm{d}t \le C\,\tau_m\left(\|\xi^-_{m-1}\|^2_{L^2(\Omega)} + \|\eta^-_{m-1}\|^2_{L^2(\Omega)} + \int_{I_m} R_m(\eta)\,\mathrm{d}t\right), \tag{6.242}$$

$$m = 1, \ldots, r,$$

where $R_m(\eta)$ is defined in (6.226), provided

$$0 < \tau_m \le C^*. \tag{6.243}$$

Proof First, let us consider $q = 1$. Then, from (6.241), (6.236) and (6.237) it follows that

$$\begin{aligned}
&\frac{L_q}{\tau_m}\int_{I_m} \|\xi\|^2_{L^2(\Omega)}\mathrm{d}t + \frac{\beta_0}{2}\int_{I_m} |||\xi|||^2_m \mathrm{d}t \\
&\le \left(C_1 + \frac{4M_q\delta_1}{\tau_m}\right)\int_{I_m} \|\xi\|^2_{L^2(\Omega)}\mathrm{d}t \\
&+ C_2\int_{I_m} R_m(\eta)\mathrm{d}t + 2\frac{\|\eta^-_{m-1}\|^2_{L^2(\Omega)}}{\delta_1} + 2\frac{\|\xi^-_{m-1}\|^2_{L^2(\Omega)}}{\delta_1}.
\end{aligned}$$

If we set

$$\delta_1 = \frac{L_q}{8M_q}, \quad C_3 = \frac{2}{\delta_1},$$

then, under the condition

$$0 < \tau_m \leq C^* := \frac{L_q}{4C_1}, \tag{6.244}$$

we get

$$\frac{L_q}{4\tau_m} \int_{I_m} \|\xi\|^2_{L^2(\Omega)} \mathrm{d}t + \frac{\beta_0}{2} \int_{I_m} |||\xi|||^2_m \mathrm{d}t$$
$$\leq C_2 \int_{I_m} R_m(\eta) \mathrm{d}t + C_3 \left\| \eta^-_{m-1} \right\|^2_{L^2(\Omega)} + C_3 \left\| \xi^-_{m-1} \right\|^2_{L^2(\Omega)}, \tag{6.245}$$

which already implies (6.242).

Further, let $q \geq 2$, $l \in \{1, ..., q-1\}$ and $\tilde{\xi}_l = \zeta_{t_{m-1+l/q}}$, where $\zeta_{t_{m-1+l/q}}$ is the discrete characteristic function of the function ξ at the point $t_{m-1+l/q}$ defined by conditions (6.146)–(6.147). Then, these conditions and (6.164) imply that

$$\int_{I_m} (\tilde{\xi}_l, \xi')\mathrm{d}t = \int_{t_{m-1}}^{t_{m-1+l/q}} (\xi, \xi')\mathrm{d}t, \quad \xi(t^+_{m-1}) = \tilde{\xi}_l(t^+_{m-1}), \tag{6.246}$$

$$\int_{I_m} |||\tilde{\xi}_l|||^2_m \mathrm{d}t \leq C_{CH} \int_{I_m} |||\xi|||^2_m \mathrm{d}t. \tag{6.247}$$

We proceed again from identity (6.206), where we set $\varphi := \tilde{\xi}_l$. Then we get

$$\int_{I_m} (\xi', \tilde{\xi}_l)\mathrm{d}t + \left(\xi^+_{m-1}, (\tilde{\xi}_l)^+_{m-1}\right) \tag{6.248}$$
$$= \int_{I_m} \left(-a_{h,m}(U, U, \tilde{\xi}_l) + a_{h,m}(U, \pi u, \tilde{\xi}_l) - \beta_0 J^\sigma_{h,m}(\xi, \tilde{\xi}_l)\right) \mathrm{d}t$$
$$+ \int_{I_m} \left(-a_{h,m}(U, \pi u, \tilde{\xi}_l) + a_{h,m}(u, \pi u, \tilde{\xi}_l)\right) \mathrm{d}t$$
$$+ \int_{I_m} \left(-a_{h,m}(u, \pi u, \tilde{\xi}_l) + a_{h,m}(u, u, \tilde{\xi}_l) - \beta_0 J^\sigma_{h,m}(\eta, \tilde{\xi}_l)\right) \mathrm{d}t$$
$$+ \int_{I_m} \left(b_{h,m}(u, \tilde{\xi}_l) - b_{h,m}(U, \tilde{\xi}_l)\right) \mathrm{d}t$$
$$+ \left(\xi^-_{m-1}, (\tilde{\xi}_l)^+_{m-1}\right) - \int_{I_m} (\eta', \tilde{\xi}_l)\mathrm{d}t - \left(\{\eta\}_{m-1}, (\tilde{\xi}_l)^+_{m-1}\right).$$

By (6.246),

$$\begin{aligned}
&\int_{I_m} (\xi', \tilde{\xi}_l)\mathrm{d}t + \left(\xi^+_{m-1}, (\tilde{\xi}_l)^+_{m-1}\right) = \int_{t_{m-1}}^{t_{m-1+l/q}} (\xi, \xi') + \left(\xi^+_{m-1}, \xi^+_{m-1}\right) \\
&= \frac{1}{2}\int_{t_{m-1}}^{t_{m-1+l/q}} \frac{d}{dt}\|\xi\|^2_{L^2(\Omega)}\mathrm{d}t + \left\|\xi^+_{m-1}\right\|^2_{L^2(\Omega)} \qquad (6.249)\\
&= \frac{1}{2}\left(\left\|\xi_{m-1+l/q}\right\|^2_{L^2(\Omega)} + \left\|\xi^+_{m-1}\right\|^2_{L^2(\Omega)}\right).
\end{aligned}$$

Then (6.91), (6.249) and (6.248) imply that

$$\begin{aligned}
&\frac{1}{2}\left(\left\|\xi_{m-1+l/q}\right\|^2_{L^2(\Omega)} + \left\|\xi^+_{m-1}\right\|^2_{L^2(\Omega)}\right) \qquad (6.250)\\
&\leq \int_{I_m}\left(\left|a_{h,m}(U,U,\tilde{\xi}_l) - a_{h,m}(U,\pi u,\tilde{\xi}_l) + \beta_0 J^\sigma_{h,m}(\xi,\tilde{\xi}_l)\right|\right)\mathrm{d}t\\
&\quad + \int_{I_m}\left(\left|a_{h,m}(U,\pi u,\tilde{\xi}_l) - a_{h,m}(u,\pi u,\tilde{\xi}_l)\right|\right)\mathrm{d}t\\
&\quad + \int_{I_m}\left(\left|a_{h,m}(u,\pi u,\tilde{\xi}_l) - a_{h,m}(u,u,\tilde{\xi}_l)\right| + \left|\beta_0 J^\sigma_{h,m}(\eta,\tilde{\xi}_l)\right|\right)\mathrm{d}t\\
&\quad + \int_{I_m}\left(\left|b_{h,m}(U,\tilde{\xi}_l) - b_{h,m}(u,\tilde{\xi}_l)\right|\right)\mathrm{d}t\\
&\quad + \left|\left(\xi^-_{m-1}, \xi^+_{m-1}\right)\right| + \left|\left(\eta^-_{m-1}, \xi^+_{m-1}\right)\right|.
\end{aligned}$$

Using Lemmas 6.39, 6.40, where we set $\varphi = \tilde{\xi}_l$, $k_a = k_c = 1$, inequalities (6.32), (6.231) (with $k_b = 1$) and the Young inequality, for an arbitrary $\delta_2 > 0$ we get

$$\begin{aligned}
&\frac{1}{2}\left(\left\|\xi_{m-1+l/q}\right\|^2_{L^2(\Omega)} + \left\|\xi^+_{m-1}\right\|^2_{L^2(\Omega)}\right)\\
&\leq \int_{I_m}\left(C\left(|||\xi|||^2_m + |||\tilde{\xi}_l|||^2_m\right) + \beta_0|||\tilde{\xi}_l|||^2_m + C_a\|\xi\|^2_{L^2(\Omega)} + C_a R_m(\eta)\right)\mathrm{d}t\\
&\quad + \int_{I_m}\left(\beta_0|||\tilde{\xi}_l|||^2_m + C_c\tilde{R}_m(\eta) + \beta_0 J^\sigma_{h,m}(\eta,\eta) + \beta_0 J^\sigma_{h,m}(\tilde{\xi}_l,\tilde{\xi}_l)\right)\mathrm{d}t\\
&\quad + \int_{I_m}\left(\beta_0|||\tilde{\xi}_l|||^2_m + C_b\|\xi\|^2_{L^2(\Omega)} + C_b R_m(\eta)\right)\mathrm{d}t\\
&\quad + \frac{\left\|\xi^-_{m-1}\right\|^2_{L^2(\Omega)}}{\delta_2} + \delta_2\left\|\xi^+_{m-1}\right\|^2_{L^2(\Omega)} + \frac{\left\|\eta^-_{m-1}\right\|^2_{L^2(\Omega)}}{\delta_2} + \delta_2\left\|\xi^+_{m-1}\right\|^2_{L^2(\Omega)}.
\end{aligned}$$

This implies that

$$\left\|\xi_{m-1+l/q}\right\|^2_{L^2(\Omega)} + \left\|\xi^+_{m-1}\right\|^2_{L^2(\Omega)} \qquad (6.251)$$

$$
\begin{aligned}
&\leq \tilde{C} \int_{I_m} \left(|||\xi|||_m^2 + \|\xi\|_{L^2(\Omega)}^2 + R_m(\eta) \right) \mathrm{d}t \\
&\quad + 2 \frac{\left\|\xi_{m-1}^-\right\|_{L^2(\Omega)}^2}{\delta_2} + 2 \frac{\left\|\eta_{m-1}^-\right\|_{L^2(\Omega)}^2}{\delta_2} + 4\delta_2 \left\|\xi_{m-1}^+\right\|_{L^2(\Omega)}^2
\end{aligned}
$$

with a constant $\tilde{C} > 0$ independent of ξ, η, h and τ.

Further, let us multiply (6.251) by $\frac{\beta_0}{4\tilde{C}(q-1)}$, sum over all $l \in \{1, ..., q-1\}$ and add inequality (6.241) to the result. Then we obtain the inequality

$$
\begin{aligned}
&\tilde{C}_1 \left(\left\|\xi_m^-\right\|_{L^2(\Omega)}^2 + \sum_{l=1}^{q-1} \left\|\xi_{m-1+l/q}\right\|_{L^2(\Omega)}^2 + \left\|\xi_{m-1}^+\right\|_{L^2(\Omega)}^2 \right) + \frac{\beta_0}{2} \int_{I_m} |||\xi|||_m^2 \mathrm{d}t \\
&\leq \int_{I_m} \left(\frac{\beta_0}{4} |||\xi|||_m^2 + \tilde{C}_2 \|\xi\|_{L^2(\Omega)}^2 + \tilde{C}_3 R_m(\eta) \right) \mathrm{d}t \\
&\quad + \left(\frac{2}{\delta_1} + \frac{\beta_0}{2\tilde{C}\delta_2} \right) \left(\left\|\xi_{m-1}^-\right\|_{L^2(\Omega)}^2 + \left\|\eta_{m-1}^-\right\|_{L^2(\Omega)}^2 \right) \\
&\quad + \left(\frac{\beta_0\delta_2}{\tilde{C}} + 4\delta_1 \right) \left\|\xi_{m-1}^+\right\|_{L^2(\Omega)}^2,
\end{aligned}
$$

where

$$
\tilde{C}_1 = \min\left\{ \frac{\beta_0}{4\tilde{C}(q-1)}, 1 \right\}, \quad \tilde{C}_2 = \frac{\beta_0}{4} + C_1, \quad \tilde{C}_3 = \frac{\beta_0}{4} + C_2.
$$

Hence, by Lemma 6.42,

$$
\begin{aligned}
&\frac{\tilde{C}_1 L_q}{\tau_m} \int_{I_m} \|\xi\|_{L^2(\Omega)}^2 \mathrm{d}t + \frac{\beta_0}{4} \int_{I_m} |||\xi|||_m^2 \mathrm{d}t \\
&\leq \left(\frac{\beta_0 M_q \delta_2}{\tilde{C}\tau_m} + \frac{4M_q\delta_1}{\tau_m} + \tilde{C}_2 \right) \int_{I_m} \|\xi\|_{L^2(\Omega)}^2 \mathrm{d}t + \tilde{C}_3 \int_{I_m} R_m(\eta) \mathrm{d}t \\
&\quad + \left(\frac{2}{\delta_1} + \frac{\beta_0}{2\tilde{C}\delta_2} \right) \left(\left\|\xi_{m-1}^-\right\|_{L^2(\Omega)}^2 + \left\|\eta_{m-1}^-\right\|_{L^2(\Omega)}^2 \right).
\end{aligned}
$$

This, and the choice

$$
\delta_1 = \frac{\tilde{C}_1 L_q}{16 M_q}, \quad \delta_2 = \frac{\tilde{C}\tilde{C}_1 L_q}{4\beta_0 M_q}, \quad \tilde{C}_4 = \frac{2}{\delta_1} + \frac{\beta_0}{2\tilde{C}\delta_2},
$$

lead to the inequality

$$
\left(\frac{\tilde{C}_1 L_q}{2\tau_m} - \tilde{C}_2 \right) \int_{I_m} \|\xi\|_{L^2(\Omega)}^2 \mathrm{d}t + \frac{\beta_0}{4} \int_{I_m} |||\xi|||_m^2 \mathrm{d}t
$$

$$\leq \tilde{C}_3 \int_{I_m} R_m(\eta)\mathrm{d}t + \tilde{C}_4 \left(\|\xi_{m-1}^-\|_{L^2(\Omega)}^2 + \|\eta_{m-1}^-\|_{L^2(\Omega)}^2 \right).$$

If the condition

$$0 < \tau_m \leq C^* := \frac{\tilde{C}_1 L_q}{4\tilde{C}_2} \tag{6.252}$$

is satisfied, then

$$\frac{\tilde{C}_1 L_q}{2\tau_m} - \tilde{C}_2 \geq \frac{\tilde{C}_1 L_q}{4\tau_m},$$

and hence,

$$\begin{aligned} &\frac{\tilde{C}_1 L_q}{4\tau_m} \int_{I_m} \|\xi\|_{L^2(\Omega)}^2 \mathrm{d}t + \frac{\beta_0}{4} \int_{I_m} |||\xi|||_m^2 \mathrm{d}t \\ &\leq \tilde{C}_3 \int_{I_m} R_m(\eta)\mathrm{d}t + \tilde{C}_4 \left(\|\xi_{m-1}^-\|_{L^2(\Omega)}^2 + \|\eta_{m-1}^-\|_{L^2(\Omega)}^2 \right). \end{aligned} \tag{6.253}$$

This already implies that (6.242) holds under condition (6.243) with C^* defined by (6.252). □

Now we finish the derivation of the *abstract error estimate* of the ST-DGM.

Theorem 6.44 *Let (6.193), (6.194) and (6.243) hold. Then there exists a constant $C_{AE} > 0$ such that the error $e = U - u$ satisfies the following estimates:*

$$\begin{aligned} &\|e_m^-\|_{L^2(\Omega)}^2 + \frac{\beta_0}{2} \sum_{j=1}^{m} \int_{I_j} |||e|||_j^2 \,\mathrm{d}t \\ &\leq C_{AE}\Big(\sum_{j=1}^{m} \|\eta_{j-1}^-\|_{L^2(\Omega)}^2 + \sum_{j=1}^{m} \int_{I_j} R_j(\eta)\,\mathrm{d}t \Big) + 2\|\eta_m^-\|_{L^2(\Omega)}^2 + \beta_0 \sum_{j=1}^{m} \int_{I_j} |||\eta|||_j^2 \,\mathrm{d}t, \\ &\quad m = 1, \ldots, r, \quad h \in (0, \bar{h}), \end{aligned} \tag{6.254}$$

and

$$\begin{aligned} \|e\|_{L^2(Q_T)}^2 \leq C_{AE} \sum_{m=1}^{r} \tau_m \Bigg(&\|\eta_{m-1}^-\|_{L^2(\Omega)}^2 + \int_{I_m} R_m(\eta)\,\mathrm{d}t \\ &+ \sum_{j=1}^{r} \|\eta_{j-1}^-\|_{L^2(\Omega)}^2 + \sum_{j=1}^{r} \int_{I_j} R_j(\eta)\,\mathrm{d}t \Bigg) + 2\|\eta\|_{L^2(Q_T)}^2, \quad h \in (0, \bar{h}), \end{aligned} \tag{6.255}$$

where $R_m(\eta)$ is defined by (6.226).

Proof (i) Substituting (6.242) in (6.235), we get

$$
\begin{aligned}
&\left\|\xi_j^-\right\|_{L^2(\Omega)}^2 + \frac{\beta_0}{2}\int_{I_j} |||\xi|||_j^2 \mathrm{d}t \qquad (6.256)\\
&\le (1+C\tau_j)\left\|\xi_{j-1}^-\right\|_{L^2(\Omega)}^2 + C\left(\left\|\eta_{j-1}^-\right\|_{L^2(\Omega)}^2 + \int_{I_j} R_j(\eta)\mathrm{d}t\right), \quad j=1,\dots,r.
\end{aligned}
$$

(As usual, by C we denote a positive constant independent of h, τ, m, attaining different values at different places.) Let $m \ge 1$. If we sum (6.256) over $j = 1, \dots, m$, and take into account that $\xi_0^- = 0$, as follows from (6.81), we get

$$
\begin{aligned}
&\|\xi_m^-\|_{L^2(\Omega)}^2 + \frac{\beta_0}{2}\sum_{j=1}^m \int_{I_j} |||\xi|||_j^2 \,\mathrm{d}t\\
&\le C\sum_{j=1}^m \tau_j \|\xi_{j-1}^-\|_{L^2(\Omega)}^2 + C\sum_{j=1}^m \left(\|\eta_{j-1}^-\|_{L^2(\Omega)}^2 + \int_{I_j} R_j(\eta)\,\mathrm{d}t\right).
\end{aligned}
$$

Now the discrete Gronwall Lemma 1.11, with

$$
\begin{aligned}
x_0 &= a_0 = c_0 = 0,\\
x_m &= \|\xi_m^-\|_{L^2(\Omega)}^2,\\
c_m &= \frac{\beta_0}{2}\sum_{j=1}^m \int_{I_j} |||\xi|||_j^2\,\mathrm{d}t,\\
a_m &= C\sum_{j=1}^m \left(\|\eta_{j-1}^-\|_{L^2(\Omega)}^2 + \int_{I_j} R_j(\eta)\,\mathrm{d}t\right),\\
b_j &= C\,\tau_{j+1}, \quad j = 0, 1, \dots, m-1,
\end{aligned}
$$

implies that

$$
\begin{aligned}
&\|\xi_m^-\|_{L^2(\Omega)}^2 + \frac{\beta_0}{2}\sum_{j=1}^m \int_{I_j} |||\xi|||_j^2\,\mathrm{d}t\\
&\le C\sum_{j=1}^m \left(\|\eta_{j-1}^-\|_{L^2(\Omega)}^2 + \int_{I_j} R_j(\eta)\,\mathrm{d}t\right)\prod_{j=0}^{m-1}(1 + C\,\tau_{j+1}).
\end{aligned}
$$

Taking into account that $1 + C\,\tau_{j+1} \le \exp(C\,\tau_{j+1})$, and hence,

$$
\prod_{j=0}^{m-1}(1+C\,\tau_{j+1}) = \prod_{j=1}^{m}(1+C\,\tau_j) \le \exp\left(C\sum_{j=1}^m \tau_j\right) = \exp(C\,t_m) \le \tilde{C} := \exp(C\,T),
$$

we get

$$\|\xi_m^-\|_{L^2(\Omega)}^2 + \frac{\beta_0}{2}\sum_{j=1}^{m}\int_{I_j} |||\xi|||_j^2\,\mathrm{d}t \tag{6.257}$$

$$\leq \tilde{C}\Big(\sum_{j=1}^{m}\|\eta_{j-1}^-\|_{L^2(\Omega)}^2 + \sum_{j=1}^{m}\int_{I_j} R_j(\eta)\,\mathrm{d}t\Big), \quad m = 1,\dots,r.$$

If we use the relation $e = \xi + \eta$ and the inequalities

$$\begin{aligned} \|e_m^-\|_{L^2(\Omega)}^2 &\leq 2(\|\xi_m^-\|_{L^2(\Omega)} + \|\eta_m^-\|_{L^2(\Omega)}^2), \\ \|e\|_{L^2(\Omega)}^2 &\leq 2(\|\xi\|_{L^2(\Omega)}^2 + \|\eta\|_{L^2(\Omega)}^2), \\ |||e|||_j^2 &\leq 2(|||\xi|||_j^2 + |||\eta|||_j^2), \end{aligned} \tag{6.258}$$

from (6.257) we immediately get (6.254).

(ii) It follows from (6.242) and (6.258) that

$$\|e\|_{L^2(Q_T)}^2 = \sum_{m=1}^{r}\int_{I_m} \|e\|_{L^2(\Omega)}^2\,\mathrm{d}t \tag{6.259}$$

$$\leq C\sum_{m=1}^{r}\tau_m\Big(\|\xi_{m-1}^-\|_{L^2(\Omega)}^2 + \|\eta_{m-1}^-\|_{L^2(\Omega)}^2 + \int_{I_m} R_m(\eta)\,\mathrm{d}t\Big) + 2\int_0^T \|\eta\|_{L^2(\Omega)}^2\,\mathrm{d}t.$$

Now we use (6.257) with $m := m-1 < r$ for the estimate of $\|\xi_{m-1}^-\|_{L^2(\Omega)}^2$, take into account that $\xi_0^- = 0$, $\eta_0^- = \Pi_{h,0}u^0 - u^0$, and get inequality (6.255). □

Remark 6.45 A detailed analysis shows that the constant C_{AE} from the abstract error estimate (6.254) behaves in dependence on β_0 as $\exp(C/\beta_0)$, which means that this constant blows up for $\beta_0 \to 0+$, and the obtained error estimates cannot be applied to the case of nonlinear singularly perturbed convection-diffusion problems with degenerated diffusion. Uniform error estimates with respect to diffusion tending to zero were obtained, e.g., in [129] for the space-time DG approximations of linear convection-diffusion-reaction problems. This will be treated in Sect. 6.4.

6.2.4 Main Result

Here we present the final error estimate of the ST-DGM applied to the nonlinear convection-diffusion equation. We assume that the exact solution satisfies the regularity conditions (6.194) and

$$u \in H^{q+1}\big(0, T; H^1(\Omega)\big) \cap C([0, T]; H^s(\Omega)) \tag{6.260}$$

with integers $s \geq 2$ and $q \geq 1$. We set $\mu = \min(p + 1, s)$. Obviously, $C([0, T]; H^s(\Omega)) \subset L^2(0, T; H^s(\Omega))$ and condition (6.193) is also satisfied.

Moreover, we assume that

$$\tau_m \geq C_S h_m^2, \quad m = 1, \dots, r. \tag{6.261}$$

Let us note that it will be shown in Remark 6.48 that this assumption is not necessary, if the meshes are not time-dependent, i.e., if all meshes $\mathscr{T}_{h,m}$, $m = 1, \dots, r$, are identical.

We recall that the meshes are assumed to satisfy the *shape-regularity assumption* (6.25) and the *equivalence condition* (6.26).

Now we can formulate the main results in our analysis of error estimates for the ST-DGM.

Theorem 6.46 *Let u be the exact solution of problem (6.191) satisfying the regularity conditions (6.194) and (6.260). Let the system of triangulation satisfy the shape-regularity assumption (6.25) and the equivalence condition (6.26) and the time steps τ_m, $m = 1, \dots, r$, satisfy conditions (6.243) and (6.261). Let U be the approximate solution to problem (6.191) obtained by scheme (6.203). Then there exists a constant $C > 0$ independent of h, τ, m, r, u, U such that*

$$\begin{aligned} &\|e_m^-\|^2_{L^2(\Omega)} + \frac{\beta_0}{2} \sum_{j=1}^{m} \int_{I_j} |||e|||_j^2 \,\mathrm{d}t \\ &\leq C \left(h^{2(\mu-1)} |u|^2_{C([0,T]; H^\mu(\Omega))} + \tau^{2(q+\gamma)} |u\|_{H^{q+1}(0,T;H^1(\Omega))} \right), \quad m = 1, \dots, r, \ h \in (0, \bar{h}), \end{aligned} \tag{6.262}$$

and

$$\|e\|^2_{L^2(Q_T)} \leq C \left(h^{2(\mu-1)} |u|^2_{L^2(0,T;H^\mu(\Omega))} + \tau^{2(q+\gamma)} |u\|^2_{H^{q+1}(0,T;H^1(\Omega))} \right), \quad h \in (0, \bar{h}). \tag{6.263}$$

Here $\gamma = 0$, if (6.118) holds and the function u_D from the boundary condition (6.191b) has a general behaviour with respect to t. If u_D is defined by (6.116), then $\gamma = 1$ and condition (6.118) is not required. (The symbol $| \cdot \|$ is defined by (6.105).)

Proof Let $j \in \{1, \dots, r\}$. By virtue of (6.27), (6.106), (6.109), (6.112), (6.114), (6.115) and Lemma 6.19, we have

$$\int_{I_j} |||\eta|||_j^2 \,\mathrm{d}t \leq C \sum_{K \in \mathscr{T}_{h,j}} \left(h_K^{2(\mu-1)} |u|^2_{L^2(I_j; H^\mu(K))} + \tau_j^{2(q+\gamma)} |u\|^2_{H^{q+1}(I_j; H^1(K))} \right), \tag{6.264}$$

with γ defined in the theorem. This, and the inequality $h_K \le h_j$, valid for $K \in \mathscr{T}_{h,j}$ imply that

$$\int_{I_j} |||\eta|||_j^2 \, \mathrm{d}t \le C\Big(h_j^{2(\mu-1)}|u|^2_{L^2(I_j;H^\mu(\Omega))} + \tau_j^{2(q+\gamma)}|u\|^2_{H^{q+1}(I_j;H^1(\Omega))}\Big). \quad (6.265)$$

Similarly, in view of (6.265), (6.226) and (6.106)–(6.113), we get

$$\int_{I_j} R_j(\eta) \, \mathrm{d}t \le C\Big(h_j^{2(\mu-1)}|u|^2_{L^2(I_j;H^\mu(\Omega))} + \tau_j^{2(q+\gamma)}|u\|^2_{H^{q+1}(I_j;H^1(\Omega))}\Big). \quad (6.266)$$

Further, by (6.142),

$$\sum_{j=0}^{m} \|\eta_j^-\|^2_{L^2(\Omega)} \le C\,(T+\bar{h}^2)h^{2(\mu-1)}|u|^2_{C([0,T];H^\mu(\Omega))}. \quad (6.267)$$

Taking into account that $|\cdot|^2_{L^2(0,T;H^\mu(\Omega))} \le T|\cdot|^2_{C([0,T];H^\mu(\Omega))}$ and using (6.254) and (6.265)–(6.267), we arrive at estimate (6.262).

Similarly as above, estimating the individual terms in (6.255) depending on η, with the aid of (6.107), (6.108), (6.111), (6.265)–(6.267) and the relation $\sum_{m=1}^{r} \tau_m = T$, we obtain (6.263). □

Exercise 6.47 Prove estimate (6.263) in detail.

Remark 6.48 The case of identical meshes on all time levels. Similarly as in Sect. 6.1.10, assumption (6.140) can be avoided, if all meshes $\mathscr{T}_{h,m}$, $m = 1\ldots,r$, are identical. Then relations (6.179) and (6.180) are valid, and it is possible to show that the expression $\sum_{j=1}^{m} \|\eta_{j-1}^-\|^2_{L^2(\Omega)}$ does not appear in estimate (6.257). We find that instead of (6.254) we get the abstract error estimate in the form

$$\|e_m^-\|^2_{L^2(\Omega)} + \frac{\beta_0}{2}\sum_{j=1}^{m}\int_{I_j} |||e|||_j^2 \, \mathrm{d}t \quad (6.268)$$

$$\le C\sum_{j=1}^{m}\int_{I_j} R_j(\eta)\,\mathrm{d}t + 2\|\eta_m^-\|^2_{L^2(\Omega)} + \beta_0\sum_{j=1}^{m}\int_{I_j} |||\eta|||_j^2\,\mathrm{d}t,$$

$$m = 1,\ldots,r,\quad h \in (0,\bar{h}).$$

Then Theorem 6.44 holds without assumption (6.261).

Remark 6.49 The error estimate (6.263) in the L^2-norm is of order $O(h^{\mu-1})$ with respect to h, which is suboptimal by comparison with the interpolation error estimate (6.108), and one would expect the error estimate in the L^2-norm of order $O(h^\mu)$. This is a well-known phenomenon in the finite element method as well as in the DGM. In several discontinuous Galerkin techniques, similarly as in conforming finite elements (cf. [52]), it is possible to prove the optimal error estimate in the L^2-norm in the case of the SIPG version with the aid of the Nitsche method, as for example in [7, 96, 125, 206]. See also Sects. 2.7.2 and 4.5. The case, when the space-time DGM is applied to the nonlinear convection-diffusion problem, remains to be solved.

Remark 6.50 Similarly, as in Remark 6.45, it is possible to show that in the above error estimates, the constants C depend on β_0 as $\exp(c/\beta_0)$, which means that these constants blow up for $\beta_0 \to 0+$. Error estimates that are uniform with respect to the diffusion coefficient, will be proven in Sect. 6.4 in the case of a linear convection-diffusion problem. The case with a nonlinear convection and linear diffusion was analyzed recently in [207] in the case, when backward Euler time discretization was used.

6.2.5 Numerical Examples

In the following, we present numerical experiments demonstrating accuracy of the space-time discontinuous Galerkin method. Namely, we investigate the *experimental order of convergence* in the $L^\infty((0,T);L^2(\Omega))$-norm, the $L^2((0,T);L^2(\Omega))$-norm and the $L^2((0,T);H^1(\Omega,\mathscr{T}_h))$-seminorm.

Let $\Omega = (0,1)^2$ and $T = 1/2$. We consider the scalar nonlinear convection-diffusion equation

$$\frac{\partial u}{\partial t} - \nabla\cdot(\beta(u)\nabla u) - \frac{\partial u^2}{\partial x_1} - \frac{\partial u^2}{\partial x_2} = g \quad \text{in } \Omega\times(0,T), \tag{6.269}$$

where $\beta(u) = 0.1(2 + \arctan(u))$. We prescribe a Dirichlet boundary condition on $\partial\Omega$ and set the source term g such that the exact solution has the form

$$u(x_1,x_2,t) = (\delta + e^{\alpha t})x_1x_2(1-x_1)(1-x_2), \tag{6.270}$$

where $\delta = 0.1$ and $\alpha = 10$.

We discretized (6.269) by the ST-DG method, cf. (6.203). The resulting nonlinear algebraic system was solved by a Newton-like method, where the Jacobian matrix is replaced by the *flux matrix* arising from a linearization of nonlinear terms. See Sect. 8.4.2, where more general problem is treated.

Since the exact solution is a quartic function with respect to space, we use a coarse triangular grid (having 512 elements) and P_4 polynomial approximation in space. We carried out computations with linear, quadratic and cubic polynomial approximation in time and with the fixed time steps $\tau = 1/10, 1/20, 1/40$ and $1/80$. Table 6.1 shows computational errors and the corresponding experimental orders of convergence (EOC).

Table 6.1 ST-DGM: computational errors in the $L^\infty(0, T; L^2(\Omega))$-norm, the $L^2(0, T; L^2(\Omega))$-norm and the $L^2(0, T; H^1(\Omega, \mathscr{T}_h))$-norm and experimental orders of convergence for $q = 1, 2, 3$

τ	q	$\|e_h\|_{L^\infty(L^2)}$	EOC	$\|e_h\|_{L^2(L^2)}$	EOC	$\|e_{h\tau}\|_{L^2(H^1)}$	EOC
1.00E–01	1	1.63E–01	–	5.057E–02	–	2.567E–01	–
5.00E–02	1	5.05E–02	1.7	1.337E–02	1.9	6.395E–02	2.0
2.50E–02	1	1.38E–02	1.9	3.360E–03	2.0	1.555E–02	2.0
1.25E–02	1	3.59E–03	1.9	8.369E–04	2.0	3.835E–03	2.0
1.00E–01	2	1.32E–02	–	4.130E–03	–	2.019E–02	–
5.00E–02	2	2.08E–03	2.7	5.292E–04	3.0	2.490E–03	3.0
2.50E–02	2	2.90E–04	2.8	6.544E–05	3.0	3.018E–04	3.0
1.25E–02	2	3.81E–05	2.9	8.116E–06	3.0	3.793E–05	3.0
1.00E–01	3	8.23E–04	–	2.483E–04	–	1.187E–03	–
5.00E–02	3	6.26E–05	3.7	1.582E–05	4.0	7.382E–05	4.0
2.50E–02	3	4.31E–06	3.9	9.865E–07	4.0	8.829E–06	3.1
1.25E–02	3	9.02E–07	2.3	1.022E–07	3.3	7.595E–06	0.2

Except the last calculation, where we observe a decrease of the EOC, we get the order of convergence $O(\tau^{q+1})$ in all investigated norms. This phenomenon of the accuracy can be caused by the limits of the nonlinear iterative solvers in the finite precision arithmetic.

6.3 Extrapolated Space-Time Discontinuous Galerkin Method for Nonlinear Convection-Diffusion Problems

The realization of space-time discontinuous Galerkin discretization of a convection-diffusion problem with nonlinear convection and nonlinear diffusion, introduced and analyzed in Sect. 6.2, requires solving a nonlinear algebraic system at each time step. In the case when the diffusion terms are linear, there is a natural question, as to whether it is possible to construct a similar technique for the solution of the discrete problem as in Sect. 5.2, using the implicit discretization of linear diffusion terms and applying a suitable extrapolation from the previous time interval in the nonlinear convective terms. Then the discrete problem will be equivalent to a linear algebraic system on each time level. We call such method *extrapolated space-time discontinuous Galerkin method* (EST-DGM). The subject of this section will be the construction and analysis of this method.

Let $\Omega \subset \mathbb{R}^d (d = 2, 3)$ be a bounded polygonal domain. We consider the *initial-boundary value problem* to find $u : Q_T \to \mathbb{R}$ such that

$$\frac{\partial u}{\partial t} + \sum_{s=1}^{d} \frac{\partial f_s(u)}{\partial x_s} = \varepsilon \Delta u + g \quad \text{in } Q_T, \tag{6.271a}$$

$$u\big|_{\partial\Omega\times(0,T)} = u_D, \tag{6.271b}$$

$$u(x,0) = u^0(x), \quad x \in \Omega. \tag{6.271c}$$

We assume that the data satisfy conditions (4.2), i.e.,

$$f = (f_1, \ldots, f_d),\ f_s \in C^1(\mathbb{R}),\ f_s' \text{ are bounded},\ f_s(0) = 0,\ s = 1, \ldots, d,$$

$$u_D = \text{trace of some } u^* \in C([0,T]; H^1(\Omega)) \cap L^\infty(Q_T) \text{ on } \partial\Omega_D \times (0,T),$$

$$\varepsilon > 0,\ g \in C([0,T]; L^2(\Omega)),\ u^0 \in L^2(\Omega).$$

For the same reasons mentioned in the previous section in Remark 6.33, we consider here only the Dirichlet boundary condition.

We assume that there exists a weak solution u of (6.271) that is sufficiently regular, namely,

$$u \in C([0,T]; H^s(\Omega)) \cap W^{q+1,\infty}(0,T; H^1(\Omega)), \tag{6.272}$$

where $s \geq 2$ is an integer. By integers $p \geq 1$ and $q \geq 1$ we denote given degrees of polynomial approximations in space and time, respectively. As usual, we set $\mu = \min(p+1, s)$. The solution satisfying the regularity condition (6.272) fulfills problem (6.271) pointwise. We use here a stronger regularity assumption than in Sect. 6.2 because of a further theoretical analysis.

6.3.1 Discretization of the Problem

We use the same notation as in Sect. 6.1.1.1 and, therefore we do not recall it. In the case of problem (6.271) with linear diffusion, the diffusion, penalty, convection and right-hand side forms (6.197)–(6.201) reduce to

$$a_{h,m}(w,\varphi) = \sum_{K\in\mathscr{T}_{h,m}} \int_K \nabla w \cdot \nabla \varphi \,\mathrm{d}x - \sum_{\Gamma\in\mathscr{F}_{h,m}^I} \int_\Gamma (\langle\nabla w\rangle \cdot \boldsymbol{n}[\varphi] + \Theta\langle\nabla\varphi\rangle \cdot \boldsymbol{n}\,[w])\mathrm{d}S$$

$$- \sum_{\Gamma\in\mathscr{F}_{h,m}^B} \int_\Gamma (\nabla w \cdot \boldsymbol{n}\,\varphi + \Theta\,\nabla\varphi\cdot\boldsymbol{n}\,w)\,\mathrm{d}S, \tag{6.273}$$

$$J_{h,m}^\sigma(w,\varphi) = C_W \sum_{\Gamma\in\mathscr{F}_{h,m}^I} h_\Gamma^{-1} \int_\Gamma [w]\,[\varphi]\,\mathrm{d}S + C_W \sum_{\Gamma\in\mathscr{F}_{h,m}^B} h_\Gamma^{-1} \int_\Gamma w\,\varphi\,\mathrm{d}S, \tag{6.274}$$

$$A_{h,m}(w,\varphi) = \varepsilon a_{h,m}(w,\varphi) + \varepsilon J_{h,m}^\sigma(w,\varphi), \tag{6.275}$$

$$b_{h,m}(w,\varphi) = -\sum_{K\in\mathscr{T}_{h,m}} \int_K \sum_{s=1}^d f_s(w)\frac{\partial\varphi}{\partial x_s}\,\mathrm{d}x + \sum_{\Gamma\in\mathscr{F}_{h,m}^I} \int_\Gamma H\left(w_\Gamma^{(L)}, w_\Gamma^{(R)}, \boldsymbol{n}\right)[\varphi]\,\mathrm{d}S$$

$$+\sum_{\Gamma\in\mathscr{F}_{h,m}^B}\int_\Gamma H\left(w_\Gamma^{(L)},w_\Gamma^{(L)},\boldsymbol{n}\right)\varphi\,\mathrm{d}S, \tag{6.276}$$

$$\ell_{h,m}(\varphi)=(g,\varphi)-\varepsilon\Theta\sum_{\Gamma\in\mathscr{F}_{h,m}^B}\int_\Gamma\nabla\varphi\cdot\boldsymbol{n}u_D\,\mathrm{d}S+\varepsilon C_W\sum_{\Gamma\in\mathscr{F}_{h,m}^B}h_\Gamma^{-1}\int_\Gamma u_D\varphi\,\mathrm{d}S, \tag{6.277}$$

making sense for $v,w,\varphi\in H^2(\Omega,\mathscr{T}_{h,m})$. Here $C_W>0$ is a fixed constant, $\Theta=-1,0,1$ and H is a numerical flux with the properties (4.18)–(4.20) introduced in Sect. 4.2.

In what follows, we use the notation $u'=\partial u/\partial t$. Obviously, the exact solution u with property (6.272) satisfies the identity

$$\left(u'(t),\varphi(t)\right)+A_{h,m}(u(t),\varphi(t))+b_{h,m}(u(t),\varphi(t))=\ell_{h,m}(\varphi)\,(t) \tag{6.278}$$

for all $\varphi\in S_{h,\tau}^{p,q}$, $t\in I_m$ and $m=1,\dots,r$. This leads us, in agreement with Definition 6.34, to the possible definition of an approximate solution as a function $U\in S_{h,\tau}^{p,q}$ such that

$$\int_{I_m}\left((U',\varphi)+A_{h,m}(U,\varphi)+b_{h,m}(U,\varphi)\right)\mathrm{d}t+(\{U\}_{m-1},\varphi_{m-1}^+)=\int_{I_m}\ell_{h,m}(\varphi)\,\mathrm{d}t$$

$$\forall\varphi\in S_{h,\tau}^{p,q},\ m=1,\dots,r, \tag{6.279a}$$

$$(U_0^-,\varphi)=(u^0,\varphi)\quad\forall\varphi\in S_{h,0}. \tag{6.279b}$$

The presence of the nonlinear convection form $b_{h,m}$ causes that (6.279) represents systems of nonlinear algebraic equations, which have to be solved at each time level. In order to avoid difficulties with solving the nonlinear discrete problem, we employ an explicit extrapolation in the first argument of the form $b_{h,m}$.

In what follows, we use the following *Lagrange extrapolation*. By $\mathscr{L}_i^m(t)\in P_q(I_m)$, $i=0,\dots,q$, $m=1,\dots,r$, we denote the Lagrange interpolation basis functions associated with the Lagrangian nodes $t_{m,j}=t_{m-1}+\frac{j}{q}\tau_m$, $j=0,\dots,q$:

$$\mathscr{L}_i^m(t):=\prod_{j=0,j\neq i}^q\frac{t-(t_{m-1}+\frac{j}{q}\tau_m)}{\frac{i-j}{q}\tau_m},\quad i=0,\dots,q,\ t\in I_m,\ m=1,\dots,r. \tag{6.280}$$

Under the assumption that $\frac{\tau_m}{\tau_{m-1}}\leq C_\theta$ with a constant C_θ independent of m, using the notation $s=t-t_{m-1}\in(0,\tau_m)$, we obtain the estimate

$$|\mathscr{L}_i^{m-1}(t)|=\prod_{j=0,j\neq i}^q\left|\frac{t-(t_{m-2}+\frac{j}{q}\tau_{m-1})}{\frac{i-j}{q}\tau_{m-1}}\right|=\prod_{j=0,j\neq i}^q\left|\frac{s+\frac{q-j}{q}\tau_{m-1}}{\frac{i-j}{q}\tau_{m-1}}\right| \tag{6.281}$$

$$\leq \prod_{j=0, j\neq i}^{q} q \frac{s+\tau_{m-1}}{\tau_{m-1}} \leq \prod_{j=0, j\neq i}^{q} q \frac{\tau_m+\tau_{m-1}}{\tau_{m-1}}$$
$$\leq \prod_{j=0, j\neq i}^{q} (qC_\theta + q) = (qC_\theta + q)^q, \quad \forall t \in I_m.$$

Now if $w \in C(\cup_{m=1}^{r} I_m; X)$, where X is a Banach space, and $w|_{I_m}$ is continuously extendable on $\overline{I}_m$ for each $m = 1, \ldots, r$, we define the extrapolation $\widehat{w}\colon \cup_{m=1}^{r} I_m \to \mathbb{R}$ of w in the following way:

$$\widehat{w}(t) = Ew(t) := w(t), \quad \forall t \in I_1, \tag{6.282}$$
$$\widehat{w}(t_{m-1}) = Ew(t_{m-1}) := w(t_{m-1}), \quad m = 2, \ldots, r,$$
$$\widehat{w}(t) = Ew(t) := \sum_{i=0}^{q} \mathscr{L}_i^{m-1}(t) w(t_{m-2} + \frac{i}{q}\tau_{m-1}), \quad t \in I_m, \ m = 2, \ldots, r.$$

See Fig. 6.2. This defines the *extrapolation operator* E.

We can see that the operator E reproduces polynomials of degree $\leq q$:

$$E\varphi = \varphi \quad \forall \varphi \in P_q(0, T), \tag{6.283}$$

where $P_q(0, T)$ denotes the space of all polynomials on $(0, T)$ of degree $\leq q$. Moreover, if $w = w(x, t) \in C([0, T]; H^1(\Omega))$, then

$$\frac{\partial}{\partial x_j}(Ew(\cdot, t)) = E\left(\frac{\partial}{\partial x_j} w(\cdot, t)\right) \in L^2(\Omega), \quad t \in I_m, \ m = 2, \ldots, r. \tag{6.284}$$

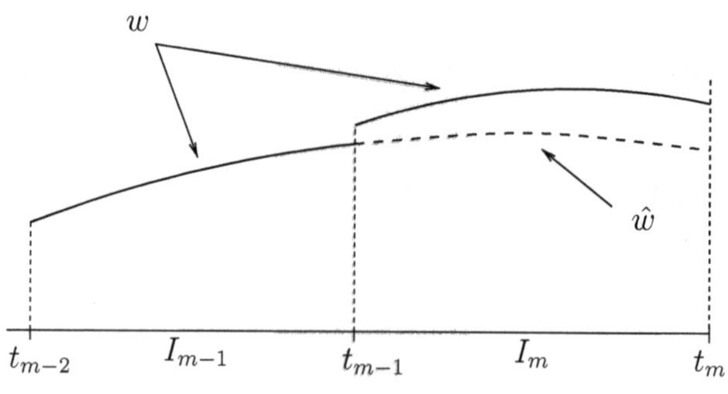

Fig. 6.2 Extrapolation of a piecewise polynomial function

If $w \in S_{h,\tau}^{p,q}$, then it may happen that in general for non-identical meshes $\widehat{w} \notin S_{h,\tau}^{p,q}$. Therefore, we apply the extrapolation operator in combination with the $L^2(\Omega)$-projection $\Pi_{h,m}$ on $S_{h,m}^p$. This means that instead of $\widehat{w}$, we use $\Pi_{h,m}\widehat{w} \in S_{h,\tau}^{p,q}$. It holds that

$$(\Pi_{h,m}\widehat{w}(t), \varphi_h) = (\widehat{w}(t), \varphi_h) \qquad \forall \varphi_h \in S_{h,m}^p,\ t \in I_m. \tag{6.285}$$

Now we are ready to introduce the *extrapolated space-time discontinuous Galerkin method* (EST-DGM).

Definition 6.51 We say that the function $U \in S_{h,\tau}^{p,q}$ is an *EST-DG approximate solution* of problem (6.271), if

$$\int_{I_m} (U', \varphi) + A_{h,m}(U, \varphi) + b_{h,m}(\Pi_{h,m}\widehat{U}, \varphi)\,\mathrm{d}t + (\{U\}_{m-1}, \varphi_{m-1}^+) \tag{6.286a}$$
$$= \int_{I_m} \ell_{h,m}(\varphi)\,\mathrm{d}t \quad \forall \varphi \in S_{h,\tau}^{p,q}, \quad m = 2, \dots, r,$$
$$(U_0^-, \varphi) = (u^0, \varphi) \quad \forall \varphi \in S_{h,0}. \tag{6.286b}$$

Remark 6.52 The existence and uniqueness of the approximate solution follows from an analogue of Theorem 6.5 since the term $b_{h,m}(\Pi_{h,m}\widehat{U}, \varphi)$ does not depend on the approximate solution on I_m.

Scheme (6.286) is not self-started; we need the information from the previous time interval for constructing $\widehat{U}$. Therefore, it is necessary to define the approximate solution on the time interval I_1 by a suitable self-started method. One possibility is to apply an explicit Runge–Kutta method with time step τ_1/q on the interval I_1 and an interpolation of the obtained approximate values at time instants $j\tau_1/q,\ j = 0, \dots, q$. For Runge–Kutta schemes, see Sect. 5.2.1.1.

For piecewise constant approximation in time ($q = 0$), we obtain the backward Euler method linearized in the convective term by extrapolating from the previous time level, treated in Sect. 5.1.

6.3.2 Auxiliary Results

In what follows, we will be concerned with the analysis of error estimates of the EST-DGM (6.286). We assume that the regularity assumption (6.272) is valid. We consider a system of triangulations $\{\mathscr{T}_{h\tau}\}_{h\in(0,\bar{h}),\tau\in(0,\bar{\tau})}$, $\bar{h} > 0,\ \bar{\tau} > 0$ satisfying the *shape-regularity assumption* (6.25) and the *equivalence condition* (6.26). We also assume that the constant C_W from the definition (6.274) of the penalty form $J_{h,m}^\sigma$ satisfies the conditions in Corollary 2.42. We recall several auxiliary results presented in previous sections. By (6.28) and (6.29),

$$
\begin{aligned}
\|\Pi_{h,m}v - v\|_{L^2(K)} &\le C_A h_K^{\mu} |v|_{H^{\mu}(K)}, \\
|\Pi_{h,m}v - v|_{H^1(K)} &\le C_A h_K^{\mu-1} |v|_{H^{\mu}(K)}, \\
\|\Pi_{h,m}v\|_{L^2(K)} &\le \|v\|_{L^2(K)},
\end{aligned}
\tag{6.287}
$$

for all $v \in H^{\mu}(K),\ K \in \mathscr{T}_{h,m},\ m = 0, \ldots, r,\ h \in (0, \bar{h})$, where $C_A > 0$ is a constant independent of v, m and h. Moreover, the definition of the norm $|||\cdot|||_m$ and estimates (2.119) and (6.287) give

$$
|||\Pi_{h,m}v - v|||_m \le c_M C_A h^{\mu-1} |v|_{H^{\mu}(\Omega, \mathscr{T}_{h,m})}, \tag{6.288}
$$

for all $v \in H^{\mu}(\Omega, \mathscr{T}_{h,m}),\ m = 0, \ldots, r,\ h \in (0, \bar{h})$, where $C_A > 0$ is the constant in (6.287) and $c_M = (1 + 4C_W C_M C_T^{-1})^{1/2}$.

Further, by Lemma 4.6 and the multiplicative inequality, there exist constants $C_b, C_{b^*}, C_{b^{**}} > 0$ such that

$$
|b_{h,m}(v, \varphi) - b_{h,m}(\bar{v}, \varphi)| \le C_b \left(\|v - \bar{v}\|_{L^2(\Omega)} + |v - \bar{v}|_{H^1(\Omega, \mathscr{T}_{h,m})} \right) |||\varphi|||_m, \quad v,\ \bar{v},\ \varphi \in H^1(\Omega, \mathscr{T}_{h,m}), \tag{6.289}
$$

$$
|b_{h,m}(v_h, \varphi_h) - b_{h,m}(\bar{v}_h, \varphi_h)| \le C_b^* \|v_h - \bar{v}_h\|_{L^2(\Omega)} |||\varphi_h|||_m, \quad v_h,\ \bar{v}_h,\ \varphi_h \in S_{h,m}^p, \tag{6.290}
$$

$$
|b_{h,m}(u, \varphi_h) - b_{h,m}(\Pi_{h,m}u, \varphi_h)| \le C_b^{**} h^{\mu} |u|_{H^{\mu}(\Omega)} |||\varphi_h|||_m, \quad u \in H^{\mu}(\Omega),\ \varphi_h \in S_{h,m}^p. \tag{6.291}
$$

Now, (2.128), (2.129) and (6.287) imply that

$$
\left| A_{h,m}(u(t) - \Pi_{h,m}u(t), \varphi_h) \right| \le c_a \varepsilon h^{\mu-1} |u(t)|_{H^{\mu}(\Omega)} |||\varphi_h|||_m, \quad \text{a.e. } t \in (0, T),\ u \in L^2(0, T; H^{\mu}(\Omega)),\ \varphi_h \in S_{h,m}^p, \tag{6.292}
$$

$$
|A_{h,m}(v_h, \varphi_h)| \le c_a \varepsilon |||v_h|||_m\, |||\varphi_h|||_m, \qquad v_h, \varphi_h \in S_{h,m}^p, \tag{6.293}
$$

where $c_a = \max(C_B, \sqrt{3} C_A \tilde{C}_B)$ is a constant independent of $h, m, r, \varepsilon, u, v_h$ and φ_h.

In the error analysis we will use the concept of the *discrete characteristic functions* ζ_y introduced in Definition 6.25 to the function ξ and every $y \in I_m$. It satisfies inequality (6.164), i.e.,

$$
\int_{I_m} |||\zeta_y|||_m^2 \, \mathrm{d}t \le C_{CH} \int_{I_m} |||\xi|||_m^2 \, \mathrm{d}t, \tag{6.294}
$$

where the constant C_{CH} depends only on q.

As before, for $\varphi = \varphi(x, \vartheta)$ and $\psi = \psi(x, t)$ we use the notation $\partial_\vartheta \varphi = \partial\varphi/\partial\vartheta$ and $\partial_t \psi = \partial\psi/\partial t$, respectively.

In the sequel, we will assume that the time partition formed by time instants t_m, $m = 0, \dots, r$, is locally quasiuniform. This means that there exist positive constants $\underline{C}_\theta$ and C_θ, independent of h, τ, τ_m, r, such that

$$\underline{C}_\theta \tau_{m-1} \le \tau_m \le C_\theta \tau_{m-1}, \quad m = 2, \dots, r. \tag{6.295}$$

The following Lemma characterizes approximation properties of the extrapolation operator E defined in (6.282).

Lemma 6.53 *Let u be the solution of (6.271) satisfying (6.272) and let condition (6.295) be valid for $m = 2, \dots, r$. Then*

$$\|u(t) - \widehat{u}(t)\|_{L^2(\Omega)} \le C_E \tau^{q+1} |u|_{W^{q+1,\infty}(0,T;L^2(\Omega))}, \quad t \in I_m, \quad m = 2, \dots, r, \tag{6.296}$$

$$\int_{I_m} |u - \widehat{u}|^2_{H^1(K)}\, dt \le C_E^*\, \tau^{2(q+1)} |u|^2_{H^{q+1}(J_m;H^1(K))}, \quad K \in \mathscr{T}_{h,m}, \ m = 2, \dots, r, \tag{6.297}$$

$$\int_{I_m} \|u - \widehat{u}\|^2_{L^2(\Omega)}\, dt \le C_E\, \tau_m \tau^{2(q+1)} |u|^2_{W^{q+1,\infty}(0,T;L^2(\Omega))}, \quad m = 2, \dots, r, \tag{6.298}$$

$$\int_{I_m} |u - \widehat{u}|^2_{H^1(\Omega)} dt \le C_E^*\, (1 + \underline{C}_\theta^{-1}) \tau_m \tau^{2(q+1)} |u|^2_{W^{q+1,\infty}(0,T;H^1(\Omega))}, \quad m = 2, \dots, r, \tag{6.299}$$

where $J_m = (t_{m-2}, t_m)$ and the constants C_E, $C_E^* > 0$ are independent of t, m, τ, h, K and u.

Proof (a) In order to establish (6.296), we can proceed in a similar way as in the proof of Lemma 6.14. We introduce the transformation

$$F(\vartheta) = \vartheta(\tau_m + \tau_{m-1}) + t_{m-2}, \ \vartheta \in (0, 1), \tag{6.300}$$

of the interval $(0, 1)$ onto $J_m := I_{m-1} \cup I_m \cup \{t_{m-1}\} = (t_{m-2}, t_m)$. Further, we set

$$Z_m(x, t) = u(x, t) - Eu(x, t) = u(x, t) - \widehat{u}(x, t), \quad x \in \Omega, \ t \in J_m, \tag{6.301}$$

and

$$z(x, \vartheta) = Z_m(x, F(\vartheta)), \quad x \in \Omega, \ \vartheta \in (0, 1). \tag{6.302}$$

(The variable x stands here as a parameter only.)

Now we introduce the operator $\tilde{E}$ obtained by the transformation of the operator E. If $\tilde{u} \in H^{q+1}(0, 1)$, then we set

$$\tilde{E}(\tilde{u}) = (Eu) \circ F, \tag{6.303}$$

where $u(t) = \tilde{u}(F^{-1}(t)),\ t \in J_m$.

The application of the same process as in the proof of Lemma 6.14 requires establishing the continuity of the operator $\tilde{E}$, considered as a mapping of the space $H^{q+1}(0, 1)$ into the spaces $L^2(0, 1)$ and $H^1(0, 1)$.

(i) Continuity of the operator $\tilde{E} : H^{q+1}(0, 1) \to L^2(0, 1)$:

We have the continuous embedding $H^{q+1}(0, 1) \hookrightarrow L^\infty(0, 1)$. Hence, there exists a constant $c_1 > 0$ such that

$$\|\varphi\|_{L^\infty(0,1)} \le c_1 \|\varphi\|_{H^{q+1}(0,1)} \quad \forall \varphi \in H^{q+1}(0, 1). \tag{6.304}$$

Now let $\tilde{u} \in H^{q+1}(0, 1)$ and $u \in H^{q+1}(J_m)$ satisfy the relation $\tilde{u} = u \circ F$. Then, by (6.282), (6.281), (6.303) and (6.304),

$$\begin{aligned}\|\tilde{E}\tilde{u}\|_{L^\infty(0,1)} &= \operatorname{ess\,sup}_{(0,1)} |\tilde{E}\tilde{u}| = \operatorname{ess\,sup}_{J_m} |Eu| \\ &\le \|u\|_{L^\infty(J_m)} \sup_{t \in J_m} \sum_{i=0}^{q} |\mathscr{L}_i^{m-1}(t)| \\ &\le (q+1)\,(C_\theta q + q)^q \|\tilde{u}\|_{L^\infty(0,1)} \le c_2 \|\tilde{u}\|_{H^{q+1}(0,1)},\end{aligned}$$

where $c_2 = c_1(q+1)\,(C_\theta q + q)^q$. This and the continuous embedding $L^\infty(0, 1) \hookrightarrow L^2(0, 1)$ imply the continuity of the operator $\tilde{E}$ as a mapping of $H^{q+1}(0, 1)$ into $L^2(0, 1)$. Hence, there exists a constant $c_3 > 0$ such that

$$\|\tilde{E}\tilde{u}\|_{L^2(0,1)} \le c_3 \|\tilde{u}\|_{H^{q+1}(0,1)} \quad \forall \tilde{u} \in H^{q+1}(0, 1). \tag{6.305}$$

(ii) Continuity of the operator $\tilde{E} : H^{q+1}(0, 1) \to H^1(0, 1)$:

In view of (6.305), it is sufficient to establish the existence of a constant $c_4 > 0$ such that

$$\left\| \frac{\mathrm{d}(\tilde{E}\tilde{u})}{\mathrm{d}\vartheta} \right\|_{L^2(0,1)} \le c_4 \|\tilde{u}\|_{H^{q+1}(0,1)} \quad \forall \tilde{u} \in H^{q+1}(0, 1). \tag{6.306}$$

By a simple calculation we find that

$$\frac{\mathrm{d}}{\mathrm{d}t} \mathscr{L}_i^{m-1}(t) = \sum_{\substack{n=0 \\ n \ne i}}^{q} \frac{1}{\frac{i-n}{q}\tau_{m-1}} \prod_{\substack{j=0 \\ j \ne i \\ j \ne n}}^{q} \frac{t - (t_{m-2} + \frac{j}{q}\tau_{m-1})}{\frac{i-j}{q}\tau_{m-1}}.$$

This relation and similar estimates as in (6.281) imply that

$$\left| \frac{\mathrm{d}}{\mathrm{d}t} \mathscr{L}_i^{m-1}(t) \right| \le \frac{q^2}{\tau_{m-1}} (C_\theta q + q)^q. \tag{6.307}$$

Further, by virtue of (6.303) and (6.300),

$$\frac{\mathrm{d}(\tilde{E}\tilde{u})(\vartheta)}{\mathrm{d}\theta} = (\tau_m + \tau_{m-1}) \frac{\mathrm{d}(Eu)}{\mathrm{d}t} \big(\vartheta(\tau_m + \tau_{m-1}) + t_{m-2}\big).$$

This relation, (6.282), (6.307) and the assumption $\tau_m/\tau_{m-1} \le C_\theta$ imply that

$$\begin{aligned}
\left\| \frac{\mathrm{d}}{\mathrm{d}\vartheta} (\tilde{E}\tilde{u}) \right\|_{L^\infty(0,1)} &\le \sup_{t \in J_m} \left| \frac{\mathrm{d}(Eu)(t)}{\mathrm{d}t} \right| (\tau_m + \tau_{m-1}) \\
&\le \|u\|_{L^\infty(J_m)} \sup_{t \in J_m} \sum_{i=0}^{q} \left| \frac{\mathrm{d}}{\mathrm{d}t} \mathscr{L}_i^{m-1}(t) \right| (\tau_m + \tau_{m-1}) \\
&\le \|u\|_{L^\infty(J_m)} (\tau_m + \tau_{m-1}) \frac{q^2(q+1)}{\tau_{m-1}} (C_\theta q + q)^q \\
&\le q^2(q+1)\,(C_\theta q + q)^{q+1}\, \|\tilde{u}\|_{L^\infty(0,1)}.
\end{aligned}$$

Finally, using this result and the inequalities

$$\left\| \frac{\mathrm{d}}{\mathrm{d}\vartheta} (\tilde{E}\tilde{u}) \right\|_{L^2(0,1)} \le \left\| \frac{\mathrm{d}}{\mathrm{d}\vartheta} (\tilde{E}\tilde{u}) \right\|_{L^\infty(0,1)}$$

and (6.304), we obtain (6.306).

Now it is already possible to proceed quite analogously as in the proof of Lemma 6.14. Actually, we have

$$\begin{aligned}
\partial_\vartheta z(x, \vartheta) &= (\tau_m + \tau_{m-1}) \partial_t Z_m(x,\ (\tau_m + \tau_{m-1})\vartheta + t_{m-2}), && (6.308)\\
\partial_\vartheta^{q+1} z(x, \vartheta) &= (\tau_m + \tau_{m-1})^{q+1}\, \partial_t^{q+1} Z_m(x,\ (\tau_m + \tau_{m-1})\vartheta + t_{m-2}), \quad \vartheta \in (0, 1).
\end{aligned}$$

This fact and (6.301) imply that

$$\|z(x,\cdot)\|^2_{L^2(0,1)} = \frac{1}{(\tau_m + \tau_{m-1})} \|Z_m(x,\cdot)\|^2_{L^2(J_m)}, \tag{6.309a}$$

$$\|\partial_\vartheta z(x,\cdot)\|^2_{L^2(0,1)} = (\tau_m + \tau_{m-1}) \|\partial_t Z_m(x,\cdot)\|^2_{L^2(J_m)}, \tag{6.309b}$$

$$\|\partial_\vartheta^{q+1} z(x,\cdot)\|^2_{L^2(0,1)} = (\tau_m + \tau_{m-1})^{2q+1} \|\partial_t^{q+1} Z_m(x,\cdot)\|^2_{L^2(J_m)}. \tag{6.309c}$$

Taking into account the continuity of the operator $\tilde{E}$, it follows from Theorem 2.16 that there exists a constant $C > 0$ independent of z such that

$$\|z(x,\cdot)\|^2_{L^2(0,1)} \le C\|\partial_\vartheta^{q+1} z(x,\cdot)\|^2_{L^2(0,1)} \qquad (6.310)$$
$$\|\partial_\vartheta z(x,\cdot)\|^2_{L^2(0,1)} \le C\|\partial_\vartheta^{q+1} z(x,\cdot)\|^2_{L^2(0,1)}.$$

Now, by (6.302) and (6.300),

$$\|z\|_{L^\infty(0,1;L^2(\Omega))} = \operatorname{ess\,sup}_{\vartheta\in(0,1)} \left(\int_\Omega |z(x,\vartheta)|^2\,\mathrm{d}x\right)^{1/2} \qquad (6.311)$$
$$= \operatorname{ess\,sup}_{t\in J_m} \left(\int_\Omega |Z_m(x,t)|^2\mathrm{d}x\right)^{1/2} = \|Z_m\|_{L^\infty(J_m;L^2(\Omega))}.$$

Further, by (6.309c) and Fubini's theorem we have

$$\|\partial_\vartheta^{q+1} z\|^2_{L^2(0,1;L^2(\Omega))} = \int_0^1 \left(\int_\Omega |\partial_\vartheta^{q+1} z|^2\,\mathrm{d}x\right)\mathrm{d}\vartheta = \int_\Omega \left(\int_0^1 |\partial_\vartheta^{q+1} z|^2\,\mathrm{d}\vartheta\right)\mathrm{d}x$$
$$= \int_\Omega \|\partial_\vartheta^{q+1} z(x,\cdot)\|^2_{L^2(0,1)}\mathrm{d}x = (\tau_m+\tau_{m-1})^{2q+1}\int_\Omega \|\partial_t^{q+1} Z_m(x,\cdot)\|^2_{L^2(J_m)}\,\mathrm{d}x$$
$$= (\tau_m+\tau_{m-1})^{2q+1}\|\partial_t^{q+1} Z_m\|^2_{L^2(J_m;L^2(\Omega))}. \qquad (6.312)$$

We will also use relations (6.67) and (6.68), i.e.,

$$\|v\|_{L^\infty(0,1;L^2(\Omega))} \le C\|v\|_{H^1(0,1;L^2(\Omega))} \quad \forall v\in H^1(0,1;L^2(\Omega)), \qquad (6.313)$$

where

$$\|v\|^2_{H^1(0,1;L^2(\Omega))} = \|v\|^2_{L^2(0,1;L^2(\Omega))} + \|\partial_\vartheta v\|^2_{L^2(0,1;L^2(\Omega))}. \qquad (6.314)$$

Now, taking into account (6.313), (6.314), (6.310)–(6.312), we obtain

$$\|Z_m\|^2_{L^\infty(J_m;L^2(\Omega))} = \|z\|^2_{L^\infty(0,1;L^2(\Omega))} \le C\left(\|z\|^2_{L^2(0,1;L^2(\Omega))} + \|\partial_\vartheta z\|^2_{L^2(0,1;L^2(\Omega))}\right)$$
$$\le C\|\partial_\vartheta^{q+1} z\|^2_{L^2(0,1;L^2(\Omega))} = C\,(\tau_m+\tau_{m-1})^{2q+1}\|\partial_t^{q+1} Z_m\|^2_{L^2(J_m;L^2(\Omega))}.$$

Finally, this estimate, (6.312) and (6.301) yield (6.296).

(b) In what follows, we prove estimate (6.297). Let $m\in\{2,\dots,r\}$, $K\in\mathscr{T}_{h,m}$, $x\in K$. Taking into account the continuity of the operator $\tilde{E}: H^{q+1}(0,1)\to L^2(0,1)$ and relation (6.284), we can apply Theorem 2.16, where we set $p := q$, $\alpha=\beta := 2$, $m := 0$, $\omega := J_m$, $h_\omega = \rho_\omega := \tau_m+\tau_{m-1}$, $\Pi := E$, $\hat{\Pi} := \tilde{E}$. Taking into account that in the part (a) we proved that the operator $\tilde{E}: H^{q+1}(0,1)\to L^2(0,1)$ is continuous, by (2.75) we find that for $i=1,\dots,d$ we have

$$\left\| \frac{\partial}{\partial x_i}(u(x,\cdot) - \hat{u}(x,\cdot)) \right\|^2_{L^2(J_m)} = \left\| \frac{\partial}{\partial x_i}(u(x,\cdot) - Eu(x,\cdot)) \right\|^2_{L^2(J_m)} \tag{6.315}$$
$$= \left\| \frac{\partial u}{\partial x_i}(x,\cdot) - E\Big(\frac{\partial u}{\partial x_j}(x,\cdot)\Big) \right\|^2_{L^2(J_m)} \leq \tilde{C}_E\,(\tau_m + \tau_{m-1})^{2(q+1)} \left| \frac{\partial u(x,\cdot)}{\partial x_i} \right|^2_{H^{q+1}(J_m)}.$$

Further, (6.315) and Fubini's theorem imply that

$$\int_{I_m} |u - \hat{u}|^2_{H^1(K)}\, \mathrm{d}t \leq \int_{J_m} |u - \hat{u}|^2_{H^1(K)}\, \mathrm{d}t \tag{6.316}$$
$$= \int_{J_m} \left(\int_K \sum_{i=1}^d \left| \frac{\partial}{\partial x_i}(u(x,t) - \hat{u}(x,t)) \right|^2 \mathrm{d}x \right) \mathrm{d}t$$
$$= \int_K \left(\int_{J_m} \sum_{i=1}^d \left| \frac{\partial}{\partial x_i}(u(x,t) - \hat{u}(x,t)) \right|^2 \mathrm{d}t \right) \mathrm{d}x$$
$$= \int_K \sum_{i=1}^d \left\| \frac{\partial}{\partial x_i}(u(x,\cdot) - \hat{u}(x,\cdot)) \right\|^2_{L^2(J_m)} \mathrm{d}x$$
$$\leq \tilde{C}_E\,(\tau_m + \tau_{m-1})^{2(q+1)} \int_K \sum_{i=1}^d \left| \frac{\partial u(x,\cdot)}{\partial x_i} \right|^2_{H^{q+1}(J_m)} \mathrm{d}x$$
$$\leq C_E^* \tau^{2(q+1)} \int_K \sum_{i=1}^d \left| \frac{\partial u(x,\cdot)}{\partial x_i} \right|^2_{H^{q+1}(J_m)} \mathrm{d}x,$$

which follows from the inequality $\tau_{m-1}, \tau_m \leq \tau$ and, thus, $C_E^* = 2^{2(q+1)}\tilde{C}_E$.

Now we show that

$$\int_K \sum_{i=1}^d \left| \frac{\partial u(x,\cdot)}{\partial x_i} \right|^2_{H^{q+1}(J_m)} \mathrm{d}t = |u|^2_{H^{q+1}(J_m;H^1(K))}. \tag{6.317}$$

Actually, Fubini's theorem, the relation $\partial_t^{q+1}\big(\frac{\partial u}{\partial x_i}\big) = \frac{\partial}{\partial x_i}(\partial_t^{q+1} u)$ in the sense of distributions and (1.40) imply that

$$\int_K \sum_{i=1}^d \left| \frac{\partial u(x,\cdot)}{\partial x_i} \right|^2_{H^{q+1}(J_m)} = \int_K \left(\int_{J_m} \sum_{i=1}^d \left| \partial_t^{q+1} \Big(\frac{\partial u(x,\cdot)}{\partial x_i} \Big) \right|^2 \mathrm{d}t \right) \mathrm{d}x \tag{6.318}$$
$$= \int_{J_m} \left(\int_K \sum_{i=1}^d \left| \partial_t^{q+1} \Big(\frac{\partial u}{\partial x_i} \Big) \right|^2 \mathrm{d}x \right) \mathrm{d}t = \int_{J_m} \left(\int_K \sum_{i=1}^d \left| \frac{\partial}{\partial x_i}(\partial_t^{q+1} u) \right|^2 \mathrm{d}x \right) \mathrm{d}t$$
$$= \int_{J_m} |\partial_t^{q+1} u|^2_{H^1(K)}\, \mathrm{d}t = |u|^2_{H^{q+1}(J_m;H^1(K))}.$$

Now (6.316) and (6.317) immediately imply estimate (6.297).

(c) The proof of (6.298) is a consequence of (6.296).

(d) The proof of (6.299) is a consequence of the inequalities (6.295), (6.297) and

$$
\begin{aligned}
|u|_{H^{q+1}(J_m;H^1(\Omega))} &\le (\tau_m + \tau_{m-1})|u|_{W^{q+1,\infty}(0,T;H^1(\Omega))} \\
&\le \tau_m(1 + \underline{C}_\theta^{-1})|u|_{W^{q+1,\infty}(0,T;H^1(\Omega))}.
\end{aligned}
$$

□

Remark 6.54 The above results can also be proven with the use of a general approximation theory in Bochner spaces derived in [279]. The technique used in the proofs of Lemmas 6.14 and 6.53, based on the continuous embedding (6.67), was proposed by K. Najzar.

Exercise 6.55 (a) Prove that $\partial_t^{q+1}\left(\frac{\partial u}{\partial x_i}\right) = \frac{\partial}{\partial x_i}(\partial_t^{q+1} u)$ in the sense of distributions.
(b) Using the properties of operator $\tilde{E}$, apply the technique from the proof of Lemma 6.14 and prove estimates (6.296)–(6.299) in detail.

6.3.3 Error Estimates

As usual, we express the error $e = U - u$ of the EST-DGM in the form $e = \xi + \eta$, where $\xi = U - \pi u$ and $\eta = \pi u - u$. For further considerations we put

$$
C_u = \max(|u|_{W^{q+1,\infty}(0,T;L^2(\Omega))}, |u|_{C([0,T];H^\mu(\Omega))}, |u|_{W^{q+1,\infty}(0,T;H^1(\Omega))}), \tag{6.319}
$$

which makes sense by virtue of assumption (6.272) and the inequality $\mu \le s$.

Lemma 6.56 *Let u be the solution of (6.271) satisfying (6.272) and let U be its approximation given by (6.286). Let (6.140), i.e., $\tau_m \ge C_S h_m^2$ hold for $m = 1, \dots, r$ with C_S independent of h and τ and let (6.295) be satisfied. Then there exists a constant $C_1 > 0$ such that*

$$
\begin{aligned}
&\int_{I_m} \Big((\xi', \varphi) + A_{h,m}(\xi, \varphi)\Big)\, \mathrm{d}t + (\{\xi\}_{m-1}, \varphi_{m-1}^+) \qquad (6.320) \\
&\le \tau_m Q(h,\tau) + \frac{\varepsilon}{8}\int_{I_m} |||\varphi|||_m^2\, \mathrm{d}t + \frac{1}{8}\|\{\varphi\}_{m-1}\|_{L^2(\Omega)}^2 + \tau_m \frac{C_1}{\varepsilon} \sup_{t\in I_{m-1}} \|\xi(t)\|_{L^2(\Omega)}^2, \\
&\qquad \forall \varphi \in S_{h,\tau}^{p,q},\ \forall m = 2, \dots, r,
\end{aligned}
$$

where $Q(h,\tau) = O(h^{2(\mu-1)} + \tau^{2(q+\gamma)})$ and γ is specified in Corollary 6.21.

Proof Integrating (6.278) over I_m and subtracting it from (6.286), we obtain

$$\int_{I_m} \big((\xi', \varphi) + A_{h,m}(\xi, \varphi)\big)\, \mathrm{d}t + (\{\xi\}_{m-1}, \varphi_{m-1}^+) \tag{6.321}$$
$$= -\Big(\int_{I_m} (\eta', \varphi)\, \mathrm{d}t + (\{\eta\}_{m-1}, \varphi_{m-1}^+)\Big) - \int_{I_m} A_{h,m}(\eta, \varphi)\, \mathrm{d}t$$
$$+ \int_{I_m} \big(b_{h,m}(u, \varphi) - b_{h,m}(\Pi_{h,m}\widehat{U}, \varphi)\big)\, \mathrm{d}t, \qquad \varphi \in S_{h,\tau}^{p,q}.$$

We estimate the terms on the right-hand side of (6.321). Using relation (6.87) with $\delta = 2$, approximation property (6.287), the Young inequality, and assumption (6.140), we find that

$$\int_{I_m} (\eta', \varphi)\, \mathrm{d}t + (\{\eta\}_{m-1}, \varphi_{m-1}^+) \tag{6.322}$$
$$\le 2\|\eta_{m-1}^-\|_{L^2(\Omega)}^2 + \frac{1}{8}\|\{\varphi\}_{m-1}\|_{L^2(\Omega)}^2 \le C_2\tau_m h^{2(\mu-1)} + \frac{1}{8}\|\{\varphi\}_{m-1}\|_{L^2(\Omega)}^2,$$

where $C_2 = 2C_A^2 C_u C_S^{-1}$. Further, the second term on the right-hand side of (6.321) can be estimated with the aid of (6.292) and (6.293) by

$$-\int_{I_m} A_{h,m}(\eta, \varphi)\, \mathrm{d}t = \int_{I_m} \big(A_{h,m}(u - \Pi_{h,m}u, \varphi) + A_{h,m}(\Pi_{h,m}u - \pi u, \varphi)\big)\, \mathrm{d}t$$
$$\le \varepsilon \int_{I_m} \Big(c_a C_u h^{\mu-1} |||\varphi|||_m + c_a |||\Pi_{h,m}u - \pi u|||_m\, |||\varphi|||_m\Big)\, \mathrm{d}t.$$

Now we use the Young inequality and the relation

$$\int_{I_m} |||\Pi_{h,m}u - \pi u|||_m^2\, \mathrm{d}t \le 2\int_{I_m} |||\Pi_{h,m}u - u|||_m^2\, \mathrm{d}t + 2\int_{I_m} |||u - \pi u|||_m^2\, \mathrm{d}t.$$

Then, by (6.288), in a similar way as in (6.264), we get

$$-\int_{I_m} A_{h,m}(\eta, \varphi)\, \mathrm{d}t \le \frac{\varepsilon}{16}\int_{I_m} |||\varphi|||_m^2\, \mathrm{d}t + C_3\tau_m\big(h^{2(\mu-1)} + \tau^{2(q+\gamma)}\big), \tag{6.323}$$

where $C_3 = C_3(c_a, C_A, C_u, \varepsilon)$.

Finally, we estimate the third term on the right-hand side of (6.321) containing the nonlinear form $b_{h,m}$. By the triangle inequality,

$$|b_{h,m}(u, \varphi) - b_{h,m}(\Pi_{h,m}\widehat{U}, \varphi)| \le |b_{h,m}(u, \varphi) - b_{h,m}(\widehat{u}, \varphi)| \tag{6.324}$$
$$+ |b_{h,m}(\widehat{u}, \varphi) - b_{h,m}(\Pi_{h,m}\widehat{u}, \varphi)| + |b_{h,m}(\Pi_{h,m}\widehat{u}, \varphi) - b_{h,m}(\Pi_{h,m}\widehat{U}, \varphi)|.$$

In what follows, we apply several times the Cauchy and Young inequalities, which will not be mentioned any more. By (6.289),

$$\begin{aligned}
&\int_{I_m} \left|b_{h,m}(u,\varphi) - b_{h,m}(\hat{u},\varphi)\right| \,\mathrm{d}t \\
&\le C_b \int_{I_m} \left(\|u-\hat{u}\|_{L^2(\Omega)} + |u-\hat{u}|_{H^1(\Omega,\mathscr{T}_{h,m})}\right) |||\varphi|||_m \,\mathrm{d}t \\
&\le \frac{C_{b1}}{\varepsilon} \int_{I_m} \left(\|u-\hat{u}\|^2_{L^2(\Omega)} + |u-\hat{u}|^2_{H^1(\Omega,\mathscr{T}_{h,m})}\right) \mathrm{d}t + \frac{\varepsilon}{48}\int_{I_m} |||\varphi|||_m^2 \,\mathrm{d}t,
\end{aligned}$$

where $C_{b1} = C_{b1}(C_b)$. Now, this inequality, (6.298) and (6.299) imply that

$$\int_{I_m} \left|b_{h,m}(u,\varphi) - b_{h,m}(\hat{u},\varphi)\right| \,\mathrm{d}t \le \frac{C_{b2}}{\varepsilon}\tau_m \tau^{2(q+1)} + \frac{\varepsilon}{48}\int_{I_m} |||\varphi|||_m^2 \,\mathrm{d}t, \tag{6.325}$$

where $C_{b2} = C_{b2}(C_{b1}, C_u, C_E, C_E^*)$.

From the definition (6.282) of $\hat{u}$ and the assumption that $u \in C([0,T]; H^s(\Omega)) \subset C([0,T]; H^\mu(\Omega))$ it follows that $\hat{u}(t) \in H^\mu(\Omega)$ for every $t \in I_m$. Moreover, by virtue of (6.281) and (6.282),

$$\int_{I_m} |\hat{u}|^2_{H^\mu(\Omega)} \,\mathrm{d}t \le (qC_\Theta + q)^q \tau_m |u|^2_{C([0,T];H^\mu(\Omega))}. \tag{6.326}$$

Taking into account (6.291) and using (6.326), we find that

$$\begin{aligned}
\int_{I_m} \left|b_{h,m}(\hat{u},\varphi) - b_{h,m}(\Pi_{h,m}\hat{u},\varphi)\right| \,\mathrm{d}t &\le C_b^{**} h^\mu \int_{I_m} |\hat{u}|_{H^\mu(\Omega)} |||\varphi|||_m \,\mathrm{d}t \qquad (6.327)\\
&\le \frac{C_{b3}}{\varepsilon}\tau_m h^{2\mu} + \frac{\varepsilon}{48}\int_{I_m} |||\varphi|||_m^2 \,\mathrm{d}t
\end{aligned}$$

with $C_{b3} = C_{b3}(C_b^{**}, C_u)$.

From (6.290) and (6.287) we get

$$\begin{aligned}
&\int_{I_m} \left|b_{h,m}(\Pi_{h,m}\hat{u},\varphi) - b_{h,m}(\Pi_{h,m}\hat{U},\varphi)\right| \,\mathrm{d}t \qquad (6.328)\\
&\le C_b^* \int_{I_m} \|\Pi_{h,m}(\hat{u}-\hat{U})\|_{L^2(\Omega)} |||\varphi|||_m \,\mathrm{d}t \le C_b^* \int_{I_m} \|\hat{u}-\hat{U}\|_{L^2(\Omega)} |||\varphi|||_m \,\mathrm{d}t.
\end{aligned}$$

In order to estimate $\|\widehat{u} - \widehat{U}\|_{L^2(\Omega)}$, we use definition (6.282), inequality (6.281) and (6.136). Then

$$\|\widehat{u} - \widehat{U}\|_{L^2(\Omega)} \le \sum_{j=0}^{q} \left(|\mathscr{L}_j^{m-1}(t)|\, \|u(t_{m-2} + \frac{j}{q}\tau_{m-1}) - U(t_{m-2} + \frac{j}{q}\tau_{m-1})\|_{L^2(\Omega)}\right)$$

$$
\begin{aligned}
&\le (qC_\theta + q)^q \sum_{j=0}^{q} \|u(t_{m-2} + \frac{j}{q}\tau_{m-1}) - \pi u(t_{m-2} + \frac{j}{q}\tau_{m-1})\|_{L^2(\Omega)} \\
&+ (qC_\theta + q)^q \sum_{j=0}^{q} \|\pi u(t_{m-2} + \frac{j}{q}\tau_{m-1}) - U(t_{m-2} + \frac{j}{q}\tau_{m-1})\|_{L^2(\Omega)} \\
&\le C_4(\tau^{q+1} + h^\mu) + C_5 \sup_{t\in I_{m-1}} \|\xi(t)\|_{L^2(\Omega)},
\end{aligned}
\tag{6.329}
$$

where $C_4 = C_4(q, C_\theta, C_\pi, C_u)$ and $C_5 = C_5(q, C_\theta)$. By virtue of (6.328), (6.329) and the relation

$$
\left(\sup_{t\in I_{m-1}} \|\xi(t)\|_{L^2(\Omega)}\right)^2 = \sup_{t\in I_{m-1}} \|\xi(t)\|^2_{L^2(\Omega)},
\tag{6.330}
$$

we have

$$
\begin{aligned}
&\int_{I_m} \left| b_{h,m}(\Pi_{h,m}\hat{u}, \varphi) - b_{h,m}(\Pi_{h,m}\hat{U}, \varphi)\right| \mathrm{d}t \\
&\le \frac{C_{b4}}{\varepsilon}\tau_m \left(\tau^{2(q+1)} + h^{2\mu} + \sup_{t\in I_{m-1}} \|\xi(t)\|^2_{L^2(\Omega)}\right) + \frac{\varepsilon}{48}\int_{I_m} |||\varphi|||_m^2 \,\mathrm{d}t.
\end{aligned}
\tag{6.331}
$$

Due to (6.324), (6.325), (6.327) and (6.331), there exists a constant C_1 depending on $C_{b1}, \ldots, C_{b4}, C_u, C_E, C_4, C_5$ such that

$$
\begin{aligned}
&\int_{I_m} |b_{h,m}(u, \varphi) - b_{h,m}(\Pi_{h,m}\widehat{U}, \varphi)| \,\mathrm{d}t \\
&\le \frac{C_1}{\varepsilon}\tau_m \left(h^{2\mu} + \tau^{2q+2} + \sup_{t\in I_{m-1}} \|\xi(t)\|^2_{L^2(\Omega)}\right) + \frac{\varepsilon}{16}\int_{I_m} |||\varphi|||_m^2 \,\mathrm{d}t.
\end{aligned}
\tag{6.332}
$$

Finally, summing (6.332) with (6.322) and (6.323), and using (6.321), we get (6.320) with

$$
Q(h, \tau) := C_6(h^{2(\mu-1)} + h^{2(\mu-1)} + \tau^{2(q+\gamma)}) + \frac{C_1}{\varepsilon}(h^{2\mu} + \tau^{2q+2}), \quad C_6 > 0. \tag{6.333}
$$

□

Now we are ready to formulate the *main result* on the error estimates of the EST-DGM.

Theorem 6.57 *Let u be the solution of problem (6.271) satisfying (6.272) and let $U \in S^{p,q}_{h,\tau}$ be its approximation obtained by scheme (6.286). Let (6.140) hold for all $m = 1, \ldots, r$ and let (6.295) hold. Let the shape-regularity assumption (6.25) and the equivalence condition (6.26) be satisfied. Then there exists a constant $C > 0$ independent of h and τ such that*

$$\sup_{t\in I_m} \|u(t) - U(t)\|^2_{L^2(\Omega)} \tag{6.334}$$
$$\le C\left(h^{2(\mu-1)} + \tau^{2(q+\gamma)} + \sup_{t\in I_1} \|u(t) - U(t)\|^2_{L^2(\Omega)} \right) e^{Ct_m/\varepsilon}, \quad m = 1, \ldots, r,$$

$$\varepsilon \sum_{m=1}^{r} \int_0^T |||u(t) - U(t)|||^2_m \, dt \tag{6.335}$$
$$\le C\left(h^{2(\mu-1)} + \tau^{2(q+\gamma)} + \sup_{t\in I_1} \|u(t) - U(t)\|^2_{L^2(\Omega)} + \int_{I_1} |||u(t) - U(t)|||^2_1 \, dt \right) e^{CT/\varepsilon}.$$

Here $\gamma = 0$, if (6.118) holds and the function u_D from the boundary condition (6.271b) has a general behaviour. If u_D is defined by (6.116), then $\gamma = 1$.

Proof By (6.233),

$$\int_{I_m} 2(\xi', \xi)\, dt + 2(\{\xi\}_{m-1}, \xi^+_{m-1}) = \|\xi^-_m\|^2_{L^2(\Omega)} - \|\xi^-_{m-1}\|^2_{L^2(\Omega)} + \|\{\xi\}_{m-1}\|^2_{L^2(\Omega)}. \tag{6.336}$$

Putting $\varphi := 2\xi$ in (6.320) and using (6.336) together with coercivity of the form $A_{h,m}$, i.e., $A_{h,m}(\xi, \xi) \ge \varepsilon|||\xi|||^2_m/2$ (cf. (6.95)), we obtain

$$\|\xi^-_m\|^2_{L^2(\Omega)} - \|\xi^-_{m-1}\|^2_{L^2(\Omega)} + \frac{\varepsilon}{2}\int_{I_m} |||\xi|||^2_m \, dt \le \tau_m Q(h, \tau) + \tau_m \frac{C_1}{\varepsilon} \sup_{t\in I_{m-1}} \|\xi(t)\|^2_{L^2(\Omega)}. \tag{6.337}$$

Now we proceed in a similar way as in the proof of Theorem 6.30. Let $y = \arg\sup_{t\in \bar{I}_m} \|\xi(t)\|_{L^2(\Omega)}$, where as the values $\xi(t_{m-1})$ and $\xi(t_m)$ we take the limits ξ^+_{m-1} and ξ^-_m, respectively. By $\tilde{\xi}$ we denote the discrete characteristic function ζ_y to ξ at the point y, introduced in Definition 6.25. Then, by (6.146) and (6.147),

$$\int_{I_m} (\xi', \tilde{\xi})\, dt = \int_{t_{m-1}}^{y} (\xi', \xi)\, dt, \quad \tilde{\xi}^+_{m-1} = \xi^+_{m-1}, \quad \{\tilde{\xi}\}_{m-1} = \{\xi\}_{m-1}, \tag{6.338}$$

since ξ' is a polynomial of degree $q - 1$ with respect to time. Then, similarly as in (6.336), from (6.338) we have

$$\int_{I_m} 2(\xi', \tilde{\xi})\, dt + 2(\{\xi\}_{m-1}, \tilde{\xi}^+_{m-1}) = \|\xi(y)\|^2_{L^2(\Omega)} - \|\xi^-_{m-1}\|^2_{L^2(\Omega)} + \|\{\xi\}_{m-1}\|^2_{L^2(\Omega)}. \tag{6.339}$$

If we set $\varphi := 2\tilde{\xi}$ in (6.320), use (6.339) and the last equality in (6.338) and take into account that $\|\xi(y)\|^2_{L^2(\Omega)} = \sup_{t\in I_m} \|\xi(t)\|^2_{L^2(\Omega)}$, we obtain the estimate

$$\sup_{t\in I_m} \|\xi(t)\|^2_{L^2(\Omega)} - \|\xi^-_{m-1}\|^2_{L^2(\Omega)} + 2\int_{I_m} A_{h,m}(\xi,\tilde{\xi})\,\mathrm{d}t \tag{6.340}$$
$$\leq \tau_m Q(h,\tau) + \varepsilon \int_{I_m} |||\tilde{\xi}|||^2_m\,\mathrm{d}t + \tau_m \frac{C_1}{\varepsilon} \sup_{t\in I_{m-1}} \|\xi(t)\|^2_{L^2(\Omega)}.$$

Since $-\sup_{t\in I_{m-1}} \|\xi(t)\|_{L^2(\Omega)} \leq -\|\xi^-_{m-1}\|_{L^2(\Omega)}$, by (6.294), (6.293) and the Young inequality, from (6.340) we get

$$\sup_{t\in I_m} \|\xi(t)\|^2_{L^2(\Omega)} - \sup_{t\in I_{m-1}} \|\xi(t)\|^2_{L^2(\Omega)} \tag{6.341}$$
$$\leq 2\int_{I_m} |A_{h,m}(\xi,\tilde{\xi})|\,\mathrm{d}t + \tau_m Q(h,\tau) + \varepsilon \int_{I_m} |||\tilde{\xi}|||^2_m\,\mathrm{d}t + \tau_m \frac{C_1}{\varepsilon} \sup_{t\in I_{m-1}} \|\xi(t)\|^2_{L^2(\Omega)}$$
$$\leq c_a\varepsilon \int_{I_m} |||\xi|||^2_m\,\mathrm{d}t + (1+c_a)\varepsilon \int_{I_m} |||\tilde{\xi}|||^2_m\,\mathrm{d}t + \tau_m Q(h,\tau) + \tau_m \frac{C_1}{\varepsilon} \sup_{t\in I_{m-1}} \|\xi(t)\|^2_{L^2(\Omega)}$$
$$\leq K\varepsilon \int_{I_m} |||\xi|||^2_m\,\mathrm{d}t + \tau_m Q(h,\tau) + \tau_m \frac{C_1}{\varepsilon} \sup_{t\in I_{m-1}} \|\xi(t)\|^2_{L^2(\Omega)},$$

where $K = c_a + (1+c_a)C_{CH}$. Multiplying inequality (6.337) by $2K$ and summing with (6.341), we get

$$2K\|\xi^-_m\|^2_{L^2(\Omega)} + \sup_{t\in I_m} \|\xi(t)\|^2_{L^2(\Omega)} - 2K\|\xi^-_{m-1}\|^2_{L^2(\Omega)} - \sup_{t\in I_{m-1}} \|\xi(t)\|^2_{L^2(\Omega)} \tag{6.342}$$
$$\leq \tau_m(2K+1)Q(h,\tau) + \tau_m \frac{C_1(2K+1)}{\varepsilon} \sup_{t\in I_{m-1}} \|\xi(t)\|^2_{L^2(\Omega)}.$$

Then, under the notation $X_m := 2K\|\xi^-_m\|^2_{L^2(\Omega)} + \sup_{t\in I_m} \|\xi(t)\|^2_{L^2(\Omega)}$, $Y := \frac{C_1(2K+1)}{\varepsilon}$ and $Z := (2K+1)Q(h,\tau)$, the inequality (6.342) implies that

$$X_m - X_{m-1} \leq \tau_m Z + \tau_m Y X_{m-1} \quad m = 2,\ldots,r. \tag{6.343}$$

Summing (6.343) over $m = 2,\ldots,n \leq r$ yields

$$X_n \leq X_1 + TZ + Y\sum_{s=2}^{n} \tau_s X_{s-1} \quad n = 2,\ldots,r. \tag{6.344}$$

Then the discrete Gronwall Lemma 1.11 implies that

$$2K\|\xi^-_n\|^2_{L^2(\Omega)} + \sup_{t\in I_n} \|\xi(t)\|^2_{L^2(\Omega)} = X_n \tag{6.345}$$
$$\leq \left(T(2K+1)Q(h,\tau) + 2K\|\xi^-_1\|^2_{L^2(\Omega)} + \sup_{t\in I_1} \|\xi(t)\|^2_{L^2(\Omega)}\right)\exp(C_1(2K+1)t_n/\varepsilon).$$

The estimate $\sup_{t\in I_n} \|\eta(t)\|_{L^2(\Omega)} \le C_u C_\pi (h^\mu + \tau^{q+1})$, which follows from (6.136), together with inequalities (6.177) and

$$\sup_{t\in I_1} \|\xi(t)\|^2_{L^2(\Omega)} \le 2\left(\sup_{t\in I_1} \|u(t) - U(t)\|^2_{L^2(\Omega)} + C_u^2 C_\pi^2 (h^{2\mu} + \tau^{2(q+1)})\right), \tag{6.346}$$

yield (6.334).

In order to prove (6.335), we sum (6.337) over $m = 2, \dots, r$ and get

$$\begin{aligned} &\varepsilon \sum_{m=1}^{r} \int_{I_m} |||\xi|||_m^2 \, \mathrm{d}t \qquad (6.347)\\ &\le 2\left(\frac{\varepsilon}{2}\int_{I_1} |||\xi|||_1^2 \, \mathrm{d}t + TQ(h,\tau) + \|\xi_1^-\|^2_{L^2(\Omega)} + \frac{C_1}{\varepsilon}\sum_{m=2}^{r} \tau_m \sup_{t\in I_{m-1}} \|\xi(t)\|^2_{L^2(\Omega)}\right). \end{aligned}$$

Now we take into account that

$$|||u - U|||_m^2 \le 2(|||\xi|||_m^2 + |||\eta|||_m^2), \tag{6.348}$$

and

$$\|\xi_1^-\|^2_{L^2(\Omega)} \le \sup\nolimits_{t\in I_1} \|\xi(t)\|^2_{L^2(\Omega)}. \tag{6.349}$$

Moreover, by (6.265),

$$\sum_{m=1}^{r} \int_{I_m} |||\eta|||_m^2 \, \mathrm{d}t \le C(h^{2(\mu-1)} + \tau^{2(q+\gamma)}). \tag{6.350}$$

It follows from (6.346)–(6.349) that

$$\begin{aligned} &\varepsilon \sum_{m=1}^{r} \int_{I_m} |||u - U|||_m^2 \, \mathrm{d}t \le 2\left(\varepsilon \sum_{m=1}^{r} \int_{I_m} |||\xi|||_m^2 \, \mathrm{d}t + \varepsilon \sum_{m=1}^{r} \int_{I_m} |||\eta|||_m^2 \, \mathrm{d}t\right) \qquad (6.351)\\ &\le C\Bigg(\varepsilon \int_{I_1} |||u - U|||_1^2 \, \mathrm{d}t + \varepsilon \sum_{m=1}^{r} \int_{I_m} |||\eta|||_m^2 \, \mathrm{d}t\\ &\quad + \sup_{t\in I_1} \|u(t) - U(t)\|^2_{L^2(\Omega)} + h^{2\mu} + \tau^{2(q+\gamma)}\Bigg). \end{aligned}$$

This and (6.350) already imply (6.335), which we wanted to prove. □

Remark 6.58 (The case of identical meshes on all time levels) Similarly as in Sect. 6.1.10, assumption (6.140) can be avoided, if all meshes $\mathscr{T}_{h,m}$, $m = 1 \ldots, r$, are identical. In this case the term $\|\eta_{m-1}^{-}\|_{L^2(\Omega)}^2$ does not appear in estimate (6.322) and then relation (6.320) is valid without assumption (6.140).

Exercise 6.59 Prove (6.335) in detail.

6.3.4 Numerical Examples

6.3.4.1 Solution Procedure

In what follows, we are concerned with the numerical realization of the EST-DGM discrete problem (6.286). It is obvious that there is no need to solve the problem simultaneously on all time intervals, because the problem on the interval I_m depends only on information from I_{m-1}. Hence, we proceed by solving problem (6.286) step-by-step for $m = 2, \ldots, r$. From (6.286a) we see that the restriction $U|_{I_m}$ of the approximate solution and test functions φ can be considered as elements of the space $S_{h,\tau,m}^{p,q} = \{v;\ v = w|_{I_m},\ w \in S_{h,\tau}^{p,q}\}$, see (6.22). Since the discrete problem is linear on each time interval I_m, it can be solved by a method for the solution of large sparse linear systems, such as GMRES (see, e.g., [249]). The order of the linear algebraic system, equivalent to the discrete problem on an individual time level, is equal to the dimension of the space $S_{h,\tau,m}^{p,q}$, i.e., $(q + 1)N$, where $N = N_m$ is the dimension of the space $S_{h,m}^{p}$. This means that the application of the EST-DGM requires solving a system of $(q + 1)N$ linear equations on each time level.

This can be considered as a drawback by comparison with other higher-order time discretizations, as the BDF schemes from Sect. 5.2, since the BDF methods do not require solving sparse systems with increasing size depending on the increasing order of time accuracy. Fortunately, in the special case when the diffusion is linear and independent of time, it is possible to split the solution of the system of order $(q + 1)N$ into smaller systems of order either N or $2N$. This procedure is described in the sequel.

In the following let $m \in \{2, \ldots, r\}$ be arbitrary but fixed. Let us introduce a suitable basis of $S_{h,m}^{p}$ consisting of $\phi_1, \ldots, \phi_N$, where $\phi_i : \Omega \to \mathbb{R}$, $i = 1, \ldots, N$. In the space $P_q(I_m)$ formed by all polynomials of degree $\leq q$ over I_m, we consider a basis consisting of polynomials $\varphi_1, \ldots, \varphi_{q+1} \colon I_m \to \mathbb{R}$. We assume that φ_i are $L^2(I_m)$-orthonormal on I_m. This means that

$$\int_{I_m} \varphi_i\, \varphi_j \,\mathrm{d}t = \delta_{ij}, \quad \int_{I_m} |\varphi_i|^2 \,\mathrm{d}t = 1, \tag{6.352}$$

where δ_{ij} is the Kronecker symbol. The tensor products

$$\psi_{ij}(x, t) = \varphi_j(t)\phi_i(x), \quad i = 1, \ldots, N,\ j = 1, \ldots, q + 1, \tag{6.353}$$

form the basis of the space $S_{h,\tau,m}^{p,q}$.

If we express the approximate solution U as a linear combination of the basis functions and use test functions in (6.286a) equal to the basis functions, i.e.,

$$U = \sum_{j=1}^{N} \sum_{l=1}^{q+1} x_{jl} \psi_{jl}, \qquad \varphi = \psi_{ik}, \tag{6.354}$$

from (6.286a) we get

$$\sum_{j=1}^{N} \sum_{l=1}^{q+1} x_{jl} f_{ijkl} = b_{ik} y_{ik}, \quad i = 1, \ldots, N, \quad k = 1, \ldots, q+1, \tag{6.355}$$

where

$$\begin{aligned} f_{ijkl} &= \int_{I_m} ((\psi'_{jl}, \psi_{ik}) + A_{h,m}(\psi_{jl}, \psi_{ik}))\, dt + (\psi_{jl}(t_{m-1}), \psi_{ik}(t_{m-1})), \\ b_{ik} &= \int_{I_m} (\ell_{h,m}(\psi_{ik}) - b_{h,m}(\Pi_{h,m}\widehat{U}, \psi_{ik}))\, dt + (U_{m-1}^{-}, \psi_{ik}) \end{aligned} \tag{6.356}$$

for $i, j = 1, \ldots, N$ and $k, l = 1, \ldots, q+1$. Now, substituting (6.353) into (6.356), using the time invariance of $A_{h,m}$ and carrying out a simple manipulation, we obtain the formula

$$f_{ijkl} = a_{ij} \delta_{lk} + m_{ij} s_{lk}, \quad i, j = 1, \ldots, N, \quad k, l = 1, \ldots, q+1, \tag{6.357}$$

where

$$\begin{aligned} m_{ij} &= (\phi_j, \phi_i), \quad i, j = 1, \ldots, N, \\ a_{ij} &= A_{h,m}(\phi_j, \phi_i), \quad i, j = 1, \ldots, N, \\ s_{lk} &= \int_{I_m} \varphi'_l(t) \varphi_k(t)\, dt + \varphi_l(t_{m-1}) \varphi_k(t_{m-1}), \quad k, l = 1, \ldots, q+1. \end{aligned}$$

Hence, we get the system

$$\sum_{j=1}^{N} \sum_{l=1}^{q+1} a_{ij} x_{jl} \delta_{lk} + m_{ij} x_{jl} s_{lk} = b_{ik}, \quad i = 1, \ldots, N, \quad k = 1, \ldots, q+1, \tag{6.358}$$

which can be written in the simplified form

$$\sum_{j=1}^{N} a_{ij}x_{jk} + \sum_{j=1}^{N}\sum_{l=1}^{q+1} m_{ij}x_{jl}s_{lk} = b_{ik}, \quad i = 1, \ldots, N, \quad k = 1, \ldots, q+1. \tag{6.359}$$

Let us introduce the matrices $\mathbb{A} = (a_{ij})_{i,j=1}^{N}$, $\mathbb{M} = (m_{ij})_{i,j=1}^{N}$, $\mathbb{X} = (x_{jl})_{\substack{j=1,\ldots,N \\ l=1,\ldots,q+1}}$, $\mathbb{B} = (b_{ik})_{\substack{i=1,\ldots,N \\ k=1,\ldots,q+1}}$ and $\mathbb{S} = (s_{lk})_{i,j=1}^{q+1}$. Hence, $\mathbb{A}$, $\mathbb{M} \in \mathbb{R}^{N\times N}$, $\mathbb{X}$, $\mathbb{B} \in \mathbb{R}^{N\times(q+1)}$ and $\mathbb{S} \in \mathbb{R}^{(q+1)\times(q+1)}$ (the first subscript denotes the row and the second denotes the column). Then (6.359) can be written as the matrix equation

$$\mathbb{A}\mathbb{X} + \mathbb{M}\mathbb{X}\mathbb{S} = \mathbb{B}, \tag{6.360}$$

where $\mathbb{X}$ is an unknown $N \times (q+1)$ matrix. We call this the mixed sparse-full Sylvester equation or just mixed Sylvester equation. This is a form of the original equation that alone would make solution by iterative methods more efficient—we now have the system matrix in implicit form, expressed using the actions of matrices $\mathbb{A}$, $\mathbb{M}$ and $\mathbb{S}$. Nevertheless, it is still a linear system of order $(q+1)N$ that is not separable, because the matrix $\mathbb{S}$ is, in general, full. To overcome this, let us introduce the Schur factorization of $\mathbb{S}$:

$$\mathbb{S} = \mathbb{Z}\mathbb{E}\mathbb{Z}^{\mathrm{T}} \tag{6.361}$$

with $\mathbb{Z}$ orthogonal and $\mathbb{E}$ upper quasi-triangular (see, e.g., [281, Theorem 5.4.22]). Substituting this into (6.360) and multiplying by $\mathbb{Z}$ from the right, we get

$$\mathbb{A}\mathbb{X}\mathbb{Z} + \mathbb{M}\mathbb{X}\mathbb{Z}\mathbb{E} = \mathbb{B}\mathbb{Z} \tag{6.362}$$

and, writing $\mathbb{Y} = \mathbb{X}\mathbb{Z}$ and $\mathbb{C} = \mathbb{B}\mathbb{Z}$ we arrive at

$$\mathbb{A}\mathbb{Y} + \mathbb{M}\mathbb{Y}\mathbb{E} = \mathbb{C}. \tag{6.363}$$

This is again the mixed Sylvester equation, but this time with a block upper triangular matrix $\mathbb{E} = \{e_{kl}\}_{k,l=1}^{q+1}$. The structure of $\mathbb{E}$ allows a more efficient solution procedure: denoting $y_1, \ldots, y_{q+1}$ the columns of $\mathbb{Y}$ and $c_1, \ldots, c_{q+1}$ the columns of $\mathbb{C}$, we evaluate vectors y_k, $k = 1, \ldots, q+1$ sequentially solving either the system

$$(\mathbb{A} + e_{kk}\mathbb{M})y_k = c_k - \sum_{l=1}^{k-1} e_{ik}\mathbb{M}y_i, \tag{6.364}$$

if e_{kk} is a 1×1 block corresponding to a real eigenvalue of $\mathbb{S}$, or the system

$$\left\{ \begin{array}{l} (\mathbb{A} + e_{kk}\mathbb{M}) y_k + e_{k\,k+1}\mathbb{M} y_{k+1} = c_k - \sum_{i=1}^{k-1} e_{ik}\mathbb{M} y_i \\ e_{k+1\,k}\mathbb{M} y_k + (\mathbb{A} + e_{kk}\mathbb{M}) y_{k+1} = c_{k+1} - \sum_{i=1}^{k-1} e_{ik+1}\mathbb{M} y_i \end{array} \right\}, \qquad (6.365)$$

if $(e_{ij})_{i,j=k}^{k+1}$ forms a diagonal 2×2 block of $\mathbb{E}$ corresponding to a complex eigenvalue pair of $\mathbb{S}$. It should be noted that by the properties of the Schur decomposition, $e_{k+1\,k+1} = e_{kk}$. The expressions (6.364)–(6.365) allow us to solve problem (6.363) in a recurrent manner, splitting it into subproblems of order either N or $2N$. The matrix $\mathbb{X}$ can be recovered from $\mathbb{Y}$ as $\mathbb{X} = \mathbb{Y}\mathbb{Z}^{\mathrm{T}}$.

In the following, we present several numerical experiments demonstrating accuracy and efficiency of the proposed extrapolated space-time discontinuous Galerkin finite element method. Namely, we investigate the *experimental order of convergence* and compare EST-DGM with the BDF-DGM from Sect. 5.2.

6.3.4.2 Experimental Order of Convergence

The first example is a simple test case for investigating the experimental order of convergence (EOC) with respect to time. Similarly as in Sect. 5.2, we solve the nonlinear convection-diffusion problem (6.271) with $\Omega = (0, 1) \times (0, 1)$, $T = 1$, $\varepsilon = 0.02$, $f_s(u) = u^2/2$, $s = 1, 2$, and functions g, u_D and u_0 chosen in such a way that the exact solution has the form

$$u(x_1, x_2, t) = 16 \frac{\exp(10t) - 1}{\exp(10) - 1} x_1 x_2 (1 - x_1)(1 - x_2). \qquad (6.366)$$

The order of accuracy of the time discretization is investigated by using an *overkill* in the spatial resolution. We use the uniform triangular grid of the form shown in Fig. 2.4 from Sect. 2.9, with $h = 1/48$ (4608 elements), and the polynomial approximation degree $p = 4$ in space. The time step τ is successively diminished and for each pair of successive steps τ_1, τ_2 and the corresponding errors, e_1, e_2, the experimental order of convergence is calculated according to the formula

$$\mathrm{EOC} = \frac{\log e_2 - \log e_1}{\log \tau_2 - \log \tau_1}. \qquad (6.367)$$

The EOC is calculated for both the $L^\infty(0, T; L^2(\Omega))$-norm of the error, defined by (6.334), and the $L^2(0, T; H^1(\Omega, \mathscr{T}_h))$-norm defined by (6.335).

Table 6.2 shows computational errors, EOC and computational (CPU) time for the EST-DGM with $q = 0, 1, 2, 3$. We see that the EOC in time is approximately $q + 1$ in both norms.

Table 6.3 shows computational errors, EOC and CPU time for the n-step BDF-DGM from Sect. 5.2 with $n = 1, 2, 3$. We observe that the EST-DGM needs a several

Table 6.2 EST-DGM: computational errors in the $L^\infty(0, T; L^2(\Omega))$-norm and the $L^2(0, T; H^1(\Omega, \mathscr{T}_h))$-norm, experimental order of convergence and computational time for $q = 0, 1, 2, 3$

	τ	$\|e_h\|_{L^\infty(0,T;L^2(\Omega))}$	EOC	$\|e_h\|_{L^2(0,T;H^1(\Omega,\mathscr{T}_h))}$	EOC	CPU(s)
$q = 0$	2.000E–01	1.034E–01	–	4.824E–01	–	18.3
	1.000E–01	4.143E–02	1.319	2.387E–01	1.015	31.6
	5.000E–02	2.122E–02	0.965	1.438E–01	0.732	57.6
	2.500E–02	1.137E–02	0.901	8.086E–02	0.830	106.6
	1.250E–02	5.843E–03	0.960	4.262E–02	0.924	208.8
	6.250E–03	2.990E–03	0.966	2.206E–02	0.950	409.9
$q = 1$	2.000E–01	3.650E–02	–	2.649E–01	–	131.4
	1.000E–01	1.899E–02	0.943	1.385E–01	0.936	179.5
	5.000E–02	6.421E–03	1.564	4.718E–02	1.554	266.5
	2.500E–02	1.802E–03	1.833	1.331E–02	1.826	395.1
	1.250E–02	4.717E–04	1.934	3.487E–03	1.932	603.6
	6.250E–03	1.203E–04	1.971	8.891E–04	1.972	1030.2
$q = 2$	2.000E–01	2.805E–02	–	2.043E–01	–	189.8
	1.000E–01	7.672E–03	1.870	5.650E–02	1.855	276.4
	5.000E–02	1.296E–03	2.565	9.732E–03	2.537	345.7
	2.500E–02	1.822E–04	2.831	1.375E–03	2.823	551.7
	1.250E–02	2.423E–05	2.911	1.844E–04	2.899	892.4
	6.250E–03	3.174E–06	2.932	2.451E–05	2.911	1564.2
$q = 3$	2.000E–01	1.667E–02	–	1.217E–01	–	185.0
	1.000E–01	2.350E–03	2.826	1.782E–02	2.772	246.0
	5.000E–02	1.980E–04	3.569	1.559E–03	3.515	396.2
	2.500E–02	1.368E–05	3.855	1.064E–04	3.873	567.7
	1.250E–02	8.929E–07	3.938	6.785E–06	3.971	999.4
	6.250E–03	5.696E–08	3.970	4.267E–07	3.991	1779.5

times higher computational time than the BDF-DG method having the same order of accuracy. On the other hand, for a given magnitude of τ, the EST-DGM gives approximately a ten times smaller error than the BDF-DGM.

6.3.4.3 Moving Interior Layer

The second example is more challenging. We consider the nonlinear convection-diffusion problem (6.271) with $\Omega = (-1, 1) \times (-1, 1)$, $T = 2$, $\varepsilon = 0.02$, $f_s(u) = u^2/2$, $s = 1, 2$, $g = 0$. The functions u_D and u_0 are chosen in such a way that the exact solution has the form

$$u(x_1, x_2, t) = \left(1 + \exp\left(\frac{x_1 + x_2 + 1 - t}{2\varepsilon}\right)\right)^{-1}. \tag{6.368}$$

This function contains an interior layer propagating in the direction $(1, 1)$.

Table 6.3 n-step BDF-DGM: computational errors in the $L^\infty(0, T; L^2(\Omega))$-norm and the $L^2(0, T; H^1(\Omega, \mathscr{T}_h))$-norm, experimental order of convergence and the computational time for n = 1, 2, 3

	τ	$\|e_h\|_{L^\infty(0,T;L^2(\Omega))}$	EOC	$\|e_h\|_{L^2(0,T;H^1(\Omega,\mathscr{T}_h))}$	EOC	CPU(s)
$n = 1$	2.000E–01	6.594E–01	–	3.062E+00	–	15.1
	1.000E–01	3.011E–01	1.131	1.398E+00	1.131	23.3
	5.000E–02	1.417E–01	1.087	6.545E–01	1.095	40.0
	2.500E–02	6.842E–02	1.051	3.148E–01	1.056	69.3
	1.250E–02	3.372E–02	1.021	1.546E–01	1.026	133.6
	6.250E–03	1.669E–02	1.015	7.641E–02	1.017	265.4
$n = 2$	2.000E–01	3.251E–01	–	1.556E+00	–	13.3
	1.000E–01	1.098E–01	1.566	5.307E–01	1.552	21.6
	5.000E–02	3.398E–02	1.693	1.643E–01	1.692	38.2
	2.500E–02	9.810E–03	1.792	4.735E–02	1.794	65.8
	1.250E–02	2.658E–03	1.884	1.285E–02	1.881	130.7
	6.250E–03	6.943E–04	1.937	3.365E–03	1.933	260.1
$n = 3$	2.000E–01	2.088E–01	–	1.027E+00	–	13.1
	1.000E–01	5.228E–02	1.998	2.636E–01	1.962	20.9
	5.000E–02	1.044E–02	2.324	5.313E–02	2.311	37.0
	2.500E–02	1.732E–03	2.592	8.865E–03	2.583	66.4
	1.250E–02	2.517E–04	2.782	1.299E–03	2.771	130.6
	6.250E–03	3.402E–05	2.887	1.764E–04	2.880	268.8

Table 6.4 Moving interior layer: computational errors in the $L^2(\Omega)$-norm and the $H^1(\Omega, \mathscr{T}_h)$-norm at time $t = 2$ and computational time; results on the mesh with $h = 1/48$ for $p = 1, 2, 3$ and $q = 1, 2$ in comparison with mesh adaptation

Method	Mesh	$\|e_h\|_{L^2(\Omega)}$	$\|e_h\|_{H^1(\Omega,\mathscr{T}_h)}$	CPU(s)
$p = 1, q = 1$	$h = 1/48$	1.188E–03	3.990E–02	584.7
$p = 2, q = 1$	$h = 1/48$	5.668E–04	2.954E–02	1201.4
$p = 3, q = 1$	$h = 1/48$	6.663E–05	3.453E–03	2316.0
$p = 1, q = 2$	$h = 1/48$	1.125E–03	3.887E–02	970.1
$p = 2, q = 2$	$h = 1/48$	5.594E–04	2.963E–02	1876.7
$p = 3, q = 2$	$h = 1/48$	2.243E–05	2.978E–03	3486.9
$p = 3, q = 2$	Adapt	4.690E–05	2.099E–03	1256.7

We carried out computations on a uniform triangular grid with $h = 1/24$ (4608 elements), the equidistant time step $\tau = 0.005$ and used the polynomial approximation in space with $p = 1, 2, 3$ and in time with $q = 1, 2$. Table 6.4 shows the computational errors in the $L^2(\Omega)$-norm and the $H^1(\Omega, \mathscr{T}_h)$-seminorm at $t = 2$. We simply observe a significant reduction of the error for increasing p but an almost negligible decrease of the error for increasing q. (It is probably caused by the fact that the error in space is dominant.)

The space-time discontinuous Galerkin methods can simply treat with different grids at different time levels. Therefore, we employ an automatic mesh adaptation strategy, which detects elements with a high error and each of these detected elements is split into four daughter elements. Then hanging nodes arise. On the other hand, four daughter elements can be derefined back to a mother element if the error indication is small.

The following adaptive strategy is applied. We start from the uniform initial grid with $h = 1/3$ and the mesh refinement/derefinement is applied after each 5 time steps. We allow 4 levels of mesh refinement. Moreover, in the beginning of the computation, we carry out 4 mesh refinements with a very small time step and then the solution is restarted. This strategy produces a reasonable mesh for the first time step.

Similarly as in the previous case, we employ the equidistant time step $\tau = 0.005$, piecewise cubic approximation with respect to space ($p = 3$) and piecewise quadratic approximation with respect to time ($q = 2$).

Figure 6.3 shows the adaptively refined grids at time instants $t = 0, 0.5, 1, 1.5$ and 2. The last frame of Fig. 6.3 shows graphs of the solution along the line (s, s), $s \in (-1, 1)$, at the time levels mentioned above. We can observe the propagation of a moving front, which is in good agreement with the exact solution.

The last line of Table 6.4 shows the errors in the L^2-norm and the H^1-seminorm at $t = 2$ achieved by computation on adaptively refined grids. We see that the error is comparable with results obtained on the fine uniform grid, using the best approximation with $p = 3$ and $q = 2$. However, the computational time is almost three times smaller for the computation realized with mesh adaptation.

Exercise 6.60 Show that function (6.368) is a solution of the equation

$$\frac{\partial u}{\partial t} + \frac{1}{2}\sum_{s=1}^{2} \frac{\partial u^2}{\partial x_s} = \varepsilon \Delta u$$

for any $\varepsilon > 0$.

6.4 Uniform Error Estimates with Respect to the Diffusion Coefficient for the ST-DGM

Section 4.6 was devoted to the derivation of error estimates, uniform with respect to the diffusion coefficient for a linear nonstationary convection-diffusion-reaction initial-boundary value problem solved by the method of lines. Now we extend these results to the full space-time discontinuous Galerkin method (ST-DGM). Under some assumptions on shape regularity of the meshes and a certain regularity of the exact solution, we prove error estimates in $L^2(0, T; L^2(\Omega))$-and $\sqrt{\varepsilon}L^2(0, T; H^1(\Omega, \mathscr{T}_{h,m}))$-norms, uniform with respect to the diffusion coefficient $\varepsilon \geq 0$. The estimates hold true even in the hyperbolic case when $\varepsilon = 0$.

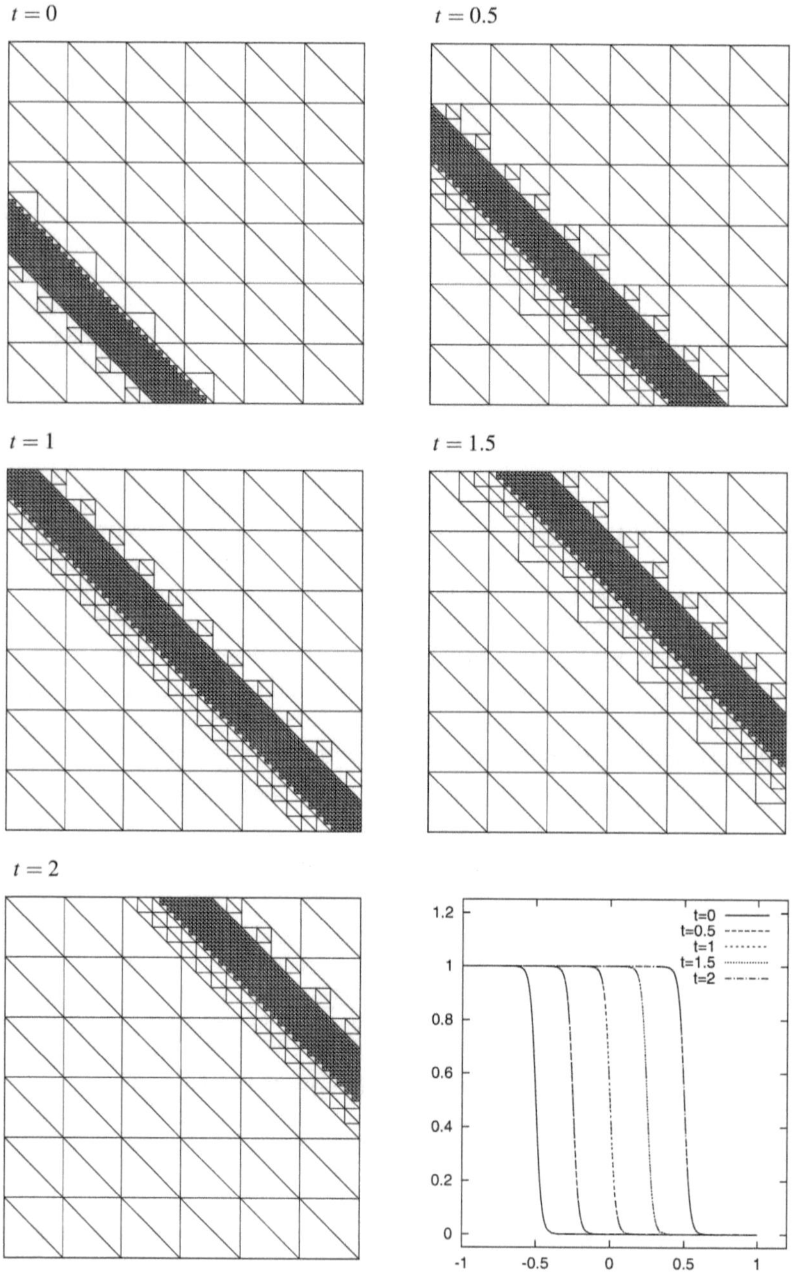

Fig. 6.3 Adaptively refined grids and the propagation of the solution along the diagonal cut (s, s), $s \in (-1, 1)$ (*right bottom*) at times $t = 0,\ 0.5,\ 1,\ 1.5,\ 2$

6.4.1 Formulation of the Problem and Some Assumptions

Similarly as in Sect. 4.6, we consider the following initial-boundary value problem: Find $u : Q_T = \Omega \times (0, T) \to \mathbb{R}$ such that

$$\frac{\partial u}{\partial t} + \boldsymbol{v} \cdot \nabla u - \varepsilon \Delta u + cu = g \text{ in } Q_T, \tag{6.369a}$$

$$u = u_D \text{ on } \partial\Omega^- \times (0, T), \tag{6.369b}$$

$$\varepsilon \boldsymbol{n} \cdot \nabla u = g_N \text{ on } \partial\Omega^+ \times (0, T), \tag{6.369c}$$

$$u(x, 0) = u^0(x), \quad x \in \Omega. \tag{6.369d}$$

We assume that $\partial\Omega = \partial\Omega^- \cup \partial\Omega^+$, and

$$\boldsymbol{v}(x, t) \cdot \boldsymbol{n}(x) < 0 \text{ on } \partial\Omega^-, \quad \forall t \in [0, T],$$
$$\boldsymbol{v}(x, t) \cdot \boldsymbol{n}(x) \geq 0 \text{ on } \partial\Omega^+ \quad \forall t \in [0, T].$$

Here $\boldsymbol{n}$ is the outer unit normal to the boundary $\partial\Omega$ of Ω, $\partial\Omega^-$ is the inflow boundary and $\partial\Omega^+$ is the outflow boundary. In the case $\varepsilon = 0$ we put $g_N = 0$ and ignore the Neumann condition (6.369c).

We assume that the data satisfy conditions (4.159), i.e.,

$$\begin{aligned}
&\varepsilon \geq 0, \ g \in C([0, T); L^2(\Omega)), \ g_N \in C([0, T); L^2(\partial\Omega^+)), \ u_0 \in L^2(\Omega), \\
&u_D \text{ is the trace of some } u^* \in C([0, T); H^1(\Omega)) \cap L^\infty(Q_T) \text{ on } \partial\Omega^- \times (0, T), \\
&\boldsymbol{v} \in C([0, T); W^{1,\infty}(\Omega)), \ |\nabla \boldsymbol{v}| \leq C_v \text{ a.e. in } Q_T, \ |\boldsymbol{v}| \leq C_v \text{ in } \overline{\Omega} \times [0, T]. \\
&c \in C([0, T); L^\infty(\Omega)), \ |c(x, t)| \leq C_c \text{ a.e. in } Q_T, \ c - \frac{1}{2}\nabla \cdot \boldsymbol{v} \geq \gamma_0 > 0 \text{ in } Q_T.
\end{aligned}$$

We assume that the exact solution u of problem (6.369) satisfies the regularity condition

$$u \in H^{q+1}(0, T; H^1(\Omega)) \cap C([0, T]; H^s(\Omega)) \tag{6.370}$$

with integers $q \geq 1$ and $s \geq 2$.

6.4.2 Discretization of the Problem

We use the notation from Sect. 6.1.1.1. We consider a system (6.24) of *conforming triangulations* $\mathscr{T}_{h,m}$ (cf. condition (MA4) from Sect. 2.3.2), satisfying the conditions of the *shape regularity* (6.25) and of *equivalence* (6.26). Let us recall that similarly as in Sect. 4.6.2 for $K \in \mathscr{T}_{h,m}$ and $t \in I_m$ we set

$$\partial K^-(t) = \{x \in \partial K;\ \boldsymbol{v}(x,t) \cdot \boldsymbol{n}(x) < 0\}, \tag{6.371}$$
$$\partial K^+(t) = \{x \in \partial K;\ \boldsymbol{v}(x,t) \cdot \boldsymbol{n}(x) \geq 0\}, \tag{6.372}$$

where $\boldsymbol{n}$ denotes the outer unit normal to ∂K.

The space-time discretization can be carried out in a similar way as in Sects. 6.2 and 4.6.2. Hence, we introduce the diffusion, convection, reaction, penalty and right-hand side forms, defined for $u, \varphi \in H^2(\Omega, \mathscr{T}_{h,m})$ in a similar way as in Sect. 4.6.2:

$$\begin{aligned} a_{h,m}(u,\varphi) = & \sum_{K\in\mathscr{T}_{h,m}} \int_K \nabla u \cdot \nabla\varphi \, \mathrm{d}x \\ & - \sum_{\Gamma\in\mathscr{F}^I_{h,m}} \int_\Gamma (\langle\nabla u\rangle \cdot \boldsymbol{n}_\Gamma [\varphi] + \Theta \langle\nabla\varphi\rangle \cdot \boldsymbol{n}_\Gamma [u]) \, \mathrm{d}S \\ & - \sum_{K\in\mathscr{T}_{h,m}} \int_{\partial K^-\cap\partial\Omega} ((\nabla u\cdot\boldsymbol{n})\varphi + \Theta(\nabla\varphi\cdot\boldsymbol{n})u) \, \mathrm{d}S, \end{aligned} \tag{6.373}$$

$$\begin{aligned} b_{h,m}(u,\varphi) = & \sum_{K\in\mathscr{T}_{h,m}} \int_K (\boldsymbol{v}\cdot\nabla u)\varphi \, \mathrm{d}x \\ & - \sum_{K\in\mathscr{T}_{h,m}} \int_{\partial K^-\cap\partial\Omega} (\boldsymbol{v}\cdot\boldsymbol{n}) u\varphi \, \mathrm{d}S - \sum_{K\in\mathscr{T}_{h,m}} \int_{\partial K^-\setminus\partial\Omega} (\boldsymbol{v}\cdot\boldsymbol{n})[u]\varphi \, \mathrm{d}S, \end{aligned} \tag{6.374}$$

$$c_{h,m}(u,\varphi) = \int_\Omega c u\varphi \, \mathrm{d}x, \tag{6.375}$$

$$J^\sigma_{h,m}(u,\varphi) = \sum_{\Gamma\in\mathscr{F}^I_{h,m}} \int_\Gamma \sigma\, [u]\,[\varphi] \, \mathrm{d}S + \sum_{K\in\mathscr{T}_{h,m}} \int_{\partial K^-\cap\partial\Omega} \sigma u\varphi \, \mathrm{d}S, \tag{6.376}$$

$$\begin{aligned} \ell_{h,m}(\varphi)(t) = & \int_\Omega g(t)\varphi \, \mathrm{d}x + \sum_{K\in\mathscr{T}_{h,m}} \int_{\partial K^+\cap\partial\Omega} g_N(t)\varphi \, \mathrm{d}S \\ & + \varepsilon \sum_{K\in\mathscr{T}_{h,m}} \int_{\partial K^-\cap\partial\Omega} \sigma u_D(t)\varphi \, \mathrm{d}S \\ & - \varepsilon\Theta \sum_{K\in\mathscr{T}_{h,m}} \int_{\partial K^-\cap\partial\Omega} u_D(t)(\nabla\varphi\cdot\boldsymbol{n}) \, \mathrm{d}S \\ & - \sum_{K\in\mathscr{T}_{h,m}} \int_{\partial K^-\cap\partial\Omega} (\boldsymbol{v}\cdot\boldsymbol{n}) u_D(t)\varphi \, dS, \end{aligned} \tag{6.377}$$

$$\mathscr{A}_{h,m}(u,\varphi) = \varepsilon a_{h,m}(u,\varphi) + \varepsilon J_{h,m}(u,\varphi) + b_{h,m}(u,\varphi) + c_{h,m}(u,\varphi). \tag{6.378}$$

In the diffusion form $a_{h,m}(u, \varphi)$ we use the nonsymmetric (NIPG) formulation for $\Theta = -1$, incomplete (IIPG) formulation for $\Theta = 0$ or symmetric formulation (SIPG) for $\Theta = 1$, and the weight σ is defined by (2.104), where the constant $C_W > 0$ is arbitrary for NIPG version, and satisfies condition (2.132) or (2.139) for SIPG or IIPG version, respectively.

In what follows, the symbols U' and u' will denote the time derivative of U and u, respectively.

Definition 6.61 We say that the function $U \in S_{h,\tau}^{p,q}$ is the *approximate solution* of problem (6.369), if it satisfies the identity

$$\sum_{m=1}^{r} \int_{I_m} \big((U', \varphi) + \mathscr{A}_{h,m}(U, \varphi)\big)\, dt + \sum_{m=2}^{r} (\{U\}_{m-1}, \varphi_{m-1}^{+}) + (U_0^{+}, \varphi_0^{+}) \tag{6.379}$$
$$= \sum_{m=1}^{r} \int_{I_m} \ell_{h,m}(\varphi)\, dt + (u_0, \varphi_0^{+}) \quad \forall \varphi \in S_{h,\tau}^{p,q}.$$

It is easy to see that this scheme can be written in a similar way as (6.203), namely,

$$\int_{I_m} \big((U', \varphi) + \mathscr{A}_{h,m}(U, \varphi)\big)\, dt + (\{U\}_{m-1}, \varphi_{m-1}^{+}) = \int_{I_m} \ell_{h,m}(\varphi)\, dt \tag{6.380}$$
$$\forall \varphi \in S_{h,m}^{p,q}, \quad m = 1, \ldots, M, \quad U_0^{-} = \Pi_{h,0} u^0.$$

If we denote

$$\mathscr{B}(U, \varphi) = \sum_{m=1}^{r} \int_{I_m} \big((U', \varphi) + \mathscr{A}_{h,m}(U, \varphi)\big)\, dt + \sum_{m=2}^{r} (\{U\}_{m-1}, \varphi_{m-1}^{+}) + (U_0^{+}, \varphi_0^{+}),$$
$$L(\varphi) = \sum_{m=1}^{r} \int_{I_m} \ell_{h,m}(v)\, dt + (u_0, \varphi_0^{+}), \tag{6.381}$$

we can write (6.379) as

$$\mathscr{B}(U, \varphi) = L(\varphi) \qquad \forall \varphi \in S_{h,\tau}^{p,q}. \tag{6.382}$$

It is possible to show that the regular exact solution u satisfies the identity $\mathscr{B}(u, \varphi) = L(\varphi)$ for all $\varphi \in S_{h,\tau}^{p,q}$, and, thus we have

$$\mathscr{B}(U, \varphi) = \mathscr{B}(u, \varphi) \qquad \forall \varphi \in S_{h,\tau}^{p,q}. \tag{6.383}$$

Our goal is the analysis of the estimate of the error $e = U - u$. To this end, as we have already mentioned, in the sequel we consider *conforming* triangulations satisfying the *shape-regularity assumption* (6.25) and assume the *equivalence condition* (6.26) holds.

6.4.3 Properties of the Discrete Problem

In this section we prove some basic properties of $\mathscr{A}_{h,m}$ and $\mathscr{B}$.
Let $u \in C^1([t_{m-1}, t_m]; H^2(\Omega, \mathscr{T}_{h,m}))$ for all $m = 1, \ldots, r$ and $v \in S_{h,\tau}^{p,q}$.

Lemma 6.62 *We can express $\mathscr{B}$ as*

$$\mathscr{B}(u, v) = \sum_{m=1}^{r} \int_{I_m} \big((-u, v') + \mathscr{A}_{h,m}(u, v)\big)\, \mathrm{d}t - \sum_{m=1}^{r-1} (u_m^-, \{v\}_m) + (u_r^-, v_r^-). \tag{6.384}$$

Proof The integration by parts yields

$$\begin{aligned}
&\sum_{m=1}^{r} \int_{I_m} (u', v)\, \mathrm{d}t + \sum_{m=2}^{r} (\{u\}_{m-1}, v_{m-1}^+) + (u_0^+, v_0^+) \\
&= \sum_{m=1}^{r} \int_{I_m} (-u, v')\, \mathrm{d}t + \sum_{m=1}^{r} \big((u_m^-, v_m^-) - (u_{m-1}^+, v_{m-1}^+)\big) \\
&\quad + \sum_{m=2}^{r} (u_{m-1}^+ - u_{m-1}^-, v_{m-1}^+) + (u_0^+, v_0^+) \\
&= \sum_{m=1}^{r} \int_{I_m} (-u, v')\, \mathrm{d}t + \sum_{m=1}^{r-1} (u_m^-, v_m^- - v_m^+) + (u_r^-, v_r^-).
\end{aligned}$$

□

Lemma 6.63 *We have*

$$\mathscr{B}(v, v) = \sum_{m=1}^{r} \int_{I_m} \mathscr{A}_{h,m}(v, v)\, \mathrm{d}t + \|v\|_T^2, \tag{6.385}$$

where

$$\|v\|_T^2 = \frac{1}{2} \|v_0^+\|_{L^2(\Omega)}^2 + \frac{1}{2} \sum_{m=1}^{r-1} \|\{v\}_m\|_{L^2(\Omega)}^2 + \frac{1}{2} \|v_r^-\|_{L^2(\Omega)}^2. \tag{6.386}$$

Proof By (6.381) and (6.384), we have

$$\mathscr{B}(v, v) = \sum_{m=1}^{r} \int_{I_m} \big((v', v) + \mathscr{A}_{h,m}(v, v)\big)\, \mathrm{d}t + \sum_{m=1}^{r-1} (\{v\}_m, v_m^+) + (v_0^+, v_0^+),$$

$$\mathscr{B}(v, v) = \sum_{m=1}^{r} \int_{I_m} \left((-v, v') + \mathscr{A}_{h,m}(v, v)\right) \mathrm{d}t + \sum_{m=1}^{r-1} (-v_m^-, \{v\}_m) + (v_r^-, v_r^-).$$

We arrive at (6.385) by adding these identities and dividing by two. □

In the sequel we use the notation

$$\|v\|_{v,\Gamma}^2 = \int_{\Gamma} |\boldsymbol{v} \cdot \boldsymbol{n}| v^2 \, \mathrm{d}S \quad \text{for } \Gamma \subset \partial K, \; K \in \mathscr{T}_{h,m}.$$

First, we will be concerned with the coercivity property.

Lemma 6.64 *The forms $\mathscr{A}_{h,m}$ are coercive:*

$$\mathscr{A}_{h,m}(v, v) \geq \|v\|_{E,m}^2, \qquad v \in H^2(\Omega, \mathscr{T}_{h,m}), \tag{6.387}$$

where

$$\|v\|_{E,m}^2 = \frac{\varepsilon}{2} |||v|||_m^2 + \gamma_0 \|v\|_{L^2(\Omega)}^2 + \frac{1}{2} \sum_{K \in \mathscr{T}_{h,m}} \left(\|v\|_{v,\partial K \cap \partial\Omega}^2 + \|[v]\|_{v,\partial K^- \setminus \partial\Omega}^2\right). \tag{6.388}$$

Proof Using a similar process as in (4.202)–(4.204), we find that

$$\begin{aligned} \mathscr{A}_{h,m}(v, v) \geq & \frac{\varepsilon}{2} |||v|||_m^2 + \int_{\Omega} (c - \frac{1}{2} \nabla \cdot \boldsymbol{v}) v^2 \, \mathrm{d}x \\ & + \frac{1}{2} \sum_{K \in \mathscr{T}_{h,m}} \left(\|v\|_{v,\partial K \cap \partial\Omega}^2 + \|[v]\|_{v,\partial K^- \setminus \partial\Omega}^2\right), \quad v \in H^2(\Omega, \mathscr{T}_{h,m}), \end{aligned} \tag{6.389}$$

which together with assumption (4.159f) yields (6.387). □

6.4.4 Abstract Error Estimate

In deriving error estimates we make use of the space time interpolation of the exact solution, introduced in Sect. 6.2.2:

$$\pi u \in S_{h,\tau}^{p,q}, \tag{6.390a}$$

$$\int_{I_m} (\pi u - u, \varphi) \, \mathrm{d}t = 0 \quad \forall \varphi \in S_{h,\tau}^{p,q-1}, \tag{6.390b}$$

$$\pi u(t_m^-, x) = \Pi_{h,m} u(t_m^-, x), \quad x \in \Omega, \tag{6.390c}$$

for $m = 1, \ldots, r$, where $\Pi_{h,m}$ is $L^2(\Omega)$-projection on $S^p_{h,m}$. This means that taking $v \in L^2(\Omega)$, we have $\Pi_{h,m}v \in S^p_{h,m}$ and $(\Pi_{h,m}v - v, \varphi) = 0$ for all $\varphi \in S^p_{h,m}$.

At this point, we derive error estimates in terms of the π-interpolation error.

Lemma 6.65 *We have*

$$\mathscr{B}(U - \pi u, U - \pi u) \tag{6.391}$$
$$= \sum_{m=1}^{r} \int_{I_m} \mathscr{A}_{h,m}(u - \pi u, U - \pi u)\, \mathrm{d}t - \sum_{m=1}^{r-1} ((u - \pi u)^-_m, \{U - \pi u\}_m).$$

Proof From (6.383) and (6.384) we get

$$\mathscr{B}(U - \pi u, U - \pi u) = \mathscr{B}(u - \pi u, U - \pi u)$$
$$= -\sum_{m=1}^{r} \int_{I_m} (u - \pi u, (U - \pi u)')\, \mathrm{d}t + \sum_{m=1}^{r} \int_{I_m} \mathscr{A}_{h,m}(u - \pi u, U - \pi u)\, \mathrm{d}t$$
$$- \sum_{m=1}^{r-1} ((u - \pi u)^-_m, \{U - \pi u\}_m) + ((u - \pi u)^-_r, (U - \pi u)^-_r).$$

The first term on the second line vanishes due to (6.390b). The second term on the last line is also zero, because we have

$$((u - \pi u)^-_m, \varphi) = ((u - \Pi_{h,m}u)^-_m, \varphi) + ((\Pi_{h,m}u - \pi u)^-_m, \varphi) \tag{6.392}$$

for $\varphi \in S^p_{h,m}$, and both terms on the right-hand side of (6.392) vanish (the first term equals zero because of the properties of the $L^2(\Omega)$-projection and the second one due to (6.390c)). □

The sum on the last line in relation (6.391) does not vanish because, in general, $\{U - \pi u\}_m \notin S^p_{h,m}$, as we use different triangulations of Ω on different time levels.

Under the notation

$$\xi = U - \pi u \in S^{p,q}_{h,\tau}, \quad \eta = \pi u - u, \tag{6.393}$$

we have $e = \xi + \eta$ and (6.391) can be rewritten as

$$\mathscr{B}(\xi, \xi) = -\sum_{m=1}^{r} \int_{I_m} \mathscr{A}_{h,m}(\eta, \xi)\, \mathrm{d}t + \sum_{m=1}^{r-1} (\eta^-_m, \{\xi\}_m). \tag{6.394}$$

Lemma 6.66 *Let us denote*

$$R_m(\eta) = \|\eta\|_{E,m} + \sqrt{\varepsilon}\left(\sum_{K\in\mathscr{T}_{h,m}} h_K^2|\eta|^2_{H^2(K)}\right)^{1/2} \tag{6.395}$$
$$+\left(\sum_{K\in\mathscr{T}_{h,m}} h_K^{-2}\|\eta\|^2_{L^2(K)}\right)^{1/2} + \left(\sum_{K\in\mathscr{T}_{h,m}} \|\eta^-\|^2_{\boldsymbol{v},\partial K^-\setminus\partial\Omega}\right)^{1/2}.$$

Then there exists a constant C_a independent of u, U, h, m and ε such that

$$|\mathscr{A}_{h,m}(\eta,\xi)| \le C_a\|\xi\|_{E,m}R_m(\eta). \tag{6.396}$$

Proof We proceed similarly as in Sect. 2.5, where instead of $\mathscr{T}_h$ we write $\mathscr{T}_{h,m}$, instead of a_h we consider $\varepsilon a_{h,m}$, and set $v_h := \xi$. We begin with the form $a_{h,m}$. Using the first inequality from (2.117), (2.123) and (6.376), we have

$$\begin{aligned}
\left|a_{h,m}(\eta,\xi)\right| &\le |\eta|_{H^1(\Omega,\mathscr{T}_{h,m})}|\xi|_{H^1(\Omega,\mathscr{T}_{h,m})} \\
&\quad+\left(\sum_{\Gamma\in\mathscr{F}_h^{ID}}\int_\Gamma \sigma^{-1}(\boldsymbol{n}\cdot\langle\nabla\eta\rangle)^2\,\mathrm{d}S\right)^{1/2}\left(\sum_{\Gamma\in\mathscr{F}_h^{ID}}\int_\Gamma \sigma[\xi]^2\,\mathrm{d}S\right)^{1/2} \\
&\quad+\left(\sum_{\Gamma\in\mathscr{F}_h^{ID}}\int_\Gamma \sigma^{-1}(\boldsymbol{n}\cdot\langle\nabla\xi\rangle)^2\,\mathrm{d}S\right)^{1/2}\left(\sum_{\Gamma\in\mathscr{F}_h^{ID}}\int_\Gamma \sigma[\eta]^2\,\mathrm{d}S\right)^{1/2} \\
&\le |\eta|_{H^1(\Omega,\mathscr{T}_{h,m})}|\xi|_{H^1(\Omega,\mathscr{T}_{h,m})} \\
&\quad+C\left(\sum_{K\in\mathscr{T}_{h,m}}\left(h_K|\eta|_{H^1(K)}|\eta|_{H^2(K)}+|\eta|^2_{H^1(K)}\right)\right)^{1/2} J^\sigma_{h,m}(\xi,\xi)^{1/2} \\
&\quad+C\left(\sum_{K\in\mathscr{T}_{h,m}}\left(h_K|\xi|_{H^1(K)}|\xi|_{H^2(K)}+|\xi|^2_{H^1(K)}\right)\right)^{1/2} J^\sigma_{h,m}(\eta,\eta)^{1/2}.
\end{aligned} \tag{6.397}$$

Moreover, Youngs's inequality gives

$$\begin{aligned}
&\sum_{K\in\mathscr{T}_{h,m}}\left(h_K|\eta|_{H^1(K)}|\eta|_{H^2(K)}+|\eta|^2_{H^1(K)}\right) \\
&\le \sum_{K\in\mathscr{T}_{h,m}}\frac{1}{2}\left(3|\eta|^2_{H^1(K)}+h_K^2|\eta|^2_{H^2(K)}\right) = \frac{3}{2}|\eta|^2_{H^1(\Omega,\mathscr{T}_h)} + \sum_{K\in\mathscr{T}_{h,m}}\frac{1}{2}h_K^2|\eta|^2_{H^2(K)}
\end{aligned} \tag{6.398}$$

and the inverse inequality implies that

$$\sum_{K\in\mathscr{T}_{h,m}} \left(h_K|\xi|_{H^1(K)}|\xi|_{H^2(K)} + |\xi|^2_{H^1(K)}\right) \le (1+C_I)|\xi|^2_{H^1(\Omega,\mathscr{T}_h)}. \qquad (6.399)$$

Hence, inserting (6.398) and (6.399) into (6.397), using the discrete Cauchy inequality and the inequality $(c_1^2+\cdots+c_n^2)^{1/2} \le |c_1|+\cdots+|c_n|$, we obtain

$$\begin{aligned} &\varepsilon|a_{h,m}(\eta,\xi)| \qquad (6.400)\\ &\le C\left(\sqrt{\varepsilon}\,|\eta|_{H^1(\Omega,\mathscr{T}_{h,m})} + \sqrt{\varepsilon}\,J_{h,m}(\eta,\eta)^{1/2} + \sqrt{\varepsilon}\left(\sum_{K\in\mathscr{T}_h} h_K^2|\eta|^2_{H^2(K)}\right)^{1/2}\right)\\ &\quad\times\left(\sqrt{\varepsilon}\,|\xi|_{H^1(\Omega,\mathscr{T}_{h,m})} + \sqrt{\varepsilon}\,J_{h,m}(\xi,\xi)^{1/2}\right)\\ &\le C\|\xi\|_{E,m}\left(\|\eta\|_{E,m} + \sqrt{\varepsilon}\left(\sum_{K\in\mathscr{T}_{h,m}} h_K^2|\eta|^2_{H^2(K)}\right)^{1/2}\right). \end{aligned}$$

Due to (4.188), where we write $b_{h,m}$ instead of b_h, we have

$$\begin{aligned} |b_{h,m}(\eta,\xi)| &\le \left|\sum_{K\in\mathscr{T}_h}\int_K \eta(\boldsymbol{v}\cdot\nabla\xi)\,\mathrm{d}x\right| + \left|\sum_{K\in\mathscr{T}_h}\int_K \eta\xi\,\nabla\cdot\boldsymbol{v}\,\mathrm{d}x\right| \qquad (6.401)\\ &+\left|\sum_{K\in\mathscr{T}_h}\left(\int_{\partial K}(\boldsymbol{v}\cdot\boldsymbol{n})\xi\eta\,\mathrm{d}S - \int_{\partial K^-\cap\partial\Omega}(\boldsymbol{v}\cdot\boldsymbol{n})\xi\eta\,\mathrm{d}S - \int_{\partial K^-\setminus\partial\Omega}(\boldsymbol{v}\cdot\boldsymbol{n})\xi[\eta]\,\mathrm{d}S\right)\right|. \end{aligned}$$

The first term in (6.401) is estimated with the aid of assumption (4.159d), the Cauchy inequality and the inverse inequality (2.86):

$$\begin{aligned} &\left|\sum_{K\in\mathscr{T}_h}\int_K \eta(\boldsymbol{v}\cdot\nabla\xi)\,\mathrm{d}x\right| \le C_{\boldsymbol{v}}\sum_{K\in\mathscr{T}_{h,m}}\|\eta\|_{L^2(K)}|\xi|_{H^1(K)} \qquad (6.402)\\ &\le C_{\boldsymbol{v}}C_I\sum_{K\in\mathscr{T}_{h,m}} h_K^{-1}\|\eta\|_{L^2(K)}\,\|\xi\|_{L^2(K)} \le C\left(\sum_{K\in\mathscr{T}_{h,m}} h_K^{-2}\|\eta\|^2_{L^2(K)}\right)^{1/2}\|\xi\|_{L^2(\Omega)}. \end{aligned}$$

The second term is estimated by

$$\left|\sum_{K\in\mathscr{T}_h}\int_K \eta\xi\,\nabla\cdot\boldsymbol{v}\,\mathrm{d}x\right| \le C_{\boldsymbol{v}}\|\eta\|_{L^2(\Omega)}\|\xi\|_{L^2(\Omega)}. \qquad (6.403)$$

The third term is estimated as in (4.191):

$$\left| \sum_{K \in \mathscr{T}_h} \left(\int_{\partial K^+} (\boldsymbol{v} \cdot \boldsymbol{n}) \xi \eta \, \mathrm{d}S + \int_{\partial K^- \setminus \partial \Omega} ((\boldsymbol{v} \cdot \boldsymbol{n}) \xi \eta - (\boldsymbol{v} \cdot \boldsymbol{n}) \xi [\eta]) \, \mathrm{d}S \right) \right| \tag{6.404}$$
$$= \left| \sum_{K \in \mathscr{T}_h} \left(\int_{\partial K^+ \cap \partial \Omega} (\boldsymbol{v} \cdot \boldsymbol{n}) \xi \eta \, \mathrm{d}S + \int_{\partial K^- \setminus \partial \Omega} (\boldsymbol{v} \cdot \boldsymbol{n}) \eta^- [\xi] \, \mathrm{d}S \right) \right|$$
$$\leq \sum_{K \in \mathscr{T}_h} \|\xi\|_{\boldsymbol{v}, \partial K^+ \cap \partial \Omega} \|\eta\|_{\boldsymbol{v}, \partial K^+ \cap \partial \Omega} + \sum_{K \in \mathscr{T}_h} \|[\xi]\|_{\boldsymbol{v}, \partial K^- \setminus \partial \Omega} \|\eta^-\|_{\boldsymbol{v}, \partial K^- \setminus \partial \Omega}.$$

Summarizing (6.401)–(6.404), we obtain

$$|b_{h,m}(\eta, \xi)| \leq C \left(\|\eta\|_{L^2(\Omega)} + \left(\sum_{K \in \mathscr{T}_{h,m}} h_K^{-2} \|\eta\|^2_{L^2(K)} \right)^{1/2} \right) \|\xi\|_{L^2(\Omega)} \tag{6.405}$$
$$+ \sum_{K \in \mathscr{T}_h} \|\xi\|_{\boldsymbol{v}, \partial K^+ \cap \partial \Omega} \|\eta\|_{\boldsymbol{v}, \partial K^+ \cap \partial \Omega} + \sum_{K \in \mathscr{T}_h} \|[\xi]\|_{\boldsymbol{v}, \partial K^- \setminus \partial \Omega} \|\eta^-\|_{\boldsymbol{v}, \partial K^- \setminus \partial \Omega}$$
$$\leq C \left(\|\eta\|_{L^2(\Omega)} + \left(\sum_{K \in \mathscr{T}_{h,m}} h_K^{-2} \|\eta\|^2_{L^2(K)} \right)^{1/2} \right.$$
$$\left. + \left(\sum_{K \in \mathscr{T}_{h,m}} \|\eta\|^2_{\boldsymbol{v}, \partial K^+ \cap \partial \Omega} \right)^{1/2} + \left(\sum_{K \in \mathscr{T}_h} \|\eta^-\|^2_{\boldsymbol{v}, \partial K^- \setminus \partial \Omega} \right)^{1/2} \right)$$
$$\times \left(\|\xi\|_{L^2(\Omega)} + \left(\sum_{K \in \mathscr{T}_h} \|\xi\|^2_{\boldsymbol{v}, \partial K^+ \cap \partial \Omega} \right)^{1/2} + \left(\sum_{K \in \mathscr{T}_h} \|[\xi]\|^2_{\boldsymbol{v}, \partial K^- \setminus \partial \Omega} \right)^{1/2} \right)$$
$$\leq C \|\xi\|_{E,m} \left(\|\eta\|_{E,m} + \left(\sum_{K \in \mathscr{T}_h} \|\eta^-\|^2_{\boldsymbol{v}, \partial K^- \setminus \partial \Omega} \right)^{\frac{1}{2}} + \left(\sum_{K \in \mathscr{T}_{h,m}} h_K^{-2} \|\eta\|^2_{L^2(\Omega)} \right)^{\frac{1}{2}} \right).$$

Further, we have

$$|c_{h,m}(\eta, \xi)| < C_c \|\eta\|_{L^2(\Omega)} \|\xi\|_{L^2(\Omega)}, \tag{6.406}$$
$$\varepsilon |J_{h,m}(\eta, \xi)| \leq \varepsilon \sqrt{J_{h,m}(\eta, \eta)} \sqrt{J_{h,m}(\xi, \xi)} < \varepsilon |||\eta||| \, |||\xi|||. \tag{6.407}$$

Now, the above estimates for $a_{h,m}$, $b_{h,m}$, $c_{h,m}$ and $J_{h,m}$ imply (6.396). □

From the estimates for $b_{h,m}$ and $c_{h,m}$ we see that it is necessary to have $\gamma_0 > 0$ as assumed in (4.159f). However, this assumption is not restrictive, as shown in Sect. 4.6.1.

Lemma 6.67 *The following estimate holds:*

$$\sum_{m=1}^{r} \int_{I_m} \|\xi\|_{E,m}^2 \, dt + \|\xi\|_T^2 \le 4C_a^2 \sum_{m=1}^{r} \int_{I_m} R_m^2(\eta) \, dt + 8 \sum_{m=1}^{r-1} \|\eta_m^-\|_{L^2(\Omega)}^2, \tag{6.408}$$

where C_a is the constant from Lemma 6.66.

Proof From Lemma 6.63 and relation (6.394) we get

$$\begin{aligned} &\sum_{m=1}^{r} \int_{I_m} \mathscr{A}_{h,m}(\xi,\xi)\, dt + \|\xi\|_T^2 \\ &= \mathscr{B}(\xi,\xi) = \sum_{m=1}^{r} \int_{I_m} \mathscr{A}_{h,m}(\eta,\xi)\, dt - \sum_{m=1}^{r-1} (\eta_m^-, \{\xi\}_m). \end{aligned} \tag{6.409}$$

By Lemma 6.64,

$$\mathscr{A}_{h,m}(\xi,\xi) \ge \|\xi\|_{E,m}^2, \tag{6.410}$$

and by Lemma 6.66,

$$|\mathscr{A}_{h,m}(\eta,\xi)| \le C_a \|\xi\|_{E,m} R_m(\eta). \tag{6.411}$$

The Cauchy inequality implies that

$$\begin{aligned} &\left| \sum_{m=1}^{r-1} (\eta_m^-, \{\xi\}_m) \right| \le \sum_{m=1}^{r-1} \|\eta_m^-\|_{L^2(\Omega)} \|\{\xi\}_m\|_{L^2(\Omega)} \\ &\le \left(2 \sum_{m=1}^{r-1} \|\eta_m^-\|_{L^2(\Omega)}^2 \right)^{1/2} \left(\frac{1}{2} \sum_{m=1}^{r-1} \|\{\xi\}_m\|_{L^2(\Omega)}^2 \right)^{1/2}, \\ &\sum_{m=1}^{r} \int_{I_m} \|\xi\|_{E,m} R_m(\eta)\, dt \le \left(\sum_{m=1}^{r} \int_{I_m} R_m^2(\eta)\, dt \right)^{1/2} \left(\sum_{m=1}^{r} \int_{I_m} \|\xi\|_{E,m}^2\, dt \right)^{1/2}. \end{aligned}$$

From the above estimates, the definition (6.386) of the norm $\|\cdot\|_T$ and (6.409) we get

$$\sum_{m=1}^{r}\int_{I_m}\|\xi\|_{E,m}^2\,\mathrm{d}t+\|\xi\|_T^2\le\sqrt{2}\left(\sum_{m=1}^{r}\int_{I_m}\|\xi\|_{E,m}^2\,\mathrm{d}t+\|\xi\|_T^2\right)^{1/2}$$
$$\times\left(C_a\left(\sum_{m=1}^{r}\int_{I_m}R_m^2(\eta)\,\mathrm{d}t\right)^{1/2}+\left(2\sum_{m=1}^{r-1}\|\eta_m^-\|_{L^2(\Omega)}^2\right)^{1/2}\right).$$

This fact and the inequality $(a+b)^2\le 2(a^2+b^2)$ already imply (6.408). □

From the above inequality with $a=\|\xi\|_{E,m}, b=\|\eta\|_{E,m}$ and $a=\|\xi\|_T, b=\|\eta\|_T$, Lemma 6.67 and definitions (6.388), (6.395) of $\|\cdot\|_{E,m}$ and $R_m(\eta)$ we deduce the following *abstract error estimate.*

Theorem 6.68 *There exists a constant $C_{AE}>0$ independent of $u, U, h, \tau, r, m, \varepsilon$ such that*

$$\sum_{m=1}^{r}\int_{I_m}\|e\|_{E,m}^2\,\mathrm{d}t+\|e\|_T^2 \tag{6.412}$$
$$\le C_{AE}\sum_{m=1}^{r}\int_{I_m}R_m^2(\eta)\,\mathrm{d}t+C_{AE}\sum_{m=1}^{r-1}\|\eta_m^-\|_{L^2(\Omega)}^2+2\|\eta\|_T^2,$$

$$\sum_{m=1}^{r}\int_{I_m}\|e\|_{E,m}^2\,\mathrm{d}t\le C_{AE}\sum_{m=1}^{r}\int_{I_m}R_m^2(\eta)\,\mathrm{d}t+C_{AE}\sum_{m=1}^{r-1}\|\eta_m^-\|_{L^2(\Omega)}^2. \tag{6.413}$$

6.4.4.1 Error Estimate in Terms of h and τ

The estimation of error $e=U-u$ of the ST-DGM (6.379) in terms of h and τ is based on the abstract error estimate (6.413) and the approximation properties of the interpolation operator π derived in Sect. 6.1.6. We again write

$$\eta|_{I_m}=(\pi u-u)|_{I_m}=\eta^{(1)}+\eta^{(2)}, \tag{6.414}$$
$$\eta^{(1)}=(\Pi_{h,m}u-u)|_{I_m},\quad \eta^{(2)}=(\pi u-\Pi_{h,m}u)|_{I_m}=(\pi(\Pi_{h,m}u)-\Pi_{h,m}u)|_{I_m},$$

because, by (6.39), $\pi u|_{I_m}=\pi(\Pi_{h,m}u)|_{I_m}$. We assume that the exact solution u satisfies the regularity condition (6.370). Then, in a similar way as in Sect. 6.2.4, we can prove the error estimate. It will be derived under several assumptions. We consider assumptions on the *shape regularity* (2.19) and *local quasi-uniformity* (2.21) of the space grids $\mathscr{T}_{h,m}$ and, moreover, some relations between the time steps and space sizes. Namely, we have in mind the condition (6.261), i.e., $\tau_m \geq C\, h_m^2$, $m = 1, \ldots, r$. Instead of condition (6.118), i.e., $\tau_m \leq \overline{C}\, h_{K_\Gamma^{(L)}}$, $\Gamma \in \mathscr{F}_{h,m}^B$, $m = 1, \ldots, r$, $h \in (0, \bar{h})$, we consider the stronger condition

$$\tau_m \leq \overline{C}^*\, h_K, \quad K \in \mathscr{T}_{h,m},\ m = 1, \ldots, r,\ h \in (0, \bar{h}). \tag{6.415}$$

Theorem 6.69 *Let u be the exact solution of problem (6.369) and let assumptions (4.159) and (6.370) be satisfied. Let U be the approximate solution of this problem obtained by method (6.379) under assumptions (2.19), (2.21), (6.261) and (6.415). Then there exists a contact $C > 0$ independent of h, τ, m, r, u, U and ε such that*

$$\sum_{m=1}^{r} \int_{I_m} \|e\|_{E,m}^2 \, \mathrm{d}t \leq C\Big(h^{2(\mu-1)}\, |u|^2_{C([0,T];H^\mu(\Omega))} + \tau^{2q} |u\|^2_{H^{q+1}(0,T;H^1(\Omega))}\Big). \tag{6.416}$$

Estimate (6.416) holds for every $\varepsilon \geq 0$. (The symbol $|\cdot\|$ is defined by (6.105).)

Proof By the abstract error estimate (6.413), the definition (6.395) of the expression $R_m(\eta)$ and the definition (6.388) of the norm $\|\cdot\|_{E,m}$, we have

$$\begin{aligned}
&\sum_{m=1}^{r} \int_{I_m} \|e\|_{E,m}^2 \, \mathrm{d}t \qquad (6.417)\\
&\leq C \sum_{m=1}^{r} \int_{I_m} \Big(\frac{\varepsilon}{2}\big(|\eta|^2_{H^1(\Omega,\mathscr{T}_{h,m})} + J_{h,m}(\eta,\eta)\big) + \gamma_0 \|\eta\|^2_{L^2(\Omega)} \\
&\quad + \varepsilon \sum_{K\in\mathscr{T}_{h,m}} h_K^2 |\eta|^2_{H^2(K)} + \sum_{K\in\mathscr{T}_{h,m}} h_K^{-2} \|\eta\|^2_{L^2(K)} \\
&\quad + \frac{1}{2} \sum_{K\in\mathscr{T}_{h,m}} \big(\|\eta\|^2_{\nu,\partial K\cap\partial\Omega} + \|[\eta]\|^2_{\nu,\partial K\setminus\partial\Omega} + \|\eta^-\|^2_{\nu,\partial K^-\setminus\partial\Omega}\big)\Big) \mathrm{d}t \\
&\quad + C \sum_{m=1}^{r-1} \|\eta_m^-\|^2_{L^2(\Omega)}.
\end{aligned}$$

Let us estimate the individual terms depending on η. By (6.109), (6.112), (6.115) and Lemma 6.19,

$$\begin{aligned}
&\sum_{m=1}^{r} \int_{I_m} \frac{\varepsilon}{2}\Big(|\eta|^2_{H^1(\Omega,\mathscr{T}_{h,m})} + J_{h,m}(\eta,\eta)\Big) \mathrm{d}t \qquad (6.418)\\
&\leq C\,\varepsilon \Big(h^{2(\mu-1)}\, |u|^2_{L^2(0,T;H^\mu(\Omega))} + \tau^{2q}\, |u|^2_{H^{q+1}(0,T;H^1(\Omega))}\Big).
\end{aligned}$$

By virtue of (6.108) and (6.111),

$$\sum_{m=1}^{r} \int_{I_m} \|\eta\|^2_{L^2(\Omega)} \, \mathrm{d}t \le C\Big(h^{2\mu}\, |u^2|^2_{L^2(0,T;H^\mu(\Omega))} + \tau^{2(q+1)}\, |u|_{H^{q+1}(0,T;L^2(\Omega))}\Big). \tag{6.419}$$

Moreover, taking into account assumption (6.415), we find that

$$\sum_{m=1}^{r} \int_{I_m} \sum_{K\in\mathscr{T}_{h,m}} h_K^{-2} \|\eta\|^2_{L^2(K)} \, \mathrm{d}t \le C\left(h^{2(\mu-1)} |u|^2_{L^2(0,T;H^\mu(\Omega))} + \tau^{2q} |u|^2_{H^{q+1}(0,T;L^2(\Omega))}\right).$$

In view of (6.110) and (6.113),

$$\begin{aligned}\sum_{m=1}^{r} \int_{I_m} \sum_{K\in\mathscr{T}_{h,m}} h_K^2\, |\eta|^2_{H^2(K)} \\ \le C\left(h^{2(\mu-1)}\, |u|^2_{L^2(0,T;H^\mu(\Omega))} + \tau^{2(q+1)}\, |u|_{H^{q+1}(0,T;H^1(\Omega))}\right).\end{aligned} \tag{6.420}$$

Further, under condition (6.261), estimate (6.107) and the relation $\sum_{m=1}^{r} \tau_m = T$ imply that

$$\begin{aligned}\sum_{m=1}^{r-1} \|\eta_m^-\|^2_{L^2(\Omega)} &\le C \sum_{m=1}^{r-1} \tau_m\, h_m^{2\mu-1} |u(t_m)|^2_{H^\mu(\Omega)} \\ &\le C\, T\, h^{2\mu-1} |u|^2_{C([0,T];H^\mu(\Omega))}.\end{aligned} \tag{6.421}$$

Finally, we estimate the expression

$$R_b(\eta) := \sum_{m=1}^{r} \int_{I_m} \sum_{K\in\mathscr{T}_{h,m}} \left(\|\eta\|^2_{\nu,\partial K\cap\partial\Omega} + \|[\eta]\|^2_{\nu,\partial K\setminus\partial\Omega} + \|\eta^-\|^2_{\nu,\partial K^-\setminus\partial\Omega}\right) \mathrm{d}t. \tag{6.422}$$

From the multiplicative trace inequality (2.78) and the Young inequality we get

$$\begin{aligned}\|\eta\|^2_{\nu,\partial K\cap\partial\Omega} + \|[\eta]\|^2_{\nu,\partial K\setminus\partial\Omega} + \|\eta^-\|^2_{\nu,\partial K^-\setminus\partial\Omega} \\ \le C\|\eta\|^2_{L^2(\partial K)} \le C\Big(h_K^{-1} \|\eta\|^2_{L^2(K)} + h_K \|\eta\|^2_{H^1(K)}\Big).\end{aligned} \tag{6.423}$$

By virtue of condition (6.415), $h_K^{-1} \tau_m^{2(q+1)} \le C\, \tau_m^{2q+1}$ and, thus by (6.111), (6.112), (6.108) and (6.109),

$$R_b(\eta) \leq C\, h^{2\mu-1}\, |u|^2_{L^2(0,T;H^\mu(\Omega))} + C \sum_{m=1}^{r} \Big(\tau_m^{2q+1} \sum_{K\in\mathscr{T}_{h,m}} |u|^2_{H^{q+1}(I_m;H^1(K))} \Big) \tag{6.424}$$
$$\leq C\Big(h^{2\mu-1}\, |u|^2_{L^2(0,T;H^\mu(\Omega))} + \tau^{2q+1}\, |u|^2_{H^{q+1}(0,T;H^1(\Omega))}\Big).$$

Taking into account estimates (6.417)–(6.424) and the relation

$$|u|_{L^2(0,T;H^\mu(\Omega))} \leq T|u|_{C([0,T];H^\mu(\Omega))},$$

we find that (6.416) holds. □

Remark 6.70 (a) In our theoretical error analysis, it was necessary to use assumption (6.415), which reminds us of the CFL-like stability condition.[1] It is an interesting subject to investigate the stability of the ST-DGM (6.379) in order to find out, if the method is unconditionally stable.

(b) If we compare Theorems 6.69, 4.28 and 6.46, we see that the uniform error estimates with respect to ε derived for the ST-DGM are of a lower-order both in space and in time. A detailed analysis shows that better estimates can be obtained in this case under the assumption that the transport velocity $\boldsymbol{v}$ behaves in time as a polynomial of degree $\leq q-1$ and the function u_D behaves in time as a polynomial of degree $\leq q$. Then it is possible to avoid the expression $\sum_{K\in\mathscr{T}_{h,m}} h_K^{-2}\|\eta\|_{L^2(K)}$ in the estimate of the term $b_{h,m}(\eta,\xi)$ and thereby to improve the error estimate. We leave the detailed analysis to the reader.

6.4.5 Numerical Examples

The error estimates derived above will be demonstrated by numerical experiments. We solve equation (6.369a) in $Q_T = (0,1)^2 \times (0,1)$ with $v_1 = v_2 = 1,\ c = 0.5$ and two choices of the diffusion coefficient: $\varepsilon = 0.005$ (parabolic case) and $\varepsilon = 0$ (hyperbolic case). The right-hand side g and the boundary and initial conditions are chosen in such way that they conform to the exact solution

$$u_{ex}(x_1,x_2,t) = (1-e^{-t})\Big(2x_1+2x_2-x_1x_2+2(1-e^{v_1(x_1-1)/\nu})(1-e^{v_2(x_2-1)/\nu})\Big), \tag{6.425}$$

where $\nu = 0.05$ is a constant determining the steepness of the boundary layer in the exact solution.

[1] CFL condition means the Courant–Friedrichs–Lewy condition, which guarantees the stability of explicit numerical schemes for the solution of first-order hyperbolic equations. See, e.g., [127, Sect. 3.2.8] or [245, Sect. 3.1.3].

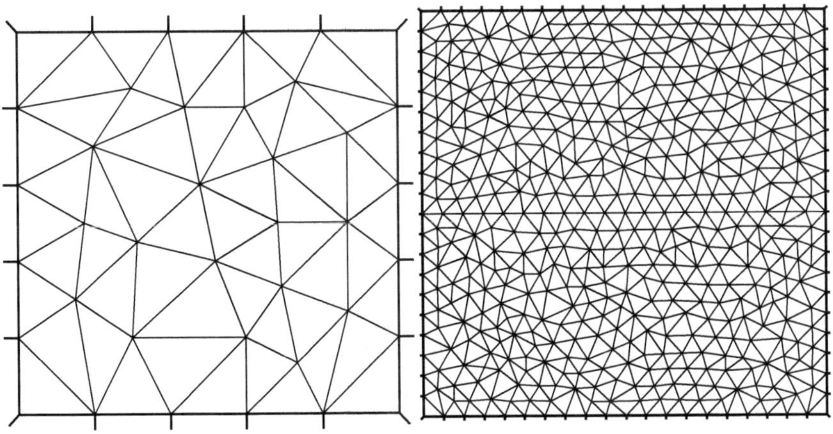

Fig. 6.4 Coarsest mesh $\mathscr{T}_{h_1}$ (*left*) and the fine mesh $\mathscr{T}_{h_4}$ (*right*)

The problem is solved on a sequence of non-nested nonuniform space meshes $\mathscr{T}_{h_1}, \mathscr{T}_{h_2}, \ldots,$ which is kept unchanged on all time levels. Figure 6.4 shows the coarsest mesh $\mathscr{T}_{h_1}$ and the mesh $\mathscr{T}_{h_4}$. We inspect the experimental order of convergence (EOC) with respect to τ and h, which are simultaneously varied. For successive pairs (τ, h) and (τ', h') we evaluate the experimental order of convergence (EOC) in space and time defined as

$$\mathrm{EOC}_{\mathrm{sp}} = \frac{\log\left(\|e_{\tau'h'}\|_{L^2(Q_T)}\right) - \log\left(\|e_{\tau h}\|_{L^2(Q_T)}\right)}{\log h' - \log h},$$

$$\mathrm{EOC}_{\mathrm{ti}} = \frac{\log\left(\|e_{\tau'h'}\|_{L^2(Q_T)}\right) - \log\left(\|e_{\tau h}\|_{L^2(Q_T)}\right)}{\log \tau' - \log \tau},$$

where $e_{\tau h} = U - u$ is the error of the method, when the exact solution u_{ex} is approximated by the DG approximate solution U that is computed with the aid of a space triangulation of size h and a time interval partition of size τ. Moreover, we compute the *global experimental order of convergence* (GEOC) with the aid of additional data sets with halved time step and fitting a general nonlinear model of the form

$$\|e_{\tau h}\|_{L^2(Q_T)} \approx C_1\, h^r + C_2\, \tau^s$$

through the data via the method of nonlinear least squares, using the MINPACK package [225]. The results are shown in Tables 6.5 and 6.6.

Table 6.5 Computational errors and the corresponding EOC: $\varepsilon = 0.005$ (parabolic case): $p = 1$, $q = 1$ (top) and $p = 2$, $q = 2$ (bottom)

h	τ	$\|e_{\tau h}\|_{L^2(Q_T)}$	$\mathrm{EOC}_{\mathrm{sp}}$	$\mathrm{EOC}_{\mathrm{ti}}$
0.2838	0.2500	4.5853E–02	–	–
0.2172	0.2000	3.5474E–02	0.96	1.15
0.1540	0.1667	2.2387E–02	1.34	2.52
0.1035	0.1000	1.2945E–02	1.38	1.07
0.0768	0.0769	5.3557E–03	2.95	3.36
0.0532	0.0526	2.3742E–03	2.22	2.14
0.0398	0.0400	1.3345E–03	1.98	2.10
0.0270	0.0270	5.2577E–04	2.40	2.38
0.0223	0.0222	2.7946E–04	3.30	3.23
0.0144	0.0145	1.1835E–04	1.98	2.01
GEOC			2.07	2.11
h	τ	$\|e_{\tau h}\|_{L^2(Q_T)}$	$\mathrm{EOC}_{\mathrm{sp}}$	$\mathrm{EOC}_{\mathrm{ti}}$
0.2838	0.2500	2.0470E–02	–	–
0.2172	0.2000	1.0103E–02	2.64	3.16
0.1540	0.1667	4.3992E–03	2.42	4.56
0.1035	0.1000	1.6821E–03	2.42	1.88
0.0768	0.0769	4.9668E–04	4.08	4.65
0.0532	0.0526	1.6550E–04	3.00	2.90
0.0398	0.0400	7.7630E–05	2.61	2.76
0.0270	0.0270	2.7654E–05	2.66	2.63
GEOC			2.89	2.78

The numerical experiments were carried out with the aid of the `FreeFEM++` modelling environment from [169], which was adapted to the space-time DGM discretization. The time integrals were evaluated by exact quadrature formulae for polynomials of degree 5 and 9 in the case of elements linear in time and quadratic in time, respectively. The quadrature formulae used for the integration over triangles and their sides were exact for polynomials of degree 5 both for linear and quadratic elements. The nonsymmetric linear problem was solved in each time step by the multifrontal direct solver `UMFPACK` [72].

It follows from these examples that the error estimate (6.416) is suboptimal both in space and time (cf. Remark 6.70b). On the other hand, a positive property of this estimate is the uniformity with respect to the diffusion coefficient $\varepsilon \to 0+$. The derivation of optimal error estimates remains open.

Table 6.6 Computational errors and the corresponding EOC: $\varepsilon = 0$ (hyperbolic case): $p = 1$, $q = 1$ (top) and $p = 2$, $q = 2$ (bottom)

h	τ	$\|e_{\tau h}\|_{L^2(Q_T)}$	$\mathrm{EOC_{sp}}$	$\mathrm{EOC_{ti}}$
0.2838	0.2500	4.9212E–02	–	–
0.2172	0.2000	3.8843E–02	0.89	1.06
0.1540	0.1667	2.5997E–02	1.17	2.20
0.1035	0.1000	1.5581E–02	1.29	1.00
0.0768	0.0769	6.9089E–03	2.72	3.10
0.0532	0.0526	3.2904E–03	2.02	1.95
0.0398	0.0400	1.8620E–03	1.96	2.07
0.0270	0.0270	7.5458E–04	2.32	2.30
0.0223	0.0222	4.1924E–04	3.07	3.00
0.0144	0.0145	1.7556E–04	2.01	2.04
GEOC			1.95	1.99
h	τ	$\|e_{\tau h}\|_{L^2(Q_T)}$	$\mathrm{EOC_{sp}}$	$\mathrm{EOC_{ti}}$
0.2838	0.2500	2.3451E–02	–	–
0.2172	0.2000	1.2484E–02	2.36	2.83
0.1540	0.1667	6.1746E–03	2.05	3.86
0.1035	0.1000	2.6342E–03	2.14	1.67
0.0768	0.0769	8.0848E–04	3.95	4.50
0.0532	0.0526	2.6400E–04	3.05	2.95
0.0398	0.0400	1.0761E–04	3.09	3.27
0.0270	0.0270	2.7962E–05	3.47	3.44
GEOC			2.87	2.98

Chapter 7
Generalization of the DGM

The aim of this chapter is to present some advanced aspects and special techniques of the discontinuous Galerkin method. First, we present the hp-discontinuous Galerkin method. Then the DGM over nonstandard nonsimplicial meshes will be treated. Finally, the effect of numerical integration in the DGM will be analyzed in the case of a nonstationary convection-diffusion problem with nonlinear convection.

7.1 hp-Discontinuous Galerkin Method

Since the DGM is based on discontinuous piecewise polynomial approximations, it is possible to use different polynomial degrees on different elements in a simple way. Then we speak of the hp-discontinuous Galerkin method (hp-DGM). A suitable adaptive mesh refinement combined with the choice of the polynomial approximation degrees, representing the hp-adaptation, can significantly increase the *efficiency* of the computational process. It allows us to achieve the given error tolerance with the aid of the low number of degrees of freedom. The origins of hp finite element methods date back to the pioneering work of Ivo Babuška et al., see the survey paper [16]. Based on several theoretical works as, e.g., monographs [253, 258] or papers [16, 78, 259], it is possible to expect that the error decreases to zero at an exponential rate with respect to the number of degrees of freedom.

We present here the analysis of error estimates for the hp-DGM in the case of a model of the Poisson boundary value problem. We underline the similarity and differences with analysis of the h-version of the DGM presented in Chap. 2. Mostly the same notation is used for several constants appearing also in Chap. 2, but some constants may have slightly different meaning. However, we suppose that there is no danger of misunderstanding. On the contrary, it helps us to adapt the techniques from Chap. 2 to this section.

The analysis of the hp-DGM can be directly extended to nonstationary convection-diffusion equations from Chaps. 4 and 5. See, e.g., [87, 178, 186].

V. Dolejší and M. Feistauer, *Discontinuous Galerkin Method*,
Springer Series in Computational Mathematics 48,
DOI 10.1007/978-3-319-19267-3_7

7.1.1 Formulation of a Model Problem

Similarly, as in Sect. 2.1, let Ω be a bounded polygonal or polyhedral domain in $\mathbb{R}^d$, $d = 2, 3$, with Lipschitz boundary $\partial\Omega$. We denote by $\partial\Omega_D$ and $\partial\Omega_N$ parts of the boundary $\partial\Omega$ such that $\partial\Omega = \partial\Omega_D \cup \partial\Omega_N$, $\partial\Omega_D \cap \partial\Omega_N = \emptyset$ and $\mathrm{meas}_{d-1}(\partial\Omega_D) > 0$.

We consider the Poisson problem (2.1) to find a function $u : \Omega \to \mathbb{R}$ such that

$$-\Delta u = f \quad \text{in } \Omega, \tag{7.1a}$$

$$u = u_D \quad \text{on } \partial\Omega_D, \tag{7.1b}$$

$$\boldsymbol{n} \cdot \nabla u = g_N \quad \text{on } \partial\Omega_N, \tag{7.1c}$$

where f, u_D and g_N are given functions. The weak solution of problem (7.1) is given by Definition 2.1.

7.1.2 Discretization

In this section we introduce the *hp*-DGM numerical solution of problem (7.1). We start from the generalization of the function spaces defined in Chap. 2.

7.1.2.1 Function Spaces

Let $\mathscr{T}_h$ $(h > 0)$ be a triangulation of Ω. In the same way as in Chap. 2, by the symbols $\mathscr{F}_h$, $\mathscr{F}_h^I$, $\mathscr{F}_h^B$, $\mathscr{F}_h^D$ and $\mathscr{F}_h^{ID}$ we denote sets of faces of elements $K \in \mathscr{T}_h$. To each $K \in \mathscr{T}_h$, we assign a positive integer s_K—*local Sobolev index* and a positive integer p_K–*local polynomial degree*. Then we define the sets

$$\boldsymbol{s} = \{s_K, K \in \mathscr{T}_h\}, \qquad \boldsymbol{p} = \{p_K, K \in \mathscr{T}_h\}. \tag{7.2}$$

Over the triangulation $\mathscr{T}_h$, we define (instead of (2.29)) the *broken Sobolev space* corresponding to the vector $\boldsymbol{s}$

$$H^{\boldsymbol{s}}(\Omega, \mathscr{T}_h) = \{v;\, v|_K \in H^{s_K}(K)\ \forall\, K \in \mathscr{T}_h\} \tag{7.3}$$

with the norm

$$\|v\|_{H^{\boldsymbol{s}}(\Omega,\mathscr{T}_h)} = \left(\sum_{K\in\mathscr{T}_h} \|v\|^2_{H^{s_K}(K)} \right)^{1/2} \tag{7.4}$$

and the seminorm

$$|v|_{H^s(\Omega,\mathscr{T}_h)} = \left(\sum_{K\in\mathscr{T}_h} |v|^2_{H^{s_K}(K)} \right)^{1/2}, \tag{7.5}$$

where $\|\cdot\|_{H^{s_K}(K)}$ and $|\cdot|_{H^{s_K}(K)}$ denotes the norm and seminorm in the Sobolev space $H^{s_K}(K) = W^{s_K,2}(K)$, respectively. If $s_K = q \geq 1$ for all $K \in \mathscr{T}_h$, then we use the notation $H^q(\Omega, \mathscr{T}_h) = H^s(\Omega, \mathscr{T}_h)$. Obviously,

$$H^{\bar{s}}(\Omega, \mathscr{T}_h) \subset H^s(\Omega, \mathscr{T}_h) \subset H^{\underline{s}}(\Omega, \mathscr{T}_h), \tag{7.6}$$

where $\bar{s} = \max\{s_K,\ s_K \in \boldsymbol{s}\}$ and $\underline{s} = \min\{s_K,\ s_K \in \boldsymbol{s}\}$.

Furthermore, we define (instead of (2.34)) the space of discontinuous piecewise polynomial functions associated with the vector $\boldsymbol{p}$ by

$$S_{h\boldsymbol{p}} = \{v \in L^2(\Omega);\ v|_K \in P_{p_K}(K)\ \forall\, K \in \mathscr{T}_h\}, \tag{7.7}$$

where $P_{p_K}(K)$ denotes the space of all polynomials on K of degree $\leq p_K$. In the hp-error analysis we assume that there exists a constant $C_P \geq 1$ such that

$$\frac{p_K}{p_{K'}} \leq C_P \quad \forall\, K,\ K' \in \mathscr{T}_h \text{ such that } K \text{ and } K' \text{ are neighbours.} \tag{7.8}$$

Assumption (7.8) may seem rather restrictive. However, it appears that the application of the hp-methods to practical problems is efficient and accurate, if the polynomial degrees of approximation on neighbouring elements do not differ too much.

7.1.2.2 *hp*-Variant of the Penalty Parameter

In Sect. 2.6.1 we introduced the penalty parameter $\sigma : \cup_{\Gamma\in\mathscr{F}_h^{ID}} \to \mathbb{R}$, which was proportional to $\mathrm{diam}(\Gamma)^{-1} \sim h_K^{-1}$ where $\Gamma \subset \partial K$, $\Gamma \in \mathscr{F}_h^{ID}$. However, the following numerical analysis shows that for the hp-DGM, the penalty parameter σ has to depend also on the degree of the polynomial approximation (see also [180]). To this end, for each $K \in \mathscr{T}_h$ we define the parameter

$$d(K) = \frac{h_K}{p_K^2}, \quad K \in \mathscr{T}_h. \tag{7.9}$$

Now for each $\Gamma \in \mathscr{F}_h^{ID}$ we introduce the hp-analogue to the quantity h_Γ from Sect. 2.6.1, which is now denoted by $d(\Gamma)$. In the theoretical analysis, we require that the quantity $d(\Gamma)$, $\Gamma \in \mathscr{F}_h$, $h \in (0, \bar{h})$, satisfies the *equivalence condition* with $d(K)$, i.e., there exist constants $C_T, C_G > 0$ independent of h, K and Γ such that

$$C_T\, d(K) \le d(\Gamma) \le C_G\, d(K), \quad K \in \mathscr{T}_h,\ \Gamma \in \mathscr{F}_h,\ \Gamma \subset \partial K. \tag{7.10}$$

Let $K_\Gamma^{(L)}$ and $K_\Gamma^{(R)}$ be the neighbouring elements sharing the face $\Gamma \in \mathscr{F}_h^I$. There are several possibilities how to define the parameter $d(\Gamma)$ for all interior faces $\Gamma \in \mathscr{F}_h^I$:

(i)

$$d(\Gamma) = \frac{2\,\mathrm{diam}(\Gamma)}{(p_{K_\Gamma^{(L)}})^2 + (p_{K_\Gamma^{(R)}})^2}, \quad \Gamma \in \mathscr{F}_h^I, \tag{7.11}$$

(ii)

$$d(\Gamma) = \max(d(K_\Gamma^{(L)}), d(K_\Gamma^{(R)})), \quad \Gamma \in \mathscr{F}_h^I, \tag{7.12}$$

(iii)

$$d(\Gamma) = \min(d(K_\Gamma^{(L)}), d(K_\Gamma^{(R)})), \quad \Gamma \in \mathscr{F}_h^I. \tag{7.13}$$

Moreover, for the boundary faces $\Gamma \in \mathscr{F}_h^D$, we put

$$d(\Gamma) = d(K_\Gamma^{(L)}), \tag{7.14}$$

where $K_\Gamma^{(L)}$ is the element adjacent to Γ.

In the sequel we consider a system $\{\mathscr{T}_h\}_{h\in(0,\bar{h})}$ of triangulations of the domain Ω satisfying the shape-regularity assumption (2.19), i.e.,

$$\frac{h_K}{\rho_K} \le C_R, \quad K \in \mathscr{T}_h, \quad h \in (0, \bar{h}). \tag{7.15}$$

The following lemma characterizes the mesh assumptions and the choices of $d(\Gamma)$, which guarantees the equivalence condition (7.10).

Lemma 7.1 *Let $\{\mathscr{T}_h\}_{h\in(0,\bar{h})}$ be a system of triangulations of the domain Ω satisfying assumption (7.15). Moreover, let $\mathbf{p}$ be the polynomial degree vector given by (7.2), satisfying assumption (7.8). Then condition (7.10) is satisfied in the following cases:*

(a) The triangulations $\mathscr{T}_h$, $h \in (0, \bar{h})$, are conforming (i.e., assumption (MA4) from Sect. 2.3.2 is satisfied) and $d(\Gamma)$ is defined by (7.11) or (7.12) or (7.13).

(b) The triangulations $\mathscr{T}_h$, $h \in (0, \bar{h})$, are, in general, nonconforming, assumption (A2) (i.e., (2.22) is satisfied and $d(\Gamma)$ is defined by (7.11).

(c) The triangulations $\mathscr{T}_h$, $h \in (0, \bar{h})$, are, in general, nonconforming, assumption (A1) is satisfied (i.e., the system $\{\mathscr{T}_h\}_{h\in(0,\bar{h})}$ is locally quasi-uniform) and $d(\Gamma)$ is defined by (7.12) or (7.13).

Exercise 7.2 Prove the above lemma and determine the constants C_T and C_G.

Remark 7.3 If $p_K = p \in \mathbb{N}$ for all $K \in \mathscr{T}_h$, then the constants C_T and C_G from (7.10) are identical with the constants from (2.20).

7.1.2.3 Approximate Solution

Now we are ready to introduce the hp-DGM approximate solution. Using the same process as in Chap. 2, we arrive at the definition of the following forms. For $u, v \in H^s(\Omega, \mathscr{T}_h)$, where $s_K \geq 2$ for all $K \in \mathscr{T}_h$, we put

$$a_h(u,v) = \sum_{K\in\mathscr{T}_h} \int_K \nabla u \cdot \nabla v \,\mathrm{d}x - \sum_{\Gamma\in\mathscr{F}_h^{ID}} \int_\Gamma (\langle\nabla u\rangle \cdot \boldsymbol{n}[v] + \Theta\langle\nabla v\rangle \cdot \boldsymbol{n}[u]) \,\mathrm{d}S, \tag{7.16}$$

$$J_h^\sigma(u,v) = \sum_{\Gamma\in\mathscr{F}_h^{ID}} \int_\Gamma \sigma[u]\,[v]\,\mathrm{d}S, \tag{7.17}$$

$$\ell_h(v) = \int_\Omega gv\,\mathrm{d}x - \Theta \sum_{\Gamma\in\mathscr{F}_h^{D}} \int_\Gamma u_D\,(\nabla v\cdot\boldsymbol{n})\,\mathrm{d}S + \sum_{\Gamma\in\mathscr{F}_h^{D}} \int_\Gamma \sigma u_D\, v\,\mathrm{d}S + \int_{\partial\Omega_N} g_N v\,\mathrm{d}S, \tag{7.18}$$

where the *penalty parameter* σ is given by

$$\sigma|_\Gamma = \sigma_\Gamma = \frac{C_W}{d(\Gamma)}, \quad \Gamma \in \mathscr{F}_h^{ID}, \tag{7.19}$$

with $d(\Gamma)$ introduced in (7.11)–(7.14), and a suitable constant $C_W > 0$. In contrast to the penalty parameter σ defined in Sect. 2.6.1, we have $\sigma|_\Gamma \sim p^2h^{-1}$, where h and p correspond to the diameter of Γ and the degree of the polynomial approximation, respectively, in the vicinity of Γ.

Similarly as in Sect. 2.4, for $\Theta = -1$, $\Theta = 0$ and $\Theta = 1$ the form a_h (together with the form J_h^σ) represents the nonsymmetric variant (NIPG), incomplete variant (IIPG) and symmetric variant (SIPG), respectively, of the approximation of the diffusion term. Moreover, we put

$$A_h(u,v) = a_h(u,v) + J_h^\sigma(u,v), \quad u, v \in H^s(\Omega, \mathscr{T}_h). \tag{7.20}$$

Now we define an approximate solution of problem (7.1).

Definition 7.4 A function $u_h \in S_{hp}$ is called an *hp-DG approximate solution* of problem (7.1), if it satisfies the identity

$$A_h(u_h, v_h) = \ell_h(v_h) \quad \forall v_h \in S_{hp}. \tag{7.21}$$

From the construction of the forms A_h and ℓ_h one can see that the exact solution $u \in H^2(\Omega)$ of problem (7.1) satisfies the identity

$$A_h(u, v) = \ell_h(v) \quad \forall v \in H^2(\Omega, \mathscr{T}_h), \tag{7.22}$$

which represents the *consistency* of the method. Identities (7.21) and (7.22) imply the *Galerkin orthogonality* of the error $e_h = u_h - u$ of the method:

$$A_h(e_h, v_h) = 0 \quad \forall v_h \in S_{hp}, \tag{7.23}$$

which will be used in the analysis of error estimates. (Compare with (2.57).)

7.1.3 Theoretical Analysis

This section is devoted to the error analysis of the hp-DGM introduced above. Namely, an error estimate in the analogue to the DG-norm introduced by (2.103) will be derived. We follow the analysis of the abstract method from Sect. 2.2 and present several "hp-variants" of results from Chap. 2. We use the same notation for constants, although they attain different values in Chap. 2 and Sect. 7.1.3.

7.1.3.1 Auxiliary Results

Similarly as in Sect. 2.5, the numerical analysis is based on three fundamental results: the multiplicative trace inequality, the inverse inequality and the approximation properties.

The *multiplicative trace inequality* presented in Lemma 2.19 remains the same. This means that under the shape-regularity assumption (7.15), there exists a constant $C_M > 0$ independent of v, h and K such that

$$\|v\|^2_{L^2(\partial K)} \le C_M \left(\|v\|_{L^2(K)} \, |v|_{H^1(K)} + h_K^{-1} \|v\|^2_{L^2(K)} \right), \tag{7.24}$$
$$K \in \mathscr{T}_h, \; v \in H^1(K), \; h \in (0, \bar{h}).$$

The proof of Lemma 2.21 gives us the hp-version of the *inverse inequality*: Let the shape-regularity assumption (7.15) be satisfied. Then there exists a constant $C_I > 0$ independent of v, h, p_K, and K such that

$$|v|_{H^1(K)} \le C_I p_K^2 h_K^{-1} \|v\|_{L^2(K)}, \quad v \in P_{p_K}(K), \; K \in \mathscr{T}_h, \; h \in (0, \bar{h}). \tag{7.25}$$

Finally, we introduce the hp-version of approximation properties of spaces S_{hp}. We present the results from [14]. Since the proof is very technical, we skip it and refer to the original work.

Lemma 7.5 (Approximation properties) *There exists a constant $C_A > 0$ independent of v, h, K and p_K and a mapping $\pi^K_{p_K} : H^{s_K}(K) \to P_{p_K}(K)$, $s_K \geq 1$, such that the inequality*

$$\|\pi^K_{p_K} v - v\|_{H^q(K)} \leq C_A \frac{h_K^{\mu_K - q}}{p_K^{s_K - q}} \|v\|_{H^{s_K}(K)} \tag{7.26}$$

holds for all $v \in H^{s_K}(K)$, $K \in \mathscr{T}_h$ and $h \in (0, \bar{h})$ with $\mu_K = \min(p_K+1, s_K)$, $0 \leq q \leq s_K$,

Proof See Lemma 4.5 in [14] for the case $d = 2$. If $d = 3$, the arguments are analogous. □

Definition 7.6 Let $\boldsymbol{s}$ and $\boldsymbol{p}$ be the vectors introduced in (7.2). We define the mapping $\Pi_{h\boldsymbol{p}} : H^{\boldsymbol{s}}(\Omega, \mathscr{T}_h) \to S_{h\boldsymbol{p}}$ by

$$\left(\Pi_{h\boldsymbol{p}} u\right)|_K = \pi^K_{p_K}(u|_K) \quad \forall\, K \in \mathscr{T}_h, \tag{7.27}$$

where $\pi^K_{p_K} : H^{s_K}(K) \to P_{p_K}(K)$ is the mapping introduced in Lemma 7.5.

Lemma 7.7 *Let $\boldsymbol{s}$ and $\boldsymbol{p}$ be the vectors introduced in (7.2) and $\Pi_{h\boldsymbol{p}} : H^{\boldsymbol{s}}(\Omega, \mathscr{T}_h) \to S_{h\boldsymbol{p}}$ the corresponding mapping defined by (7.27). If $v \in H^{\boldsymbol{s}}(\Omega, \mathscr{T}_h)$, then*

$$\left\|\Pi_{h\boldsymbol{p}} v - v\right\|^2_{H^q(\Omega, \mathscr{T}_h)} \leq C_A^2 \sum_{K \in \mathscr{T}_h} \frac{h_K^{2(\mu_K - q)}}{p_K^{2s_K - 2q}} \|v\|^2_{H^{s_K}(K)}, \tag{7.28}$$

where $\mu_K = \min(p_K + 1, s_K)$, $K \in \mathscr{T}_h$ and $0 \leq q \leq \min_{s_K \in \boldsymbol{s}} s_K$ and C_A is the constant from Lemma 7.5.

Proof Using definition (7.27) and the approximation properties (7.26), we obtain (7.28). □

Moreover, using the previous results, we prove some technical inequalities analogous to Lemma 2.27.

Lemma 7.8 *Let (7.10) be valid and let σ be defined by (7.19). Then for each $v \in H^1(\Omega, \mathscr{T}_h)$ we have*

$$\sum_{\Gamma \in \mathscr{F}_h^{ID}} d(\Gamma)^{-1} \int_\Gamma [v]^2 \,\mathrm{d}S \leq \frac{2}{C_T} \sum_{K \in \mathscr{T}_h} d(K)^{-1} \int_{\partial K} |v|^2 \,\mathrm{d}S, \tag{7.29}$$

$$\sum_{\Gamma \in \mathscr{F}_h^{ID}} d(\Gamma) \int_\Gamma \langle v \rangle^2 \,\mathrm{d}S \leq C_G \sum_{K \in \mathscr{T}_h} d(K) \int_{\partial K} |v|^2 \,\mathrm{d}S. \tag{7.30}$$

Hence,

$$\sum_{\Gamma\in\mathscr{F}_h^{ID}} \sigma_\Gamma \|[v]\|^2_{L^2(\Gamma)} \le \frac{2C_W}{C_T} \sum_{K\in\mathscr{T}_h} d(K)^{-1}\|v\|^2_{L^2(\partial K)}, \tag{7.31}$$

$$\sum_{\Gamma\in\mathscr{F}_h^{ID}} \frac{1}{\sigma_\Gamma} \|\langle v\rangle\|^2_{L^2(\Gamma)} \le \frac{C_G}{C_W} \sum_{K\in\mathscr{T}_h} d(K)\|v\|^2_{L^2(\partial K)}, \tag{7.32}$$

where the penalty parameter σ is given by (7.19).

Proof (a) By definition (2.32), (2.33), inequality (2.110) and assumption (7.10), we have

$$\begin{aligned}
&\sum_{\Gamma\in\mathscr{F}_h^{ID}} d(\Gamma)^{-1}\int_\Gamma [v]^2\,\mathrm{d}S\\
&= \sum_{\Gamma\in\mathscr{F}_h^{I}} d(\Gamma)^{-1}\int_\Gamma \left|v_\Gamma^{(L)} - v_\Gamma^{(R)}\right|^2 \mathrm{d}S + \sum_{\Gamma\in\mathscr{F}_h^{D}} d(\Gamma)^{-1}\int_\Gamma \left|v_\Gamma^{(L)}\right|^2 \mathrm{d}S\\
&\le 2\sum_{\Gamma\in\mathscr{F}_h^{I}} d(\Gamma)^{-1}\int_\Gamma \left(\left|v_\Gamma^{(L)}\right|^2 + \left|v_\Gamma^{(R)}\right|^2\right) \mathrm{d}S + \sum_{\Gamma\in\mathscr{F}_h^{D}} d(\Gamma)^{-1}\int_\Gamma \left|v_\Gamma^{(L)}\right|^2 \mathrm{d}S\\
&\le 2C_T^{-1}\sum_{\Gamma\in\mathscr{F}_h^{ID}} d(K_\Gamma^{(L)})^{-1}\int_\Gamma \left|v_\Gamma^{(L)}\right|^2 \mathrm{d}S + 2C_T^{-1}\sum_{\Gamma\in\mathscr{F}_h^{I}} d(K_\Gamma^{(R)})^{-1}\int_\Gamma \left|v_\Gamma^{(R)}\right|^2 \mathrm{d}S\\
&\le 2C_T^{-1}\sum_{K\in\mathscr{T}_h} d(K)^{-1}\int_{\partial K} |v|^2\,\mathrm{d}S,
\end{aligned}$$

which proves (7.29). Moreover, using (7.19) we immediately obtain (7.31).

(b) In the proof of (7.30), we proceed in a similar way, using (2.32), (7.10) and (2.110). Inequality (7.32) is a direct consequence of (7.30) and (7.19). □

Analogously to Lemma 2.32, we present its hp-variant.

Lemma 7.9 *Let $v \in H^1(\Omega, \mathscr{T}_h)$. Then*

$$\begin{aligned}
J_h^\sigma(v,v) &\le \frac{2C_WC_M}{C_T}\sum_{K\in\mathscr{T}_h}\left(\frac{p_K^2}{h_K^2}\|v\|^2_{L^2(K)} + \frac{p_K^2}{h_K}\|v\|_{L^2(K)}|v|_{H^1(K)}\right) \qquad (7.33)\\
&\le \frac{C_WC_M}{C_T}\sum_{K\in\mathscr{T}_h}\left(\frac{2p_K^2}{h_K^2}\|v\|^2_{L^2(K)} + \frac{p_K^3}{h_K^2}\|v\|^2_{L^2(K)} + p_K|v|^2_{H^1(K)}\right).
\end{aligned}$$

Proof If $v \in H^1(\Omega, \mathscr{T}_h)$, then the definition (7.17) of the form J_h^σ, (7.31) and (7.9) imply that

$$J_h^\sigma(v,v) = \sum_{\Gamma\in\mathscr{F}_h^{ID}} \int_\Gamma \sigma[v]^2\,\mathrm{d}S = \sum_{\Gamma\in\mathscr{F}_h^{ID}} \sigma_\Gamma \|[v]^2\|_{L^2(\Gamma)}$$
$$\le \frac{2C_W}{C_T}\sum_{K\in\mathscr{T}_h} d(K)^{-1}\|v\|^2_{L^2(\partial K)} = \frac{2C_W}{C_T}\sum_{K\in\mathscr{T}_h}\frac{p_K^2}{h_K}\|v\|^2_{L^2(\partial K)}.$$

Now, using the multiplicative trace inequality (7.24), we get

$$J_h^\sigma(v,v) \le \frac{2C_W C_M}{C_T}\sum_{K\in\mathscr{T}_h}\left(\frac{p_K^2}{h_K^2}\|v\|^2_{L^2(K)} + \frac{p_K^2}{h_K}\|v\|_{L^2(K)}|v|_{H^1(K)}\right),$$

which gives the first inequality in (7.33). Moreover, the application of the Young inequality yields the second one. □

Finally, we introduce the *hp*-variant of Lemma 2.34.

Lemma 7.10 *Under assumptions (7.15) and (7.10), for any* $v \in H^2(\Omega,\mathscr{T}_h)$ *the following estimate holds:*

$$\sum_{\Gamma\in\mathscr{F}_h^{ID}}\int_\Gamma \sigma^{-1}(\boldsymbol{n}\cdot\langle\nabla v\rangle)^2\,\mathrm{d}S \le \frac{C_G C_M}{C_W}\sum_{K\in\mathscr{T}_h}\frac{h_K}{p_K^2}\left(|v|_{H^1(K)}|v|_{H^2(K)} + h_K^{-1}|v|^2_{H^1(K)}\right)$$
$$\le \frac{C_G C_M}{2C_W}\sum_{K\in\mathscr{T}_h}\left(\frac{3}{p_K^2}|v|^2_{H^1(K)} + \frac{h_K^2}{p_K^2}|v|^2_{H^2(K)}\right). \quad (7.34)$$

Moreover, for $v_h \in S_{hp}$ *we have*

$$\sum_{\Gamma\in\mathscr{F}_h^{ID}}\int_\Gamma \sigma^{-1}(\boldsymbol{n}\cdot\langle\nabla v_h\rangle)^2\,\mathrm{d}S \le \frac{C_G C_M}{C_W}(C_I+1)|v_h|^2_{H^1(\Omega,\mathscr{T}_h)}. \quad (7.35)$$

Proof Using (7.32), the multiplicative trace inequality (7.24) and notation (7.9), we find that

$$\sum_{\Gamma\in\mathscr{F}_h^{ID}}\int_\Gamma \sigma^{-1}(\boldsymbol{n}\cdot\langle\nabla v\rangle)^2\,\mathrm{d}S \le \frac{C_G}{C_W}\sum_{K\in\mathscr{T}_h} d(K)\|\nabla v\|^2_{L^2(\partial K)}$$
$$\le \frac{C_G C_M}{C_W}\sum_{K\in\mathscr{T}_h}\frac{h_K}{p_K^2}\left(\|\nabla v\|_{L^2(K)}|\nabla v|_{H^1(K)} + h_K^{-1}\|\nabla v\|^2_{L^2(K)}\right),$$
$$= \frac{C_G C_M}{C_W}\sum_{K\in\mathscr{T}_h}\frac{h_K}{p_K^2}\left(|v|_{H^1(K)}|v|_{H^2(K)} + h_K^{-1}|v|^2_{H^1(K)}\right),$$

which is the first inequality in (7.34). The second one is obtained by the application of the Young inequality.

Further, for $v_h \in S_{hp}$, estimate (7.34), the inverse inequality (7.25) and the inequality $1/p_K^2 \le 1$ give

$$\begin{aligned}
&\sum_{\Gamma \in \mathscr{F}_h^{ID}} \int_\Gamma \sigma^{-1}(\boldsymbol{n} \cdot \langle \nabla v_h \rangle)^2 \,\mathrm{d}S \\
&\quad \le \frac{C_G C_M}{C_W} \sum_{K \in \mathscr{T}_h} \frac{h_K}{p_K^2} \left(\|\nabla v_h\|_{L^2(K)} |\nabla v_h|_{H^1(K)} + h_K^{-1} \|\nabla v_h\|_{L^2(K)}^2 \right), \\
&\quad \le \frac{C_G C_M}{C_W} \sum_{K \in \mathscr{T}_h} \frac{h_K}{p_K^2} \left(C_I p_K^2 h_K^{-1} \|\nabla v_h\|_{L^2(K)} \|\nabla v_h\|_{L^2(K)} + h_K^{-1} \|\nabla v_h\|_{L^2(K)}^2 \right), \\
&\quad \le \frac{C_G C_M}{C_W} (C_I + 1) \sum_{K \in \mathscr{T}_h} \|\nabla v_h\|_{L^2(K)}^2 = \frac{C_G C_M}{C_W} (C_I + 1) |v_h|_{H^1(\Omega, \mathscr{T}_h)}^2,
\end{aligned}$$

which implies (7.35). □

7.1.3.2 Continuity of the Bilinear Forms

Now, we prove the continuity of the bilinear form A_h defined by (7.20). In the space S_{hp} we again employ the DG-norm

$$|||u||| = \left(|u|_{H^1(\Omega, \mathscr{T}_h)}^2 + J_h^\sigma(u, u) \right)^{1/2}. \tag{7.36}$$

Comparing (7.36) with (2.103), both relations are formally identical. However, the norm in (7.36) is p-dependent, because σ depends on the polynomial degrees p_K, $K \in \mathscr{T}_h$.

Exercise 7.11 Prove that $|||\cdot|||$ is a norm in the spaces $H^s(\Omega, \mathscr{T}_h)$ and S_{hp}.

Furthermore, due to (2.122), we have

$$|A_h(u, v)| \le 2\|u\|_{1,\sigma} \|v\|_{1,\sigma} \quad \forall u, v \in H^2(\Omega, \mathscr{T}_h), \tag{7.37}$$

where

$$\begin{aligned}
\|v\|_{1,\sigma}^2 &= |||v|||^2 + \sum_{\Gamma \in \mathscr{F}_h^{ID}} \int_\Gamma \sigma^{-1}(\boldsymbol{n} \cdot \langle \nabla v \rangle)^2 \,\mathrm{d}S \\
&= |v|_{H^1(\Omega, \mathscr{T}_h)}^2 + J_h^\sigma(v, v) + \sum_{\Gamma \in \mathscr{F}_h^{ID}} \int_\Gamma \sigma^{-1}(\boldsymbol{n} \cdot \langle \nabla v \rangle)^2 \mathrm{d}S.
\end{aligned} \tag{7.38}$$

Now, we derive the hp-estimate of the $\|\cdot\|_{1,\sigma}$-norm, compare with Lemma 2.35.

Lemma 7.12 *Let (7.10) be valid and let σ be defined by (7.19). Then, there exist constants C_σ, $\tilde{C}_\sigma > 0$ such that*

$$J_h^\sigma(u,u)^{1/2} \le |||u||| \le \|u\|_{1,\sigma} \le C_\sigma\, R_a(u) \quad \forall\, u \in H^2(\Omega, \mathscr{T}_h),\ h \in (0,\bar{h}), \tag{7.39}$$

$$J_h^\sigma(v_h,v_h)^{1/2} \le |||v_h||| \le \|v_h\|_{1,\sigma} \le \tilde{C}_\sigma |||v_h||| \quad \forall\, v_h \in S_{hp},\ h \in (0,\bar{h}), \tag{7.40}$$

where

$$R_a(u) = \left(\sum_{K\in\mathscr{T}_h} \left(\frac{p_K^3}{h_K^2}\|u\|^2_{L^2(K)} + p_K|u|^2_{H^1(K)} + \frac{h_K^2}{p_K^2}|u|^2_{H^2(K)} \right) \right)^{1/2}, \quad u \in H^2(\Omega,\mathscr{T}_h). \tag{7.41}$$

Proof The first two inequalities in (7.39) as well as in (7.40) follow immediately from the definition of the DG-norm (7.36) and $\|\cdot\|_{1,\sigma}$-norm (7.38). Moreover, in view of (7.38), (7.4), (7.33) and (7.34), for $u \in H^2(\Omega, \mathscr{T}_h)$, we have

$$\begin{aligned}
\|u\|^2_{1,\sigma} &= |u|^2_{H^1(\Omega,\mathscr{T}_h)} + J_h^\sigma(u,u) + \sum_{\Gamma\in\mathscr{F}_h^{ID}} \int_\Gamma \sigma^{-1}(\boldsymbol{n}\cdot\langle\nabla u\rangle)^2 \mathrm{d}S \\
&\le \sum_{K\in\mathscr{T}_h} |u|^2_{H^1(K)} + \frac{C_W C_M}{C_T} \sum_{K\in\mathscr{T}_h} \left(\frac{2p_K^2}{h_K^2}\|u\|^2_{L^2(K)} + \frac{p_K^3}{h_K^2}\|u\|^2_{L^2(K)} + p_K|u|^2_{H^1(K)} \right) \\
&\quad + \frac{C_G C_M}{2C_W} \sum_{K\in\mathscr{T}_h} \left(\frac{3}{p_K^2}|u|^2_{H^1(K)} + \frac{h_K^2}{p_K^2}|u|^2_{H^2(K)} \right).
\end{aligned}$$

Now, using the inequalities $p_k \ge 1$ and $1/p_K \le 1$, we get

$$\begin{aligned}
\|u\|^2_{1,\sigma} \le \sum_{K\in\mathscr{T}_h} \Bigg(&\left(1 + \frac{3C_G C_M}{2C_W} + \frac{C_W C_M}{C_T}\right) p_K |u|^2_{H^1(K)} \\
&+ \frac{C_G C_M}{2C_W}\frac{h_K^2}{p_K^2}|u|^2_{H^2(K)} + \frac{3C_W C_M}{C_T}\frac{p_K^3}{h_k^2}\|u\|^2_{L^2(K)} \Bigg).
\end{aligned}$$

Hence, (7.39) holds with

$$C_\sigma = \left(\max\left(1 + \frac{3C_G C_M}{2C_W} + \frac{C_W C_M}{C_T}, \frac{C_G C_M}{2C_W}, \frac{3C_W C_M}{C_T} \right) \right)^{1/2}.$$

Further, if $v_h \in S_{hp}$, then (7.38) and (7.35) immediately imply (7.40) with $\tilde{C}_\sigma = (1 + C_G C_M (C_I + 1)/C_W)^{1/2}$. $\square$

Lemma 7.12 directly implies the continuity of the form A_h.

Corollary 7.13 *Let (7.10) be valid and let σ be defined by (7.19). Then there exist constants $C_B > 0$ and $\tilde{C}_B > 0$ such that the form A_h defined by (7.20) satisfies the estimates*

$$|A_h(u_h, v_h)| \le C_B |||u_h||| \, |||v_h||| \quad \forall u_h, v_h \in S_{hp}, \tag{7.42}$$

$$|A_h(u, v_h)| \le \tilde{C}_B \, R_a(u) \, |||v_h||| \quad \forall u \in H^2(\Omega, \mathscr{T}_h) \, \forall v_h \in S_{hp} \, \forall h(0, \bar{h}), \tag{7.43}$$

where R_a is defined by (7.41).

Proof Estimates (7.37), (7.39) and (7.40) give (7.42) with $C_B = 2\tilde{C}_\sigma^2$. Moreover, by (7.37) and (7.39),

$$|A_h(u, v_h)| \le 2\|u\|_{1,\sigma} \|v_h\|_{1,\sigma} \le 2C_\sigma \tilde{C}_\sigma R_a(u) |||v_h|||,$$

which is (7.43) with $\tilde{C}_B = 2C_\sigma \tilde{C}_\sigma$. □

7.1.3.3 Coercivity of the Bilinear Forms

In order to derive error estimates of the approximate solution (7.21), we need the coercivity of the form A_h. To this end, we present here the generalization of the results from Sect. 2.6.3.

Lemma 7.14 (NIPG coercivity) *For any $C_W > 0$ the bilinear form A_h defined by (7.20) with $\Theta = -1$ in (7.16) satisfies the coercivity condition*

$$A_h(v, v) \ge |||v|||^2 \quad \forall v \in H^2(\Omega, \mathscr{T}_h). \tag{7.44}$$

Proof If $\Theta = -1$, then from (7.16) and (7.20) it immediately follows that

$$A_h(v, v) = a_h(v, v) + J_h^\sigma(v, v) = |v|^2_{H^1(\Omega, \mathscr{T}_h)} + J_h^\sigma(v, v) = |||v|||^2, \tag{7.45}$$

which we wanted to prove. □

The proof of coercivity of the symmetric bilinear form A_h defined by (7.16) with $\Theta = 1$ is more complicated.

Lemma 7.15 (SIPG coercivity) *Let assumptions (7.15) and (7.10) be satisfied, let*

$$C_W \ge 4C_G C_M (1 + C_I), \tag{7.46}$$

where C_M, C_I and C_G are the constants from (7.24), (7.25) and (7.10), respectively, and let the penalty parameter σ be given by (7.19) for all $\Gamma \in \mathscr{F}_h^{ID}$. Then the bilinear form A_h defined by (7.20) and (7.16) with $\Theta = 1$ satisfies the coercivity condition

$$A_h(v_h, v_h) \ge \frac{1}{2} |||v_h|||^2 \quad \forall v_h \in S_{hp}, \ \forall h \in (0, \bar{h}).$$

Proof Let $\delta > 0$. Then (7.17), (7.19), (7.16) with $\Theta = 1$ and the Cauchy and Young inequalities imply that

$$
\begin{aligned}
&a_h(v_h, v_h) \qquad (7.47)\\
&= |v_h|^2_{H^1(\Omega,\mathcal{T}_h)} - 2 \sum_{\Gamma\in\mathcal{F}_h^{ID}} \int_\Gamma \boldsymbol{n}\cdot\langle\nabla v_h\rangle[v_h]\mathrm{d}S\\
&\geq |v_h|^2_{H^1(\Omega,\mathcal{T}_h)} - 2\left\{\frac{1}{\delta}\sum_{\Gamma\in\mathcal{F}_h^{ID}} \int_\Gamma d(\Gamma)(\boldsymbol{n}\cdot\langle\nabla v_h\rangle)^2\mathrm{d}S\right\}^{\frac{1}{2}} \left\{\delta \sum_{\Gamma\in\mathcal{F}_h^{ID}} \int_\Gamma \frac{[v_h]^2}{d(\Gamma)}\mathrm{d}S\right\}^{\frac{1}{2}}\\
&\geq |v_h|^2_{H^1(\Omega,\mathcal{T}_h)} - \omega - \frac{\delta}{C_W} J_h^\sigma(v_h, v_h),
\end{aligned}
$$

where

$$
\omega = \frac{1}{\delta}\sum_{\Gamma\in\mathcal{F}_h^{ID}} \int_\Gamma d(\Gamma)|\langle\nabla v_h\rangle|^2\,\mathrm{d}S. \qquad (7.48)
$$

Further, from (7.9), assumption (7.10), inequality (7.30), the multiplicative trace inequality (7.24), the inverse inequality (7.25) and the inequality $p_K^{-2} \leq 1$, we get

$$
\begin{aligned}
\omega &\leq \frac{C_G}{\delta}\sum_{K\in\mathcal{T}_h} \frac{h_K}{p_K^2}\|\nabla v_h\|^2_{L^2(\partial K)} \qquad (7.49)\\
&\leq \frac{C_G C_M}{\delta}\sum_{K\in\mathcal{T}_h} \frac{h_K}{p_K^2}\left(|v_h|_{H^1(K)}|\nabla v_h|_{H^1(K)} + h_K^{-1}|v_h|^2_{H^1(K)}\right)\\
&\leq \frac{C_G C_M}{\delta}\sum_{K\in\mathcal{T}_h} \frac{h_K}{p_K^2}\left(C_I p_K^2 h_K^{-1}|v_h|^2_{H^1(K)} + h_K^{-1}|v_h|^2_{H^1(K)}\right)\\
&\leq \frac{C_G C_M(1+C_I)}{\delta}|v_h|^2_{H^1(\Omega,\mathcal{T}_h)}.
\end{aligned}
$$

Now let us choose

$$
\delta = 2C_G C_M(1+C_I). \qquad (7.50)
$$

Then it follows from (7.46) and (7.47)–(7.50) that

$$
\begin{aligned}
a_h(v_h, v_h) &\geq \frac{1}{2}\left(|v_h|^2_{H^1(\Omega,\mathcal{T}_h)} - \frac{4C_G C_M(1+C_I)}{C_W} J_h^\sigma(v_h, v_h)\right) \qquad (7.51)\\
&\geq \frac{1}{2}\left(|v_h|^2_{H^1(\Omega,\mathcal{T}_h)} - J_h^\sigma(v_h, v_h)\right).
\end{aligned}
$$

Finally, from the definition (7.20) of the form A_h and from (7.51) we have

$$A_h(v_h, v_h) = a_h(v_h, v_h) + J_h^\sigma(v_h, v_h) \tag{7.52}$$
$$\geq \frac{1}{2}\left(|v_h|^2_{H^1(\Omega,\mathscr{T}_h)} + J_h^\sigma(v_h, v_h)\right) = \frac{1}{2}|||v_h|||^2,$$

□

Lemma 7.16 (IIPG coercivity) *Let assumptions (7.15) and (7.10) be satisfied, let*

$$C_W \geq C_G C_M(1 + C_I), \tag{7.53}$$

where C_M, C_I and C_G are constants from (7.24), (7.25) and (7.10), respectively, and let the penalty parameter σ be given by (7.19) for all $\Gamma \in \mathscr{F}_h^{ID}$. Then the bilinear form A_h defined by (7.20) and (7.16) with $\Theta = 0$ satisfies the coercivity condition

$$A_h(v_h, v_h) \geq \frac{1}{2}|||v_h|||^2 \quad \forall v_h \in S_{hp}.$$

Proof The proof is almost identical with the proof of the previous lemma. □

Corollary 7.17 *We can summarize the above results in the following way. We have*

$$A_h(v_h, v_h) \geq C_C|||v_h|||^2 \quad \forall v_h \in S_{hp}, \tag{7.54}$$

with

$$\begin{array}{lll} C_C = 1 & \textit{for } \Theta = -1, & \textit{if } C_W > 0, \\ C_C = 1/2 & \textit{for } \Theta = 1, & \textit{if } C_W \geq 4C_G C_M(1 + C_I), \\ C_C = 1/2 & \textit{for } \Theta = 0, & \textit{if } C_W \geq C_G C_M(1 + C_I). \end{array}$$

Corollary 7.18 *By virtue of Corollary 1.7, the coercivity of the form A_h implies the existence and uniqueness of the solution of the discrete problem.*

7.1.3.4 Error Estimates in the DG-Norm

In this section we will be concerned with the derivation of the error estimates of the hp-discontinuous Galerkin method (7.21). Let u and u_h denote the exact solution of problem (7.1) and the approximate solution obtained by method (7.21), respectively. The error $e_h = u_h - u$ can be written in the form

$$e_h = \xi + \eta, \quad \text{with } \xi = u_h - \Pi_{hp}u \in S_{hp}, \ \eta = \Pi_{hp}u - u, \tag{7.55}$$

where Π_{hp} is the S_{hp}-interpolation defined by (7.27). The estimation of the error e_h will be carried out in several steps.

We suppose that the system of triangulations $\{\mathscr{T}_h\}_{h\in(0,\bar{h})}$ satisfies the shape-regularity assumption (7.15) and that the relations (7.10) between $d(\Gamma)$ and $d(K)$ are valid.

First, we prove the *abstract error estimate*, representing a bound of the error in terms of the S_{hp}-interpolation error η, cf. Theorem 2.43.

Theorem 7.19 *Let (7.10) be valid, let σ be defined by (7.19) and let the exact solution of problem (7.1) satisfy the condition $u \in H^2(\Omega)$. Then there exists a constant $C_{\mathrm{AE}} > 0$ such that*

$$|||e_h||| \le C_{\mathrm{AE}}\, R_a(\eta) = C_{\mathrm{AE}}\, R_a(\Pi_{hp}u - u) \quad \forall\, h \in (0,\bar{h}), \tag{7.56}$$

where $R_a(u)$ is given by (7.41).

Proof The proof is completely identical with the proof of Theorem 2.43. We obtain again $C_{\mathrm{AE}} = C_\sigma + \tilde{C}_B/C_C$, where C_σ and $\tilde{C}_B$ and C_C are constants from (7.39) and (7.43) and (7.54). □

The abstract error estimate is the basis for the estimation of the error e_h in terms of the mesh size h.

Theorem 7.20 (DG-norm error estimate) *Let $\{\mathscr{T}_h\}_{h\in(0,\bar{h})}$ be a system of triangulations of the domain Ω satisfying the shape-regularity assumption (7.15). Let* **s** *and* **p** *be the vectors (7.2) such that $s_K \ge 2$, $p_K \ge 1$ and $\mu_K = \min(p_K+1, s_K)$ for each $K \in \mathscr{T}_h$. Let the condition of equivalence (7.10) between $d(\Gamma)$ and $d(K)$ be valid (cf. Lemma 7.1). Let u be the solution of problem (7.1) such that $u \in H^2(\Omega)\cap H^{\mathbf{s}}(\Omega,\mathscr{T}_h)$ for any $h \in (0,\bar{h})$. Moreover, let the penalty constant C_W satisfy the conditions from Corollary 7.17. Let $u_h \in S_{h\mathbf{p}}$ be the approximate solution obtained by means of method (7.21). Then the error $e_h = u_h - u$ satisfies the estimate*

$$|||e_h||| \le \tilde{C}\left(\sum_{K\in\mathscr{T}_h} \frac{h_K^{2(\mu_K-1)}}{p_K^{2s_K-3}}\|u\|^2_{H^{s_K}(K)}\right)^{\frac{1}{2}}, \quad h\in(0,\bar{h}), \tag{7.57}$$

where $\tilde{C}$ is a constant independent of h and **p**.

Proof It is enough to use the abstract error estimate (7.56), where the expressions $|\eta|_{H^1(K)}$, $|\eta|_{H^2(K)}$ and $\|\eta\|_{L^2(K)}$, $K \in \mathscr{T}_h$, are estimated on the basis of the approximation properties (7.26), rewritten for $\eta|_K = (\Pi_{hp}u - u)|_K = \pi_{K,p}(u|_K) - u|_K$ and $K \in \mathscr{T}_h$:

$$\begin{aligned} \|\eta\|_{L^2(K)} &\le C_A \frac{h_K^{\mu_K}}{p_K^{s_K}}\|u\|_{H^{\mu}(K)}, \\ |\eta|_{H^1(K)} &\le C_A \frac{h_K^{\mu_K-1}}{p_K^{s_K-1}}\|u\|_{H^{\mu}(K)}, \end{aligned} \tag{7.58}$$

$$|\eta|_{H^2(K)} \leq C_A \frac{h_K^{\mu_K - 2}}{p_K^{s_K - 2}} \|u\|_{H^{\mu_K}(K)}.$$

The above, the definition (7.41) of the expression R_a and the inequalities $1/p_K^{2s-2} \leq 1/p_K^{2s-3}$, $p_K \geq 1$ imply

$$\begin{aligned}
R_a(\eta)^2 &= \sum_{K\in\mathscr{T}_h} \left(\frac{p_K^3}{h_K^2} \|\eta\|^2_{L^2(K)} + p_K |\eta|^2_{H^1(K)} + \frac{h_K^2}{p_K^2} |\eta|^2_{H^2(K)} \right) \\
&\leq C_A^2 \sum_{K\in\mathscr{T}_h} \left(\frac{p_K^3}{h_K^2} \frac{h_K^{2\mu_K}}{p_K^{2s_K}} + p_K \frac{h_K^{2(\mu_K-1)}}{p_K^{2s_K-2}} + \frac{h_K^2}{p_K^2} \frac{h_K^{2\mu_K-4}}{p_K^{2s_K-4}} \right) \|u\|^2_{H^{\mu_K}(K)} \\
&\leq C_A^2 \sum_{K\in\mathscr{T}_h} \left(\frac{h_K^{2(\mu_K-1)}}{p_K^{2s_K-3}} + \frac{h_K^{2(\mu_K-1)}}{p_K^{2s_K-3}} + \frac{h_K^{2(\mu_K-1)}}{p_K^{2s_K-3}} \right) \|u\|^2_{H^{\mu_K}(K)} \\
&= 3C_A^2 \sum_{K\in\mathscr{T}_h} \frac{h_K^{2(\mu_K-1)}}{p_K^{2s_K-3}} \|u\|^2_{H^{\mu_K}(K)}.
\end{aligned}$$

Together with (7.56) this gives (7.57) with the constant $\tilde{C} = \sqrt{3} C_{\mathrm{AE}}\, C_A$. □

Comparing error estimate (7.57) with the approximation property (7.28) with $q = 1$, we see that (7.57) is suboptimal with respect to the polynomial degrees p_K, $K \in \mathscr{T}_h$. This is caused by the presence of the interior penalty form J_h^σ, see the last two terms in the second inequality in (7.33), namely the terms

$$\frac{p_K^3}{h_K^2} \|v\|^2_{L^2(K)} + p_K |v|^2_{H^1(K)} = p_K \left(\frac{p_K^2}{h_K^2} \|v\|^2_{L^2(K)} + |v|^2_{H^1(K)} \right), \quad K \in \mathscr{T}_h.$$

The error estimates optimal with respect to p were derived in [150] using an augmented Sobolev space.

As for the analysis of further subjects concerned with the hp-DGM, we refer to several works, namely [181, 185] dealing with the hp-DGM for quasilinear elliptic problems, [148, 149] dealing with the hp-DGM on anisotropic meshes, [285] proving the exponential rate of the convergence of the hp-DGM, [44, 183] dealing with the hp-DGM for convection-diffusion problems and [252, 270] analyzing the hp-DGM for the Stokes problem.

7.1.4 Computational Performance of the hp-DGM

In the previous sections we analyzed the hp-DGM, where the mesh $\mathscr{T}_h$ and the approximation polynomial degrees p_K, $K \in \mathscr{T}_h$, were given in advance. In practice, the hp-DGM can be applied in the combination with an adaptive algorithm, where the size h_K of the elements $K \in \mathscr{T}_h$ as well as the polynomial degrees p_K on elements $K \in \mathscr{T}_h$ are adaptively determined. The aim of this section is to demonstrate the ability of the hp-DGM to deal with refined grids and with different polynomial degrees on different $K \in \mathscr{T}_h$. We present one numerical example showing the efficiency and a possible potential of the hp-DGM.

7.1.4.1 Mesh Adaptation—An Overview

Numerical examples presented in Sect. 2.9.2 show that if the exact solution of the given problem is not sufficiently regular, then the experimental order of convergence of the DGM is low for any polynomial approximation degree. Therefore, a high number of *degrees of freedom* (DOF) (=dim S_{hp}) has to be used in order to achieve a given accuracy. A significant reduction of the number of DOF can be achieved by a local mesh refinement of the given grid $\mathscr{T}_h$, in which we look for elements $K \in \mathscr{T}_h$, for which the computational error is too large. Then these marked elements are refined. In practice, for each element $K \in \mathscr{T}_h$ we define an *error estimator* η_K such that

$$\|u - u_h\|_K \approx \eta_K, \tag{7.59}$$

where $\|\cdot\|_K$ denotes a suitable norm of functions defined on $K \in \mathscr{T}_h$. The elements, where η_K is larger than a prescribed tolerance, are split into several daughter elements. E.g., for $d = 2$, by connecting the mid points of edges of the triangle marked for refinement, new four daughter triangles arise in place of the original one. This refinement strategy leads to hanging nodes, see Sect. 2.3.1. Figure 7.2 shows a sequence of adaptively refined triangular grids.

There exist a number of works dealing with strategies for the error estimation and the corresponding mesh adaptive techniques. Since a posteriori error analysis and mesh adaption are out of the scope of this book, we refer only to [114], where an introduction to adaptive methods for partial differential equations can be found. Moreover, an overview of standard approaches was presented in [274, 276, 280].

Here we use the *residual error estimator* η_K, $K \in \mathscr{T}_h$, developed in [90], which is based on the approximation of the computational error measured in the dual norm. We suppose that similar results can be obtained by any other reasonable error estimator. However, a single error estimator η_K cannot simultaneously decide whether it is better to accomplish h or p refinement. Several strategies for making this decision have been proposed. See, e.g., [184] or [113] for a survey.

In the following numerical examples, we employ the approach from [90], where the regularity indicator is based on measuring the interelement jumps of the DG solution.

7.1.4.2 Numerical Example

We illustrate the efficiency of the hp-discontinuous Galerkin method by the following example. Let $\Omega = (0, 1) \times (0, 1)$, $\partial\Omega_D := \partial\Omega$. We consider the Poisson problem (7.1), where the right-hand side f and the Dirichlet boundary condition u_D are chosen so that the exact solution has the form

$$u(x_1, x_2) = 2(x_1^2 + x_2^2)^{-3/4} x_1 x_2 (1 - x_1)(1 - x_2), \tag{7.60}$$

cf. Sect. 2.9.2. The function u has a singularity at the origin and, hence, $u \in H^1(\Omega)$ but $u \notin H^2(\Omega)$. Numerical examples presented in Sect. 2.9.2 showed that the experimental order of convergence of DGM in the $H^1(\Omega, \mathscr{T}_h)$-seminorm is approximately $O(h^{1/2})$ for any tested polynomial approximation degree.

In order to study the computational properties of the hp-DGM, we carried out three types of calculations:

- fix-DGM: P_p, $p = 1, 3, 5$, *approximations on uniformly refined grids*, i.e., the computation with fixed polynomial approximation degree ($p_K = p$ for all $K \in \mathscr{T}_h$) on uniform triangular grids with $h_\ell = 1/2^{2+\ell}$, $\ell = 0, 1, \ldots$. Figure 7.1 shows the uniformly refined grids for $\ell = 0, 2, 4$.
- h-DGM: *h-adaptive DGM for P_p, $p = 1, 3, 5$, polynomial approximations*, i.e., the computation with fixed polynomial approximation degree ($p_K = p$ for all $K \in \mathscr{T}_h$) on adaptively (locally) refined grids. Figure 7.2 shows the example of the sequence of meshes generated by the h-refinement algorithm for $p = 3$ together with details at the singularity corner.
- hp-DGM: *hp-adaptive DGM*, i.e., the computation with adaptively chosen polynomial approximation degree p_K, $K \in \mathscr{T}_h$, on adaptively (locally) refined grids using the algorithm from [90]. Figure 7.3 shows the hp-grids generated by this algorithm for selected levels of adaptation. Each $K \in \mathscr{T}_h$ is marked by the colour corresponding to the used polynomial approximation degree.

Our aim is to identify the *experimental order of convergence* (EOC), similarly as in Sect. 2.9. Since we employ locally adaptive grids and possible different polynomial approximation degrees on $K \in \mathscr{T}_h$, it does not make sense to use formula (2.198)

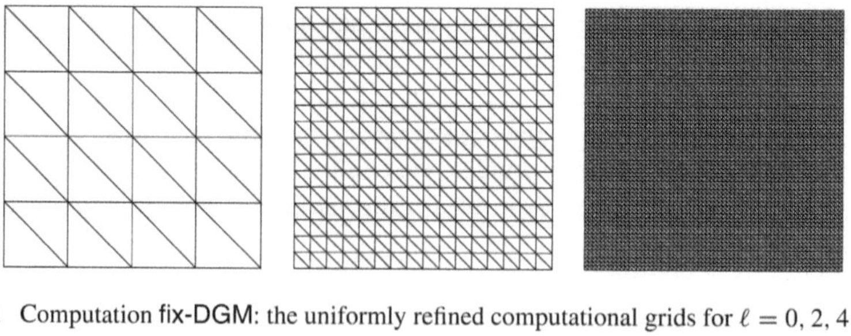

Fig. 7.1 Computation fix-DGM: the uniformly refined computational grids for $\ell = 0, 2, 4$

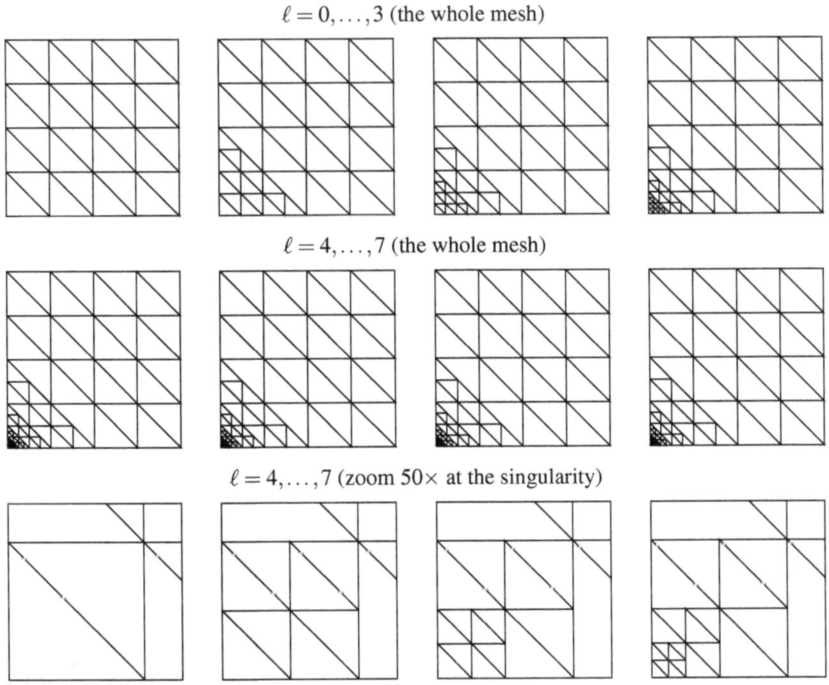

Fig. 7.2 Computation *h*-DGM: example of the sequence of the meshes $\ell = 0, \dots, 7$, generated by the *h*-refinement algorithm for $p = 3$; the last row shows the details at the singularity corner

and to define the EOC by (2.199). Therefore, we expect that the computational error $e_h = u_h - u$ behaves according to the formula

$$\|e_h\| \approx C N_h^{-\frac{\mathrm{EOC}}{d}}, \tag{7.61}$$

where $\|e_h\|$ is the computational error in the (semi-)norm of interest, $d = 2$ is the space dimension, $C > 0$ is a constant, $\mathrm{EOC} \in \mathbb{R}$ is the experimental order of convergence and N_h is the number of degrees of freedom given by (cf., e.g., [37, Chap. 3] or [52])

$$N_h = \dim S_{hp} = \sum_{K \in \mathscr{T}_h} \frac{1}{d!} \prod_{j=1}^{d} (p_K + j). \tag{7.62}$$

Obviously, if the mesh $\mathscr{T}_h$ is quasi-uniform (cf. Remark 2.3) and $p_K = p$ for all $K \in \mathscr{T}_h$, then the experimental orders of convergence defined by (7.61) and by (2.198) are identical.

Since the exact solution is known and, therefore, $\|e_h\|$ can be exactly evaluated, it is possible to determine the EOC in the following way. Let $\|e_{h_1}\|$ and $\|e_{h_2}\|$ be the

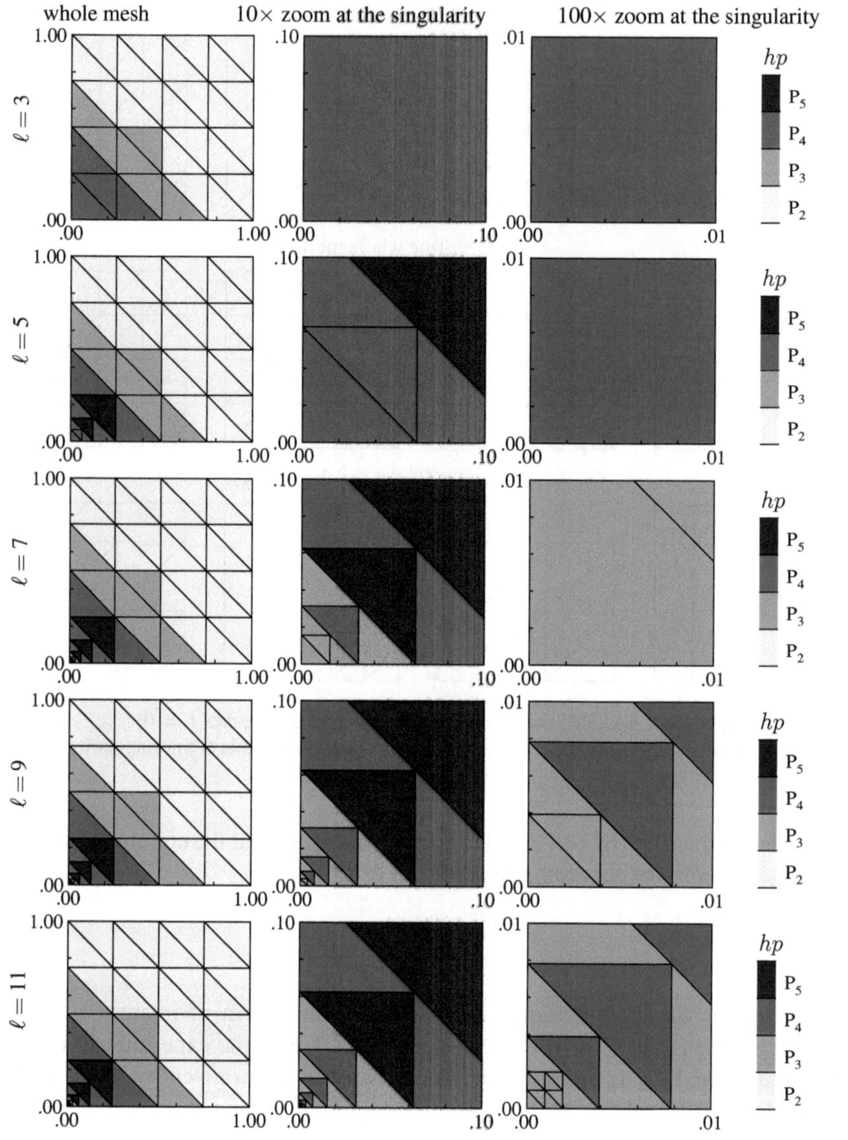

Fig. 7.3 Computation *h*p-DGM: the *hp*-meshes for the levels of adaptation $\ell = 3, 5, 7, 9, 11$; each $K \in \mathscr{T}_h$ is marked by the colour corresponding to the used polynomial approximation degree; the whole domain (*left*), zooms 10× and 100× at the singularity corner (*center* and *right*), respectively

computational errors of numerical solutions obtained on two different meshes $\mathscr{T}_{h_1}$ and $\mathscr{T}_{h_2}$ having the numbers of degrees of freedom N_{h_1} and N_{h_2}, respectively. Then eliminating the constant C from (7.61), we come to the definition of the EOC in the form

$$\mathrm{EOC} = -\frac{\log(\|e_{h_1}\|/\|e_{h_2}\|)}{\log((N_{h_1}/N_{h_2})^{1/d})}. \tag{7.63}$$

Table 7.1 shows the results of all types of computations (fix-DGM, h-DGM, hp-DGM), namely, the computational errors in the $L^\infty(\Omega)$-norm, the $L^2(\Omega)$-norm and the $H^1(\Omega, \mathscr{T}_h)$-seminorm and the corresponding EOC together with the computational time in seconds. The results with the error in the $H^1(\Omega, \mathscr{T}_h)$-seminorm are visualized in Fig. 7.4. We observe that the fix-DGM computations give a low experimental order of convergence in agreement with results in Tables 2.5 and 2.6. Moreover, the h-mesh refinements h-DGM achieve the same error level with smaller number of DOF. Namely, for P_3 and P_5 approximation the decrease of the number of DOF is essential. Finally, the hp-adaptive strategy hp-DGM leads to the lower number of DOF (and a shorter computational time) in comparison to h-DGM.

We observe that in some cases EOC is negative for the hp-DGM. The relation (7.63) gives EOC < 0 in two situations:

- The adaptive algorithm increases the number of degrees of freedom N_h but the computational error e_h increases too. This is the usual property of hp-adaptive methods, when at the beginning of the adaptation algorithm we use high polynomial degrees on coarse grids. The polynomial approximation oscillates and thus e_h is large.
- The adaptive algorithm reduces the number of degrees of freedom N_h together with a decrease of the computational error e_h (see level 7 of hp-DGM in Table 7.1). This is in fact a positive property of the used algorithm.

Furthermore, from Table 7.1, we find out that for the hp-DGM computations, the error in the $L^2(\Omega)$-norm is almost constant for the levels $\ell = 8, 9, 10$ and 11, whereas the errors in the $L^\infty(\Omega)$-norm and in the $H^1(\Omega, \mathscr{T}_h)$-seminorm are decreasing. This is caused be the fact that the piecewise constant function $F^0 : \Omega \to \mathbb{R}$ given by

$$F^0|_K = \|u - u_h\|_{L^2(K)}, \quad K \in \mathscr{T}_h$$

attains the maximal values for K far from the singularity (if the mesh is already sufficiently refined), whereas the piecewise constant function $F^1 : \Omega \to \mathbb{R}$ given by

$$F^1|_K = |u - u_h|_{H^1(K)}, \quad K \in \mathscr{T}_h$$

attains the maximal values for K near the singularity even for sufficiently refined grids. Figure 7.3 shows that for $\ell \geq 5$ only elements near the singularity are adapted, and hence the error in the $L^2(\Omega)$-norm cannot be further decreased.

Table 7.1 Computational errors in the $L^\infty(\Omega)$-norm, the $L^2(\Omega)$-norm and the $H^1(\Omega, \mathscr{T}_h)$-seminorm, the corresponding EOC and the CPU time for all types of computations

Level	p	$\#\mathscr{T}_h$	DOF	$\Vert e_h\Vert_{L^\infty(\Omega)}$	EOC	$\Vert e_h\Vert_{L^2(\Omega)}$	EOC	$\Vert e_h\Vert_{H^1(\Omega,\mathscr{T}_h)}$	EOC	CPU(s)
fix-DGM										
0	1	32	96	2.47E−01	–	4.22E−02	–	7.01E−01	–	0.3
1	1	128	384	1.99E−01	0.3	1.83E−02	1.2	5.61E−01	0.3	0.5
2	1	512	1536	1.50E−01	0.4	7.28E−03	1.3	4.26E−01	0.4	1.4
3	1	2048	6144	1.09E−01	0.5	2.77E−03	1.4	3.14E−01	0.4	6.3
4	1	8192	24,576	7.84E−02	0.5	1.02E−03	1.4	2.27E−01	0.5	38.9
0	3	32	320	1.51E−01	–	5.79E−03	–	4.63E−01	–	0.4
1	3	128	1280	1.07E−01	0.5	2.13E−03	1.4	3.34E−01	0.5	1.0
2	3	512	5120	7.55E−02	0.5	7.71E−04	1.5	2.39E−01	0.5	3.9
3	3	2048	20,480	5.34E−02	0.5	2.76E−04	1.5	1.70E−01	0.5	16.8
4	3	8192	81,920	3.78E−02	0.5	9.83E−05	1.5	1.20E−01	0.5	82.2
0	5	32	672	2.29E−01	–	5.09E−03	–	3.85E−01	–	0.6
1	5	128	2688	1.62E−01	0.5	1.81E−03	1.5	2.75E−01	0.5	2.2
2	5	512	10,752	1.15E−01	0.5	6.42E−04	1.5	1.95E−01	0.5	9.2
3	5	2048	43,008	8.12E−02	0.5	2.28E−04	1.5	1.38E−01	0.5	41.2
4	5	8192	1,72,032	5.74E−02	0.5	8.05E−05	1.5	9.80E−02	0.5	235.3
h-DGM										
0	1	32	96	2.47E−01	–	4.22E−02	–	7.01E−01	–	0.3
1	1	128	384	1.99E−01	0.3	1.83E−02	1.2	5.61E−01	0.3	0.5
2	1	410	1230	1.50E−01	0.5	7.34E−03	1.6	4.26E−01	0.5	1.3
3	1	959	2877	1.09E−01	0.7	2.89E−03	2.2	3.15E−01	0.7	3.2
4	1	1952	5856	7.84E−02	0.9	1.15E−03	2.6	2.30E−01	0.9	8.2

(continued)

Table 7.1 (continued)

Level	p	$\#\mathscr{T}_h$	DOF	$\|e_h\|_{L^\infty(\Omega)}$	EOC	$\|e_h\|_{L^2(\Omega)}$	EOC	$\|e_h\|_{H^1(\Omega,\mathscr{T}_h)}$	EOC	CPU(s)
5	1	3491	10,473	5.59E−02	1.2	5.28E−04	2.7	1.67E−01	1.1	21.1
6	1	5567	16,701	3.96E−02	1.5	3.11E−04	2.3	1.21E−01	1.4	47.8
7	1	7922	23,766	2.81E−02	2.0	2.40E−04	1.5	8.95E−02	1.7	86.0
8	1	11,387	34,161	1.99E−02	1.9	1.77E−04	1.7	6.73E−02	1.6	168.6
0	3	32	320	1.51E−01	–	5.79E−03	–	4.63E−01	–	0.4
1	3	44	440	1.07E−01	2.2	2.14E−03	6.3	3.34E−01	2.0	0.7
2	3	56	560	7.55E−02	2.9	7.99E−04	8.2	2.39E−01	2.8	1.0
3	3	68	680	5.34E−02	3.6	3.42E−04	8.7	1.70E−01	3.5	1.3
4	3	80	800	3.78E−02	4.3	2.23E−04	5.3	1.21E−01	4.2	1.8
5	3	86	860	2.67E−02	9.5	2.03E−04	2.6	8.67E−02	9.3	2.2
6	3	92	920	1.89E−02	10.3	2.00E−04	0.4	6.25E−02	9.7	2.7
7	3	98	980	1.34E−02	11.0	2.00E−04	0.1	4.57E−02	9.9	3.1
0	5	32	672	2.29E−01	–	5.09E−03	–	3.85E−01	–	0.6
1	5	38	798	1.62E−01	4.0	1.81E−03	12.0	2.75E−01	3.9	1.0
2	5	44	924	1.15E−01	4.7	6.43E−04	14.1	1.95E−01	4.7	1.5
3	5	50	1050	8.12E−02	5.4	2.29E−04	16.1	1.38E−01	5.4	2.1
4	5	56	1176	5.74E−02	6.1	8.53E−05	17.5	9.80E−02	6.1	2.8
5	5	62	1302	4.06E−02	6.8	3.99E−05	15.0	6.94E−02	6.8	3.6

(continued)

Table 7.1 (continued)

Level	p	$\#\mathscr{T}_h$	DOF	$\|e_h\|_{L^\infty(\Omega)}$	EOC	$\|e_h\|_{L^2(\Omega)}$	EOC	$\|e_h\|_{H^1(\Omega,\mathscr{T}_h)}$	EOC	CPU(s)
hp-DGM										
0	–	32	96	2.47E−01	–	4.22E−02	–	7.01E−01	–	0.3
1	–	32	192	1.14E−01	2.2	8.68E−03	4.6	4.14E−01	1.5	0.4
2	–	32	232	1.51E−01	−3.0	5.86E−03	4.1	4.63E−01	−1.2	0.5
3	–	32	252	2.01E−01	−7.0	5.98E−03	−0.5	3.77E−01	5.0	0.6
4	–	35	303	1.43E−01	3.7	2.25E−03	10.6	2.71E−01	3.6	0.8
5	–	38	354	1.01E−01	4.4	1.03E−03	10.0	1.95E−01	4.3	1.0
6	–	44	424	5.34E−02	7.1	7.40E−04	3.7	1.72E−01	1.3	1.2
7	–	44	420	3.78E−02	−81.5	6.93E−04	−15.4	1.25E−01	−76.1	1.4
8	–	47	455	2.67E−02	8.7	6.86E−04	0.2	9.15E−02	7.7	1.7
9	–	50	490	1.89E−02	9.4	6.86E−04	0.0	6.91E−02	7.6	1.9
10	–	53	525	1.34E−02	10.1	6.85E−04	0.0	5.46E−02	6.9	2.2
11	–	59	585	9.45E−03	6.4	6.85E−04	0.0	4.55E−02	3.3	2.4

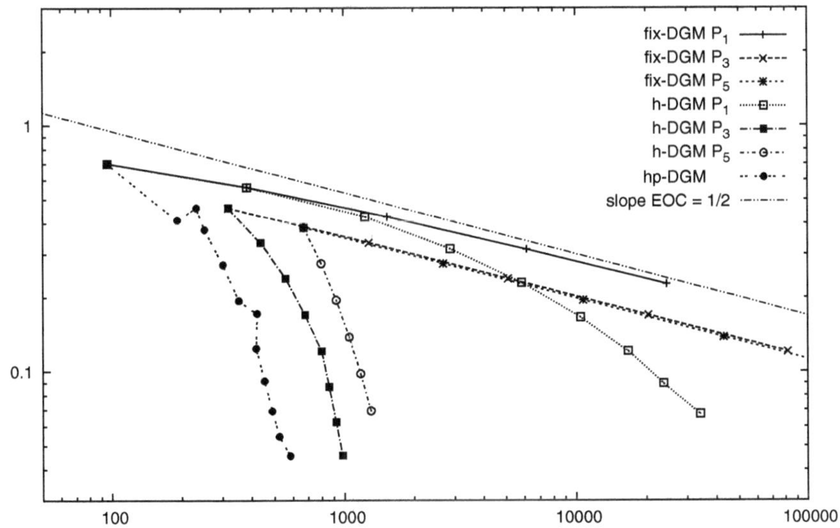

Fig. 7.4 Convergence of errors in the $H^1(\Omega, \mathscr{T}_h)$-seminorm with respect to the number of DOF for fix-DGM, h-DGM, hp-DGM computations. Moreover the slope corresponding to EOC $= 1/2$ is plotted

The presented numerical experiments show that the hp-DGM can treat locally refined grids with hanging nodes and different approximation polynomial degrees generated by an hp-adaptive technique. This approach allows us to achieve the given error tolerance with the aid of a low number of DOF.

7.2 DGM on General Elements

The versatility of the DGM can also be exhibited by the possible use of general nonsimplicial partitions of the computational domain. Hence, for $d = 2$, not only triangles and quadrilaterals, but also even nonconvex polygons can be treated. Let us mention, e.g., the *dual elements* (dual finite volumes) considered in [126] or the Voronoi cells from [108, 223] widely used in the *finite volume methods*. The use of such nonstandard elements can reduce the number of degrees of freedom of the given problem. On the other hand, the numerical implementation is more complicated, and therefore this approach has not yet been used in practical applications.

In this section we present the analysis of the DGM applied to the Poisson problem (7.1) on star-shaped elements. In this way we demonstrate that the DGM can really be considered as a generalization of both the finite element as well as finite volume methods. Similarly as in Sect. 7.1, we derive the basic tools for the theoretical analysis, namely the multiplicative trace inequality, the inverse inequality and the approximation properties of a suitable interpolation operator. We see that then the error analysis will be straightforward. Of course, the DGM applied on general elements can be used for the numerical approximations of more complex problems. See, for example [99], where the DGM on star-shaped elements is analyzed in the case of the nonstatinary convection-diffusion problem (4.1).

7.2.1 Assumptions on the Domain Partition

Let us consider a system $\{\mathscr{T}_h\}_{h\in(0,\bar{h})}$, $\bar{h} > 0$, of partitions of the domain Ω, formed by general polygonal (for $d = 2$) or polyhedral (for $d = 3$) elements K. Naturally, we require that $\cup_{K\in\mathscr{T}_h} K = \overline{\Omega}$ and that elements from $\mathscr{T}_h$ have disjoint interiors.

We use the same notation as in Chap. 2 and introduce the concept of interfaces between neighbouring elements and the symbols $\mathscr{F}_h$, $\mathscr{F}_h^I$, $\mathscr{F}_h^D$ for the sets of all faces of all elements $K \in \mathscr{T}_h$, all inner faces and all Dirichlet boundary faces. Now the faces between neighbouring elements may be formed by several straight lines (if $d = 2$) or parts of planes (if $d = 3$). Each face Γ is associated with a normal $\boldsymbol{n}_\Gamma$, which may be piecewise constant in general.

We assume that the system $\{\mathscr{T}_h\}_{h\in(0,\bar{h})}$, $\bar{h} > 0$, of partitions of Ω has the following properties:

Assumptions 7.21 (*Star-shaped elements*) Each element $K \in \mathscr{T}_h$, $h \in (0, \bar{h})$, is a *star-shaped* polygonal (or polyhedral) domain with respect to at least one point $x_K = (x_{K1}, \dots, x_{Kd}) \in K^\circ$, where K° is the interior of K. (This means that the straight segment connecting x_K with any $x \in K$ is a part of K.) We assume that

(i) there exists a constant $\kappa > 0$ independent of K and h such that

$$\frac{\max_{x\in\partial K} |x - x_K|}{\min_{x\in\partial K} |x - x_K|} \le \kappa \quad \forall\, K \in \mathscr{T}_h,\ h \in (0, \bar{h}), \tag{7.64}$$

(ii) element K can be divided into a finite number of closed simplexes:

$$K = \cup_{S\in\mathscr{S}_K} S, \tag{7.65}$$

there exists a positive constant C_S independent of K, S and h such that

$$\frac{h_S}{\rho_S} \le C_S \quad \forall\, S \in \mathscr{S}_K \qquad \text{(shape regularity)}, \tag{7.66}$$

where h_S is the diameter of S, ρ_S is the radius of the largest d-dimensional ball inscribed into S and, moreover,

$$1 \le \frac{h_K}{h_S} \le \tilde{\kappa} < \infty \quad \forall\, S \in \mathscr{S}_K, \tag{7.67}$$

where $\tilde{\kappa}$ is a constant independent of K, S and h.

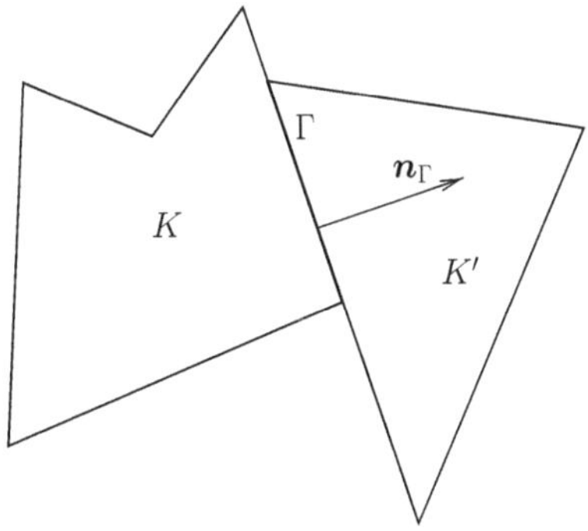

Fig. 7.5 Admissible elements K and K', the common face Γ and the corresponding normal $\boldsymbol{n}_\Gamma$

Remark 7.22 Assumptions 7.21 represent a generalization of the shape-regularity assumption (2.19).

Moreover, similarly as in (2.21), we assume that elements are *locally quasi-uniform*, which means that there exists a constant $C_Q > 0$ such that

$$h_K \leq C_Q h_{K'} \quad \forall\, K,\ K' \in \mathscr{T}_h,\ K,\ K' \text{ are neighbours.} \tag{7.68}$$

Elements satisfying Assumptions 7.21 and condition (7.68) are called *admissible*. Figure 7.5 shows a possible situation with a nonconvex element and a hanging node.

Moreover, let $h_\Gamma,\ \Gamma \in \mathscr{F}_h$, be given by (2.25) or (2.26). If Assumptions 7.21 and condition (7.68) are satisfied, then the *equivalence condition*

$$C_T h_K \leq h_\Gamma \leq C_G h_K, \quad K \in \mathscr{T}_h,\ \Gamma \in \mathscr{F}_h^{ID},\ \Gamma \subset \partial K, \tag{7.69}$$

is also valid in the case of star-shaped elements (cf. (2.20)).

Exercise 7.23 Verify relation (7.69) for meshes satisfying Assumptions 7.21 and condition (7.68).

7.2.2 Function Spaces

Let $\mathscr{T}_h$ be a triangulation consisting of star-shaped elements. In the same way as in Sect. 2.3.3, for a positive integer k we define the *broken Sobolev space*

$$H^k(\Omega, \mathscr{T}_h) = \{v \in L^2(\Omega),\ v|_K \in H^k(K)\ \forall\, K \in \mathscr{T}_h\}, \quad k \in \mathbb{N}, \tag{7.70}$$

with the norm $\|v\|_{H^k(\Omega,\mathscr{T}_h)} = \left(\sum_{K\in\mathscr{T}_h} \|v\|^2_{H^k(K)}\right)^{1/2}$ and the seminorm $|v|_{H^k(\Omega,\mathscr{T}_h)} = \left(\sum_{K\in\mathscr{T}_h} |v|^2_{H^k(K)}\right)^{1/2}$.

Similarly, the symbols $[v]_\Gamma$ and $\langle v\rangle_\Gamma$ denote the jump and the mean value of a function $v \in H^k(\Omega, \mathscr{T}_h)$ on $\Gamma \in \mathscr{F}_h$.

Finally, for an integer $p \geq 1$ we define the space of discontinuous piecewise polynomial functions

$$S_{hp} = \{v \in L^2(\Omega); v|_K \in P_p(K)\ \forall\, K \in \mathscr{T}_h\}, \tag{7.71}$$

where $P_p(K)$ denotes the space of all polynomials on element K of degree $\leq p$. We call the number p the degree of polynomial approximation.

7.2.3 Approximate Solution

In Sect. 2.4 we derived the DGM for the problem (2.1) on simplicial meshes. This process can be applied in the same way also in the case of general meshes introduced above. Hence, for $u, v \in H^2(\Omega, \mathscr{T}_h)$, we define the diffusion, penalty and right-hand side forms

$$a_h(u, v) = \sum_{K\in\mathscr{T}_h} \int_K \nabla u \cdot \nabla v \,\mathrm{d}x - \sum_{\Gamma\in\mathscr{F}_h^{ID}} \int_\Gamma (\langle \nabla u\rangle \cdot \boldsymbol{n}[v] + \Theta\langle\nabla v\rangle \cdot \boldsymbol{n}[u])\,\mathrm{d}S, \tag{7.72}$$

$$J_h^\sigma(u, v) = \sum_{\Gamma\in\mathscr{F}_h^{ID}} \int_\Gamma \sigma[u]\,[v]\,\mathrm{d}S, \tag{7.73}$$

$$\ell_h(v)\,(t) = \int_\Omega g(t)v\,\mathrm{d}x - \Theta \sum_{\Gamma\in\mathscr{F}_h^D} \int_\Gamma u_D(t)\,(\nabla v\cdot\boldsymbol{n})\,\mathrm{d}S \tag{7.74}$$
$$+ \sum_{\Gamma\in\mathscr{F}_h^D} \int_\Gamma \sigma u_D(t)\,v\,\mathrm{d}S + \int_{\partial\Omega_N} g_N(t)v\,\mathrm{d}S,$$

$$A_h(u, v) = a_h(u, v) + J_h^\sigma(u.v). \tag{7.75}$$

In the above integrals over each face $\Gamma \in \mathscr{F}_h$ the symbol $\boldsymbol{n}$ denotes the normal $\boldsymbol{n}_\Gamma$. The *penalty parameter* σ is given by

$$\sigma|_\Gamma = \sigma_\Gamma = \frac{C_W}{h_\Gamma}, \quad \Gamma \in \mathscr{F}_h^{ID}, \tag{7.76}$$

where h_Γ is given either by (2.25) or by (2.26) and $C_W > 0$ is a suitable constant. Similarly as in Sect. 2.4, for $\Theta = -1$, $\Theta = 0$ and $\Theta = 1$ the form A_h represents the nonsymmetric variant (NIPG), incomplete variant (IIPG) and symmetric variant (SIPG), respectively, of the diffusion form.

Now we are ready to introduce the approximate solution of problem (7.1).

Definition 7.24 A function $u_h \in S_{hp}$ is called an *approximate solution on a non-simplicial mesh* of problem (7.1) if it satisfies the identity

$$A_h(u_h, v_h) = \ell_h(v_h) \quad \forall v_h \in S_{hp}. \tag{7.77}$$

From the construction of the forms A_h and ℓ_h one can see that the strong solution $u \in H^2(\Omega)$ of problem (7.1) satisfies the identity

$$A_h(u, v) = \ell_h(v) \quad \forall v \in H^2(\Omega, \mathscr{T}_h), \tag{7.78}$$

which represents the *consistency* of the method. The expressions (7.77) and (7.78) imply the *Galerkin orthogonality* of the error $e_h = u_h - u$ of the method:

$$A_h(e_h, v_h) = 0 \quad \forall v_h \in S_{hp}, \tag{7.79}$$

which will be used in the analysis of error estimates (compare with (2.57)).

7.2.4 Auxiliary Results

In what follows we consider a system $\{\mathscr{T}_h\}_{h\in(0,\bar{h})}$, $\bar{h} > 0$, of partitions of the domain Ω satisfying Assumptions 7.21 and condition (7.68). The derivation of the error of the DGM solution computed on these meshes is based on analyzing the abstract numerical method in Sect. 2.2. Hence, we need to introduce analogies to the multiplicative trace inequality (Lemma 2.19), the inverse inequality (Lemma 2.21), S_{hp}-interpolation operator and its approximation properties (Lemmas 2.22 and 2.24).

Lemma 7.25 *Let Assumptions 7.21 be valid. Then for each $K \in \mathscr{T}_h$, $h \in (0, \bar{h})$, there exist axiparallel boxes B_0, B_1 such that*

(i) $B_0 \subset K \subset B_1$,
(ii) centers of B_0 and B_1 are identical and equal to x_K,
(iii) we have

$$1 \le \frac{h_{B_1}}{h_{B_0}} \le \bar{\kappa} < \infty, \tag{7.80}$$

where $\bar{\kappa}$ is a constant independent of K and h, and the numbers h_{B_0} and h_{B_1} are the half-sizes of B_0 and B_1, respectively, i.e.,

$$B_l = \prod_{s=1}^{d} [x_{Ks} - h_{B_l}, x_{Ks} + h_{B_l}], \qquad l = 0, 1, \tag{7.81}$$

and

$$h_K \le 2h_{B_1} \qquad (\Rightarrow h_K \le 2\bar{\kappa} h_{B_0}). \tag{7.82}$$

Proof Condition (7.64) implies that there exists a d-dimensional ball A_0 with center x_K inscribed into K and there exists a d-dimensional ball A_1 with the center x_K containing K. The ratio of diameters of A_0 and A_1 satisfies

$$1 \le \frac{r_{A_1}}{r_{A_0}} = \frac{\max_{x\in\partial K} |x - x_K|}{\min_{x\in\partial K} |x - x_K|} \le \kappa. \tag{7.83}$$

This implies the existence of axiparallel boxes B_0, B_1 satisfying conditions (i)–(ii) from Lemma 7.25. Obviously,

$$h_{B_1} = r_{A_1}, \quad h_{B_0}\sqrt{d} = r_{A_0}, \tag{7.84}$$

see Fig. 7.6. Then (7.83) and (7.84) imply that

$$\frac{h_{B_1}}{\sqrt{d} h_{B_0}} = \frac{r_{A_1}}{r_{A_0}} \le \kappa \tag{7.85}$$

and condition (iii) is valid with $\bar{\kappa} = \sqrt{d}\kappa$. □

Now, we introduce the variant of the multiplicative trace inequality over star-shaped elements.

Lemma 7.26 (Multiplicative trace inequality) *Let Assumptions 7.21 be valid. Then there exists a constant $C_M > 0$ independent of v, h and K such that*

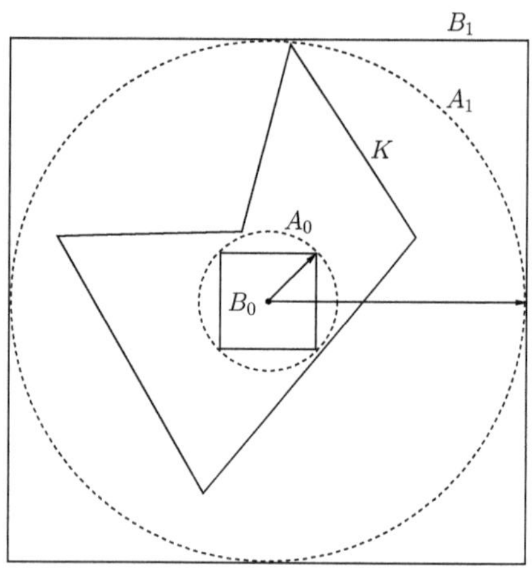

Fig. 7.6 Balls A_0, A_1 and axiparallel boxes B_0, B_1 used in the proof of Lemma 7.25

$$\|v\|^2_{L^2(\partial K)} \le C_M \left(\|v\|_{L^2(K)} \, |v|_{H^1(K)} + h_K^{-1} \|v\|^2_{L^2(K)} \right), \tag{7.86}$$
$$K \in \mathscr{T}_h, \ v \in H^1(K), \ h \in (0, \bar{h}).$$

Proof We prove the lemma in two steps.

(i) Let $K \in \mathscr{T}_h$ and let $S \in \mathscr{S}_K$ be an arbitrary but fixed simplex from (7.65). Then, applying (2.78) to the simplex S, we have

$$\|v\|^2_{L^2(\partial S)} \le \tilde{C}_M \left[2\|v\|_{L^2(S)} |v|_{H^1(S)} + \frac{d}{h_S} \|v\|^2_{L^2(S)} \right], \tag{7.87}$$

where $\tilde{C}_M$ is the constant from (7.66). (Cf. the proof of Lemma 2.19.)

(ii) Let $K \in \mathscr{T}_h$ and $S \in \mathscr{S}_K$. Then (7.87) yields

$$\|v\|^2_{L^2(\partial S)} \le C \left(\|v\|_{L^2(S)} \, |v|_{H^1(S)} + h_S^{-1} \|v\|^2_{L^2(S)} \right), \quad S \in \mathscr{S}_K, \ v \in H^1(K), \tag{7.88}$$

where $C = \tilde{C}_M \max\{2, d\}$. Using the inclusion $\partial K \subset \cup_{S \in \mathscr{S}_K} \partial S$, relations (7.88) and (7.67) and the Cauchy inequality, we find that

$$\begin{aligned}
\|v\|^2_{L^2(\partial K)} &\le \sum_{S \in \mathscr{S}_K} \|v\|^2_{L^2(\partial S)} \\
&\le C \left\{ \sum_{S \in \mathscr{S}_K} \left(\|v\|_{L^2(S)} \, |v|_{H^1(S)} + h_S^{-1} \|v\|^2_{L^2(S)} \right) \right\} \\
&\le C \left\{ \left(\sum_{S \in \mathscr{S}_K} \|v\|^2_{L^2(S)} \right)^{\frac{1}{2}} \left(\sum_{S \in \mathscr{S}_K} |v|^2_{H^1(S)} \right)^{\frac{1}{2}} + \tilde{\kappa} h_K^{-1} \|v\|^2_{L^2(K)} \right\} \\
&\le C_M \left(\|v\|_{L^2(K)} \, |v|_{H^1(K)} + h_K^{-1} \|v\|^2_{L^2(K)} \right),
\end{aligned} \tag{7.89}$$

where $C_M = \tilde{\kappa} C = \tilde{\kappa} \tilde{C}_M \max\{2, d\}$ and $\tilde{\kappa}$ is the constant from (7.67). □

We proceed with the generalization of the inverse inequality over star-shaped elements.

Lemma 7.27 (Inverse inequality) *Let Assumptions 7.21 be valid. Then there exists a constant $C_I > 0$ independent of v, h and K such that*

$$|v|_{H^1(K)} \le C_I h_K^{-1} \|v\|_{L^2(K)}, \quad v \in P_p(K), \ K \in \mathscr{T}_h, \ h \in (0, \bar{h}). \tag{7.90}$$

Proof Let $K \in \mathscr{T}_h$ and let B_0 and B_1 be the boxes from Lemma 7.25 satisfying (7.80). Without loss of generality we suppose that the center of B_0 and B_1 is the origin of the coordinate system. Then obviously

$$[-1, 1]^d \subset h_{B_0}^{-1} K \subset h_{B_0}^{-1} B_1, \tag{7.91}$$

where

$$h_{B_0}^{-1} K = \left\{ y = (y_1, \dots, y_d) \in \mathbb{R}^d;\ y_l = h_{B_0}^{-1} x_l,\ l = 1, \dots, d,\ x = (x_1, \dots, x_d) \in K \right\}. \tag{7.92}$$

Similarly $h_{B_0}^{-1} B_1$ is defined. Further, we define the reference boxes

$$\hat{B}_0 = [-1, 1]^d, \quad \hat{B}_1 = [-\bar{\kappa}, \bar{\kappa}]^d, \tag{7.93}$$

where $\bar{\kappa}$ is the constant from (7.80). It can be verified that $\|.\|_{L^2(\hat{B}_0)}$ is a norm in the space $P_p(\hat{B}_1)$. The equivalence of norms on the finite-dimensional space $P_p(\hat{B}_1)$ implies that there exists a constant $C(p, \bar{\kappa})$ such that

$$\|v\|_{L^2(\hat{B}_1)} \leq C(p, \bar{\kappa}) \|v\|_{L^2(\hat{B}_0)} \quad \forall v \in P_p(\hat{B}_1). \tag{7.94}$$

In what follows by $\hat{v}$ we denote the function obtained by the scaling of $v \in P_p$ defined in (7.92). Furthermore, the transformation to the reference element and the equivalence of the norms $\|\cdot\|_{L^2(\hat{B}_0)}$ and $\|\cdot\|_{H^1(\hat{B}_0)}$ on the finite-dimensional space $P_p(\hat{B}_0)$ imply that

$$\begin{aligned} \|\nabla v\|_{L^2(B_0)} &= h_{B_0}^{d/2-1} \|\hat{\nabla} \hat{v}\|_{L^2(\hat{B}_0)} = h_{B_0}^{d/2-1} \|\hat{v}\|_{H^1(\hat{B}_0)} \\ &\leq \tilde{C}(p) h_{B_0}^{d/2-1} \|\hat{v}\|_{L^2(\hat{B}_0)} = \tilde{C}(p) h_{B_0}^{-1} \|v\|_{L^2(B_0)} \quad \forall v \in P_p(B_0). \end{aligned} \tag{7.95}$$

where $\tilde{C}(p)$ is a constant independent of h and v.

Let $v \in P_p(B_1)$. Then the transformation to the reference box $\hat{B}_0$ and relations (7.91)–(7.95), (7.67) and (7.82) yield

$$\begin{aligned} \|\nabla v\|_{L^2(K)} &\leq \|\nabla v\|_{L^2(B_1)} = (h_{B_1}/\bar{\kappa})^{d/2-1} \|\hat{\nabla} \hat{v}\|_{L^2(\hat{B}_1)} \\ &\leq (h_{B_1}/\bar{\kappa})^{d/2-1} C(p, \bar{\kappa}) \|\hat{\nabla} \hat{v}\|_{L^2(\hat{B}_0)} \\ &= (h_{B_1}/\bar{\kappa})^{d/2-1} h_{B_0}^{1-d/2} C(p, \bar{\kappa}) \|\nabla v\|_{L^2(B_0)} \\ &\leq C(p, \bar{\kappa}) \tilde{C}(p) h_{B_0}^{-1} \|v\|_{L^2(B_0)} \leq 2\bar{\kappa} C(p, \bar{\kappa}) \tilde{C}(p) h_K^{-1} \|v\|_{L^2(K)}, \end{aligned}$$

which proves the lemma with $C_I = 2\bar{\kappa} C(p, \bar{\kappa}) \tilde{C}(p)$. □

Finally, we introduce an interpolation operator over star-shaped elements and its approximation properties, which generalizes Lemma 2.22.

Lemma 7.28 (Approximation properties) *Let Assumptions 7.21 be valid and integers p, s, q be given satisfying the conditions $p \geq 0, 0 \leq q \leq s$. Then there exist a* *mapping* $\pi_{K,p} : H^1(K) \to P_p(K)$ *and a constant* $C_A > 0$ *independent of* v *and* h *such that*

$$|\pi_{K,p}v - v|_{H^q(K)} \le C_A h_K^{\mu-q} |v|_{H^\mu(K)} \quad \forall\, v \in H^s(K),\ K \in \mathscr{T}_h, \tag{7.96}$$

where $\mu = \min(p+1, s)$.

Proof The existence of the mapping $\pi_{K,p}$ is based on the results of Verfürth from [275]. In the same way as in Sect. 1.3.1 we denote by $\alpha = (\alpha_1, \dots, \alpha_d)$ a multi-index with integers $\alpha_i \ge 0,\ i = 1, \dots, d$. For $x = (x_1, \dots, x_d)$ we set $x^\alpha = x_1^{\alpha_1} \cdots x_d^{\alpha_d}$.

Now for any function φ integrable over K we define its mean value

$$\pi_K \varphi = \frac{1}{|K|} \int_K \varphi \,\mathrm{d}x. \tag{7.97}$$

For each $v \in H^s(K)$ we recursively define polynomials $\rho_{p,K}(v), \rho_{p-1,K}(v), \dots, \rho_{0,K}(v)$ in $P_p(K)$. First, we set

$$\rho_{p,K}(v) = \sum_{|\alpha|=p} \frac{1}{p!} x^\alpha \pi_K(D^\alpha v). \tag{7.98}$$

Then, for $k = p, p-1, \dots, 1$ we put

$$\rho_{k-1,K}(v) = \rho_{k,K}(v) + \sum_{|\alpha|=k-1} \frac{1}{p!} x^\alpha \pi_K \left(D^\alpha(v - \rho_{k,K}(u))\right). \tag{7.99}$$

Finally, we put

$$\pi_{K,p}v = \rho_{0,K}(v). \tag{7.100}$$

For the proof of the approximation properties (7.96) we refer to [275]. □

Definition 7.29 Let $\mathscr{T}_h$ be a triangulation and let $p > 0$. We define the mapping $\Pi_{hp} : H^1(\Omega, \mathscr{T}_h) \to S_{hp}$ by

$$\left(\Pi_{hp}u\right)|_K = \pi_{K,p}(u|_K) \quad \forall\, K \in \mathscr{T}_h, \tag{7.101}$$

where $\pi_{K,p} : H^1(K) \to P_p(K)$ is the mapping introduced in Lemma 7.28.

Lemma 7.30 *Let* $\mathscr{T}_h$ *be a triangulation, let integers* p, s, q *satisfy assumptions from Lemma 7.28 and let* $\Pi_{hp} : H^1(\Omega, \mathscr{T}_h) \to S_{hp}$ *be the mapping given by (7.101). Then*

$$\left|\Pi_{hp}v - v\right|_{H^q(\Omega,\mathscr{T}_h)} \le C_A h^{\mu-q} |v|_{H^\mu(\Omega,\mathscr{T}_h)}, \quad v \in H^s(\Omega, \mathscr{T}_h), \tag{7.102}$$

where $\mu = \min(p+1, s)$ *and* C_A *is the constant from (7.96).*

Proof Using (7.101), the definition of a seminorm in a broken Sobolev space and the approximation properties (7.96), we obtain (7.102). □

7.2.5 Error Analysis

Now we present the error estimate in the DG-norm (cf. (2.103)) of the DGM applied on star-shaped elements. The error analysis follows the technique from Chap. 2, where Lemmas 2.19, 2.21, 2.22 and 2.24 have to be replaced by Lemmas 7.26, 7.27, 7.28 and 7.30, respectively.

Moreover, in order to ensure the coercivity of form A_h we have to choose the penalization constant C_W sufficiently large depending on the used variants of DGM ($\Theta = -1,\ 0,\ 1$ in (7.72)). Particularly, C_W is specified in Corollary 2.41, where C_M, C_I and C_G are the constants from (7.86), (7.90) and (7.69).

Then, the error estimate is formulated in the following way.

Theorem 7.31 (DG-norm error estimate) *Let us assume that $s \geq 2,\ p \geq 1$ are integers, $u \in H^s(\Omega)$ is the solution of problem (7.1), $\{\mathscr{T}_h\}_{h\in(0,\bar{h})}$, $h \in (0, \bar{h})$, is a system of triangulations of the domain Ω satisfying Assumptions 7.21 and condition (7.68). Moreover, let the penalization constant C_W satisfy the conditions from Corollary 2.41 with constants C_M, C_I and C_G from (7.86), (7.90) and (7.69). Let $u_h \in S_{hp}$ be the approximate solution obtained by means of method (7.77). Then the error $e_h = u_h - u$ satisfies the estimate*

$$|||e_h||| \leq \tilde{C} h^{\mu-1} |u|_{H^{\mu}(\Omega)}, \quad h \in (0, \bar{h}), \tag{7.103}$$

where $\mu = \min(p+1, s)$ and $\tilde{C}$ is a constant independent of h.

Exercise 7.32 Prove Theorem 7.31 in details. Hints: Adapt the technique from Sect. 2.7 to this case.

7.2.6 Numerical Examples

In order to demonstrate the theoretical error estimate obtained in the previous section, we present examples of the solution of problem (7.1) computed by the DGM applied on general meshes satisfying Assumptions 7.21 and condition (7.68). Namely, we consider nonconforming meshes containing triangular and nonconvex quadrilateral star-shaped elements constructed by the following algorithm presented in [95]:

1. We start from a vertically oriented structured triangular grid, see Fig. 7.7a.
2. We apply a vertical shift to some vertices, which creates a triangular mesh with hanging nodes, shown in Fig. 7.7b.

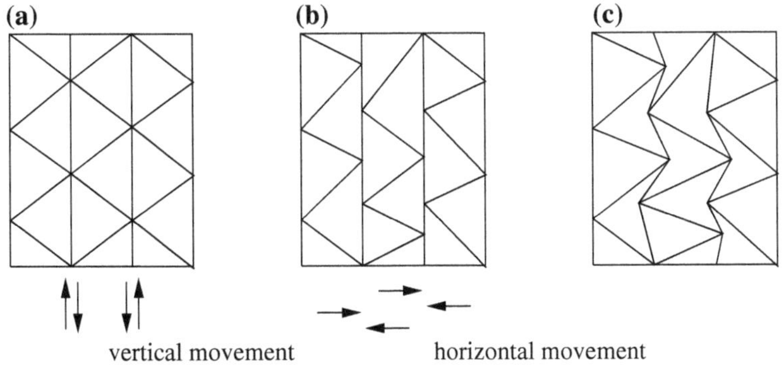

Fig. 7.7 Algorithm generating meshes with nonconvex quadrilateral elements

3. We apply a horizontal shift to some vertices, which creates nonconvex quadrilaterals shown in Fig. 7.7c.

This algorithm allows us to construct meshes with a prescribed constant C_d given by

$$C_d = \max_{K \in \mathscr{T}_h} \max_{\substack{\Gamma \in \mathscr{F}_h \\ \Gamma \subset \partial K}} \frac{h_K}{\mathrm{diam}(\Gamma)}. \tag{7.104}$$

(Compare with condition (2.22).) The parameter C_d characterizes the nonconformity of the mesh. Figure 7.8 shows meshes with different numbers $\#\mathscr{T}_h$ of elements and different values of C_d. Of course, these types of meshes are artificial and not used in practice. We only want to demonstrate that our scheme is robust with respect to rather rough meshes.

We solve the Poisson problem (7.1) in $\Omega = (0, 1) \times (0, 1)$ with $\partial\Omega_N = \emptyset$. We prescribe the right-hand side f and the Dirichlet boundary condition in such a way that the exact solution has the form

$$u(x_1, x_2) = (1 - x_1^2)^2(1 - x_2^2)^2. \tag{7.105}$$

The solution of the problem was carried out with the aid of piecewise linear elements on 6 grids $\mathscr{T}_{h_l}$, $l = 1, \ldots, 6$, having different numbers of elements and different parameters C_d, see Table 7.2. Some of the meshes are shown in Fig. 7.8.

The computational error $e_h = u_h - u$ of the solution is evaluated in the $L^2(\Omega)$-norm and we compute the experimental order of convergence (EOC). The *global experimental order of convergence* (GEOC) $\bar{\alpha}$ is obtained by the least squares method.

Table 7.2 (left) shows the computational error e_h in the $L^2(\Omega)$-norm and the experimental order of convergence together with the global experimental order of convergence, see Sect. 2.9.1. Moreover, Table 7.2 (right) shows the error e_h obtained

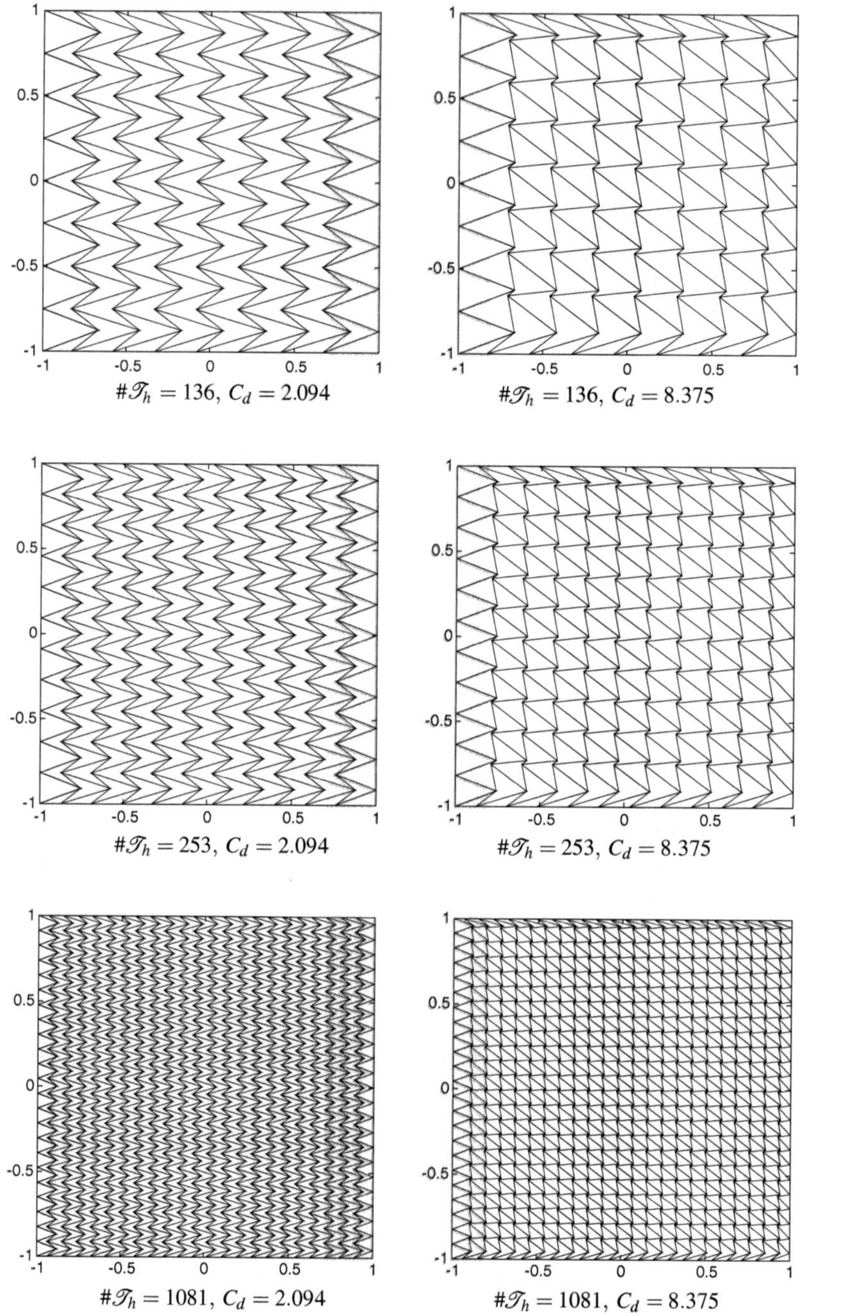

Fig. 7.8 Example of meshes formed by triangular and nonconvex quadrilateral elements with different numbers $\#\mathscr{T}_h$ of elements and different values of C_d

Table 7.2 Error e_h in $L^2(\Omega)$-norm and EOC for meshes with $C_d = 2.094$ and $C_d = 8.375$ (left) and dependence of e_h on the value of C_d for $\#T_h = 528$ (right)

			$C_d = 2.094$		$C_d = 8.375$	
l	$\#T_{h_l}$	h_l	$\|e_h\|_{L^2(\Omega)}$	EOC	$\|e_h\|_{L^2(\Omega)}$	EOC
1	136	4.334E-01	1.9775E-02	-	1.5393E-02	-
2	253	3.152E-01	1.0404E-02	2.017	7.9873E-03	2.060
3	528	2.167E-01	4.9109E-03	2.004	3.6525E-03	2.088
4	1081	1.508E-01	2.3905E-03	1.986	1.7223E-03	2.073
5	2080	1.084E-01	1.2450E-03	1.976	8.8145E-04	2.029
6	4095	7.705E-02	6.1307E-04	2.075	4.3596E-04	2.062
GEOC				2.005		2.064

l	C_d	e_h
1	2.094	4.9109E-03
2	4.188	3.8514E-03
3	8.375	3.6525E-03
4	16.106	3.6403E-03
5	29.911	3.6550E-03
6	52.344	3.6676E-03

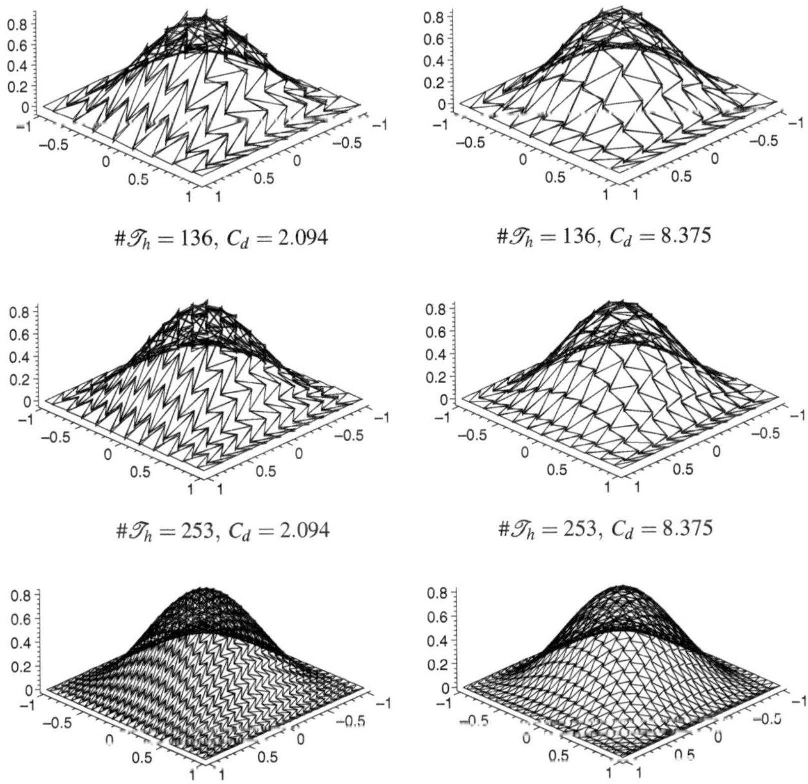

$\#\mathscr{T}_h = 136,\ C_d = 2.094$ $\quad$ $\#\mathscr{T}_h = 136,\ C_d = 8.375$

$\#\mathscr{T}_h = 253,\ C_d = 2.094$ $\quad$ $\#\mathscr{T}_h = 253,\ C_d = 8.375$

$\#\mathscr{T}_h = 1081,\ C_d = 2.094$ $\quad$ $\#\mathscr{T}_h = 1081,\ C_d = 8.375$

Fig. 7.9 Numerical solution on meshes with nonconvex quadrilateral elements with different numbers $\#\mathscr{T}_h$ of elements and different values of C_d

on meshes with 528 elements and different values of the parameter C_d. We see that the dependence of the error on C_d is not significant.

Figure 7.9 shows the numerical solution obtained on meshes with $\#\mathscr{T}_h = 136, 253, 1081$ and $C_d = 2.094,\ 8.375$. We observe here the convergence of the approximate solution to the exact solution for $h \to 0$.

7.3 The Effect of Numerical Integration

In practical applications of the finite element method, integrals appearing in the definition of the discrete problem have to be evaluated with the aid of numerical integration. According to the terminology introduced by G. Strang [262], the use of numerical integration belongs to finite element variational crimes. The theory of numerical integration was developed by P.G. Ciarlet and P.A. Raviart in [53]. Their ideas and results were extended to nonlinear elliptic problems, e.g., in [132, 136, 138, 140, 287].

Here we pay attention to the effect of numerical integration used in the DGM for the numerical solution of nonlinear convection-diffusion problems equipped with initial condition and Dirichlet boundary condition.

7.3.1 *Continuous Problem*

Let us consider a bounded polygonal domain $\Omega \subset \mathbb{R}^2$ with Lipschitz boundary $\partial\Omega$ and time interval $[0, T]$ with $T > 0$. Similarly as in Sect. 6.2, we consider the problem with only a Dirichlet boundary condition for the same reasons as explained in Remark 6.33. We will be concerned with the nonstationary nonlinear convection-diffusion problem (4.1): Find $u : Q_T = \Omega \times (0, T) \to \mathbb{R}$ such that

$$\frac{\partial u}{\partial t} + \sum_{s=1}^{2} \frac{\partial f_s(u)}{\partial x_s} = \varepsilon \Delta u + g \quad \text{in } Q_T, \tag{7.106a}$$

$$u\big|_{\partial\Omega\times(0,T)} = u_D, \tag{7.106b}$$

$$u(x, 0) = u^0(x), \quad x \in \Omega. \tag{7.106c}$$

We assume that the data satisfy conditions (4.2), i.e.,

$$\begin{aligned}
&f_s \in C^1(\mathbb{R}),\ f_s' \text{ are bounded},\ f_s(0) = 0,\ s = 1, \dots, d,\\
&u_D = \text{trace of some } u^* \in C([0, T]; H^1(\Omega)) \cap L^\infty(Q_T) \text{ on } \partial\Omega \times (0, T),\\
&\varepsilon > 0,\ g \in C([0, T]; L^2(\Omega)),\ u^0 \in L^2(\Omega).
\end{aligned}$$

We assume the existence of a solution u of problem (7.106) satisfying the following *regularity conditions*:

$$u \in L^2(0, T; H^2(\Omega)), \quad \frac{\partial u}{\partial t} \in L^2(0, T; H^1(\Omega)). \tag{7.107}$$

Then $u \in C([0, T]; H^1(\Omega))$ satisfies (7.106) pointwise (almost everywhere). Later we will introduce a stronger assumption on the regularity of u.

7.3.2 Space Semidiscretization

The space semidiscretization is introduced in the same way as in Sect. 4.2 with the notation from Sects. 2.3 and 4.2. We get the forms defined in (4.10), (4.23), (4.11) and 4.13:

$$a_h(u, v) = \sum_{K \in \mathscr{T}_h} \int_K \nabla u \cdot \nabla v \, dx - \sum_{\Gamma \in \mathscr{F}_h^{ID}} \int_\Gamma (\langle \nabla u \rangle \cdot \boldsymbol{n}[v] + \Theta \langle \nabla v \rangle \cdot \boldsymbol{n}[u]) \, dS, \tag{7.108}$$

$$b_h(u, v) = \sum_{\Gamma \in \mathscr{F}_h} \int_\Gamma H(u_\Gamma^{(L)}, u_\Gamma^{(R)}, \boldsymbol{n}) \, [v] \, dS - \sum_{K \in \mathscr{T}_h} \int_K f(u) \cdot \nabla v \, dx, \tag{7.109}$$

$$J_h^\sigma(u, v) = \sum_{\Gamma \in \mathscr{F}_h^{ID}} \int_\Gamma \sigma [u] \, [v] \, dS, \tag{7.110}$$

$$\ell_h(v)\,(t) = \int_\Omega g(t) v \, dx - \varepsilon \Theta \sum_{\Gamma \in \mathscr{F}_h^D} \int_\Gamma u_D(t) \, (\nabla v \cdot \boldsymbol{n}) \, dS + \varepsilon \sum_{\Gamma \in \mathscr{F}_h^D} \int_\Gamma \sigma u_D(t) \, v \, dS. \tag{7.111}$$

As usual, we set $(\psi, \varphi) = \int_\Omega \psi \varphi \, dx$. We assume that the numerical flux H satisfies conditions (4.18)–(4.20).

Definition 7.33 We say that a function u_h is the *approximate solution* of problem (7.106), if it satisfies the conditions

$$u_h \in C^1([0, T]; S_{hp}), \tag{7.112a}$$

$$\left(\frac{\partial u_h(t)}{\partial t}, v_h \right) + \varepsilon a_h(u_h(t), v_h) + b_h(u_h(t), v_h) + \varepsilon J_h^\sigma(u_h(t), v_h) = \ell_h(v_h)\,(t)$$
$$\forall v_h \in S_{hp}, \ \forall t \subset [0, T], \tag{7.112b}$$

$$u_h(0) = u_h^0 \text{ and } (u_h(0), v_h) = (u^0, v_h) \quad \forall v_h \in S_{hp}, \tag{7.112c}$$

where $p \geq 1$ is an integer and $S_{hp} = \{v_h; v_h|K \in P_p(K) \ \forall K \in \mathscr{T}_h\}$. The weight σ in the penalty form J_h^σ is defined as

$$\sigma|_\Gamma = \frac{C_W}{h_\Gamma}, \quad \Gamma \in \mathscr{F}_h, \tag{7.113}$$

where h_Γ is introduced in Sect. 2.6 and $C_W > 0$ is a suitable constant.

7.3.3 Numerical Integration

In practical computations the integrals appearing in (7.108)–(7.111) are evaluated with the aid of numerical integration. This means that for functions $F \in C(K)$ and $G \in C(\Gamma)$, where $K \in \mathscr{T}_h$ and Γ is a face of K, we use the approximations

$$\int_K F\,\mathrm{d}x \approx |K| \sum_{\alpha=1}^{n_K} \omega_\alpha^K F(x_\alpha^K), \tag{7.114}$$

$$\int_\Gamma G\,\mathrm{d}S \approx |\Gamma| \sum_{\alpha=1}^{m_\Gamma} \beta_\alpha^\Gamma G(x_\alpha^\Gamma), \tag{7.115}$$

where $\omega_\alpha^K, \beta_\alpha^\Gamma \in \mathbb{R}$ are integration weights and $x_\alpha^K \in K$, $x_\alpha^\Gamma \in \Gamma$ are integration points. (We recall that $|K|$ denotes the area of K and $|\Gamma|$ denotes the length of Γ.) Examples of the volume and face quadrature formulae are the Dunavant [111] and the Gauss formulae, respectively. For more formulae in the context of finite element approximations, see, e.g., [260].

Using quadrature formulae (7.114) and (7.115), we obtain the approximations $(\cdot,\cdot)_h, \tilde{a}_h, \tilde{J}_h^\sigma, \tilde{b}_h, \tilde{\ell}_h$ of the forms $(\cdot,\cdot), a_h, J_h^\sigma, b_h, \ell_h$ defined by (7.108)–(7.111).

Definition 7.34 A function $\tilde{u}_h$ is called the *approximate solution of the discrete problem with numerical integration* if it satisfies the conditions

$$\tilde{u}_h \in C^1([0,T]; S_{hp}), \tag{7.116a}$$

$$\left(\frac{\partial \tilde{u}_h(t)}{\partial t}, \varphi_h\right)_h + \varepsilon \tilde{a}_h(\tilde{u}_h(t), \varphi_h) + \tilde{b}_h(\tilde{u}_h(t), \varphi_h) + \varepsilon \tilde{J}_h^\sigma(\tilde{u}_h(t), \varphi_h)$$
$$= \tilde{\ell}_h(\varphi_h)(t) \quad \forall \varphi_h \in S_{hp},\ \forall t \in (0,T), \tag{7.116b}$$

$$\tilde{u}_h(0) = \tilde{u}_h^0. \tag{7.116c}$$

By $\tilde{u}_h^0$ we denote a suitable S_{hp}-approximation of the initial condition u^0. One possibility is to assume that $\tilde{u}_h^0 \in S_{hp}$ satisfies the condition

$$(\tilde{u}_h^0 - u^0, \varphi_h)_h = 0 \quad \forall \varphi_h \in S_{hp}. \tag{7.117}$$

(Another possibility is to define the function $\tilde{u}_h^0$ on each $K \in \mathscr{T}_h$ as the Lagrange interpolation, provided $u^0 \in C(\overline{\Omega})$.)

Our goal is to evaluate the effect of numerical integration and to derive error estimates of method (7.116). We assume that Assumptions 4.5 from the beginning of Sect. 4.3 are valid, and the equivalence condition (2.20), i.e.,

$$C_T h_K \le h_\Gamma \le C_G h_K, \quad K \in \mathscr{T}_h, \quad \Gamma \subset \partial K, \tag{7.118}$$

holds. Then, by Corollary 2.41, the forms a_h and J_h^σ have the coercivity property

$$a_h(\varphi_h, \varphi_h) + \varepsilon J_h^\sigma(\varphi_h, \varphi_h) \ge \frac{\varepsilon}{2} |||\varphi_h|||^2 \quad \forall \varphi_h \in S_{hp}. \tag{7.119}$$

Moreover, let us assume that there exist constants $\omega,\ \beta > 0$ such that

$$\sum_{\alpha=1}^{n_K} |\omega_\alpha^K| \le \omega, \ \sum_{\alpha=1}^{m_\Gamma} |\beta_\alpha^\Gamma| \le \beta, \quad \forall K \in \mathscr{T}_h, \quad \forall \Gamma \in \mathscr{F}_h, \quad \forall h \in (0, \bar{h}). \tag{7.120}$$

7.3.4 Some Important Results

In analyzing the effect of numerical integration we assume that the exact solution satisfies the regularity condition

$$\frac{\partial u}{\partial t} \in L^2(0, T; H^{p+1}(\Omega)). \tag{7.121}$$

Then $u \in C([0, T]; H^{p+1}(\Omega))$. Because of simplicity, here we consider the regularity of the exact solution u related to degree p of the polynomial approximation.

Important tools are the multiplicative trace inequality (2.78), the inverse inequalities and approximation properties of the operator Π_{hp}. By Lemma 2.21, we have

$$|v|_{H^1(K)} \le C_I h_K^{-1} \|v\|_{L^2(K)}. \tag{7.122}$$

valid for all $v \in P_p(K)$, $K \in \mathscr{T}_h$, and $h \in (0, \bar{h})$, with a constant $C_I > 0$ independent of v, h, and K. By virtue of Lemma 2.22, there exists a constant $C_A > 0$ independent of v and h such that

$$\|\Pi_{hp} v - v\|_{L^2(K)} \le C_A h_K^{p+1} |v|_{H^{p+1}(K)}, \tag{7.123}$$

$$|\Pi_{hp} v - v|_{H^1(K)} \le C_A h_K^{p} |v|_{H^{p+1}(K)}, \tag{7.124}$$

$$|\Pi_{hp} v - v|_{H^2(K)} \le C_A h_K^{p-1} |v|_{H^{p+1}(K)} \tag{7.125}$$

for all $v \in H^{p+1}(K)$, $K \in \mathscr{T}_h$, and $h \in (0, \bar{h})$.

Moreover, we use the following results.

Lemma 7.35 *If $\{\mathscr{T}_h\}_{h\in(0,\bar{h})}$ is a shape-regular system of triangulations of the domain Ω, then there exist constants $C_I > 0$ and $C_A > 0$ independent of v, h, and K such that*

$$\|v\|_{L^\infty(K)} \le C_I h_K^{-1} \|v\|_{L^2(K)}, \tag{7.126}$$
$$|v|_{W^{1,\infty}(K)} \le C_I h_K^{-1} |v|_{H^1(K)}, \tag{7.127}$$
$$\|v\|_{L^\infty(\Gamma)} \le C_I |\Gamma|^{-1/2} \|v\|_{L^2(\Gamma)}, \tag{7.128}$$

hold for all $v \in P_p(K)$, $K \in \mathscr{T}_h$, $\Gamma \in \mathscr{F}_h, \Gamma \subset \partial K$, $h \in (0, \bar{h})$. Moreover,

$$\|\Pi_{hp} v - v\|_{L^\infty(K)} \le C_A |K|^{-1/2} h_K^{p+1} |v|_{H^{p+1}(K)} \tag{7.129}$$

for all $v \in H^{p+1}(K)$, $K \in \mathscr{T}_h$, and $h \in (0, \bar{h})$.

Proof Obviously, since $d = 2$, we have

$$\pi\, h_K^2 / C_R^2 \le |K| \le h_K^2/2, \tag{7.130}$$

which follows from the shape-regularity (2.19) of the system of triangulations. By $\widehat{K}$ we denote the reference triangle with vertices $(0, 0)$, $(1, 0)$, $(0, 1)$. Since the area of the reference triangle $\widehat{K}$ is equal to $|\widehat{K}| = 1/2$, by (2.65) and (7.130) we get

$$|\det(\mathbb{B}_K)| = 2|K|, \tag{7.131}$$
$$C_1\, h_K^2 \le |\det(\mathbb{B}_K)| \le h_K^2 \tag{7.132}$$

with the constant $C_1 = 2\pi / C_R^2$.

Now, let $v \in P_p(K)$. Elements K and $\widehat{K}$ are affine-equivalent via an invertible affine mapping $F_K : \widehat{K} \overset{\text{onto}}{\longrightarrow} K$, $x = F_K(\widehat{x}) = \mathbb{B}_K \widehat{x} + b_K \in K$ for $\widehat{x} \in \widehat{K}$ with a nonsingular 2×2 matrix $\mathbb{B}_K$. We set $\widehat{v} = v \circ F_K$. Further, successively we use the obvious relation $\|v\|_{L^\infty(K)} = \|\widehat{v}\|_{L^\infty(\widehat{K})}$, the equivalence of norms in the space $P_p(\widehat{K})$, relation (2.66) with $m = 0$ and $\alpha = 2$ and then relations (7.132). We get

$$\begin{aligned}\|v\|_{L^\infty(K)} = \|\widehat{v}\|_{L^\infty(\widehat{K})} &\le C_2 \|\widehat{v}\|_{L^2(\widehat{K})}\\ &\le C_3 |\det(\mathbb{B}_K)|^{-1/2}\, \|v\|_{L^2(K)}\\ &\le C_3\, C_1^{-1/2}\, h_K^{-1} \|v\|_{L^2(K)},\end{aligned}$$

which is (7.126) with $C_I = C_3\, C_1^{-1/2}$.

Inequality (7.127) is a consequence of (7.126) applied to ∇v.

The proof of (7.128) can be carried out in the same way as in the case of (7.126), replacing K and $\widehat{K}$ by Γ and $\widehat{\Gamma} = [0, 1]$ and using the relations $\|v\|_{L^\infty(\Gamma)} = \|\widehat{v}\|_{L^\infty(\widehat{\Gamma})}$ and $\|v\|_{L^2(\Gamma)} = |\Gamma|^{1/2}\, \|\widehat{v}\|_{L^2(\widehat{\Gamma})}$.

Finally, estimate (7.129) is a consequence of (2.75) with $m := 0, \beta := \infty, p := p$, $\alpha := 2$, $\omega := K$ and $\Pi_K := \Pi_{hp}$. We get

$$\|v - \Pi_{hp}\|_{L^\infty(K)} \le C_4 |K|^{-1/2} h_K^{p+1} |v|_{W^{p+1,2}(K)},$$

which is (7.129). □

It follows from Theorem 4.14 that under the mentioned assumptions the error $e_h = u_h - u$ of the approximate solution computed with exact integration satisfies the estimate

$$\max_{t\in[0,T]} \|e_h(t)\|^2_{L^2(\Omega)} + \frac{\varepsilon}{2}\int_0^T \left(|e_h(\vartheta)|^2_{H^1(\Omega,\mathscr{T}_h)} + J_h^\sigma(e_h(\vartheta), e_h(\vartheta))\right) \mathrm{d}\vartheta \le \tilde{C} h^{2p} \tag{7.133}$$

In what follows, by C we denote a generic positive constant independent of h, attaining in general different values in different places.

7.3.5 Truncation Error of Quadrature Formulae

In this section we analyze the error caused by numerical integration. If $\Gamma \in \mathscr{F}_h$, $K \in \mathscr{T}_h$, $\varphi \in C(\Gamma)$ and $\psi \in C(K)$, then the symbols $E_\Gamma(\varphi)$ and $E_K(\psi)$ will denote the error of the numerical integration of the function φ over Γ and of ψ over K, respectively:

$$E_\Gamma(\varphi) = \int_\Gamma \varphi \,\mathrm{d}S - |\Gamma| \sum_{\alpha=1}^{m_\Gamma} \beta_\alpha^\Gamma \varphi(x_\alpha^\Gamma), \tag{7.134}$$

$$E_K(\psi) = \int_K \psi \,\mathrm{d}x - |K| \sum_{\alpha=1}^{n_K} \omega_\alpha^K \psi(x_\alpha^K). \tag{7.135}$$

We start with integrating along a face $\Gamma \in \mathscr{F}_h$. In the sequel, we set $1/\infty := 0$.

Lemma 7.36 *Let $s,\ p > 0$ be integers, $r,\ q \in [1,\infty]$ and $rs > 1$. Let assumption (7.120) be satisfied and let the quadrature formula (7.115) be exact for polynomials of degree $\le s + p - 1$. Then there exists a constant $C > 0$ such that*

$$|E_\Gamma(Qv)| = \left| \int_\Gamma Qv \,\mathrm{d}S - |\Gamma| \sum_{\alpha=1}^{m_\Gamma} \beta_\alpha^\Gamma (Qv)(x_\alpha^\Gamma) \right| \tag{7.136}$$

$$\le C|\Gamma|^{s+1-1/r-1/q} |Q|_{W^{s,r}(\Gamma)} \|v\|_{L^q(\Gamma)},$$

$$Q \in W^{s,r}(\Gamma),\ v \in P_p(\Gamma),\ \Gamma \in \mathscr{F}_h,\ h \in (0,\bar{h}). \tag{7.137}$$

Proof First let us consider $r \in [1, \infty)$. Let $F_\Gamma : \widehat{\Gamma} = [0, 1] \stackrel{\text{onto}}{\longrightarrow} \Gamma$ be an affine mapping. Then for $\psi \in C(\Gamma)$, $Q \in W^{s,r}(\Gamma)$, $v \in P_p(\Gamma)$ we denote

$$\widehat{\psi} = \psi \circ F_\Gamma, \quad \widehat{Q} = Q \circ F_\Gamma, \quad \widehat{v} = v \circ F_\Gamma. \tag{7.138}$$

Thus,

$$\widehat{\psi} \in C(\widehat{\Gamma}), \quad \widehat{Q} \in W^{s,r}(\widehat{\Gamma}), \quad \widehat{v} \in P_p(\widehat{\Gamma}). \tag{7.139}$$

Let us set $\widehat{x}_\alpha^\Gamma = F_\Gamma^{-1}(x_\alpha^\Gamma)$, $\alpha = 1, \dots, m_\Gamma$, where F_Γ^{-1} is the inverse of F_Γ. Then we obtain the transformed quadrature formula

$$\int_{\widehat{\Gamma}} \widehat{\psi} \, d\widehat{x} \approx \sum_{\alpha=1}^{m_\Gamma} \beta_\alpha^\Gamma \widehat{\psi}(\widehat{x}_\alpha^\Gamma) \tag{7.140}$$

with error

$$\widehat{E}(\widehat{\psi}) = \int_{\widehat{\Gamma}} \widehat{\psi} \, d\widehat{x} - \sum_{\alpha=1}^{m_\Gamma} \beta_\alpha^\Gamma \widehat{\psi}(\widehat{x}_\alpha^\Gamma). \tag{7.141}$$

We have

$$\int_\Gamma \psi \, dS = |\Gamma| \int_{\widehat{\Gamma}} \widehat{\psi} \, d\widehat{x}, \tag{7.142}$$

which implies that

$$E_\Gamma(\psi) = |\Gamma| \, \widehat{E}(\widehat{\psi}). \tag{7.143}$$

Since $rs > 1$, by virtue of the embedding Theorem 1.3, we have the continuous embedding $W^{s,r}(\widehat{\Gamma}) \hookrightarrow C(\widehat{\Gamma})$. Hence, there exists a constant $c_1 > 0$ independent of $\widehat{Q}$ such that

$$\|\widehat{Q}\|_{C(\widehat{\Gamma})} = \max_{\widehat{\Gamma}} |\widehat{Q}| \le c_1 \|\widehat{Q}\|_{W^{s,r}(\widehat{\Gamma})}. \tag{7.144}$$

Now, by (7.141), where we put $\widehat{\psi} := \widehat{Q}\widehat{v} \in C(\widehat{\Gamma})$, and (7.120),

$$|\widehat{E}(\widehat{Q}\widehat{v})| \le \left(1 + \sum_{\alpha=1}^{m_\Gamma} |\beta_\alpha^\Gamma|\right) \max_{\widehat{\Gamma}} |\widehat{Q}| \max_{\widehat{\Gamma}} |\widehat{v}| \le c_1(1+\beta) \|\widehat{Q}\|_{W^{s,r}(\widehat{\Gamma})} \max_{\widehat{\Gamma}} |\widehat{v}|. \tag{7.145}$$

Taking into account that all norms are equivalent in the finite-dimensional space $P_p(\widehat{\Gamma})$, there exists a constant $\widehat{c} > 0$ such that

$$\max_{\widehat{\Gamma}} |\widehat{w}| \leq \widehat{c}\, \|\widehat{w}\|_{L^q(\widehat{\Gamma})} \quad \forall \widehat{w} \in P_p(\widehat{\Gamma}). \tag{7.146}$$

This inequality and (7.145) imply the estimate

$$|\widehat{E}(\widehat{Q}\widehat{v})| \leq c_2 \|\widehat{Q}\|_{W^{s,r}(\widehat{\Gamma})} \|\widehat{v}\|_{L^q(\widehat{\Gamma})}, \tag{7.147}$$

with $c_2 = c_1 \widehat{c}(1 + \beta)$.

Now let us consider $\widehat{v} \in P_p(\widehat{\Gamma})$ fixed. Then, by (7.147), the mapping

$$\widehat{Q} \in W^{s,r}(\widehat{\Gamma}) \mapsto f_{\widehat{v}}(\widehat{Q}) = \widehat{E}(\widehat{Q}\widehat{v}) \tag{7.148}$$

is a continuous linear functional on $W^{s,r}(\widehat{\Gamma})$ with the norm

$$\| f_{\widehat{v}} \|^* \leq c_2 \|\widehat{v}\|_{L^q(\widehat{\Gamma})}. \tag{7.149}$$

By virtue of the assumption that the quadrature formula (7.115) is exact for polynomials of degree $\leq s + p - 1$, by (7.143) we have

$$\widehat{E}(\widehat{\psi}) = 0 \quad \forall \widehat{\psi} \in P_{s+p-1}(\widehat{\Gamma}),$$

and from (7.148) we find that

$$f_{\widehat{v}}(\widehat{Q}) = 0 \quad \forall \widehat{Q} \in P_{s-1}(\widehat{\Gamma}).$$

Together with the Bramble–Hilbert lemma, i.e., Theorem 2.18, this implies that there exists a constant $c_3 > 0$ independent of $\widehat{Q}$ and $\widehat{v}$ such that

$$| f_{\widehat{v}}(\widehat{Q})| \leq c_3 \| f_{\widehat{v}}\|^* \, |\widehat{Q}|_{W^{s,r}(\widehat{\Gamma})}. \tag{7.150}$$

Hence, from (7.149) and (7.150) we get

$$|\widehat{E}(\widehat{Q}\widehat{v})| \leq c_4 |\widehat{Q}|_{W^{s,r}(\widehat{\Gamma})} \|\widehat{v}\|_{L^q(\widehat{\Gamma})}, \tag{7.151}$$

where $c_4 = c_2 c_3 > 0$ is a constant independent of $\widehat{Q}$, and $\widehat{v}$.

Now we carry out the transformation back to the face Γ. Using the relations

$$|F'_\Gamma| = |\Gamma|, \qquad |\widehat{Q}^{(s)}(\widehat{x})| = |\Gamma|^s \, |Q^{(s)}(F_\Gamma(\widehat{x}))|, \quad \widehat{x} \in \widehat{\Gamma},$$

(where $Q^{(s)}$ means the sth derivative of Q with respect to the arclength measured along Γ), by (7.142), for $r \in [1, \infty)$ we get

$$|\widehat{Q}|_{W^{s,r}(\widehat{\Gamma})} = \left(\int_{\widehat{\Gamma}} |\widehat{Q}^{(s)}(\widehat{x})|^r \, d\widehat{x} \right)^{1/r} \tag{7.152}$$
$$= \left(|\Gamma|^{-1} \int_{\Gamma} (|\Gamma|^s \, |Q^{(s)}|)^r dS \right)^{1/r} = |\Gamma|^{s-1/r} \, |Q|_{W^{s,r}(\Gamma)}.$$

Moreover,

$$\|\widehat{v}\|_{L^q(\widehat{\Gamma})} = |\Gamma|^{-1/q} \|v\|_{L^q(\Gamma)}. \tag{7.153}$$

Finally, from (7.142), (7.143), (7.151)–(7.153) we immediately obtain (7.136).
In the case $r = \infty$ we can proceed similarly. □

Furthermore, we consider approximation (7.114) of the volume integrals.

Lemma 7.37 *Let us consider a shape-regular system $\{\mathscr{T}_h\}_{h\in(0,\bar{h})}$ of triangulations. Moreover, let $z,\ p > 0$ be integers, $r,\ q \in [1, \infty]$, $rz > 2$, let the quadrature formula (7.114) be exact for polynomials of degree $\le z + p - 1$ and let assumption (7.120) be satisfied. Then there exists a constant $C > 0$ such that*

$$|E_K(Qv)| = \left| \int_K Qv \, dx - |K| \sum_{\alpha=1}^{n_K} \omega_\alpha^K \, (Qv) \, (x_\alpha^K) \right| \tag{7.154}$$
$$\le C h_K^{z+2-2/r-2/q} |Q|_{W^{z,r}(K)} \|v\|_{L^q(K)},$$
$$Q \in W^{z,r}(K),\ v \in P_p(K),\ K \in \mathscr{T}_h,\ h \in (0, \bar{h}).$$

Proof We can proceed analogously as in the proof of Lemma 7.36. Now we use the standard reference triangle $\widehat{K}$ with vertices (0, 0), (1, 0), (0, 1) instead of the interval $\widehat{\Gamma}$, write K instead of Γ, z instead of s and use the mapping F_K of $\widehat{K}$ onto K, $F_K(\widehat{x}) = \mathbb{B}_K \widehat{x} + \boldsymbol{b}_K$, instead of F_Γ. Here $\mathbb{B}_K$ is a nonsingular 2×2 matrix and $\boldsymbol{b}_K \in \mathbb{R}^2$. Then the standard relation

$$|\det \mathbb{B}_K| = 2|K| \tag{7.155}$$

yields the transformed integral

$$\int_K \psi \, dx = 2|K| \int_{\widehat{K}} \widehat{\psi} \, d\widehat{x} \tag{7.156}$$

and integration formula

$$\int_{\widehat{K}} \widehat{\psi} \, d\widehat{x} \approx \sum_{\alpha=1}^{n_K} \widehat{\omega}_\alpha^K \, \widehat{\psi}(\widehat{x}_\alpha^K),$$

where $\widehat{\omega}_\alpha^K = \omega_\alpha^K/2$ and $\widehat{x}_\alpha^K = F_K^{-1}(x_\alpha^K)$. Thus, the error of numerical integration is transformed in the following way:

$$E_K(\psi) = 2|K|\,\widehat{E}(\widehat{\psi}) = 2|K|\left(\int_{\widehat{K}} \widehat{\psi}\,\mathrm{d}\widehat{x} - \sum_{\alpha=1}^{n_K} \widehat{\omega}_\alpha^K \widehat{\psi}(\widehat{x}_\alpha^K)\right). \tag{7.157}$$

If we take into account that in view of (1.27), the assumption $rz > 2$ implies the continuous embedding $W^{z,r}(K) \hookrightarrow C(K)$, similarly as in the proof of Lemma 7.36 we find that

$$|\widehat{E}(\widehat{Q}\widehat{v})| \le C|\widehat{Q}|_{W^{z,r}(\widehat{K})}\|\widehat{v}\|_{L^q(\widehat{K})}, \quad \widehat{Q} \in W^{z,r}(\widehat{K}),\ \widehat{v} \in P_p(\widehat{K}). \tag{7.158}$$

Now it is necessary to pass to the original element $K \in \mathscr{T}_h$ using standard scaling arguments. By Lemmas 2.13 and 2.12,

$$|\widehat{Q}|_{W^{z,r}(\widehat{K})} \le \tilde{C}_1\|\mathbb{B}_K\|^z\,|\det \mathbb{B}_K|^{-1/r}|Q|_{W^{z,r}(K)}, \qquad \|\mathbb{B}_K\| \le \tilde{C}_2 h_K. \tag{7.159}$$

Moreover, by (7.130)–(7.132),

$$h_K^2/2 \ge |K| > h_K^2/C_R^2. \tag{7.160}$$

These inequalities, (7.156), (7.155) and (7.159) imply that

$$\|\widehat{v}\|_{L^q(\widehat{K})} \le \tilde{C}_3 h_K^{-2/q}\|v\|_{L^q(K)}, \tag{7.161}$$

$$|\widehat{Q}|_{W^{z,r}(\widehat{K})} \le \tilde{C}_4 h_K^{z-2/r}|Q|_{W^{z,r}(K)}, \tag{7.162}$$

which together with (7.157) and (7.158) yield estimate (7.154).

In the case $r = \infty$ we can again proceed similarly. □

7.3.6 Properties of the Convection Forms

The convection form b_h satisfies relations from Lemma 4.6. Namely, if $\mu = p + 1$ and $\xi = u_h - \Pi_{hp}u$, $u \in H^{p+1}(\Omega)$ and φ_h, $u_h \in S_{hp}$, then, by (4.34), (7.123) and (7.124), we have

$$|b_h(u_h, \varphi_h) - b_h(u, \varphi_h)| \tag{7.163}$$
$$\le C_B\left(|\varphi_h|_{H^1(\Omega,\mathscr{T}_h)} + J_h^\sigma(\varphi_h, \varphi_h)^{1/2}\right)\left(h^{p+1}|u|_{H^{p+1}(\Omega)} + \|u_h - \Pi_{hp}u\|_{L^2(\Omega)}\right),$$

with a constant C_B depending on C_{b4} and C_A.

The form b_h can be expressed as

$$b_h(w,\varphi) = b_h^1(w,\varphi) + b_h^2(w,\varphi), \tag{7.164}$$
$$b_h^1(w,\varphi) = -\sum_{K\in\mathscr{T}_h}\int_K \sum_{\ell=1}^{2} f_\ell(w)\frac{\partial\varphi}{\partial x_\ell}\,\mathrm{d}x,$$
$$b_h^2(w,\varphi) = \sum_{\Gamma\in\mathscr{F}_h}\int_\Gamma H(w_\Gamma^{(L)}, w_\Gamma^{(R)}, \boldsymbol{n}_\Gamma)[\varphi]\,\mathrm{d}S,$$
$$w,\ \varphi \in H^2(\Omega,\mathscr{T}_h).$$

The convection form $\tilde{b}_h$ is obtained using the quadrature formulae (7.114) and (7.115) for the evaluation of integrals appearing in (7.164). We have

$$\tilde{b}_h(w,\varphi) = \tilde{b}_h^1(w,\varphi) + \tilde{b}_h^2(w,\varphi), \tag{7.165}$$
$$\tilde{b}_h^1(w,\varphi) = -\sum_{K\in\mathscr{T}_h}|K|\sum_{\alpha=1}^{n_K}\omega_\alpha^K\sum_{\ell=1}^{2} f_\ell(w(x_\alpha^K))\frac{\partial\varphi(x_\alpha^K)}{\partial x_\ell},$$
$$\tilde{b}_h^2(w,\varphi) = \sum_{\Gamma\in\mathscr{F}_h}|\Gamma|\sum_{\alpha=1}^{m_\Gamma}\beta_\alpha^\Gamma H\left(w_\Gamma^{(L)}(x_\alpha^\Gamma),\ w_\Gamma^{(R)}(x_\alpha^\Gamma),\ \boldsymbol{n}_\Gamma\right)[\varphi(x_\alpha^\Gamma)],$$
$$w,\ \varphi \in H^2(\Omega,\mathscr{T}_h).$$

Similarly as in the case of the form b_h, we use in $\tilde{b}_h$ extrapolation for the treatment of the boundary value $w_\Gamma^{(R)}(x_\alpha^\Gamma)$, if $\Gamma\subset\partial\Omega$. This means that we set

$$w_\Gamma^{(R)}(x_\alpha^\Gamma) := w_\Gamma^{(L)}(x_\alpha^\Gamma) \quad \text{for } \Gamma\in\mathscr{F}_h^B. \tag{7.166}$$

Let us note that if $w\in H^2(\Omega,\mathscr{T}_h)$, then $w|_K\in C(K)$ for each $K\in\mathscr{T}_h$ and the definition (7.165) makes sense.

Lemma 7.38 *Let Assumptions 4.5 from Sect. 4.3 be satisfied. Then there exists a constant $C_{LB}>0$ such that*

$$|\tilde{b}_h(u_h,\varphi_h) - \tilde{b}_h(v_h,\varphi_h)| \tag{7.167}$$
$$\le C_{LB}\|u_h - v_h\|_{L^2(\Omega)}\left(|\varphi_h|_{H^1(\Omega,\mathscr{T}_h)} + J_h^\sigma(\varphi_h,\varphi_h)^{1/2}\right),$$
$$u_h,\ v_h,\ \varphi_h\in S_{hp},\ h\in(0,\bar{h}).$$

Proof Let $u_h,\ v_h,\ \varphi_h\in S_{hp}$. By (7.165), (7.160) and the Lipschitz continuity of the functions f_ℓ (with constant $L_f = 2L_H$),

$$
\begin{aligned}
&|\tilde{b}_h^1(u_h,\varphi_h)-\tilde{b}_h^1(v_h,\varphi_h)|\\
&\le L_f\sum_{K\in\mathscr{T}_h}|K|\sum_{\alpha=1}^{n_K}|\omega_\alpha^K|\sum_{\ell=1}^{2}|u_h(x_\alpha^K)-v_h(x_\alpha^K)|\left|\frac{\partial\varphi_h(x_\alpha^K)}{\partial x_\ell}\right|\\
&\le C\sum_{K\in\mathscr{T}_h}h_K^2\|u_h-v_h\|_{L^\infty(K)}|\varphi_h|_{W^{1,\infty}(K)}.
\end{aligned}
$$

This estimate, (7.126), (7.127) and the Cauchy inequality imply that

$$
\begin{aligned}
&|\tilde{b}_h^1(u_h,\varphi_h)-\tilde{b}_h^1(v_h,\varphi_h)| \qquad (7.168)\\
&\le C\sum_{K\in\mathscr{T}_h}\|u_h-v_h\|_{L^2(K)}|\varphi_h|_{H^1(K)}\le C\|u_h-v_h\|_{L^2(\Omega)}|\varphi_h|_{H^1(\Omega,\mathscr{T}_h)}.
\end{aligned}
$$

Further, by the definition of $\tilde{b}_h^2$ in (7.165), we have

$$
\begin{aligned}
&\tilde{b}_h^2(u_h,\varphi_h)-\tilde{b}_h^2(v_h,\varphi_h)\\
&=\sum_{\Gamma\in\mathscr{F}_h}|\Gamma|\sum_{\alpha=1}^{m_\Gamma}\beta_\alpha^\Gamma[\varphi_h(x_\alpha^\Gamma)]\\
&\quad\times\Big(H(u_{h\Gamma}^{(L)}(x_\alpha^\Gamma),u_{h\Gamma}^{(R)}(x_\alpha^\Gamma),\boldsymbol{n}_\Gamma)-H(v_{h\Gamma}^{(L)}(x_\alpha^\Gamma),v_{h\Gamma}^{(R)}(x_\alpha^\Gamma),\boldsymbol{n}_\Gamma)\Big).
\end{aligned}
$$

From this relation, the Lipschitz continuity of the numerical flux H and assumption (7.120) it follows that

$$
\begin{aligned}
&|\tilde{b}_h^2(u_h,\varphi_h)-\tilde{b}_h^2(v_h,\varphi_h)| \qquad (7.169)\\
&\le C\sum_{\Gamma\in\mathscr{F}_h}|\Gamma|\sum_{\alpha=1}^{m_\Gamma}|\beta_\alpha^\Gamma|\,|[\varphi_h(x_\alpha^\Gamma)]|\Big(|u_{h\Gamma}^{(L)}(x_\alpha^\Gamma)-v_{h\Gamma}^{(L)}(x_\alpha^\Gamma)|+|u_{h\Gamma}^{(R)}(x_\alpha^\Gamma)-v_{h\Gamma}^{(R)}(x_\alpha^\Gamma)|\Big)\\
&\le C\sum_{\Gamma\in\mathscr{F}_h}|\Gamma|\,\|[\varphi_h]\|_{L^\infty(\Gamma)}\|u_h-v_h\|_{L^\infty(\Gamma)}
\end{aligned}
$$

Now we use the inverse inequality (7.128), relations (7.113), (7.118) and the Cauchy inequality and get

$$
\begin{aligned}
&|\tilde{b}_h^2(u_h,\varphi_h)-\tilde{b}_h^2(v_h,\varphi_h)|\le C\sum_{\Gamma\in\mathscr{F}_h}\Big\{\|[\varphi_h]\|_{L^2(\Gamma)}\|u_h-v_h\|_{L^2(\Gamma)}\Big\}\\
&\le C\sum_{\Gamma\in\mathscr{F}_h}\Big(\int_\Gamma\sigma[\varphi_h]^2\mathrm{d}S\Big)^{1/2}\Big(\sum_{\Gamma\in\mathscr{F}_h}\int_\Gamma\sigma^{-1}|u_h-v_h|^2\mathrm{d}S\Big)^{1/2}\\
&< CJ_h^\sigma(\varphi_h,\varphi_h)^{1/2}\Big(\sum_{K\in\mathscr{T}_h}h_K\|u_h-v_h\|_{L^2(\partial K)}^2\Big)^{1/2}.
\end{aligned}
$$

This inequality, the multiplicative trace inequality (2.78) and the inverse inequality (7.122) imply that

$$|\tilde{b}_h^2(u_h, \varphi_h) - \tilde{b}_h^2(v_h, \varphi_h)| \leq C\, J_h^\sigma(\varphi_h, \varphi_h)^{1/2} \Big(\sum_{K \in \mathscr{T}_h} \|u_h - v_h\|^2_{L^2(K)} \Big)^{1/2} \quad (7.170)$$
$$= C J_h^\sigma(\varphi_h, \varphi_h)^{1/2} \|u_h - v_h\|_{L^2(\Omega)}.$$

From (7.165), (7.168) and (7.170) we immediately obtain (7.167). □

Lemma 7.39 *Let Assumptions 4.5 from Sect. 4.3 be satisfied. Then there exists a constant $C_{BB} > 0$ such that*

$$|\tilde{b}_h(u, \varphi_h) - \tilde{b}_h(\Pi_{hp}u, \varphi_h)| \leq C_{BB} h^{p+1} |u|_{H^{p+1}(\Omega)} \Big(|\varphi_h|_{H^1(\Omega, \mathscr{T}_h)} + J_h^\sigma(\varphi_h, \varphi_h)^{1/2} \Big),$$
$$u \in H^{p+1}(\Omega),\ \varphi_h \in S_{hp},\ h \in (0, \bar{h}). \quad (7.171)$$

Proof Since $H^{p+1}(\Omega) \hookrightarrow C(\overline{\Omega})$, the expression $\tilde{b}_h(u, \varphi_h)$ makes sense. Similarly as in the proof of Lemma 7.38, we find that

$$|\tilde{b}_h^1(u, \varphi_h) - \tilde{b}_h^1(\Pi_{hp}u, \varphi_h)| \leq C \sum_{K \in \mathscr{T}_h} h_K^2 \|u - \Pi_{hp}u\|_{L^\infty(K)} |\varphi_h|_{W^{1,\infty}(K)}. \quad (7.172)$$

Taking into account (7.160) and using (7.172), (7.129), (7.127) and the Cauchy inequality, we easily arrive at the estimate

$$|\tilde{b}_h^1(u, \varphi_h) - \tilde{b}_h^1(\Pi_{hp}u, \varphi_h)| \leq C \sum_{K \in \mathscr{T}_h} h_K^{p+1} |u|_{H^{p+1}(K)} |\varphi_h|_{H^1(K)} \quad (7.173)$$
$$\leq C\, h^{p+1} |u|_{H^{p+1}(\Omega)} |\varphi_h|_{H^1(\Omega, \mathscr{T}_h)}.$$

Further, we can proceed similarly as in the derivation of (7.169) and show that

$$|\tilde{b}_h^2(u, \varphi_h) - \tilde{b}_h^2(\Pi_{hp}u, \varphi_h)| \leq C \sum_{\Gamma \in \mathscr{F}_h} |\Gamma|\, \|[\varphi_h]\|_{L^\infty(\Gamma)} \|u - \Pi_{hp}u\|_{L^\infty(\Gamma)}. \quad (7.174)$$

Since $(u - \Pi_{hp}u)|_K \in C(K)$, we see that $\|u - \Pi_{hp}u\|_{L^\infty(\Gamma)} \leq \|u - \Pi_{hp}u\|_{L^\infty(K_\Gamma^{(L)})}$. By virtue of this inequality, (7.174), (7.128), (7.126), (7.113), (7.123) and the inequality $|\Gamma| \leq h_{K_\Gamma^{(L)}}$, we have

$$|\tilde{b}_h^2(u,\varphi_h) - \tilde{b}_h^2(\Pi_{hp}u,\varphi_h)| \le C \sum_{\Gamma\in\mathscr{F}_h} |\Gamma| \|[\varphi_h]\|_{L^\infty(\Gamma)} \|u - \Pi_{hp}u\|_{L^\infty(K_\Gamma^{(L)})} \tag{7.175}$$
$$\le C \sum_{\Gamma\in\mathscr{F}_h} |\Gamma|^{-1/2}|\Gamma| \|[\varphi_h]\|_{L^2(\Gamma)} h_{K_\Gamma^{(L)}}^{-1} \|u - \Pi_{hp}u\|_{L^2(K_\Gamma^{(L)})}$$
$$\le C \sum_{\Gamma\in\mathscr{F}_h} h_{K_\Gamma^{(L)}}^{1/2} h_{K_\Gamma^{(L)}}^{1/2} \|h_{K_\Gamma^{(L)}}^{-1/2}[\varphi_h]\|_{L^2(\Gamma)} h_{K_\Gamma^{(L)}}^{-1} \|u - \Pi_{hp}u\|_{L^2(K_\Gamma^{(L)})}$$
$$\le C \sum_{\Gamma\in\mathscr{F}_h} \|\sigma^{1/2}[\varphi_h]\|_{L^2(\Gamma)} h_{K_\Gamma^{(L)}}^{p+1} |u|_{H^{p+1}(K_\Gamma^{(L)})}.$$

Now we take into account that

$$\sum_{\Gamma\in\mathscr{F}_h} \|\sigma^{1/2}[\varphi_h]\|_{L^2(\Gamma)}^2 = J_h^\sigma(\varphi,\varphi), \tag{7.176}$$
$$\sum_{\Gamma\in\mathscr{F}_h} |u|_{H^{p+1}(K_\Gamma^{(L)})}^2 \le 3 \sum_{K\in\mathscr{T}_h} |u|_{H^{p+1}(K)}^2 = 3|u|_{H^{p+1}(\Omega)}^2,$$

apply the Cauchy inequality to (7.175) and get the estimate

$$|\tilde{b}_h^2(u,\varphi_h) - \tilde{b}_h^2(\Pi_{hp}u,\varphi_h)| \le C J_h^\sigma(\varphi_h,\varphi_h)^{1/2} h^{p+1} |u|_{H^{p+1}(\Omega)}. \tag{7.177}$$

Finally, (7.165), (7.173) and (7.177) yield (7.171), what we wanted to prove. □

7.3.7 The Effect of Numerical Integration in the Convection Form

In this section we will be first concerned with analyzing the effect of numerical integration in the convection form b_h. This means that our goal is to estimate the expression

$$E_b(u,\varphi_h) = \tilde{b}_h(u,\varphi_h) - b_h(u,\varphi_h) \tag{7.178}$$

for $\varphi_h \in S_{hp}$ and a sufficiently regular function u.

Taking into account the definitions (7.109) and (7.165) of the forms b_h and $\tilde{b}_h$, respectively, we see that it is necessary to investigate the properties of the superposition operators "$u \to f_\ell(u)$", $\ell = 1, 2$, defined for functions u from a suitable Sobolev space $W^{k,r}(\Omega)$. (For a general theory, see [248] or [6].) By [6], Theorem 9.7, $f_\ell(u) \in W^{k,r}(\Omega)$ for $u \in W^{k,r}(\Omega)$ ($k \ge 1$ is an integer, $r \in [1,\infty]$), provided $f_\ell \in C^k(\mathbb{R})$ and $kr > 2$. Moreover, if the derivative $f_\ell^{(k)}$ is bounded, then it is also possible to consider the case $kr = 2$. In this section we assume that the functions u and f_ℓ are such that

$$f_\ell(u) \in W^{k,\infty}(\Omega), \quad \ell = 1, 2. \tag{7.179}$$

Lemma 7.40 *Let s, $p \geq 1$ be integers, $k = s$ and let the quadrature formulae (7.114) and (7.115) be exact for polynomials of degree $\leq s + p - 2$ and $s + p - 1$, respectively. Let Assumptions 4.5 from Sect. 4.3 be satisfied. Then*

$$|b_h(u, \varphi_h) - \tilde{b}_h(u, \varphi_h)| \leq C\, h^{s-1} \max_{\ell=1,2} \left\{ |f_\ell(u)|_{W^{s,\infty}(\Omega)} \right\} \|\varphi_h\|_{L^2(\Omega)} \tag{7.180}$$

for u satisfying (7.179), $\varphi_h \in S_{hp}$ and $h \in (0, \bar{h})$, with a constant $C > 0$ independent of u, φ_h and h.

Proof By (7.164) and (7.165), $b_h = b_h^1 + b_h^2$ and $\tilde{b}_h = \tilde{b}_h^1 + \tilde{b}_h^2$. First we estimate the effect of numerical integration in the form $\tilde{b}_h^1$. In view of (7.135), we can write

$$E_b^1(u, \varphi_h) = b_h^1(u, \varphi_h) - \tilde{b}_h^1(u, \varphi_h) = \sum_{K \in \mathscr{T}_h} E_K\left(-\sum_{\ell=1}^{2} f_\ell(u) \frac{\partial \varphi_h}{\partial x_\ell} \right), \tag{7.181}$$

where

$$E_K\left(-\sum_{\ell=1}^{2} f_\ell(u) \frac{\partial \varphi_h}{\partial x_\ell} \right)$$
$$= -\left(\int_K \sum_{\ell=1}^{2} f_\ell(u) \frac{\partial \varphi_h}{\partial x_\ell}\, \mathrm{d}x - |K| \sum_{\alpha=1}^{n_K} \omega_\alpha^K \sum_{\ell=1}^{2} \left(f_\ell(u) \frac{\partial \varphi_h}{\partial x_\ell} \right)(x_\alpha^K) \right).$$

In view of (7.179), we have $f_\ell(u) \in C(\overline{\Omega})$ and all above terms make sense.

Taking into account that $\left.\frac{\partial \varphi_h}{\partial x_\ell}\right|_K \in P_{p-1}(K)$ and using Lemma 7.37 with $r = \infty$, $q = 2$, $z = s$, $p := p - 1$, and the inverse inequality (7.122), we find that

$$\left| E_K\left(-\sum_{\ell=1}^{2} f_\ell(u) \frac{\partial \varphi_h}{\partial x_\ell} \right) \right| \leq C\, h_K^{s+1} \max_{\ell=1,2} |f_\ell(u)|_{W^{s,\infty}(K)} |\varphi_h|_{H^1(K)}$$
$$\leq C\, h_K^{s} \max_{\ell=1,2} |f_\ell(u)|_{W^{s,\infty}(K)} \|\varphi_h\|_{L^2(K)}.$$

This inequality and (7.181) imply that

$$|E_b^1(u, \varphi_h)| \leq C\, h^{s-1} \max_{\ell=1,2} |f_\ell(u)|_{W^{s,\infty}(\Omega)} \sum_{K \in \mathscr{T}_h} h_K \|\varphi_h\|_{L^2(K)}.$$

By (7.160) and the Cauchy inequality,

$$\sum_{K\in\mathscr{T}_h} h_K \|\varphi_h\|_{L^2(K)} \le \left(C_R \sum_{K\in\mathscr{T}_h} |K| \right)^{1/2} \|\varphi_h\|_{L^2(\Omega)} = C_R^{1/2} |\Omega|^{1/2} \|\varphi_h\|_{L^2(\Omega)}.$$

Hence,

$$|E_b^1(u,\varphi_h)| \le C\, h^{s-1} \max_{\ell=1,2} |f_\ell(u)|_{W^{s,\infty}(\Omega)} \|\varphi_h\|_{L^2(\Omega)} \tag{7.182}$$

with a constant $C > 0$ independent of u, φ_h and h.

Now let us deal with the form $\tilde{b}_h^2$. Since $u,\ f_\ell(u) \in C(\overline{\Omega})$ $(\ell = 1, 2)$, due to the consistency (4.19) of the numerical flux and the definition (7.165) of $\tilde{b}_h^2$, we have

$$\tilde{b}_h^2(u,\varphi_h) = \sum_{\Gamma\in\mathscr{F}_h} |\Gamma| \sum_{\alpha=1}^{m_\Gamma} \beta_\alpha^\Gamma \sum_{\ell-1}^{2} f_\ell(u(x_\alpha^\Gamma))\, (n_\Gamma)_\ell\, [\varphi_h](x_\alpha^\Gamma). \tag{7.183}$$

Moreover, by (7.164) and (4.19),

$$b_h^2(u,\varphi_h) = \sum_{\Gamma\in\mathscr{F}_h} \int_\Gamma \sum_{\ell=1}^{2} f_\ell(u)\, (n_\Gamma)_\ell [\varphi_h]\, \mathrm{d}S. \tag{7.184}$$

Since $|(n_\Gamma)_\ell| \le 1$, $\ell = 1, 2$, we can write

$$|b_h^2(u,\varphi_h) - \tilde{b}_h^2(u,\varphi_h)| \le \sum_{\Gamma\in\mathscr{F}_h} \sum_{\ell=1}^{2} |E_\Gamma^\ell(f_\ell(u)\,[\varphi_h])|, \tag{7.185}$$

where, in view of (7.134),

$$E_\Gamma^\ell(f_\ell(u)\,[\varphi_h]) = \int_\Gamma f_\ell(u)\,[\varphi_h]\,\mathrm{d}S - |\Gamma| \sum_{\alpha=1}^{m_\Gamma} \beta_\alpha^\Gamma f_\ell \left(u(x_\alpha^\Gamma)\right)\, [\varphi_h](x_\alpha^\Gamma). \tag{7.186}$$

Since $f_\ell(u) \in W^{s,\infty}(\Omega)$, the derivatives of $f_\ell(u)$ of order $s-1$ are Lipschitz-continuous in $\overline{\Omega}$, and hence these derivatives are also Lipschitz-continuous on each Γ with the same Lipschitz constant. This implies that $f_\ell(u) \in W^{s,\infty}(\Gamma)$ and

$$|f_\ell(u)|_{W^{s,\infty}(\Gamma)} \le |f_\ell(u)|_{W^{s,\infty}(\Omega)}. \tag{7.187}$$

If we consider $\Gamma \subset \partial K$, employ Lemma 7.36 with $r = \infty$ and $q = 2$, relation (7.187), multiplicative trace inequality (2.78) and inverse inequality (7.122) for φ_h, we find that

$$\left|E_{\Gamma}^{\ell}(f_{\ell}(u)\,[\varphi_h])\right| \leq C|\Gamma|^{s+1/2}|f_{\ell}(u)|_{W^{s,\infty}(\Gamma)}\|[\varphi_h]\|_{L^2(\Gamma)}$$
$$\leq C\,h_K^{s+1/2}|f_{\ell}(u)|_{W^{s,\infty}(\Omega)}\left(\|[\varphi_h]\|_{L^2(K)}|[\varphi_h]|_{H^1(K)} + h_K^{-1}\|[\varphi_h]\|^2_{L^2(K)}\right)^{1/2}$$
$$\leq C\,h_K^{s}|f_{\ell}(u)|_{W^{s,\infty}(\Omega)}\|[\varphi_h]\|_{L^2(K)}.$$

This estimate and the inequality $|[\varphi_h]_\Gamma| \leq |\varphi_{hK}^{(L)}| + |\varphi_{hK}^{(R)}|$ imply that

$$|b_h^2(u,\varphi_h) - \tilde{b}_h^2(u,\varphi_h)| \leq C\,h^{s-1}\sum_{\ell=1}^{2}|f_{\ell}(u)|_{W^{s,\infty}(\Omega)}\sum_{\Gamma\in\mathscr{F}_h} h_K\,\|\varphi_h\|_{L^2(K)}. \tag{7.188}$$

By the Cauchy inequality and (7.160),

$$\sum_{K\in\mathscr{T}_h} h_K\,\|\varphi_h\|_{L^2(K)} \leq \left(\sum_{K\in\mathscr{T}_h} h_K^2\right)^{1/2}\|\varphi_h\|_{L^2(\Omega)} \leq C|\Omega|^{1/2}\|\varphi_h\|_{L^2(\Omega)}.$$

From this, (7.188) and (7.182) we immediately get (7.180). □

Now we analyze the effect of numerical integration on the right-hand side form ℓ_h (cf. (7.111)), approximated by

$$\tilde{\ell}_h(\varphi_h) = \sum_{K\in\mathscr{T}_h}|K|\sum_{\alpha=1}^{n_K}\omega_\alpha^K(g\varphi_h)(x_\alpha^K) - \varepsilon\Theta\sum_{\Gamma\in\mathscr{F}_h^B}|\Gamma|\sum_{\alpha=1}^{m_\Gamma}\beta_\alpha^\Gamma(\nabla\varphi_h\cdot\boldsymbol{n}_\Gamma\,u_D)(x_\alpha^\Gamma)$$
$$+\,\varepsilon\sum_{\Gamma\in\mathscr{F}_h^B}\sigma_\Gamma|\Gamma|\sum_{\alpha=1}^{m_\Gamma}\beta_\alpha^\Gamma(u_D\varphi_h)(x_\alpha^\Gamma). \tag{7.189}$$

Lemma 7.41 *Let us assume that $g \in H^{s_\Omega}(\Omega)$ and $u_D \in H^{s_D}(\partial\Omega)$, where $s_\Omega, s_D \geq \tilde{p} = \max\{2, p\}$ are integers. Moreover, let the system $\{\mathscr{T}_h\}_{h\in(0,\bar{h})}$ of triangulations be shape-regular, let the quadrature formulae (7.114) be exact for polynomials of degree $\leq s_\Omega + p - 2$ and let the quadrature formula (7.115) be exact for polynomials of degree $\leq s_D + p - 1$ on $\partial\Omega$. Then there exists a constant $C > 0$ such that*

$$|\ell_h(\varphi_h) - \tilde{\ell}_h(\varphi_h)| \leq C\left\{h^{s_\Omega}|g|_{H^{s_\Omega}(\Omega)} + \varepsilon h^{s_D-3/2}\|u_D\|_{H^{s_D}(\partial\Omega)}\right\}\|\varphi_h\|_{L^2(\Omega)} \tag{7.190}$$
$$+\,C\varepsilon h^{s_D-3/2}\|u_D\|_{H^{s_D}(\partial\Omega)}\|\varphi_h\|_{L^2(\Omega,\mathscr{T}_h)},\ \ \varphi_h \in S_{hp},\ h\in(0,\bar{h}).$$

Proof In view of (7.111) and (7.189),

$$|\ell_h(\varphi_h) - \tilde{\ell}_h(\varphi_h)| \le E_1 + E_2 + E_3, \tag{7.191}$$

where

$$E_1 = \left| \sum_{K \in \mathscr{T}_h} \left(\int_K g\varphi_h \,\mathrm{d}x - |K| \sum_{\alpha=1}^{n_K} \omega_\alpha^K (g\varphi_h)(x_\alpha^K) \right) \right|, \tag{7.192}$$

$$E_2 = \varepsilon \left| \sum_{\Gamma \in \mathscr{F}_h^B} \left(\int_\Gamma (\nabla\varphi_h \cdot \boldsymbol{n}_\Gamma)\, u_D \,\mathrm{d}S - |\Gamma| \sum_{\alpha=1}^{m_\Gamma} \beta_\alpha^\Gamma (\nabla\varphi_h \cdot \boldsymbol{n}_\Gamma\, u_D)(x_\alpha^\Gamma) \right) \right|,$$

$$E_3 = \varepsilon \left| \sum_{\Gamma \in \mathscr{F}_h^B} \sigma|_\Gamma \left(\int_\Gamma u_D\varphi_h \,\mathrm{d}S - |\Gamma| \sum_{\alpha=1}^{m_\Gamma} \beta_\alpha^\Gamma (u_D\varphi_h)(x_\alpha^\Gamma) \right) \right|.$$

Now we estimate expressions E_1, E_2, E_3. By Lemma 7.37 and the Cauchy inequality we have

$$E_1 \le C \sum_{K \in \mathscr{T}_h} h_K^{s_\Omega} |g|_{H^{s_\Omega}(K)} \|\varphi_h\|_{L^2(K)} \le C\, h^{s_\Omega} |g|_{H^{s_\Omega}(\Omega)} \|\varphi_h\|_{L^2(\Omega)}. \tag{7.193}$$

Further,

$$E_2 \le \varepsilon \sum_{\Gamma \in \mathscr{F}_h^B} \sum_{\ell=1}^{2} E_{2\Gamma}^\ell, \tag{7.194}$$

where

$$E_{2\Gamma}^\ell = \left| \int_\Gamma \frac{\partial\varphi_h}{\partial x_\ell} (n_\Gamma)_\ell\, u_D \,\mathrm{d}S - |\Gamma| \sum_{\alpha=1}^{m_\Gamma} \beta_\alpha^\Gamma \frac{\partial\varphi_h(x_\alpha^\Gamma)}{\partial x_\ell} (n_\Gamma)_\ell\, u_D(x_\alpha^\Gamma) \right|.$$

By Lemma 7.36, where we set $s = s_D$ and $r = q = 2$, the multiplicative trace inequality (2.78), the inverse inequality (7.122) and the inequality $|\Gamma| \le h_{K_\Gamma^{(L)}}$, we get

$$\begin{aligned} E_{2\Gamma}^\ell &\le C|\Gamma|^{s_D} \|u_D\|_{H^{s_D}(\Gamma)} \left\| \frac{\partial\varphi_h}{\partial x_\ell} \right\|_{L^2(\Gamma)} \le C h_{K_\Gamma^{(L)}}^{s_D - 1/2} \|u_D\|_{H^{s_D}(\Gamma)} |\varphi_h|_{H^1(K_\Gamma^{(L)})} \\ &\le C h_{K_\Gamma^{(L)}}^{s_D - 3/2} \|u_D\|_{H^{s_D}(\Gamma)} \|\varphi_h\|_{L^2(K_\Gamma^{(L)})}, \end{aligned}$$

which together with (7.194) and the Cauchy inequality implies that

$$E_2 \le C\varepsilon h^{s_D-3/2}\|u_D\|_{H^{s_D}(\partial\Omega)}\|\varphi_h\|_{L^2(\Omega)}. \tag{7.195}$$

Finally, we will estimate the expression E_3. We have

$$E_3 \le C\varepsilon \sum_{\Gamma\in\mathscr{F}_h^B} h_\Gamma^{-1} E_{3\Gamma}, \tag{7.196}$$

where

$$E_{3\Gamma} = \left| \int_\Gamma u_D\varphi_h \, \mathrm{d}S - |\Gamma| \sum_{\alpha=1}^{m_\Gamma} \beta_\alpha^\Gamma (u_D\varphi_h)(x_\alpha^\Gamma) \right|.$$

Again, by virtue of Lemma 7.36, the multiplicative trace inequality (2.78), the inverse inequality (7.122) and the inequality $|\Gamma| \le h_{K_\Gamma^{(L)}}$,

$$E_{3\Gamma} \le C|\Gamma|^{s_D}\|u_D\|_{H^{s_D}(\Gamma)}\|\varphi_h\|_{L^2(\Gamma)} \le C h_{K_\Gamma^{(L)}}^{s_D-1/2}\|u_D\|_{H^{s_D}(\Gamma)}\|\varphi_h\|_{L^2(K_\Gamma^{(L)})}.$$

These estimates, (7.196), the inequality $h_\Gamma^{-1} \le C_T^{-1} h_{K_\Gamma^{(L)}}^{-1}$ (which follows from relation (7.118)), and the Cauchy inequality imply that

$$E_3 \le C\varepsilon h^{s_D-3/2}\|u_D\|_{H^{s_D}(\partial\Omega)}\|\varphi_h\|_{L^2(\Omega)}. \tag{7.197}$$

Taking into account (7.191), (7.193), (7.195) and (7.197), we get (7.190). □

Corollary 7.42 *Under the assumptions of Lemma 7.41, where $s_\Omega = \tilde{p} = \max\{p, 2\}$ and $s_D = p + 2$, there exists a constant $C > 0$ such that*

$$\begin{aligned} &|\ell_h(\varphi_h) - \tilde{\ell}_h(\varphi_h)| \\ &\le C\left(h^{\tilde{p}}|g|_{H^{\tilde{p}}(\Omega)}\|\varphi_h\|_{L^2(\Omega)} + \varepsilon h^{p+1/2}\|u_D\|_{H^{p+2}(\partial\Omega)}\right)\|\varphi_h\|_{L^2(\Omega)}. \end{aligned} \tag{7.198}$$

7.3.8 Error Estimates for the Method of Lines with Numerical Integration

In this section using the obtained results we prove error estimates of the DGM with numerical integration. We apply the following version of the Gronwall lemma.

Lemma 7.43 *Let $y,\ q \in C([0,T])$, $y,\ q \ge 0$ in $[0,T]$, $Z,\ R \ge 0$ and*

$$y(t) + q(t) \le Z + R\int_0^t y(s)\,\mathrm{d}s, \quad t \in [0,T]. \tag{7.199}$$

Then

$$y(t) + q(t) \le Z \exp(Rt), \quad t \in [0, T]. \tag{7.200}$$

Exercise 7.44 Prove Lemma 7.43 on the basis of Lemma 1.9.

In the sequel we assume the regularity property (7.121) of the exact solution. Moreover, let the function u^0 from the initial condition satisfy

$$u^0 \in H^{\tilde{p}}(\Omega), \tag{7.201}$$

where $\tilde{p} = \max(p, 2)$.

We need the following estimate of the error in the initial condition.

Lemma 7.45 *Let the assumptions of the shape-regularity and (7.120) be satisfied and let the quadrature formulae (7.114) be exact for polynomials of degree $\le 2p$. Let u_h^0 be defined by (7.112c) and let $\tilde{u}_h^0$ be defined by (7.117). Then we have*

$$\|\tilde{u}_h^0 - u_h^0\|_{L^2(\Omega)} \le Ch^{\tilde{p}}|u^0|_{H^{\tilde{p}}(\Omega)}, \qquad h \in (0, \bar{h}). \tag{7.202}$$

Proof Let $\varphi_h \in S_{hp}$. Since $(u_h^0\varphi_h)|_K$ is a polynomial of degree $\le 2p$, we have $(u_h^0, \varphi_h) = (u_h^0, \varphi_h)_h$, in view of the assumption of lemma. Moreover, by (7.112) and (7.117),

$$(u_h^0 - u^0, \varphi_h) = 0, \qquad (\tilde{u}_h^0 - u^0, \varphi_h)_h = 0, \qquad \forall \varphi_h \in S_{hp}.$$

This implies that for $e_I^0 = \tilde{u}_h^0 - u_h^0 \in S_{hp}$ we have

$$\begin{aligned}(e_I^0, \varphi_h) &= (\tilde{u}_h^0, \varphi_h) - (u_h^0, \varphi_h) = (\tilde{u}_h^0, \varphi_h)_h - (u_h^0, \varphi_h)_h = (u^0, \varphi_h)_h - (u^0, \varphi_h) \\ &= \sum_{K \in \mathscr{T}_h} |K| \sum_{\alpha=1}^{n_K} \omega_\alpha^K (u^0\varphi_h)(x_\alpha^K) - \int_\Omega u^0 \varphi_h \mathrm{d}x.\end{aligned}$$

Let us put $\varphi_h := e_I^0$ and use the fact that $u^0 \in H^{\tilde{p}}(\Omega)$. Then Lemma 7.37 with $z = \tilde{p} \ge 2$, $r = q = 2$ and the Cauchy inequality imply that

$$\|e_I^0\|_{L^2(\Omega)}^2 \le C \sum_{K \in \mathscr{T}_h} h_K^{\tilde{p}} |u^0|_{H^{\tilde{p}}(K)} \|e_I^0\|_{L^2(K)} \le Ch^{\tilde{p}}|u^0|_{H^{\tilde{p}}(\Omega)} \|e_I^0\|_{L^2(\Omega)}.$$

Thus,

$$\|e_I^0\|_{L^2(\Omega)} \le Ch^{\tilde{p}}|u^0|_{H^{\tilde{p}}(\Omega)},$$

which we wanted to prove. □

Now we formulate and prove the main result on the error estimates of the DGM with numerical integration.

Theorem 7.46 *Let Assumptions 4.5 from Sect. 4.3 be satisfied. Let us assume that u is the exact solution of problem (7.106) satisfying the regularity conditions (7.121) and let $f_\ell(u) \in L^2(0,T; W^{p+1,\infty}(\Omega))$, $\ell = 1,2$. Moreover, let u_h be the approximate solution obtained by scheme (7.116) with numerical integration and let the quadrature formulae (7.114) and (7.115) be exact for polynomials of degree $\le 2p$. Let $g \in L^2(0,T; H^{\tilde{p}}(\Omega))$, $u_D \in L^2(0,T; H^{p+2}(\partial\Omega))$, $u^0 \in H^{\tilde{p}}(\Omega)$, where $\tilde{p} = \max\{p, 2\}$, and let $\tilde{u}_h^0$ and u_h^0 be defined as in Lemma 7.45. Then for the error $\tilde{e}_h = \tilde{u}_h - u$ of the method with numerical integration we have the estimate*

$$\max_{t\in[0,T]} \|\tilde{e}_h(t)\|^2_{L^2(\Omega)} + \frac{\varepsilon}{2}\int_0^T \left(|\tilde{e}_h(\vartheta)|^2_{H^1(\Omega,\mathscr{T}_h)} + \tilde{J}_h^\sigma(\tilde{e}_h(\vartheta), \tilde{e}_h(\vartheta))\right) \mathrm{d}\vartheta \le \widehat{C} h^{2p} \tag{7.203}$$

with a constant $\widehat{C}$ depending on u, g, u_D, T, ε and $\bar{h}$, but independent of h.

Proof Since $\tilde{e}_h = e_h + e_I$, where $e_h = u_h - u$ and $e_I = \tilde{u}_h - u_h$, and e_h satisfies estimate (7.133), it is enough to prove the estimate of e_I.

From the assumption on quadrature formulae and from the fact that $\tilde{u}_h(t) \in S_{hp}$ we conclude that for all $\varphi_h \in S_{hp}$ we have

$$\begin{aligned} \left(\frac{\partial \tilde{u}_h}{\partial t}, \varphi_h\right) &= \left(\frac{\partial \tilde{u}_h}{\partial t}, \varphi_h\right)_h, \\ a_h(\tilde{u}_h, \varphi_h) &= \tilde{a}_h(\tilde{u}_h, \varphi_h), \\ J_h^\sigma(\tilde{u}_h, \varphi_h) &= \tilde{J}_h^\sigma(\tilde{u}_h, \varphi_h). \end{aligned} \tag{7.204}$$

It follows from (7.204), (7.112) and (7.116) that

$$\left(\frac{\partial \tilde{u}_h}{\partial t}, \varphi_h\right) + \tilde{b}_h(\tilde{u}_h, \varphi_h) + a_h(\tilde{u}_h, \varphi_h) + \varepsilon J_h^\sigma(\tilde{u}_h, \varphi_h) = \tilde{\ell}_h(\varphi_h), \tag{7.205}$$

$$\left(\frac{\partial u_h}{\partial t}, \varphi_h\right) + b_h(u_h, \varphi_h) + a_h(u_h, \varphi_h) + \varepsilon J_h^\sigma(u_h, \varphi_h) = \ell_h(\varphi_h), \tag{7.206}$$

for all $\varphi_h \in S_{hp}$. If we subtract (7.206) from (7.205) and set $\varphi_h := e_I$, we get

$$\begin{aligned} &\frac{1}{2}\frac{\mathrm{d}}{\mathrm{d}t}\|e_I\|^2_{L^2(\Omega)} + a_h(e_I, e_I) + \varepsilon J_h^\sigma(e_I, e_I) \\ &= \left(\tilde{\ell}_h(e_I) - \ell_h(e_I)\right) + \left(b_h(u_h, e_I) - \tilde{b}_h(\tilde{u}_h, e_I)\right). \end{aligned} \tag{7.207}$$

Let us estimate the expression $b_h(u_h, e_I) - \tilde{b}_h(\tilde{u}_h, e_I)$. We can write

$$b_h(u_h, e_I) - \tilde{b}_h(\tilde{u}_h, e_I) = \sum_{k=1}^{4} \vartheta_k, \tag{7.208}$$

where

$$\vartheta_1 = b_h(u_h, e_I) - b_h(u, e_I), \qquad \vartheta_2 = b_h(u, e_I) - \tilde{b}_h(u, e_I), \tag{7.209}$$
$$\vartheta_3 = \tilde{b}_h(u, e_I) - \tilde{b}_h(\Pi_{hp}u, e_I), \qquad \vartheta_4 = \tilde{b}_h(\Pi_{hp}u, e_I) - \tilde{b}_h(\tilde{u}_h, e_I),$$

and estimate the individual terms ϑ_k. By (7.163), (7.123), (7.133) and the inequality $\|u_h - \Pi_{hp}u\|_{L^2(\Omega)} \le \|u_h - u\|_{L^2(\Omega)} + \|u - \Pi_{hp}u\|_{L^2(\Omega)}$,

$$|\vartheta_1| \le C \left(J_h^\sigma(e_I, e_I)^{1/2} + |e_I|_{H^1(\Omega, \mathscr{T}_h)} \right) \left(h^{p+1} |u|_{H^{p+1}(\Omega)} + \tilde{C}^{1/2} h^p \right). \tag{7.210}$$

Further, Lemma 7.40 implies that

$$|\vartheta_2| \le C\, h^p \max_{\ell=1,2} \left\{ |f_\ell(u)|_{W^{p+1,\infty}(\Omega)} \right\} \|e_I\|_{L^2(\Omega)}. \tag{7.211}$$

In view of Lemmas 7.39 and 7.38,

$$|\vartheta_3| \le C\, h^{p+1} |u|_{H^{p+1}(\Omega)} \left(J_h^\sigma(e_I, e_I)^{1/2} + |e_I|_{H^1(\Omega, \mathscr{T}_h)} \right) \tag{7.212}$$

and

$$|\vartheta_4| \le C \left(J_h^\sigma(e_I, e_I)^{1/2} + |e_I|_{H^1(\Omega, \mathscr{T}_h)} \right) \|\Pi_{hp}u - \tilde{u}_h\|_{L^2(\Omega)}.$$

It follows from this estimate, the inequality

$$\|\Pi_{hp}u - \tilde{u}_h\|_{L^2(\Omega)} \le \|\Pi_{hp}u - u\|_{L^2(\Omega)} + \|u - u_h\|_{L^2(\Omega)} + \|u_h - \tilde{u}_h\|_{L^2(\Omega)},$$

and estimates (7.133), (7.123) that

$$|\vartheta_4| \le C \left(J_h^\sigma(e_I, e_I)^{1/2} + |e_I|_{H^1(\Omega, \mathscr{T}_h)} \right) \left(h^{p+1} |u|_{H^{p+1}(\Omega)} + \tilde{C}^{1/2} h^p + \|e_I\|_{L^2(\Omega)} \right). \tag{7.213}$$

Summarizing (7.208)–(7.213), we find that

$$\begin{aligned} &|\tilde{b}_h(\tilde{u}_h, e_I) - b_h(u_h, e_I)| \\ &\le C \left(J_h^\sigma(e_I, e_I)^{1/2} + |e_I|_{H^1(\Omega, \mathscr{T}_h)} \right) \left(h^{p+1} |u|_{H^{p+1}(\Omega)} + \tilde{C}^{1/2} h^p + \|e_I\|_{L^2(\Omega)} \right) \\ &\quad + C\, h^p \max_{\ell=1,2} \left\{ |f_\ell(u)|_{W^{p+1,\infty}(\Omega)} \right\} \|e_I\|_{L^2(\Omega)}. \end{aligned} \tag{7.214}$$

Now, (7.119), Corollary 7.42, relation (7.207) and estimate (7.214) imply that

$$\frac{\mathrm{d}}{\mathrm{d}t}\|e_I\|^2_{L^2(\Omega)} + \varepsilon\left(|e_I|^2_{H^1(\Omega,\mathscr{T}_h)} + J^\sigma_h(e_I,e_I)\right) \tag{7.215}$$
$$\le C h^p\left(|g|_{H^{\bar p}(\Omega)} + \varepsilon h^{1/2}\|u_D\|_{H^{p+2}(\partial\Omega)}\right)\|e_I\|_{L^2(\Omega)}$$
$$+ C\left(J^\sigma_h(e_I,e_I)^{1/2} + |e_I|_{H^1(\Omega,\mathscr{T}_h)}\right)\left(h^{p+1}|u|_{H^{p+1}(\Omega)} + \tilde C^{1/2}h^p + \|e_I\|_{L^2(\Omega)}\right)$$
$$+ C h^p \max_{\ell=1,2}\left\{|f_\ell(u)|_{W^{p+1,\infty}(\Omega)}\right\}\|e_I\|_{L^2(\Omega)}.$$

With the aid of the Young inequality and integration with respect to time from 0 to $t \in [0,T]$, using the relations $e_I(0) = e^0_I = u^0_h - \tilde u^0_h$ and (7.202), we obtain

$$\|e_I(t)\|^2_{L^2(\Omega)} + \frac{\varepsilon}{2}\int_0^t\left(|e_I(\vartheta)|^2_{H^1(\Omega,\mathscr{T}_h)} + J^\sigma_h(e_I(\vartheta),e_I(\vartheta))\right)\mathrm{d}\vartheta \tag{7.216}$$
$$\le C h^{2p}\Big\{|g|^2_{L^2(0,T;H^{\bar p}(\Omega))} + h\varepsilon^2\|u_D\|^2_{L^2(0,T;H^{p+2}(\partial\Omega))}$$
$$+ \max_{\ell=1,2}\left\{|f_\ell(u)|^2_{L^2(0,T;W^{p+1,\infty}(\Omega))}\right\} + \frac{1}{\varepsilon}h^2|u|^2_{L^2(0,T;H^{p+1}(\Omega))} + \frac{\tilde C}{\varepsilon} + |u^0|^2_{H^{\bar p}(\Omega)}\Big\}$$
$$+ C\left(1+\frac{1}{\varepsilon}\right)\int_0^t\|e_I(\vartheta)\|^2_{L^2(\Omega)}\mathrm{d}\vartheta,$$
$$\le Z + C\left(1+\frac{1}{\varepsilon}\right)\int_0^t\|e_I(\vartheta)\|^2_{L^2(\Omega)}\mathrm{d}\vartheta, \tag{7.217}$$

where

$$Z = C^* h^{2p}(h\varepsilon^2 + 1 + h^2\varepsilon^{-2} + \varepsilon^{-1}) \tag{7.218}$$

with C^* depending on $\tilde C$, $|g|_{L^2(0,T;H^{\bar p}(\Omega))}$, $\|u_D\|_{L^2(0,T;H^{p+2}(\partial\Omega))}$, $|u|_{L^2(0,T;H^{p+1}(\Omega))}$, $|u^0|_{H^{\bar p}(\Omega)}$, and $\max_{\ell=1,2}\{|f_\ell(u)|^2_{L^2(0,T;W^{p+1,\infty}(\Omega))}\}$, but independent of ε and h. The constant C is independent of $u,\ g,\ u_D,\ T,\ \varepsilon,\ \bar h$, and $\tilde C$ is the constant from estimate (7.133).

The use of the Gronwall Lemma 7.43 leads to estimate

$$\max_{t\in[0,T]}\|e_I(t)\|^2_{L^2(\Omega)} + \frac{\varepsilon}{2}\int_0^T\left(|e_I(\vartheta)|^2_{H^1(\Omega,\mathscr{T}_h)} + J^\sigma_h(e_I(\vartheta),e_I(\vartheta))\right)\mathrm{d}\vartheta \tag{7.219}$$
$$\le Z\exp(CT/\varepsilon), \qquad h\in(0,\bar h).$$

Finally, from the relations

$$\begin{aligned}\tilde{e}_h &= u - \tilde{u}_h = e_I + e_h, \\ \|\tilde{e}_h\|^2_{L^2(\Omega)} &\leq 2\left(\|e_I\|^2_{L^2(\Omega)} + \|e_h\|^2_{L^2(\Omega)}\right), \\ |\tilde{e}_h|^2_{H^1(\Omega,\mathscr{T}_h)} &\leq 2\left(|e_I|^2_{H^1(\Omega,\mathscr{T}_h)} + |e_h|^2_{H^1(\Omega,\mathscr{T}_h)}\right), \\ J_h^\sigma(\tilde{e}_h, \tilde{e}_h) &\leq 2\left(J_h^\sigma(e_I, e_I) + J_h^\sigma(e_h, e_h)\right)\end{aligned} \tag{7.220}$$

and estimate (7.133) we obtain (7.203). □

Remark 7.47 (a) It follows from the presented analysis that in the definition of individual forms appearing in (7.116) one could apply quadrature formulae exact for polynomials of different degrees. For example, for evaluating the volume integrals in the form $\tilde{a}_h$, it is enough to use quadratures that are exact for polynomials of degree $\leq 2p - 2$, whereas the evaluation of $(\cdot, \cdot)_h$ and $\tilde{b}_h$ requires formulae to be exact for polynomials of degree $\leq 2p$. For the sake of simplicity, in the formulation of the main result, we consider the same quadrature formulae for all volume integrals.

(b) The effect of numerical integration has not yet been studied in other DG techniques. The approach presented in this section can be adapted, for example, to the BDF-DGM, but in some cases, as for the ST-DGM, analyzing the effect of numerical integration will require special attention.

Part II
Applications of the Discontinuous Galerkin Method

Chapter 8
Inviscid Compressible Flow

In previous chapters we introduced and analyzed the *discontinuous Galerkin method* (DGM) for the numerical solution of several scalar equations. However, many practical problems are described by systems of partial differential equations. In the second part of this book, we present the application of the DGM to solving compressible flow problems. The numerical schemes, analyzed for a scalar equation, are extended to a system of equations and numerically verified. We also deal with an efficient solution of resulting systems of algebraic equations.

One of the models used for the numerical simulation of a compressible (i.e., gas) flow is based on the assumption that the flow is inviscid and adiabatic. This means that in gas we neglect the internal friction and heat transfer. Inviscid adiabatic flow is described by the *continuity equation*, the *Euler equations* of motion and the *energy equation*, to which we add closing thermodynamical relations. See, for example, [127, Sect. 1.2]. This complete system is usually called the Euler equations.

The Euler equations, similarly as other nonlinear hyperbolic systems of conservation laws, may have discontinuous solutions. This is one of the reasons that the finite volume method (FVM) using piecewise constant approximations became very popular for the numerical solution of compressible flow. For a detailed treatment of finite volume techniques, we can refer to [119, 205]. See also [122, 127]. Moreover, the FVM is applicable on general polygonal meshes and its algorithmization is relatively easy. Therefore, many fluid dynamics codes and program packages are based on the FVM. However, the standard FVM is only of the first order, which is not sufficient in a number of applications. The increase of accuracy in finite volume schemes applied on unstructured and/or anisotropic meshes seems to be problematic and is not theoretically sufficiently justified.

As for the finite element method (FEM), the standard conforming finite element techniques were considered to be suitable for the numerical solution of elliptic and parabolic problems, linear elasticity and incompressible viscous flow, when the exact solution is sufficiently regular. Of course, there are also conforming finite element techniques applied to the solution of compressible flow, but the treatment of discontinuous solutions is rather complicated. For a survey, see [127, Sect. 4.3].

V. Dolejší and M. Feistauer, *Discontinuous Galerkin Method*,
Springer Series in Computational Mathematics 48,
DOI 10.1007/978-3-319-19267-3_8

A combination of ideas and techniques of the FV and FE methods yields the discontinuous Galerkin method using advantages of both approaches and allowing to obtain schemes with a higher-order accuracy in a natural way. In this chapter we present the application of the DGM to the Euler equations. We describe the discretization, a special attention is paid to the choice of boundary conditions and we also discuss an efficient solution of the resulting discrete problem.

8.1 Formulation of the Inviscid Flow Problem

8.1.1 Governing Equations

We consider the unsteady compressible inviscid adiabatic flow in a domain $\Omega \subset \mathbb{R}^d$ ($d = 2$ or 3) and time interval $(0, T)$, $0 < T < \infty$. In what follows, we present only the governing equations, their derivation can be found, e.g., in [127, Sect. 3.1].

We use the standard notation: ρ-density, p-pressure (symbol p denotes the degree of polynomial approximation), E-total energy, v_s, $s = 1, \dots, d$-components of the velocity vector $\boldsymbol{v} = (v_1, \dots, v_d)^{\mathrm{T}}$ in the directions x_s, θ-absolute temperature, $c_v > 0$-specific heat at constant volume, $c_{\mathrm{p}} > 0$-specific heat at constant pressure, $\gamma = c_{\mathrm{p}}/c_v > 1$-Poisson adiabatic constant, $R = c_{\mathrm{p}} - c_v > 0$-gas constant. We will be concerned with the flow of a perfect gas, for which the equation of state has the form

$$\mathrm{p} = R\rho\theta, \tag{8.1}$$

and assume that c_{p}, c_v are constants. Since the gas is light, we neglect the outer volume force.

The system of governing equations formed by the continuity equation, the Euler equations of motion and the energy equation (see [127, Sect. 3.1]) considered in the space-time cylinder $Q_T = \Omega \times (0, T)$ can be written in the form

$$\frac{\partial \rho}{\partial t} + \sum_{s=1}^{d} \frac{\partial(\rho v_s)}{\partial x_s} = 0, \tag{8.2}$$

$$\frac{\partial(\rho v_i)}{\partial t} + \sum_{s=1}^{d} \frac{\partial(\rho v_i v_s + \delta_{is}\mathrm{p})}{\partial x_s} = 0, \quad i = 1, \dots, d, \tag{8.3}$$

$$\frac{\partial E}{\partial t} + \sum_{s=1}^{d} \frac{\partial((E + \mathrm{p})v_s)}{\partial x_s} = 0. \tag{8.4}$$

To the above system, we add the thermodynamical relations defining the pressure

$$\mathrm{p} = (\gamma - 1)(E - \rho|\boldsymbol{v}|^2/2), \tag{8.5}$$

and the total energy

$$E = \rho(c_v\theta + |\boldsymbol{v}|^2/2), \tag{8.6}$$

in terms of other quantities.

We define the *speed of sound* a and the *Mach number* M by

$$a = \sqrt{\gamma \mathrm{p}/\rho}, \qquad M = \frac{|\boldsymbol{v}|}{a}. \tag{8.7}$$

The flow is called *subsonic* and *supersonic* in a region ω, if $M < 1$ and $M > 1$, respectively, in ω. If $M \gg 1$, we speak about *hypersonic flow*. If there are two subregions ω_1 and ω_2 in the flow field such that $M < 1$ in ω_1 and $M > 1$ in ω_2, the flow is called *transonic*.

Exercise 8.1 Derive (8.5) from (8.1) and (8.6).

System (8.2)–(8.4) has $m = d + 2$ equations and it can be written in the form

$$\frac{\partial \boldsymbol{w}}{\partial t} + \sum_{s=1}^{d} \frac{\partial \boldsymbol{f}_s(\boldsymbol{w})}{\partial x_s} = 0, \tag{8.8}$$

where

$$\boldsymbol{w} = (w_1, \ldots, w_m)^{\mathrm{T}} = (\rho, \rho v_1, \ldots, \rho v_d, E)^{\mathrm{T}} \in \mathbb{R}^m, \tag{8.9}$$

is the so-called *state vector*, and

$$\boldsymbol{f}_s(\boldsymbol{w}) = \begin{pmatrix} f_{s,1}(\boldsymbol{w}) \\ f_{s,2}(\boldsymbol{w}) \\ \vdots \\ f_{s,m-1}(\boldsymbol{w}) \\ f_{s,m}(\boldsymbol{w}) \end{pmatrix} = \begin{pmatrix} \rho v_s \\ \rho v_1 v_s + \delta_{1s}\mathrm{p} \\ \vdots \\ \rho v_d v_s + \delta_{ds}\mathrm{p} \\ (E + \mathrm{p})v_s \end{pmatrix} \tag{8.10}$$

$$= \begin{pmatrix} w_{s+1} \\ \frac{w_2 w_{s+1}}{w_1} + \delta_{1s}(\gamma - 1)\left(w_m - \frac{1}{2w_1}\sum_{i=2}^{m-1} w_i^2\right) \\ \vdots \\ \frac{w_{m-1} w_{s+1}}{w_1} + \delta_{ds}(\gamma - 1)\left(w_m - \frac{1}{2w_1}\sum_{i=2}^{m-1} w_i^2\right) \\ \frac{w_{s+1}}{w_1}\left(w_m - \frac{\gamma-1}{2w_1}\sum_{i=2}^{m-1} w_i^2\right) \end{pmatrix},$$

is the *flux* of the quantity $\boldsymbol{w}$ in the direction x_s, $s = 1, \ldots, d$. By δ_{ij} we denote the Kronecker symbol. Often, $\boldsymbol{f}_s$, $s = 1, \ldots, d$, are called *inviscid Euler fluxes*.

Usually, system (8.2)–(8.4), i.e., (8.8), is called the system of the Euler equations, or simply the *Euler equations*. The functions $\rho, v_1, \ldots, v_d, \mathrm{p}$ are called *primitive* (or physical) *variables*, whereas $w_1 = \rho$, $w_2 = \rho v_1, \dots, w_{m-1} = \rho v_d$, $w_m = E$ are *conservative variables*. It is easy to show that

$$\begin{aligned} v_i &= w_{i+1}/w_1,\ i = 1, \dots, d, \\ \mathrm{p} &= (\gamma - 1)\left(w_m - \sum_{i=2}^{m-1} w_i^2/(2w_1) \right), \\ \theta &= \left(w_m/w_1 - \frac{1}{2} \sum_{i=2}^{m-1} (w_i/w_1)^2 \right) /c_v. \end{aligned} \tag{8.11}$$

The domain of definition of the vector-valued functions $\boldsymbol{f}_s$, $s = 1, \dots, d$, is the open set $\mathscr{D} \subset \mathbb{R}^m$ of vectors $\boldsymbol{w} = (w_1, \dots, w_m)^{\mathrm{T}}$ such that the corresponding density and pressure are positive:

$$\mathscr{D} = \left\{ \boldsymbol{w} \in \mathbb{R}^m;\ w_1 = \rho > 0,\ w_m - \sum_{i=2}^{m-1} w_i^2/(2w_1) = \mathrm{p}/(\gamma - 1) > 0 \right\}. \tag{8.12}$$

Obviously, $\boldsymbol{f}_s \in (C^1(\mathscr{D}))^m$.

Differentiation in (8.8) and the use of the chain rule lead to a *first-order quasilinear system* of partial differential equations

$$\frac{\partial \boldsymbol{w}}{\partial t} + \sum_{s=1}^{d} \mathbb{A}_s(\boldsymbol{w}) \frac{\partial \boldsymbol{w}}{\partial x_s} = 0, \tag{8.13}$$

where $\mathbb{A}_s(\boldsymbol{w})$ is the $m \times m$ Jacobi matrix of the mapping $\boldsymbol{f}_s$ defined for $\boldsymbol{w} \in \mathscr{D}$:

$$\mathbb{A}_s(\boldsymbol{w}) = \frac{\mathrm{D}\boldsymbol{f}_s(\boldsymbol{w})}{\mathrm{D}\boldsymbol{w}} = \left(\frac{\partial f_{s,i}(\boldsymbol{w})}{\partial w_j} \right)_{i,j=1}^{m}, \quad s = 1, \dots, d. \tag{8.14}$$

Let

$$\mathrm{B}_1 = \{ \boldsymbol{n} \in \mathbb{R}^d;\ |\boldsymbol{n}| = 1 \} \tag{8.15}$$

denote the unit sphere in $\mathbb{R}^d$. For $\boldsymbol{w} \in \mathscr{D}$ and $\boldsymbol{n} = (n_1, \dots, n_d)^{\mathrm{T}} \in \mathrm{B}_1$ we denote

$$\boldsymbol{P}(\boldsymbol{w}, \boldsymbol{n}) = \sum_{s=1}^{d} \boldsymbol{f}_s(\boldsymbol{w}) n_s, \tag{8.16}$$

which is the *physical flux* of the quantity $\boldsymbol{w}$ in the direction $\boldsymbol{n}$. Obviously, the Jacobi matrix $\mathrm{D}\boldsymbol{P}(\boldsymbol{w}, \boldsymbol{n})/\mathrm{D}\boldsymbol{w}$ can be expressed in the form

$$\frac{\mathrm{D}\boldsymbol{P}(\boldsymbol{w}, \boldsymbol{n})}{\mathrm{D}\boldsymbol{w}} = \mathbb{P}(\boldsymbol{w}, \boldsymbol{n}) := \sum_{s=1}^{d} \mathbb{A}_s(\boldsymbol{w}) n_s. \tag{8.17}$$

Exercise 8.2 Let $d = 2$. Prove that the Jacobi matrices $\mathbb{A}_s,\ s = 1, 2$, have the form

$$\mathbb{A}_1(\boldsymbol{w}) = \left(\begin{array}{c|c|c|c} 0 & 1 & 0 & 0 \\ \frac{\gamma_1}{2}|\boldsymbol{v}|^2 - v_1^2 & (3-\gamma)v_1 & -\gamma_1 v_2 & \gamma_1 \\ -v_1 v_2 & v_2 & v_1 & 0 \\ v_1\left(\gamma_1|\boldsymbol{v}|^2 - \gamma\frac{E}{\rho}\right) & \gamma\frac{E}{\rho} - \gamma_1 v_1^2 - \frac{\gamma_1}{2}|\boldsymbol{v}|^2 & -\gamma_1 v_1 v_2 & \gamma v_1 \end{array}\right), \tag{8.18}$$

$$\mathbb{A}_2(\boldsymbol{w}) = \left(\begin{array}{c|c|c|c} 0 & 0 & 1 & 0 \\ -v_1 v_2 & v_2 & v_1 & 0 \\ \frac{\gamma_1}{2}|\boldsymbol{v}|^2 - v_2^2 & -\gamma_1 v_1 & (3-\gamma)v_2 & \gamma_1 \\ v_2\left(\gamma_1|\boldsymbol{v}|^2 - \gamma\frac{E}{\rho}\right) & -\gamma_1 v_1 v_2 & \gamma\frac{E}{\rho} - \gamma_1 v_2^2 - \frac{\gamma_1}{2}|\boldsymbol{v}|^2 & \gamma v_2 \end{array}\right), \tag{8.19}$$

where $\gamma_1 = \gamma - 1$.

Exercise 8.3 Let $d = 2$. With the aid of (8.18)–(8.19) show that the matrix $\mathbb{P}(\boldsymbol{w}, \boldsymbol{n})$ has the form

$$\mathbb{P}(\boldsymbol{w},\boldsymbol{n}) = \left(\begin{array}{c|c|c|c} 0 & n_1 & n_2 & 0 \\ \frac{\gamma_1}{2}|\boldsymbol{v}|^2 n_1 - v_1 \boldsymbol{v}\cdot\boldsymbol{n} & -\gamma_2 v_1 n_1 + \boldsymbol{v}\cdot\boldsymbol{n} & v_1 n_2 - \gamma_1 v_2 n_1 & \gamma_1 n_1 \\ \frac{\gamma_1}{2}|\boldsymbol{v}|^2 n_2 - v_2 \boldsymbol{v}\cdot\boldsymbol{n} & v_2 n_1 - \gamma_1 v_1 n_2 & -\gamma_2 v_2 n_2 + \boldsymbol{v}\cdot\boldsymbol{n} & \gamma_1 n_2 \\ \left(\gamma_1|\boldsymbol{v}|^2 - \frac{\gamma E}{\rho}\right)\boldsymbol{v}\cdot\boldsymbol{n} & G n_1 - \gamma_1 v_1 \boldsymbol{v}\cdot\boldsymbol{n} & G n_2 - \gamma_1 v_2 \boldsymbol{v}\cdot\boldsymbol{n} & \gamma \boldsymbol{v}\cdot\boldsymbol{n} \end{array}\right), \tag{8.20}$$

where $\boldsymbol{n} = (n_1, n_2)$, $\gamma_1 = \gamma - 1$, $\gamma_2 = \gamma - 2$ and $G = \gamma\frac{E}{\rho} - \frac{\gamma_1}{2}|\boldsymbol{v}|^2$.

Exercise 8.4 Let $d = 3$. Prove that the Jacobi matrices $\mathbb{A}_s,\ s = 1, 2, 3$, have the form

$$\mathbb{A}_1 = \left(\begin{array}{c|c|c|c|c} 0 & 1 & 0 & 0 & 0 \\ \frac{\gamma_1}{2}|\boldsymbol{v}|^2 - v_1^2 & (3-\gamma)v_1 & -\gamma_1 v_2 & -\gamma_1 v_3 & \gamma_1 \\ -v_1 v_2 & v_2 & v_1 & 0 & 0 \\ -v_1 v_3 & v_3 & 0 & v_1 & 0 \\ v_1\left(\gamma_1|\boldsymbol{v}|^2 - \gamma\frac{E}{\rho}\right) & \gamma\frac{E}{\rho} - \gamma_1 v_1^2 - \frac{\gamma_1}{2}|\boldsymbol{v}|^2 & -\gamma_1 v_1 v_2 & -\gamma_1 v_1 v_3 & \gamma v_1 \end{array}\right), \tag{8.21}$$

$$\mathbb{A}_2 = \left(\begin{array}{c|c|c|c|c} 0 & 0 & 1 & 0 & 0 \\ -v_1 v_2 & v_2 & v_1 & 0 & 0 \\ \frac{\gamma_1}{2}|\boldsymbol{v}|^2 - v_2^2 & -\gamma_1 v_1 & (3-\gamma)v_2 & -\gamma_1 v_3 & \gamma_1 \\ -v_2 v_3 & 0 & v_3 & v_2 & 0 \\ v_2\left(\gamma_1|\boldsymbol{v}|^2 - \gamma\frac{E}{\rho}\right) & -\gamma_1 v_1 v_2 & \gamma\frac{E}{\rho} - \gamma_1 v_2^2 - \frac{\gamma_1}{2}|\boldsymbol{v}|^2 & -\gamma_1 v_2 v_3 & \gamma v_2 \end{array}\right), \tag{8.22}$$

$$\mathbb{A}_3 = \begin{pmatrix} 0 & 0 & 0 & 1 & 0 \\ -v_1v_3 & v_3 & 0 & v_1 & 0 \\ -v_2v_3 & 0 & v_3 & v_2 & 0 \\ \frac{\gamma_1}{2}|\boldsymbol{v}|^2 - v_3^2 & -\gamma_1 v_1 & -\gamma_1 v_2 & (3-\gamma)v_3 & \gamma_1 \\ v_3\left(\gamma_1|\boldsymbol{v}|^2 - \gamma\frac{E}{\rho}\right) & -\gamma_1 v_1 v_3 & -\gamma_1 v_2 v_3 & \gamma\frac{E}{\rho} - \gamma_1 v_3^2 - \frac{\gamma_1}{2}|\boldsymbol{v}|^2 & \gamma v_3 \end{pmatrix}, \tag{8.23}$$

where $\gamma_1 = \gamma - 1$.

Exercise 8.5 Let $d = 3$. With the aid of (8.21)–(8.23), show that the matrix $\mathbb{P}(\boldsymbol{w}, \boldsymbol{n})$ has the form

$$\mathbb{P}(\boldsymbol{w}, \boldsymbol{n}) = \tag{8.24}$$

$$\begin{pmatrix} 0 & n_1 & n_2 & n_3 & 0 \\ \frac{\gamma_1}{2}|\boldsymbol{v}|^2 n_1 - v_1\boldsymbol{v}\cdot\boldsymbol{n} & -\gamma_2 v_1 n_1 + \boldsymbol{v}\cdot\boldsymbol{n} & v_1 n_2 - \gamma_1 v_2 n_1 & v_1 n_3 - \gamma_1 v_3 n_1 & \gamma_1 n_1 \\ \frac{\gamma_1}{2}|\boldsymbol{v}|^2 n_2 - v_2\boldsymbol{v}\cdot\boldsymbol{n} & v_2 n_1 - \gamma_1 v_1 n_2 & -\gamma_2 v_2 n_2 + \boldsymbol{v}\cdot\boldsymbol{n} & v_2 n_3 - \gamma_1 v_3 n_2 & \gamma_1 n_2 \\ \frac{\gamma_1}{2}|\boldsymbol{v}|^2 n_3 - v_3\boldsymbol{v}\cdot\boldsymbol{n} & v_3 n_1 - \gamma_1 v_1 n_3 & v_3 n_2 - \gamma_1 v_2 n_3 & -\gamma_2 v_3 n_3 + \boldsymbol{v}\cdot\boldsymbol{n} & \gamma_1 n_3 \\ \left(\gamma_1|\boldsymbol{v}|^2 - \frac{\gamma E}{\rho}\right)\boldsymbol{v}\cdot\boldsymbol{n} & G n_1 - \gamma_1 v_1 \boldsymbol{v}\cdot\boldsymbol{n} & G n_2 - \gamma_1 v_2 \boldsymbol{v}\cdot\boldsymbol{n} & G n_3 - \gamma_1 v_3 \boldsymbol{v}\cdot\boldsymbol{n} & \gamma \boldsymbol{v}\cdot\boldsymbol{n} \end{pmatrix},$$

where $\boldsymbol{n} = (n_1, n_2, n_3)$, $\gamma_1 = \gamma - 1$, $\gamma_2 = \gamma - 2$ and $G = \gamma\frac{E}{\rho} - \frac{\gamma_1}{2}|\boldsymbol{v}|^2$.

Let us summarize some important properties of the system of the Euler equations (8.8).

Lemma 8.6 *(a) The vector-valued functions $\boldsymbol{f}_s$ defined by (8.10) are homogeneous mappings of order 1:*

$$\boldsymbol{f}_s(\alpha\boldsymbol{w}) = \alpha\boldsymbol{f}_s(\boldsymbol{w}), \quad \alpha > 0. \tag{8.25}$$

Moreover, we have

$$\boldsymbol{f}_s(\boldsymbol{w}) = \mathbb{A}_s(\boldsymbol{w})\boldsymbol{w}. \tag{8.26}$$

(b) Similarly,

$$\boldsymbol{P}(\alpha\boldsymbol{w}, \boldsymbol{n}) = \alpha\boldsymbol{P}(\boldsymbol{w}, \boldsymbol{n}), \quad \alpha > 0, \tag{8.27}$$

$$\boldsymbol{P}(\boldsymbol{w}, \boldsymbol{n}) = \mathbb{P}(\boldsymbol{w}, \boldsymbol{n})\boldsymbol{w}. \tag{8.28}$$

(c) The system of the Euler equations is diagonally hyperbolic. This means that the matrix $\mathbb{P} = \sum_{j=1}^{d}\mathbb{A}_j(\boldsymbol{w})n_j$ has only real eigenvalues $\lambda_i = \lambda_i(\boldsymbol{w}, \boldsymbol{n})$, $i = 1, \ldots, m$, and is diagonalizable: there exists a nonsingular matrix $\mathbb{T} = \mathbb{T}(\boldsymbol{w}, \boldsymbol{n})$ such that

$$\mathbb{T}^{-1}\mathbb{P}\mathbb{T} = \boldsymbol{\Lambda} = \boldsymbol{\Lambda}(\boldsymbol{w}, \boldsymbol{n}) = \mathrm{diag}(\lambda_1, \dots, \lambda_m) = \begin{pmatrix} \lambda_1 & 0 & \dots & 0 & 0 \\ 0 & \lambda_2 & 0 & \dots & 0 \\ \vdots & & \ddots & & \vdots \\ 0 & \dots & 0 & \lambda_{m-1} & 0 \\ 0 & 0 & \dots & 0 & \lambda_m \end{pmatrix}. \quad (8.29)$$

The columns of the matrix $\mathbb{T}$ *are the eigenvectors of the matrix* $\mathbb{P}$.

(d) The eigenvalues of the matrix $\mathbb{P}(\boldsymbol{w}, \boldsymbol{n})$, $\boldsymbol{w} \in \mathscr{D}$, $\boldsymbol{n} \in \mathrm{B}_1$ *have the form*

$$\begin{aligned} \lambda_1(\boldsymbol{w}, \boldsymbol{n}) &= \boldsymbol{v} \cdot \boldsymbol{n} - a, \\ \lambda_2(\boldsymbol{w}, \boldsymbol{n}) &= \dots = \lambda_{d+1}(\boldsymbol{w}, \boldsymbol{n}) = \boldsymbol{v} \cdot \boldsymbol{n}, \\ \lambda_m(\boldsymbol{w}, \boldsymbol{n}) &= \boldsymbol{v} \cdot \boldsymbol{n} + a, \end{aligned} \quad (8.30)$$

where $a = \sqrt{\gamma \mathrm{p}/\rho}$ *is the speed of sound and* $\boldsymbol{v}$ *is the velocity vector given by* $\boldsymbol{v} = (w_2/w_1, w_3/w_1, \dots, w_{d+1}/w_1)^{\mathrm{T}}$.

(e) The system of the Euler equations is rotationally invariant. Namely, for $\boldsymbol{n} = (n_1, \dots, n_d) \in \mathrm{B}_1$, $\boldsymbol{w} \in \mathscr{D}$ *it holds*

$$\boldsymbol{P}(\boldsymbol{w}, \boldsymbol{n}) = \sum_{s=1}^{d} \boldsymbol{f}_s(\boldsymbol{w}) n_s = \mathbb{Q}^{-1}(\boldsymbol{n}) \boldsymbol{f}_1(\mathbb{Q}(\boldsymbol{n})\boldsymbol{w}), \quad (8.31)$$

$$\mathbb{P}(\boldsymbol{w}, \boldsymbol{n}) = \sum_{s=1}^{d} \mathbb{A}_s(\boldsymbol{w}) n_s = \mathbb{Q}^{-1}(\boldsymbol{n}) \mathbb{A}_1(\mathbb{Q}(\boldsymbol{n})\boldsymbol{w}) \mathbb{Q}(\boldsymbol{n}), \quad (8.32)$$

where $\mathbb{Q}(\boldsymbol{n})$ *is the* $m \times m$ *matrix corresponding to* $\boldsymbol{n} \in \mathrm{B}_1$ *given by*

$$\mathbb{Q}(\boldsymbol{n}) = \begin{pmatrix} 1 & \mathbf{0} & 0 \\ \mathbf{0}^{\mathrm{T}} & \mathbb{Q}_0(\boldsymbol{n}) & \mathbf{0}^{\mathrm{T}} \\ 0 & \mathbf{0} & 1 \end{pmatrix}, \quad (8.33)$$

where the $d \times d$ *rotation matrix* $\mathbb{Q}_0(\boldsymbol{n})$ *is defined for* $d = 2$ *by*

$$\mathbb{Q}_0(\boldsymbol{n}) = \begin{pmatrix} n_1 & n_2 \\ -n_2 & n_1 \end{pmatrix}, \qquad \boldsymbol{n} = (n_1, n_2), \quad (8.34)$$

and for $d = 3$ *by*

$$\mathbb{Q}_0(\boldsymbol{n}) = \begin{pmatrix} \cos\alpha\cos\beta & \sin\alpha\cos\beta & \sin\beta \\ \sin\alpha & \cos\alpha & 0 \\ -\cos\alpha\sin\beta & -\sin\alpha\sin\beta & \cos\beta \end{pmatrix}, \quad (8.35)$$

$$\boldsymbol{n} = (\cos\alpha\cos\beta, \sin\alpha\cos\beta, \sin\beta), \ \alpha \in [0, 2\pi), \ \beta \in [-\pi/2, \pi/2].$$

By $\mathbf{0}$ *we denote the vector* $(0, 0)$, *if* $d = 2$, *and* $(0, 0, 0)$, *if* $d = 3$.

Proof See [127, Lemmas 3.1 and 3.3, Theorem 3.4]. □

8.1.2 Initial and Boundary Conditions

In order to formulate the problem of inviscid compressible flow, the system of the Euler equations (8.8) has to be equipped with initial and boundary conditions. Let $\Omega \subset \mathbb{R}^d$, $d = 2, 3$, be a bounded computational domain with a piecewise smooth Lipschitz boundary $\partial\Omega$. We prescribe the *initial condition*

$$\boldsymbol{w}(x, 0) = \boldsymbol{w}^0(x), \quad x \in \Omega, \tag{8.36}$$

where $\boldsymbol{w}^0 : \Omega \to \mathscr{D}$ is a given vector-valued function. Moreover, the *boundary conditions* are given formally by

$$\mathscr{B}(\boldsymbol{w}) = 0 \quad \text{on } \partial\Omega \times (0, T), \tag{8.37}$$

where $\mathscr{B}$ is a boundary operator.

The choice of appropriate boundary conditions is a very important and delicate question in the numerical simulation of fluid flow. Determining of boundary conditions is, basically, a physical problem, but it must correspond to the mathematical character of the solved equations. Great care is required in their numerical implementation. Usually two types of boundaries are considered: *reflective* and *transparent* or *transmissive*. The reflective boundaries usually consist of fixed walls. Transmissive or transparent boundaries arise from the need to replace unbounded or rather large physical domains by bounded or sufficiently small computational domains. The corresponding boundary conditions are devised so that they allow the passage of waves without any effect on them. For 1D problems the objective is reasonably well attained. For multidimensional problems this is a substantial area of current research, usually referred to *open-end* boundary conditions, *transparent* boundary conditions, *far-field* boundary conditions, *radiation* boundary conditions or *non-reflecting* boundary conditions. Useful publications dealing with boundary conditions are [31, 151, 154–156, 160, 170, 204, 241], [152, Chap. V]. A rigorous mathematical theory of boundary conditions to conservation laws was developed only for a scalar equation in [19].

The choice of well-posed boundary conditions for the Euler equations (or, in general, of conservation laws) is a delicate question, not completely satisfactorily solved (see, e.g., the paper [19] dealing with the boundary conditions for a scalar equation). We discuss the choice of the boundary conditions in Sect. 8.3 in relation to the definition of the numerical solution of (8.8).

Let us only mention that we distinguish several disjoint parts of the boundary $\partial\Omega$, namely *inlet* $\partial\Omega_i$, *outlet* $\partial\Omega_o$ and *impermeable walls* $\partial\Omega_W$, i.e., $\partial\Omega = \partial\Omega_i \cup \partial\Omega_o \cup \partial\Omega_W$. In some situations the inlet and outlet parts are considered together.

Therefore, we speak about the inlet/outlet part of the boundary. On $\partial\Omega_W$ we prescribe the impermeability condition

$$\boldsymbol{v} \cdot \boldsymbol{n} = 0 \quad \text{on } \partial\Omega_W, \tag{8.38}$$

where $\boldsymbol{n}$ denotes the outer unit normal to $\partial\Omega_W$ and $\boldsymbol{v}$ is the velocity vector.

Concerning the inlet/outlet part of the boundary $\partial\Omega_i \cup \partial\Omega_o$, the boundary conditions are usually chosen in such a way that problem (8.8) is linearly well-posed. (See, e.g., [127, Sect. 3.3.6].) Practically it means that the number of prescribed boundary conditions is equal to the number of negative eigenvalues of the matrix $\mathbb{P}(\boldsymbol{w}, \boldsymbol{n})$ defined by (8.31). See Sect. 8.3.

8.2 DG Space Semidiscretization

In the following, we deal with the discretization of the Euler equations (8.8) by the DGM. We recall some notation introduced in Chaps. 2 and 4. Similarly as in Chap. 4, we derive the DG space semidiscretization leading to a system of ordinary differential equations. Moreover, we develop a (semi-)implicit time discretization technique which is based on a formal linearization of nonlinear terms. We will also pay attention to some further aspects of the DG discretization of the Euler equations, namely the choice of boundary conditions, the approximation of nonpolygonal boundary and the shock capturing.

8.2.1 Notation

We recall and extend notation introduced in Chaps. 2 and 4. In the finite element method, the computational domain Ω is usually approximated by a polygonal (if $d = 2$) or polyhedral (if $d = 3$) domain Ω_h, which is the domain of definition of the approximate solution. For the sake of simplicity, we assume that the domain Ω is polygonal, and thus $\Omega_h = \Omega$. By $\mathscr{T}_h$ we denote a partition of Ω consisting of closed d-dimensional simplexes with mutually disjoint interiors. We call $\mathscr{T}_h$ the triangulation of Ω.

By $\mathscr{F}_h$ we denote the set of all open $(d-1)$-dimensional faces (open edges when $d = 2$ or open faces when $d = 3$) of all elements $K \in \mathscr{T}_h$. Further, the symbol $\mathscr{F}_h^I$ stands for the set of all $\Gamma \subset \mathscr{F}_h$ that are contained in Ω (inner faces). Moreover, we define $\mathscr{F}_h^W$, $\mathscr{F}_h^i$ and $\mathscr{F}_h^o$ as the sets of all $\Gamma \in \mathscr{F}_h$ such that $\Gamma \subset \partial\Omega_W$, $\Gamma \subset \partial\Omega_i$ and $\Gamma \subset \partial\Omega_o$, respectively. In order to simplify the notation, we put $\mathscr{F}_h^{io} = \mathscr{F}_h^i \cup \mathscr{F}_h^o$ and $\mathscr{F}_h^B = \mathscr{F}_h^W \cup \mathscr{F}_h^i \cup \mathscr{F}_h^o$. Finally, for each $\Gamma \in \mathscr{F}_h$ we define a unit normal vector $\boldsymbol{n}_\Gamma = (n_{\Gamma,1}, \dots, n_{\Gamma,d})$ We assume that for $\Gamma \subset \mathscr{F}_h^B$ the vector $\boldsymbol{n}_\Gamma$ has the same orientation as the outer normal of $\partial\Omega$. For each $\Gamma \in \mathscr{F}_h^I$, the orientation of $\boldsymbol{n}_\Gamma$ is arbitrary but fixed.

Over the triangulation $\mathscr{T}_h$ we define the *broken Sobolev space* of vector-valued functions (cf. (1.1))

$$\boldsymbol{H}^1(\Omega, \mathscr{T}_h) = (H^1(\Omega, \mathscr{T}_h))^m, \tag{8.39}$$

where

$$H^1(\Omega, \mathscr{T}_h) = \{v;\ v : \Omega \to \mathbb{R},\ v|_K \in H^1(K)\ \forall\, K \in \mathscr{T}_h\} \tag{8.40}$$

is the broken Sobolev space of scalar functions introduced by (2.29).

For each $\Gamma \in \mathscr{F}_h^I$ there exist two elements $K_\Gamma^{(L)}, K_\Gamma^{(R)} \in \mathscr{T}_h$ such that $\Gamma \subset K_\Gamma^{(L)} \cap K_\Gamma^{(R)}$. We use again the convention that $K_\Gamma^{(R)}$ lies in the direction of $\boldsymbol{n}_\Gamma$ and $K_\Gamma^{(L)}$ in the opposite direction of $\boldsymbol{n}_\Gamma$, see Fig. 2.2.

In agreement with Sect. 2.3.3, for $\boldsymbol{u} \in \boldsymbol{H}^1(\Omega, \mathscr{T}_h)$ and $\Gamma \in \mathscr{F}_h^I$, we introduce the notation:

$$\boldsymbol{u}_\Gamma^{(L)} \text{ is the trace of } \boldsymbol{u}|_{K_\Gamma^{(L)}} \text{ on } \Gamma, \qquad \boldsymbol{u}_\Gamma^{(R)} \text{ is the trace of } \boldsymbol{u}|_{K_\Gamma^{(R)}} \text{ on } \Gamma \tag{8.41}$$

and

$$\langle \boldsymbol{u} \rangle_\Gamma = \frac{1}{2}\left(\boldsymbol{u}_\Gamma^{(L)} + \boldsymbol{u}_\Gamma^{(R)}\right), \tag{8.42}$$

$$[\boldsymbol{u}]_\Gamma = \boldsymbol{u}_\Gamma^{(L)} - \boldsymbol{u}_\Gamma^{(R)}. \tag{8.43}$$

In case that $[\cdot]_\Gamma$, $\langle\cdot\rangle_\Gamma$ and $\boldsymbol{n}_\Gamma$ are arguments of $\int_\Gamma \ldots \,\mathrm{d}S$, $\Gamma \in \mathscr{F}_h$, we usually omit the subscript Γ and write simply $[\cdot]$, $\langle\cdot\rangle$ and $\boldsymbol{n}$, respectively. The value $[\boldsymbol{u}]_\Gamma$ depends on the orientation of $\boldsymbol{n}_\Gamma$, but the value $[\boldsymbol{u}]_\Gamma \cdot \boldsymbol{n}_\Gamma$ is independent of this orientation.

Finally, for $\boldsymbol{u} \in \boldsymbol{H}^1(\Omega, \mathscr{T}_h)$ and $\Gamma \in \mathscr{F}_h^B$, we denote by $\boldsymbol{u}_\Gamma^{(L)}$ the trace of $\boldsymbol{u}|_{K^{(L)}}$ on Γ, where $K^{(L)} \in \mathscr{T}_h$ such that $\Gamma \subset K^{(L)} \cap \partial\Omega$.

The discontinuous Galerkin (DG) approximate solution of (8.8) is sought in a finite-dimensional subspace of $\boldsymbol{H}^1(\Omega, \mathscr{T}_h)$ which consists of piecewise polynomial functions. Hence, over the triangulation $\mathscr{T}_h$ we define the space of vector-valued discontinuous piecewise polynomial functions

$$\boldsymbol{S}_{hp} = (S_{hp})^m, \tag{8.44}$$

where

$$S_{hp} = \{v \in L^2(\Omega);\ v|_K \in P_p(K)\ \forall\, K \in \mathscr{T}_h\} \tag{8.45}$$

is the space of scalar functions defined by (2.34). Here $P_p(K)$ denotes the space of all polynomials on K of degree $\le p$, $K \in \mathscr{T}_h$. Obviously, $\boldsymbol{S}_{hp} \subset \boldsymbol{H}^1(\Omega, \mathscr{T}_h)$.

8.2.2 Discontinuous Galerkin Space Semidiscretization

In order to derive the discrete problem, we assume that there exists an exact solution $\boldsymbol{w} \in C^1([0,T]; \boldsymbol{H}^1(\Omega, \mathscr{T}_h))$ of the Euler equations (8.8). Then we multiply (8.8) by a test function $\boldsymbol{\varphi} \in \boldsymbol{H}^1(\Omega, \mathscr{T}_h)$, integrate over any element $K \in \mathscr{T}_h$, apply Green's theorem and sum over all $K \in \mathscr{T}_h$. Then we get

$$\sum_{K\in\mathscr{T}_h} \int_K \frac{\partial \boldsymbol{w}}{\partial t} \cdot \boldsymbol{\varphi} \, \mathrm{d}x - \sum_{K\in\mathscr{T}_h} \int_K \sum_{s=1}^{d} \boldsymbol{f}_s(\boldsymbol{w}) \cdot \frac{\partial \boldsymbol{\varphi}}{\partial x_s} \, \mathrm{d}x + \sum_{K\in\mathscr{T}_h} \int_{\partial K} \sum_{s=1}^{d} \boldsymbol{f}_s(\boldsymbol{w}) n_s \cdot \boldsymbol{\varphi} \, \mathrm{d}S = 0, \tag{8.46}$$

where $\boldsymbol{n} = (n_1, \dots, n_d)$ denotes the outer unit normal to the boundary of $K \in \mathscr{T}_h$. Similarly as in Sect. 2.4, we rewrite (8.46) in the form

$$\sum_{K\in\mathscr{T}_h} \int_K \frac{\partial \boldsymbol{w}}{\partial t} \cdot \boldsymbol{\varphi} \, \mathrm{d}x - \sum_{K\in\mathscr{T}_h} \int_K \sum_{s=1}^{d} \boldsymbol{f}_s(\boldsymbol{w}) \cdot \frac{\partial \boldsymbol{\varphi}}{\partial x_s} \, \mathrm{d}x \tag{8.47}$$
$$+ \sum_{\Gamma\in\mathscr{F}_h^I} \int_\Gamma \sum_{s=1}^{d} \boldsymbol{f}_s(\boldsymbol{w}) n_{\Gamma,s} \cdot [\boldsymbol{\varphi}] \, \mathrm{d}S + \sum_{\Gamma\in\mathscr{F}_h^B} \int_\Gamma \sum_{s=1}^{d} \boldsymbol{f}_s(\boldsymbol{w}) n_{\Gamma,s} \cdot \boldsymbol{\varphi} \, \mathrm{d}S = 0.$$

The crucial point of the DG approximation of conservation laws is the evaluation of the integrals over $\Gamma \in \mathscr{F}_h$ in (8.47). These integrals are approximated with the aid of the *numerical flux* $\mathbf{H} : \mathscr{D} \times \mathscr{D} \times \mathrm{B}_1 \to \mathbb{R}^m$ by

$$\int_\Gamma \sum_{s=1}^{d} \boldsymbol{f}_s(\boldsymbol{w}) \, n_{\Gamma,s} \cdot \boldsymbol{\varphi} \, \mathrm{d}S \approx \int_\Gamma \mathbf{H}(\boldsymbol{w}_\Gamma^{(L)}, \boldsymbol{w}_\Gamma^{(R)}, \boldsymbol{n}_\Gamma) \cdot \boldsymbol{\varphi} \, \mathrm{d}S, \tag{8.48}$$

where the functions $\boldsymbol{w}_\Gamma^{(L)}$ and $\boldsymbol{w}_\Gamma^{(R)}$ are defined by (8.41) and B_1 by (8.15). The meaning of $\boldsymbol{w}_\Gamma^{(R)}$ for $\Gamma \in \mathscr{F}_h^B$ will be specified later in the treatment of boundary conditions in Sect. 8.3. The numerical flux is an important concept in the finite volume method (see, e.g., [127, Sect. 3.2] or [282]). It has to satisfy some basic conditions:

- *continuity*: $\mathbf{H}(\boldsymbol{w}_1, \boldsymbol{w}_2, \boldsymbol{n})$ is locally Lipschitz continuous with respect to the variables $\boldsymbol{w}_1$ and $\boldsymbol{w}_2$,
- *consistency*:

$$\mathbf{H}(\boldsymbol{w}, \boldsymbol{w}, \boldsymbol{n}) = \sum_{s=1}^{d} \boldsymbol{f}_s(\boldsymbol{w}) \, n_s, \quad \boldsymbol{w} \in \mathscr{D}, \ \boldsymbol{n} = (n_1, \dots, n_d) \in \mathrm{B}_1, \tag{8.49}$$

- *conservativity*:

$$\mathbf{H}(\boldsymbol{w}_1, \boldsymbol{w}_2, \boldsymbol{n}) = -\mathbf{H}(\boldsymbol{w}_2, \boldsymbol{w}_1, -\boldsymbol{n}), \quad \boldsymbol{w}_1, \boldsymbol{w}_2 \in \mathscr{D}, \ \boldsymbol{n} \in \mathrm{B}_1. \tag{8.50}$$

Examples of numerical fluxes can be found, e.g., in [122, 127, 205, 269].

Now, we complete the DG space semidiscretization of (8.8). Approximating the face integrals in (8.47) by (8.48) and interchanging the derivative and integral in the first term, we obtain the identity

$$\frac{\mathrm{d}}{\mathrm{d}t}(\boldsymbol{w}(t), \boldsymbol{\varphi}) + \boldsymbol{b}_h(\boldsymbol{w}(t), \boldsymbol{\varphi}) = 0 \quad \forall \boldsymbol{\varphi} \in \boldsymbol{H}^1(\Omega, \mathscr{T}_h)\; \forall t \in (0, T), \tag{8.51}$$

where

$$(\boldsymbol{w}, \boldsymbol{\varphi}) = \int_\Omega \boldsymbol{w} \cdot \boldsymbol{\varphi}\, \mathrm{d}x, \tag{8.52}$$

$$\begin{aligned} \boldsymbol{b}_h(\boldsymbol{w}, \boldsymbol{\varphi}) = &\sum_{\Gamma \in \mathscr{F}_h^I} \int_\Gamma \mathbf{H}(\boldsymbol{w}_\Gamma^{(L)}, \boldsymbol{w}_\Gamma^{(R)}, \boldsymbol{n}_\Gamma) \cdot [\boldsymbol{\varphi}]\, \mathrm{d}S + \sum_{\Gamma \in \mathscr{F}_h^B} \int_\Gamma \mathbf{H}(\boldsymbol{w}_\Gamma^{(L)}, \boldsymbol{w}_\Gamma^{(R)}, \boldsymbol{n}_\Gamma) \cdot \boldsymbol{\varphi}\, \mathrm{d}S \\ &- \sum_{K \in \mathscr{T}_h} \int_K \sum_{s=1}^d \boldsymbol{f}_s(\boldsymbol{w}) \cdot \frac{\partial \boldsymbol{\varphi}}{\partial x_s}\, \mathrm{d}x. \end{aligned} \tag{8.53}$$

The meaning of $\boldsymbol{w}_\Gamma^{(R)}$ for $\Gamma \in \mathscr{F}_h^B$ will be specified in Sect. 8.3. We call $\boldsymbol{b}_h$ the *convection* (or inviscid) *form*. The expressions in (8.51)–(8.53) make sense for $\boldsymbol{w}, \boldsymbol{\varphi} \in \boldsymbol{H}^1(\Omega, \mathscr{T}_h)$. The approximation of the exact solution $\boldsymbol{w}(t)$ of (8.8) will be sought in the finite-dimensional space $\boldsymbol{S}_{hp} \subset \boldsymbol{H}^1(\Omega, \mathscr{T}_h)$ for each $t \in (0, T)$. Therefore, using (8.51), we immediately arrive at the definition of an approximate solution.

Definition 8.7 We say that a function $\boldsymbol{w}_h : \Omega \times (0, T) \to \mathbb{R}^m$ is the *space semidiscrete solution* of the Euler equations (8.8), if the following conditions are satisfied:

$$\boldsymbol{w}_h \in C^1([0, T]; \boldsymbol{S}_{hp}), \tag{8.54a}$$

$$\frac{\mathrm{d}}{\mathrm{d}t}\big(\boldsymbol{w}_h(t), \boldsymbol{\varphi}_h\big) + \boldsymbol{b}_h(\boldsymbol{w}_h(t), \boldsymbol{\varphi}_h) = 0 \quad \forall \boldsymbol{\varphi}_h \in \boldsymbol{S}_{hp}\; \forall t \in (0, T), \tag{8.54b}$$

$$\boldsymbol{w}_h(0) = \Pi_h \boldsymbol{w}^0, \tag{8.54c}$$

where $\Pi_h \boldsymbol{w}^0$ is the $\boldsymbol{S}_{hp}$-approximation of the function $\boldsymbol{w}^0$ from the initial condition (8.36). Usually it is defined as the L^2-projection of $\boldsymbol{w}^0$ on the space $\boldsymbol{S}_{hp}$.

Problem (8.54) represents a system of N_{hp} ordinary differential equations (ODEs), where N_{hp} is equal to the dimension of the space $\boldsymbol{S}_{hp}$. Its solution will be discussed in Sect. 8.4.

Remark 8.8 If we consider the case $p = 0$ (i.e., the approximate solution is piecewise constant on $\mathscr{T}_h$), then the numerical scheme (8.54) represents the standard *finite volume method*. See, e.g., [127, 205, 282]. Actually, for $p = 0$ we choose the basis functions of $\boldsymbol{S}_{h0}$ as characteristic functions χ_K of $K \in \mathscr{T}_h$. Let us recall that $\chi_K = 1$ on K and $\chi_K = 0$ elsewhere. Therefore, putting $\boldsymbol{\varphi}_h = \chi_K,\ K \in \mathscr{T}_h$, in (8.54b), we obtain

$$\frac{\mathrm{d}}{\mathrm{d}t}\boldsymbol{w}_K(t) + \sum_{K' \in \mathscr{N}(K)} |\Gamma_{K,K'}| \mathbf{H}(\boldsymbol{w}_K(t), \boldsymbol{w}_{K'}(t), \boldsymbol{n}_{K,K'}) = 0, \tag{8.55}$$

where

$$\boldsymbol{w}_K = \frac{1}{|K|} \int_K \boldsymbol{w}_h \, \mathrm{d}x, \quad K \in \mathscr{T}_h, \tag{8.56}$$

and $\mathscr{N}(K) = \{K', \ \partial K \cap \partial K' \in \mathscr{F}_h\}$ is the set of all elements K' having a common face $\Gamma_{K,K'}$ with K. The set $\mathscr{N}(K)$ contains also fictitious elements outside of Ω having a common face $\partial K \cap \Omega$ with $K \in \mathscr{T}_h$. In this case, the value $\boldsymbol{w}_{K'}$ in the numerical flux $\mathbf{H}$ is determined from boundary conditions. By $|\Gamma_{K,K'}|$ and $|K|$ we denote the $(d-1)$-Lebesgue measure of the common face $\Gamma_{K,K'}$ between K and K' and the d-dimensional measure of the element K, respectively. The symbol $\boldsymbol{n}_{K,K'}$ denotes the outer unit normal to ∂K on $\Gamma_{K,K'}$.

8.3 Numerical Treatment of Boundary Conditions

If $\Gamma \in \mathscr{F}_h^B$, then it is necessary to specify the boundary state $\boldsymbol{w}_\Gamma^{(R)}$ appearing in the numerical flux $\mathbf{H}$ in the definition (8.53) of the convection form $\boldsymbol{b}_h$. In what follows, we describe the treatment of the boundary conditions for impermeable walls and the inlet/outlet part of the boundary. The boundary conditions should be theoretically determined at all boundary points. In practical computations, when the integrals are evaluated with the aid of quadrature formulae, it is enough to consider the boundary conditions at only integration boundary points. Therefore, for the sake of simplicity, the symbol $\boldsymbol{w}_\Gamma^{(R)}$ will mean the value of this function at a boundary point in consideration.

8.3.1 *Boundary Conditions on Impermeable Walls*

For $\Gamma \in \mathscr{F}_h^W$ we should interpret in a suitable way the impermeability condition (8.38), i.e., $\boldsymbol{v} \cdot \boldsymbol{n} = 0$, where $\boldsymbol{v}$ is the velocity vector and $\boldsymbol{n}$ the outer unit normal to $\partial\Omega_W$. This condition has to be incorporated in some sense into the expression $\mathbf{H}(\boldsymbol{w}_\Gamma^{(L)}, \boldsymbol{w}_\Gamma^{(R)}, \boldsymbol{n}_\Gamma)$ appearing in the definition (8.53) of the form $\boldsymbol{b}_h$.

We describe two possibilities. The first one is based on the direct use of the impermeability condition in the physical flux $\boldsymbol{P}(\boldsymbol{w}, \boldsymbol{n})$. The second one applies the *mirror operator* to the state $\boldsymbol{w}$.

8.3.1.1 Direct Use of the Impermeability Condition

Let $\boldsymbol{n} = (n_1, \dots, n_d) \in \mathrm{B}_1$. Then from (8.16) and (8.10) we have

$$\boldsymbol{P}(\boldsymbol{w}, \boldsymbol{n}) = \sum_{s=1}^{d} \boldsymbol{f}_s(\boldsymbol{w}) n_s = \sum_{s=1}^{d} \begin{pmatrix} f_{s,1}(\boldsymbol{w}) \\ f_{s,2}(\boldsymbol{w}) \\ \vdots \\ f_{s,m-1}(\boldsymbol{w}) \\ f_{s,m}(\boldsymbol{w}) \end{pmatrix} n_s = \begin{pmatrix} \rho \boldsymbol{v} \cdot \boldsymbol{n} \\ \rho v_1 \boldsymbol{v} \cdot \boldsymbol{n} + \mathrm{p} n_1 \\ \vdots \\ \rho v_d \boldsymbol{v} \cdot \boldsymbol{n} + \mathrm{p} n_d \\ (E + \mathrm{p}) \boldsymbol{v} \cdot \boldsymbol{n} \end{pmatrix}. \tag{8.57}$$

Using the condition $\boldsymbol{v} \cdot \boldsymbol{n} = 0$ in (8.57), we obtain

$$\boldsymbol{P}(\boldsymbol{w}, \boldsymbol{n}) = \sum_{s=1}^{d} \boldsymbol{f}_s(\boldsymbol{w}) n_s = (0, \mathrm{p} n_1, \dots, \mathrm{p} n_d, 0)^{\mathrm{T}} =: \boldsymbol{f}_{\mathrm{W}}^1(\boldsymbol{w}, \boldsymbol{n}), \tag{8.58}$$

where the pressure satisfies the relation $\mathrm{p} = (\gamma - 1)(w_m - (w_2^2 + \dots + w_{m-1}^2)/(2w_1))$. Then, taking into account (8.48) and (8.58), for $\Gamma \in \mathscr{F}_h^W$ we can put

$$\int_\Gamma \mathbf{H}(\boldsymbol{w}_\Gamma^{(L)}, \boldsymbol{w}_\Gamma^{(R)}, \boldsymbol{n}_\Gamma) \cdot \boldsymbol{\varphi}_h \, \mathrm{d}S = \int_\Gamma \boldsymbol{f}_{\mathrm{W}}^1(\boldsymbol{w}_\Gamma^{(L)}, \boldsymbol{n}_\Gamma) \cdot \boldsymbol{\varphi}_h \, \mathrm{d}S, \quad \Gamma \in \mathscr{F}_h^W. \tag{8.59}$$

For the purpose of the solution strategy developed in Sect. 8.4, we introduce a linearization of $\boldsymbol{f}_{\mathrm{W}}^1$. By virtue of (8.28), we have

$$\sum_{s=1}^{d} \boldsymbol{f}_s(\boldsymbol{w}) n_s = \boldsymbol{P}(\boldsymbol{w}, \boldsymbol{n}) = \mathbb{P}(\boldsymbol{w}, \boldsymbol{n}) \boldsymbol{w} \quad \forall \boldsymbol{w} \in \mathscr{D} \; \forall \boldsymbol{n} = (n_1, \dots, n_d) \in \mathrm{B}_1. \tag{8.60}$$

Our aim is to introduce a matrix (denoted by $\mathbb{P}_W$ hereafter), which is the simplest possible and such that

$$\mathbb{P}(\boldsymbol{w}, \boldsymbol{n}) \boldsymbol{w} = \mathbb{P}_W(\boldsymbol{w}, \boldsymbol{n}) \boldsymbol{w} \tag{8.61}$$

provided that $\boldsymbol{w} \in \mathscr{D}$ and $\boldsymbol{n} \in \mathrm{B}_1$ satisfy the impermeability condition $\boldsymbol{v} \cdot \boldsymbol{n} = 0$, where $\boldsymbol{v}$ is the velocity vector corresponding to $\boldsymbol{w}$. Taking into account the explicit expression (8.24) for $\mathbb{P}$, we remove some of its entries and define the matrix

$$\mathbb{P}_W(\boldsymbol{w}, \boldsymbol{n}) = (\gamma - 1) \begin{pmatrix} 0 & 0 & \dots & 0 & 0 \\ |\boldsymbol{v}|^2 n_1/2 & -v_1 n_1 & \dots & -v_d n_1 & n_1 \\ \vdots & \vdots & \ddots & \vdots & \vdots \\ |\boldsymbol{v}|^2 n_d/2 & -v_1 n_d & \dots & -v_d n_d & n_d \\ 0 & 0 & \dots & 0 & 0 \end{pmatrix}, \tag{8.62}$$

where $w \in \mathscr{D}$, $n = (n_1, \ldots, n_d) \in \mathrm{B}_1$, $v_j = w_{j+1}/w_1$, $j = 1, \ldots, d$, are the components of the velocity vector and $|v|^2 = v_1^2 + \cdots + v_d^2$. We can verify by a simple calculation that (8.61) is valid.

Moreover, we define the linearized form of f_{W}^1 by

$$f_{\mathrm{W}}^{1,\mathrm{L}}(\bar{w}, w, n) = \mathbb{P}_W(\bar{w}, n)w, \quad \bar{w}, w \in \mathscr{D}, \ n \in \mathrm{B}_1, \tag{8.63}$$

which is linear with respect to the argument w. Obviously, due to (8.58), (8.61) and (8.63), we find that under the condition $v \cdot n = 0$, the linearized form $f_{\mathrm{W}}^{1,\mathrm{L}}$ is consistent with f_{W}^1, i.e.,

$$f_{\mathrm{W}}^{1,\mathrm{L}}(w, w, n) = f_{\mathrm{W}}^1(w, n) \quad \forall w \in \mathscr{D} \, \forall n \in \mathrm{B}_1 \text{ such that } v \cdot n = 0. \tag{8.64}$$

Exercise 8.9 Verify relation (8.61) for $\mathbb{P}_W$ given by (8.62), provided $v \cdot n = 0$.

8.3.1.2 Inviscid Mirror Boundary Conditions

This approach is based on the definition of the state vector $w_\Gamma^{(R)}$, $\Gamma \in \mathscr{F}_h^W$ in the form

$$w_\Gamma^{(R)} = \mathscr{M}(w_\Gamma^{(L)}), \tag{8.65}$$

where the boundary operator $\mathscr{M}$, called the *inviscid mirror operator*, is defined in the following way. If $w \in \mathscr{D}$, $w = (\rho, \rho v, E)^{\mathrm{T}}$ and $n \in \mathrm{B}_1$ is the outer unit normal to $\partial\Omega$ at a point in consideration lying on $\partial\Omega_W$, then we set

$$v^\perp = v - 2(v \cdot n)n, \tag{8.66}$$

and

$$\mathscr{M}(w) = (\rho, \rho v^\perp, E)^{\mathrm{T}}. \tag{8.67}$$

The vectors v and $v^\perp$ have the same tangential component but opposite normal components, see Fig. 8.1. Obviously, the operator $\mathscr{M}$ is linear.

Now we define the mapping $f_{\mathrm{W}}^2 : \mathscr{D} \times \mathrm{B}_1 \to \mathbb{R}^m$ by

$$f_{\mathrm{W}}^2(w, n) = \mathbf{H}(w, \mathscr{M}(w), n) \tag{8.68}$$

and, if $\Gamma \in \mathscr{F}_h^W$, then in (8.53) we have

$$\mathbf{H}(w_\Gamma^{(L)}, w_\Gamma^{(R)}, n_\Gamma) = f_{\mathrm{W}}^2(w_\Gamma^{(L)}, n_\Gamma). \tag{8.69}$$

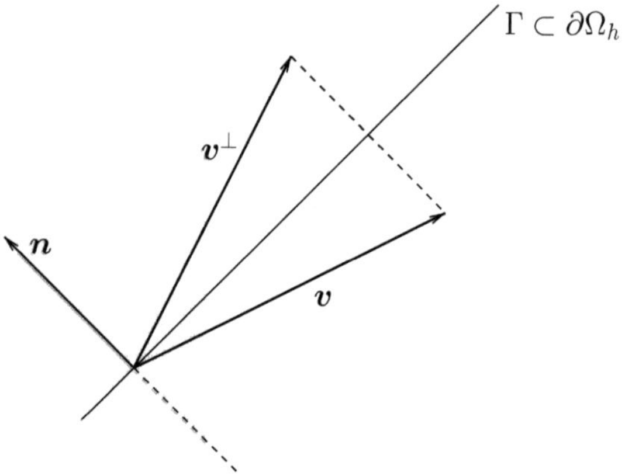

Fig. 8.1 Impermeability conditions defined by the mirror operator, vectors of velocity of $\boldsymbol{v}$ and $\boldsymbol{v}^\perp = \boldsymbol{v} - 2(\boldsymbol{v} \cdot \boldsymbol{n})\boldsymbol{n}$

8.3.2 Boundary Conditions on the Inlet and Outlet

The definition of the boundary state $\boldsymbol{w}_\Gamma^{(R)}$ in (8.53) for $\Gamma \in \mathscr{F}_h^{io} \subset \partial\Omega_i \cup \partial\Omega_o$ (i.e., $\Gamma \subset \partial\Omega_i \cup \partial\Omega_o$) is more delicate. The determination of the inlet/outlet boundary conditions is usually based on a given state vector function $\boldsymbol{w}_{\mathrm{BC}}$ prescribed on $(\partial\Omega_i \cup \partial\Omega_o) \times (0, T)$. For example, when we solve flow around an isolated profile, the state vector $\boldsymbol{w}_{\mathrm{BC}}$ corresponds to the unperturbed far-field flow (flow at infinity). For flow in a channel, the state vector $\boldsymbol{w}_{\mathrm{BC}}$ may correspond to a flow at the inlet and outlet of the channel.

However, since system (8.8) is hyperbolic, we cannot simply put $\boldsymbol{w}_\Gamma^{(R)} = \boldsymbol{w}_{\mathrm{BC}}$. As we will show later (see also [127]), for a *linear hyperbolic* system with one space variable

$$\frac{\partial \boldsymbol{q}}{\partial t} + \bar{\mathbb{A}} \frac{\partial \boldsymbol{q}}{\partial x} = 0, \quad (x, t) \in (-\infty, 0) \times (0, \infty), \tag{8.70}$$

where $\boldsymbol{q} : (-\infty, 0) \times [0, \infty) \to \mathbb{R}^m$ and $\bar{\mathbb{A}}$ is a constant $m \times m$ matrix, only some quantities defining $\boldsymbol{q}$ at $x = 0$ can be prescribed, whereas other quantities have to be extrapolated from the interior of the computational domain. We will see that the number of prescribed components of $\boldsymbol{q}$ is equal to the number of negative eigenvalues of $\bar{\mathbb{A}}$.

However, for *nonlinear hyperbolic* systems the theory is missing. Therefore, a usual approach is to choose the boundary conditions in such a way that a linearized initial-boundary value problem is well-posed, i.e., it has a unique solution. We describe this method in the following part of this section.

8.3.2.1 Approach Based on the Solution of the Linearized Riemann Problem

Let $\Gamma \in \mathscr{F}_h^{io}$ and let $x_\Gamma \in \Gamma$ be a point in consideration, at which we want to determine boundary conditions. We introduce a new coordinate system $(\tilde{x}_1, \ldots, \tilde{x}_d)$ such that the coordinate origin lies at the point x_Γ, the axis $\tilde{x}_1$ is parallel to the normal direction $\boldsymbol{n}$ to the boundary, and the coordinate axes $\tilde{x}_2, \ldots, \tilde{x}_d$ are tangential to the boundary, see Fig. 8.2. This transformation of the space coordinates is carried out by the mapping $\tilde{x} = \mathbb{Q}_0(\boldsymbol{n})(x - x_\Gamma)$, where $\mathbb{Q}_0(\boldsymbol{n})$ is the rotation matrix defined by (8.34) for $d = 2$ and (8.35) for $d = 3$.

Let $\boldsymbol{w}_\Gamma^{(L)}$ be the value of the trace of the state vector $\boldsymbol{w}$ on Γ from the interior of Ω at the point x_Γ and let

$$\boldsymbol{q}_\Gamma^{(L)} = \mathbb{Q}(\boldsymbol{n}_\Gamma)\boldsymbol{w}_\Gamma^{(L)}, \tag{8.71}$$

where $\mathbb{Q}(\boldsymbol{n}_\Gamma)$ is given by (8.33).

Using rotational invariance of the Euler equations introduced in Lemma 8.6e, we transform them to the coordinates $\tilde{x}_1, \ldots, \tilde{x}_d$, neglect the derivative with respect to $\tilde{x}_j,\ j = 2, \ldots, d$, and linearize the resulting system around the state $\boldsymbol{q}_\Gamma^{(L)}$. Then we obtain the linear system

$$\frac{\partial \boldsymbol{q}}{\partial t} + \mathbb{A}_1(\boldsymbol{q}_\Gamma^{(L)})\frac{\partial \boldsymbol{q}}{\partial \tilde{x}_1} = 0, \quad (\tilde{x}_1, t) \in (-\infty, 0) \times [0, \infty) \tag{8.72}$$

for the transformed vector-valued function $\boldsymbol{q} = \mathbb{Q}(\boldsymbol{n}_\Gamma)\boldsymbol{w}$, see Fig. 8.3, left. To this system we add the initial and boundary conditions

$$\begin{aligned} \boldsymbol{q}(\tilde{x}_1, 0) &= \boldsymbol{q}_\Gamma^{(L)}, \quad \tilde{x}_1 < 0, \\ \boldsymbol{q}(0, t) &= \boldsymbol{q}_\Gamma^{(R)}, \quad t > 0, \end{aligned} \tag{8.73}$$

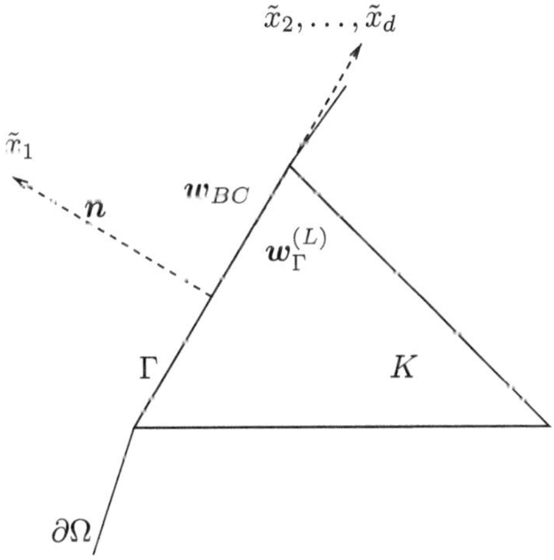

Fig. 8.2 New coordinate system $(\tilde{x}_1, \ldots, \tilde{x}_d)$

where $\boldsymbol{q}_{\Gamma}^{(L)}$ is given by (8.71) and the unknown state vector $\boldsymbol{q}_{\Gamma}^{(R)}$ should be determined in such a way that it reflects the state vector $\boldsymbol{q}_{\mathrm{BC}} = \mathbb{Q}(\boldsymbol{n}_{\Gamma})\boldsymbol{w}_{\mathrm{BC}}$ with a prescribed state $\boldsymbol{w}_{\mathrm{BC}}$, and the initial-boundary value problem (8.72) and (8.73) is well-posed, i.e., has a unique solution.

In order to find the vector $\boldsymbol{q}_{\Gamma}^{(R)}$, we consider the *linearized Riemann problem*

$$\frac{\partial \boldsymbol{q}}{\partial t} + \mathbb{A}_1(\boldsymbol{q}_{\Gamma}^{(L)})\frac{\partial \boldsymbol{q}}{\partial \tilde{x}_1} = 0, \quad (\tilde{x}_1, t) \in (-\infty, \infty) \times [0, \infty) \tag{8.74}$$

with the initial condition

$$\boldsymbol{q}(\tilde{x}_1, 0) = \begin{cases} \boldsymbol{q}_{\Gamma}^{(L)}, & \text{if } \tilde{x}_1 < 0, \\ \boldsymbol{q}_{\mathrm{BC}}, & \text{if } \tilde{x}_1 > 0, \end{cases} \tag{8.75}$$

see Fig. 8.3 (right).

The exact solution of problem (8.74) and (8.75) can be found by the method of characteristics in the following way: Let $\boldsymbol{g}_s, s = 1, \ldots, m$, be the eigenvectors corresponding to the eigenvalues $\tilde{\lambda}_s,\ s = 1, \ldots, m$, of the matrix $\mathbb{A}_1 = \mathbb{A}_1(\boldsymbol{q}_{\Gamma}^{(L)})$. Hence, $\mathbb{A}_1 \boldsymbol{g}_s = \tilde{\lambda}_s \boldsymbol{g}_s,\ s = 1, \ldots, m$.

Taking into account (8.32), we see that the eigenvalues of the matrices $\mathbb{A}_1(\boldsymbol{q}_{\Gamma}^{(L)})$ and $\mathbb{P}(\boldsymbol{w}_{\Gamma}^{(L)}, \boldsymbol{n}_{\Gamma})$ attain the same values, i.e.,

$$\tilde{\lambda}_s = \lambda_s, \quad s = 1, \ldots, m, \tag{8.76}$$

where λ_s are the eigenvalues of the matrix $\mathbb{P}(\boldsymbol{w}_{\Gamma}^{(L)}, \boldsymbol{n}_{\Gamma})$.

The explicit formulae for the eigenvectors $\boldsymbol{g}_s,\ s = 1, \ldots, m$, can be found in [127], Sect. 3.1. These eigenvectors form a basis of $\mathbb{R}^m$, and thus the exact solution of (8.74) can be written in the form

$$\boldsymbol{q}(\tilde{x}_1, t) = \sum_{s=1}^{m} \mu_s(\tilde{x}_1, t)\boldsymbol{g}_s, \quad \tilde{x}_1 \in \mathbb{R},\ t > 0, \tag{8.77}$$

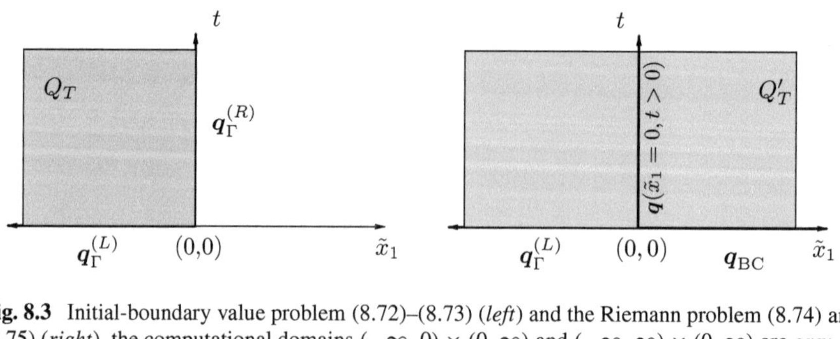

Fig. 8.3 Initial-boundary value problem (8.72)–(8.73) (*left*) and the Riemann problem (8.74) and (8.75) (*right*), the computational domains $(-\infty, 0) \times (0, \infty)$ and $(-\infty, \infty) \times (0, \infty)$ are *grey*

where μ_s, $s = 1, \ldots, m$, are unknown functions defined in $(-\infty, \infty) \times [0, \infty)$. Similarly, the initial states from (8.75) can be expressed as

$$\boldsymbol{q}_\Gamma^{(L)} = \sum_{s=1}^{m} \alpha_s \boldsymbol{g}_s, \quad \boldsymbol{q}_{\mathrm{BC}} = \sum_{s=1}^{m} \beta_s \boldsymbol{g}_s. \tag{8.78}$$

The vectors $\boldsymbol{\alpha} = (\alpha_1, \ldots, \alpha_m)$ and $\boldsymbol{\beta} = (\beta_1, \ldots, \beta_m)$ are given by the relations

$$\boldsymbol{\alpha} = \mathbb{T}^{-1}\boldsymbol{q}_\Gamma^{(L)}, \quad \boldsymbol{\beta} = \mathbb{T}^{-1}\boldsymbol{q}_{\mathrm{BC}}, \tag{8.79}$$

where $\mathbb{T}$ is the $m \times m$-matrix whose columns are the eigenvectors $\boldsymbol{g}_s$, $s = 1, \ldots, m$. The functions μ_s, $s = 1, \ldots, m$, are called the *characteristic variables*.

Substituting (8.77) into (8.74), we get

$$0 = \sum_{s=1}^{m} \left(\frac{\partial \mu_s}{\partial t} + \tilde{\lambda}_s \frac{\partial \mu_s}{\partial \tilde{x}_1} \right) \boldsymbol{g}_s, \quad s = 1, \ldots, m, \tag{8.80}$$

which holds if and only if

$$\frac{\partial \mu_s}{\partial t} + \tilde{\lambda}_s \frac{\partial \mu_s}{\partial \tilde{x}_1} = 0, \quad \tilde{x}_1 \in \mathbb{R},\ t > 0,\ s = 1, \ldots, m. \tag{8.81}$$

These equations are equipped with initial conditions following from (8.75) and (8.78)

$$\mu_s(\tilde{x}_1, 0) = \bar{\mu}_s(\tilde{x}_1) := \begin{cases} \alpha_s, & \tilde{x}_1 < 0, \\ \beta_s, & \tilde{x}_1 > 0, \end{cases} \quad s = 1, \ldots, m. \tag{8.82}$$

We can simply verify that the exact solution of (8.81) and (8.82) reads

$$\mu_s(\tilde{x}_1, t) = \bar{\mu}_s(\tilde{x}_1 - \tilde{\lambda}_s t), \quad \tilde{x}_1 \in \mathbb{R},\ t \geq 0,$$

where $\bar{\mu}_s$ is given by (8.82). This together with (8.82) gives

$$\mu_s(\tilde{x}_1, t) = \begin{cases} \alpha_s, & \text{if } \tilde{x}_1 - \tilde{\lambda}_s t < 0, \\ \beta_s, & \text{if } \tilde{x}_1 - \tilde{\lambda}_s t > 0, \end{cases} \quad s = 1, \ldots, m. \tag{8.83}$$

We define the sought state $\boldsymbol{q}_\Gamma^{(R)}$ as the solution of problem (8.74) and (8.75) at $\tilde{x}_1 = 0$. Hence, we put $\boldsymbol{q}_\Gamma^{(R)} = \boldsymbol{q}(0, t)$, and by (8.77) and (8.83), we get

$$\boldsymbol{q}_\Gamma^{(R)} = \sum_{s=1}^{m} \eta_s \boldsymbol{g}_s, \quad \eta_s = \begin{cases} \alpha_s, & \tilde{\lambda}_s \geq 0, \\ \beta_s, & \tilde{\lambda}_s < 0. \end{cases} \tag{8.84}$$

Finally, we introduce the inlet/outlet *boundary operator* based on the solution of the linearized Riemann problem

$$\mathscr{B}^{\mathrm{LRP}}(\boldsymbol{w}_\Gamma^{(L)}, \boldsymbol{w}_{\mathrm{BC}}) := \mathbb{Q}^{-1}(\boldsymbol{n}_\Gamma)\boldsymbol{q}_\Gamma^{(R)}. \tag{8.85}$$

Then we define the sought boundary state

$$\boldsymbol{w}_\Gamma^{(R)} := \mathscr{B}^{\mathrm{LRP}}(\boldsymbol{w}_\Gamma^{(L)}, \boldsymbol{w}_{\mathrm{BC}}). \tag{8.86}$$

Remark 8.10 From the above process (taking into account (8.76)) we can conclude that the sought boundary state $\boldsymbol{w}_\Gamma^{(R)}$ is determined using m_{pr} quantities characterizing the prescribed boundary state $\boldsymbol{w}_{\mathrm{BC}}$, where m_{pr} is the number of negative eigenvalues of the matrix $\mathbb{P}(\boldsymbol{w}_\Gamma^{(L)}, \boldsymbol{n}_\Gamma)$, whereas we extrapolate m_{ex} quantities defining the state $\boldsymbol{w}_\Gamma^{(L)}$, where $m_{\mathrm{ex}} = m - m_{\mathrm{pr}}$ is the number of nonnegative eigenvalues of the matrix $\mathbb{P}(\boldsymbol{w}_\Gamma^{(L)}, \boldsymbol{n}_\Gamma)$.

This observation is in agreement with the definitions of boundary conditions on impermeable walls. Taking into account that by (8.30) the eigenvalues of the matrix $\mathbb{P}(\boldsymbol{w}_\Gamma^{(L)}, \boldsymbol{n}_\Gamma)$ read

$$\lambda_1 = \boldsymbol{v}\cdot\boldsymbol{n} - a, \quad \lambda_2 = \cdots = \lambda_{d+1} = \boldsymbol{v}\cdot\boldsymbol{n}, \quad \lambda_{d+2} = \boldsymbol{v}\cdot\boldsymbol{n} + a, \tag{8.87}$$

where $\boldsymbol{v}$ and a represent the velocity vector and the speed of sound, respectively, corresponding to the state $\boldsymbol{w}_\Gamma^{(L)}$, and $\boldsymbol{n} = \boldsymbol{n}_\Gamma$. Then the impermeability condition $\boldsymbol{v}\cdot\boldsymbol{n} = 0$ implies that $\lambda_1 < 0$, $\lambda_2 = \cdots = \lambda_{d+1} = 0$, $\lambda_{d+2} > 0$. Hence, in this case we prescribe only one quantity, namely $\boldsymbol{v}\cdot\boldsymbol{n} = 0$ or the opposite normal component $-\boldsymbol{v}\cdot\boldsymbol{n}$ of the velocity vector and the remaining quantities defining the state $\boldsymbol{w}_\Gamma^{(R)}$ are obtained by extrapolation.

8.3.2.2 Approach Based on Physical Properties of the Flow

It follows from the above considerations and the form (8.87) of eigenvalues λ_s, $s = 1, \ldots, m = d+2$, that in the case of the inlet or outlet, on which $\boldsymbol{v}\cdot\boldsymbol{n} < 0$ or $\boldsymbol{v}\cdot\boldsymbol{n} > 0$, respectively, it is necessary to distinguish between the subsonic or supersonic regime, when $|\boldsymbol{v}\cdot\boldsymbol{n}| < a$ or $|\boldsymbol{v}\cdot\boldsymbol{n}| > a$, respectively. The number of prescribed and extrapolated boundary conditions for the mentioned possibilities is shown in Table 8.1.

On the basis of these results, it is possible to introduce a widely used method for determining the inlet/outlet boundary conditions based on the use of physical variables. In this approach we extrapolate or prescribe directly some physical variables. Particularly, we distinguish the following cases:

- *supersonic inlet*, $m_{\mathrm{pr}} = m$, we prescribe all components of the boundary state $\boldsymbol{w}_\Gamma^{(R)}$. Hence, we set $\boldsymbol{w}_\Gamma^{(R)} = \boldsymbol{w}_{\mathrm{BC}}$,

Table 8.1 Boundary conditions based on the well-posedness of the linearized problem: number of prescribed m_{pr} and extrapolated m_{ex} components of $\boldsymbol{w}$ for subsonic/supersonic inlet/outlet

Flow regime	m_{pr}	m_{ex}
Supersonic inlet	m	0
Subsonic inlet	$m-1$	1
Subsonic outlet	1	$m-1$
Supersonic outlet	0	m

- *subsonic inlet*, $m_{\text{pr}} = m-1$, we extrapolate the pressure from the interior of the domain, and prescribe the density and the components of the velocity on the boundary,
- *subsonic outlet*, $m_{\text{pr}} = 1$, we prescribe the pressure and extrapolate the density and the components of the velocity on the boundary,
- *supersonic outlet*, $m_{\text{pr}} = 0$, we extrapolate all components of $\boldsymbol{w}$ from the interior of Ω on the boundary. This means that we set $\boldsymbol{w}_\Gamma^{(R)} = \boldsymbol{w}_\Gamma^{(L)}$.

Hence, we define the inlet/outlet boundary operator based on physical variables:

$$\mathscr{B}^{\text{phys}}(\boldsymbol{w}_\Gamma^{(L)}, \boldsymbol{w}_{\text{BC}}) = \begin{cases} \boldsymbol{w}_{\text{BC}} & \text{if } \boldsymbol{v}\cdot\boldsymbol{n} < -a \quad \text{supersonic inlet} \\ \mathsf{Phys}(\rho_{\text{BC}}, \boldsymbol{v}_{\text{BC}}, \mathrm{p}_\Gamma^{(L)}) & \text{if } -a \le \boldsymbol{v}\cdot\boldsymbol{n} < 0 \quad \text{subsonic inlet} \\ \mathsf{Phys}(\rho_\Gamma^{(L)}, \boldsymbol{v}_\Gamma^{(L)}, \mathrm{p}_{\text{BC}}) & \text{if } 0 < \boldsymbol{v}\cdot\boldsymbol{n} \le a \quad \text{subsonic outlet} \\ \boldsymbol{w}_\Gamma^{(L)} & \text{if } \boldsymbol{v}\cdot\boldsymbol{n} > a \quad \text{supersonic outlet} \end{cases} \tag{8.88}$$

where $\rho_{\text{BC}}, \boldsymbol{v}_{\text{BC}}, \mathrm{p}_{\text{BC}}$ are the density, the velocity vector and the pressure, respectively, corresponding to the prescribed state $\boldsymbol{w}_{\text{BC}}$ and $\rho_\Gamma^{(L)}, \boldsymbol{v}_\Gamma^{(L)}, \mathrm{p}_\Gamma^{(L)}$ denote the density, the velocity vector and the pressure corresponding to $\boldsymbol{w}_\Gamma^{(L)}$. The symbol Phys denotes the transformation from the physical variables to the conservative ones, namely, for $\rho > 0$, $\mathrm{p} > 0$ and $\boldsymbol{v} \in \mathbb{R}^d$ we set

$$\mathsf{Phys}(\rho, \boldsymbol{v}, \mathrm{p}) = \left(\rho,\ \rho\boldsymbol{v},\ \mathrm{p}/(\gamma-1) + \rho|\boldsymbol{v}|^2/2\right)^{\mathrm{T}} \in \mathbb{R}^m. \tag{8.89}$$

This approach is usually used with success for the transonic flow. However, its application to low Mach number flows does not give reasonable results, because these boundary conditions are not transparent for acoustic waves coming from inside of the computational domain Ω. In numerical simulations, we observe some reflection from the inlet/outlet parts of the boundary. Therefore, in a low Mach number flow, it is suitable to apply the method based on the solution of a linearized Riemann problem. This means that the boundary state $\boldsymbol{w}_\Gamma^{(R)}$ is defined by (8.86). Another more sophisticated method will be treated in the following section.

8.3.2.3 Boundary Conditions Based on the Exact Solution of the Nonlinear Riemann Problem

The generalization of the method based on the solution of the linearized Riemann problem uses the exact solution of the nonlinear Riemann problem. The only difference is that we do not linearize the system of the Euler equations around the state $\boldsymbol{w}_\Gamma^{(L)}$, but instead of (8.72) we consider the nonlinear system

$$\frac{\partial \boldsymbol{q}}{\partial t} + \mathbb{A}_1(\boldsymbol{q})\frac{\partial \boldsymbol{q}}{\partial \tilde{x}_1} = 0, \quad (\tilde{x}_1, t) \in (-\infty, 0) \times [0, \infty) \tag{8.90}$$

with the initial and boundary conditions (8.73). This means that instead of (8.74), we consider the *Riemann problem*

$$\frac{\partial \boldsymbol{q}}{\partial t} + \mathbb{A}_1(\boldsymbol{q})\frac{\partial \boldsymbol{q}}{\partial \tilde{x}_1} = 0, \quad (\tilde{x}_1, t) \in (-\infty, \infty) \times [0, \infty) \tag{8.91}$$

equipped with the initial condition (8.75). The solution of problem (8.91) and (8.75) is much more complicated than the solution of the linearized problem (8.74) and (8.75) but for the Euler equations it can be constructed analytically, see e.g., [127, Sect. 3.1.6] or [282, Sect. 10.2]. This analytical solution contains an implicit formula for the pressure p, which has to be obtained iteratively.

When the solution $\boldsymbol{q}$ of the Riemann problem (8.91) and (8.75) is obtained, then we define the inlet/outlet boundary operator based on the solution of the nonlinear Riemann problem as

$$\mathscr{B}^{\mathrm{RP}}(\boldsymbol{w}_\Gamma^{(L)}, \boldsymbol{w}_{\mathrm{BC}}) := \mathbb{Q}^{-1}(\boldsymbol{n}_\Gamma)\boldsymbol{q}(0, t) \tag{8.92}$$

and set $\boldsymbol{w}_\Gamma^{(R)} := \mathscr{B}^{\mathrm{RP}}(\boldsymbol{w}_\Gamma^{(L)}, \boldsymbol{w}_{\mathrm{BC}})$.

Finally, based on the presented approaches to the choice of boundary conditions we specify the definition (8.53) of the form $\boldsymbol{b}_h$ by

$$\begin{aligned} \boldsymbol{b}_h(\boldsymbol{w}, \boldsymbol{\varphi}) = &- \sum_{K \in \mathscr{T}_h} \int_K \sum_{s=1}^{d} \boldsymbol{f}_s(\boldsymbol{w}) \cdot \frac{\partial \boldsymbol{\varphi}}{\partial x_s} \, \mathrm{d}x \\ &+ \sum_{\Gamma \in \mathscr{F}_h^I} \int_\Gamma \mathbf{H}(\boldsymbol{w}_\Gamma^{(L)}, \boldsymbol{w}_\Gamma^{(R)}, \boldsymbol{n}_\Gamma) \cdot [\boldsymbol{\varphi}] \, \mathrm{d}S \\ &+ \sum_{\Gamma \in \mathscr{F}_h^W} \int_\Gamma \boldsymbol{f}_{\mathrm{W}}^i(\boldsymbol{w}_\Gamma^{(L)}, \boldsymbol{n}_\Gamma) \cdot \boldsymbol{\varphi} \, \mathrm{d}S \\ &+ \sum_{\Gamma \in \mathscr{F}_h^{io}} \int_\Gamma \mathbf{H}\left(\boldsymbol{w}_\Gamma^{(L)}, \mathscr{B}(\boldsymbol{w}_\Gamma^{(L)}, \boldsymbol{w}_{\mathrm{BC}}), \boldsymbol{n}_\Gamma\right) \cdot \boldsymbol{\varphi} \, \mathrm{d}S, \end{aligned} \tag{8.93}$$

where $i = 1$ or $i = 2$, if we use the impermeability boundary condition (8.58) or (8.68), respectively. Moreover, the inlet/outlet boundary operator $\mathscr{B}$ represents $\mathscr{B}^{\text{phys}}$, $\mathscr{B}^{\text{LRP}}$ and $\mathscr{B}^{\text{RP}}$ given by (8.85), (8.88) and (8.92), respectively.

Remark 8.11 The definitions of the boundary operators $\mathscr{B}^{\text{phys}}$, $\mathscr{B}^{\text{LRP}}$ and $\mathscr{B}^{\text{RP}}$ and of the form $\boldsymbol{b}_h$ and their evaluations may seem to be rather complicated and CPU time demanding. However, it is necessary to take into account that the integrals appearing in the definition of the form $\boldsymbol{b}_h$ are computed with the aid of numerical integration and the boundary conditions have to be determined only at integration points.

8.4 Time Discretization

The space semidiscrete problem (8.54) represents a system of ordinary differential equations (ODEs), which has to be solved with the aid of suitable numerical schemes. In the framework of the finite difference and finite volume methods, the explicit Euler or Runge–Kutta time discretization is very popular for solving the Euler equations. In early works on the DGM for the Euler equations [24, 30, 62], explicit time discretization was also used. Their advantage is a simple algorithmization, but on the other hand, the size of the time step τ is strongly restricted by the *Courant–Friedrichs–Lewy (CFL) stability condition* written, for example, in the form

$$\tau \leq \text{CFL} \min_{\substack{K \in \mathscr{T}_h \\ \Gamma \subset \partial K}} \frac{|K|}{\varrho(\mathbb{P}(\boldsymbol{w}_h, \boldsymbol{n})|_\Gamma)|\Gamma|}, \tag{8.94}$$

where $\varrho(\mathbb{P}(\boldsymbol{w}_h, \boldsymbol{n})|_\Gamma)$ denotes the spectral radius of the matrix $\mathbb{P}(\boldsymbol{w}_h, \boldsymbol{n})|_\Gamma$ given by (8.17) and evaluated at the points of $\Gamma \in \mathscr{F}_h$, $|K|$ is the d-dimensional measure of $K \in \mathscr{T}_h$ and $|\Gamma|$ denotes the $(d-1)$-dimensional measure of $\Gamma \in \mathscr{F}_h$. Moreover, $0 < \text{CFL} \leq 1$ is the Courant–Friedrichs–Lewy (CFL) number. Our numerical experiments indicate that whereas the value $\text{CFL} = 0.85$ was sufficient for almost all flow regimes in finite volume computations, the P_1 discontinuous Galerkin approximation requires the value $\text{CFL} \approx 0.15$ in order to guarantee stability. Moreover, the stability condition (8.94) becomes more and more restrictive with increasing polynomial approximation degree p.

Therefore, it is suitable to consider implicit methods for numerically solving compressible flow problems, see, e.g., [20, 25, 165, 166]. It is well known that the use of implicit methods contributes to improving the efficiency of numerical schemes for solving the Euler equations in some cases, because implicit methods allow using longer time steps. In the framework of the finite volume methods, implicit schemes were used, for example in [141, 220, 261]. The drawback of the implicit schemes is having to solve a large nonlinear algebraic system on each time level. To this end, the Newton method is often applied leading to a sequence of linear discrete problems. One variant of this approach is a well known Δ-scheme by Beam and Warming [32, 33], see also [177]. This approach is often combined with multigrid techniques, see e.g., [81, 171, 199].

The application of the Newton schemes requires, of course, the differentiability of the numerical flux and the computation of its partial derivatives, which is usually rather complicated. This is the reason that some authors use artificial pseudo-time-integration, as was applied together with multigrid in [272, 273] for the DG discrete problem. Multigrid techniques usually require using structured meshes and, in the case of the mesh refinement, a sequence of nested meshes. This is not the case when the anisotropic mesh adaptation (AMA) method is used. Then the algebraic multigrid would have to be applied, but its efficiency is not high. Therefore, one often uses the Krylov subspace methods for solving linear systems in linearized schemes for the Euler equations (cf., e.g., [220]).

In the following we will be concerned with developing several numerical schemes for the full space-time discretization of the Euler equations. The presented techniques were developed on the basis of results from [92, 93, 98, 123, 124, 133].

8.4.1 Backward Euler Method

The implicit *backward Euler* time discretization of (8.54) is the simplest implicit method for numerically solving ODEs. It can be formally considered either as the first-order implicit Runge–Kutta method or as the first-order backward difference formula (BDF), or as the first-order time discontinuous Galerkin method, see [161, 268]. The higher-order time discretizations will be discussed in Sect. 8.4.5.

In what follows we consider a partition $0 = t_0 < t_1 < t_2 \cdots < t_r = T$ of the time interval $[0, T]$ and set $\tau_k = t_k - t_{k-1}$ for $k = 1, \ldots, r$. We use the symbol $\boldsymbol{w}_h^k$ for the approximation of $\boldsymbol{w}_h(t_k)$, $k = 1, \ldots, r$.

Using the backward Euler scheme for the time discretization of (8.54), we can define the following method for the numerical solution of problem (8.8).

Definition 8.12 We say that the finite sequence of functions $\boldsymbol{w}_h^k$, $k = 0, \ldots, r$, is an *approximate solution* of problem (8.8) obtained by the *backward Euler–discontinuous Galerkin method* (BE-DGM) if the following conditions are satisfied:

$$\boldsymbol{w}_h^k \in \boldsymbol{S}_{hp}, \ k = 0, 1, \ldots, r, \tag{8.95a}$$

$$\frac{1}{\tau_k}\left(\boldsymbol{w}_h^k - \boldsymbol{w}_h^{k-1}, \boldsymbol{\varphi}_h\right) + \boldsymbol{b}_h(\boldsymbol{w}_h^k, \boldsymbol{\varphi}_h) = 0 \quad \forall \boldsymbol{\varphi}_h \in \boldsymbol{S}_{hp}, \ k = 1, \ldots, r, \tag{8.95b}$$

$$\boldsymbol{w}_h^0 = \Pi_h \boldsymbol{w}^0, \tag{8.95c}$$

where $\Pi_h \boldsymbol{w}^0$ is the $\boldsymbol{S}_{hp}$-approximation (usually $L^2(\Omega)$-projection on the space $\boldsymbol{S}_{hp}$) of the function $\boldsymbol{w}^0$ from the initial condition (8.36).

Remark 8.13 The BE-DGM has formally the order of convergence $O(h^{p+1} + \tau)$ in the $L^\infty(0, T; (L^2(\Omega))^m)$-norm and the order of convergence $O(h^p + \tau)$ in the $L^2(0, T; (H^1(\Omega))^m)$-seminorm, provided that the exact solution is sufficiently regular. These results were verified numerically in [89, 93].

Problem (8.95) represents a nonlinear algebraic system for each $k = 1, \ldots, r$. Its solution will be discussed in the following sections. First, we present its solution with the aid of the standard Newton method [79]. Then we develop a Newton-like method based on the approximation of the Jacobi matrix by the flux matrix.

8.4.2 Newton Method Based on the Jacobi Matrix

In order to develop the solution strategy for the nonlinear systems (8.95b), we introduce their algebraic representation. Let N_{hp} denote the dimension of the space $\boldsymbol{S}_{hp}$ and let $\mathrm{B}_{hp} = \{\boldsymbol{\varphi}_i(x),\ i = 1, \ldots, N_{hp}\}$ denote a set of linearly independent functions forming a basis of $\boldsymbol{S}_{hp}$. It is possible to construct a basis B_{hp} as a composition of local bases constructed separately for each $K \in \mathscr{T}_h$. See Sect. 8.4.8, where one possibility is described in detail.

Any function $\boldsymbol{w}_h^k \in \boldsymbol{S}_{hp}$ can be expressed in the form

$$\boldsymbol{w}_h^k(x) = \sum_{j=1}^{N_{hp}} \xi^{k,j} \boldsymbol{\varphi}_j(x) \in \boldsymbol{S}_{hp} \quad \longleftrightarrow \quad \boldsymbol{\xi}_k = (\xi^{k,j})_{j=1}^{N_{hp}} \in \mathbb{R}^{N_{hp}},\ k = 1, \ldots, r, \tag{8.96}$$

where $\xi^{k,j} \in \mathbb{R},\ j = 1, \ldots, N_{hp},\ k = 1, \ldots, r$, are its basis coefficients. Obviously, (8.96) defines an isomorphism between $\boldsymbol{w}_h^k \in \boldsymbol{S}_{hp}$ and $\boldsymbol{\xi}_k \in \mathbb{R}^{N_{hp}}$. We call $\boldsymbol{\xi}_k$ the *algebraic representation* of $\boldsymbol{w}_h^k$.

In order to rewrite the nonlinear algebraic systems (8.95b), we define the vector-valued function $\boldsymbol{F}_h : \mathbb{R}^{N_{hp}} \times \mathbb{R}^{N_{hp}} \to \mathbb{R}^{N_{hp}}$ by

$$\boldsymbol{F}_h\left(\boldsymbol{\xi}_{k-1}; \boldsymbol{\xi}_k\right) = \left(\frac{1}{\tau_k}\left(\boldsymbol{w}_h^k - \boldsymbol{w}_h^{k-1}, \boldsymbol{\varphi}_i\right) + \boldsymbol{b}_h(\boldsymbol{w}_h^k, \boldsymbol{\varphi}_i)\right)_{i=1}^{N_{hp}},\quad k = 1, \ldots, r, \tag{8.97}$$

where $\boldsymbol{\xi}_{k-l} \in \mathbb{R}^{N_{hp}}$ is the algebraic representation of $\boldsymbol{w}_h^{k-l} \in \boldsymbol{S}_{hp}$ for $l = 0, 1$. We do not emphasize that $\boldsymbol{F}_h$ depends explicitly on τ_k. Therefore, the algebraic representation of the systems (8.95b) reads: For a given vector $\boldsymbol{\xi}_{k-1} \in \mathbb{R}^{N_{hp}}$ find $\boldsymbol{\xi}_k \in \mathbb{R}^{N_{hp}}$ such that

$$\boldsymbol{F}_h(\boldsymbol{\xi}_{k-1}; \boldsymbol{\xi}_k) = \boldsymbol{0}, \quad k = 1, \ldots, r. \tag{8.98}$$

Here $\boldsymbol{0}$ denotes a generic zero vector (i.e., all entries of $\boldsymbol{0}$ are equal to zero) and $\boldsymbol{\xi}_0$ is given by the initial condition (8.95c) and the isomorphism (8.96). Systems (8.98) are strongly nonlinear and their efficient and accurate solution is demanding.

A natural strategy is to apply the (damped) *Newton method* [79] which generates a sequence of approximations $\boldsymbol{\xi}_k^l$, $l = 0, 1, \ldots$, to the actual numerical solution $\boldsymbol{\xi}_k$ using the following algorithm. Given an iterate $\boldsymbol{\xi}_k^l \in \mathbb{R}^{N_{hp}}$, the update of $\boldsymbol{\xi}_k^l$ reads

$$\boldsymbol{\xi}_k^{l+1} = \boldsymbol{\xi}_k^l + \lambda^l \boldsymbol{\delta}^l, \tag{8.99}$$

where $\boldsymbol{\delta}^l \in \mathbb{R}^{N_{hp}}$ is defined as the solution of the system

$$\mathbb{D}_h(\boldsymbol{\xi}_k^l)\boldsymbol{\delta}^l = -\boldsymbol{F}_h(\boldsymbol{\xi}_{k-1}; \boldsymbol{\xi}_k^l). \tag{8.100}$$

Here $\lambda^l \in (0, 1]$ is the damping parameter (for its choice see Sect. 8.4.4.1) and $\mathbb{D}_h$ is the *Jacobi matrix* of the vector-valued function $\boldsymbol{F}_h$ given by (8.97), i.e.,

$$\mathbb{D}_h(\boldsymbol{\xi}_k^l) = \frac{\mathrm{D}\boldsymbol{F}_h(\boldsymbol{\xi}_{k-1}; \boldsymbol{\xi}_k^l)}{\mathrm{D}\boldsymbol{\xi}_k^l}. \tag{8.101}$$

From (8.96), (8.97) and (8.101) we obtain

$$\mathbb{D}_h(\boldsymbol{\xi}_k) = (d_{ij}(\boldsymbol{\xi}_k))_{i,j=1}^{N_{hp}}, \tag{8.102}$$

$$d_{ij}(\boldsymbol{\xi}_k) = \frac{1}{\tau_k}(\boldsymbol{\varphi}_j, \boldsymbol{\varphi}_i) + \frac{\partial \boldsymbol{b}_h\left(\sum_{l=1}^{N_{hp}} \xi^{k,l} \boldsymbol{\varphi}_l, \boldsymbol{\varphi}_i\right)}{\partial \xi^{k,j}}, \quad i, j = 1, \ldots, N_{hp}.$$

For $\lambda^l = 1$ we get the standard Newton method. This technique was also successfully applied in [25, 165] for computing viscous flow.

Evaluating of the Jacobi matrix $\mathbb{D}_h$ is not quite easy, since the form $\boldsymbol{b}_h$ depends nonlinearly on its first argument. Moreover, there are difficulties with the differentiability of the mapping $\boldsymbol{F}_h$, because the numerical flux $\mathbf{H}$ is sometimes only Lipschitz-continuous, but not differentiable.

In the following section we present an alternative approach inspired by the semi-implicit technique from [93, 133] and based on the so-called flux matrix.

8.4.3 Newton-Like Method Based on the Flux Matrix

Evaluating of the Jacobi matrix $\mathbb{D}_h$ in (8.100) can be avoided with the aid of a formal linearization of the convection form $\boldsymbol{b}_h$. The aim is to define the form $\boldsymbol{b}_h^L$: $\boldsymbol{S}_{hp} \times \boldsymbol{S}_{hp} \times \boldsymbol{S}_{hp} \to \mathbb{R}$ such that it is linear with respect to its second and third arguments and is consistent with $\boldsymbol{b}_h$, i.e.,

$$\boldsymbol{b}_h(\boldsymbol{w}_h, \boldsymbol{\varphi}_h) = \boldsymbol{b}_h^L(\boldsymbol{w}_h, \boldsymbol{w}_h, \boldsymbol{\varphi}_h) - \tilde{\boldsymbol{b}}_h(\boldsymbol{w}_h, \boldsymbol{\varphi}_h) \quad \forall \boldsymbol{w}_h, \boldsymbol{\varphi}_h \in \boldsymbol{S}_{hp}, \tag{8.103}$$

where $\tilde{\boldsymbol{b}}_h : \boldsymbol{S}_{hp} \times \boldsymbol{S}_{hp} \to \mathbb{R}$ is some "residual" form, vanishing for the majority of functions $\boldsymbol{\varphi}_h \in \boldsymbol{S}_{hp}$, see (8.121).

By (8.93), we defined the form

$$
\begin{aligned}
\boldsymbol{b}_h(\boldsymbol{w}_h, \boldsymbol{\varphi}_h) = &- \sum_{K \in \mathscr{T}_h} \int_K \sum_{s=1}^{d} \boldsymbol{f}_s(\boldsymbol{w}_h) \cdot \frac{\partial \boldsymbol{\varphi}_h}{\partial x_s} \, \mathrm{d}x && (=: \eta_1) \\
&+ \sum_{\Gamma \in \mathscr{F}_h^I} \int_\Gamma \mathbf{H}(\boldsymbol{w}_{h\Gamma}^{(L)}, \boldsymbol{w}_{h\Gamma}^{(R)}, \boldsymbol{n}_\Gamma) \cdot [\boldsymbol{\varphi}_h] \, \mathrm{d}S && (=: \eta_2) \\
&+ \sum_{\Gamma \in \mathscr{F}_h^W} \int_\Gamma \boldsymbol{f}_{\mathrm{W}}^i(\boldsymbol{w}_{h\Gamma}^{(L)}, \boldsymbol{n}) \cdot \boldsymbol{\varphi}_h \, \mathrm{d}S && (=: \eta_3) \\
&+ \sum_{\Gamma \in \mathscr{F}_h^{io}} \int_\Gamma \mathbf{H}\left(\boldsymbol{w}_{h\Gamma}^{(L)}, \mathscr{B}(\boldsymbol{w}_{h\Gamma}^{(L)}, \boldsymbol{w}_{\mathrm{BC}}), \boldsymbol{n}_\Gamma\right) \cdot \boldsymbol{\varphi}_h \mathrm{d}S && (=: \eta_4),
\end{aligned}
\tag{8.104}
$$

where $\boldsymbol{w}_{h\Gamma}^{(L)}$ and $\boldsymbol{w}_{h\Gamma}^{(R)}$ denote the traces of $\boldsymbol{w}_h$ on $\Gamma \in \mathscr{F}_h$, cf. (8.41). The individual terms $\eta_1, \ldots, \eta_4$ will be partially linearized.

For η_1 we use the property (8.26) of the Euler fluxes and define the form η_1^{L} : $\boldsymbol{S}_{hp} \times \boldsymbol{S}_{hp} \times \boldsymbol{S}_{hp} \to \mathbb{R}$ by

$$
\eta_1^{\mathrm{L}}(\bar{\boldsymbol{w}}_h, \boldsymbol{w}_h, \boldsymbol{\varphi}_h) = - \sum_{K \in \mathscr{T}_h} \int_K \sum_{s=1}^{d} \mathbb{A}_s(\bar{\boldsymbol{w}}_h) \boldsymbol{w}_h \cdot \frac{\partial \boldsymbol{\varphi}_h}{\partial x_s} \, \mathrm{d}x. \tag{8.105}
$$

Obviously, $\eta_1^{\mathrm{L}}(\boldsymbol{w}_h, \boldsymbol{w}_h, \boldsymbol{\varphi}_h) = \eta_1$ and η_1^{L} is linear with respect to its second and third arguments.

Linearizing of the term η_2 can be carried out on the basis of a suitable choice of the numerical flux $\mathbf{H}$. For example, let us use in (8.104) the Vijayasundaram numerical flux, see [277], [122, Sect. 7.3] or [127, Sect. 3.3.4]. It is defined in the following way. By (8.29), the matrix $\mathbb{P} = \mathbb{P}(\boldsymbol{w}, \boldsymbol{n})$ defined in (8.17) is diagonalizable: there exists a nonsingular matrix $\mathbb{T} = \mathbb{T}(\boldsymbol{w}, \boldsymbol{n})$ such that

$$
\mathbb{P} = \mathbb{T} \boldsymbol{\Lambda} \mathbb{T}^{-1}, \tag{8.106}
$$

where $\boldsymbol{\Lambda} = \mathrm{diag}\ (\lambda_1, \ldots, \lambda_m)$ and $\lambda_1, \ldots, \lambda_m$ are the eigenvalues of $\mathbb{P}$. The columns of the matrix $\mathbb{T}$ are the eigenvectors of the matrix $\mathbb{P}$. We define the "positive" and "negative" part of $\mathbb{P}$ by

$$
\mathbb{P}^{\pm} = \mathbb{T} \boldsymbol{\Lambda}^{\pm} \mathbb{T}^{-1}, \ \boldsymbol{\Lambda}^{\pm} = \mathrm{diag}\ (\lambda_1^{\pm}, \ldots, \lambda_m^{\pm}), \tag{8.107}
$$

where $a^+ = \max(a, 0)$ and $a^- = \min(a, 0)$ for $a \in \mathbb{R}$. Then the *Vijayasundaram numerical flux* reads as

$$\mathbf{H}_{VS}(\boldsymbol{w}_1, \boldsymbol{w}_2, \boldsymbol{n}) = \mathbb{P}^+\left(\frac{\boldsymbol{w}_1 + \boldsymbol{w}_2}{2}, \boldsymbol{n}\right)\boldsymbol{w}_1 + \mathbb{P}^-\left(\frac{\boldsymbol{w}_1 + \boldsymbol{w}_2}{2}, \boldsymbol{n}\right)\boldsymbol{w}_2. \tag{8.108}$$

We can characterize the properties of the Vijayasundaram numerical flux.

Lemma 8.14 *The Vijayasundaram numerical flux* $\mathbf{H}_{VS} = \mathbf{H}(\boldsymbol{w}_1, \boldsymbol{w}_2, \boldsymbol{n})$ *is Lipschitz-continuous with respect to* $\boldsymbol{w}_1, \boldsymbol{w}_2 \in \mathscr{D}$ *and satisfies conditions (8.49) and (8.50), i.e., it is consistent and conservative.*

Proof (a) From (8.10) it follows that the entries of the matrix $\mathbb{P}$ are continuously differentiable. This fact, the definition of the matrices $\mathbb{P}^\pm$, definition (8.108) and the Lipschitz-continuity of the functions $\lambda \in \mathbb{R} \to \lambda^+$ and $\lambda \in \mathbb{R} \to \lambda^-$ imply that the Vijayasundaram numerical flux is locally Lipschitz-continuous.

(b) The consistency of $\mathbf{H}_{VS}$ is a consequence of the relations (8.16), (8.28) and $\mathbb{P}(\boldsymbol{w}, \boldsymbol{n}) = \mathbb{P}^+(\boldsymbol{w}, \boldsymbol{n}) + \mathbb{P}^-(\boldsymbol{w}, \boldsymbol{n})$.

(c) The proof of the consistency of $\mathbf{H}_{VS}$ is more complicated. First, we show that

$$\mathbb{P}^\pm(\boldsymbol{w}, -\boldsymbol{n}) = -\mathbb{P}^\mp(\boldsymbol{w}, \boldsymbol{n}) \tag{8.109}$$

for $\boldsymbol{w} \in \mathscr{D}$ and $\boldsymbol{n} = (n_1, \ldots, n_d) \in B_1$. It follows from (8.16) that

$$\boldsymbol{P}(\boldsymbol{w}, -\boldsymbol{n}) = -\boldsymbol{P}(\boldsymbol{w}, \boldsymbol{n}).$$

By differentiation,

$$\mathbb{P}(\boldsymbol{w}, -\boldsymbol{n}) = -\mathbb{P}(\boldsymbol{w}, \boldsymbol{n}),$$

and thus

$$\mathbb{P}^\pm(\boldsymbol{w}, -\boldsymbol{n}) = (-\mathbb{P}(\boldsymbol{w}, \boldsymbol{n}))^\pm. \tag{8.110}$$

Further, by (8.106),

$$\pm\mathbb{P}(\boldsymbol{w}, \boldsymbol{n}) = \mathbb{T}(\boldsymbol{w}, \boldsymbol{n})\,(\pm\boldsymbol{\Lambda}(\boldsymbol{w}, \boldsymbol{n}))\,\mathbb{T}^{-1}(\boldsymbol{w}, \boldsymbol{n}),$$

where

$$\boldsymbol{\Lambda}(\boldsymbol{w}, \boldsymbol{n}) = \operatorname{diag}\left(\lambda_1(\boldsymbol{w}, \boldsymbol{n}), \ldots, \lambda_m(\boldsymbol{w}, \boldsymbol{n})\right).$$

Thus,

$$\mathbb{P}^\pm(\boldsymbol{w}, \boldsymbol{n}) = \mathbb{T}(\boldsymbol{w}, \boldsymbol{n})\,\boldsymbol{\Lambda}^\pm(\boldsymbol{w}, \boldsymbol{n})\,\mathbb{T}^{-1}(\boldsymbol{w}, \boldsymbol{n}) \tag{8.111}$$

and

$$(-\mathbb{P}(\boldsymbol{w}, \boldsymbol{n}))^\pm = \mathbb{T}(\boldsymbol{w}, \boldsymbol{n})\,(-\boldsymbol{\Lambda}(\boldsymbol{w}, \boldsymbol{n}))^\pm\,\mathbb{T}^{-1}(\boldsymbol{w}, \boldsymbol{n}). \tag{8.112}$$

Here

$$\boldsymbol{\Lambda}^{\pm}(\boldsymbol{w},\boldsymbol{n}) = \operatorname{diag}\left(\lambda_1^{\pm}(\boldsymbol{w},\boldsymbol{n}),\dots,\lambda_m^{\pm}(\boldsymbol{w},\boldsymbol{n})\right),$$
$$(-\boldsymbol{\Lambda}(\boldsymbol{w},\boldsymbol{n}))^{\pm} = \operatorname{diag}\left((-\lambda_1)^{\pm}(\boldsymbol{w},\boldsymbol{n}),\dots,(-\lambda_m)^{\pm}(\boldsymbol{w},\boldsymbol{n})\right),$$

It is easy to find that $(-\lambda)^{\pm} = -\lambda^{\mp}$, which implies that

$$(-\boldsymbol{\Lambda}(\boldsymbol{w},n))^{\pm} = -\boldsymbol{\Lambda}^{\mp}(\boldsymbol{w},n).$$

The above, (8.111) and (8.112) yield

$$\begin{aligned}(-\mathbb{P}(\boldsymbol{w},\boldsymbol{n}))^{\pm} &= -\mathbb{T}(\boldsymbol{w},\boldsymbol{n})\,\boldsymbol{\Lambda}^{\mp}(\boldsymbol{w},n)\,\mathbb{T}(\boldsymbol{w},\boldsymbol{n}) \\ &= -\mathbb{P}^{\mp}(\boldsymbol{w},\boldsymbol{n}).\end{aligned} \tag{8.113}$$

Now, by (8.110) and (8.113) we get (8.109).

Finally, by virtue of (8.109), for $\boldsymbol{w}_1,\boldsymbol{w}_2 \in \mathscr{D}$ and $\boldsymbol{n} \in B_1$,

$$\begin{aligned}&\mathbf{H}_{VS}(\boldsymbol{w}_1,\boldsymbol{w}_2,\boldsymbol{n}) - \mathbb{P}^{+}\left(\frac{\boldsymbol{w}_1+\boldsymbol{w}_2}{2},\boldsymbol{n}\right)\boldsymbol{w}_1 \mid \mathbb{P}^{-}\left(\frac{\boldsymbol{w}_1+\boldsymbol{w}_2}{2},\boldsymbol{n}\right)\boldsymbol{w}_2\\ &= -\mathbb{P}^{-}\left(\frac{\boldsymbol{w}_1+\boldsymbol{w}_2}{2},-\boldsymbol{n}\right)\boldsymbol{w}_1 - \mathbb{P}^{+}\left(\frac{\boldsymbol{w}_1+\boldsymbol{w}_2}{2},-\boldsymbol{n}\right)\boldsymbol{w}_2 = -\mathbf{H}_{VS}(\boldsymbol{w}_2,\boldsymbol{w}_1,-\boldsymbol{n}),\end{aligned}$$

which is what we wanted to prove. □

The form of $\mathbf{H}_{VS}$ is a way of defining the form $\eta_2^{\mathrm{L}}: \boldsymbol{S}_{hp}\times\boldsymbol{S}_{hp}\times\boldsymbol{S}_{hp}\to\mathbb{R}$ by

$$\eta_2^{\mathrm{L}}(\bar{\boldsymbol{w}}_h,\boldsymbol{w}_h,\boldsymbol{\varphi}_h) = \sum_{\Gamma\in\mathscr{F}_h^I}\int_\Gamma\left[\mathbb{P}^{+}\left(\langle\bar{\boldsymbol{w}}_h\rangle_\Gamma,\boldsymbol{n}_\Gamma\right)\boldsymbol{w}_{h\Gamma}^{(L)} + \mathbb{P}^{-}\left(\langle\bar{\boldsymbol{w}}_h\rangle_\Gamma,\boldsymbol{n}_\Gamma\right)\boldsymbol{w}_{h\Gamma}^{(R)}\right]\cdot\boldsymbol{\varphi}_h\,\mathrm{d}S, \tag{8.114}$$

where $\langle\bar{\boldsymbol{w}}_h\rangle_\Gamma$ denotes the mean value of $\bar{\boldsymbol{w}}_h$ on $\Gamma\in\mathscr{F}_h$ defined by (8.42). Obviously, $\eta_2^{\mathrm{L}}(\boldsymbol{w}_h,\boldsymbol{w}_h,\boldsymbol{\varphi}_h) = \eta_2$ and η_2^{L} is linear with respect to its second and third arguments.

Concerning the term η_3 in (8.104), we distinguish between the direct use of the impermeability condition and the inviscid mirror boundary condition presented in Sect. 8.3.1. For the former case we define the form

$$\eta_3^{\mathrm{L}}(\bar{\boldsymbol{w}}_h,\boldsymbol{w}_h,\boldsymbol{\varphi}_h) = \sum_{\Gamma\in\mathscr{F}_h^W}\int_\Gamma \boldsymbol{f}_{\mathrm{W}}^{1,\mathrm{L}}(\bar{\boldsymbol{w}}_{h\Gamma}^{(L)},\boldsymbol{w}_{h\Gamma}^{(L)},\boldsymbol{n})\cdot\boldsymbol{\varphi}_h\,\mathrm{d}S, \tag{8.115}$$

where $\boldsymbol{f}_{\mathrm{W}}^{1,\mathrm{L}}$ is defined by (8.63), i.e.,

$$\boldsymbol{f}_{\mathrm{W}}^{1,\mathrm{L}}(\bar{\boldsymbol{w}},\boldsymbol{w},\boldsymbol{n}) = \mathbb{P}_W(\bar{\boldsymbol{w}},\boldsymbol{n})\boldsymbol{w},\quad \bar{\boldsymbol{w}},\boldsymbol{w}\in\mathscr{D},\ \boldsymbol{n}\in\mathrm{B}_1, \tag{8.116}$$

with $\mathbb{P}_W$ given in (8.62).

In the case of inviscid mirror boundary conditions we use relations (8.68) and (8.108) and put

$$f_{\mathrm{W}}^{2,\mathrm{L}}(\bar{\boldsymbol{w}}_h, \boldsymbol{w}_h, \boldsymbol{n}) = \mathbb{P}^{+}(\bar{\boldsymbol{w}}_h, \boldsymbol{n})\, \boldsymbol{w}_h + \mathbb{P}^{-}(\bar{\boldsymbol{w}}_h, \boldsymbol{n})\, \mathscr{M}(\boldsymbol{w}_h), \tag{8.117}$$

where $\mathbb{P}^{\pm}$ are defined by (8.107). Now, on the basis of (8.68), (8.69) and (8.117), we put

$$\eta_3^{\mathrm{L}}(\bar{\boldsymbol{w}}_h, \boldsymbol{w}_h, \boldsymbol{\varphi}_h) = \sum_{\Gamma \in \mathscr{F}_h^W} \int_\Gamma f_{\mathrm{W}}^{2,\mathrm{L}}(\bar{\boldsymbol{w}}_{h\Gamma}^{(L)}, \boldsymbol{w}_{h\Gamma}^{(L)}, \boldsymbol{n}) \cdot \boldsymbol{\varphi}_h \,\mathrm{d}S. \tag{8.118}$$

Therefore, (8.115) and (8.118) can be written as

$$\eta_3^{\mathrm{L}}(\bar{\boldsymbol{w}}_h, \boldsymbol{w}_h, \boldsymbol{\varphi}_h) = \sum_{\Gamma \in \mathscr{F}_h^W} \int_\Gamma f_{\mathrm{W}}^{\alpha,\mathrm{L}}(\bar{\boldsymbol{w}}_{h\Gamma}^{(L)}, \boldsymbol{w}_{h\Gamma}^{(L)}, \boldsymbol{n}) \cdot \boldsymbol{\varphi}_h \,\mathrm{d}S, \tag{8.119}$$

where $\alpha = 1$ for directly using the impermeability condition and $\alpha = 2$ for the inviscid mirror boundary condition. It follows from (8.116)–(8.119) and the linearity of the operator $\mathscr{M}$ that η_3^{L} is linear with respect to its second and third arguments. Moreover, $\eta_3^{\mathrm{L}}(\boldsymbol{w}_h, \boldsymbol{w}_h, \boldsymbol{\varphi}_h) = \eta_3$.

Finally, η_4 is approximated with the aid of the forms

$$\eta_4^{\mathrm{L}}(\bar{\boldsymbol{w}}_h, \boldsymbol{w}_h, \boldsymbol{\varphi}_h) = \sum_{\Gamma \in \mathscr{F}_h^{io}} \int_\Gamma \left(\mathbb{P}^{+}(\bar{\boldsymbol{w}}_{h\Gamma}^{(L)}, \boldsymbol{n}_\Gamma)\boldsymbol{w}_{h\Gamma}^{(L)}\right) \cdot \boldsymbol{\varphi}_h \mathrm{d}S, \tag{8.120}$$

and

$$\tilde{\boldsymbol{b}}_h(\bar{\boldsymbol{w}}_h, \boldsymbol{\varphi}_h) = -\sum_{\Gamma \in \mathscr{F}_h^{io}} \int_\Gamma \left(\mathbb{P}^{-}(\bar{\boldsymbol{w}}_{h\Gamma}^{(L)}, \boldsymbol{n}_\Gamma)\mathscr{B}(\bar{\boldsymbol{w}}_\Gamma^{(L)}, \boldsymbol{w}_{\mathrm{BC}})\right) \cdot \boldsymbol{\varphi}_h \mathrm{d}S, \tag{8.121}$$

where $\mathscr{B}$ represents the boundary operators $\mathscr{B}^{\mathrm{phys}}$, $\mathscr{B}^{\mathrm{LRP}}$ and $\mathscr{B}^{\mathrm{RP}}$ given by (8.88), (8.85) and (8.92), respectively. Let us underline that in the arguments of $\mathbb{P}^{\pm}$ we do not use the mean value of the state vectors from the left- and right-hand side of Γ as in (8.108). Moreover, if $\operatorname{supp} \boldsymbol{\varphi}_h \cap (\partial\Omega_i \cup \partial\Omega_o) = \emptyset$, then $\tilde{\boldsymbol{b}}_h(\bar{\boldsymbol{w}}_h, \boldsymbol{\varphi}_h) = 0$.

Obviously, due to (8.93) and (8.120), we have

$$\eta_4^{\mathrm{L}}(\boldsymbol{w}_h, \boldsymbol{w}_h, \boldsymbol{\varphi}_h) - \tilde{\boldsymbol{b}}_h(\boldsymbol{w}_h, \boldsymbol{\varphi}_h) = \eta_4. \tag{8.122}$$

Taking into account (8.93), (8.105), (8.114), (8.119) and (8.120), we introduce the form

$$\boldsymbol{b}_h^L(\bar{\boldsymbol{w}}_h, \boldsymbol{w}_h, \boldsymbol{\varphi}_h) = -\sum_{K \in \mathscr{T}_h} \int_K \sum_{s=1}^d \mathbb{A}_s(\bar{\boldsymbol{w}}_h)\boldsymbol{w}_h \cdot \frac{\partial \boldsymbol{\varphi}_h}{\partial x_s} \,\mathrm{d}x \tag{8.123}$$

$$+ \sum_{\Gamma \in \mathscr{F}_h^I} \int_\Gamma \left[\mathbb{P}^+ \left(\langle \bar{\boldsymbol{w}}_h \rangle_\Gamma , \boldsymbol{n}_\Gamma \right) \boldsymbol{w}_{h\Gamma}^{(L)} + \mathbb{P}^- \left(\langle \bar{\boldsymbol{w}}_h \rangle_\Gamma , \boldsymbol{n}_\Gamma \right) \boldsymbol{w}_{h\Gamma}^{(R)} \right] \cdot \boldsymbol{\varphi}_h \mathrm{d}S$$

$$+ \sum_{\Gamma \in \mathscr{F}_h^W} \int_\Gamma \boldsymbol{f}_\mathrm{W}^{\alpha,\mathrm{L}} (\bar{\boldsymbol{w}}_{h\Gamma}^{(L)}, \boldsymbol{w}_{h\Gamma}^{(L)}, \boldsymbol{n}) \cdot \boldsymbol{\varphi}_h \, \mathrm{d}S$$

$$+ \sum_{\Gamma \in \mathscr{F}_h^{io}} \int_\Gamma \left(\mathbb{P}^+ \left(\bar{\boldsymbol{w}}_{h\Gamma}^{(L)}, \boldsymbol{n}_\Gamma \right) \boldsymbol{w}_{h\Gamma}^{(L)} + \mathbb{P}^- (\bar{\boldsymbol{w}}_{h\Gamma}^{(L)}, \boldsymbol{n}_\Gamma) \mathscr{B}(\bar{\boldsymbol{w}}_\Gamma^{(L)}, \boldsymbol{w}_\mathrm{BC}) \right) \cdot \boldsymbol{\varphi}_h \mathrm{d}S.$$

From the definitions (8.93) of $\boldsymbol{b}_h$, (8.123) of $\boldsymbol{b}_h^L$ and (8.121) of $\tilde{\boldsymbol{b}}_h$ we can see that relation (8.103) is valid. Moreover, the form $\boldsymbol{b}_h^L$ is linear with respect to the arguments $\boldsymbol{w}_h$ and $\boldsymbol{\varphi}_h$.

Now we introduce the Newton-like method for solving systems (8.98) based on the flux matrix. We again return to the algebraic representation of the method. Using notation from Sect. 8.4.2, we define the $N_{hp} \times N_{hp}$ *flux matrix*

$$\mathbb{C}_h \left(\bar{\boldsymbol{\xi}} \right) = \left(\frac{1}{\tau_k} \left(\boldsymbol{\varphi}_j , \boldsymbol{\varphi}_i \right) + \boldsymbol{b}_h^L (\bar{\boldsymbol{w}}_h, \boldsymbol{\varphi}_j, \boldsymbol{\varphi}_i) \right)_{i,j=1}^{N_{hp}} \tag{8.124}$$

and the vector

$$\boldsymbol{d}_h \left(\boldsymbol{\xi}_{k-1}, \bar{\boldsymbol{\xi}} \right) = \left(\frac{1}{\tau_k} \left(\boldsymbol{w}_h^{k-1}, \boldsymbol{\varphi}_i \right) + \tilde{\boldsymbol{b}}_h (\bar{\boldsymbol{w}}_h, \boldsymbol{\varphi}_i) \right)_{i=1}^{N_{hp}}, \tag{8.125}$$

where $\boldsymbol{\varphi}_i \in \mathrm{B}_{hp}$, $i = 1, \ldots, N_{hp}$, are the basis functions in the space $\boldsymbol{S}_{hp}$, $\bar{\boldsymbol{\xi}} \in \mathbb{R}^{N_{hp}}$ and $\boldsymbol{\xi}_{k-l} \in \mathbb{R}^{N_{hp}}$, $l = 0, 1$, are the algebraic representations of $\bar{\boldsymbol{w}}_h \in \boldsymbol{S}_{hp}$ and $\boldsymbol{w}_h^{k-l} \in \boldsymbol{S}_{hp}$, $l = 0, 1$, respectively. (We do not emphasize that $\mathbb{C}_h$ and $\boldsymbol{d}_h$ depend explicitly on τ_k.) Finally, using (8.97), (8.103), (8.124) and (8.125), we have

$$\boldsymbol{F}_h(\boldsymbol{\xi}_{k-1}; \boldsymbol{\xi}_k) = \mathbb{C}_h(\boldsymbol{\xi}_k)\boldsymbol{\xi}_k - \boldsymbol{d}_h(\boldsymbol{\xi}_{k-1}, \boldsymbol{\xi}_k), \quad k = 1, \ldots, r. \tag{8.126}$$

Obviously, the sparsity of $\mathbb{C}_h$ is identical with the sparsity of the Jacobi matrix $\mathbb{D}_h$ introduced in (8.101). Therefore, in the following Newton-like method for solving systems (8.98), we use $\mathbb{C}_h$ as the approximation of $\mathbb{D}_h$ in the definition of our iterative Newton-like method, which is represented as the following algorithm.

If the approximate solution $\boldsymbol{w}_h^{k-1} \in \boldsymbol{S}_{hp}$, represented by $\boldsymbol{\xi}_{k-1}$, was already computed, then we set $\boldsymbol{\xi}_k^0 = \boldsymbol{\xi}_{k-1}$ and apply the iterative process

$$\boldsymbol{\xi}_k^{l+1} = \boldsymbol{\xi}_k^l + \lambda^l \boldsymbol{\delta}^l, \quad l = 0, 1, \ldots, \tag{8.127}$$

with $\boldsymbol{\delta}^l$ defined by

$$\mathbb{C}_h(\boldsymbol{\xi}_k^l)\boldsymbol{\delta}^l = -\boldsymbol{F}_h(\boldsymbol{\xi}_{k-1}; \boldsymbol{\xi}_k^l). \tag{8.128}$$

The term $\lambda^l \in (0, 1]$ is a damping parameter which should ensure convergence of (8.127) and (8.128) in case when the initial guess $\boldsymbol{\xi}_k^0$ is far from the solution of (8.98). The initial guess $\boldsymbol{\xi}_k^0$ can be defined as

$$\boldsymbol{\xi}_k^0 = \boldsymbol{\xi}_{k-1}, \quad k = 1, \ldots, r, \tag{8.129}$$

where $\boldsymbol{\xi}_{k-1}$ corresponds to the approximate solution $\boldsymbol{w}_h^{k-1}$.

In the following section we discuss several aspects of the iterative method (8.127) and (8.128).

Remark 8.15 Let us note that if we carry out only one Newton-like iteration at each time level, put $\lambda^0 = 1$, and the matrix $\mathbb{C}_h$ is updated at each time step, then the implicit method (8.95) reduces to the *semi-implicit time discretization* approach presented in [93, 133]. It can be formulated in the following way: We seek the finite sequence of functions $\boldsymbol{w}_h^k$, $k = 0, 1, \ldots, r$, such that

$$\boldsymbol{w}_h^k \in \boldsymbol{S}_{hp}, \ k = 0, 1, \ldots, r, \tag{8.130a}$$

$$\frac{1}{\tau_k}\left(\boldsymbol{w}_h^k - \boldsymbol{w}_h^{k-1}, \boldsymbol{\varphi}_h\right) + \hat{\boldsymbol{b}}_h(\boldsymbol{w}_h^{k-1}, \boldsymbol{w}_h^k, \boldsymbol{\varphi}_h) = 0 \quad \forall \boldsymbol{\varphi}_h \in \boldsymbol{S}_{hp}, \ k = 1, \ldots, r, \tag{8.130b}$$

$$\boldsymbol{w}_h^0 = \Pi_h \boldsymbol{w}^0, \tag{8.130c}$$

where $\Pi_h \boldsymbol{w}^0$ is the $\boldsymbol{S}_{hp}$-approximation of $\boldsymbol{w}^0$ from the initial condition (8.36) and

$$\hat{\boldsymbol{b}}_h(\bar{\boldsymbol{w}}_h, \boldsymbol{w}_h, \boldsymbol{\varphi}_h) = \boldsymbol{b}_h^L(\bar{\boldsymbol{w}}_h, \boldsymbol{w}_h, \boldsymbol{\varphi}_h) - \tilde{\boldsymbol{b}}_h(\bar{\boldsymbol{w}}_h, \boldsymbol{\varphi}_h) \tag{8.131}$$

with $\boldsymbol{b}_h^L$ and $\tilde{\boldsymbol{b}}_h$ given by (8.121) and (8.123), respectively.

8.4.4 Realization of the Iterative Algorithm

In this section we mention some aspects of the Newton-like iterative process (8.127) and (8.128).

8.4.4.1 Choice of Damping Parameters

The damping parameters λ^l, $l = 0, 1, \ldots$, should guarantee convergence of the iterative process (8.127) and (8.128). Following the analysis presented in [79], we start from the value $\lambda^l = 1$ and evaluate a monitoring function

$$\kappa^l = \frac{\left\| \boldsymbol{F}_h(\boldsymbol{\xi}_{k-1}; \boldsymbol{\xi}_k^{l+1}) \right\|}{\left\| \boldsymbol{F}_h(\boldsymbol{\xi}_{k-1}; \boldsymbol{\xi}_k^{l}) \right\|}, \tag{8.132}$$

where $\|\cdot\|$ is a norm in the space $\mathbb{R}^{N_{hp}}$. If $\kappa^l < 1$, we proceed to the next Newton-like iteration. Otherwise, we put $\lambda^l = \lambda^l/2$ and repeat the actual lth Newton-like iteration.

8.4.4.2 Update of the Flux Matrix

As numerical experiments show, in the iterative process it is not necessary to update the flux matrix $\mathbb{C}_h(\boldsymbol{\xi}_k^l)$ at each Newton-like iteration $l = 1, 2, \ldots$ and each time level $k = 1, \ldots, r$. Computational costs of the evaluation of $\boldsymbol{F}_h$ are much smaller than the evaluation of $\mathbb{C}_h$. For simplicity, let us consider the case $d = 2$ and assume that $\mathscr{T}_h$ is a conforming triangulation. By $\#\mathscr{T}_h$ we denote the number of elements of $\mathscr{T}_h$. Then $\boldsymbol{F}_h$ has $N_{hp} = \#\mathscr{T}_h(p+1)(p+1)/2$ components and $\mathbb{C}_h$ has approximately $4\#\mathscr{T}_h((p+1)(p+1)/2)^2$ non-vanishing components. Hence, the evaluation of $\boldsymbol{F}_h$ is approximately $2(p+1)(p+2)$ times cheaper than the evaluation of $\mathbb{C}_h$.

Therefore, it is more efficient to perform more Newton-like iterations than to update the matrix $\mathbb{C}_h$. In practice, we update $\mathbb{C}_h$, when either the damping parameter λ achieves a minimal prescribed value (using the algorithm described in Sect. 8.4.4.1) or the prescribed maximal number of Newton-like iterations is achieved.

8.4.4.3 Termination of the Iterative Process

The iterative process (8.127) and (8.128) is terminated if a suitable *algebraic stopping criterion* is achieved. The standard approach is based on the condition

$$\left\| \boldsymbol{F}_h(\boldsymbol{\xi}_{k-1}; \boldsymbol{\xi}_k^{l}) \right\| \leq \text{TOL}, \tag{8.133}$$

where $\|\cdot\|$ is a norm in $\mathbb{R}^{N_{hp}}$ and TOL is a given tolerance. However, it is difficult to choose TOL in order to guarantee the accuracy of the solution and to avoid a too long iterative process. The optimal stopping criterion, which balances the accuracy and efficiency, should be derived from a posteriori estimates taking into account algebraic errors. This is out of the scope of this monograph and we refer, for example, to [5, 49], dealing with this subject. In [89] a heuristic stopping criterion solving this problem was proposed.

8.4.4.4 Solution of the Linear Algebraic Systems (8.128)

The linear algebraic systems (8.128) can be solved by a direct solver (e.g., UMFPACK [72]) in case that the number of unknowns is not high (the limit value is usually 10^5). In general, iterative solvers are more efficient, because a good initial approximation is obtained from the previous Newton-like iteration or the previous time level. Usually it is necessary to compute only a few iterations. Among the iterative solvers, very efficient are the Krylov subspace methods, see [215].

It is possible to apply, e.g., the GMRES method [249] with block diagonal or block ILU(0) preconditioning [102]. Usually, the GMRES iterative process is stopped, when the preconditioned residuum is two times smaller than the initial one. This criterion may seem to be too weak, but numerical experiments show that it is sufficient in a number of applications.

8.4.5 Higher-Order Time Discretization

In Sect. 8.4.1, we have introduced the space-time discretization of the Euler equations (8.8) with the aid of the backward Euler—discontinuous Galerkin method (BE-DGM). However, by virtue of Remark 8.13, this method is only of the first order in time. In solving nonstationary flows, it is necessary to apply schemes that are sufficiently accurate in space as well as in time. There are several possibilities (see, e.g., [161, 268]) how to obtain a higher-order time discretizations.

We mention three techniques having the order n with respect to the time discretization, i.e., the error is of order $O(\tau^n)$:

- *backward difference formula* (BDF) method,which is a multistep method using computed approximate solutions from n previous time levels. On each time level, it is necessary to solve one nonlinear algebraic system with N_{hp} equations, where N_{hp} is the dimension of the space S_{hp}. Hence, the BDF method has (approximately) the same computational costs as the backward Euler method.
- *implicit Runge–Kutta* (IRK) method,which is a one-step method and it evaluates several (at least n) stages within one time step. This means that we solve (atleast) n-nonlinear algebraic systems with N_{hp} equations at each time level. Hence, the IRK method has approximately n-times higher computational cost than the backward Euler method.
- *time discontinuous Galerkin* (TDG) method, which is based on a polynomial approximation of degree $n-1$ with respect to time. The TDG method was introduced in Sect. 6.2 for a scalar equation. We solve one nonlinear algebraic system with nN_{hp} equations at each time level. As we see, the TDG method has approximately n^2-times higher computational cost than the backward Euler method or the BDF method.

The BDF, IRK and TDG time discretizations reduce to backward Euler method for the limit case $n = 1$. An overview of theoretical aspects of the higher-order time discretization in combination with the DG space discretization can be found in [278].

It follows from the above discussion that the cheapest approach is the BDF technique, which will be described in this section. Again let $0 = t_0 < t_1 < t_2 < \cdots t_r = T$ be a partition of the time interval $[0, T]$, $\tau_k = t_k - t_{k-1}$, $k = 1, \ldots r$, and let

$\boldsymbol{w}_h^k \in \boldsymbol{S}_{hp}$ denote a piecewise polynomial approximation of $\boldsymbol{w}_h(t_k),\ k = 0, 1, \ldots, r$. We define the following scheme.

Definition 8.16 We say that the finite sequence of functions $\boldsymbol{w}_h^k,\ k = 0, \ldots, r$, is the *approximate solution* of (8.8) computed by the *n-step backward difference formula—discontinuous Galerkin method* (BDF-DGM) if the following conditions are satisfied:

$$\boldsymbol{w}_h^k \in \boldsymbol{S}_{hp},\ k = 0, 1, \ldots, r, \tag{8.134a}$$

$$\frac{1}{\tau_k}\left(\sum_{l=0}^{n} \alpha_{n,l} \boldsymbol{w}_h^{k-l}, \boldsymbol{\varphi}_h\right) + \boldsymbol{b}_h\left(\boldsymbol{w}_h^k, \boldsymbol{\varphi}_h\right) = 0 \quad \forall \boldsymbol{\varphi}_h \in \boldsymbol{S}_{hp},\ k = n, \ldots, r, \tag{8.134b}$$

$\boldsymbol{w}_h^0$ is the $\boldsymbol{S}_{hp}$-approximation (usually $L^2(\Omega)$-projection on $\boldsymbol{S}_{hp}$) of the initial condition $\boldsymbol{w}^0$, (8.134c)

$\boldsymbol{w}_h^l \in \boldsymbol{S}_{hp},\ l = 1, \ldots, n-1,$ are determined by a suitable q-step method with $q \le l$ or by an explicit Runge–Kutta method. (8.134d)

Some Runge–Kutta schemes can be found in Sect. 5.2.1.1. Their application to a system of partial differential equations can be written in the same form.

The BDF coefficients $\alpha_{n,l},\ l = 0, \ldots, n$, depend on time steps $\tau_{k-l},\ l = 0, \ldots, n$. They can be derived from the Lagrange interpolation of pairs $(t_{k-l}, \boldsymbol{w}_h^{k-l}),\ l = 0, \ldots, n$, see, e.g. [161]. Table 8.2 shows their values in the case of constant and variable time steps for $n = 1, 2, 3$. Obviously, the one-step BDF-DGM is identical with the BE-DGM defined by (8.95). In Table 8.3 these coefficients are expressed directly in terms of the time steps τ_j.

Remark 8.17 (*Stability of the BDF-DGM*) The n-step BDF method is unconditionally stable for $n = 1$ and $n = 2$, and for increasing n the region of stability decreasing. For $n > 7$ this method is unconditionally unstable, see [161, Sect. 3.5]. In practice, the n-BDF-DGM with $n = 1, 2, 3$ is usually used.

Table 8.2 Values of $\alpha_{n,l},\ l = 0, \ldots, n$, for $n = 1, 2, 3$ for constant and variable time steps, $\theta_k = \tau_k/\tau_{k-1},\ k = 1, 2, \ldots, r$

	Constant time step			Variable time step		
	$n = 1$	$n = 2$	$n = 3$	$n = 1$	$n = 2$	$n = 3$
$\alpha_{n,0}$	1	$\frac{3}{2}$	$\frac{11}{6}$	1	$\frac{2\theta_k+1}{\theta_k+1}$	$\frac{\theta_k\theta_{k-1}}{\theta_k\theta_{k-1}+\theta_{k-1}+1} + \frac{2\theta_k+1}{\theta_k+1}$
$\alpha_{n,1}$	-1	-2	-3	1	$-(\theta_k + 1)$	$-\frac{(\theta_k+1)(\theta_k\theta_{k-1}+\theta_{k-1}+1)}{\theta_{k-1}+1}$
$\alpha_{n,2}$		$\frac{1}{2}$	$\frac{3}{2}$		$\frac{\theta_k^2}{\theta_k+1}$	$\frac{\theta_k^2(\theta_k\theta_{k-1}+\theta_{k-1}+1)}{\theta_k+1}$
$\alpha_{n,3}$			$-\frac{1}{3}$			$-\frac{(\theta_k+1)\theta_k^2\theta_{k-1}^3}{(\theta_{k-1}+1)(\theta_k\theta_{k-1}+\theta_{k-1}+1)}$

Table 8.3 Values of the coefficients $\alpha_{n,l}$ expressed in terms of the time steps

	$n=1$	$n=2$	$n=3$
$\alpha_{n,0}$	1	$\frac{2\tau_k+\tau_{k-1}}{\tau_k+\tau_{k-1}}$	$\frac{(2\tau_k+\tau_{k-1})(2\tau_k+\tau_{k-1}+\tau_{k-2})-\tau_k^2}{(\tau_k+\tau_{k-1})(\tau_k+\tau_{k-1}+\tau_{k-2})}$
$\alpha_{n,1}$	-1	$-\frac{\tau_k+\tau_{k-1}}{\tau_{k-1}}$	$-\frac{(\tau_k+\tau_{k-1})(\tau_k+\tau_{k-1}+\tau_{k-2})}{\tau_{k-1}(\tau_{k-1}+\tau_{k-2})}$
$\alpha_{n,2}$		$\frac{\tau_k^2}{\tau_{k-1}(\tau_k+\tau_{k-1})}$	$\frac{\tau_k^2(\tau_k+\tau_{k-1}+\tau_{k-2})}{\tau_{k-1}\tau_{k-2}(\tau_k+\tau_{k-1})}$
$\alpha_{n,3}$			$-\frac{\tau_k^2(\tau_k+\tau_{k-1})}{\tau_{k-2}(\tau_k+\tau_{k-1}+\tau_{k-2})(\tau_{k-1}+\tau_{k-2})}$

Remark 8.18 (*Accuracy of the BDF-DGM*) The n-step BDF-DGM has formally the order of convergence $O(h^{p+1}+\tau^n)$ in the $L^\infty(0,T;(L^2(\Omega))^m)$-norm and $O(h^p+\tau^n)$ in the $L^2(0,T;(H^1(\Omega))^m)$-seminorm, provided that the exact solution is sufficiently regular. These orders of convergence were numerically verified for a scalar equation.

Problem (8.134) represents a nonlinear algebraic system for each $k=1,\dots,r$, which can be solved with the strategy presented in Sect. 8.4.3.

Again, let N_{hp} denote the dimension of the space $\boldsymbol{S}_{hp}$ of the piecewise polynomial functions and let $\mathrm{B}_{hp}=\{\boldsymbol{\varphi}_i(x),\ i=1,\dots,N_{hp}\}$ be a basis of $\boldsymbol{S}_{hp}$. Using the isomorphism (8.96) between $\boldsymbol{w}_h^k\in\boldsymbol{S}_{hp}$ and $\boldsymbol{\xi}_k\in\mathbb{R}^{N_{hp}}$, we define the vector-valued function $\boldsymbol{F}_h:(\mathbb{R}^{N_{hp}})^n\times\mathbb{R}^{N_{hp}}\to\mathbb{R}^{N_{hp}}$ by

$$\boldsymbol{F}_h\left(\{\boldsymbol{\xi}_{k-l}\}_{l=1}^n;\boldsymbol{\xi}_k\right)=\left(\frac{1}{\tau_k}\left(\sum_{l=0}^{n}\alpha_{n,l}\boldsymbol{w}_h^{k-l},\boldsymbol{\varphi}_i\right)+\boldsymbol{b}_h(\boldsymbol{w}_h^k,\boldsymbol{\varphi}_i)\right)_{i=1}^{N_{hp}},\quad k=1,\dots,r, \tag{8.135}$$

where $\boldsymbol{\xi}_{k-l}\in\mathbb{R}^{N_{hp}}$ is the algebraic representation of $\boldsymbol{w}_h^{k-l}\in\boldsymbol{S}_{hp}$ for $l=1,\dots,n$. Then scheme (8.134) has the following algebraic representation. If $\boldsymbol{\xi}_{k-l}$, $l=1,\dots,n$, $(k=1,\dots,r)$ are given vectors, then we want to find $\boldsymbol{\xi}_k\in\mathbb{R}^{N_{hp}}$ such that

$$\boldsymbol{F}_h(\{\boldsymbol{\xi}_{k-l}\}_{l=1}^n;\boldsymbol{\xi}_k)=\boldsymbol{0}. \tag{8.136}$$

System (8.136) is strongly nonlinear. It can be solved with the aid of the Newton-like method based on the flux matrix, presented in Sect. 8.4.3. Let $\boldsymbol{b}_h^L$ and $\tilde{\boldsymbol{b}}_h$ be the forms defined by (8.121) and (8.123), respectively. Then (8.103) implies the consistency

$$\boldsymbol{b}_h(\boldsymbol{w}_h,\boldsymbol{\varphi}_h)=\boldsymbol{b}_h^L(\boldsymbol{w}_h,\boldsymbol{w}_h,\boldsymbol{\varphi}_h)-\tilde{\boldsymbol{b}}_h(\boldsymbol{w}_h,\boldsymbol{\varphi}_h)\quad\forall\,\boldsymbol{w}_h,\boldsymbol{\varphi}_h\in\boldsymbol{S}_{hp}, \tag{8.137}$$

where the form $\boldsymbol{b}_h^L$ is defined in (8.123).

We see that instead of (8.124) and (8.125), we define the *flux matrix* $\mathbb{C}_h$ and the vector $\boldsymbol{d}_h$ by

$$\mathbb{C}_h\left(\bar{\boldsymbol{\xi}}\right) = \left(\frac{\alpha_{n,0}}{\tau_k}\left(\boldsymbol{\varphi}_j, \boldsymbol{\varphi}_i\right) + \boldsymbol{b}_h^L(\bar{\boldsymbol{w}}_h, \boldsymbol{\varphi}_j, \boldsymbol{\varphi}_i)\right)_{i,j=1}^{N_{hp}} \tag{8.138}$$

and

$$\boldsymbol{d}_h\left(\{\boldsymbol{\xi}_{k-l}\}_{l=1}^n, \bar{\boldsymbol{\xi}}\right) = \left(\frac{1}{\tau_k}\left(\sum_{i=1}^n \alpha_{n,i}\boldsymbol{w}_h^{k-l}, \boldsymbol{\varphi}_i\right) + \tilde{\boldsymbol{b}}_h(\bar{\boldsymbol{w}}_h, \boldsymbol{\varphi}_i)\right)_{i=1}^{N_{hp}}, \tag{8.139}$$

respectively. Here $\boldsymbol{\varphi}_i \in \mathrm{B}_{hp}$, $i = 1, \dots, N_{hp}$, are the basis functions, $\bar{\boldsymbol{\xi}} \in \mathbb{R}^{N_{hp}}$ and $\boldsymbol{\xi}_{k-l} \in \mathbb{R}^{N_{hp}}$, $l = 1, \dots, n$, are the algebraic representations of $\bar{\boldsymbol{w}}_h \in \boldsymbol{S}_{hp}$ and $\boldsymbol{w}_h^{k-l} \in \boldsymbol{S}_{hp}$, $l = 1, \dots, n$, respectively. Finally, using (8.135) and (8.137)–(8.139), we have

$$\boldsymbol{F}_h(\{\boldsymbol{\xi}_{k-l}\}_{l=1}^n; \boldsymbol{\xi}_k) = \mathbb{C}_h(\boldsymbol{\xi}_k)\boldsymbol{\xi}_k - \boldsymbol{d}_h(\{\boldsymbol{\xi}_{k-l}\}_{l=1}^n, \boldsymbol{\xi}_k), \quad k = 1, \dots, r. \tag{8.140}$$

Let us note that the flux matrix $\mathbb{C}_h$ given by (8.138) has the same block structure as the matrix $\mathbb{C}_h$ defined by (8.124). The sequence of nonlinear algebraic systems can be solved by the damped Newton-like iterative process (8.127) and (8.128) treated in Sect. 8.4.4.

Concerning the initial guess $\boldsymbol{\xi}_k^0$ for the iterative process (8.127) and (8.128), we use either the value known from the previous time level given by (8.129), i.e., $\boldsymbol{\xi}_k^0 = \boldsymbol{\xi}_{k-1}$, $k = 1, \dots, r$, or it is possible to apply a higher-order extrapolation from previous time levels similarly as in the high-order semi-implicit time discretization from [88]. Hence, we put

$$\boldsymbol{\xi}_k^0 = \sum_{l=1}^n \beta_{n,l}\boldsymbol{\xi}_{k-l}, \quad k = 1, \dots, r, \tag{8.141}$$

where $\boldsymbol{\xi}_{k-l}$, $l = 1, \dots, n$, correspond to the solution $\boldsymbol{w}_h^{k-l}$ at the time level t_{k-l} and $\beta_{n,l}$, $l = 1, \dots, n$, are coefficients depending on time steps τ_{k-l}, $l = 0, \dots, n$. Table 8.4 shows the values of $\beta_{n,l}$, $l = 1, \dots, n$, for $n = 1, 2, 3$. In Table 8.5, these coefficients are expressed in terms of the time steps.

Remark 8.19 Similarly as in Remark 8.15, if we carry out only one Newton-like iteration at each time level, put $\lambda^0 = 1$, the matrix $\mathbb{C}$ is updated at each time step and use the extrapolation (8.141); then the implicit method (8.134) reduces to the

Table 8.4 Values of $\beta_{n,l}$, $l = 0, \dots, n$, for $n = 1, 2, 3$ for constant and variable time steps, $\theta_k = \tau_k/\tau_{k-1}$, $k = 1, 2, \dots, r$

	Constant time step			Variable time step		
	$n = 1$	$n = 2$	$n = 3$	$n = 1$	$n = 2$	$n = 3$
$\beta_{n,1}$	1	2	3	1	$1+\theta_k$	$(1+\theta_k)\frac{\theta_k\theta_{k-1}+\theta_{k-1}+1}{\theta_{k-1}+1}$
$\beta_{n,2}$		-1	-3		$-\theta_k$	$-\theta_k(\theta_k\theta_{k-1}+\theta_{k-1}+1)$
$\beta_{n,3}$			1			$\theta_k\theta_{k-1}\frac{\theta_k\theta_{k-1}+\theta_{k-1}}{\theta_{k-1}+1}$

Table 8.5 Values of $\beta_{n,l}$ expressed in terms of time steps

	$n = 1$	$n = 2$	$n = 3$
$\beta_{n,1}$	1	$\frac{\tau_k+\tau_{k-1}}{\tau_{k-1}}$	$\frac{(\tau_k+\tau_{k-1}+\tau_{k-2})(\tau_k+\tau_{k-1})}{\tau_{k-1}(\tau_{k-1}+\tau_{k-2})}$
$\beta_{n,2}$		$-\frac{\tau_k}{\tau_{k-1}}$	$-\frac{\tau_k(\tau_k+\tau_{k-1}+\tau_{k-2})}{\tau_{k-1}\tau_{k-2}}$
$\beta_{n,3}$			$\frac{\tau_k(\tau_k+\tau_{k-1})}{\tau_{k-2}(\tau_{k-1}+\tau_{k-2})}$

high-order semi-implicit time discretization approach presented in [93, 133], which can be formulated in the following way: We seek the finite sequence of functions $\{\boldsymbol{w}_h^k\}_{k=0}^r$ such that

$$\boldsymbol{w}_h^k \in \boldsymbol{S}_{hp},\ k = 0, 1, \dots, r, \tag{8.142a}$$

$$\frac{1}{\tau_k}\left(\sum_{l=0}^{n} \alpha_{n,l}\boldsymbol{w}_h^{k-l}, \boldsymbol{\varphi}_h\right) + \hat{\boldsymbol{b}}_h\left(\sum_{l=1}^{n} \beta_{n,l}\boldsymbol{w}_h^{k-l}, \boldsymbol{w}_h^k, \boldsymbol{\varphi}_h\right) = 0 \tag{8.142b}$$

$$\forall \boldsymbol{\varphi}_h \in \boldsymbol{S}_{hp},\ k = 1, \dots, r.$$

Similarly as in (8.134), $\boldsymbol{w}_h^0, \dots, \boldsymbol{w}_h^{n-1}$ are defined by (8.134c) and (8.134d). Here, $\beta_{n,l}$, $l = 1, \dots, n$, are coefficients introduced above and $\hat{\boldsymbol{b}}_h$ is the form given by (8.131), i.e.,

$$\hat{\boldsymbol{b}}_h(\bar{\boldsymbol{w}}_h, \boldsymbol{w}_h, \boldsymbol{\varphi}_h) = \boldsymbol{b}_h^L(\bar{\boldsymbol{w}}_h, \boldsymbol{w}_h, \boldsymbol{\varphi}_h) - \tilde{\boldsymbol{b}}_h(\bar{\boldsymbol{w}}_h, \boldsymbol{\varphi}_h), \quad \boldsymbol{w}_h, \boldsymbol{\varphi}_h \in \boldsymbol{S}_{hp}.$$

Obviously, $\hat{\boldsymbol{b}}_h$ is consistent with $\boldsymbol{b}_h$ because $\boldsymbol{b}_h(\boldsymbol{w}_h, \boldsymbol{\varphi}_h) = \hat{\boldsymbol{b}}_h(\boldsymbol{w}_h, \boldsymbol{w}_h, \boldsymbol{\varphi}_h)$ for all $\boldsymbol{w}_h, \boldsymbol{\varphi}_h \in \boldsymbol{S}_{hp}$. Problem (8.142) represents a sequence of systems of linear algebraic equations.

8.4.6 Choice of the Time Step

The choice of the time step has a great influence on the efficiency of the BDF-DGM. We already mentioned that the implicit time discretization allows us to choose the time step many times larger than an explicit scheme. Too large time step causes the loss of accuracy and too small time step reduces the efficiency of the computation.

On the other hand, in the beginning of the computation, we usually start from a nonphysical initial condition and a large time step may cause failure of the computational process. Therefore, the aim is to develop a sufficiently robust algorithm which automatically increases the time step from small values in the beginning of the computation to larger values, but which also ensures accuracy with respect to time.

The standard ODE strategy chooses the size of the time step so that the corresponding *local discretization error* is below a given tolerance, see, e.g., [161]. Very often, the local discretization error is estimated by a difference of two numerical solutions obtained by two time integration methods. However, we have to solve two nonlinear algebraic systems at each time level which leads to higher computational costs, see [103].

In this section we present a strategy, which is based on a very low cost estimation of the local discretization error. For simplicity, we deal only with the first-order method, but these considerations can be simply extended to higher-order schemes. Let us consider the ordinary differential equation

$$y' := \frac{\mathrm{d}y}{\mathrm{d}t} = f(y), \quad y(0) = y_0, \tag{8.143}$$

where $y : [0, T] \to \mathbb{R}$, $f : \mathbb{R} \to \mathbb{R}$ and $y_0 \in \mathbb{R}$. We assume that problem (8.143) has a unique solution $y \in C^2([0, T])$. Moreover, let $0 = t_0 < t_1 < t_2 < \cdots < t_r = T$ be a partition of $[0, T]$. We denote by $y_k \approx y(t_k)$ an approximation of the solution y at t_k, $k = 1, \ldots, r$. The *backward Euler method* reads as

$$y_k = y_{k-1} + \tau_k f(y_k), \qquad k = 1, 2, \ldots, r, \tag{8.144}$$

where $\tau_k = t_k - t_{k-1}$. By the Taylor theorem, there exists $\theta_k \in [t_{k-1}, t_k]$ such that the corresponding local discretization error L_k has the form

$$L_k = \frac{1}{2}\tau_k^2 y''(\theta_k), \qquad \theta_k \in (t_{k-1}, t_k), \tag{8.145}$$

where y'' denotes the second-order derivative of y.

Our idea is the following. We define the quadratic function $\tilde{y}_k : [t_{k-2}, t_k] \to \mathbb{R}$ such that $\tilde{y}_k(t_{k-l}) = y_{k-l}$, $l = 0, 1, 2$. The second-order derivative of $\tilde{y}_k$ is constant on (t_{k-2}, t_k). We use the approximation

$$|L_k| \approx L_k^{\mathrm{app}} = \frac{1}{2}\tau_k^2 |\tilde{y}_k''|. \tag{8.146}$$

Let $\omega > 0$ be a given tolerance for the local discretization error. Our aim is to choose the time step as large as possible but guaranteeing the condition $L_k^{\mathrm{app}} \le \omega,\ k = 1, \dots, r$. On the basis of (8.146), we assume that

$$\omega \approx \frac{1}{2}(\tau_k^{\mathrm{opt}})^2 |\tilde{y}_k''|, \tag{8.147}$$

where τ_k^{opt} denotes the optimal size of τ_k. We express $|\tilde{y}_k''|$ from (8.146), insert it in (8.147) and express τ_k^{opt} as

$$\tau_k^{\mathrm{opt}} := \tau_k \left(\frac{\omega}{L_k^{\mathrm{app}}} \right)^{1/2}. \tag{8.148}$$

On the basis of the above considerations, we define the following

Adaptive time step algorithm

(1) let $\omega > 0$, $k > 1$, $y_{k-1}, y_{k-2} \in \mathbb{R}$ and $\tau_k > 0$ be given,
(2) compute y_k by (8.144),
(3) from $[t_{k-l}, y_{k-l}]$, $l = 0, 1, 2$, construct $\tilde{y}_k$,
(4) compute τ_k^{opt} by (8.146) and (8.148),
(5) *if* $\tau_k^{\mathrm{opt}} \ge \tau_k$
then

  (i) put $\tau_{k+1} = \min(\tau_k^{\mathrm{opt}}, c_1\tau_k, \tau^{\max})$,
  (ii) put $k = k + 1$
  (iii) go to step 2)
  else

  (i) put $\tau_k = \tau_k^{\mathrm{opt}}$,
  (ii) go to step 2).

The constant $c_1 > 1$ restricts the maximal ratio of two successive time steps. It is possible to use the value $c_1 = 2.5$. The value $\tau^{\max}$ restricts the maximal size of the time step for practical reasons. For example, $\tau^{\max} = 2\tau_0 10^{12}$, but any sufficiently large value yields similar results. If the *else* branch in step (5) of the algorithm is reached, then on each time level we solve more than one algebraic problem, which is expensive. However, this branch is reached very rarely in practice. It may occur only if the initial time step τ_0 or the constant c_1 are chosen too large.

This approach is extended to a system of ODEs in the following way. Let $\boldsymbol{y}_k \in \mathbb{R}^N$ be an approximation of the solution of the system of ODEs at $t_k,\ k = 0, 1, \dots$. For each time level t_k, we define a vector-valued quadratic function $\tilde{\boldsymbol{y}}_k(t) : [t_{k-2}, t_k] \to \mathbb{R}^N$ such that $\tilde{\boldsymbol{y}}_k(t_{k-l}) = \boldsymbol{y}_{k-l}$, $l = 0, 1, 2$. Then the optimal time step is given by (8.148) with the approximation of the local discretization error

$$L_k^{\mathrm{app}} = \frac{1}{2}\tau_k^2 |\tilde{\boldsymbol{y}}_k''|, \tag{8.149}$$

where $\tilde{\boldsymbol{y}}_k'' \in \mathbb{R}^N$ denotes the second-order derivative of $\tilde{\boldsymbol{y}}_k(t)$ with respect to t. The adaptive time stepping algorithm remains the same, $\tilde{y}_k$ is replaced by $\tilde{\boldsymbol{y}}_k$ and (8.146) is replaced by (8.149).

Concerning the choice of the first two time steps in the case of the solution of the Euler equations, we use the relation (8.94), namely

$$\tau_k = \mathrm{CFL} \min_{K \in \mathscr{T}_h} \frac{|K|}{\max_{\Gamma \subset \partial K} \varrho(\mathbb{P}(\boldsymbol{w}_h^k|_\Gamma))|\Gamma|}, \qquad k = 0, 1, \tag{8.150}$$

where $\varrho(\mathbb{P}(\boldsymbol{w}_h^k|_\Gamma))$ is the spectral radius of the matrix $\mathbb{P}(\boldsymbol{w}_h^k|_\Gamma, \boldsymbol{n}_\Gamma)$ given by (8.17) on $\Gamma \in \mathscr{F}_h$ and the value CFL is the initial Courant–Friedrichs–Lewy number. In order to avoid drawback resulting from a nonphysical initial condition (which is the usual case), we put $\mathrm{CFL} = 0.5$. Thus τ_0 and τ_1 correspond to the time steps used for the explicit time discretization with this CFL value. This choice may be underestimated in some cases, but based on our numerical experiments, it is robust with respect to the flow regime.

Remark 8.20 The presented technique can be simply extended to n-step BDF-DGM. For $n \geq 1$ we derive (instead of (8.146)) the relation $L_k^{\mathrm{app}} = \gamma_n \tau_k^{n+1} |\tilde{y}_k^{(n+1)}|$, where $\gamma_n > 0$. Then relations (8.147) and (8.148) have to be modified.

Remark 8.21 In order to accelerate the convergence to the steady state solutions, it is possible to apply local time stepping. However, our aim is to develop a scheme which can also be applied to nonstationary problems. Therefore, we consider only global time stepping.

8.4.7 Structure of the Flux Matrix

The flux matrix $\mathbb{C}_h$ given by (8.124) can be written in the form

$$\mathbb{C}_h(\boldsymbol{\xi}) = \frac{1}{\tau_k}\mathbb{M}_h + \mathbb{B}_h(\boldsymbol{\xi}), \tag{8.151}$$

where

$$\mathbb{M}_h = \left((\boldsymbol{\varphi}_j, \boldsymbol{\varphi}_i)\right)_{i,j=1}^{N_{hp}}, \qquad \mathbb{B}_h(\boldsymbol{\xi}) = \left(\boldsymbol{b}_h^L(\bar{\boldsymbol{w}}_h, \boldsymbol{\varphi}_j, \boldsymbol{\varphi}_i)\right)_{i,j=1}^{N_{hp}}. \tag{8.152}$$

The matrix $\mathbb{M}_h$ is called the *mass matrix*. If the basis in $\boldsymbol{S}_{hp}$ is constructed elementwise (i.e., the support of each basis function is just one simplex from $\mathscr{T}_h$), then $\mathbb{M}_h$ is block diagonal. Similarly, the matrices $\mathbb{B}_h$ and therefore $\mathbb{C}_h$ have a block structure. By virtue of (8.123), we easily find that each block-row of $\mathbb{B}_h$ corresponds to one element $K \in \mathscr{T}_h$ and contains a diagonal block and several off-diagonal blocks. Each off-diagonal block corresponds to one face $\Gamma \in \mathscr{F}_h$. See Fig. 8.4, where an illustrative mesh and the corresponding block structures of matrices $\mathbb{M}_h$ and $\mathbb{C}_h$ are shown.

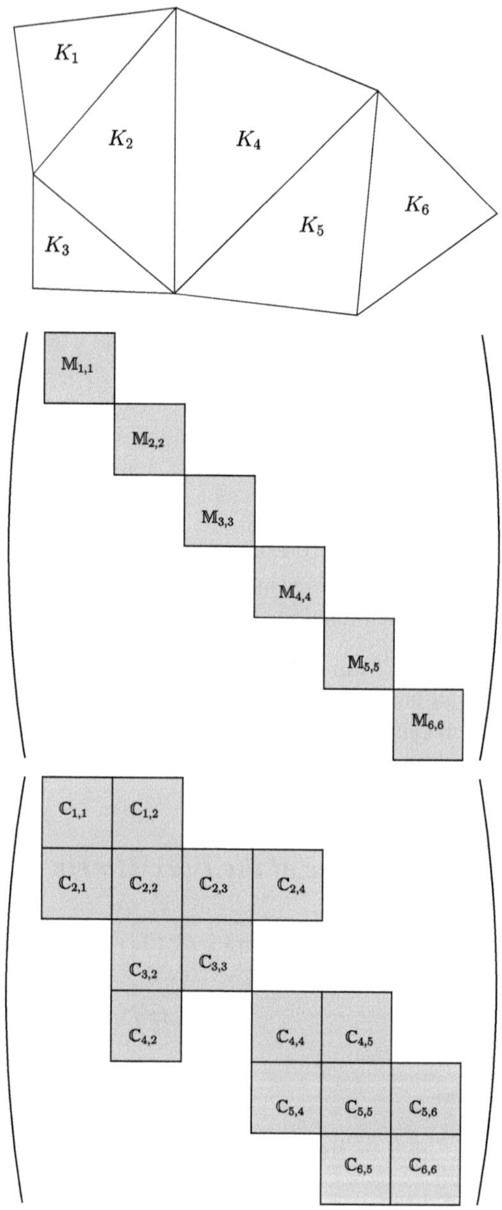

Fig. 8.4 Example of a triangular mesh with elements $K_\mu,\ \mu = 1, \ldots, 6$ (*top*) and the corresponding block structure of the matrices $\mathbb{M}_h$ (*center*) and $\mathbb{C}_h$ (*bottom*)

Similarly, the vector $\boldsymbol{d}_h$ from (8.125) can be written as

$$\boldsymbol{d}_h(\boldsymbol{\xi}_{k-1}, \bar{\boldsymbol{\xi}}) = \frac{1}{\tau_k}\boldsymbol{m}_h(\boldsymbol{\xi}_{k-1}) + \boldsymbol{u}_h(\bar{\boldsymbol{\xi}}), \tag{8.153}$$

where

$$\boldsymbol{m}_h(\boldsymbol{\xi}_{k-1}) = \left(\left(\boldsymbol{w}_h^{k-l}, \boldsymbol{\varphi}_i\right)\right)_{i=1}^{N_{hp}}, \quad \boldsymbol{u}_h(\bar{\boldsymbol{\xi}}) = \left(\tilde{\boldsymbol{b}}_h(\bar{\boldsymbol{w}}_h, \boldsymbol{\varphi}_i)\right)_{i=1}^{N_{hp}}. \tag{8.154}$$

If the time step τ_k in (8.151) is small enough, then the matrix $\mathbb{M}_h/\tau_k$ dominates over $\mathbb{B}_h$. Hence, if we construct a basis of $\boldsymbol{S}_{hp}$ which is orthonormal with respect to the L^2-scalar product, then $\mathbb{M}_h$ is the identity matrix and the linear algebraic problems (8.128) is solved easily for small τ_k.

Remark 8.22 On the other hand, there exists a limit value $\tau^\infty \gg 1$, such that for any $\tau_k \geq \tau^\infty$ we have

$$\mathbb{C}_h(\bar{\boldsymbol{\xi}}) \doteq \mathbb{B}_h(\bar{\boldsymbol{\xi}}), \quad \bar{\boldsymbol{\xi}} \in \mathbb{R}^{N_{hp}}, \tag{8.155}$$

where the symbol $\doteq$ denotes the equality in the finite precision arithmetic. Similarly, for any $\tau_k \geq \tau^\infty$ from (8.153) and (8.154) we obtain the relation

$$\boldsymbol{d}_h(\boldsymbol{\xi}_{k-1}, \bar{\boldsymbol{\xi}}) \doteq \boldsymbol{u}_h(\bar{\boldsymbol{\xi}}). \tag{8.156}$$

This means that $\mathbb{C}_h$ as well as $\boldsymbol{d}_h$ are independent of the size of τ_k. Moreover, by virtue of (8.126), problem (8.98)

$$\begin{aligned} \boldsymbol{0} = \boldsymbol{F}_h(\boldsymbol{\xi}_{k-1}; \boldsymbol{\xi}_k) =& \mathbb{C}_h(\boldsymbol{\xi}_k)\boldsymbol{\xi}_k - \boldsymbol{d}_h(\boldsymbol{\xi}_{k-1}, \boldsymbol{\xi}_k) \\ \doteq& \mathbb{B}_h(\boldsymbol{\xi}_k)\boldsymbol{\xi}_k - \boldsymbol{u}_h(\boldsymbol{\xi}_k), \quad k = 1, \ldots, r, \end{aligned} \tag{8.157}$$

is independent (in the finite precision arithmetic) on the size of τ_k provided that $\tau_k \geq \tau^\infty$. Our numerical experiments indicated that limit value $\tau_\infty \approx 10^{12}$ in the double precision arithmetic.

8.4.8 Construction of the Basis in the Space S_{hp}

In this section we present one possibility, how to construct a basis $\mathrm{B}_{hp} = \{\varphi_i(x),\ i = 1, \ldots, N_{hp}\}$ in the space $\boldsymbol{S}_{hp}$, in order to solve efficiently the Euler equations with the aid of the DGM. Obviously, it is advantageous to use functions from B_{hp} with small supports. Since $\boldsymbol{S}_{hp}$ consists of discontinuous functions, for each element $K \subset \mathscr{T}_h$ it is possible to define a local basis

$$\mathrm{B}_K = \left\{\boldsymbol{\psi}_{K,i} \in \boldsymbol{S}_{hp};\ \mathrm{supp}\,(\boldsymbol{\psi}_{K,i}) \subset K,\ i = 1, \ldots, \hat{N}\right\}, \tag{8.158}$$

with $\boldsymbol{\psi}_{K,i} \in (P_p(K))^m$ (= the space of vector-valued polynomials of degree $\leq p$ on $K \in \mathscr{T}_h$), where $\hat{N} = \frac{d+2}{d!}\Pi_{j=1}^d(p+j)$ is its dimension. Then the basis B_{hp} will be a composition of the local bases $\mathrm{B}_K,\ K \in \mathscr{T}_h$.

Let

$$\hat{K} = \{(\hat{x}_1, \ldots, \hat{x}_d);\ \hat{x}_i \geq 0,\ i = 1, \ldots, d,\ \sum_{i=1}^d \hat{x}_i \leq 1\} \tag{8.159}$$

be the *reference simplex*. We consider affine mappings

$$F_K : \hat{K} \to \mathbb{R}^d, \qquad F_K(\hat{K}) = K, \quad K \in \mathscr{T}_h. \tag{8.160}$$

(In Sect. 8.6 we deal with curved elements. In this case F_K is a polynomial mapping of degree > 1.)

On the reference element $\hat{K}$ we define a basis in the space of vector-valued polynomials of degree $\leq p$ by

$$\hat{\boldsymbol{S}}_p = (\hat{S}_p)^m, \tag{8.161}$$

$$\hat{S}_p = \left\{\phi_{n_1,\ldots,n_d}(\hat{x}_1, \ldots, \hat{x}_d) = \Pi_{i=1}^d(\hat{x}_i - \hat{x}_i^c)^{n_i};\quad n_1, \ldots, n_d \geq 0,\ \sum_{j=1}^d n_j \leq p\right\},$$

where $(\hat{x}_1^c, \ldots, \hat{x}_d^c)$ is the barycenter of $\hat{K}$. The dimension of the space spanned over the set $\hat{\boldsymbol{S}}_p$ is $\hat{N} = \frac{d+2}{d!}\Pi_{j=1}^d(p+j)$. By the Gram–Schmidt $L^2(\hat{K})$-orthonormalization process applied to $\hat{\boldsymbol{S}}_p$ we obtain the orthonormal system $\{\hat{\phi}_j,\ j = 1, \ldots, \hat{N}\}$. The Gram–Schmidt orthonormalization on the reference element can be easily computed, because $\hat{N}$ is small (moreover, the orthonormalization can be done for each component of $\boldsymbol{S}_{hp}$ independently). Hence, this orthonormalization does not cause any essential loss of accuracy.

Furthermore, let F_K, $K \in \mathscr{T}_h$, be the mapping introduced in (8.160). We put

$$\mathrm{B}_K = \{\boldsymbol{\psi}_{K,j};\ \boldsymbol{\psi}_{K,j}(x) = \hat{\phi}_j(F_K^{-1}(x)),\ x \in K,\ j = 1, \ldots, \hat{N}\}, \tag{8.162}$$

which defines a local basis B_K for each element $K \in \mathscr{T}_h$ separately. For an affine mapping F_K the basis B_K is $L^2(K)$-orthogonal with respect to the L^2-scalar product and the blocks $\mathbb{M}_{K,K}$ of the mass matrix $\mathbb{M}$ given by (8.152) are diagonal. If F_K is not afine, then the orthogonality of B_K is violated. However, in practical applications, the curved face $K_K \cap \partial\Omega$ is close to a straight (polygonal) one (see Sect. 8.6), and thus the matrix block $\mathbb{M}_{K,K}$ is strongly diagonally dominant.

Finally, a composition of the local bases $\mathrm{B}_K,\ K \in \mathscr{T}_h$, defines a basis of $\boldsymbol{S}_{hp}$, i.e.,

$$\mathrm{B}_{hp} = \{\boldsymbol{\psi}_{K,j};\ \boldsymbol{\psi}_{K,j} \in \mathrm{B}_K,\ j = 1, \ldots, \hat{N},\ K \in \mathscr{T}_h\}, \tag{8.163}$$

which is, for affine mappings F_K, $K \in \mathscr{T}_h$, the L^2-orthogonal basis of S_{hp}. In case that F_K is not an affine mapping for some $K \in \mathscr{T}_h$, the L^2-orthogonality is violated, i.e., $(\psi_{K,i}, \psi_{K,j}) \neq 0$ for $i, j = 1, \ldots, \hat{N}$, $i \neq j$. However, since F_K is usually close to an affine mapping, we have $|(\psi_{K,i}, \psi_{K,j})| \ll |(\psi_{K,i}, \psi_{K,i})|$ for $i, j = 1, \ldots, \hat{N}, i \neq j$.

Remark 8.23 It is possible to find that every entry of $\boldsymbol{F}_h$ and/or $\mathbb{C}_h$ depends on $\boldsymbol{w}_h$ on at most two neighbouring elements. This is a favourable property which simplifies the parallelization of the algorithm.

8.4.9 Steady-State Solution

Very often, we are interested in the solution of the *stationary Euler equations*, i.e., we seek $\boldsymbol{w} : \Omega \to \mathscr{D}$ ($\mathscr{D}$ is given by (8.12)) such that

$$\sum_{s=1}^{d} \frac{\partial \boldsymbol{f}_s(\boldsymbol{w})}{\partial x_s} = 0, \tag{8.164}$$

where $\boldsymbol{w}$ is the steady-state vector and $\boldsymbol{f}_s$, $s = 1, \ldots, d$, are the Euler fluxes defined in (8.9) and (8.10), respectively. This system is equipped with boundary conditions (8.37), discussed in detail in Sect. 8.3.

The stationary Euler equations can be discretized in the same way as the non-stationary ones, omitting only the approximation of the time derivative.

Definition 8.24 We say that $\boldsymbol{w}_h \in \boldsymbol{S}_{hp}$ is a DG approximate solution of (8.164) if

$$\boldsymbol{b}_h(\boldsymbol{w}_h, \boldsymbol{\varphi}_h) = 0 \quad \forall \boldsymbol{\varphi}_h \in \boldsymbol{S}_{hp}, \tag{8.165}$$

where $\boldsymbol{b}_h$ is given by (8.93). We call $\boldsymbol{w}_h$ the *steady-state solution* of the Euler equations.

With the aid of the notation introduced in Sect. 8.4.2, we can formulate (8.165) as the algebraic problem to find $\boldsymbol{\xi} \in \mathbb{R}^{N_{hp}}$ such that

$$\boldsymbol{F}_h^{\mathrm{SS}}(\boldsymbol{\xi}) = \boldsymbol{0}, \tag{8.166}$$

where $\boldsymbol{\xi}$ is the algebraic representation of $\boldsymbol{w}_h$ by the isomorphism (8.96) and

$$\boldsymbol{F}_h^{\mathrm{SS}}(\boldsymbol{\xi}) = \left(\boldsymbol{b}_h \left(\sum_{j=1}^{N_{hp}} \xi_j \boldsymbol{\varphi}_j, \boldsymbol{\varphi}_i \right) \right)_{i=1}^{N_{hp}} \in \mathbb{R}^{N_{hp}}. \tag{8.167}$$

By virtue of (8.137), (8.152) and (8.154), we have

$$\boldsymbol{F}_h^{\mathrm{SS}}(\boldsymbol{\xi}) = \mathbb{B}(\boldsymbol{\xi})\boldsymbol{\xi} - \boldsymbol{u}_h(\boldsymbol{\xi}), \quad \boldsymbol{\xi} \in \mathbb{R}^{N_{hp}}. \tag{8.168}$$

Problem (8.166) represents a system of nonlinear algebraic equations. It can be solved directly by the (damped) Newton method, see [164]. Another very often used possibility is to apply the *time-marching* (or *time stabilization*) method based on the solution of the nonstationary Euler equations (8.8) and to seek the *steady-state* solution as a limit of the nonstationary solution for $t \to \infty$. This means that the methods for solving unsteady flow are applied as iterative processes, assuming that $\boldsymbol{w}_h = \lim_{k\to\infty}\boldsymbol{w}_h^k$. The nonstationary computational process is stopped, when a suitable *steady-state criterion* is achieved.

The usual steady-state criterion often used for explicit time discretization reads (for an orthonormal basis) as

$$\left\| \frac{\partial \boldsymbol{w}_h}{\partial t} \right\|_{L^2(\Omega)} \approx \eta_k = \frac{1}{\tau_k}\|\boldsymbol{w}_h^k - \boldsymbol{w}_h^{k-1}\|_{L^2(\Omega)} = \frac{1}{\tau_k}|\boldsymbol{\xi}_k - \boldsymbol{\xi}_{k-1}| \le \mathrm{TOL}, \tag{8.169}$$

where $\boldsymbol{w}_h^{k-l}$, $l = 0, 1$, denote the values of the approximate solution at time levels t_{k-l}, $l = 0, 1$, $\boldsymbol{\xi}_{k-l}$, $l = 0, 1$, are their algebraic representations given by the isomorphism (8.96) and TOL is a given tolerance.

Criterion (8.169) is not suitable for the implicit time discretization, when very large time steps are used, see [102, Sect. 4.3.1.]. Then it is suitable to use the *steady-state residual criterion*

$$|\boldsymbol{F}_h^{\mathrm{SS}}(\boldsymbol{\xi}_k)| = |\mathbb{B}(\boldsymbol{\xi}_k)\boldsymbol{\xi}_k - \boldsymbol{u}_h(\boldsymbol{\xi}_k)| \le \mathrm{TOL}, \tag{8.170}$$

which is independent of τ_k and measures the residuum of the nonlinear algebraic system (8.167).

However, it is an open question as to how to choose the tolerance TOL in (8.170), since the residuum depends on the size of the computational domain Ω, on the magnitude of components of $\boldsymbol{w}_h^k$, etc. Therefore, from the practical reasons, we use the *relative residuum steady-state criterion*

$$\mathrm{SSres}(k) := \frac{|\boldsymbol{F}_h^{\mathrm{SS}}(\boldsymbol{\xi}_k)|}{|\boldsymbol{F}_h^{\mathrm{SS}}(\boldsymbol{\xi}_0)|} \le \mathrm{TOL}, \tag{8.171}$$

which already does not suffer from the mentioned drawbacks. Here $\boldsymbol{\xi}_0$ is the algebraic representation of the initial state $\boldsymbol{w}_h^0$.

Another possibility are the stopping criteria which follow from the physical nature of the considered problem. E.g., in aerodynamics, when we solve flow around a 2D profile, we are often interested in the *aerodynamic coefficients* of the considered flow, namely coefficients of *drag* (c_D), *lift* (c_L) and *momentum* (c_M). In the 2D case, the coefficients c_D and c_L are defined as the first and second components of the vector

$$\frac{1}{\frac{1}{2}\rho_\infty |\boldsymbol{v}_\infty|^2 L_{\text{ref}}} \int_{\Gamma_{\text{prof}}} \text{p}\boldsymbol{n}\, \mathrm{d}S, \tag{8.172}$$

where ρ_∞ and $\boldsymbol{v}_\infty$ are the far-field density and velocity, respectively, L_{ref} is the reference length, Γ_{prof} is the profile, $\boldsymbol{n}$ is outer unit normal to the profile pointing into the profile and p is the pressure. Moreover, c_M is given by

$$\frac{1}{\frac{1}{2}\rho_\infty |\boldsymbol{v}_\infty|^2 L_{\text{ref}}^2} \int_{\Gamma_{\text{prof}}} (x - x_{\text{ref}}) \times \text{p}\boldsymbol{n}\, \mathrm{d}S, \tag{8.173}$$

where x_{ref} is the moment reference point. We adopt the notation $x \times y = x_1 y_2 - x_2 y_1$ for $x = (x_1, x_2)$, $y = (y_1, y_2) \in \mathbb{R}^2$.

Then it is natural to stop the computation when these coefficients achieve a given tolerance tol, e.g.,

$$\Delta c_\alpha(k) \leq \text{tol}, \qquad \Delta c_\alpha(k) = \max_{l=\overline{k},\dots,k} c_\alpha(l) - \min_{l=\overline{k},\dots,k} c_\alpha(l), \tag{8.174}$$

where $\alpha = D$, L and M (for the drag, lift and momentum), $c_\alpha(k)$ is the value of the corresponding aerodynamic coefficient at the kth-time level and $\overline{k}$ is the entire part of the number 0.9 k. This means that the minimum and maximum in (8.174) are taken over the last 10 % of the number of time levels.

In contrast to the tolerance TOL in (8.171), which has to be chosen empirically, the tolerance tol in (8.174) can be chosen only on the basis of our accuracy requirements (without any previous numerical experiments). Since the absolute values of aerodynamic coefficient are (usually) less than one, the stopping criterion (8.174) with tolerance, e.g., tol $= 10^{-4}$, gives accuracy of the aerodynamic coefficients for 3 decimal digits.

Finally, let us note that since we seek only the steady-state solution, we do not need to take care of an accurate approximation of the evolution process. Therefore, we can choose the time step τ_k relatively large. Hence, the tolerance ω appearing in (8.148) can also be large.

8.5 Shock Capturing

In higher-order numerical methods, applied to the solution of high speed flows with shock waves and contact discontinuities, we can observe the Gibbs phenomenon manifested by spurious (nonphysical) oscillations in computed quantities propagating from discontinuities. In the standard Galerkin finite element methods, these oscillations propagate far into the computational domain. However, in DG numerical solutions the Gibbs phenomenon is manifested only by spurious overshoots and undershoots appearing in the vicinity of discontinuities. These phenomena do not occur in low Mach number regimes, when the exact solution is regular, but in the high-speed flow they cause instabilities in the numerical solution and collapse of the computational process.

In order to cure this undesirable feature, in the framework of higher-order finite volume methods one uses suitable limiting procedures. They should preserve the higher-order accuracy of the method in regions where the solution is regular, and decrease the order to 1 in a neighbourhood of discontinuities or steep gradients. These methods are based on the use of the flux limiter. See e.g., [127] and citations therein. In [57, 62], the finite volume limiting procedures were generalized also to DGM.

Here we present another technique, based on the concept of artificial viscosity applied locally on the basis of a suitable *jump (discontinuity) indicator.*

8.5.1 Jump Indicators

Approximate solutions obtained by the DGM are, in general, discontinuous on interfaces between neighbouring elements. If the exact solution is sufficiently regular, then the jumps in the approximate solution are small and, as follows from the theory as well as numerical experiments, tend to zero if $h \to 0$.

The DG solution of inviscid flow can contain large inter-element jumps in subdomains, where the solution is not sufficiently smooth, i.e., in areas with discontinuities (shock waves or contact discontinuities). Numerical experiments show that the inter-element jumps in the approximate solution are $[\boldsymbol{w}_h]_\Gamma = O(1)$ on discontinuities, but $[\boldsymbol{w}_h]_\Gamma = O(h^{p+1})$ in the areas where the solution is regular. This inspires us to define a *jump indicator*, which evaluates the inter-element jumps of the approximate solution. On general unstructured grids, it appears to be suitable to measure the magnitude of inter-element jumps in the integral form by

$$\int_{\partial K \cap \Omega} [w_{h,1}]^2 \, \mathrm{d}S, \quad K \in \mathscr{T}_h \tag{8.175}$$

on interior faces $\Gamma \in \mathscr{F}_h^I$, where $w_{h,1}$ denotes the first component, i.e., the density ρ_h corresponding to the state $\boldsymbol{w}_h$. (Here we take into account that the density is discontinuous both on shock waves and contact discontinuities.)

This leads us to the definition of the *jump indicator* in the form

$$g_K(\boldsymbol{w}_h) = \frac{\int_{\partial K \cap \Omega} [w_{h,1}]^2 \, \mathrm{d}S}{|K| \sum_{\Gamma \subset \partial K \cap \Omega} \mathrm{diam}(\Gamma)}, \quad K \in \mathscr{T}_h, \tag{8.176}$$

where $|K|$ denotes the d-dimensional measure of K and $\mathrm{diam}(\Gamma)$ is the diameter of Γ. We see that we have

$$g_K(\boldsymbol{w}_h) = \begin{cases} O(h^{2p}) & \text{for } K \in \mathscr{T}_h \text{, where the solution is smooth,} \\ O(h^{-2}) & \text{for } K \in \mathscr{T}_h \text{ near discontinuities.} \end{cases} \tag{8.177}$$

Thus, $g_K \to 0$ for $h \to 0$ in the case when $K \in \mathscr{T}_h$ is in a subdomain where the solution is regular, and $g_K \to \infty$ for $h \to 0$ in the case when $K \in \mathscr{T}_h$ is in the vicinity of a discontinuity.

There are various modifications of this indicator, as for example,

$$g_K(\boldsymbol{w}_h) = \int_{\partial K} [w^k_{h,1}]^2 \,\mathrm{d}S / (h_K |K|^{3/4}), \quad K \in \mathscr{T}_h, \tag{8.178}$$

in the 2D case, proposed in [98] and applied in [133]. The indicator g_K was constructed in such a way that it takes an anisotropy of the computational mesh into account. It was shown in [98] that the indicator $g_K(\boldsymbol{w}_h)$ identifies discontinuities safely on unstructured and anisotropic meshes.

Now we introduce the *discrete jump (discontinuity) indicator*

$$G_K(\boldsymbol{w}_h) = 0, \quad \text{if } g_K(\boldsymbol{w}_h) < 1, \quad G_K(\boldsymbol{w}_h) = 1, \quad \text{if } g_K(\boldsymbol{w}_h) \geq 1, \quad K \in \mathscr{T}_h. \tag{8.179}$$

Numerical experiments show that under the assumption that the mesh space size $h < 1$, it is possible to indicate the areas without discontinuities checking the condition $G_K(\boldsymbol{w}_h) < 1$. On the other hand, if $G_K(\boldsymbol{w}_h) > 1$, the element K is lying in a neighbourhood of a discontinuity.

However, it appears that the above discrete discontinuity indicators and the artificial viscosity forms (8.181) and (8.182) introduced in the following section are too strict. Particularly, it may happen in some situations that the value of g_K in (8.176) is close to 1 and then during the computational process the value G_K from (8.179) oscillates between 1 and 0. This can disable to achieve a steady-state solution. Therefore, it is suitable to introduce some "smoothing" of the discrete indicator (8.179). Namely we set

$$G_K(\boldsymbol{w}_h) = \begin{cases} 0, & \text{if } g_K(\boldsymbol{w}_h) < \xi_{\min}, \\ \frac{1}{2}\sin\left(\pi \frac{g_K(\boldsymbol{w}_h) - (\xi_{\max} - \xi_{\min})}{2(\xi_{\max} - \xi_{\min})}\right) + \frac{1}{2}, & \text{if } g_K(\boldsymbol{w}_h) \in [\xi_{\min}; \xi_{\max}), \\ 1, & \text{if } g_K(\boldsymbol{w}_h) \geq \xi_{\max}, \end{cases} \tag{8.180}$$

where $0 \leq \xi_{\min} < \xi_{\max}$. In practical applications, it is suitable to set $\xi_{\min} = 0.5$ and $\xi_{\max} = 1.5$.

8.5.2 Artificial Viscosity Shock Capturing

On the basis of the discrete discontinuity indicator we introduce local artificial viscosity forms, which are included in the numerical schemes for solving inviscid compressible flow. For example, we define the artificial viscosity form $\boldsymbol{\beta}_h : \boldsymbol{S}_{hp} \times \boldsymbol{S}_{hp} \times \boldsymbol{S}_{hp} \to \mathbb{R}$ by

$$\boldsymbol{\beta}_h(\bar{\boldsymbol{w}}_h, \boldsymbol{w}_h, \boldsymbol{\varphi}_h) = \nu_1 \sum_{K \in \mathscr{T}_h} h_K G_K(\bar{\boldsymbol{w}}_h) \int_K \nabla \boldsymbol{w}_h \cdot \nabla \boldsymbol{\varphi}_h \, \mathrm{d}\boldsymbol{x} \tag{8.181}$$

with $\nu_1 = O(1)$. Since this artificial viscosity form is rather local, we propose to augment it by the form $\boldsymbol{\gamma}_h : \boldsymbol{S}_{hp} \times \boldsymbol{S}_{hp} \times \boldsymbol{S}_{hp} \to \mathbb{R}$ defined as

$$\boldsymbol{\gamma}_h(\bar{\boldsymbol{w}}_h, \boldsymbol{w}_h, \boldsymbol{\varphi}_h) = \nu_2 \sum_{\Gamma \in \mathscr{F}_h^I} \frac{1}{2}\big(G_{K_\Gamma^{(L)}}(\bar{\boldsymbol{w}}_h) + G_{K_\Gamma^{(R)}}(\bar{\boldsymbol{w}}_h)\big) \int_\Gamma [\boldsymbol{w}_h] \cdot [\boldsymbol{\varphi}_h] \, \mathrm{d}S, \tag{8.182}$$

where $\nu_2 = O(1)$ and $K_\Gamma^{(L)}, K_\Gamma^{(R)} \in \mathscr{T}_h$ are the elements sharing the inner face $\Gamma \in \mathscr{F}_h^I$. This form allows strengthening the influence of neighbouring elements and improves the behaviour of the method in the case, when strongly unstructured and/or anisotropic meshes are used. These artificial viscosity forms were introduced in [133], where the indicator (8.179) was used.

Because of the reasons mentioned already above, using the discontinuity indicator (8.180), we also introduce more sophisticated artificial viscosity forms $\boldsymbol{\beta}_h, \boldsymbol{\gamma}_h : \boldsymbol{S}_{hp} \times \boldsymbol{S}_{hp} \times \boldsymbol{S}_{hp} \to \mathbb{R}$, defined as

$$\boldsymbol{\beta}_h(\bar{\boldsymbol{w}}_h, \boldsymbol{w}_h, \boldsymbol{\varphi}_h) = \nu_1 \sum_{K \in \mathscr{T}_h} G_K(\bar{\boldsymbol{w}}_h)\, h_K^{\alpha_1} \int_K \nabla \boldsymbol{w}_h \cdot \nabla \boldsymbol{\varphi}_h \, \mathrm{d}x, \tag{8.183}$$

and

$$\boldsymbol{\gamma}_h(\bar{\boldsymbol{w}}_h, \boldsymbol{w}_h, \boldsymbol{\varphi}_h) = \nu_2 \sum_{\Gamma \in \mathscr{F}_h^I} \frac{1}{2}\big(G_{K_\Gamma^{(L)}}(\bar{\boldsymbol{w}}_h) + G_{K_\Gamma^{(R)}}(\bar{\boldsymbol{w}}_h)\big)\, h_\Gamma^{\alpha_2} \int_\Gamma [\boldsymbol{w}_h] \cdot [\boldsymbol{\varphi}_h] \, \mathrm{d}S, \tag{8.184}$$

with the parameters $\alpha_1, \alpha_2, \nu_1, \nu_2 = O(1)$.

The described approach was partly motivated by the theoretical paper [188]. However, the artificial viscosity was applied there in the whole domain, which can lead to a nonphysical entropy production. In our case, it is important that the discrete indicators G_K vanish in regions where the solution is regular and the artificial viscosity acts only locally in the vicinity of discontinuities. Therefore, the scheme does not produce any nonphysical entropy in regions where the exact solution is regular.

The artificial viscosity forms $\boldsymbol{\beta}_h$ and $\boldsymbol{\gamma}_h$ are added to the left-hand side of the numerical schemes presented in previous sections. For example, the backward Euler—discontinuous Galerkin method with shock capturing now reads as

$$\frac{1}{\tau_k}\left(\boldsymbol{w}_h^k - \boldsymbol{w}_h^{k-1}, \boldsymbol{\varphi}_h\right) + \boldsymbol{b}_h(\boldsymbol{w}_h^k, \boldsymbol{\varphi}_h) + \boldsymbol{\beta}_h(\boldsymbol{w}_h^k, \boldsymbol{w}_h^k, \boldsymbol{\varphi}_h) + \boldsymbol{\gamma}_h(\boldsymbol{w}_h^k, \boldsymbol{w}_h^k, \boldsymbol{\varphi}_h) = 0$$
$$\forall \boldsymbol{\varphi}_h \in \boldsymbol{S}_{hp}, \ k = 1, \ldots, r. \tag{8.185}$$

Equalities (8.185) represent a system on nonlinear algebraic equations. In the case when the artificial viscosity forms $\boldsymbol{\beta}_h$ and $\boldsymbol{\gamma}_h$ are defined with the aid of the jump indicator (8.180), the discrete problem can be solved by the Newton-like method, presented in Sect. 8.4.3. Namely, in (8.97), we replace $\boldsymbol{b}_h(\boldsymbol{w}_h^k, \boldsymbol{\varphi}_i)$ by

$$\boldsymbol{b}_h(\boldsymbol{w}_h^k, \boldsymbol{\varphi}_i) + \boldsymbol{\beta}_h(\boldsymbol{w}_h^k, \boldsymbol{w}_h^k, \boldsymbol{\varphi}_i) + \boldsymbol{\gamma}_h(\boldsymbol{w}_h^k, \boldsymbol{w}_h^k, \boldsymbol{\varphi}_i),$$

and, in (8.124), we replace $\boldsymbol{b}_h^L(\bar{\boldsymbol{w}}_h, \boldsymbol{\varphi}_j, \boldsymbol{\varphi}_i)$ by

$$\boldsymbol{b}_h^L(\bar{\boldsymbol{w}}_h, \boldsymbol{\varphi}_j, \boldsymbol{\varphi}_i) + \boldsymbol{\beta}_h(\bar{\boldsymbol{w}}_h, \boldsymbol{\varphi}_j, \boldsymbol{\varphi}_i) + \boldsymbol{\gamma}_h(\bar{\boldsymbol{w}}_h, \boldsymbol{\varphi}_j, \boldsymbol{\varphi}_i).$$

Also in other schemes we proceed in a similar way. The discrete problem with higher-order time discretization and shock capturing reads as

$$\boldsymbol{w}_h^k \in \boldsymbol{S}_{hp}, \ k = 0, 1, \ldots, r, \tag{8.186a}$$

$$\frac{1}{\tau_k}\left(\sum_{l=0}^{n} \alpha_{n,l}\boldsymbol{w}_h^{k-l}, \boldsymbol{\varphi}_h\right) + \boldsymbol{b}_h\left(\boldsymbol{w}_h^k, \boldsymbol{\varphi}_h\right) + \boldsymbol{\beta}_h(\boldsymbol{w}_h^k, \boldsymbol{w}_h^k, \boldsymbol{\varphi}_h) + \boldsymbol{\gamma}_h(\boldsymbol{w}_h^k, \boldsymbol{w}_h^k, \boldsymbol{\varphi}_h) = 0$$
$$\forall \boldsymbol{\varphi}_h \in \boldsymbol{S}_{hp}, \ k = n, \ldots, r, \tag{8.186b}$$

where $\boldsymbol{w}_h^0, \ldots, \boldsymbol{w}_h^{n-1}$ are defined by (8.134c) and (8.134d).

Similarly we formulate the higher-order semi-implicit scheme with shock capturing:

$$\boldsymbol{w}_h^k \in \boldsymbol{S}_{hp}, \ k = 0, 1, \ldots, r, \tag{8.187a}$$

$$\frac{1}{\tau_k}\left(\sum_{l=0}^{n} \alpha_{n,l}\boldsymbol{w}_h^{k-l}, \boldsymbol{\varphi}_h\right) + \hat{\boldsymbol{b}}_h\left(\sum_{l=1}^{n} \beta_{n,l}\boldsymbol{w}_h^{k-l}, \boldsymbol{w}_h^k, \boldsymbol{\varphi}_h\right) + \boldsymbol{\beta}_h\left(\sum_{l=1}^{n} \beta_{n,l}\boldsymbol{w}_h^{k-l}, \boldsymbol{w}_h^k, \boldsymbol{\varphi}_h\right)$$
$$+ \boldsymbol{\gamma}_h\left(\sum_{l=1}^{n} \beta_{n,l}\boldsymbol{w}_h^{k-l}, \boldsymbol{w}_h^k, \boldsymbol{\varphi}_h\right) = 0 \quad \forall \boldsymbol{\varphi}_h \in \boldsymbol{S}_{hp}, \ k = n, \ldots, r, \tag{8.187b}$$

where $\boldsymbol{w}_h^0, \ldots, \boldsymbol{w}_h^{n-1}$ are defined by (8.134c) and (8.134d). Problem (8.187) represents again a sequence of systems of linear algebraic equations. In this case the artificial viscosity can be defined by any jump indicator introduced in Sect. 8.5.1.

8.5.3 Numerical Examples

In this section we present the solution of some test problems showing the performance of the shock capturing technique introduced above.

We consider transonic inviscid flow past the profile NACA 0012 given by the parametrization

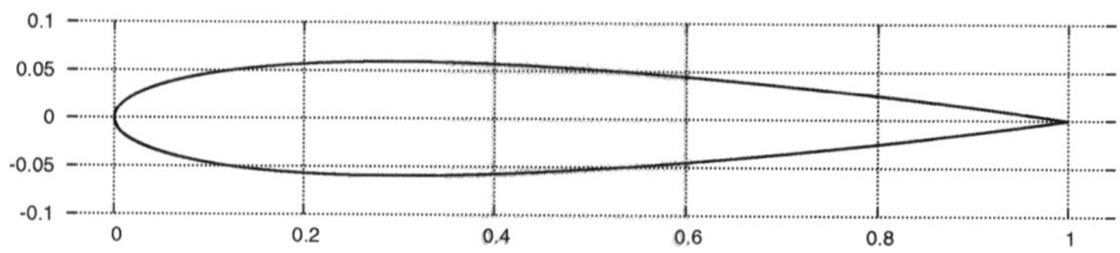

Fig. 8.5 Geometry of the NACA 0012 profile

$$\left[x, \pm\frac{0.12}{0.6}(0.2969\sqrt{x} - 0.126x - 0.3516x^2 + 0.2843x^3 - 0.1015x^4)\right], \; x \in [0, 1],$$

see Fig. 8.5. We consider the far-field Mach number $M_\infty = 0.8$ (see (8.7)) and the angle of attack $\alpha = 1.25°$. (Let us note that $\tan\alpha = v_2/v_1$, where (v_1, v_2) is the far-field velocity vector.) This flow regime leads to two shock waves (discontinuities in the solution). The shock wave on the upper side of the profile is stronger than the shock wave on the lower side.

We seek the steady-state solution of the Euler equations (8.8) with the aid of the time stabilization technique described in Sect. 8.4.9, using the backward Euler—discontinuous Galerkin method (BE-DGM) (8.95). The nonlinear algebraic systems are solved by the Newton-like iterative process (8.127)–(8.128).

We employ two unstructured triangular grids with piecewise polynomial approximation of the boundary described in Sect. 8.6. The first grid is formed by 2120 triangles and is not adapted. The second one with 2420 elements was adaptively refined along the shock waves by ANGENER code [84] developed in papers [83, 85, 100]. See Fig. 8.6. The problem was solved by the DGM using the P_p polynomial approximations with $p = 1, 2, 3$.

Figure 8.7 shows the Mach number isolines and the distribution of the Mach number along the profile in dependence on the horizontal component obtained with the aid of the P_1 and P_2 approximation on the non-adapted mesh without the shock

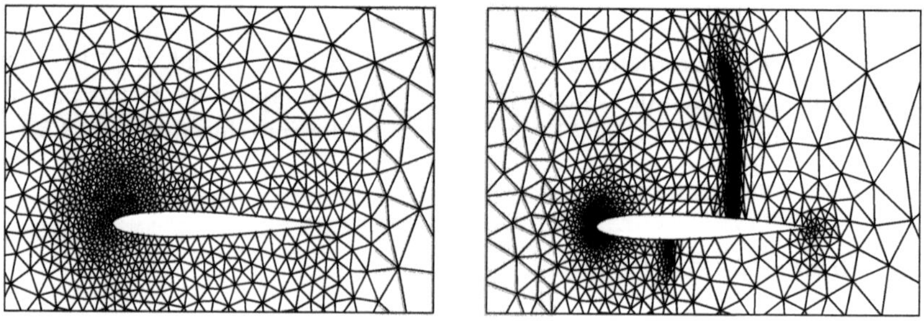

Fig. 8.6 Transonic inviscid flow around the NACA 0012 profile ($M_\infty = 0.8$, $\alpha = 1.25°$): the non-adapted (*left*) and the adapted (*right*) computational meshes

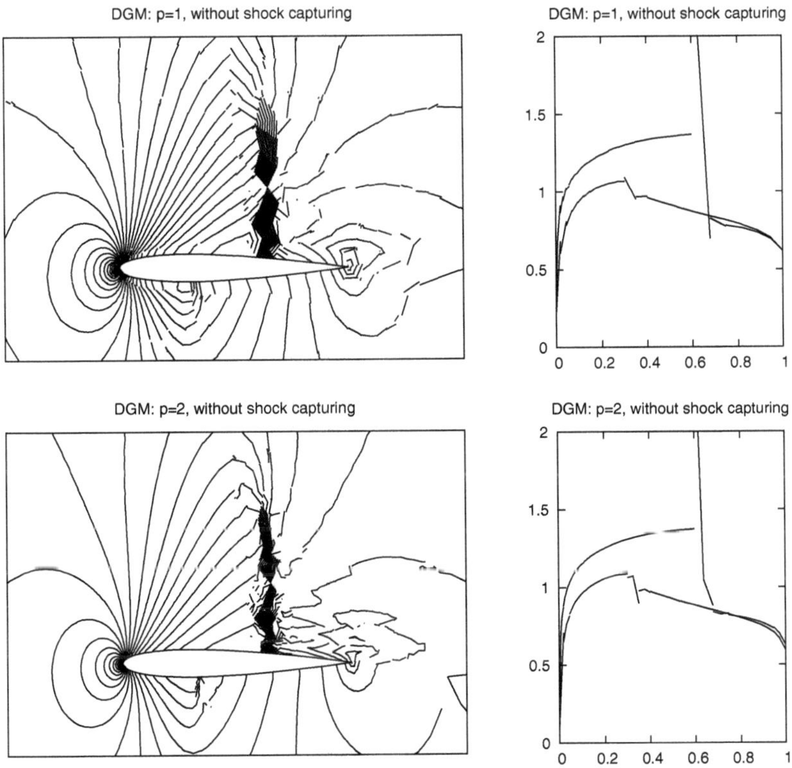

Fig. 8.7 Transonic inviscid flow around the NACA 0012 profile ($M_\infty = 0.8$, $\alpha = 1.25°$): DGM with P_1 approximation (*top*) and P_2 approximation (*bottom*), Mach number isolines (*left*) and the distribution of the Mach number along the profile (*right*) on a non-adapted mesh without the shock capturing technique

capturing technique. We observe overshoots and undershoots in the approximate solution near the shock waves. Let us note that the P_3 computation failed.

Figure 8.8 shows the results obtained with the aid of the P_1, P_2 and P_3 approximations on the non-adapted mesh with the shock capturing technique. We can see that the nonphysical overshoots and undershoots are mostly suppressed. Finally, Fig. 8.9 shows the results for P_1, P_2 and P_3 approximations on the adapted mesh with the shock capturing technique. We see that a very good resolution of the shock waves was obtained.

Further numerical experiment can be found in Sect. 8.7.4, where an example of the supersonic flow past the NACA 0012 profile is presented.

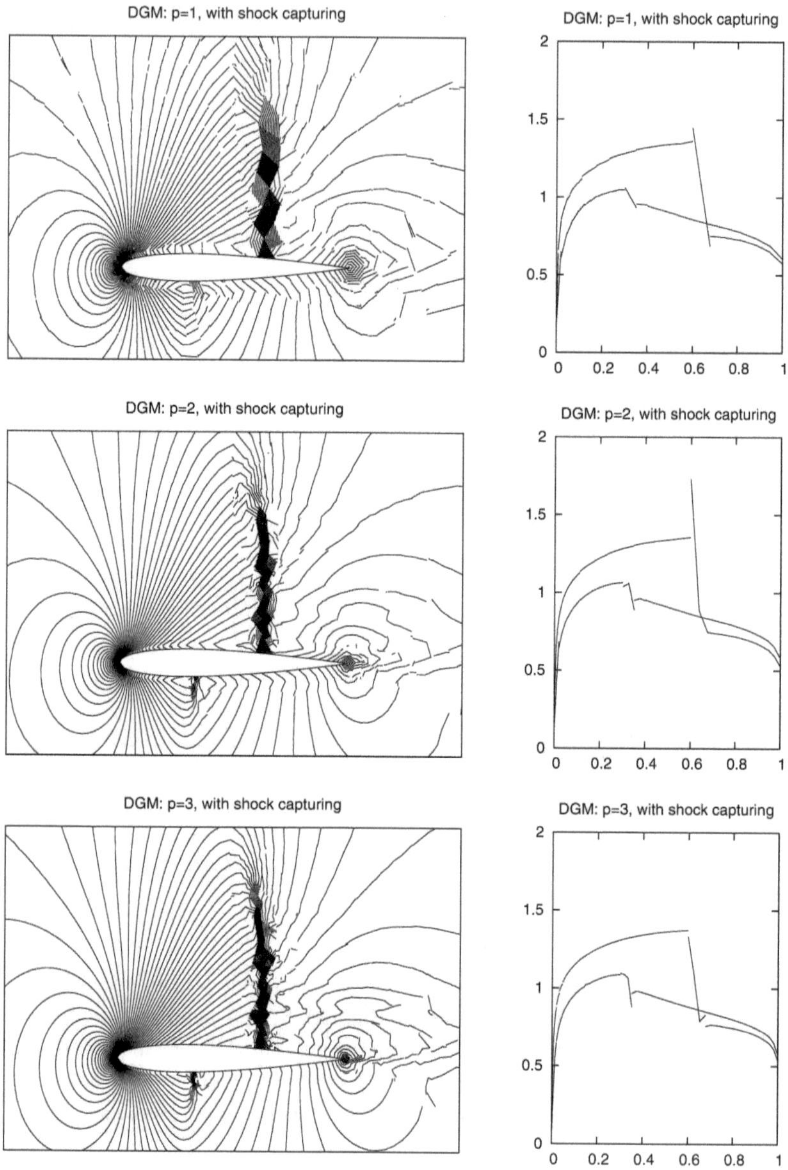

Fig. 8.8 Transonic inviscid flow around the NACA 0012 profile ($M_\infty = 0.8, \alpha = 1.25°$): DGM with P_1 approximation (*top*), P_2 approximation (*center*) and P_3 approximation (*bottom*), Mach number isolines (*left*) and the distribution of the Mach number along the profile (*right*) on a **non-adapted** mesh with the shock capturing technique

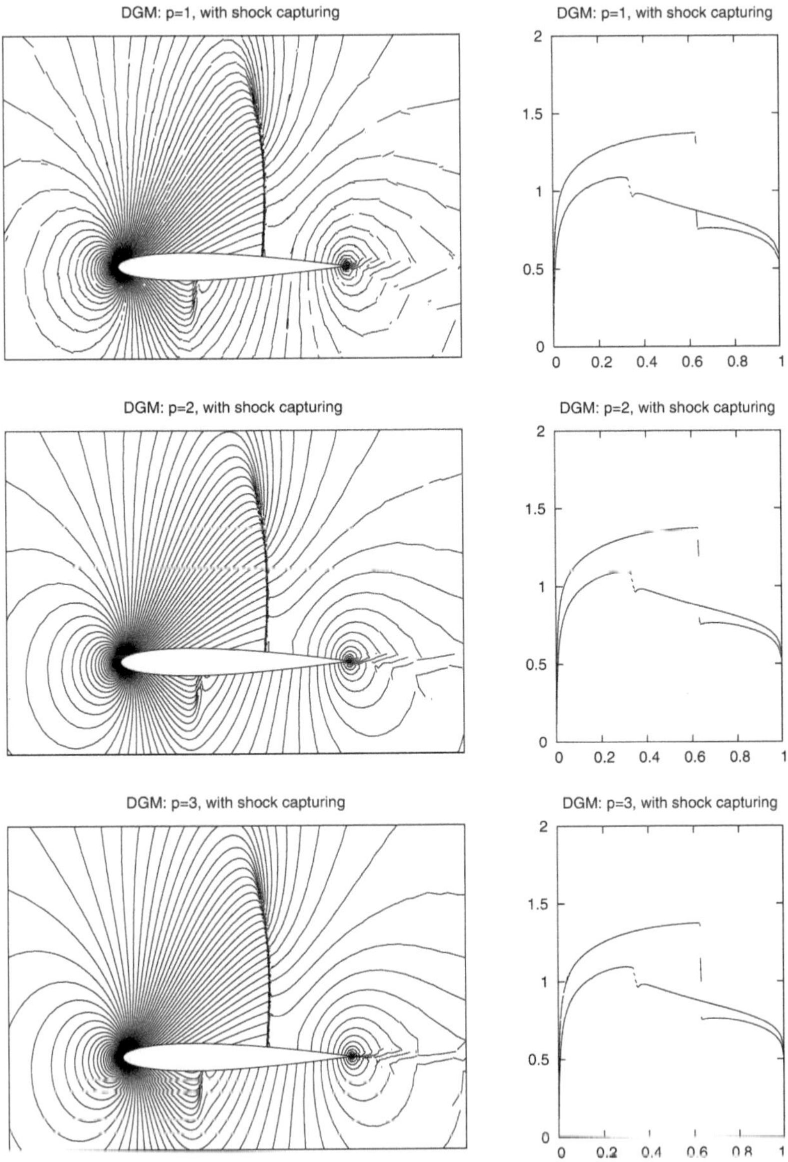

Fig. 8.9 Transonic inviscid flow around the NACA 0012 profile ($M_\infty = 0.8$, $\alpha = 1.25°$): DGM with P_1 approximation (*top*), P_2 approximation (*center*) and P_3 approximation (*bottom*) and with boundary approximation, Mach number isolines (*left*) and the distribution of the Mach number along the profile (*right*) on an **adapted** mesh with the shock capturing technique

8.6 Approximation of a Nonpolygonal Boundary

In practical applications, the computational domain Ω is usually nonpolygonal, and thus its boundary has to be approximated in some way. In [25], Bassi and Rebay showed that a piecewise linear approximation of $\partial\Omega$ can lead to a nonphysical production of entropy and expansion waves at boundary corner points, leading to incorrect numerical solutions. In order to obtain an accurate and physically admissible solution, it is necessary to use a higher-order approximation of the boundary. We proceed in such a way that a reference triangle is transformed by a polynomial mapping onto the approximation of a curved triangle adjacent to the boundary $\partial\Omega$.

8.6.1 Curved Elements

Here we describe only the two dimensional ($d = 2$) situation, the case $d = 3$ has to be generalized in a suitable way. Let K be a triangle with vertices P_K^l, $l = 1, 2, 3$, numbered in a such way that P_K^1 and P_K^2 lie on a curved part of $\partial\Omega$ and P_K^3 lies in the interior of Ω. By Γ we denote the edge $P_K^1 P_K^2$. Moreover, we assume that P_K^1 and P_K^2 are oriented in such a way that Ω is on the left-hand side of the oriented edge from P_K^1 to P_K^2, see Fig. 8.10. We consider elements having at most one curved edge. The generalization to the case with elements having more curved edges is straightforward.

Let $q \geq 2$ be an integer denoting the *polynomial degree of the boundary approximation*. We define $q - 1$ nodes $P_K^{C,j}$, $j = 1, \ldots, q - 1$, lying on $\partial\Omega$ between P_K^1 and P_K^2 in such a way that nodes $P_K^{C,j}$, $j = 1, \ldots, q - 1$, divide the curved segment of $\partial\Omega$ between P_K^1 and P_K^2 into q parts having (approximately) the same length. We assume that $P_K^{C,j}$, $j = 1, \ldots, q - 1$, are ordered with an increasing index on the

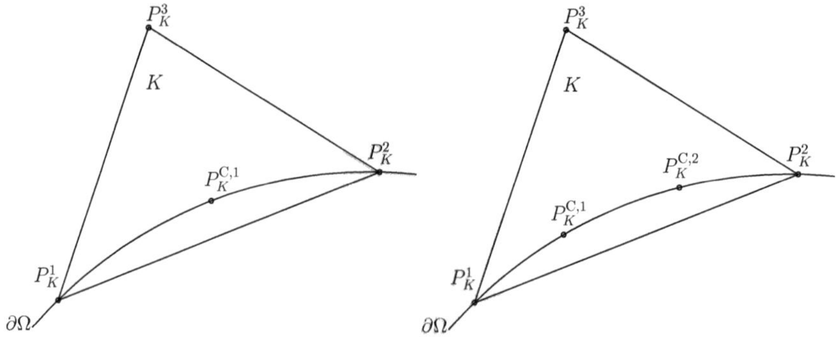

Fig. 8.10 Triangle K with vertices P_k^1 and P_K^2 lying on a nonpolygonal part of $\partial\Omega$; adding one (*left*) and two (*right*) nodes on $\partial\Omega$

path along $\partial\Omega$ from P_K^1 to P_K^2. See Fig. 8.10 showing a possible situation for $q = 2$ and $q = 3$.

Let

$$\hat{K} = \{(\hat{x}_1, \hat{x}_2);\ \hat{x}_i \geq 0,\ i = 1, 2,\ \hat{x}_1 + \hat{x}_2 \leq 1\} \tag{8.188}$$

be the *reference triangle*. In $\hat{K}$, we define the Lagrangian nodes of degree q by

$$\hat{P}^{\frac{i}{q};\frac{j}{q}} = [i/q;\ j/q],\quad 0 \leq i \leq q,\ 0 \leq j \leq q,\ 0 \leq i + j \leq q, \tag{8.189}$$

i.e., the vertices of $\hat{K}$ are the points $\hat{P}^{0;0}$, $\hat{P}^{0;1}$ and $\hat{P}^{1;0}$.

Let K be the triangle with vertices P_K^l, $l = 1, 2, 3$, and let $P_K^{C,j} \in \partial\Omega$, $j = 1, \ldots, q-1$, be the points lying on $\partial\Omega$ between P_K^1 and P_K^2 as described above. We define the Lagrangian nodes of degree q of K by

$$P_K^{\frac{i}{q};\frac{j}{q}} = \frac{i}{q} P_K^1 + \frac{j}{q} P_K^2 + \frac{1-i-j}{q} P_K^3,\quad 0 \leq i \leq q,\ 0 \leq j \leq q,\ 0 \leq i + j \leq q. \tag{8.190}$$

Obviously, $P_K^{0;0} = P_K^1$, $P_K^{1;0} = P_K^2$ and $P_K^{0;1} = P_K^3$.

Then, there exists a unique polynomial mapping $F_K : \hat{K} \to \mathbb{R}^2$ of degree $\leq q$ such that

$$\begin{aligned}
&F_K(\hat{P}^{0;0}) = P_K^1,\quad F_K(\hat{P}^{1;0}) = P_K^2,\quad F_K(\hat{P}^{0;1}) = P_K^3 \quad \text{are vertices,}\\
&F_K(\hat{P}^{\frac{i}{q};0}) = P_K^{C,i},\quad i = 1, \ldots, q-1,\quad \text{are nodes on the curved edge,}\\
&F_K(\hat{P}^{\frac{i}{q};\frac{j}{q}}) = P_K^{\frac{i}{q};\frac{j}{q}},\quad 0 \leq i \leq q,\ 1 \leq j \leq q-1,\ 0 \leq i + j \leq q,\quad \text{are other nodes.}
\end{aligned} \tag{8.191}$$

The existence and uniqueness of the mapping F_K follows from the fact that a polynomial mapping of degree q from $\mathbb{R}^2$ to $\mathbb{R}^2$ has $(q+1)(q+2)$ degrees of freedom equal to the number of conditions in (8.191). Then we obtain a linear algebraic system, which is regular, since the Lagrangian nodes on $\hat{K}$ are mutually different and at most q nodes belong to any straight line.

Then the triangle K will be replaced by the *curved triangle*

$$\tilde{K} = F_K(\hat{K}). \tag{8.192}$$

The set $\tilde{K}$ is a plane figure having two straight sides and one curved side $\tilde{\Gamma}$, which is an image of the *reference edge* $\hat{P}^{0;0}\,\hat{P}^{1;0}$, see Fig. 8.11.

Using the described procedure, we get a partition $\tilde{\mathscr{T}}_h$ associated with the triangulation $\mathscr{T}_h$. The partition $\tilde{\mathscr{T}}_h$, called the *curved triangulation*, consists of triangles $K \in \mathscr{T}_h$ and curved elements $\tilde{K}$, associated with triangles $K \in \mathscr{T}_h$ with one edge approximating a curved part of $\partial\Omega$.

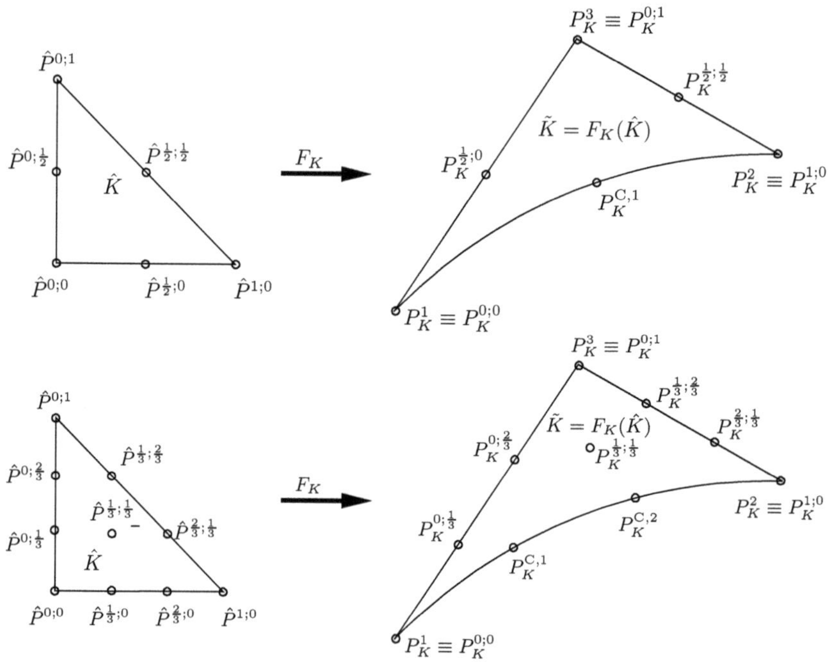

Fig. 8.11 Mapping $F_K : \hat{K} \to \tilde{K}$: quadratic (*top*) and cubic (*bottom*)

Remark 8.25 Let us note that the considerations presented in this section make sense also for $q = 1$. In this case, any node $P_K^{C,i}$ that is not inserted on $\partial\Omega$, mapping F_K given by (8.191) is linear and $\tilde{K} = F_K(\hat{K}) = K$ is the triangle with straight edges.

Remark 8.26 The concept of the curved element can be extended also to 3D by defining a polynomial mapping F_K from a reference tetrahedron $\hat{K}_{3D}$ into $\mathbb{R}^3$ for each tetrahedron K with one face approximating a curved part of $\partial\Omega$. Then K is replaced by $F_K(\hat{K}_{3D})$.

8.6.2 DGM Over Curved Elements

Let $\tilde{\mathscr{T}}_h$ be a curved triangulation consisting of (non-curved) simplexes K as well as possible curved elements $\tilde{K}$. By virtue of Remark 8.25, a non-curved element can be considered as a special curved simplex obtained by a linear ($q = 1$) mapping F_K. Therefore, we do not distinguish between curved and non-curved elements in the following and we use the symbol K also for curved elements. Moreover, instead of $\tilde{\mathscr{T}}_h$, we write $\mathscr{T}_h$.

Since $\mathscr{T}_h$ may contain curved elements, we have to modify the definition (8.44) of the space $\boldsymbol{S}_{hp}$. For an integer $p \geq 0$, over the triangulation $\mathscr{T}_h$ we define the finite-dimensional function space

$$\boldsymbol{S}_{hp} = (S_{hp})^m, \quad S_{hp} = \{v;\ v \in L^2(\Omega),\ v|_K \circ F_K \in P_p(\hat{K})\ \forall\, K \in \mathscr{T}_h\}, \quad (8.193)$$

where $P_p(\hat{K})$ denotes the space of all polynomials of degree $\leq p$ on the reference element $\hat{K}$ and the symbol $\circ$ denotes the composition of mappings. Hence, instead of (8.44) and (8.45), we employ definition (8.193).

Remark 8.27 The definition (8.193) of the space $\boldsymbol{S}_{hp}$ implies that for a curved element K, the function $\boldsymbol{w}_h|_K$ is not a polynomial of degree $\leq p$. Moreover, if all $K \in \mathscr{T}_h$ are non-curved (i.e., F_K are linear for all $K \in \mathscr{T}_h$), then the spaces defined by (8.193) are identical with the spaces defined by (8.44) and (8.45).

Now let us describe how to evaluate the volume and boundary integrals over elements K and their sides Γ. We denote by

$$J_{F_K}(\hat{x}) = \frac{D\,F_K}{D\,\hat{x}}(\hat{x}), \quad \hat{x} \in \hat{K}, \quad (8.194)$$

the Jacobian matrix of the mapping F_K. Since F_K is a polynomial mapping of degree q, J_{F_K} is a polynomial mapping of degree $q-1$ in the variable $\hat{x} = (\hat{x}_1, \hat{x}_2)$. The components of the vector-valued test functions $\boldsymbol{\varphi}_h \in \boldsymbol{S}_{hp}$ from (8.193) are defined on the curved elements K (adjacent to the boundary $\partial\Omega$) with the aid of the mapping F_K. Hence, for each $\boldsymbol{\varphi}_h \in \boldsymbol{S}_{hp}$ and each $K \in \mathscr{T}_h$ there exists a function $\hat{\boldsymbol{\varphi}}_K \in (P_p(\hat{K}))^m$ such that

$$\hat{\boldsymbol{\varphi}}_K(\hat{x}) = \boldsymbol{\varphi}_h(F_K(\hat{x})), \quad \hat{x} \in \hat{K}. \quad (8.195)$$

In the following, we describe how to evaluate the volume and face integrals appearing in the definition of the forms $\boldsymbol{b}_h$ and $\boldsymbol{b}_h^L$ given by (8.93) and (8.123), respectively. Evaluating the integrals is based on the transformation to the reference element (or reference edge) with the aid of the substitution theorem.

8.6.2.1 Volume Integrals

The volume integral of a product of two (or more) functions is simply expressed as

$$\int_K \boldsymbol{w}_h(x,t) \cdot \boldsymbol{\varphi}_h(x)\,\mathrm{d}x = \int_{\hat{K}} \hat{\boldsymbol{w}}_K(\hat{x},t) \cdot \hat{\boldsymbol{\varphi}}_K(\hat{x}) |\det J_{F_K}(\hat{x})|\,\mathrm{d}\hat{x}, \quad K \in \mathscr{T}_h,\ t \in (0,T), \quad (8.196)$$

where $\hat{\boldsymbol{w}}_K(\hat{x},t) = \boldsymbol{w}_h|_K(F_K(\hat{x},t))$ and $\hat{\boldsymbol{\varphi}}_K$ is given by (8.195).

Moreover, the evaluation of the volume integral of a product of a function and the gradient of a function requires a transformation of the gradient with respect to the variable x to the gradient with respect to $\hat{x}$. Hence, we obtain

$$\int_K \sum_{s=1}^{d} f_s(w_h(x,t)) \cdot \frac{\partial \varphi_h(x)}{\partial x_s} \, dx \tag{8.197}$$

$$= \int_{\hat{K}} \sum_{s=1}^{d} f_s(\hat{w}_K(\hat{x},t)) \cdot \sum_{j=1}^{d} \frac{\partial \hat{\varphi}_K(\hat{x})}{\partial \hat{x}_j} \frac{\partial F_{K,j}^{-1}(F_K(\hat{x}))}{\partial x_s} |\det J_{F_K}(\hat{x})| \, d\hat{x}, \ K \in \mathscr{T}_h, \ t \in (0,T),$$

where $F_{K,j}^{-1}$ denotes the jth component of the inverse mapping F_K^{-1}. In order to compute the inverse mapping F_K^{-1}, we use the following relation written in the matrix form:

$$\frac{D F_K^{-1}}{Dx}(F_K(\hat{x})) = \left(\frac{D F_K}{D\hat{x}}(\hat{x}) \right)^{-1} \tag{8.198}$$

following from the identity $x = F_K(F_K^{-1}(x))$. The computation of the inverse matrix in (8.198) is simpler than the evaluation of F_K^{-1}.

8.6.2.2 Face Integrals

Finally, we describe the evaluation of face integrals along a curved edge in $\mathbb{R}^2$. The three-dimensional case can be generalized in a natural way. Let $\Gamma \in \mathscr{F}_h$ be a (possibly curved) edge of $K \in \mathscr{T}_h$. Our aim is to evaluate the integrals

$$\int_\Gamma f(x) \, dS, \qquad \int_\Gamma \boldsymbol{f}(x) \cdot \boldsymbol{n}(x) \, \varphi(x) \, dS, \tag{8.199}$$

where $\boldsymbol{n}$ is the normal vector to Γ and $f : \Gamma \to \mathbb{R}$, $\boldsymbol{f} : \Gamma \to \mathbb{R}^2$ are given functions. Such type of integral appears in (8.93) in terms containing the numerical flux. Let us recall the definition of the face integral. If $\psi = (\psi_1, \psi_2) : [0,1] \to \Gamma$ is a parameterization of the edge Γ, then

$$\int_\Gamma f(x) \, dS = \int_0^1 f(\psi(\xi)) \sqrt{(\psi_1'(\xi))^2 + (\psi_2'(\xi))^2} \, d\xi, \tag{8.200}$$

where $\psi_i'(\xi)$, $i = 1, 2$, denotes the derivative of $\psi_i(\xi)$ with respect to ξ.

Integrals (8.199) are evaluated with the aid of a transformation to the reference element. Let $\hat{\Gamma}$ be an edge of the reference element $\hat{K}$ such that $K = F_K(\hat{K})$ and $\Gamma = F_K(\hat{\Gamma})$. We call $\hat{\Gamma}$ the *reference edge*. Let

$$x_{\hat{\Gamma}}(\xi) = (x_{\hat{\Gamma},1}(\xi), x_{\hat{\Gamma},2}(\xi)) : [0,1] \to \hat{\Gamma} \tag{8.201}$$

be a parametrization of the reference edge $\hat{\Gamma}$ preserving the counterclockwise orientation of the element boundary. Namely, the reference triangle given by (8.159) (with $d = 2$) has three reference edges parametrized by

$$\begin{aligned} x_{\hat{\Gamma}_1}(\xi) &= (\xi, 0), \quad \xi \in [0, 1], \\ x_{\hat{\Gamma}_2}(\xi) &= (1 - \xi, \xi), \quad \xi \in [0, 1], \\ x_{\hat{\Gamma}_3}(\xi) &= (0, 1 - \xi), \quad \xi \in [0, 1]. \end{aligned} \tag{8.202}$$

Moreover, we use the notation $\dot{x}_{\hat{\Gamma}}(\xi) = \frac{\mathrm{d}}{\mathrm{d}\xi} x_{\hat{\Gamma}}(\xi) \in \mathbb{R}^2$ and have

$$\begin{aligned} \dot{x}_{\hat{\Gamma}_1} &= (1, 0), \quad \xi \in [0, 1], \\ \dot{x}_{\hat{\Gamma}_2} &= (-1, 1), \quad \xi \in [0, 1], \\ \dot{x}_{\hat{\Gamma}_3} &= (0, -1), \quad \xi \in [0, 1]. \end{aligned} \tag{8.203}$$

Therefore, the edge Γ is parameterized by

$$\begin{aligned} x = F_K(x_{\hat{\Gamma}}(\xi)) &= \left(F_{K,1}(x_{\hat{\Gamma}}(\xi)), F_{K,2}(x_{\hat{\Gamma}}(\xi))\right) \\ &= \left(F_{K,1}(\hat{x}_{\hat{\Gamma},1}(\xi), \hat{x}_{\hat{\Gamma},2}(\xi)), F_{K,2}(\hat{x}_{\hat{\Gamma},1}(\xi), \hat{x}_{\hat{\Gamma},2}(\xi))\right), \quad \xi \in [0, 1]. \end{aligned} \tag{8.204}$$

The first integral in (8.199) is transformed by

$$\begin{aligned} \int_\Gamma f(x)\,\mathrm{d}S &= \int_0^1 f(F_K(x_{\hat{\Gamma}}(\xi))) \left(\sum_{i=1}^2 \left(\frac{\mathrm{d}}{\mathrm{d}\xi} F_{K,i}(x_{\hat{\Gamma}}(\xi))\right)^2\right)^{1/2} \mathrm{d}\xi \\ &= \int_0^1 f(F_K(x_{\hat{\Gamma}}(\xi))) \left(\sum_{i,j=1}^2 \left(\frac{\partial F_{K,i}(x_{\hat{\Gamma}}(\xi))}{\partial \hat{x}_j} \dot{\hat{x}}_{\hat{\Gamma},j}(\xi)\right)^2\right)^{1/2} \mathrm{d}\xi \\ &= \int_0^1 f(F_K(x_{\hat{\Gamma}}(\xi))) \left|J_{F_K}(x_{\hat{\Gamma}}(\xi))\dot{x}_{\hat{\Gamma}}\right| \mathrm{d}\xi, \end{aligned} \tag{8.205}$$

where J_{F_K} is the Jacobian matrix of the mapping F_K multiplied by the vector $\dot{x}_{\hat{\Gamma}}$ given by (8.203) and $|\cdot|$ is the Euclidean norm of the vector. Let us note that if F_K is a linear mapping, then e is a straight edge and $\left|J_{F_K}(x_{\hat{\Gamma}}(\xi))\dot{x}_{\hat{\Gamma}}(\xi)\right|$ is equal to its length.

Now, we focus on the second integral from (8.199). Let $\boldsymbol{t}_\Gamma$ be the tangential vector to Γ defined by

$$\begin{aligned} \boldsymbol{t}_\Gamma(x(\xi)) &= (t_{\Gamma,1}(x(\xi)), t_{\Gamma,2}(x(\xi))) \\ &= \frac{\mathrm{d}}{\mathrm{d}\xi} F_K(x_{\hat{\Gamma}}(\xi)) = \left(J_{F_{K,1}}(x_{\hat{\Gamma}}(\xi))\dot{x}_{\hat{\Gamma}}(\xi), J_{F_{K,2}}(x_{\hat{\Gamma}}(\xi))\dot{x}_{\hat{\Gamma}}(\xi)\right). \end{aligned} \tag{8.206}$$

(If Γ is a straight line, then $\boldsymbol{t}_\Gamma$ is constant on Γ, it has the orientation of Γ and $|\boldsymbol{t}_\Gamma| = |\Gamma|$.) Now, by the rotation we obtain the normal vector $\boldsymbol{n}_\Gamma$ pointing outside of K, namely

$$\boldsymbol{n}_\Gamma(x(\xi)) = (n_{\Gamma,1}(x(\xi)), n_{\Gamma,2}(x(\xi))), \qquad (8.207)$$
$$n_{\Gamma,1}(x(\xi)) = t_{\Gamma,2}(x(\xi)), \;\; n_{\Gamma,2}(x(\xi)) = -t_{\Gamma,1}(x(\xi)).$$

Here it is important that the counter-clockwise orientation of the elements is considered. Therefore, from (8.206) and (8.207), we have

$$\boldsymbol{n}_\Gamma(x(\xi)) = \left(J_{F_{K,2}}(x_{\hat{\Gamma}}(\xi))\dot{x}_{\hat{\Gamma}}(\xi), -J_{F_{K,1}}(x_{\hat{\Gamma}}(\xi))\dot{x}_{\hat{\Gamma}}(\xi)\right). \qquad (8.208)$$

Let us note that because $\boldsymbol{n}_\Gamma(x(\xi))$ is not normalized, it is necessary to divide it by $|\boldsymbol{n}_\Gamma(x(\xi))| = |J_{F_K}(x_{\hat{\Gamma}}(\xi))\dot{x}_{\hat{\Gamma}}(\xi)|$. Finally, similarly as in (8.205), we obtain

$$\int_\Gamma \boldsymbol{f}(x) \cdot \boldsymbol{n}(x)\, \boldsymbol{\varphi}(x)\, \mathrm{d}S \qquad (8.209)$$
$$= \int_0^1 \boldsymbol{f}(F_K(x_{\hat{\Gamma}}(\xi))) \cdot \frac{\boldsymbol{n}_\Gamma(x(\xi))}{|\boldsymbol{n}_\Gamma(x(\xi))|} \left|J_{F_K}(x_{\hat{\Gamma}}(\xi))\dot{x}_{\hat{\Gamma}}(\xi)\right| \boldsymbol{\varphi}(F_K(x_{\hat{\Gamma}}(\xi)))\, \mathrm{d}t\xi$$
$$= \int_0^1 \boldsymbol{f}(F_K(x_{\hat{\Gamma}}(\xi))) \cdot \boldsymbol{n}_\Gamma(x(\xi))\, \hat{\boldsymbol{\varphi}}(x_{\hat{\Gamma}}(\xi))\, \mathrm{d}\xi,$$

where $\boldsymbol{n}_\Gamma(x(\xi))$ is given by (8.208) and $\hat{\boldsymbol{\varphi}}$ was obtained by transformation of the function $\boldsymbol{\varphi}$: $\hat{\boldsymbol{\varphi}}(\hat{x}) = \boldsymbol{\varphi}(F_K(\hat{x}))$. Let us note that if F_K is a linear mapping, then Γ is a straight edge and $|\boldsymbol{n}_\Gamma(x(\xi))|$ is equal to its length.

8.6.2.3 Implementation Aspects of Curved Elements

The integrals over the reference triangle $\hat{K}$ and over the reference edge $\hat{\Gamma}$ in (8.196), (8.197), (8.205) and (8.209) are evaluated with the aid of suitable numerical quadratures. For the volume integrals we can employ the *Dunavant quadrature rules* [111], which give the optimal order of accuracy of the numerical integration. For face integrals the well-known *Gauss quadrature rules*, having the maximal degree of approximation for the given number of integration nodes, can be used. For other possibilities, we refer to [260].

Finally, let us mention the data structure in the implementation. Let $\hat{p}$ be an integer denoting the maximal implemented degree of the polynomial approximation in the DGM. We put $\hat{N} = (\hat{p}+1)(\hat{p}+2)/2$ denoting the corresponding maximal number of degrees of freedom for one element and one component of $\boldsymbol{w}$ for $d = 2$. Hence, in order to evaluate integrals appearing in (8.93) and (8.123) with the aid of the techniques presented above and with the aid of numerical quadratures, it is enough to evaluate (and store) the following quantities:

- for each $K \in \mathscr{T}_h$, the determinant $\det J_{F_K}$ of the Jacobi matrix and the transposed matrix to the inversion of the Jacobi matrix $J_{J_{F_K}}$ evaluated at the used *edge and volume quadrature nodes*,
- the reference basis functions $\hat{\varphi}_i(\hat{x})$, $i = 1, \ldots, \hat{N}$, with their partial derivatives $\partial\hat{\varphi}_i(\hat{x})/\partial\hat{x}_j$, $j = 1, 2$, $i = 1, \ldots \hat{N}$, on $\hat{K}$ evaluated at the used *edge and volume quadrature nodes*.

8.6.3 Numerical Examples

In this section we present the results of numerical experiments demonstrating the influence of higher-order approximation of the nonpolygonal boundary. We consider an inviscid flow around the NACA 0012 profile with the far-field Mach number $M_\infty = 0.5$ (see (8.7)) and the angle of attack $\alpha = 2°$. We seek the steady-state solution of the Euler equations (8.8) with the aid of the time stabilization described in Sect. 8.4.9, using the BE-DGM (8.95) combined with the Newton-like iterations (8.127)–(8.128).

The computation was performed on a coarse unstructured triangular grid having 507 elements, refined around the leading edge of the profile by the ANGENER code [84] (see Fig. 8.12). The polynomial approximations P_p, $p = 1, 3, 5$, in the DGM and the polynomial approximations P_q, $q = 1, 2, 3$, of the boundary described in Sect. 8.6 were used. Figures 8.13, 8.14 and 8.15 show results of these computations, namely Mach number isolines and the Mach number distribution along the profile.

We observe that the P_1 approximation of the boundary produces nonphysical oscillations in the solution. This unpleasant behaviour disappears for P_2 or P_3 approximation of the boundary. There is almost no difference between P_2 and P_3. Finally, it is possible to see that the high-order DG approximation (P_5) gives very smooth isolines even on a coarse grid.

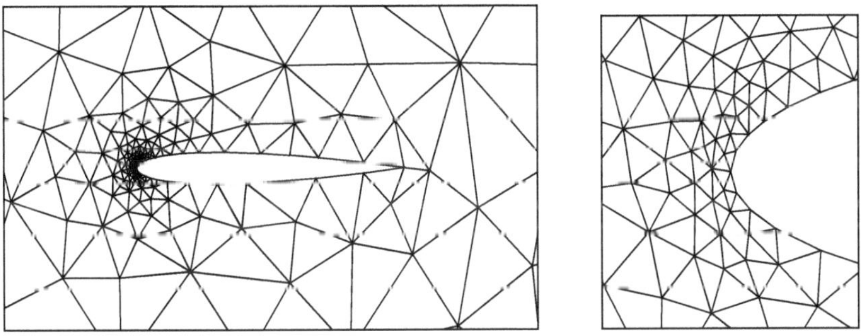

Fig. 8.12 Subsonic inviscid flow around the NACA 0012 profile ($M_\infty = 0.5$, $\alpha = 2°$): computational mesh, detail around the whole profile (*left*) and around the leading edge (*right*)

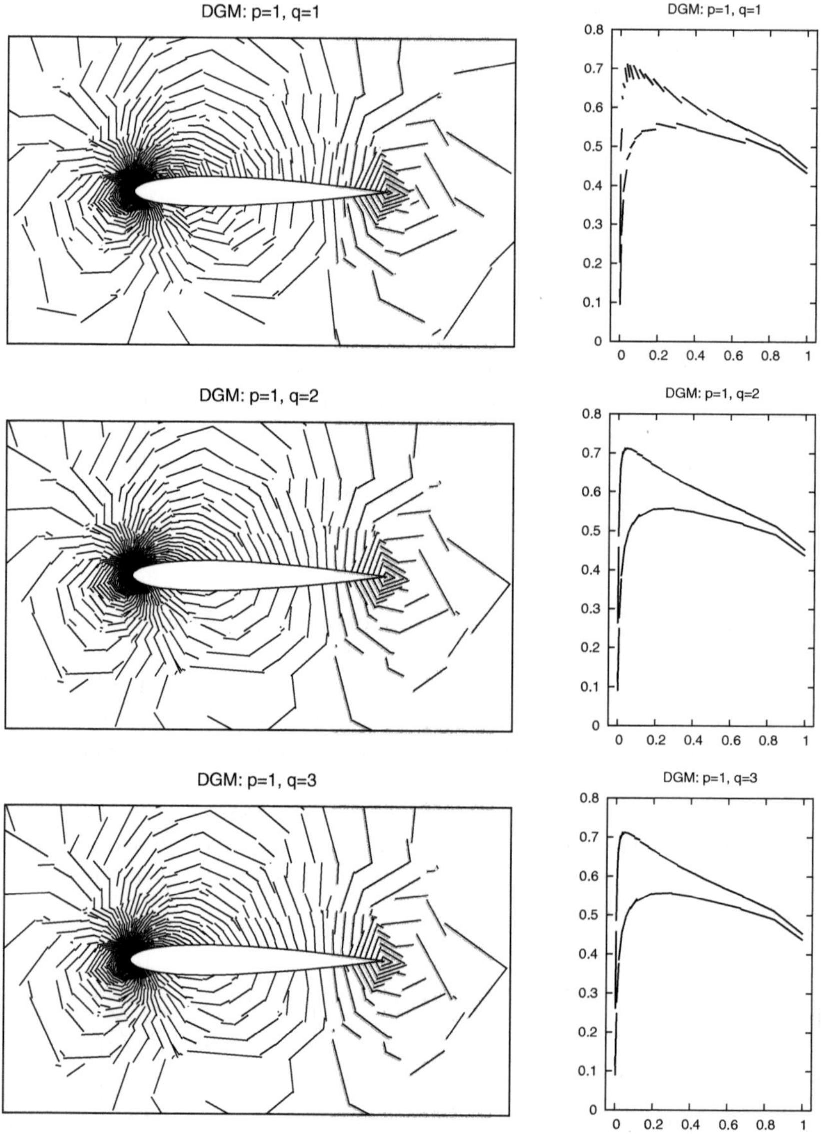

Fig. 8.13 Subsonic inviscid flow around the NACA 0012 profile ($M_\infty = 0.5, \alpha = 2°$): DGM with polynomial approximation with $p = 1$, boundary approximation with $q = 1$ (*top*), $q = 2$ (*center*) and $q = 3$ (*bottom*), Mach number isolines (*left*) and the Mach number distribution around the profile (*right*)

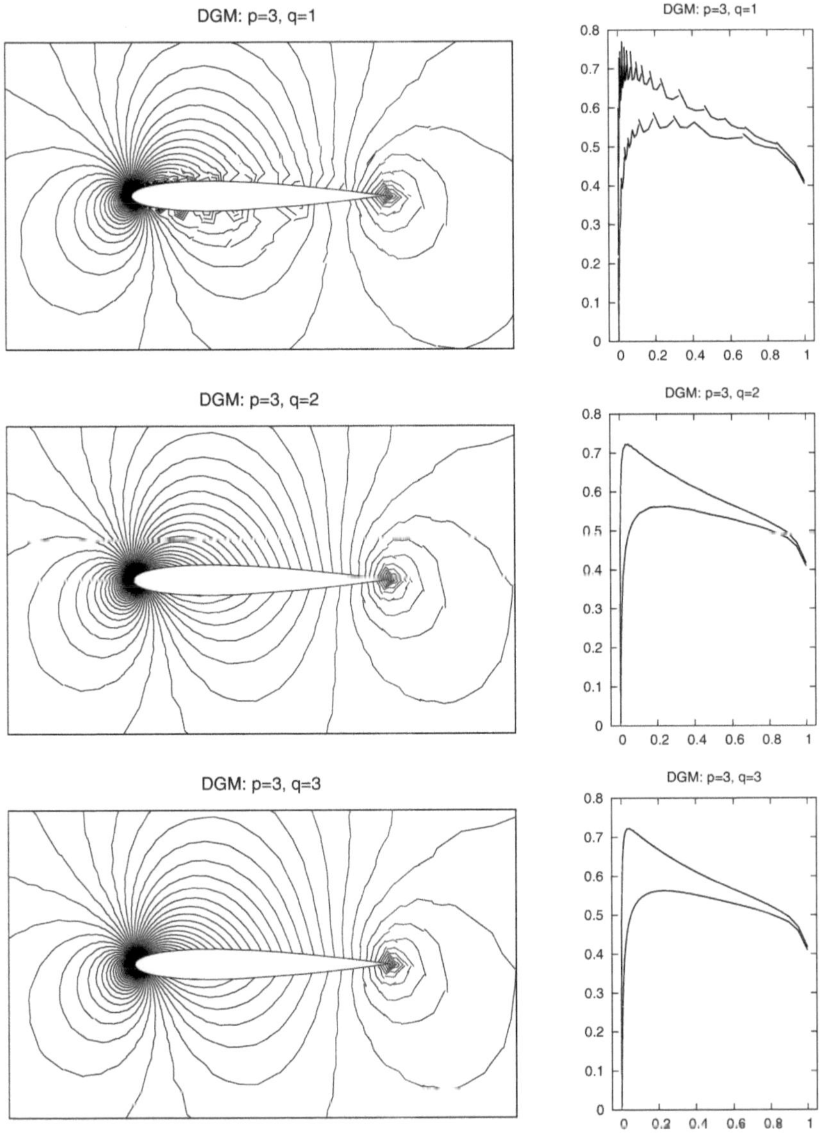

Fig. 8.14 Subsonic inviscid flow around the NACA 0012 profile ($M_\infty = 0.5, \alpha = 2°$): DGM with polynomial approximation with $p = 3$, boundary approximation with $q = 1$ (*top*), $q = 2$ (*center*) and $q = 3$ (*bottom*), Mach number isolines (*left*) and the Mach number distribution around the profile (*right*)

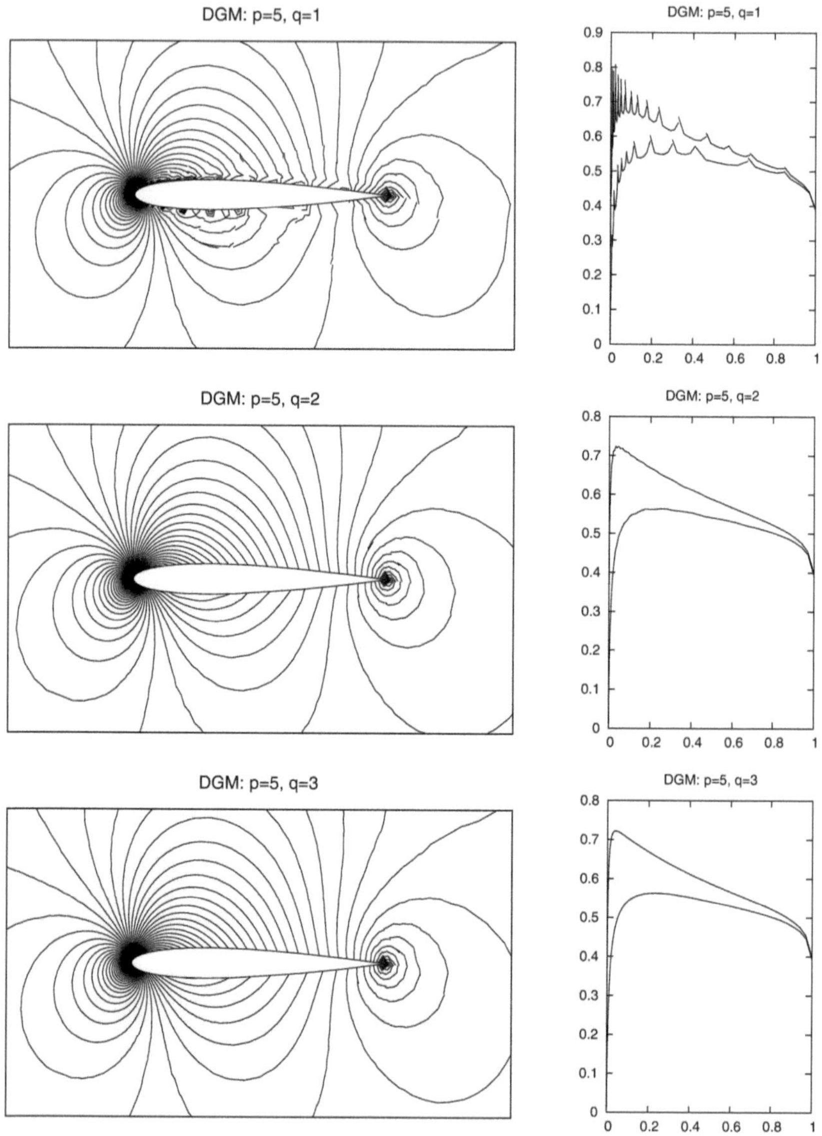

Fig. 8.15 Subsonic inviscid flow around the NACA 0012 profile ($M_\infty = 0.5, \alpha = 2°$): DGM with polynomial approximation with $p = 5$, boundary approximation with $q = 1$ (*top*), $q = 2$ (*center*) and $q = 3$ (*bottom*), Mach number isolines (*left*) and the Mach number distribution around the profile (*right*)

8.7 Numerical Verification of the BDF-DGM

In this section we present computational results demonstrating the robustness and accuracy of the BDF-DGM for solving the Euler equations.

8.7.1 Inviscid Low Mach Number Flow

It is well-known that the numerical solution of low Mach number compressible flow is rather difficult. This is caused by the stiff behaviour of numerical schemes and acoustic phenomena appearing in low Mach number flows at incompressible limit. In this case, standard finite volume and finite element methods fail. This led to the development of special finite volume techniques allowing for the simulation of compressible flow at incompressible limit, which are based on modifications of the Euler or Navier–Stokes equations. We can mention works by Klein, Munz, Meister, Wesseling and their collaborators (see e.g. [198, 242], [222, Chap. 5], or [282, Chap. 14]). However, these techniques could not be applied to the solution of high speed flow. Therefore, further attempts were concentrated on extending these methods to solving flows at all speeds. A success in this direction was achieved by several authors. Let us mention, for example, the works by Wesseling et al. (e.g., [175]), Parker and Munz [231], Meister [221] and Darwish et al. [71]. The main ingredients of these techniques are finite volume schemes applied on staggered grids, combined with multigrid, the use of the pressure-correction, multiple pressure variables and flux preconditioning.

In 2007, in paper [133], it was discovered that the DG method described above allows the solution of compressible flow with practically all Mach numbers, without any modification of the governing equations, written in the conservative form with conservative variables. The robustness with respect to the magnitude of the Mach number of this method is based on the following ingredients:

- the application of the discontinuous Galerkin method for space discretization,
- special treatment of boundary conditions,
- (semi-)implicit time discretization,
- limiting of the order of accuracy in the vicinity of discontinuities based on the locally applied artificial viscosity,
- the use of curved elements near curved parts of the boundary.

In this section we present results of numerical examples showing that the described DG method allows for the low Mach number flow, nearly at incompressible limit. First, we solve stationary inviscid low Mach number flow around the NACA 0012 profile similarly as in [20]. The angle of attack is equal to zero and the far field Mach number M_∞ is equal to 10^{-1}, 10^{-2}, 10^{-3} and 10^{-4}. The computation was carried out on a grid having 3587 elements (see Fig. 8.16, bottom) with the aid of the 3-steps BDF-DGM with $P_p,\ p = 1, 2, 3, 4$, polynomial approximation in space.

The computations are stop when the relative residuum steady-state criterion (8.171) is achieved for TOL $= 10^{-5}$.

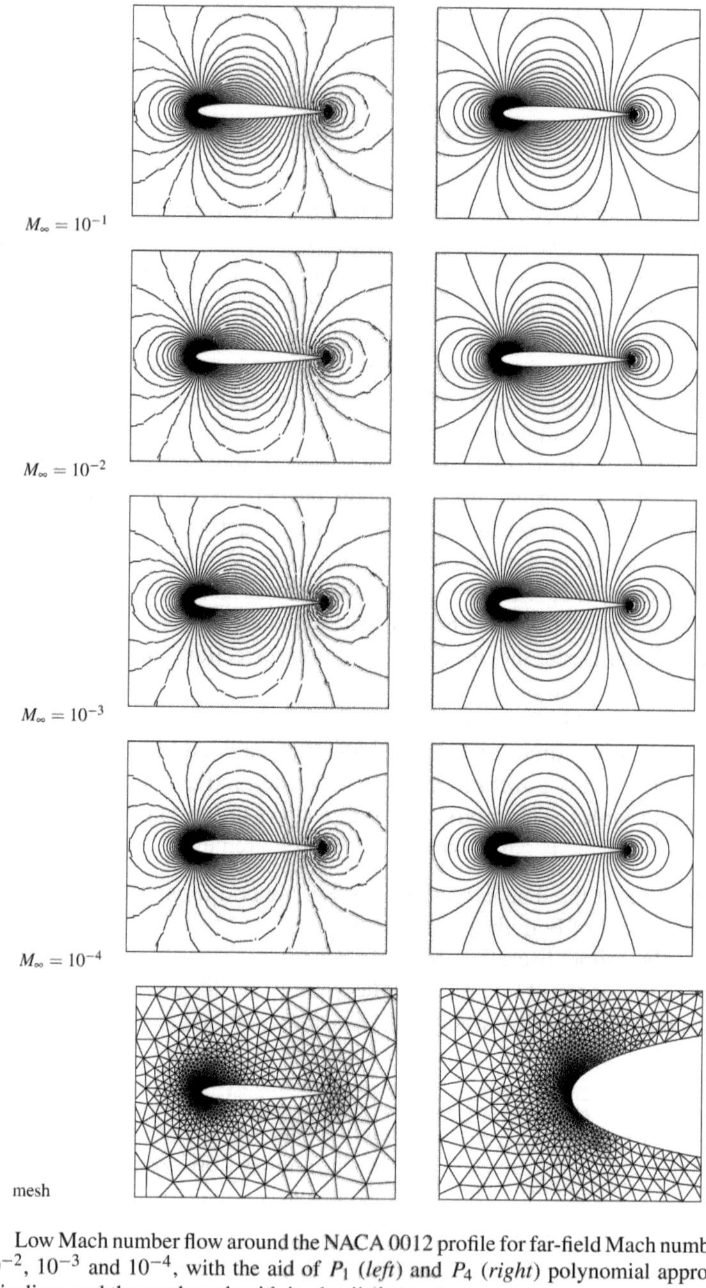

Fig. 8.16 Low Mach number flow around the NACA 0012 profile for far-field Mach number $M_\infty = 10^{-1}$, 10^{-2}, 10^{-3} and 10^{-4}, with the aid of P_1 (*left*) and P_4 (*right*) polynomial approximation: pressure isolines and the used mesh with its detail (*bottom*)

Table 8.6 Low Mach number flow around the NACA 0012 profile for far-field Mach number $M_\infty = 10^{-1}, 10^{-2}, 10^{-3}$ and 10^{-4}, with the aid of P_p, $p = 1, \ldots, 4$, polynomial approximation: ratios $(\mathrm{p}_{\max} - \mathrm{p}_{\min})/\mathrm{p}_{\max}$, $(\rho_{\max} - \rho_{\min})/\rho_{\max}$, drag coefficient c_D and lift coefficient c_L

M_∞	p	$\frac{\mathrm{p}_{\max}-\mathrm{p}_{\min}}{\mathrm{p}_{\max}}$	$\frac{\rho_{\max}-\rho_{\min}}{\rho_{\max}}$	c_D	c_L
10^{-1}	1	9.89E−03	7.08E−03	2.57E−04	1.46E−03
10^{-1}	2	9.87E−03	7.09E−03	6.63E−05	1.20E−03
10^{-1}	3	9.87E−03	7.06E−03	4.26E−05	7.97E−04
10^{-1}	4	9.87E−03	7.06E−03	1.90E−05	6.83E−04
10^{-2}	1	9.92E−05	7.10E−05	3.80E−04	1.80E−03
10^{-2}	2	9.91E−05	7.11E−05	9.63E−05	1.22E−03
10^{-2}	3	9.90E−05	7.65E−05	4.68E−05	1.11E−03
10^{-2}	4	9.91E−05	7.13E−05	−5.73E−05	3.01E−04
10^{-3}	1	9.92E−07	7.11E−07	3.95E−04	1.57E−03
10^{-3}	2	9.93E−07	7.56E−07	3.74E−05	4.75E−04
10^{-3}	3	9.90E−07	7.08E−07	5.70E−05	8.96E−04
10^{-3}	4	9.90E−07	7.08E−07	3.69E−05	6.64E−04
10^{-4}	1	9.88E−09	4.84E−08	−1.69E−05	5.42E−04
10^{-4}	2	9.91E−09	8.29E−08	1.17E−04	1.10E−03
10^{-4}	3	9.90E−09	2.51E−08	−9.56E−06	5.02E−04
10^{-4}	4	9.93E−09	3.32E−08	−2.80E−04	3.17E−04

Table 8.6 shows the relative maximum pressure and density variations $(\mathrm{p}_{\max} - \mathrm{p}_{\min})/\mathrm{p}_{\max}$ and $(\rho_{\max} - \rho_{\min})/\rho_{\max}$, respectively, the *drag coefficient* c_D and the *lift coefficient* c_L, see (8.172). Let us note that

$$\mathrm{p}_{\max} = \max_{x\in\Omega} \mathrm{p}_h(x), \quad \mathrm{p}_{\min} = \min_{x\in\Omega} \mathrm{p}_h(x), \quad \rho_{\max} = \max_{x\in\Omega} \rho_h(x), \quad \rho_{\min} = \min_{x\in\Omega} \rho_h(x),$$

where $\mathrm{p}_h(x)$ and $\rho_h(x)$ are the numerical approximations of the pressure and the density, respectively, evaluated from $\boldsymbol{w}_h$.

Both the pressure and density maximum variations are of order M_∞^2, which is in agreement with theoretical results in the analysis of compressible flow at incompressible limit. One can also see that the drag and lift coefficients attain small values, which correspond to the fact that in inviscid flow around a symmetric airfoil with zero angle of attack these quantities vanish. Figure 8.16 shows the pressure isolines obtained with the aid of P_1 and P_4 approximations.

8.7.2 Low Mach Number Flow at Incompressible Limit

It is well-known that compressible flow with a very low Mach number is very close to incompressible flow. This fact allows us to test the quality of numerical schemes for solving compressible low Mach number flow using a comparison with exact solutions of the corresponding incompressible flow, which are available in some cases. Here we present two examples of stationary compressible flow compared with incompressible flow. The steady-state solution was obtained with the aid of the time stabilization using the backward Euler linearized semi-implicit scheme (8.130). The computational grids were constructed with the aid of the anisotropic mesh adaptation technique by the ANGENER code [84]. In both examples quadratic elements ($p = 2$) were applied.

8.7.2.1 Irrotational Flow Around a Joukowski Profile

We consider flow around a negatively oriented Joukowski profile given by parameters $\Delta = 0.07$, $a = 0.5$, $h = 0.05$ (under the notation from [122], Sect. 2.2.68) with zero angle of attack. The far-field quantities are constant, which implies that the flow is irrotational and homoentropic. Using the complex function method from [122], we can obtain the exact solution of incompressible inviscid irrotational flow satisfying the Kutta–Joukowski trailing condition, provided the velocity circulation around the profile, related to the magnitude of the far-field velocity, $\gamma_{\text{ref}} = 0.7158$. We assume that the far-field Mach number of compressible flow $M_\infty = 10^{-4}$. The computational domain is of the form of a square with side of the length equal to 10 chords of the profile from which the profile is removed. The mesh (in the whole computational domain) was formed by 5418 triangular elements and refined towards the profile. Figure 8.17 (top) shows a detail near the profile of the velocity isolines for the exact solution of incompressible flow and for the approximate solution of compressible flow. Further, in Fig. 8.17 (bottom), the distribution of the velocity related to the far-field velocity and the pressure coefficient distribution around the profile is plotted in the direction from the leading edge to the trailing edge. The pressure coefficient was defined as $10^7 \cdot (\mathrm{p} - \mathrm{p}_\infty)$, where p_∞ denotes the far-field pressure.

The maximum density variation is $1.04 \cdot 10^{-8}$. The computed velocity circulation related to the magnitude of the far-field velocity is $\gamma_{\text{refcomp}} = 0.7205$, which gives the relative error 0.66 % with respect to the theoretical value γ_{ref} obtained for incompressible flow.

In order to establish the quality of the computed pressure of the low Mach compressible flow in a quantitative way, we introduce the function

$$B = \frac{\mathrm{p}}{\rho} + \frac{1}{2}|\boldsymbol{v}|^2, \tag{8.210}$$

which is constant for incompressible, inviscid, irrotational flow, as follows from the Bernoulli equation. In the considered compressible case, the relative variation of the function B, i.e., $(B_{\max} - B_{\min})/B_{\max} = 3.84 \cdot 10^{-6}$, where $B_{\max} = \max_{x \in \Omega} B(x)$ and $B_{\min} = \min_{x \in \Omega} B(x)$. This means that the Bernoulli equation is satisfied with a small error in the case of the compressible low Mach number flow computed by the developed method.

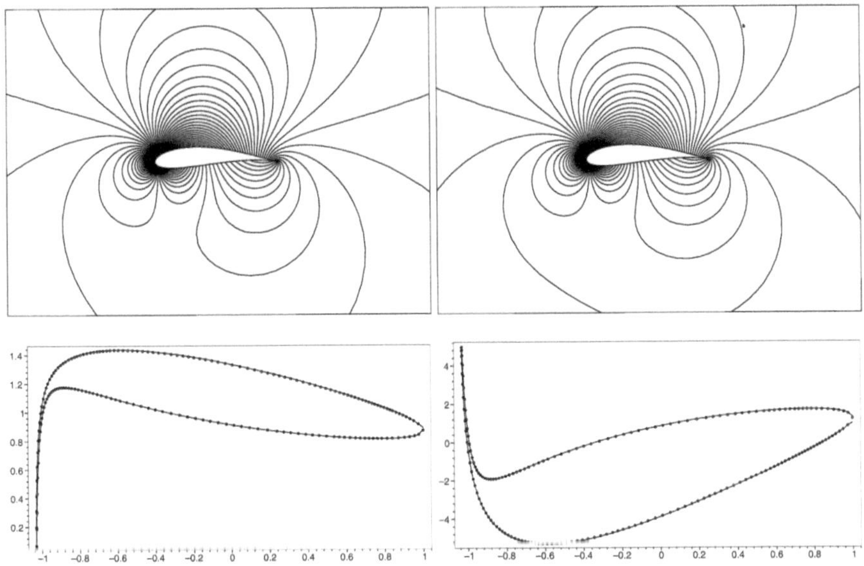

Fig. 8.17 Flow around a Joukowski airfoil, velocity isolines for the exact solution of incompressible flow (*top left*) and approximate solution of compressible low Mach number flow (*top right*), velocity (*left bottom*) and pressure coefficient (*right bottom*) distribution along the profile: exact solution of incompressible flow (*dots*) and the approximate solution of compressible flow (*full line*)

8.7.2.2 Rotational Flow Past a Circular Half-Cylinder

In the second example we present the comparison of the exact solution of incompressible inviscid rotational flow past a circular half-cylinder, with center at the origin and diameter equal to one, and with an approximate solution of compressible flow. The far-field Mach number is 10^{-4} and the far-field velocity has the components $v_1 = x_2, v_2 = 0$. The analytical exact solution was obtained in [142]. This flow is interesting for its corner vortices. The computational domain was chosen in the form of a rectangle with length 10 and width 5, from which the half-cylinder was cut off. The mesh was formed by 3541 elements. We present here computational results in the vicinity of the half-cylinder. Figure 8.18 shows streamlines of incompressible and compressible flow. Figure 8.18 (bottom) shows the velocity distribution along the half-cylinder in dependence on the variable $\vartheta - \pi/2$, where $\vartheta \in [0, \pi]$ is the angle from cylindrical coordinates. The maximum density variation is $3.44 \cdot 10^{-9}$.

8.7.2.3 Accuracy of the Method

An interesting question is the order of accuracy of the semi-implicit DG method. We tested numerically the accuracy of the piecewise quadratic DG approximations of the stationary inviscid flow past a circular cylinder with the far-field velocity

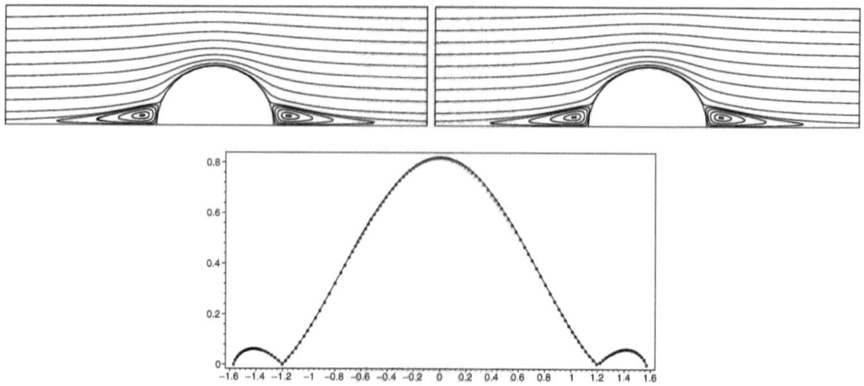

Fig. 8.18 Flow past a half-cylinder, streamlines of rotational incompressible (*top left*) and compressible (*top right*) flows and the velocity distribution (*bottom*) on the half-cylinder incompressible flow (*dots*) and compressible flow (*full line*)

parallel to the axis x_1 and the Mach number $M_\infty = 10^{-4}$. The problem was solved in a computational domain in the form of a square with sides of length equal to 20 diameters of the cylinder. Table 8.7 presents the behaviour of the error in the magnitude of the velocity related to the far-field velocity and experimental order of convergence (EOC) for approximating of the exact incompressible solution by compressible low Mach number flow on successively refined meshes measured in the $L^\infty(\Omega)$-norm.

We see that the experimental order of convergence is close to 2.5, which is comparable to theoretical error estimate (in the $L^\infty(0, T; L^2(\Omega))$-norm) obtained in Sect. 4.6.

8.7.3 Isentropic Vortex Propagation

We consider the propagation of an isentropic vortex in a compressible inviscid flow, analyzed numerically in [255]. This example is suitable for demonstrating the order of accuracy of the BDF-DGM, because the regular exact solution is known, and thus we can simply evaluate the computational error.

Table 8.7 Error in the $L^\infty(\Omega)$-norm and corresponding experimental order of convergence for approximating incompressible flow by low Mach number compressible flow with respect to $h \to 0$

$\#\mathscr{T}_h$	$\|error\|_{L^\infty(\Omega)}$	EOC
1251	5.05E−01	–
1941	4.23E−01	0.41
5031	2.77E−02	2.86
8719	6.68E−03	2.59

The computational domain is taken as $[0, 10] \times [0, 10]$ and extended periodically in both directions. The mean flow is $\bar{\rho} = 1, \bar{\boldsymbol{v}} = (1, 1)$ (diagonal flow) and $\bar{\mathrm{p}} = 1$. To this mean flow we add an *isentropic vortex*, i.e., perturbation in $\boldsymbol{v}$ and the temperature $\theta = \mathrm{p}/\rho$, but no perturbation in the entropy $\eta = \mathrm{p}/\rho^\gamma$:

$$\delta \boldsymbol{v} = \frac{\varepsilon}{2\pi} \exp[(1 - r^2)/2](-\bar{x}_2, \bar{x}_1), \quad \delta\theta = -\frac{(\gamma - 1)\varepsilon^2}{8\gamma\pi^2} \exp[1 - r^2], \quad \delta\eta = 0, \tag{8.211}$$

where $(-\bar{x}_2, \bar{x}_1) = (x_1 - 5, x_2 - 5)$, $r^2 = x_1^2 + x_2^2$, and the vortex strength $\varepsilon = 5$. The perturbations $\delta\rho$ and $\delta\mathrm{p}$ are obtained from the above relations according to

$$\begin{aligned} &\bar{\eta} = \bar{\mathrm{p}}/\bar{\rho}^\gamma, && \bar{\theta} = \bar{\mathrm{p}}/\bar{\rho}, \\ &\delta\rho = \left(\tfrac{\bar{\theta}+\delta\theta}{\bar{\eta}}\right)^{1/(\gamma-1)} && - \bar{\rho},\ \delta\mathrm{p} = (\bar{\rho} + \delta\rho)(\bar{\theta} + \delta\theta) - \bar{\mathrm{p}}. \end{aligned}$$

It is possible to see that the exact solution of the Euler equations with the initial conditions

$$\rho(x, 0) = \bar{\rho} + \delta\rho, \quad \boldsymbol{v}(x, 0) = \bar{\boldsymbol{v}} + \delta\boldsymbol{v}, \quad \mathrm{p}(x, 0) = \bar{\mathrm{p}} + \delta\mathrm{p}, \tag{8.212}$$

and periodic boundary conditions is just the passive convection of the vortex with the mean velocity. Therefore, we are able to evaluate the computational error $\|\boldsymbol{w} - \boldsymbol{w}_{h\tau}\|$ over the space-time domain $Q_T := \Omega \times (0, T)$, where $\boldsymbol{w}$ is the exact solution and $\boldsymbol{w}_{h\tau}$ is the approximate solution obtained by the time interpolation of the approximate solution computed by the n-step BDF-DGM with the discretization parameters h and τ. This means that the function $\boldsymbol{w}_{h\tau}$ is defined by

$$\begin{aligned} &\boldsymbol{w}_{h\tau}(x, t_k) = \boldsymbol{w}_h^k(x),\ x \in \Omega,\ k = 0, \ldots, r, \\ &\boldsymbol{w}_{h\tau}(x, t)|_{\Omega \times I_k} = \mathscr{L}^n(\boldsymbol{w}_h^{k+1}, \boldsymbol{w}_h^k, \ldots, \boldsymbol{w}_k^{k-n+1})|_{\Omega \times I_k}, \end{aligned} \tag{8.213}$$

where $I_k = (t_{k-1}, t_k)$ and $\mathscr{L}^n$ is the Lagrange interpolation of degree n in the space $\mathbb{R} \times \boldsymbol{S}_{hp}$ constructed over the pairs

$$(t_{k-n+1}, \boldsymbol{w}_h^{k-n+1}),\ (t_{k-n+2}, \boldsymbol{w}_h^{k-n+2}),\ \ldots,\ (t_k, \boldsymbol{w}_h^k),\ (t_{k+1}, \boldsymbol{w}_h^{k+1}).$$

In our computations we evaluate the following errors:

- $\|e_h(T)\|_{(L^2(\Omega))^m}$—error over Ω at the final time T,
- $|e_h(T)|_{(H^1(\Omega))^m}$—error over Ω at the final time T,
- $\|e_{h\tau}\|_{(L^2(Q_T))^m}$—error over the space-time cylinder $\Omega \times (0, T)$,
- $\|e_{h\tau}\|_{(L^2(0,T;\, H^1(\Omega)))^m}$—error over the space-time cylinder $\Omega \times (0, T)$.

We perform the computation on unstructured quasi-uniform triangular grids having 580, 2484 and 10008 elements, which corresponds to the average element size

Table 8.8 Isentropic vortex propagation: computational errors and the corresponding EOC

h	τ	$k = n$	$\|e_h(T)\|_{L^2(\Omega)}$	$\|e_h(T)\|_{H^1(\Omega)}$	$\|e_{h\tau}\|_{L^2(Q_T)}$	$\|e_{h\tau}\|_{L^2(0,T;\, H^1(\Omega))}$
5.87E−01	1.00E−02	1	8.54E−01	1.69E+00	1.71E+00	4.01E+00
2.84E−01	5.00E−03	1	3.30E−01	7.56E−01	6.27E−01	1.81E+00
		EOC	(1.31)	(1.11)	(1.38)	(1.09)
1.41E−01	2.50E−03	1	1.50E−01	3.51E−01	2.82E−01	8.66E−01
		EOC	(1.13)	(1.10)	(1.15)	(1.06)
5.87E−01	1.00E−02	2	3.93E−02	2.40E−01	9.64E−02	7.10E−01
2.84E−01	5.00E−03	2	3.84E−03	5.05E−02	1.02E−02	1.61E−01
		EOC	(3.20)	(2.14)	(3.09)	(2.04)
1.41E−01	2.50E−03	2	6.69E−04	1.26E−02	1.55E−03	3.96E−02
		EOC	(2.51)	(1.99)	(2.70)	(2.01)
5.87E−01	1.00E−02	3	3.97E−03	3.75E−02	1.19E−02	1.30E−01
2.84E−01	5.00E−03	3	4.89E−04	5.04E−03	1.47E−03	1.56E−02
		EOC	(2.88)	(2.76)	(2.88)	(2.91)
1.41E−01	2.50E−03	3	1.14E−04	7.38E−04	3.45E−04	2.87E−03
		EOC	(2.09)	(2.76)	(2.08)	(2.43)

$h = 0.587$, $h = 0.284$ and $h = 0.141$, respectively. For each grid, we employ the k-step BDF-DGM with P_k polynomial approximation, $k = 1, 2, 3$. We use a fixed time step $\tau = 0.01$ on the coarsest mesh, $\tau = 0.005$ on the middle one and $\tau = 0.0025$ on the finest one. It means that the ratio h/τ is almost fixed for all computations. The final time was set $T = 10$.

Table 8.8 shows the computational errors in the norms mentioned above for each case and also the corresponding *experimental orders of convergence* (EOC). We observe that EOC measured in the H^1-seminorm is roughly $O(h^k)$ for $k = 1, 2, 3$, cf. Remarks 8.13 and 8.18. On the other hand, EOC measured in the L^2-norms are higher for $k = 2$ than for $k = 3$. However, the size of the error is smaller for $k = 3$ than for $k = 2$.

Moreover, Fig. 8.19 shows the isolines of the Mach number for P_1 polynomial approximation on the coarsest mesh and for P_3 polynomial approximation on the finest mesh.

8.7.4 Supersonic Flow

In order to demonstrate the applicability of the described DG schemes to the solution of supersonic flow with high Mach numbers, we present an inviscid supersonic flow around the NACA 0012 profile with the far-field Mach number $M_\infty = 2$ and the angle of attack $\alpha = 2°$. This flow produces a strong oblique shock wave in front the leading edge of the profile. The computation was performed on the anisotropically refined grid by the ANGENER code [84] shown in Fig. 8.20. We observe a strong refinement along shock waves. Some elements in front of the oblique shock wave are

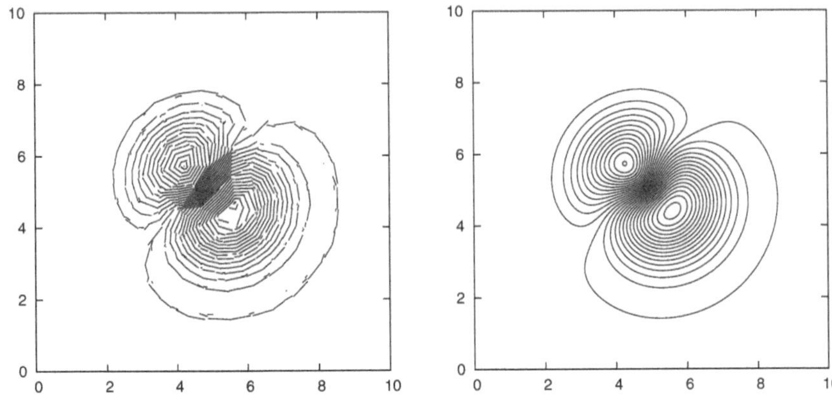

Fig. 8.19 Isentropic vortex propagation: the isolines of the Mach number computed with the aid of P_1 approximation on the coarsest mesh (*left*) and P_3 approximation of the finest one (*right*)

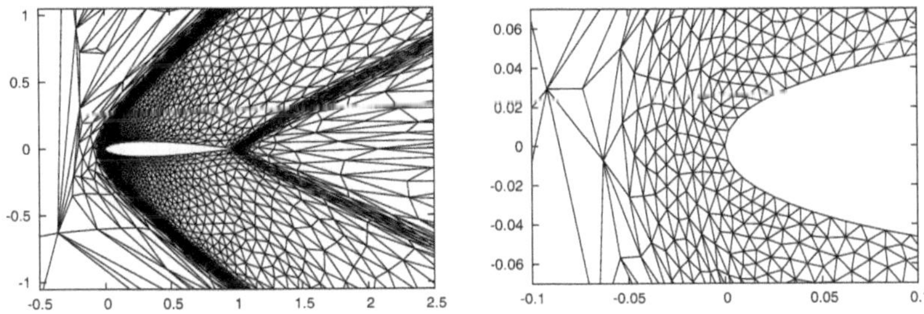

Fig. 8.20 Supersonic flow around the NACA 0012 profile ($M_\infty = 2, \alpha = 2°$): the grid used, details around the profile (*left*) and the leading edge (*right*)

very obtuse, however the DGM was able to overcome this annoyance. Figure 8.21 shows the Mach number obtained with the aid of the P_3 approximation. Due to the applied shock capturing technique presented in Sect. 8.5 (with the same setting of all parameters α_1, α_2, ν_1 and ν_2), a good resolution of the shock waves is obtained.

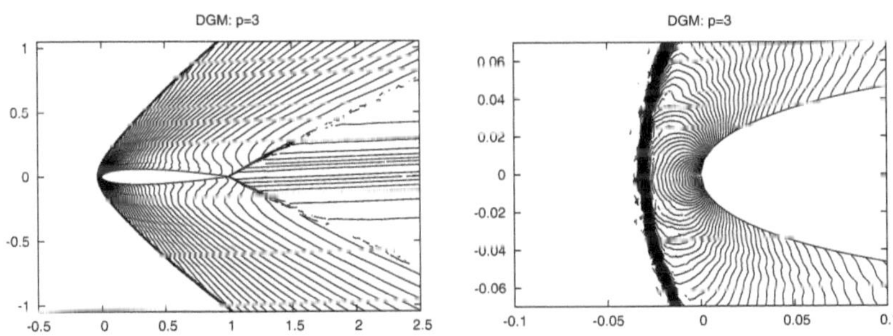

Fig. 8.21 Supersonic flow around the NACA 0012 profile ($M_\infty = 2$, $\alpha = 2°$): Mach number isolines, around of the profile (*left*) and at the leading edge (*right*)

Chapter 9
Viscous Compressible Flow

This chapter is devoted to the numerical simulation of viscous compressible flow. The methods treated here represent the generalization of techniques for solving inviscid flow problems contained in Chap. 8. Viscous compressible flow is described by the continuity equation, the Navier–Stokes equations of motion and the energy equation, to which we add closing thermodynamical relations.

In the following, we introduce the DG space semidiscretization of the compressible Navier–Stokes equations with the aid of the interior penalty Galerkin (IPG) techniques. Since the convective terms were treated in detail in Chap. 8, we focus on discretization of viscous diffusion terms. We extend heuristically the approach developed in Chap. 2. Semidiscretization leads to a system of ordinary differential equations (ODEs), which is solved by the approach presented in Chap. 8 for the Euler equations. We demonstrate the accuracy, robustness and efficiency of the DG method in the solution of several flow problems.

9.1 Formulation of the Viscous Compressible Flow Problem

9.1.1 Governing Equations

We consider unsteady compressible viscous flow in a domain $\Omega \subset \mathbb{R}^d$ ($d = 2$ or 3) and time interval $(0, T)$ $(0 < T < \infty)$. In what follows, we present the governing equations. Their derivation can be found, e.g., in [127, Sect. 1.2].

We use the standard notation: ρ-density, p-pressure (symbol p denotes the degree of polynomial approximation), E-total energy, v_s-components of the velocity vector $v = (v_1, \dots, v_d)^T$ in the directions x_s, $s = 1, \dots, d$, θ—absolute temperature, $c_v > 0$—specific heat at constant volume, $c_\text{p} > 0$—specific heat at constant pressure, $\gamma = c_\text{p}/c_v > 1$—Poisson adiabatic constant, $R = c_\text{p} - c_v > 0$—gas constant, τ_{ij}^V, $i, j = 1, \dots, d$—components of the viscous part of the stress tensor,

V. Dolejší and M. Feistauer, *Discontinuous Galerkin Method*,
Springer Series in Computational Mathematics 48,
DOI 10.1007/978-3-319-19267-3_9

$\boldsymbol{q} = (q_1, \ldots, q_d)$—heat flux. We will be concerned with the flow of a perfect gas, for which the equation of state (8.1) reads as

$$\mathrm{p} = R\rho\theta, \tag{9.1}$$

and assume that $c_\mathrm{p}, c_\mathrm{v}$ are constants. Since the gas is light, we neglect the outer volume force and heat sources.

The system of governing equations formed by the continuity equation, the Navier–Stokes equations of motion and the energy equation (see [127, Sect. 3.1]) considered in the space-time cylinder $Q_T = \Omega \times (0, T)$ can be written in the form

$$\frac{\partial \rho}{\partial t} + \sum_{s=1}^{d} \frac{\partial(\rho v_s)}{\partial x_s} = 0, \tag{9.2}$$

$$\frac{\partial(\rho v_i)}{\partial t} + \sum_{s=1}^{d} \frac{\partial(\rho v_i v_s + \delta_{is}\mathrm{p})}{\partial x_s} = \sum_{s=1}^{d} \frac{\partial \tau_{is}^{\mathrm{V}}}{\partial x_s}, \quad i = 1, \ldots, d, \tag{9.3}$$

$$\frac{\partial E}{\partial t} + \sum_{s=1}^{d} \frac{\partial((E + \mathrm{p})v_s)}{\partial x_s} = \sum_{s,j=1}^{d} \frac{\partial(\tau_{sj}^{\mathrm{V}} v_j)}{\partial x_s} - \sum_{s=1}^{d} \frac{\partial q_s}{\partial x_s}, \tag{9.4}$$

$$\mathrm{p} = (\gamma - 1)(E - \rho|\boldsymbol{v}|^2/2). \tag{9.5}$$

As we see, system (9.2)–(9.4) consists of $m = d + 2$ partial differential equations. This whole system is usually simply called compressible Navier–Stokes equations. The total energy is defined by the relation

$$E = \rho(c_\mathrm{v}\theta + |\boldsymbol{v}|^2/2). \tag{9.6}$$

This relation allows us to express the absolute temperature θ in terms of the quantities E, ρ and $|\boldsymbol{v}|^2$. The heat flux $\boldsymbol{q} = (q_1, \ldots, q_d)$ satisfies the *Fourier law*

$$\boldsymbol{q} = -k\nabla\theta, \tag{9.7}$$

where $k > 0$ is the *heat conductivity* assumed here to be constant. Furthermore, we consider the Newtonian type of fluid, i.e., the viscous part of the stress tensor has the form

$$\tau_{sk}^{\mathrm{V}} = \mu\left(\frac{\partial v_s}{\partial x_k} + \frac{\partial v_k}{\partial x_s}\right) + \lambda\nabla\cdot\boldsymbol{v}\,\delta_{sk}, \quad s, k = 1, \ldots, d, \tag{9.8}$$

where δ_{sk} is the Kronecker symbol and $\mu > 0$ and λ are the viscosity coefficients. We assume that $\lambda = -\frac{2}{3}\mu$. It is valid, for example, for a monoatomic gas, but very often it is also used for more complicated gases.

Moreover, we recall the definition of the *speed of sound* a and the *Mach number* M by

$$a = \sqrt{\gamma \mathrm{p}/\rho}, \qquad M = |\boldsymbol{v}|/a. \tag{9.9}$$

It appears suitable to write and solve numerically the Navier–Stokes equations describing viscous compressible flow in a *dimensionless form*. We introduce the following positive *reference* (scalar) *quantities*: a reference length L^*, a reference velocity U^*, a reference density ρ^*. All other reference quantities can be derived from these basic ones: we choose L^*/U^* for t, $\rho^* U^{*2}$ for both p and E, U^{*3}/L^* for heat sources q, U^{*2}/c_v for θ. Then we can define the dimensionless quantities denoted here by primes:

$$\begin{aligned} &x_i' = x_i/L^*, \quad v_i' = v_i/U^*, \quad \boldsymbol{v}' = \boldsymbol{v}/U^*, \quad \rho' = \rho/\rho^*, \\ &\mathrm{p}' = \mathrm{p}/(\rho^* U^{*2}), \quad E' = E/(\rho^* U^{*2}), \quad \theta' = \frac{c_v \theta}{U^{*2}}, \quad t' = tU^*/L^*. \end{aligned} \tag{9.10}$$

Moreover, we introduce the *Reynolds number* Re and the *Prandtl number* Pr defined as

$$\mathrm{Re} = \rho^* U^* L^*/\mu, \quad \mathrm{Pr} = c_p\, \mu/k. \tag{9.11}$$

In the sequel we denote the dimensionless quantities by the same symbols as the original dimensional quantities. This means that $\boldsymbol{v}$ will denote the dimensionless velocity, p will denote the dimensionless pressure, etc. Then system (9.2)–(9.4) can be written in the dimensionless form (cf. [127])

$$\frac{\partial \boldsymbol{w}}{\partial t} + \sum_{s=1}^{d} \frac{\partial \boldsymbol{f}_s(\boldsymbol{w})}{\partial x_s} = \sum_{s=1}^{d} \frac{\partial \boldsymbol{R}_s(\boldsymbol{w}, \nabla \boldsymbol{w})}{\partial x_s} \quad \text{in } Q_T, \tag{9.12}$$

where

$$\boldsymbol{w} = (w_1, \ldots, w_{d+2})^{\mathrm{T}} = (\rho, \rho v_1, \ldots, \rho v_d, E)^{\mathrm{T}} \tag{9.13}$$

is the *state vector*,

$$\boldsymbol{f}_s(\boldsymbol{w}) = \begin{pmatrix} f_{s,1}(\boldsymbol{w}) \\ f_{s,2}(\boldsymbol{w}) \\ \vdots \\ f_{s,m-1}(\boldsymbol{w}) \\ f_{s,m}(\boldsymbol{w}) \end{pmatrix} = \begin{pmatrix} \rho v_s \\ \rho v_1 v_s + \delta_{1,s}\mathrm{p} \\ \vdots \\ \rho v_d v_s + \delta_{d,s}\mathrm{p} \\ (E + \mathrm{p}) v_s \end{pmatrix}, \quad s = 1, \ldots, d, \tag{9.14}$$

are the *inviscid (Euler) fluxes* introduced already in (8.10). The expressions

$$\boldsymbol{R}_s(\boldsymbol{w}, \nabla \boldsymbol{w}) = \begin{pmatrix} R_{s,1}(\boldsymbol{w}, \nabla \boldsymbol{w}) \\ R_{s,2}(\boldsymbol{w}, \nabla \boldsymbol{w}) \\ \vdots \\ R_{s,m-1}(\boldsymbol{w}, \nabla \boldsymbol{w}) \\ R_{s,m}(\boldsymbol{w}, \nabla \boldsymbol{w}) \end{pmatrix} = \begin{pmatrix} 0 \\ \tau^{\mathrm{V}}_{s1} \\ \vdots \\ \tau^{\mathrm{V}}_{sd} \\ \sum_{k=1}^{d} \tau^{\mathrm{V}}_{sk} v_k + \frac{\gamma}{\mathrm{Re}\,\mathrm{Pr}} \frac{\partial \theta}{\partial x_s} \end{pmatrix}, \quad s = 1, \dots, d, \tag{9.15}$$

represent the *viscous* and *heat conduction* terms, and

$$\tau^{\mathrm{V}}_{sk} = \frac{1}{\mathrm{Re}} \left(\frac{\partial v_s}{\partial x_k} + \frac{\partial v_k}{\partial x_s} - \frac{2}{3} \nabla \cdot \boldsymbol{v}\, \delta_{sk} \right), \quad s, k = 1, \dots, d, \tag{9.16}$$

are the dimensionless components of the viscous part of the stress tensor. The dimensionless pressure and temperature are defined by

$$\mathrm{p} = (\gamma - 1)(E - \rho|\boldsymbol{v}|^2/2), \quad \theta = E/\rho - |\boldsymbol{v}|^2/2. \tag{9.17}$$

Of course, the set Q_T is obtained by the transformation of the original space-time cylinder using the relations for t' and x'_i.

The domain of definition of the vector-valued functions $\boldsymbol{f}_s$ and $\boldsymbol{R}_s$, $s = 1, \dots, d$, is the open set $\mathscr{D} \subset \mathbb{R}^m$ of vectors $\boldsymbol{w} = (w_1, \dots, w_m)^T$ such that the corresponding density and pressure are positive:

$$\mathscr{D} = \left\{ \boldsymbol{w} \in \mathbb{R}^m;\, w_1 = \rho > 0,\ w_m - \sum_{i=2}^{m-1} w_i^2/(2w_1) = \mathrm{p}/(\gamma - 1) > 0 \right\}. \tag{9.18}$$

Obviously, $\boldsymbol{f}_s, \boldsymbol{R}_s \in (C^1(\mathscr{D}))^m$, $s = 1, \dots, d$.

Similarly as in (8.13)–(8.17), the differentiation of the second term on the left-hand side of (9.12) and using the chain rule give

$$\sum_{s=1}^{d} \frac{\partial \boldsymbol{f}_s(\boldsymbol{w})}{\partial x_s} = \sum_{s=1}^{d} \mathbb{A}_s(\boldsymbol{w}) \frac{\partial \boldsymbol{w}}{\partial x_s}, \tag{9.19}$$

where $\mathbb{A}_s(\boldsymbol{w})$ is the $m \times m$ Jacobi matrix of the mapping $\boldsymbol{f}_s$ defined for $\boldsymbol{w} \in \mathscr{D}$:

$$\mathbb{A}_s(\boldsymbol{w}) = \frac{\mathrm{D}\boldsymbol{f}_s(\boldsymbol{w})}{\mathrm{D}\boldsymbol{w}} = \left(\frac{\partial f_{s,i}(\boldsymbol{w})}{\partial w_j} \right)_{i,j=1}^{m}, \quad s = 1, \dots, d. \tag{9.20}$$

Moreover, let

$$\mathrm{B}_1 = \{ \boldsymbol{n} \in \mathbb{R}^d;\ |\boldsymbol{n}| = 1 \} \tag{9.21}$$

denote the unit sphere in $\mathbb{R}^d$. Then, for $\boldsymbol{w} \in \mathscr{D}$ and $\boldsymbol{n} = (n_1, \ldots, n_d)^T \in \mathrm{B}_1$ we denote

$$\boldsymbol{P}(\boldsymbol{w}, \boldsymbol{n}) = \sum_{s=1}^{d} \boldsymbol{f}_s(\boldsymbol{w}) n_s, \tag{9.22}$$

which is the *physical flux* of the quantity $\boldsymbol{w}$ in the direction $\boldsymbol{n}$. Obviously, the Jacobi matrix $\mathrm{D}\boldsymbol{P}(\boldsymbol{w}, \boldsymbol{n})/\mathrm{D}\boldsymbol{w}$ can be expressed in the form

$$\frac{\mathrm{D}\boldsymbol{P}(\boldsymbol{w}, \boldsymbol{n})}{\mathrm{D}\boldsymbol{w}} = \mathbb{P}(\boldsymbol{w}, \boldsymbol{n}) = \sum_{s=1}^{d} \mathbb{A}_s(\boldsymbol{w}) n_s. \tag{9.23}$$

The explicit form of the matrices $\mathbb{A}_s$, $s = 1, \ldots, d$, and $\mathbb{P}$ is given in Exercises 8.2–8.5.

Furthermore, the viscous terms $\boldsymbol{R}_s(\boldsymbol{w}, \nabla \boldsymbol{w})$ can be expressed in the form

$$\boldsymbol{R}_s(\boldsymbol{w}, \nabla \boldsymbol{w}) = \sum_{k=1}^{d} \mathbb{K}_{s,k}(\boldsymbol{w}) \frac{\partial \boldsymbol{w}}{\partial x_k}, \quad s = 1, \ldots, d, \tag{9.24}$$

where $\mathbb{K}_{s,k}(\cdot)$ are $m \times m$ matrices ($m = d + 2$) dependent on $\boldsymbol{w}$. These matrices $\mathbb{K}_{s,k} := (K_{s,k}^{(\alpha,\beta)})_{\alpha,\beta=1}^{d+2}$, $s, k = 1, \ldots, d$, have for $d = 3$ the following form:

$$\mathbb{K}_{1,1}(\boldsymbol{w}) = \begin{pmatrix} 0 & 0 & 0 & 0 & 0 \\ -\frac{4}{3}\frac{w_2}{\mathrm{Re}\, w_1^2} & \frac{4}{3}\frac{1}{\mathrm{Re}\, w_1} & 0 & 0 & 0 \\ -\frac{w_3}{\mathrm{Re}\, w_1^2} & 0 & \frac{1}{\mathrm{Re}\, w_1} & 0 & 0 \\ -\frac{w_4}{\mathrm{Re}\, w_1^2} & 0 & 0 & \frac{1}{\mathrm{Re}\, w_1} & 0 \\ K_{1,1}^{(5,1)} & \frac{1}{\mathrm{Re}}(\frac{4}{3} - \frac{\gamma}{\mathrm{Pr}})\frac{w_2}{w_1^2} & \frac{1}{\mathrm{Re}}(1 - \frac{\gamma}{\mathrm{Pr}})\frac{w_3}{w_1^2} & \frac{1}{\mathrm{Re}}(1 - \frac{\gamma}{\mathrm{Pr}})\frac{w_4}{w_1^2} & \frac{\gamma}{\mathrm{Re\,Pr}}\frac{1}{w_1} \end{pmatrix}, \tag{9.25}$$

with $K_{1,1}^{(5,1)} = -\frac{1}{\mathrm{Re}}\left(\frac{4}{3}w_2^2 + w_3^2 + w_4^2\right)/w_1^3 + \frac{\gamma}{\mathrm{Re\,Pr}}\left(-w_5/w_1^2 + (w_2^2 + w_3^2 + w_4^2)/w_1^3\right)$,

$$\mathbb{K}_{2,2}(\boldsymbol{w}) = \begin{pmatrix} 0 & 0 & 0 & 0 & 0 \\ -\frac{w_2}{\mathrm{Re}\, w_1^2} & \frac{1}{\mathrm{Re}\, w_1} & 0 & 0 & 0 \\ -\frac{4}{3}\frac{w_3}{\mathrm{Re}\, w_1^2} & 0 & \frac{4}{3}\frac{1}{\mathrm{Re}\, w_1} & 0 & 0 \\ -\frac{w_4}{\mathrm{Re}\, w_1^2} & 0 & 0 & \frac{1}{\mathrm{Re}\, w_1} & 0 \\ K_{2,2}^{(5,1)} & \frac{1}{\mathrm{Re}}(1 - \frac{\gamma}{\mathrm{Pr}})\frac{w_2}{w_1^2} & \frac{1}{\mathrm{Re}}(\frac{4}{3} - \frac{\gamma}{\mathrm{Pr}})\frac{w_3}{w_1^2} & \frac{1}{\mathrm{Re}}(1 - \frac{\gamma}{\mathrm{Pr}})\frac{w_4}{w_1^2} & \frac{\gamma}{\mathrm{Re\,Pr}}\frac{1}{w_1} \end{pmatrix}, \tag{9.26}$$

with $K_{2,2}^{(5,1)} = -\frac{1}{\mathrm{Re}}\left(w_2^2 + \frac{4}{3}w_3^2 + w_4^2\right)/w_1^3 + \frac{\gamma}{\mathrm{Re\,Pr}}\left(-w_5/w_1^2 + (w_2^2+w_3^2+w_4^2)/w_1^3\right)$,

$$\mathbb{K}_{3,3}(\boldsymbol{w}) = \begin{pmatrix} 0 & 0 & 0 & 0 & 0 \\ -\frac{w_2}{\mathrm{Re}\,w_1^2} & \frac{1}{\mathrm{Re}\,w_1} & 0 & 0 & 0 \\ -\frac{w_3}{\mathrm{Re}\,w_1^2} & 0 & \frac{1}{\mathrm{Re}\,w_1} & 0 & 0 \\ -\frac{4}{3}\frac{w_4}{\mathrm{Re}\,w_1^2} & 0 & 0 & \frac{4}{3}\frac{1}{\mathrm{Re}\,w_1} & 0 \\ K_{3,3}^{(5,1)} & \frac{1}{\mathrm{Re}}(1-\frac{\gamma}{\mathrm{Pr}})\frac{w_2}{w_1^2} & \frac{1}{\mathrm{Re}}(1-\frac{\gamma}{\mathrm{Pr}})\frac{w_3}{w_1^2} & \frac{1}{\mathrm{Re}}(\frac{4}{3}-\frac{\gamma}{\mathrm{Pr}})\frac{w_4}{w_1^2} & \frac{\gamma}{\mathrm{Re\,Pr}}\frac{1}{w_1} \end{pmatrix}, \tag{9.27}$$

with $K_{3,3}^{(5,1)} = -\frac{1}{\mathrm{Re}}\left(w_2^2 + w_3^2 + \frac{4}{3}w_4^2\right)/w_1^3 + \frac{\gamma}{\mathrm{Re\,Pr}}\left(-w_5/w_1^2 + (w_2^2+w_3^2+w_4^2)/w_1^3\right)$,

$$\mathbb{K}_{1,2}(\boldsymbol{w}) = \begin{pmatrix} 0 & 0 & 0 & 0 & 0 \\ \frac{2}{3}\frac{w_3}{\mathrm{Re}\,w_1^2} & 0 & -\frac{2}{3}\frac{1}{\mathrm{Re}\,w_1} & 0 & 0 \\ -\frac{w_2}{\mathrm{Re}\,w_1^2} & \frac{1}{\mathrm{Re}\,w_1} & 0 & 0 & 0 \\ 0 & 0 & 0 & 0 & 0 \\ -\frac{1}{3}\frac{w_2 w_3}{\mathrm{Re}\,w_1^3} & \frac{w_3}{\mathrm{Re}\,w_1^2} & -\frac{2}{3}\frac{w_2}{\mathrm{Re}\,w_1^2} & 0 & 0 \end{pmatrix}, \tag{9.28}$$

$$\mathbb{K}_{1,3}(\boldsymbol{w}) = \begin{pmatrix} 0 & 0 & 0 & 0 & 0 \\ \frac{2}{3}\frac{w_4}{\mathrm{Re}\,w_1^2} & 0 & 0 & -\frac{2}{3}\frac{1}{\mathrm{Re}\,w_1} & 0 \\ 0 & 0 & 0 & 0 & 0 \\ -\frac{w_2}{\mathrm{Re}\,w_1^2} & \frac{1}{\mathrm{Re}\,w_1} & 0 & 0 & 0 \\ -\frac{1}{3}\frac{w_2 w_4}{\mathrm{Re}\,w_1^3} & \frac{w_4}{\mathrm{Re}\,w_1^2} & 0 & -\frac{2}{3}\frac{w_2}{\mathrm{Re}\,w_1^2} & 0 \end{pmatrix}, \tag{9.29}$$

$$\mathbb{K}_{2,1}(\boldsymbol{w}) = \begin{pmatrix} 0 & 0 & 0 & 0 & 0 \\ -\frac{w_3}{\mathrm{Re}\,w_1^2} & 0 & \frac{1}{\mathrm{Re}\,w_1} & 0 & 0 \\ \frac{2}{3}\frac{w_2}{\mathrm{Re}\,w_1^2} & -\frac{2}{3}\frac{1}{\mathrm{Re}\,w_1} & 0 & 0 & 0 \\ 0 & 0 & 0 & 0 & 0 \\ -\frac{1}{3}\frac{w_2 w_3}{\mathrm{Re}\,w_1^3} & -\frac{2}{3}\frac{w_3}{\mathrm{Re}\,w_1^2} & \frac{w_2}{\mathrm{Re}\,w_1^2} & 0 & 0 \end{pmatrix}, \tag{9.30}$$

$$\mathbb{K}_{2,3}(\boldsymbol{w}) = \begin{pmatrix} 0 & 0 & 0 & 0 & 0 \\ 0 & 0 & 0 & 0 & 0 \\ \frac{2}{3}\frac{w_4}{\mathrm{Re}\,w_1^2} & 0 & 0 & -\frac{2}{3}\frac{1}{\mathrm{Re}\,w_1} & 0 \\ -\frac{w_3}{\mathrm{Re}\,w_1^2} & 0 & \frac{1}{\mathrm{Re}\,w_1} & 0 & 0 \\ -\frac{1}{3}\frac{w_3 w_4}{\mathrm{Re}\,w_1^3} & 0 & \frac{w_4}{\mathrm{Re}\,w_1^2} & -\frac{2}{3}\frac{w_2}{\mathrm{Re}\,w_1^2} & 0 \end{pmatrix}, \tag{9.31}$$

$$\mathbb{K}_{3,1}(\boldsymbol{w}) = \begin{pmatrix} 0 & 0 & 0 & 0 & 0 \\ -\frac{w_4}{\mathrm{Re}\, w_1^2} & 0 & 0 & \frac{1}{\mathrm{Re}\, w_1} & 0 \\ 0 & 0 & 0 & 0 & 0 \\ \frac{2}{3}\frac{w_2}{\mathrm{Re}\, w_1^2} & -\frac{2}{3}\frac{1}{\mathrm{Re}\, w_1} & 0 & 0 & 0 \\ -\frac{1}{3}\frac{w_2 w_4}{\mathrm{Re}\, w_1^3} & -\frac{2}{3}\frac{w_4}{\mathrm{Re}\, w_1^2} & 0 & \frac{w_2}{\mathrm{Re}\, w_1^2} & 0 \end{pmatrix}, \tag{9.32}$$

$$\mathbb{K}_{3,2}(\boldsymbol{w}) = \begin{pmatrix} 0 & 0 & 0 & 0 & 0 \\ 0 & 0 & 0 & 0 & 0 \\ -\frac{w_4}{\mathrm{Re}\, w_1^2} & 0 & 0 & \frac{1}{\mathrm{Re}\, w_1} & 0 \\ \frac{2}{3}\frac{w_3}{\mathrm{Re}\, w_1^2} & 0 & -\frac{2}{3}\frac{1}{\mathrm{Re}\, w_1} & 0 & 0 \\ -\frac{1}{3}\frac{w_3 w_4}{\mathrm{Re}\, w_1^3} & 0 & -\frac{2}{3}\frac{w_4}{\mathrm{Re}\, w_1^2} & \frac{w_3}{\mathrm{Re}\, w_1^2} & 0 \end{pmatrix}. \tag{9.33}$$

Exercise 9.1 Verify the form of $\mathbb{K}_{s,k}$, $s, k = 1, 2, 3$, given by (9.25)–(9.33).

Exercise 9.2 Derive the form of $\mathbb{K}_{s,k}$, $s, k = 1, 2$, for $d = 2$.

9.1.2 Initial and Boundary Conditions

In order to formulate the problem of viscous compressible flow, the system of the Navier–Stokes equations (9.12) has to be equipped with initial and boundary conditions. Let $\Omega \subset \mathbb{R}^d$, $d = 2, 3$, be a bounded computational domain with a piecewise smooth boundary $\partial\Omega$. We prescribe the *initial condition*

$$\boldsymbol{w}(x, 0) = \boldsymbol{w}^0(x), \quad x \in \Omega, \tag{9.34}$$

where $\boldsymbol{w}^0 : \Omega \to \mathscr{D}$ is a given vector-valued function.

Concerning the *boundary conditions*, we distinguish (as in Chap. 8) the following disjoint parts of the boundary $\partial\Omega$: *inlet* $\partial\Omega_i$, *outlet* $\partial\Omega_o$ and *impermeable walls* $\partial\Omega_W$, i.e., $\partial\Omega = \partial\Omega_i \cup \partial\Omega_o \cup \partial\Omega_W$. We prescribe the following boundary conditions on individual parts of the boundary:

$$\rho = \rho_D, \quad \boldsymbol{v} = \boldsymbol{v}_D, \quad \sum_{k=1}^{d}\left(\sum_{l=1}^{d} \tau_{lk}^{\mathrm{V}} n_l\right) v_k + \frac{\gamma}{\mathrm{Re}\ \mathrm{Pr}}\frac{\partial\theta}{\partial \boldsymbol{n}} = 0 \quad \text{on } \partial\Omega_i, \tag{9.35}$$

$$\sum_{k=1}^{d} \tau_{sk}^{\mathrm{V}} n_k = 0, \ s = 1, \ldots, d, \quad \frac{\partial\theta}{\partial \boldsymbol{n}} = 0 \quad \text{on } \partial\Omega_o, \tag{9.36}$$

$$\boldsymbol{v} = 0, \quad \frac{\partial\theta}{\partial \boldsymbol{n}} = 0 \quad \text{on } \partial\Omega_W, \tag{9.37}$$

where ρ_D and $\boldsymbol{v}_D$ are given functions and $\boldsymbol{n} = (n_1, \dots, n_d)$ is the outer unit normal to $\partial\Omega$. Another possibility is to replace the *adiabatic boundary condition* (9.37) by

$$\boldsymbol{v} = 0, \quad \theta = \theta_D \quad \text{on } \partial\Omega_W, \tag{9.38}$$

with a given function θ_D defined on $\partial\Omega_W$. Moreover, in the sequel we apply also boundary conditions in the discretization of the convective terms, similarly as in Sect. 8.3.

Finally, we introduce two relations, which we employ in the DG discretization. If $\boldsymbol{w}$ is the state vector satisfying the outlet boundary condition (9.36), then, using (9.15) and (9.24), on $\partial\Omega_o$ we have

$$\sum_{s=1}^{d} \boldsymbol{R}_s(\boldsymbol{w}, \nabla\boldsymbol{w})\, n_s\Big|_{\partial\Omega_o} = \begin{pmatrix} 0 \\ \sum_{s=1}^{d} \tau_{s1}^{\mathrm{V}} n_s \\ \vdots \\ \sum_{s=1}^{d} \tau_{sd}^{\mathrm{V}} n_s \\ \sum_{k,s=1}^{d} \tau_{sk}^{\mathrm{V}} n_k v_s + \frac{\gamma}{\mathrm{Re\,Pr}} \sum_{s=1}^{d} \frac{\partial\theta}{\partial x_s} n_s \end{pmatrix} = 0. \tag{9.39}$$

Therefore, condition (9.36) represents the so-called “do-nothing” boundary condition.

Moreover, if $\boldsymbol{w}$ is the state vector satisfying the no-slip wall boundary condition (9.37), then using (9.15) we have

$$\sum_{s,k=1}^{d} \mathbb{K}_{s,k}(\boldsymbol{w}) \frac{\partial \boldsymbol{w}}{\partial x_k} n_s\Big|_{\partial\Omega_W} = \begin{pmatrix} 0 \\ \sum_{s=1}^{d} \tau_{1s}^{\mathrm{V}} n_s \\ \vdots \\ \sum_{s=1}^{d} \tau_{ds}^{\mathrm{V}} n_s \\ 0 \end{pmatrix} =: \sum_{s,k=1}^{d} \mathbb{K}_{s,k}^{W}(\boldsymbol{w}) \frac{\partial \boldsymbol{w}}{\partial x_k} n_s\Big|_{\partial\Omega_W}, \tag{9.40}$$

where τ_{ks}^{V} are the components of the viscous part of the stress tensor and $\mathbb{K}_{s,k}^{W}$, $s, k = 1, \dots, d$, are the matrices that have the last row equal to zero and the other rows are identical with the rows of $\mathbb{K}_{s,k}$, $s, k = 1, \dots, d$, i.e.,

$$\mathbb{K}_{s,k}^{W} = (\mathbb{K}_{s,k}^{W,(i,j)})_{i,j=1}^{m}, \quad \text{where} \tag{9.41}$$

$$\mathbb{K}_{s,k}^{W,(i,j)} = \begin{cases} \mathbb{K}_{s,k}^{(i,j)} & \text{for } i = 1, \dots, m-1,\ j = 1, \dots, m, \\ 0 & \text{for } i = m,\ j = 1, \dots, m, \end{cases} \quad s, k = 1, \dots, d,$$

where $\mathbb{K}_{s,k}$ are given by (9.24).

9.2 DG Space Semidiscretization

In the following, we describe the discretization of the Navier–Stokes equations (9.12) by the DGM. Similarly as in Chap. 8, we derive the DG space semidiscretization leading to a system of ordinary differential equations.

9.2.1 Notation

We use the same notation as in Sect. 8.2.1. It means that we assume that the domain Ω is polygonal (if $d = 2$) or polyhedral (if $d = 3$), $\mathscr{T}_h$ is a triangulation of Ω and $\mathscr{F}_h$ denotes the set of all faces of elements from $\mathscr{T}_h$. Further, $\mathscr{F}_h^I$, $\mathscr{F}_h^i$, $\mathscr{F}_h^o$ and $\mathscr{F}_h^W$ denote the set of all interior, inlet, outlet and wall faces, respectively. Moreover, we put $\mathscr{F}_h^B = \mathscr{F}_h^W \cup \mathscr{F}_h^i \cup \mathscr{F}_h^o$. Each face $\Gamma \in \mathscr{F}_h$ is associated with a unit normal $\boldsymbol{n}_\Gamma$, which is the outer unit normal to $\partial\Omega$ on $\Gamma \in \mathscr{F}_h{}^B$.

Further, over $\mathscr{T}_h$ we define the *broken Sobolev space* of vector-valued functions

$$\boldsymbol{H}^2(\Omega, \mathscr{T}_h) = (H^2(\Omega, \mathscr{T}_h))^m, \tag{9.42}$$

where

$$H^2(\Omega, \mathscr{T}_h) = \{v : \Omega \to \mathbb{R};\ v|_K \in H^2(K)\ \forall\, K \in \mathscr{T}_h\} \tag{9.43}$$

is the broken Sobolev space of scalar functions introduced by (2.29) (cf. (8.39) and (8.40)). The symbols $[\boldsymbol{u}]_\Gamma$ and $\langle \boldsymbol{u} \rangle_\Gamma$ denote the jump and the mean value of $\boldsymbol{u} \in \boldsymbol{H}^2(\Omega, \mathscr{T}_h)$ on $\Gamma \in \mathscr{F}_h^I$ and $[\boldsymbol{u}]_\Gamma = \langle \boldsymbol{u} \rangle_\Gamma = \boldsymbol{u}|_\Gamma$ for $\Gamma \in \mathscr{F}_h^B$. The approximate solution is sought in the *space of piecewise polynomial functions*

$$\boldsymbol{S}_{hp} = (S_{hp})^m, \tag{9.44}$$

where

$$S_{hp} = \left\{ v \in L^2(\Omega);\ v|_K \in P_p(K)\ \forall\, K \in \mathscr{T}_h \right\}. \tag{9.45}$$

Finally, let us note that the inviscid Euler fluxes $\boldsymbol{f}_s$, $s = 1, \ldots, d$, are discretized (including the boundary conditions) with the same approach as presented in Sect. 8.2.2. Therefore, we will pay attention here mainly to the discretization of the viscous terms.

9.2.2 DG Space Semidiscretization of Viscous Terms

In order to derive the discrete problem, we assume that there exists an exact solution $\boldsymbol{w} \in C^1([0,T]; \boldsymbol{H}^2(\Omega, \mathscr{T}_h))$ of the Navier–Stokes equations (9.12). We multiply (9.12) by a test function $\boldsymbol{\varphi} \in \boldsymbol{H}^2(\Omega, \mathscr{T}_h)$, integrate over an element $K \in \mathscr{T}_h$, apply Green's theorem and sum over all $K \in \mathscr{T}_h$. Then we can formally write

$$\sum_{K \in \mathscr{T}_h} \int_K \frac{\partial \boldsymbol{w}}{\partial t} \cdot \boldsymbol{\varphi} \, \mathrm{d}x + \mathrm{Inv} + \mathrm{Vis} = 0, \tag{9.46}$$

where

$$\mathrm{Inv} = \sum_{K \in \mathscr{T}_h} \int_{\partial K} \sum_{s=1}^{d} \boldsymbol{f}_s(\boldsymbol{w}) n_s \cdot \boldsymbol{\varphi} \, \mathrm{d}S - \sum_{K \in \mathscr{T}_h} \int_K \sum_{s=1}^{d} \boldsymbol{f}_s(\boldsymbol{w}) \cdot \frac{\partial \boldsymbol{\varphi}}{\partial x_s} \, \mathrm{d}x \tag{9.47}$$

$$\mathrm{Vis} = - \sum_{K \in \mathscr{T}_h} \int_{\partial K} \sum_{s=1}^{d} \boldsymbol{R}_s(\boldsymbol{w}, \nabla \boldsymbol{w}) n_s \cdot \boldsymbol{\varphi} \, \mathrm{d}S + \sum_{K \in \mathscr{T}_h} \int_K \sum_{s=1}^{d} \boldsymbol{R}_s(\boldsymbol{w}, \nabla \boldsymbol{w}) \cdot \frac{\partial \boldsymbol{\varphi}}{\partial x_s} \, \mathrm{d}x \tag{9.48}$$

represent the inviscid and viscous terms and $(n_1, \ldots, n_d)$ is the outer unit normal to ∂K.

The inviscid terms Inv are discretized by the technique presented in Chap. 8, namely, by (8.53). Hence,

$$\mathrm{Inv} \approx \boldsymbol{b}_h(\boldsymbol{w}, \boldsymbol{\varphi}), \tag{9.49}$$

where $\boldsymbol{b}_h$ is the *convection form*, given by (8.93). Let us mention that now the inviscid mirror boundary condition (8.68) is replaced by the *viscous mirror boundary condition* with the *viscous mirror operator*

$$\mathscr{M}(\boldsymbol{w}) = (\rho, -\rho \boldsymbol{v}, E)^{\mathrm{T}}, \tag{9.50}$$

replacing (8.67).

Here, we focus on the discretization of the viscous terms Vis. Similarly as in (2.36), we rearrange the first term in (9.48) according to the type of faces Γ, i.e.,

$$\sum_{K \in \mathscr{T}_h} \int_{\partial K} \sum_{s=1}^{d} \boldsymbol{R}_s(\boldsymbol{w}, \nabla \boldsymbol{w}) n_s \cdot \boldsymbol{\varphi} \, \mathrm{d}S \tag{9.51}$$

$$= \sum_{\Gamma \in \mathscr{F}_h^I} \int_\Gamma \sum_{s=1}^{d} \langle \boldsymbol{R}_s(\boldsymbol{w}, \nabla \boldsymbol{w}) \rangle \, n_s \cdot [\boldsymbol{\varphi}] \, \mathrm{d}S + \sum_{\Gamma \in \mathscr{F}_h^B} \int_\Gamma \sum_{s=1}^{d} \boldsymbol{R}_s(\boldsymbol{w}, \nabla \boldsymbol{w}) \, n_s \cdot \boldsymbol{\varphi} \, \mathrm{d}S.$$

Let us deal with treating of the boundary conditions on the outlet, where only the "Neumann" boundary conditions are prescribed. With the aid of (9.39), we immediately get the relation

$$\sum_{\Gamma\in\mathscr{F}_h^o}\int_\Gamma\sum_{s=1}^d \boldsymbol{R}_s(\boldsymbol{w},\nabla\boldsymbol{w})\,n_s\cdot\boldsymbol{\varphi}\,\mathrm{d}S=0. \tag{9.52}$$

Concerning the boundary conditions on the inlet and fixed walls, the situation is more complicated, because both the Dirichlet and Neumann boundary conditions are prescribed there. However, using (9.48), (9.51), (9.52) and (9.24) we obtain

$$\begin{aligned}\mathrm{Vis}=&\sum_{K\in\mathscr{T}_h}\int_K\sum_{s=1}^d \boldsymbol{R}_s(\boldsymbol{w},\nabla\boldsymbol{w})\cdot\frac{\partial\boldsymbol{\varphi}}{\partial x_s}\,\mathrm{d}x \\ &-\sum_{\Gamma\in\mathscr{F}_h^I}\int_\Gamma\sum_{s=1}^d\left\langle\sum_{k=1}^d \mathbb{K}_{s,k}(\boldsymbol{w})\frac{\partial\boldsymbol{w}}{\partial x_k}\right\rangle n_s\cdot[\boldsymbol{\varphi}]\,\mathrm{d}S \\ &-\sum_{\Gamma\in\mathscr{F}_h^i}\int_\Gamma\sum_{s=1}^d\sum_{k=1}^d \mathbb{K}_{s,k}(\boldsymbol{w})\frac{\partial\boldsymbol{w}}{\partial x_k}n_s\cdot\boldsymbol{\varphi}\,\mathrm{d}S \\ &-\sum_{\Gamma\in\mathscr{F}_h^W}\int_\Gamma\sum_{s=1}^d\sum_{k=1}^d \mathbb{K}_{s,k}(\boldsymbol{w})\frac{\partial\boldsymbol{w}}{\partial x_k}n_s\cdot\boldsymbol{\varphi}\,\mathrm{d}S.\end{aligned} \tag{9.53}$$

In the last term of (9.53), we use relation (9.40) following from the wall boundary condition (9.37). Hence, we obtain

$$\begin{aligned}\mathrm{Vis}=&\sum_{K\in\mathscr{T}_h}\int_K\sum_{s=1}^d \boldsymbol{R}_s(\boldsymbol{w},\nabla\boldsymbol{w})\cdot\frac{\partial\boldsymbol{\varphi}}{\partial x_s}\,\mathrm{d}x \\ &-\sum_{\Gamma\in\mathscr{F}_h^I}\int_\Gamma\sum_{s=1}^d\left\langle\sum_{k=1}^d \mathbb{K}_{s,k}(\boldsymbol{w})\frac{\partial\boldsymbol{w}}{\partial x_k}\right\rangle n_s\cdot[\boldsymbol{\varphi}]\,\mathrm{d}S \\ &-\sum_{\Gamma\in\mathscr{F}_h^i}\int_\Gamma\sum_{s=1}^d\sum_{k=1}^d \mathbb{K}_{s,k}(\boldsymbol{w})\frac{\partial\boldsymbol{w}}{\partial x_k}n_s\cdot\boldsymbol{\varphi}\,\mathrm{d}S \\ &-\sum_{\Gamma\in\mathscr{F}_h^W}\int_\Gamma\sum_{s=1}^d\sum_{k=1}^d \mathbb{K}_{s,k}^W(\boldsymbol{w})\frac{\partial\boldsymbol{w}}{\partial x_k}n_s\cdot\boldsymbol{\varphi}\,\mathrm{d}S.\end{aligned} \tag{9.54}$$

Similarly as in Sect. 2.4, relation (2.44), we have to add to the relation (9.54) a *stabilization term*, which vanishes for a smooth solution satisfying the Dirichlet boundary conditions. Analogous to scalar problems, by the formal exchange of arguments $\boldsymbol{w}$ and $\boldsymbol{\varphi}$ in the second term of (9.54), for the interior faces we obtain the expression

$$-\Theta \sum_{\Gamma \in \mathscr{F}_h^I} \int_\Gamma \sum_{s=1}^{d} \left(\sum_{k=1}^{d} \mathbb{K}_{s,k}(\boldsymbol{w}) \frac{\partial \boldsymbol{\varphi}}{\partial x_k} \right) n_s \cdot [\boldsymbol{w}] \, \mathrm{d}S \tag{9.55}$$

with $\Theta = -1$ or 1 depending on the type of stabilization, i.e., NIPG or SIPG variants. If we do not consider this stabilization, i.e., if $\Theta = 0$, we get the simple IIPG variant. However, numerical experiments indicate that this choice of stabilization is not suitable. It is caused by the fact that for $\boldsymbol{\varphi} = (\varphi_1, 0, \ldots, 0)^{\mathrm{T}}$, $\varphi_1 \in H^2(\Omega, \mathscr{T}_h)$, $\varphi_1 \neq$ const, we obtain a nonzero term (9.55), whereas all terms in (9.54) are equal to zero, because the first rows of $\boldsymbol{R}_s, \mathbb{K}_{s,k}, s, k = 1, \ldots, d$, vanish, see (9.15) and (9.25)–(9.33). This means that we would get nonzero additional terms on the right-hand side of the continuity equation, which is zero in the continuous problem. Therefore, in [30, 165, 166] the stabilization term

$$-\Theta \sum_{\Gamma \in \mathscr{F}_h^I} \int_\Gamma \sum_{s=1}^{d} \left(\sum_{k=1}^{d} \mathbb{K}_{s,k}^{\mathrm{T}}(\boldsymbol{w}) \frac{\partial \boldsymbol{\varphi}}{\partial x_k} \right) n_s [\boldsymbol{w}] \, \mathrm{d}S \tag{9.56}$$

was proposed. This avoids the drawback mentioned above. Here, $\mathbb{K}_{s,k}^{\mathrm{T}}$ denotes the matrix transposed to $\mathbb{K}_{s,k}, s, k = 1, \ldots, d$. Obviously, expression (9.56) vanishes for $\boldsymbol{w}(t) \in (H^2(\Omega))^m$, $t \in (0, T)$.

Moreover, similarly as in Sect. 2.4, we consider an extra stabilization term for the boundary faces, where at least one Dirichlet boundary condition is prescribed. Particularly, for the inlet part of the boundary, we add

$$-\Theta \sum_{\Gamma \in \mathscr{F}_h^i} \int_\Gamma \sum_{s,k=1}^{d} \mathbb{K}_{s,k}^{\mathrm{T}}(\boldsymbol{w}) \frac{\partial \boldsymbol{\varphi}}{\partial x_k} n_s (\boldsymbol{w} - \boldsymbol{w}_B) \, \mathrm{d}S, \tag{9.57}$$

where $\boldsymbol{w}_B$ is a *boundary state*. It is defined on the basis of the prescribed density ρ and the velocity $\boldsymbol{v}$ in condition (9.35) and the extrapolation of the absolute temperature. This yields the boundary state

$$\boldsymbol{w}_B|_\Gamma := \left(\rho_D, \rho_D v_{D,1}, \ldots, \rho_D v_{D,d}, \rho_D \theta_\Gamma^{(L)} + \frac{1}{2}\rho_D |\boldsymbol{v}_D|^2\right)^{\mathrm{T}}, \quad \Gamma \in \mathscr{F}_h^i, \tag{9.58}$$

where $\theta_\Gamma^{(L)}$ is the trace of the temperature on $\Gamma \in \mathscr{F}_h^i$ from the interior of Ω, and ρ_D and $\boldsymbol{v}_D = (v_{D,1}, \ldots, v_{D,d})$ are the prescribed density and velocity from (9.35), respectively.

In the case of the flow past an airfoil, when usually the far-field state vector $\boldsymbol{w}_{\mathrm{BC}}$ is prescribed, it is possible to define $\boldsymbol{w}_B$ to agree with the inviscid boundary conditions introduced in Sect. 8.3.2. In this case, we put

$$\boldsymbol{w}_B|_\Gamma := \mathscr{B}(\boldsymbol{w}_\Gamma^{(L)}, \boldsymbol{w}_{\mathrm{BC}}), \quad \Gamma \in \mathscr{F}_h^i, \tag{9.59}$$

where the inlet/outlet boundary operator $\mathscr{B}$ represents $\mathscr{B}^{\mathrm{phys}}$, $\mathscr{B}^{\mathrm{LRP}}$ and $\mathscr{B}^{\mathrm{RP}}$ given by (8.88), (8.85) and (8.92), respectively, and $\boldsymbol{w}_\Gamma^{(L)}$ is the trace of the state vector on $\Gamma \in \mathscr{F}_h^i$ from the interior of Ω.

The last term in (9.54) is stabilized by the expression

$$-\Theta \sum_{\Gamma \in \mathscr{F}_h^W} \int_\Gamma \sum_{s,k=1}^{d} (\mathbb{K}_{s,k}^W(\boldsymbol{w}))^{\mathrm{T}} \frac{\partial \boldsymbol{\varphi}}{\partial x_k} n_s (\boldsymbol{w} - \boldsymbol{w}_B)\, \mathrm{d}S, \tag{9.60}$$

where $(\mathbb{K}_{s,k}^W(\boldsymbol{w}))^{\mathrm{T}}$ is the transposed matrix to $\mathbb{K}_{s,k}^W(\boldsymbol{w})$, $s, k = 1, \ldots, m$, and $\boldsymbol{w}_B$ is the prescribed boundary state vector. In the case of the adiabatic boundary condition (9.37), we define the *boundary state* as

$$\boldsymbol{w}_B|_\Gamma := (\rho_\Gamma^{(L)}, 0, \ldots, 0, \rho_\Gamma^{(L)} \theta_\Gamma^{(L)})^{\mathrm{T}}, \quad \Gamma \in \mathscr{F}_h^W, \tag{9.61}$$

where $\rho_\Gamma^{(L)}$ and $\theta_\Gamma^{(L)}$ are the traces of the density and temperature on $\Gamma \in \mathscr{F}_h^W$ from the interior of Ω, respectively. In the case of the boundary condition (9.38), we put

$$\boldsymbol{w}_B|_\Gamma := (\rho_\Gamma^{(L)}, 0, \ldots, 0, \rho_\Gamma^{(L)} \theta_D)^{\mathrm{T}}, \quad \Gamma \in \mathscr{F}_h^W, \tag{9.62}$$

where $\rho_\Gamma^{(L)}$ is the trace of the density on $\Gamma \in \mathscr{F}_h^W$ and θ_D is the prescribed temperature on the solid wall $\partial\Omega^W$.

As we see, the boundary state $\boldsymbol{w}_B$ depends partly on the unknown solution $\boldsymbol{w}$ and partly on the prescribed Dirichlet boundary conditions. Hence, we can write

$$\boldsymbol{w}_B = BC(\boldsymbol{w}, \boldsymbol{u}_D), \tag{9.63}$$

where $\boldsymbol{u}_D$ represents the Dirichlet boundary data and BC represents the definitions of boundary states (9.58), (9.59), (9.61) and (9.62).

Analogous to the DG discretization of the model problem in Sect. 2.4, for $\boldsymbol{w}, \boldsymbol{\varphi} \in \boldsymbol{H}^2(\Omega, \mathscr{T}_h)$ we define the *viscous form*

$$\begin{aligned} \boldsymbol{a}_h(\boldsymbol{w}_h, \boldsymbol{\varphi}_h) = & \sum_{K \subset \mathscr{T}_h} \int_K \sum_{s,k=1}^{d} \left(\mathbb{K}_{s,k}(\boldsymbol{w}_h) \frac{\partial \boldsymbol{w}_h}{\partial x_k} \right) \cdot \frac{\partial \boldsymbol{\varphi}_h}{\partial x_s}\, \mathrm{d}x \\ & - \sum_{\Gamma \in \mathscr{F}_h^I} \int_\Gamma \sum_{s=1}^{d} \left\langle \sum_{k=1}^{d} \mathbb{K}_{s,k}(\boldsymbol{w}_h) \frac{\partial \boldsymbol{w}_h}{\partial x_k} \right\rangle n_s \cdot [\boldsymbol{\varphi}_h]\, \mathrm{d}S \end{aligned} \tag{9.64}$$

$$
\begin{aligned}
&- \sum_{\Gamma \in \mathscr{F}_h^i} \int_\Gamma \sum_{s,k=1}^{d} \mathbb{K}_{s,k}(\boldsymbol{w}_h) \frac{\partial \boldsymbol{w}_h}{\partial x_k} n_s \cdot \boldsymbol{\varphi}_h \, \mathrm{d}S \\
&- \sum_{\Gamma \in \mathscr{F}_h^W} \int_\Gamma \sum_{s,k=1}^{d} \mathbb{K}^W_{s,k}(\boldsymbol{w}_h) \frac{\partial \boldsymbol{w}_h}{\partial x_k} n_s \cdot \boldsymbol{\varphi}_h \, \mathrm{d}S \\
&- \Theta \left(\sum_{\Gamma \in \mathscr{F}_h^I} \int_\Gamma \sum_{s,k=1}^{d} \left\langle \mathbb{K}^{\mathrm{T}}_{s,k}(\boldsymbol{w}_h) \frac{\partial \boldsymbol{\varphi}_h}{\partial x_k} \right\rangle n_s \cdot [\boldsymbol{w}_h] \, \mathrm{d}S \right. \\
&\quad + \sum_{\Gamma \in \mathscr{F}_h^i} \int_\Gamma \sum_{s,k=1}^{d} \mathbb{K}^{\mathrm{T}}_{s,k}(\boldsymbol{w}_h) \frac{\partial \boldsymbol{\varphi}_h}{\partial x_k} n_s \cdot (\boldsymbol{w}_h - \boldsymbol{w}_B) \, \mathrm{d}S \\
&\quad \left. + \sum_{\Gamma \in \mathscr{F}_h^W} \int_\Gamma \sum_{s,k=1}^{d} \left(\mathbb{K}^W_{s,k}(\boldsymbol{w}_h) \right)^{\mathrm{T}} \frac{\partial \boldsymbol{\varphi}_h}{\partial x_k} n_s \cdot (\boldsymbol{w}_h - \boldsymbol{w}_B) \, \mathrm{d}S \right).
\end{aligned}
$$

We consider $\Theta = -1, 0, 1$ and get the NIPG, IIPG and SIPG variant of the viscous form, respectively.

Similarly as in Sect. 2.4, relations (2.41) and (2.42), in the scheme we include *interior and boundary penalty* terms, vanishing for the smooth solution satisfying the boundary conditions. Here we define the form

$$
\begin{aligned}
J_h^\sigma(\boldsymbol{w}_h, \boldsymbol{\varphi}_h) :=& \sum_{\Gamma \in \mathscr{F}_h^I} \int_\Gamma \sigma [\boldsymbol{w}_h] \cdot [\boldsymbol{\varphi}_h] \, \mathrm{d}S + \sum_{\Gamma \in \mathscr{F}_h^i} \int_\Gamma \sigma (\boldsymbol{w}_h - \boldsymbol{w}_B) \cdot \boldsymbol{\varphi}_h \, \mathrm{d}S \\
&+ \sum_{\Gamma \in \mathscr{F}_h^W} \int_\Gamma \sigma (\boldsymbol{w}_h - \boldsymbol{w}_B) \cdot \mathscr{V}(\boldsymbol{\varphi}_h) \, \mathrm{d}S, \qquad (9.65)
\end{aligned}
$$

where, in view of (9.63), $\boldsymbol{w}_B = BC(\boldsymbol{w}_h, \boldsymbol{u}_D)$ is the boundary state vector (given either by (9.58) or (9.59) for $\Gamma \in \mathscr{F}_h^i$ and either by (9.61) or (9.62) for $\Gamma \in \mathscr{F}_h^W$). The operator $\mathscr{V} : \mathbb{R}^{d+2} \to \mathbb{R}^{d+2}$ is defined as

$$
\mathscr{V}(\boldsymbol{\varphi}) := (0, \varphi_2, \ldots, \varphi_{d+1}, 0)^{\mathrm{T}} \quad \text{for } \boldsymbol{\varphi} = (\varphi_1, \varphi_2, \ldots, \varphi_{d+1}, \varphi_{d+2})^{\mathrm{T}}. \qquad (9.66)
$$

The role of $\mathscr{V}$ is to penalize only the components of $\boldsymbol{w}$, for which the Dirichlet boundary conditions are prescribed on fixed walls. Let us mention that we penalize all components of $\boldsymbol{w}$ on the inlet. It would also be possible to define a similar operator $\mathscr{V}$ for $\Gamma \in \mathscr{F}_h^i$. However, numerical experiments show that it is not necessary.

The penalty weight σ is chosen as

$$
\sigma|_\Gamma = \frac{C_W}{\mathrm{diam}(\Gamma)\,\mathrm{Re}}, \quad \Gamma \in \mathscr{F}_h, \qquad (9.67)
$$

where Re is the Reynolds number of the flow, and $C_W > 0$ is a suitable constant which guarantees the stability of the method. Its choice depends on the variant of the DG method used (NIPG, IIPG or SIPG), see Sect. 9.4.1.1, where the choice of C_W is investigated with the aid of numerical experiments. The expression $\mathrm{diam}(\Gamma)$ can be replaced by the value h_Γ defined in Sect. 2.6. (Another possibility was used in [165].)

We conclude that if $\boldsymbol{w}$ is a sufficiently regular exact solution of (9.12) satisfying the boundary conditions (9.35)–(9.37), then the viscous expression Vis from (9.48) can be rewritten in the form

$$\mathrm{Vis} = \boldsymbol{a}_h(\boldsymbol{w}, \boldsymbol{\varphi}) + \boldsymbol{J}_h^\sigma(\boldsymbol{w}, \boldsymbol{\varphi}) \quad \forall \boldsymbol{\varphi} \in \boldsymbol{H}^2(\Omega, \mathscr{T}_h). \tag{9.68}$$

9.2.3 Semidiscrete Problem

Now, we complete the DG space semidiscretization of (9.12). By $(\cdot, \cdot)$ we denote the scalar product in the space $(L^2(\Omega))^{d+2}$:

$$(\boldsymbol{w}, \boldsymbol{\varphi}) = \int_\Omega \boldsymbol{w} \cdot \boldsymbol{\varphi} \, \mathrm{d}x, \quad \boldsymbol{w}, \ \boldsymbol{\varphi} \in (L^2(\Omega))^{d+2}. \tag{9.69}$$

From (9.46), where we interchange the time derivative and integral in the first term, (9.47) and (9.68) we obtain the identity

$$\frac{\mathrm{d}}{\mathrm{d}t}(\boldsymbol{w}(t), \boldsymbol{\varphi}) + \boldsymbol{b}_h(\boldsymbol{w}(t), \boldsymbol{\varphi}) + \boldsymbol{a}_h(\boldsymbol{w}(t), \boldsymbol{\varphi}) + \boldsymbol{J}_h^\sigma(\boldsymbol{w}(t), \boldsymbol{\varphi}) = 0 \tag{9.70}$$
$$\forall \boldsymbol{\varphi} \in \boldsymbol{H}^2(\Omega, \mathscr{T}_h) \ \forall t \in (0, T),$$

In the discrete problem, because of the solution of high-speed flow containing discontinuities (shock waves and contact discontinuities, slightly smeared by the viscosity and heat conduction), we also consider the artificial viscosity forms $\boldsymbol{\beta}_h$ and $\boldsymbol{\gamma}_h$ introduced in (8.183) and (8.184), respectively. Therefore, we set

$$\begin{aligned} \boldsymbol{c}_h(\boldsymbol{w}, \boldsymbol{\varphi}) = {} & \boldsymbol{b}_h(\boldsymbol{w}, \boldsymbol{\varphi}) + \boldsymbol{a}_h(\boldsymbol{w}, \boldsymbol{\varphi}) + \boldsymbol{J}_h^\sigma(\boldsymbol{w}, \boldsymbol{\varphi}) \\ & + \boldsymbol{\beta}_h(\boldsymbol{w}, \boldsymbol{w}, \boldsymbol{\varphi}) + \boldsymbol{\gamma}_h(\boldsymbol{w}, \boldsymbol{w}, \boldsymbol{\varphi}), \quad \boldsymbol{w}, \boldsymbol{\varphi} \in \boldsymbol{H}^2(\Omega, \mathscr{T}_h), \end{aligned} \tag{9.71}$$

with the forms $\boldsymbol{b}_h$, $\boldsymbol{a}_h$, $\boldsymbol{J}_h^\sigma$, $\boldsymbol{\beta}_h$ and $\boldsymbol{\gamma}_h$ defined by (8.93), (9.64), (9.65), (8.183) and (8.184), respectively. The expressions in (9.70) and (9.71) make sense for $\boldsymbol{w}, \boldsymbol{\varphi} \in \boldsymbol{H}^2(\Omega, \mathscr{T}_h)$. For each $t \in [0, T]$ the approximation of $\boldsymbol{w}(t)$ will be sought in the finite-dimensional space $\boldsymbol{S}_{hp} \subset \boldsymbol{H}^2(\Omega, \mathscr{T}_h)$ defined by (9.44) and (9.45). Using (9.70), we immediately arrive at the definition of an approximate solution.

Definition 9.3 We say that a function $\boldsymbol{w}_h$ is the *space semidiscrete solution* of the compressible Navier–Stokes equations (9.12), if the following conditions are satisfied:

$$\boldsymbol{w}_h \in C^1([0,T]; \boldsymbol{S}_{hp}), \tag{9.72a}$$

$$\frac{\mathrm{d}}{\mathrm{d}t}\left(\boldsymbol{w}_h(t), \boldsymbol{\varphi}_h\right) + \boldsymbol{c}_h(\boldsymbol{w}_h(t), \boldsymbol{\varphi}_h) = 0 \quad \forall \boldsymbol{\varphi}_h \in \boldsymbol{S}_{hp}\ \forall t \in (0,T), \tag{9.72b}$$

$$\boldsymbol{w}_h(0) = \Pi_h \boldsymbol{w}^0, \tag{9.72c}$$

where $\Pi_h \boldsymbol{w}^0$ is an $\boldsymbol{S}_{hp}$-approximation of $\boldsymbol{w}^0$ from the initial condition (9.34). Usually it is defined as the $L^2(\Omega)$-projection on the space $\boldsymbol{S}_{hp}$.

9.3 Time Discretization

The space semidiscrete problem (9.72) represents a system of N_{hp} ordinary differential equations (ODEs), where N_{hp} is equal to the dimension of the space $\boldsymbol{S}_{hp}$. This system has to be solved with the aid of a suitable numerical scheme. Often the Runge–Kutta methods are used. (See e.g., Sect. 5.2.1.1.) However, they are conditionally stable and the CFL stability condition represents a strong restriction of the time step. This is the reason that we will be concerned with using implicit or semi-implicit time discretizations. We follow the approach developed in Sect. 8.4.1 and introduce the backward Euler and the BDF discretization of the ODE system (9.72). Then we develop the solution strategy of the corresponding nonlinear algebraic systems with the aid of the Newton-like method based on the flux matrix. In Chap. 10, the full space-time discontinuous Galerkin method will be described and applied to the solution of flows in time-dependent domains and fluid-structure interaction problems.

9.3.1 Time Discretization Schemes

In what follows, we consider a partition $0 = t_0 < t_1 < t_2 \ldots < t_r = T$ of the time interval $[0,T]$ and set $\tau_k = t_k - t_{k-1},\ k = 1,\ldots,r$. We use the symbol $\boldsymbol{w}_h^k$ for the approximation of $\boldsymbol{w}_h(t_k)$, $k = 1,\ldots,r$.

Similarly as in Definitions 8.12 and 8.16, we define the following methods for the time discretization of (9.72).

Definition 9.4 We say that the finite sequence of functions $\boldsymbol{w}_h^k,\ k = 0,\ldots,r$, is an *approximate solution* of problem (9.12) obtained by the *backward Euler-discontinuous Galerkin* method (BE-DGM), if the following conditions are satisfied:

$$\boldsymbol{w}_h^k \in \boldsymbol{S}_{hp},\ k = 0, 1, \ldots, r, \tag{9.73a}$$

$$\frac{1}{\tau_k}\left(\boldsymbol{w}_h^k - \boldsymbol{w}_h^{k-1}, \boldsymbol{\varphi}_h\right) + c_h(\boldsymbol{w}_h^k, \boldsymbol{\varphi}_h) = 0 \quad \forall \boldsymbol{\varphi}_h \in \boldsymbol{S}_{hp},\ k = 1, \ldots, r, \tag{9.73b}$$

$$\boldsymbol{w}_h^0 = \Pi_h \boldsymbol{w}^0, \tag{9.73c}$$

where $\Pi_h \boldsymbol{w}^0$ is the $\boldsymbol{S}_{hp}$-approximation of $\boldsymbol{w}^0$.

Definition 9.5 We say that the finite sequence of functions $\boldsymbol{w}_h^k,\ k = 0, \ldots, r$, is the *approximate solution* of (9.12) computed by the *n-step backward difference formula-discontinuous Galerkin method* (BDF-DGM) if the following conditions are satisfied:

$$\boldsymbol{w}_h^k \in \boldsymbol{S}_{hp},\ k = 0, 1, \ldots, r, \tag{9.74a}$$

$$\frac{1}{\tau_k}\left(\sum_{l=0}^{n} \alpha_{n,l} \boldsymbol{w}_h^{k-l}, \boldsymbol{\varphi}_h\right) + c_h\left(\boldsymbol{w}_h^k, \boldsymbol{\varphi}_h\right) = 0 \quad \forall \boldsymbol{\varphi}_h \in \boldsymbol{S}_{hp},\ k = n, \ldots, r, \tag{9.74b}$$

$$\boldsymbol{w}_h^0 = \Pi_h \boldsymbol{w}^0, \tag{9.74c}$$

$\boldsymbol{w}_h^l \in \boldsymbol{S}_{hp},\ l = 1, \ldots, n-1$, are determined by a suitable q-step method with $q < n$ or by an explicit Runge–Kutta method—cf. Sect. 5.2.1.1. (9.74d)

The BDF coefficients $\alpha_{n,l},\ l = 0, \ldots, n$, depend on time steps $\tau_{k-l},\ l = 0, \ldots, n$. They can be derived from the Lagrange interpolation of pairs $[t_{k-l}, \boldsymbol{w}_h^{k-l}],\ l = 0, \ldots, n$, see e.g. [161]. Tables 8.2 and 8.3 show their values in the case of constant and variable time steps for $n = 1, 2, 3$. One-step BDF-DGM is identical with BE-DGM defined by (9.73).

Remark 9.6 By virtue of Remark 8.18 and Chaps. 2–5, we expect that the n-step BDF-DGM has formally the order of accuracy $O(h^p + \tau^n)$ in the $L^\infty(0, T; L^2(\Omega))$-norm as well as in the $L^2(0, T; H^1(\Omega))$-seminorm, provided that the exact solution is sufficiently regular. Concerning the stability of the BDF-DGM, we refer to Remark 8.17.

Schemes (9.73) and (9.74) represent nonlinear algebraic systems for each time level $t_k,\ k = 1, \ldots, r$, which should be solved by a suitable technique. It will be discussed in the following sections.

9.3.2 Solution Strategy

Since the backward Euler method (9.73) is a special case of the BDF discretization (9.74), we deal here only with the latter case. The nonlinear algebraic system arising from (9.74) for each $k = n, \ldots, r$ will be solved by the Newton-like method based on the approximation of the Jacobi matrix by the flux matrix, which was developed in Sects. 8.4.3–8.4.5.

Again, let N_{hp} denote the dimension of the piecewise polynomial space $\boldsymbol{S}_{hp}$ and let $\mathrm{B}_{hp} = \{\boldsymbol{\varphi}_i(x),\ i = 1,\dots,N_{hp}\}$ be a basis of $\boldsymbol{S}_{hp}$, see Sect. 8.4.8. Using the isomorphism (8.96) between $\boldsymbol{w}_h^k \in \boldsymbol{S}_{hp}$ and $\boldsymbol{\xi}_k \in \mathbb{R}^{N_{hp}}$, we define the vector-valued function $\boldsymbol{F}_h : (\mathbb{R}^{N_{hp}})^n \times \mathbb{R}^{N_{hp}} \to \mathbb{R}^{N_{hp}}$ by

$$\boldsymbol{F}_h\left(\{\boldsymbol{\xi}_{k-l}\}_{l=1}^n ; \boldsymbol{\xi}_k\right) = \left(\frac{1}{\tau_k}\left(\sum_{l=0}^n \alpha_{n,l}\boldsymbol{w}_h^{k-l}, \boldsymbol{\varphi}_i\right) + \boldsymbol{c}_h(\boldsymbol{w}_h^k, \boldsymbol{\varphi}_i)\right)_{i=1}^{N_{hp}}, \quad k = n,\dots,r, \tag{9.75}$$

where $\boldsymbol{\xi}_{k-l} \in \mathbb{R}^{N_{hp}}$ is the algebraic representation of $\boldsymbol{w}_h^{k-l} \in \boldsymbol{S}_{hp}$ for $l = 1,\dots,n$. We do not emphasize that $\boldsymbol{F}_h$ depends explicitly on τ_k. Then scheme (9.74) has the following algebraic representation. If $\boldsymbol{\xi}_{k-l}$, $l = 1,\dots,n$, $(k = 1,\dots,r)$ are given vectors, then we want to find $\boldsymbol{\xi}_k \in \mathbb{R}^{N_{hp}}$ such that

$$\boldsymbol{F}_h(\{\boldsymbol{\xi}_{k-l}\}_{l=1}^n ; \boldsymbol{\xi}_k) = \boldsymbol{0}. \tag{9.76}$$

System (9.76) is strongly nonlinear. In order to solve (9.76) with the aid of the Newton-like method based on the flux matrix, presented in Sect. 8.4.3, we have to linearize the form $\boldsymbol{c}_h$ similarly as the form $\boldsymbol{b}_h$ was linearized in (8.137).

To this end, on the basis of (9.64) we introduce the forms

$$\begin{aligned}
\boldsymbol{a}_h^L(\bar{\boldsymbol{w}}_h, \boldsymbol{w}_h, \boldsymbol{\varphi}_h) = &\sum_{K\in\mathscr{T}_h}\int_K \sum_{s,k=1}^d \left(\mathbb{K}_{s,k}(\bar{\boldsymbol{w}}_h)\frac{\partial \boldsymbol{w}_h}{\partial x_k}\right)\cdot\frac{\partial \boldsymbol{\varphi}_h}{\partial x_s}\,\mathrm{d}x \qquad (9.77)\\
&- \sum_{\Gamma\in\mathscr{F}_h^I}\int_\Gamma \sum_{s=1}^d \left\langle \sum_{k=1}^d \mathbb{K}_{s,k}(\bar{\boldsymbol{w}}_h)\frac{\partial \boldsymbol{w}_h}{\partial x_k}\right\rangle n_s\cdot[\boldsymbol{\varphi}_h]\,\mathrm{d}S\\
&- \sum_{\Gamma\in\mathscr{F}_h^i}\int_\Gamma \sum_{s,k=1}^d \mathbb{K}_{s,k}(\bar{\boldsymbol{w}}_h)\frac{\partial \boldsymbol{w}_h}{\partial x_k}n_s\cdot\boldsymbol{\varphi}_h\,\mathrm{d}S\\
&- \sum_{\Gamma\in\mathscr{F}_h^W}\int_\Gamma \sum_{s,k=1}^d \mathbb{K}_{s,k}^W(\bar{\boldsymbol{w}}_h)\frac{\partial \boldsymbol{w}_h}{\partial x_k}n_s\cdot\boldsymbol{\varphi}_h\,\mathrm{d}S\\
&- \Theta\left(\sum_{\Gamma\in\mathscr{F}_h^I}\int_\Gamma \sum_{s,k=1}^d \left\langle \mathbb{K}_{s,k}^{\mathrm{T}}(\bar{\boldsymbol{w}}_h)\frac{\partial \boldsymbol{\varphi}_h}{\partial x_k}\right\rangle n_s\cdot[\boldsymbol{w}_h]\,\mathrm{d}S\right.\\
&+ \sum_{\Gamma\in\mathscr{F}_h^i}\int_\Gamma \sum_{s,k=1}^d \mathbb{K}_{s,k}^{\mathrm{T}}(\bar{\boldsymbol{w}}_h)\frac{\partial \boldsymbol{\varphi}_h}{\partial x_k}n_s\cdot\boldsymbol{w}_h\,\mathrm{d}S\\
&\left.+ \sum_{\Gamma\in\mathscr{F}_h^W}\int_\Gamma \sum_{s,k=1}^d \left(\mathbb{K}_{s,k}^W(\bar{\boldsymbol{w}}_h)\right)^{\mathrm{T}}\frac{\partial \boldsymbol{\varphi}_h}{\partial x_k}n_s\cdot\boldsymbol{w}_h\,\mathrm{d}S\right),
\end{aligned}$$

$$\tilde{a}_h(\bar{w}_h, \varphi_h) = -\Theta\left(\sum_{\Gamma\in\mathscr{F}_h^i}\int_\Gamma \sum_{s,k=1}^{d} \mathbb{K}_{s,k}^{\mathrm{T}}(\bar{w}_h)\frac{\partial\varphi_h}{\partial x_k}n_s\cdot\bar{w}_B\,\mathrm{d}S\right. \tag{9.78}$$
$$\left.+\sum_{\Gamma\in\mathscr{F}_h^W}\int_\Gamma \sum_{s,k=1}^{d}\left(\mathbb{K}_{s,k}^{W}(\bar{w}_h)\right)^{\mathrm{T}}\frac{\partial\varphi_h}{\partial x_k}n_s\cdot\bar{w}_B\,\mathrm{d}S\right),$$

where $\bar{w}_B = BC(\bar{w}_h, u_D)$ is the boundary state vector given either by (9.58) or (9.59) for $\Gamma\in\mathscr{F}_h^i$ and either by (9.61) or (9.62) for $\Gamma\in\mathscr{F}_h^W$. The above forms are consistent with the form a_h:

$$a_h(w_h,\varphi_h) = a_h^L(w_h,w_h,\varphi_h) - \tilde{a}_h(w_h,\varphi_h)\qquad \forall w_h,\varphi_h\in S_{hp}. \tag{9.79}$$

The form a_h^L is linear with respect to the second and third variables.

Furthermore, because of the penalty form J_h^σ given by (9.65), we introduce the forms

$$J_h^{\sigma,L}(w_h,\varphi_h) = \sum_{\Gamma\in\mathscr{F}_h^I}\int_\Gamma \sigma[w_h]\cdot[\varphi_h]\,\mathrm{d}S + \sum_{\Gamma\in\mathscr{F}_h^i}\int_\Gamma \sigma w_h\cdot\varphi_h\,\mathrm{d}S \tag{9.80}$$
$$+\sum_{\Gamma\in\mathscr{F}_h^W}\int_\Gamma \sigma w_h\cdot\mathscr{V}(\varphi_h)\,\mathrm{d}S,$$

$$\tilde{J}_h^\sigma(\bar{w}_h,\varphi_h) = \sum_{\Gamma\in\mathscr{F}_h^i}\int_\Gamma \sigma\bar{w}_B\cdot\varphi_h\,\mathrm{d}S + \sum_{\Gamma\in\mathscr{F}_h^W}\int_\Gamma \sigma\bar{w}_B\cdot\mathscr{V}(\varphi_h)\,\mathrm{d}S, \tag{9.81}$$

where $\bar{w}_B = BC(\bar{w}_h, u_D)$ is the boundary state vector corresponding to the function $\bar{w}_h$. Obviously,

$$J_h^\sigma(w_h,\varphi_h) = J_h^{\sigma,L}(w_h,\varphi_h) - \tilde{J}_h^\sigma(w_h,\varphi_h)\qquad \forall w_h,\varphi_h\in S_{hp}. \tag{9.82}$$

Finally, let b_h, b_h^L and $\tilde{b}_h$ be the forms defined by (8.93), (8.123) and (8.121) and respectively. By virtue of (9.71), we define the forms

$$c_h^L(\bar{w}_h,w_h,\varphi_h) = b_h^L(\bar{w}_h,w_h,\varphi) + a_h^L(\bar{w}_h,w_h,\varphi_h) + J_h^{\sigma,L}(w_h,\varphi_h) \tag{9.83}$$
$$+\beta_h(\bar{w}_h,w_h,\varphi_h) + \gamma_h(\bar{w}_h,w_h,\varphi_h),\quad w_h,w_h,\varphi_h\in S_{hp},$$
$$\tilde{c}_h(\bar{w}_h,\varphi_h) = \tilde{b}_h(\bar{w}_h,\varphi_h) + \tilde{a}_h(\bar{w}_h,\varphi_h) + \tilde{J}_h^\sigma(\bar{w}_h,\varphi_h),\quad \bar{w}_h,\varphi_h\in S_{hp},$$

which together with (8.137), (9.79) and (9.82) imply consistency:

$$c_h(w_h,\varphi_h) = c_h^L(w_h,w_h,\varphi_h) - \tilde{c}_h(w_h,\varphi_h),\quad w_h,\varphi_h\in S_{hp}. \tag{9.84}$$

Following directly the approach from Sect. 8.4.5, we transform problem (9.88b) into a system of algebraic equations. Instead of (8.138) and (8.139), for $k = n, \ldots, r$ we define the *flux matrix* $\mathbb{C}_h$ and the vector $\boldsymbol{d}_h$ by

$$\mathbb{C}_h\left(\bar{\boldsymbol{\xi}}\right) = \left(\frac{\alpha_{n,0}}{\tau_k}\left(\boldsymbol{\varphi}_j, \boldsymbol{\varphi}_i\right) + \boldsymbol{c}_h^L\left(\bar{\boldsymbol{w}}_h, \boldsymbol{\varphi}_j, \boldsymbol{\varphi}_i\right)\right)_{i,j=1}^{N_{hp}} \tag{9.85}$$

and

$$\boldsymbol{d}_h\left(\{\boldsymbol{\xi}_{k-l}\}_{l=1}^{n}, \bar{\boldsymbol{\xi}}\right) = \left(\frac{1}{\tau_k}\left(\sum_{i=1}^{n} \alpha_{n,i}\boldsymbol{w}_h^{k-l}, \boldsymbol{\varphi}_i\right) + \tilde{\boldsymbol{c}}_h\left(\bar{\boldsymbol{w}}_h, \boldsymbol{\varphi}_i\right)\right)_{i=1}^{N_{hp}}, \tag{9.86}$$

respectively. Here $\boldsymbol{\varphi}_i \in \mathrm{B}_{hp}$, $i = 1, \ldots, N_{hp}$, are the basis functions in the space $\boldsymbol{S}_{hp}$, $\bar{\boldsymbol{\xi}} \in \mathbb{R}^{N_{hp}}$ and $\boldsymbol{\xi}_{k-l} \in \mathbb{R}^{N_{hp}}$, $l = 1, \ldots, n$, are the algebraic representations of $\bar{\boldsymbol{w}}_h \in \boldsymbol{S}_{hp}$ and $\boldsymbol{w}_h^{k-l} \in \boldsymbol{S}_{hp}$, $l = 1, \ldots, n$, respectively. Then problem (9.74) is equivalent to the nonlinear systems (compare with (8.126))

$$\boldsymbol{F}_h(\{\boldsymbol{\xi}_{k-l}\}_{l=1}^{n}; \boldsymbol{\xi}_k) = \mathbb{C}_h(\boldsymbol{\xi}_k)\boldsymbol{\xi}_k - \boldsymbol{d}_h(\{\boldsymbol{\xi}_{k-l}\}_{l=1}^{n}, \boldsymbol{\xi}_k) = 0, \quad k = n, \ldots, r. \tag{9.87}$$

Let us note that the flux matrix $\mathbb{C}_h$ given by (9.85) has the same block structure as the matrix $\mathbb{C}_h$ given by (8.124). The sequence of nonlinear algebraic systems (9.87) can be solved by the damped Newton-like iterative process (8.127) and (8.128) treated in Sect. 8.4.4.

Concerning the initial guess $\boldsymbol{\xi}_k^0$ for the iterative process (8.127) and (8.128), we use either the value known from the previous time level given by (8.129), i.e., $\boldsymbol{\xi}_k^0 = \boldsymbol{\xi}_{k-1}$, $k = 1, \ldots, r$, or it is possible to apply a higher-order extrapolation from previous time levels given by (8.141).

Remark 9.7 Similarly as in Remarks 8.15 and 8.19, if we carry out only one Newton iteration ($l = 0$) at each time level, put $\lambda^0 = 1$ and use the extrapolation (8.141), then the implicit method (9.74) reduces to the *BDF-DG higher-order semi-implicit method* of the viscous compressible flow including the shock capturing, which can be formulated in the following way: We seek the finite sequence of functions $\{\boldsymbol{w}_h^k\}_{k=0}^r$ such that

$$\boldsymbol{w}_h^k \in \boldsymbol{S}_{hp}, \ k = 0, 1, \ldots, r, \tag{9.88a}$$

$$\frac{1}{\tau_k}\left(\sum_{l=0}^{n} \alpha_{n,l}\boldsymbol{w}_h^{k-l}, \boldsymbol{\varphi}_h\right) + \boldsymbol{c}_h^L\left(\sum_{l=1}^{n} \beta_{n,l}\boldsymbol{w}_h^{k-l}, \boldsymbol{w}_h^k, \boldsymbol{\varphi}_h\right) = \tilde{\boldsymbol{c}}_h\left(\sum_{l=1}^{n} \beta_{n,l}\boldsymbol{w}_h^{k-l}, \boldsymbol{\varphi}_h\right)$$
$$\forall \boldsymbol{\varphi}_h \in \boldsymbol{S}_{hp}, \ k = n, \ldots, r, \tag{9.88b}$$

$$\boldsymbol{w}_h^0 = \Pi_h \boldsymbol{w}^0, \tag{9.88c}$$

$\boldsymbol{w}_h^l \in \boldsymbol{S}_{hp}$, $l = 1, \ldots, n-1$, are determined by a suitable q-step method with $q < n$ or by an explicit Runge–Kutta method—cf. Sect. 5.2.1.1. (9.88d)

Here $\Pi_h \boldsymbol{w}^0$ is the $\boldsymbol{S}_{hp}$-approximation of $\boldsymbol{w}^0$, $\alpha_{n,l},\ l = 0, \ldots, n$, are the BDF coefficients and $\beta_{n,l},\ l = 0, \ldots, n$, are the coefficients of the extrapolation (8.141). (See Tables 8.2, 8.3, 8.4 and 8.5, for $n = 1, 2, 3$.)

Setting

$$\bar{\boldsymbol{w}}_h^k = \sum_{l=1}^{n} \beta_{n,l} \boldsymbol{w}_h^{k-l}, \quad \bar{\boldsymbol{\xi}}_k = \sum_{l=1}^{n} \beta_{n,l} \boldsymbol{\xi}_{k-l}, \tag{9.89}$$

problem (9.88) is equivalent to the linear algebraic systems

$$\boldsymbol{F}_h(\{\boldsymbol{\xi}_{k-l}\}_{l=1}^{n}; \boldsymbol{\xi}_k) = \mathbb{C}_h(\bar{\boldsymbol{\xi}}_k)\boldsymbol{\xi}_k - \boldsymbol{d}_h(\{\boldsymbol{\xi}_{k-l}\}_{l=1}^{n}, \boldsymbol{\xi}_k) = 0, \quad k = n, \ldots, r. \tag{9.90}$$

Finally, because of our considerations in Chap. 10, we introduce the notation

$$\hat{\boldsymbol{a}}_h(\bar{\boldsymbol{w}}_h, \boldsymbol{w}_h, \boldsymbol{\varphi}_h) = \boldsymbol{a}_h^L(\bar{\boldsymbol{w}}_h, \boldsymbol{w}_h, \boldsymbol{\varphi}_h) - \tilde{\boldsymbol{a}}_h(\bar{\boldsymbol{w}}_h, \boldsymbol{\varphi}_h), \tag{9.91}$$

$$\hat{\boldsymbol{J}}_h^{\sigma}(\bar{\boldsymbol{w}}_h, \boldsymbol{w}_h, \boldsymbol{\varphi}_h) = \boldsymbol{J}_h^{\sigma,L}(\bar{\boldsymbol{w}}_h, \boldsymbol{w}_h, \boldsymbol{\varphi}_h) - \tilde{\boldsymbol{J}}_h^{\sigma}(\bar{\boldsymbol{w}}_h, \boldsymbol{\varphi}_h), \tag{9.92}$$

for the viscous and penalty forms. Then (9.88b), can be replaced by the identity

$$\frac{1}{\tau_k}\left(\sum_{l=0}^{n} \alpha_{n,l} \boldsymbol{w}_h^{k-l}, \boldsymbol{\varphi}_h\right) + \hat{\boldsymbol{b}}_h(\bar{\boldsymbol{w}}_h^k, \boldsymbol{w}_h^k, \boldsymbol{\varphi}_h) + \hat{\boldsymbol{a}}_h(\bar{\boldsymbol{w}}_h^k, \boldsymbol{w}_h^k, \boldsymbol{\varphi}_h) + \hat{\boldsymbol{J}}_h^{\sigma}(\bar{\boldsymbol{w}}_h^k, \boldsymbol{w}_h^k, \boldsymbol{\varphi}_h) \tag{9.93}$$

$$+ \boldsymbol{\beta}_h(\bar{\boldsymbol{w}}_h^k, \boldsymbol{w}_h^k, \boldsymbol{\varphi}_h) + \boldsymbol{\gamma}_h(\bar{\boldsymbol{w}}_h^k, \boldsymbol{w}_h^k, \boldsymbol{\varphi}_h) = 0, \ \forall \boldsymbol{\varphi}_h \in \boldsymbol{S}_{hp}, \ k = n, \ldots, r,$$

where $\hat{\boldsymbol{b}}_h$ is given by (8.131) and $\bar{\boldsymbol{w}}_h^k$ is defined in (9.89).

9.4 Numerical Examples

This section is devoted to applications of the presented BDF-DG schemes to the numerical solution of several test problems for the compressible Navier–Stokes equations. First, we consider a low Mach number flow past an adiabatic flat plate, where the analytical solution of incompressible flow is known. This example shows that the developed method is sufficiently accurate and stable even for compressible flow at an incompressible limit. Further, we present several flow regimes around the NACA 0012 profile, demonstrate the high accuracy of the DG discretization and mention some possible problems in the simulation of unsteady flows with the aid of implicit time discretization. Finally, we present a simulation of the viscous shock-vortex interaction by high-order methods. For the steady-state problems, the backward Euler method is used for the time discretization.

9.4.1 Blasius Problem

The so-called Blasius problem represents the well-known test case, when a low-speed laminar flow along an adiabatic flat plate is considered. In this case the exact analytical solution is known for incompressible flow, see [35]. Since the flow speed is low, similarly as in Sect. 8.7.2, we compare the compressible numerical solution with the exact solution of the corresponding incompressible flow.

We consider the laminar flow past the adiabatic flat plate $\{(x_1, x_2);\ 0 \leq x_1 \leq 1,\ x_2 = 0\}$ characterized by the freestream Mach number $M = 0.1$ and the Reynolds number $\mathrm{Re} = 10^4$. The computational domain is shown in Fig. 9.1, where two used triangular grids are plotted together with their details around the leading edge. We prescribe the adiabatic boundary conditions (9.37) on the flat plate, the outflow boundary conditions (9.36) at $\{(x_1, x_2);\ x_1 = 1,\ -1.5 \leq x_2 \leq 1.5\}$ and the inflow boundary conditions (9.35) on the rest of the boundary.

We seek the steady-state solution by the time stabilization approach, in which the computational process is carried out for "$t \to \infty$". As a stopping criterion we use condition (8.171) (adapted to the viscous flow problem) with $\mathrm{TOL} = 10^{-6}$.

In the following, we investigate two items:

- the *stability of the method*, namely the influence of the value of the constant C_W in (9.67) on the convergence of the numerical scheme to the stationary solution,

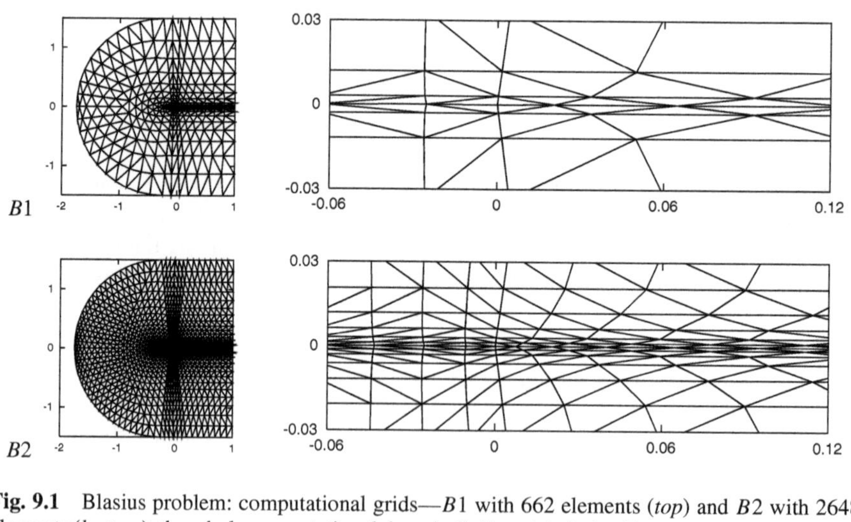

Fig. 9.1 Blasius problem: computational grids—$B1$ with 662 elements (*top*) and $B2$ with 2648 elements (*bottom*), the whole computational domain (*left*) and their details around the leading edge (*right*)

- the *accuracy of the method*, namely the comparison of the numerical solutions with the exact solution of the incompressible flow.

Exercise 9.8 Modify the stop criterion (8.171) to the viscous flow problem.

9.4.1.1 Stability of the Method

We compare the NIPG, IIPG, SIPG variants of the DGM using piecewise linear, quadratic and cubic space approximations. Our aim is to find a suitable value of the constant C_W in (9.67), which ensures the stability of the method and the convergence to the steady-state solution. First, we carried out computations for the values $C_W = 1, 5, 25, 125, 625, 3\,125$ and consequently, several additional values of C_W were chosen in order to find the limit value of C_W. These results obtained on the grid $B1$ are shown in Table 9.1, where an indication of the convergence of the appropriate variant of the DGM with a given value C_W is marked, namely,

- "convergence" (C): the stopping condition (8.171) was achieved after less than 200 time steps,
- "quasiconvergence" (qC): the stopping condition (8.171) was achieved after more than 200 time steps,
- "no-convergence" (NC): the stopping condition (8.171) was not achieved after 500 time steps.

Table 9.1 Blasius problem: the convergence (C), non-convergence (NC) or quasiconvergence (qC) of the NIPG, IIPG and SIPG variants of the DGM for P_1, P_2 and P_3 approximations for different values of C_W (symbol "–" means that the corresponding case was not tested)

	NIPG			IIPG			SIPG		
C_W	P_1	P_2	P_3	P_1	P_2	P_3	P_1	P_2	P_3
1	C	C	C	C	NC	NC	NC	NC	NC
5	C	C	C	C	C	NC	NC	NC	NC
10	–	–	–	–	C	C	–	–	–
25	C	C	C	C	C	C	NC	NC	NC
100	–	–	–	–	–	–	NC	–	–
125	C	C	C	C	C	C	C	NC	NC
150	–	–	–	–	–	–	C	–	–
250	–	–	–	–	–	–	–	NC	–
300	–	–	–	–	–	–		qC	–
400	–	–	–	–	–	–	–	C	NC
500		–	–	–	–	–	–	C	NC
625	C	C	C	C	C	C	C	C	qC
1 000	–	–	–	–	–	–	–	–	C
3 125	C	C	C	C	C	C	C	C	C

The “quasiconvergence” in fact means that the appropriate value C_W is just under the limit value ensuring the convergence to the steady-state solution.

From Table 9.1 we can find that

- NIPG variant converges for any $C_W \geq 1$ independently of the degree of polynomial approximation,
- IIPG variant requires higher values of C_W for P_2 and P_3 approximations, namely $C_W = 5$ and $C_W = 10$ are sufficient, respectively. On the other hand, P_1 approximation converges for any $C_W \geq 1$.
- SIPG variant requires significantly higher values of C_W. We observe that $C_W \geq 125$ for P_1, $C_W \geq 400$ for P_1 and $C_W \geq 1\,000$ for P_3. This is in a good agreement with theoretical results from [180] carried out for a scalar quasilinear elliptic problem, where the dependence $C_W = cp^2$ with a constant $c > 0$ is derived (p denotes the degree of the polynomial approximation).

Figure 9.2 shows the convergence history to the steady-state solution (i.e., the dependence of the steady-state residuum defined as in (8.170) on the number of time steps) for some interesting cases from Table 9.1.

9.4.1.2 Accuracy of the Method

In order to analyze the accuracy of the method at incompressible limit, we compare the numerical solution of the Blasius problem for viscous compressible flow with its incompressible analytical solution. To this end, we introduce the dimensionless velocities in the streamwise direction and in the direction orthogonal to the stream by

$$v_1^\star := \frac{v_1(\eta)}{|\boldsymbol{v}_\infty|} \quad \text{and} \quad v_2^\star := \sqrt{\mathrm{Re}_x}\,\frac{v_2(\eta)}{|\boldsymbol{v}_\infty|}, \tag{9.94}$$

respectively, where

$$\eta := \sqrt{\mathrm{Re}_x}\,\frac{x_2}{x_1}, \qquad \mathrm{Re}_x := |\boldsymbol{v}_\infty|\,\mathrm{Re}\; x_1, \tag{9.95}$$

Re is the Reynolds number and $\boldsymbol{v}_\infty$ is the freestream velocity.

Figures 9.3, 9.4, 9.5 and 9.6 show the velocity profiles $v_1^\star$ and $v_2^\star$ obtained by P_1, P_2 and P_3 approximations on the meshes $B1$ and $B2$ at $x_1 = 0.1$, $x_1 = 0.3$ and $x_1 = 0.5$ in comparison with the exact solution. We present here results obtained by the NIPG method with $C_W = 25$. (The difference between the results obtained by the NIPG, SIPG and IIPG variants are negligible.) We observe a very accurate capturing of the $v_1^\star$-profile and a reasonable capturing of the $v_2^\star$-profile. An increase of accuracy for an increasing degree of approximation and a decreasing mesh size is evident.

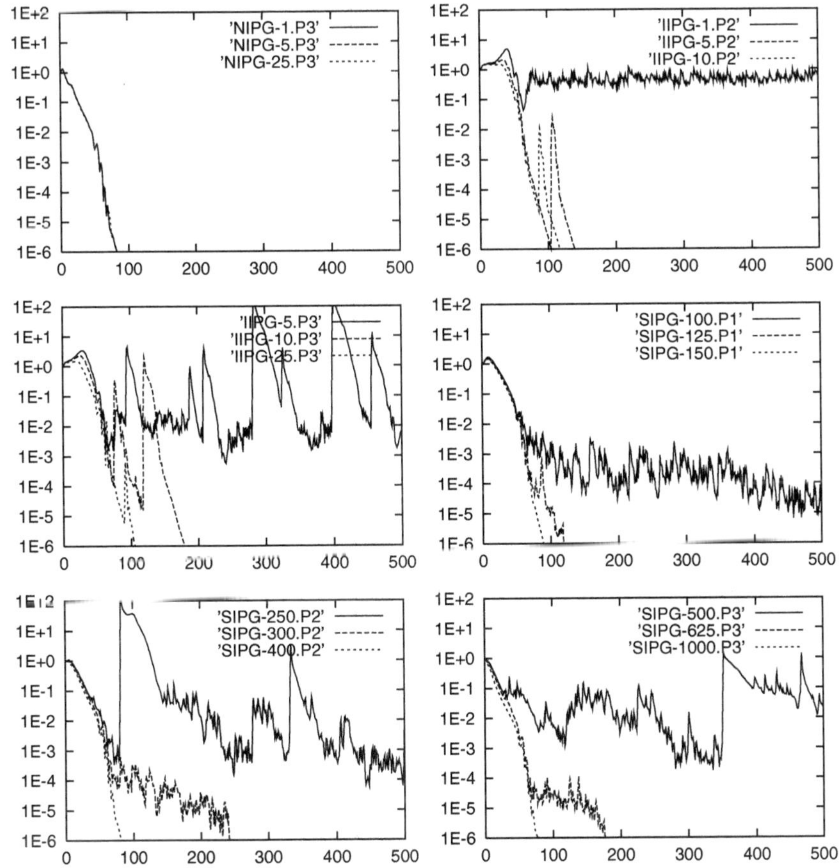

Fig. 9.2 Blasius problem: the convergence of the steady-state residuum (8.170) in the logarithmic scale on the number of time steps for some computations from Table 9.1, (e.g., 'NIPG-625.P3' means the NIPG variant of the DGM with $C_W = 625$ and P_3 approximation)

Moreover, Fig. 9.7 shows the comparison of the skin friction coefficient c_f computed by P_1, P_2 and P_3 approximations on the meshes $B1$ and $B2$ with the exact solution given by the Blasius formula. The *skin friction coefficient* is defined by

$$c_f = \frac{2\boldsymbol{t} \cdot (\mathbb{T}^{\mathrm{V}} \boldsymbol{n})}{\rho_\infty |\boldsymbol{v}_\infty|^2 L_{\mathrm{ref}}}, \tag{9.96}$$

where ρ_∞ and $\boldsymbol{v}_\infty$ are the freestream density and velocity, respectively, L_{ref} is the reference length, $\boldsymbol{n}$ and $\boldsymbol{t}$ are the unit normal and tangential vectors to the wall and $\mathbb{T}^{\mathrm{V}} = (\tau_{ij}^{\mathrm{V}})_{i,j=1}^{2}$ is the viscous part of the stress tensor. (The components τ_{ij}^{V} are defined in (9.8).)

We observe good agreement with the Blasius solution. The P_2 and P_3 approximations give the same value of c_f at the first element on the flat plate. Similar results

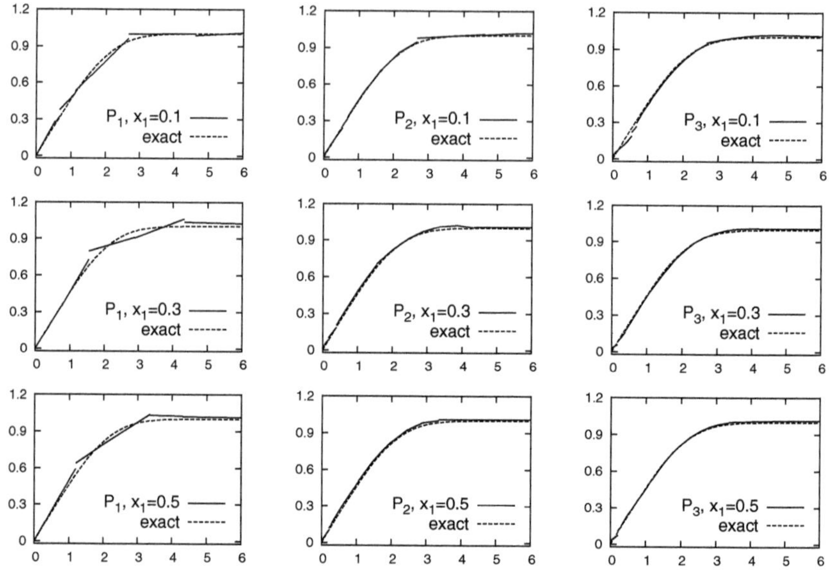

Fig. 9.3 Blasius problem: mesh $B1$, velocity profiles $v_1^\star = v_1^\star(\eta)$ for P_1, P_2 and P_3 approximations at $x_1 = 0.1$, $x_1 = 0.3$ and $x_1 = 0.5$ in comparison with the exact solution

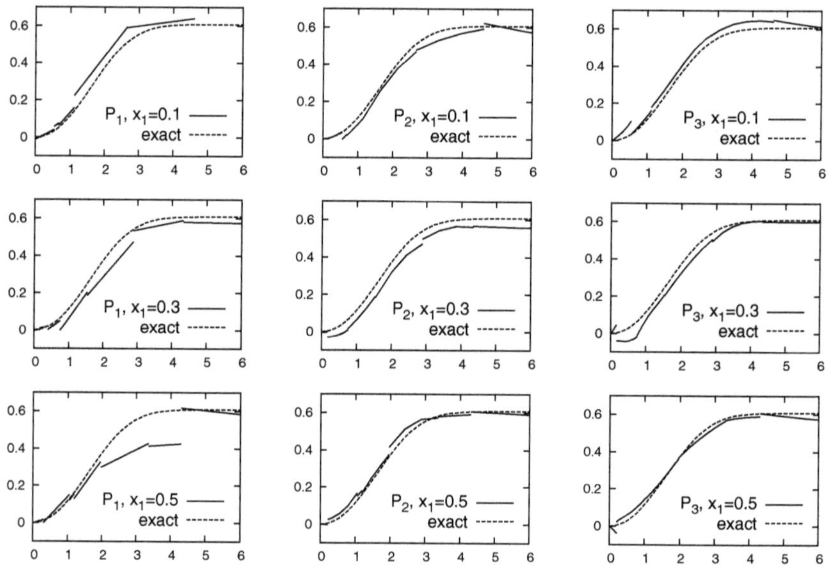

Fig. 9.4 Blasius problem: mesh $B1$, velocity profiles $v_2^\star = v_2^\star(\eta)$ for P_1, P_2 and P_3 approximations at $x_1 = 0.1$, $x_1 = 0.3$ and $x_1 = 0.5$ in comparison with the exact solution

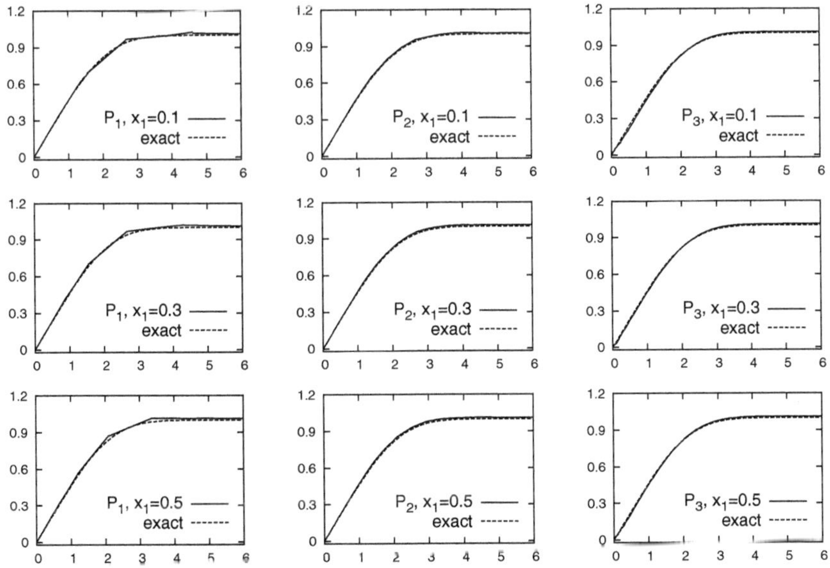

Fig. 9.5 Blasius problem: mesh $B2$, velocity profiles $v_1^\star = v_1^\star(\eta)$ for P_1, P_2 and P_3 approximations at $x_1 = 0.1$, $x_1 = 0.3$ and $x_1 = 0.5$ in comparison with the exact solution

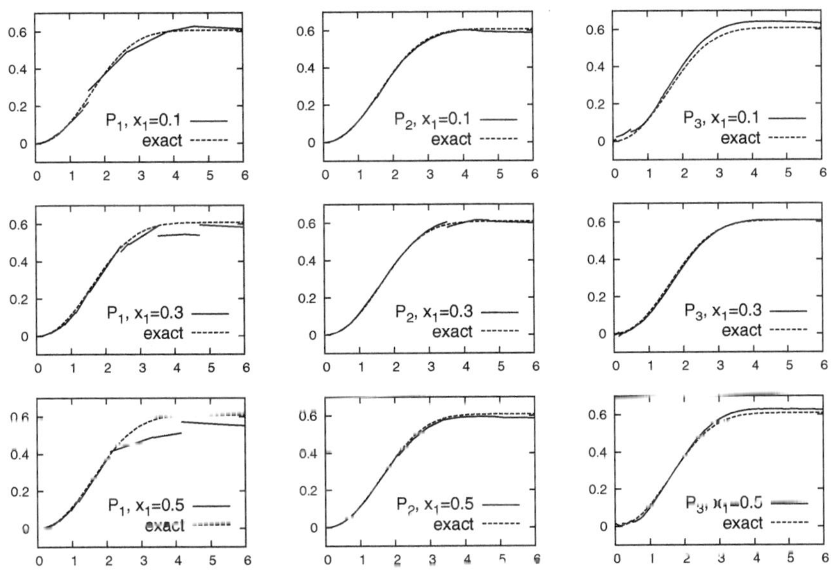

Fig. 9.6 Blasius problem: mesh $B2$, velocity profiles $v_2^\star = v_2^\star(\eta)$ for P_1, P_2 and P_3 approximations at $x_1 = 0.1$, $x_1 = 0.3$ and $x_1 = 0.5$ in comparison with the exact solution

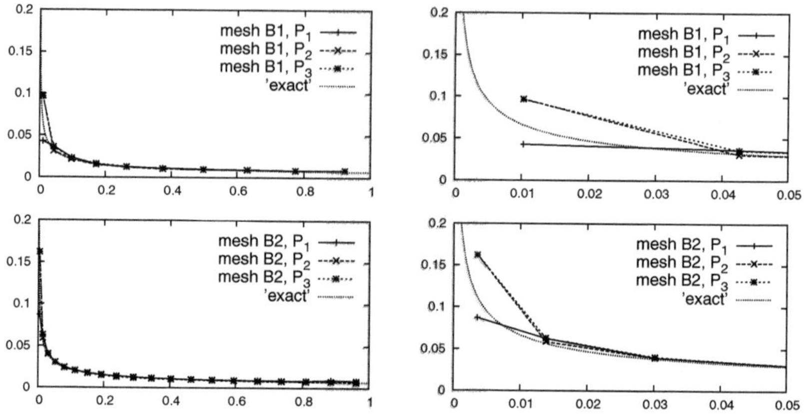

Fig. 9.7 Blasius problem: skin friction coefficient computed on the meshes $B1$ (*top*) and $B2$ (*bottom*) by P_1, P_2 and P_3 approximation in comparison with the Blasius formula (exact), distributions along the whole plate (*left*), their details around $x_1 = 0$ (*right*)

were obtained in [23, Fig. 2], where the improvement of the quality of the approximate solution on the first cell of the flat plate obtained by increasing the polynomial degree $p = 1, 2, 3$ is almost negligible. It is caused by the singularity in the solution at the leading edge of the flat plate at the point $(x_1, x_2) = (0, 0)$, which causes the decrease of the local order of accuracy of the DG method. This phenomenon was numerically verified also for a scalar nonlinear equation in Chap. 2.

9.4.2 Stationary Flow Around the NACA 0012 Profile

We consider laminar steady-state viscous subsonic flow around the NACA 0012 profile for three different flow regimes characterized by the far-field Mach number M_∞, angle of attack α and the Reynolds number Re:

$$\begin{aligned}
&(C1)\ M_\infty = 0.50,\ \alpha = 2^\circ,\ \mathrm{Re} = 500,\\
&(C2)\ M_\infty = 0.50,\ \alpha = 2^\circ,\ \mathrm{Re} = 2\,000,\\
&(C3)\ M_\infty = 0.85,\ \alpha = 2^\circ,\ \mathrm{Re} = 2\,000.
\end{aligned}$$

We carried out computations on four triangular grids $N1$–$N4$. Figure 9.8 shows these grids around the NACA 0012 profile and their zooms around the trailing and leading edges.

We evaluate the *aerodynamic coefficients drag* (c_D), *lift* (c_L) and *moment* (c_M). The coefficients c_D and c_L are defined as the first and the second components of the vector

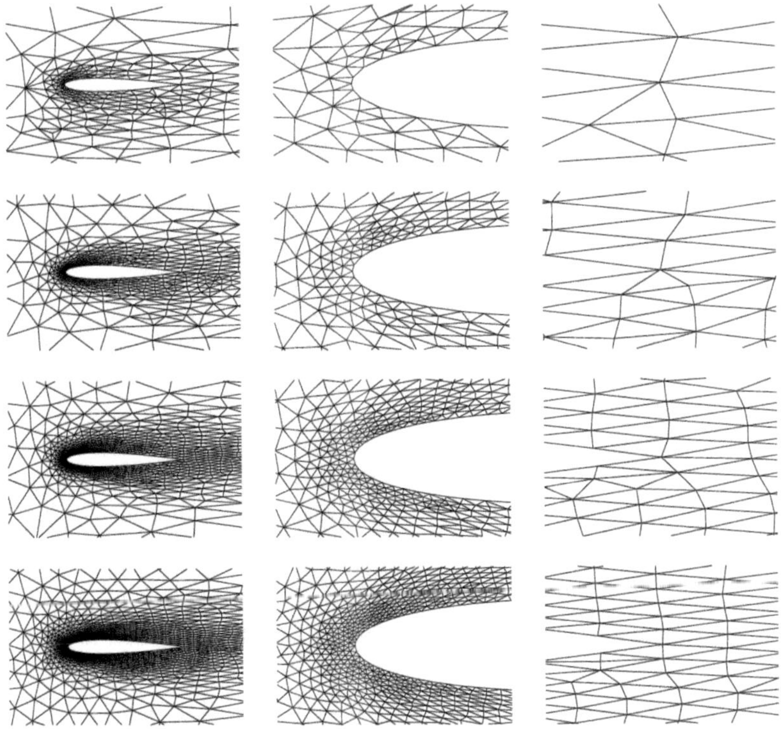

$N1$

$N2$

$N3$

$N4$

Fig. 9.8 Computational grids $N1$–$N4$ around the NACA 0012 profile (*left*) with details around the leading (*middle*) and trailing edges (*right*) used for steady-state examples

$$\frac{1}{\frac{1}{2}\rho_\infty|\boldsymbol{v}_\infty|^2 L_{\text{ref}}}\int_{\Gamma_{\text{prof}}}(\mathrm{p}\mathbb{I}-T^{\mathrm{V}})\boldsymbol{n}\,\mathrm{d}S, \tag{9.97}$$

where ρ_∞ and $\boldsymbol{v}_\infty$ are the far-field density and velocity, respectively, L_{ref} is the reference length, Γ_{prof} is the profile, p is the pressure, $\mathbb{I}$ is the identity matrix and T^{V} is the viscous part of the stress tensor given by (9.8). Moreover, c_M is given by

$$\frac{1}{\frac{1}{2}\rho_\infty|\boldsymbol{v}_\infty|^2 L_{\text{ref}}^2}\int_{\Gamma_{\text{prof}}}(x-x_{\text{ref}})\times\left((\mathrm{p}\mathbb{I}-T^{\mathrm{V}})\boldsymbol{n}\right)\mathrm{d}S, \tag{9.98}$$

where $x_{\text{ref}} = (\frac{1}{4}, 0)$ is the moment reference point. We use the notation $x \times y = x_1 y_2 - x_2 y_1$ for $x = (x_1, x_2)$, $y = (y_1, y_2) \in \mathbb{R}^2$.

For each flow regime C1, C2 and C3, we carried out computations with polynomial approximation P_p, $p = 1, 3, 5$, on grids $N1$–$N4$. We apply the stopping criterion (8.174) with tolerance tol $= 10^{-4}$.

Tables 9.2, 9.3 and 9.4 show the values of the corresponding drag, lift and moment coefficients for each computation. These tables show also the number N_h of elements

Table 9.2 NACA 0012 ($M_\infty = 0.5$, $\alpha = 0°$, Re $= 500$): the values of the drag, lift and moment coefficient obtained by the BDF-DGM for P_p, $p = 1, 3, 5$, polynomial approximations on grids $N1$–$N4$

p	N_h	N_{hp}	c_D	c_L	c_M
1	782	9384	1.7416E–01	1.0260E–01	−3.3278E-03
1	1442	17304	1.7632E–01	1.1225E–01	−2.8440E–03
1	2350	28 200	1.7767E–01	1.1291E–01	−2.8089E–03
1	3681	44 172	1.7775E–01	1.1338E–01	−2.8734E–03
3	782	31 280	1.8086E–01	1.1283E–01	−3.1439E–03
3	1442	57680	1.8093E–01	1.1284E–01	−3.1186E–03
3	2350	94 000	1.8080E–01	1.1322E–01	−3.0036E–03
3	3681	147 240	1.8085E–01	1.1302E–01	−3.0590E–03
5	782	65688	1.8077E–01	1.1269E–01	−3.1054E–03
5	1442	121 128	1.8085E–01	1.1299E–01	−3.0896E–03
5	2350	197 400	1.8087E–01	1.1310E–01	−3.0601E–03
5	3681	309 204	1.8088E–01	1.1304E–01	−3.0719E–03

Table 9.3 NACA 0012 ($M_\infty = 0.5$, $\alpha = 0°$, Re $= 2\,000$): the values of the drag, lift and moment coefficient obtained by the BDF-DGM for P_p, $p = 1, 3, 5$, polynomial approximations on grids $N1$–$N4$

p	N_h	N_{hp}	c_D	c_L	c_M
1	782	9384	8.5405E–02	9.0263E–02	−6.7673E–03
1	1442	17304	8.5231E–02	8.2415E–02	−9.7498E–03
1	2350	28 200	8.6387E–02	8.0999E–02	−1.0283E–02
1	3681	44 172	8.6219E–02	8.2633E–02	−1.0149E–02
3	782	31 280	8.7319E–02	8.5077E–02	−1.0116E–02
3	1442	57 680	8.8193E–02	8.4048E–02	−1.0124E–02
3	2350	94 000	8.8148E–02	8.4091E–02	−1.0079E–02
3	3681	147 240	8.8264E–02	8.4082E–02	−1.0094E–02
5	782	65 688	8.8124E–02	8.4008E–02	−1.0048E–02
5	1442	121 128	8.8281E–02	8.4201E–02	−1.0091E–02
5	2350	197 400	8.8283E–02	8.4290E–02	−1.0075E–02
5	3681	309 204	8.8284E–02	8.4317E–02	−1.0068E–02

of each mesh and corresponding number of degrees of freedom N_{hp}. We observe that the high degree polynomial approximation gives a sufficiently accurate solution even on coarse grids. On the other hand, P_1 polynomial approximation is not sufficiently accurate even for the finest mesh.

Further, Figs. 9.9, 9.10, 9.11, 9.12, 9.13 and 9.14 show Mach number isolines and the distribution of the skin friction coefficient (9.96) obtained for each flow regime on the meshes $N1$ and $N4$.

Table 9.4 NACA 0012 ($M_\infty = 0.85, \alpha = 0°$, Re = 2 000): the values of the drag, lift and moment coefficient obtained by the BDF-DGM for P_p, $p = 1, 3, 5$, polynomial approximations on grids $N1$–$N4$

p	N_h	N_{hp}	c_D	c_L	c_M
1	782	9384	1.1610E–01	4.4091E–02	−1.4702E–02
1	1442	17 304	1.1444E–01	3.8107E–02	−1.5934E–02
1	2350	28 200	1.1605E–01	3.4837E–02	−1.6923E–02
1	3681	44 172	1.1566E–01	3.3338E–02	−1.7027E–02
3	782	31 280	1.1809E–01	3.1726E–02	−1.7463E–02
3	1442	57 680	1.1892E–01	3.1212E–02	−1.7163E–02
3	2350	94 000	1.1887E–01	3.0834E–02	−1.7164E–02
3	3681	147 240	1.1898E-01	3.0918E-02	−1.7142E-02
5	782	65 688	1.1885E–01	3.1034E–02	−1.7048E–02
5	1442	121 128	1.1899E–01	3.1056E–02	−1.7128E–02
5	2350	197 400	1.1899E–01	3.0971E–02	−1.7154E–02
5	3681	309 204	1.1899E–01	3.0981E–02	−1.7148E–02

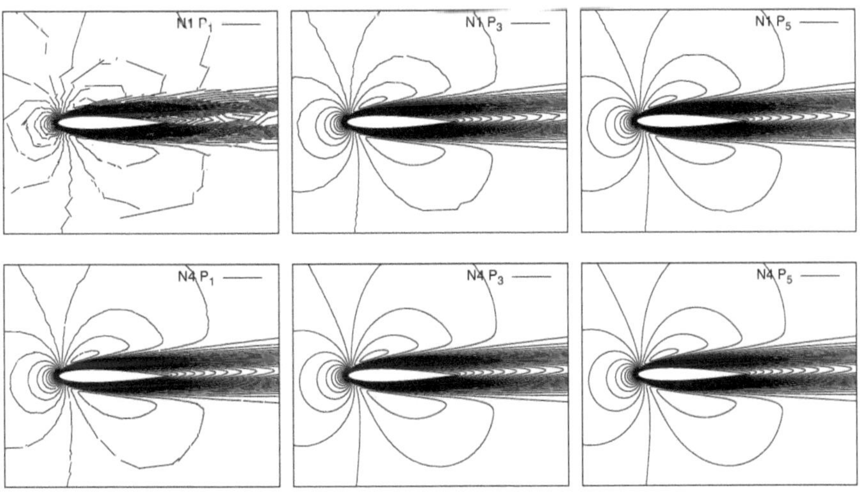

Fig. 9.9 NACA 0012 ($M_\infty = 0.5, \alpha = 2°$, Re = 500). Mach number isolines for P_1, P_3 and P_5 polynomial approximations on grids $N1$ and $N4$

The presented numerical results of examples C1, C2 and C3 show that the high-order DG method is suitable for the numerical solution of the compressible viscous flow. With the aid of the P_5 polynomial approximation we obtain the aerodynamic coefficients with sufficient accuracy even on the coarsest grid.

Finally, we demonstrate the stability of the time discretization schemes with respect to the size of the time steps. According to (8.150), we define the value

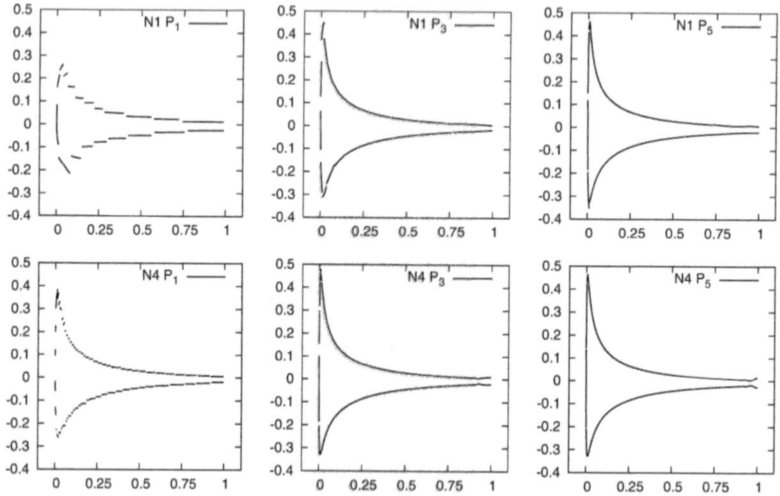

Fig. 9.10 NACA 0012 ($M_\infty = 0.5, \alpha = 2°, \mathrm{Re} = 500$): distribution of the skin friction coefficient for P_1, P_3 and P_5 polynomial approximations on grids $N1$ and $N4$

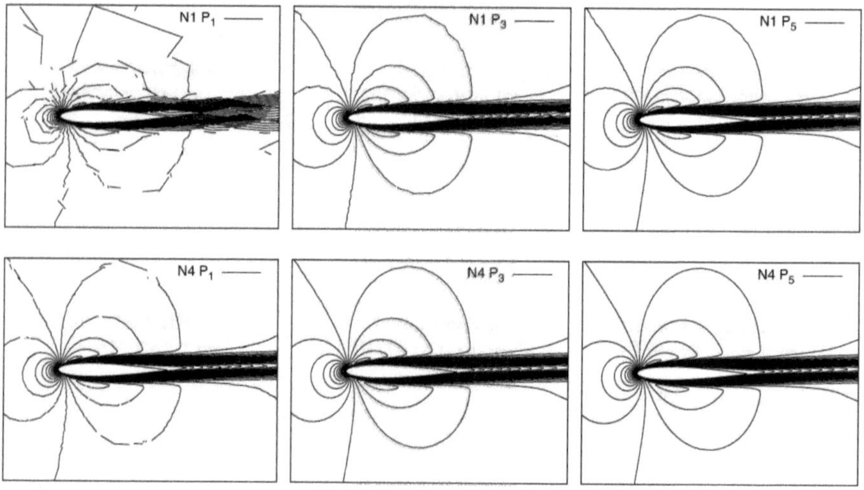

Fig. 9.11 NACA 0012 ($M_\infty = 0.5, \alpha = 2°, \mathrm{Re} = 2000$): Mach number isolines for P_1, P_3 and P_5 polynomial approximations on grids $N1$ and $N4$

$$\mathrm{CFL}_k = \frac{\tau_k}{\min_{K\in\mathscr{T}_h}\left(|K|^{-1}\max_{\Gamma\in\partial K}\varrho(\mathbb{P}(\boldsymbol{w}_h^k|_\Gamma))|\Gamma|\right)}, \quad k = 0, 1, \ldots, r, \tag{9.99}$$

which measures how many times the time step is larger in comparison to the time step for an explicit time discretization. Here $\varrho(\mathbb{P}(\boldsymbol{w}_h^k|_\Gamma))$ denotes the spectral radius of the matrix $\mathbb{P}(\boldsymbol{w}_h^k|_\Gamma)$ defined by (9.23). Figure 9.15 shows the dependence of CFL_k on the

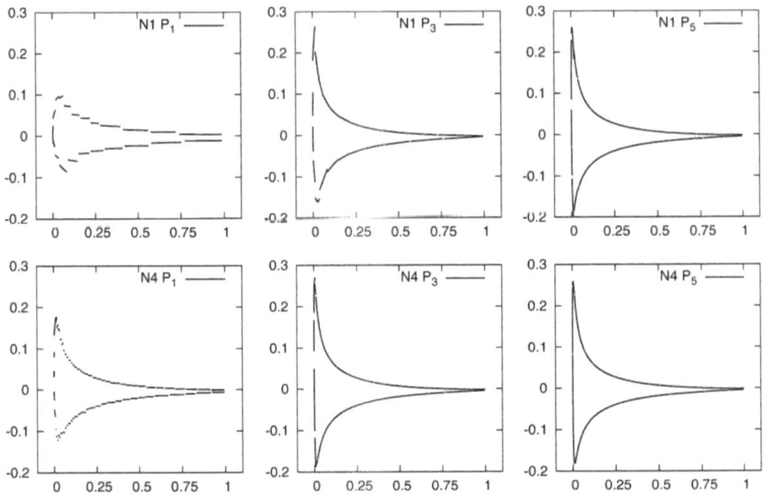

Fig. 9.12 NACA 0012 ($M_\infty = 0.5, \alpha = 2°, \mathrm{Re} = 2000$): distribution of the skin friction coefficient for P_1, P_3 and P_5 polynomial approximations on grids $N1$ and $N4$

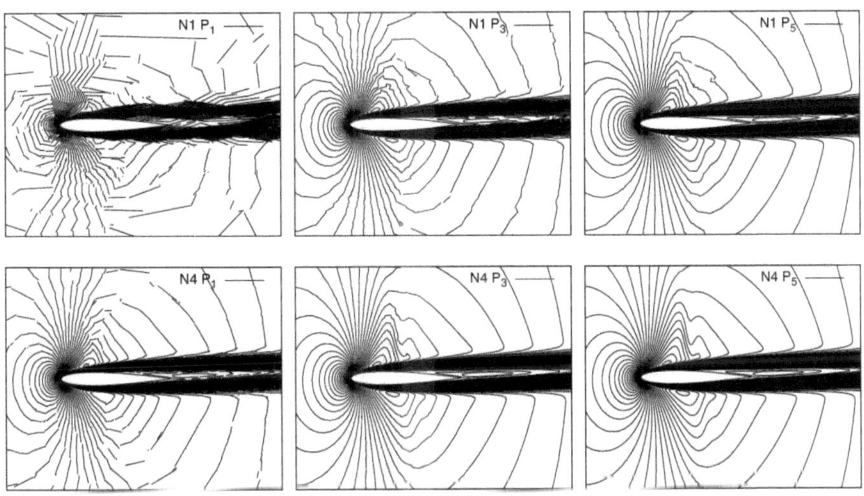

Fig. 9.13 NACA 0012 ($M_\infty = 0.85$, $\alpha = 2°$, $\mathrm{Re} = 2000$): Mach number isolines for P_1, P_3 and P_5 polynomial approximations on grids $N1$ and $N4$

parameter k for the flow regime C1, C2 and C3 using P_1 polynomial approximation on grid $N4$. We observe that very large values CFL_k are attained, and hence the BDF-DGFE method is practically unconditionally stable.

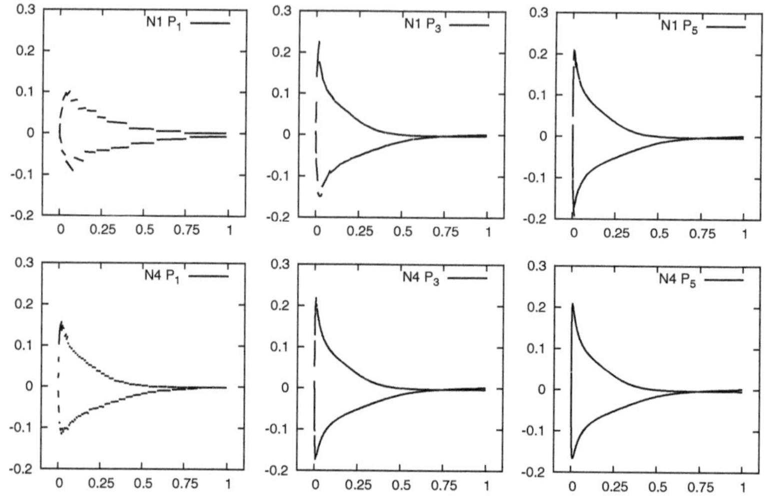

Fig. 9.14 NACA 0012 ($M_\infty = 0.85$, $\alpha = 2°$, Re $= 2000$): distribution of the skin friction coefficient for P_1, P_3 and P_5 polynomial approximations on grids $N1$ and $N4$

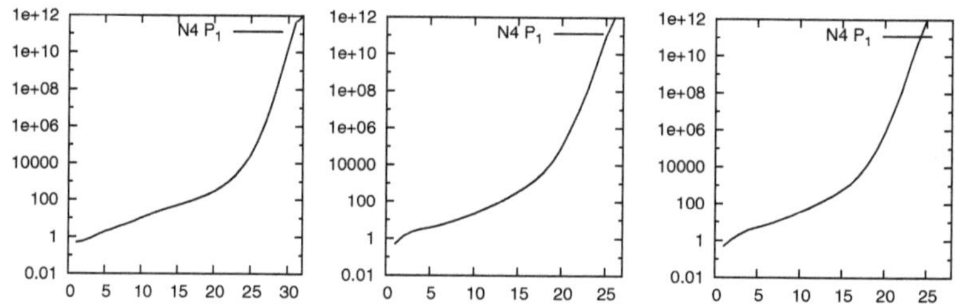

Fig. 9.15 Dependence of the value CFL_k on the parameter k for the flow regimes C1 (*left*), C2 (*center*) and C3 (*right*)

9.4.3 Unsteady Flow

We consider a transonic flow around the NACA 0012 profile with the far-field Mach number $M_\infty = 0.85$, angle of attack $\alpha = 0°$ and the Reynolds number Re $= 10\,000$. In this case the flow is unsteady with a periodic propagation of vortices behind the profile, see [224].

In the numerical simulation of nonstationary processes, it is necessary to use a sufficiently small time step in order to guarantee accuracy with respect to time. In our computations the time step was chosen adaptively with the aid of the adaptive algorithm presented in Sect. 8.4.6 with the tolerance $\omega = 10^{-2}$ in (8.148).

We applied the 3-step BDF-DGM with the P_2 polynomial approximation on the mesh from Fig. 9.16. The computation was carried out for the dimensionless time $t \in (0, 90)$. Figure 9.17 shows the dependence of the lift, drag and moment coefficients

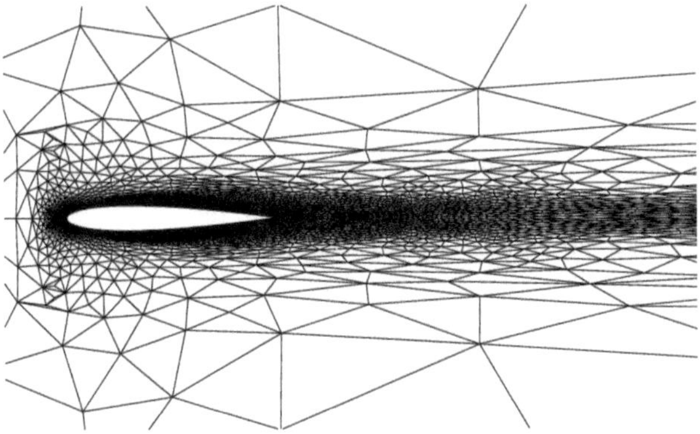

Fig. 9.16 NACA 0012, $M_\infty = 0.85$, $\alpha = 0°$ and $\mathrm{Re} = 10\,000$: triangular grid

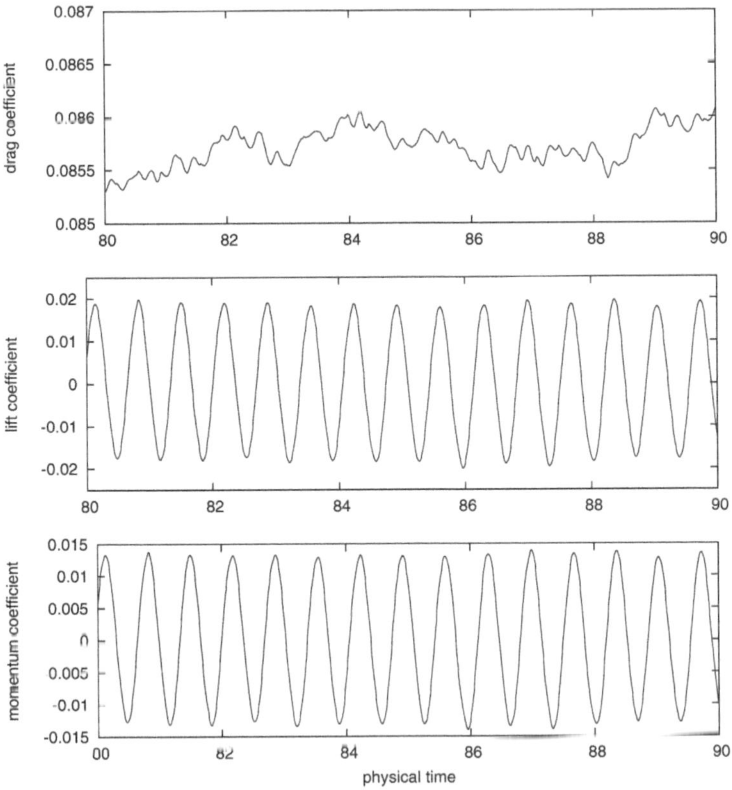

Fig. 9.17 NACA 0012, $M_\infty = 0.85$, $\alpha = 0°$ and $\mathrm{Re} = 10\,000$: dependence of the drag coefficient c_D, lift coefficient c_L and moment coefficient c_M on the dimensionless time $t \in (80, 90)$

on time $t \in (80, 90)$. We observe periodic oscillations of c_L and c_M with period $\Delta t \approx 0.7$. Figure 9.18 shows the Mach number isolines at time instants $t_i = 89.3 + i\Delta t/7,\ i = 1, 2, \ldots, 7$, demonstrating the periodic propagation of vortices behind the profile. These results are in a good agreement with results from [88, 224].

This example demonstrates that the presented BDF-DGM is able to resolve steady as well as unsteady flow without any modification of the scheme. It is very important in the case, when it is not a priori known, whether the considered flow is steady or unsteady.

9.4.4 Steady Versus Unsteady Flow

The numerical examples presented in the previous sections lead us to the conclusion that the presented BDF-DGM is robust with respect to the magnitude of the Mach number and is practically unconditionally stable. This means that large time steps can be used, cf. Fig. 9.15. However, there is a danger that the use of too long time steps can lead to qualitatively different results.

As an example we consider a laminar viscous subsonic flow around the NACA 0012 profile with the far-field Mach number $M_\infty = 0.5$, angle of attack $\alpha = 2°$ and the Reynolds number $\mathrm{Re} = 5\,000$. This flow is close to a limit between the steady and unsteady flow regimes. In [88, 102], we presented steady-state solutions for this flow regime computed using several degrees of polynomial approximation and several grids.

Here we present computations carried out by the 3-step BDF-DGM with P_3 and P_4 polynomial approximation, applied on an unstructured mesh shown in Fig. 9.19. The time steps were chosen adaptively with the aid of the adaptive algorithm presented in Sect. 8.4.6 with two different tolerances $\omega = 1$ and $\omega = 10^{-4}$ in (8.148). This means that in the former case we do not take care of the accuracy with respect to time. In the latter case, the problem was solved with a high accuracy with respect to time. Of course, the computation needs much longer CPU time.

Figure 9.20 shows the convergence of the steady-state residuum (cf. the criterion (8.171) adapted to the viscous flow problem) and the corresponding value CFL_k (cf. (9.99)) for both settings $\omega = 1$ and $\omega = 10^{-4}$.

It can be seen that for $\omega = 1$ a steady-state solution is obtained. On the other hand, for $\omega = 10^{-4}$ the resolution in time is much more accurate and an unsteady solution is obtained. Moreover, Fig. 9.21 shows the dependence of the lift coefficient c_L on the dimensionless time for P_3 and P_4 polynomial approximations with $\omega = 10^{-4}$ in (8.148). The constant value c_L-'steady' was obtained with the same method but with $\omega = 1$. Finally, Fig. 9.22 shows Mach number isolines for P_3 and P_4 polynomial approximations and for $\omega = 1$ and $\omega = 10^{-4}$.

These experiments indicate that an insufficiently accurate resolution with respect to time can lead to different flow regimes (steady versus unsteady). These results are in agreement with [203], where this example was solved by several research groups. They achieved mostly the steady state solution using steady-state solvers or implicit

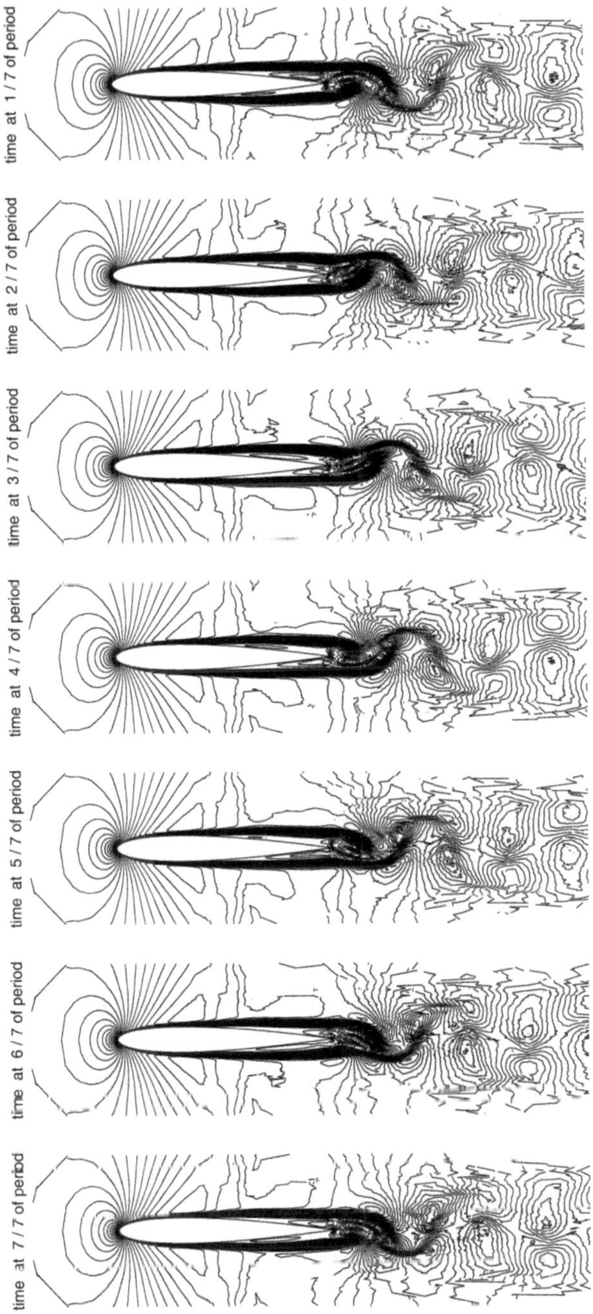

Fig. 9.18 NACA 0012, $M_\infty = 0.85$, $\alpha = 0°$ and Re = 10 000: Mach number isolines at the time instants $t_i = 89.3 + i\,\Delta t/7$, $i = 1, \ldots, 7$, in one period

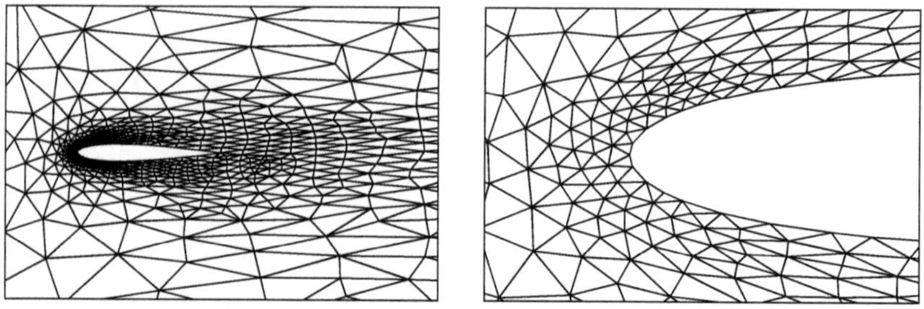

Fig. 9.19 NACA 0012, $M_\infty = 0.5, \alpha = 0°$and Re = 5 000: computational grid, around the profile (*left*) and a detail at the leading edge (*right*)

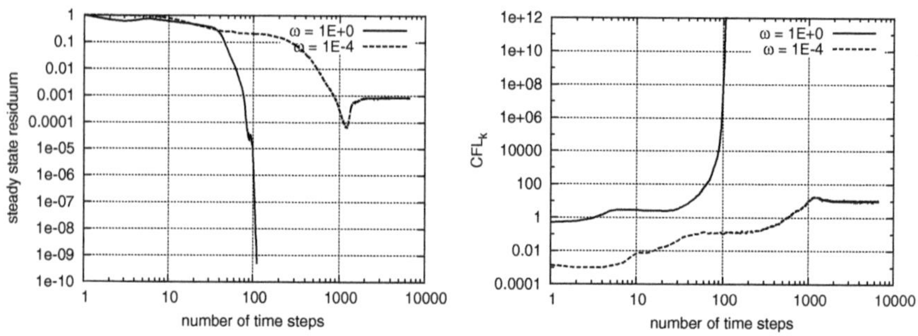

Fig. 9.20 NACA 0012, $M_\infty = 0.5$, $\alpha = 0°$and Re $= 5\,000$, P_4 approximation, $\omega = 1$ and $\omega = 10^{-4}$: steady-state residuum (*left*) and the value CFL_k (*right*) with respect to the number of time steps

time discretizations with large time steps. Only a sufficiently accurate (explicit) time discretization (carried out at the University of Stuttgart) gave the unsteady flow regime, see [203, Chap. 5].

9.4.5 Viscous Shock-Vortex Interaction

This example represents a challenging unsteady viscous flow simulation. Similarly as in [70, 143, 267], we consider the viscous interaction of a plane weak shock wave with a single isentropic vortex. During the interaction, acoustic waves are produced, and we investigate the ability of the numerical scheme to capture these waves. The computational domain is $\Omega = (0, 2) \times (0, 2)$ with the periodic extension in the x_2-direction. A stationary plane shock wave is located at $x_1 = 1$. The prescribed pressure jump through the shock is $\mathrm{p}_R - \mathrm{p}_L = 0.4$, where p_L and p_R are the pressure values from the left and right of the shock wave, respectively, corresponding to the inlet (left) Mach number $M_L = 1.1588$. The reference density and velocity are those of the free uniform flow at infinity. In particular, we define the initial density, x_1-component of velocity and pressure by

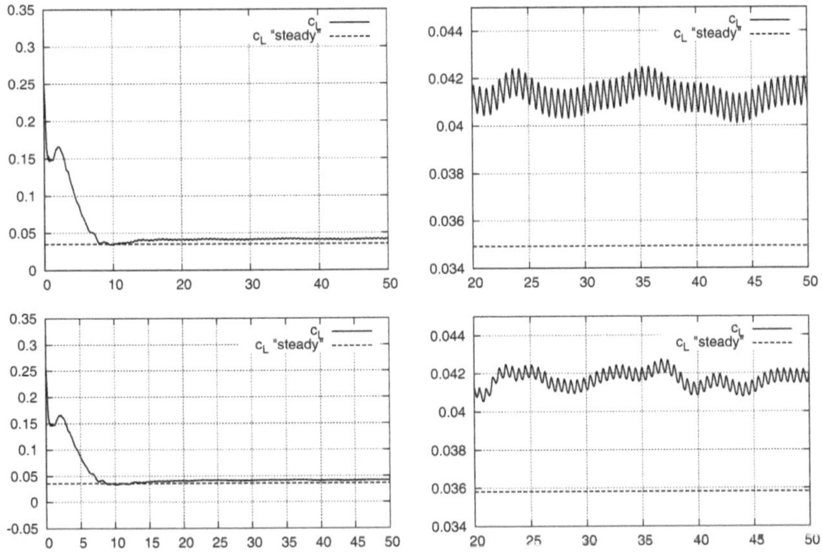

Fig. 9.21 NACA 0012, $M_\infty = 0.5$, $\alpha = 0°$and Re $= 5\,000$: P_3 (*top*) and P_4 (*bottom*) approximation, time evolution of the lift coefficient c_L with respect to the physical time for the setting $\omega = 10^{-4}$ (*left*) and its detail (*right*), the value c_L "steady" was obtain with $\omega = 1$

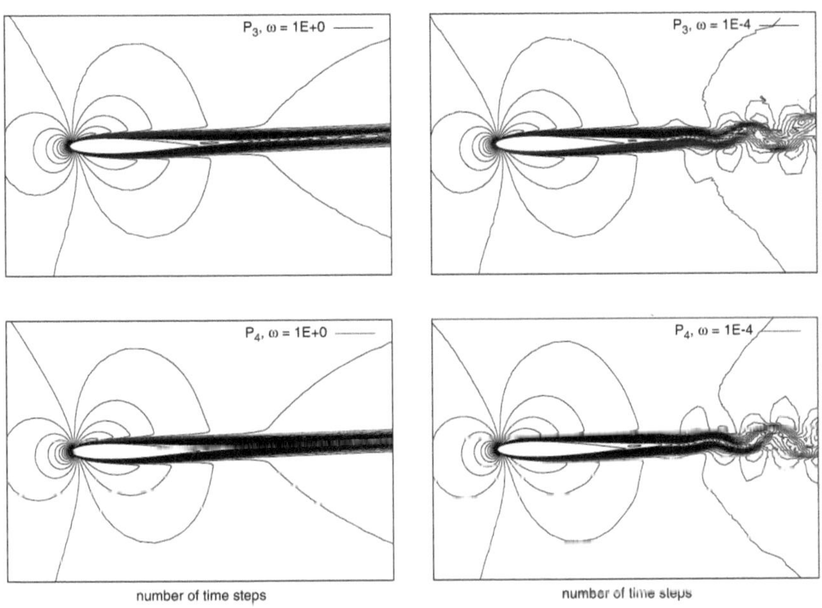

Fig. 9.22 NACA 0012, $M_\infty = 0.5$, $\alpha = 0°$and Re $= 5\,000$ for P_3 and P_4 polynomial approximations and for $\omega = 1$ and $\omega = 10^{-4}$: Mach number isolines

$$\rho_L = 1,\ u_L = M_L \gamma^{1/2},\ \mathrm{p}_L = 1,\quad \rho_R = \rho_L K_1,\ u_R = u_L K_1^{-1},\ \mathrm{p}_R = \mathrm{p}_1 K_2, \tag{9.100}$$

where

$$K_1 = \frac{\gamma+1}{2}\frac{M_L^2}{1+\frac{\gamma-1}{2}M_L^2},\quad K_2 = \frac{2}{\gamma+1}\left(\gamma M_L^2 - \frac{\gamma-1}{2}\right). \tag{9.101}$$

Here, the subscripts $_L$ and $_R$ denote the quantities at $x < 1$ and $x > 1$, respectively, $\gamma = 1.4$ is the Poisson constant. The Reynolds number is 2 000. An isolated isentropic vortex centered at (0.5, 1) is added to the basic flow. The angular velocity in the vortex is given by

$$v_\theta = c_1 r \exp(-c_2 r^2),\quad c_1 = u_c/r_c,\quad c_2 = r_c^{-2}/2, \qquad r = ((x_1 - 0.5)^2 - (x_2 - 1)^2)^{1/2}, \tag{9.102}$$

where we set $r_c = 0.075$ and $u_c = 0.5$. Computations are stopped at the dimensionless time $T = 0.7$.

We solved this problem with the aid of the 3-steps BDF-DGM (9.74) with P_4 polynomial approximation in space. The computational grid with 3 072 triangles was a priori refined in the vicinity of the stationary shock wave, see Fig. 9.23. This figure shows also the initial setting of the shock wave and the isentropic vortex with their details.

Figures 9.24 and 9.25 show the results of the simulation of viscous shock-vortex interaction, namely, the isolines of the pressure and the pressure distribution along $x_2 = 1$ at several time instants. We observe a capturing of the shock-vortex interaction with the appearance of incident and reflected acoustic waves. These results are in agreement with results presented in [70, 143, 267]. Hence, we can conclude that the DGM is able to capture such complicated physical phenomena as shock-vortex interaction.

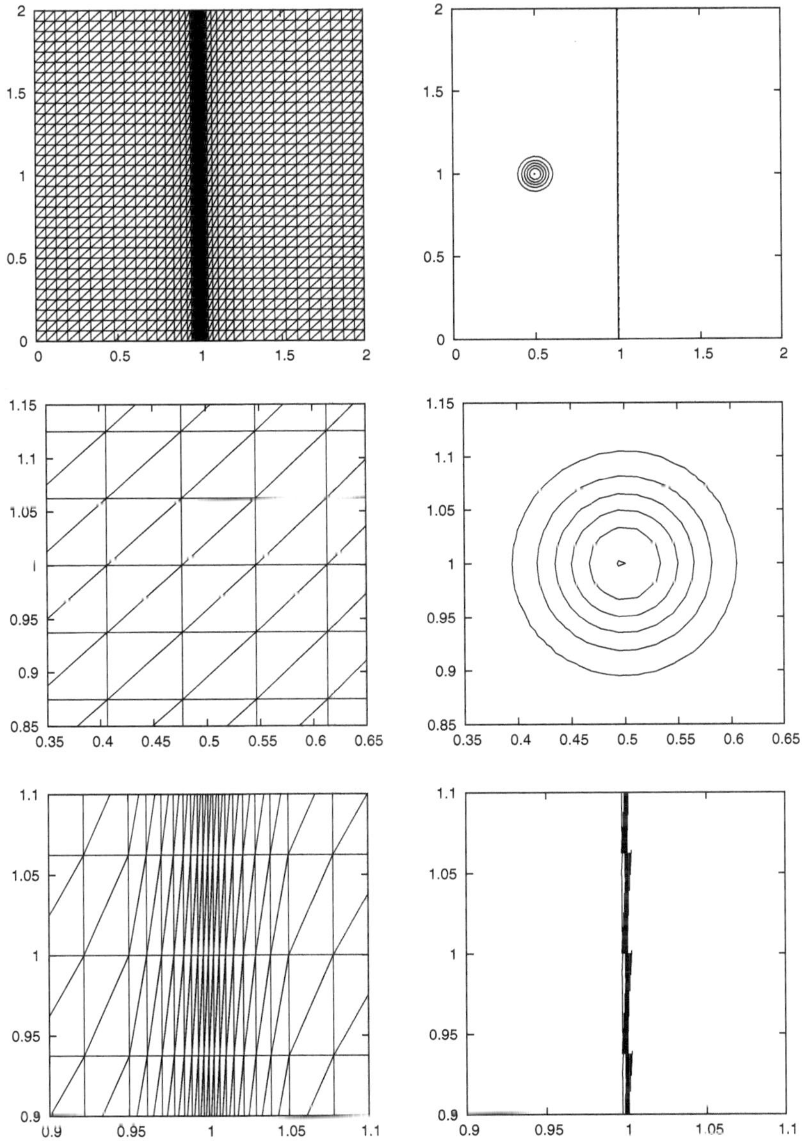

Fig. 9.23 Viscous shock vortex interaction: the used grid (*left*) and pressure isolines (*right*) at $t = 0$, the total view (*top*), its details near the vortex (*center*) and the shock wave (*bottom*)

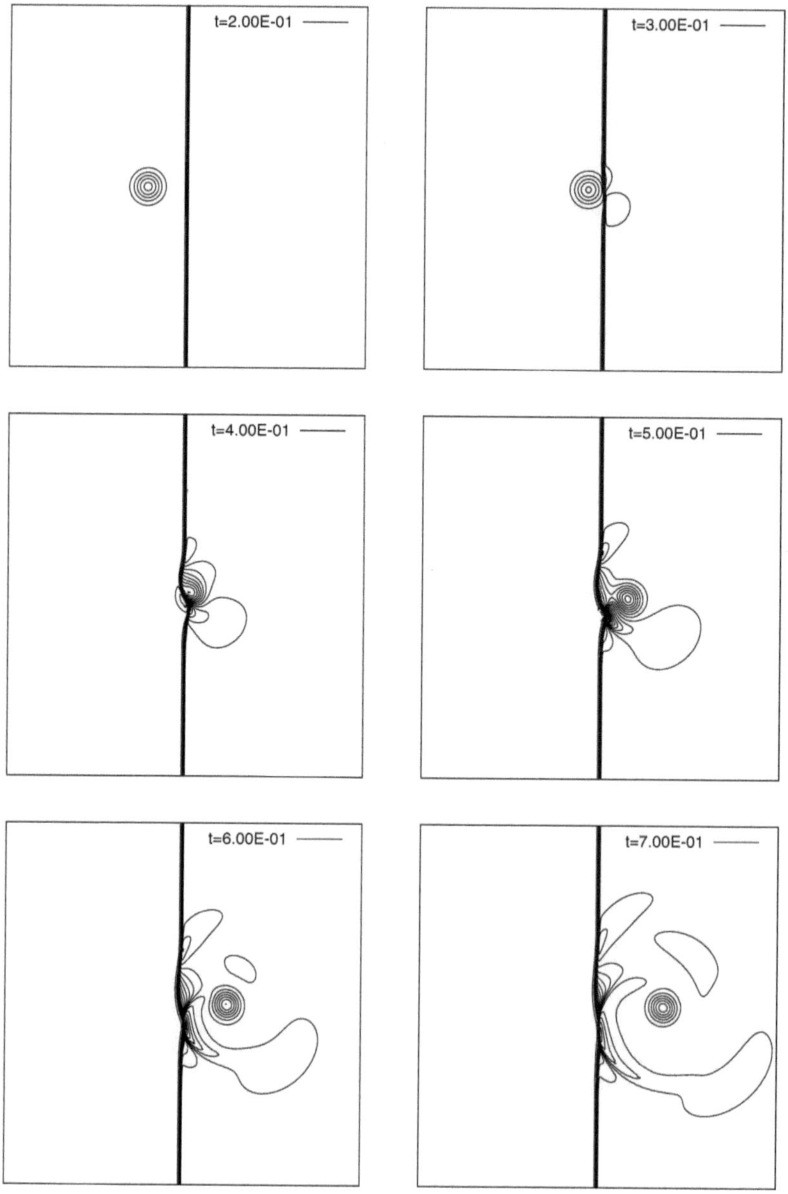

Fig. 9.24 Viscous shock-vortex interaction: pressure isolines at $t = 0.2, 0.3, 0.4, 0.5, 0.6$ and 0.7

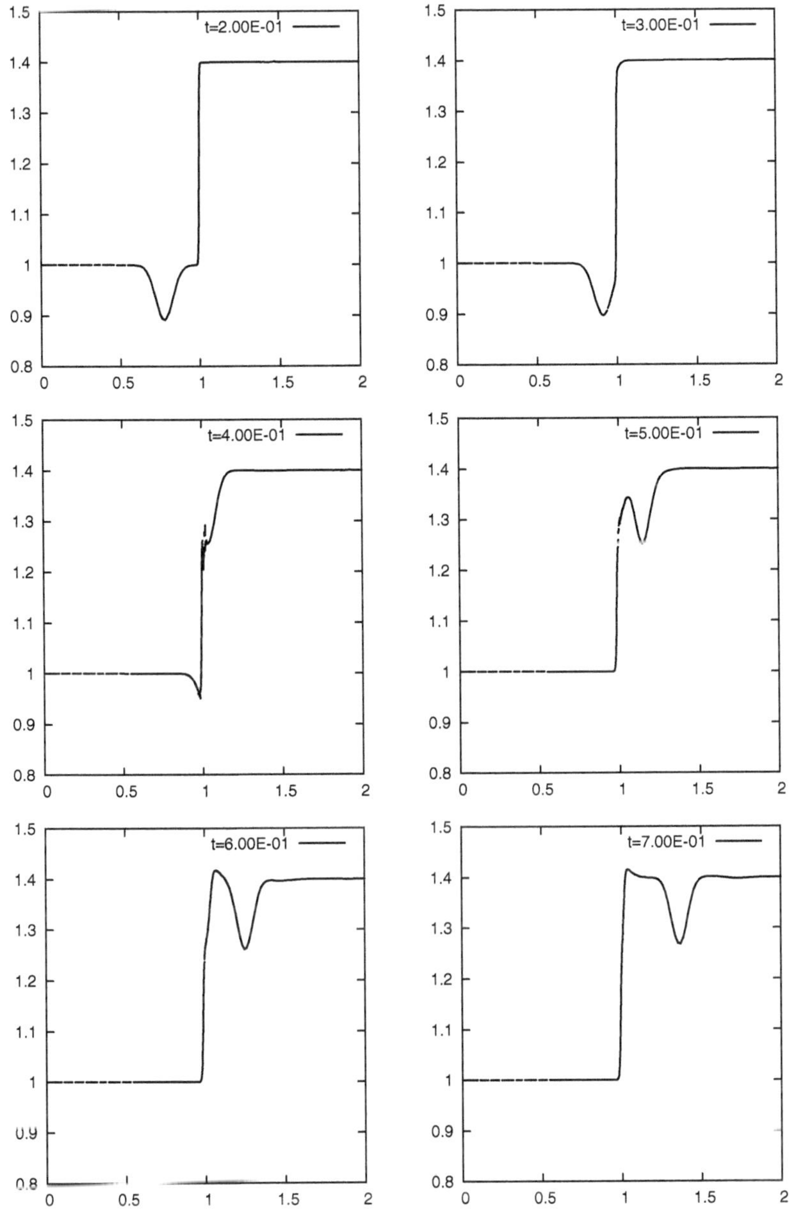

Fig. 9.25 Viscous shock-vortex interaction: the pressure distribution along the line $x_2 = 1$ at $t = 0.2, 0.3, 0.4, 0.5, 0.6$ and 0.7

Chapter 10
Fluid-Structure Interaction

Simulating a flow in time dependent domains is a significant part of fluid-structure interaction. It plays an important role in many disciplines. We mention, for example, construction of airplanes (vibrations of wings) or turbines (vibrations of blades), some problems in civil engineering (interaction of wind with constructions of bridges, TV towers or cooling towers of power stations), car industry (vibrations of various elements of a coachwork), but also in medicine (haemodynamics or flow of air in vocal folds). In a number of these examples the moving medium is a gas and the flow is compressible. For low Mach number flows, incompressible models are often used (as e.g., in [266]), but in some cases compressibility plays an important role.

In this chapter we describe the discontinuous Galerkin method applied to the numerical solution of compressible flow in time dependent domains and present some applications to problems in fluid-structure interaction. The main ingredient of this technique is the ALE (arbitrary Lagrangian–Eulerian) formulation of the compressible Navier–Stokes equations, which is discretized by modifying the DGM described in the previous chapter.

10.1 Formulation of Flow in a Time-Dependent Domain

We will be concerned with the numerical solution of a compressible flow in a bounded domain $\Omega_t \subset \mathbb{R}^d$ ($d = 2$ or 3) depending on time $t \in [0, T]$, $0 < T < \infty$. We start from the system of the compressible Navier–Stokes equations written in the dimensionless form (9.12), i.e.,

$$\frac{\partial \boldsymbol{w}}{\partial t} + \sum_{s=1}^{d} \frac{\partial \boldsymbol{f}_s(\boldsymbol{w})}{\partial x_s} = \sum_{s=1}^{d} \frac{\partial \boldsymbol{R}_s(\boldsymbol{w}, \nabla \boldsymbol{w})}{\partial x_s} \quad \text{in } Q_T, \tag{10.1}$$

V. Dolejší and M. Feistauer, *Discontinuous Galerkin Method*,
Springer Series in Computational Mathematics 48,
DOI 10.1007/978-3-319-19267-3_10

where we set $Q_T = \{(x,t);\ x \in \Omega_t,\ t \in (0,T)\}$. We use the notation (9.13)–(9.33) from Chap. 9.

In order to take into account the time dependence of the domain, we use the *arbitrary Lagrangian-Eulerian (ALE) method*, proposed, e.g., in [229]. We define a *reference domain* Ω_0 (also called reference configuration) and introduce a regular one-to-one *ALE mapping* of Ω_0 onto the *current configuration* Ω_t (cf. [229], [266] and [272])

$$\mathrm{A}_t : \overline{\Omega}_0 \longrightarrow \overline{\Omega}_t, \text{ i.e., } X \in \overline{\Omega}_0 \longmapsto x = x(X,t) = \mathrm{A}_t(X) \in \overline{\Omega}_t.$$

Here we use the notation X for points in $\overline{\Omega}_0$ and $x = x(X,t)$ for points in $\overline{\Omega}_t$.

Further, we define the *domain velocity*:

$$\begin{aligned} \tilde{z}(X,t) &= \frac{\partial}{\partial t}\mathrm{A}_t(X), \quad t \in [0,T],\ X \in \Omega_0, \\ z(x,t) &= \tilde{z}(\mathrm{A}_t^{-1}(x),t), \quad t \in [0,T],\ x \in \Omega_t, \end{aligned}$$

and the *ALE derivative* of a function $f = f(x,t)$ defined for $x \in \Omega_t$ and $t \in [0,T]$:

$$\frac{\mathrm{D}^{\mathrm{A}}}{\mathrm{D}t} f(x,t) = \frac{\partial \tilde{f}}{\partial t}(X,t), \tag{10.2}$$

where

$$\tilde{f}(X,t) = f(\mathrm{A}_t(X),t),\ X \in \Omega_0,\ x = \mathrm{A}_t(X). \tag{10.3}$$

As a direct consequence of the chain rule we get the relation

$$\frac{\mathrm{D}^{\mathrm{A}} f}{\mathrm{D}t} = \frac{\partial f}{\partial t} + \nabla \cdot (z f) - f\, \nabla \cdot z.$$

This leads to the *ALE formulation of the Navier–Stokes equations*

$$\frac{\mathrm{D}^{\mathrm{A}} \boldsymbol{w}}{\mathrm{D}t} + \sum_{s=1}^{d} \frac{\partial \boldsymbol{g}_s(\boldsymbol{w})}{\partial x_s} + \boldsymbol{w}\, \nabla \cdot z = \sum_{s=1}^{d} \frac{\partial \boldsymbol{R}_s(\boldsymbol{w}, \nabla \boldsymbol{w})}{\partial x_s}, \tag{10.4}$$

where

$$\boldsymbol{g}_s(\boldsymbol{w}) := \boldsymbol{f}_s(\boldsymbol{w}) - z_s \boldsymbol{w}, \quad s = 1, \ldots, d, \tag{10.5}$$

are the *ALE modified inviscid fluxes*. We see that the partial time derivative $\partial/\partial t$ in (10.1) is replaced by the ALE derivative $\mathrm{D}^{\mathrm{A}}/\mathrm{D}t$, the inviscid Euler fluxes $\boldsymbol{f}_s(\boldsymbol{w})$ are replaced by the ALE modified inviscid fluxes $\boldsymbol{g}_s(\boldsymbol{w})$ and new linear reaction term $\boldsymbol{w}\, \nabla \cdot z$ appears in (10.4).

System (10.4) is equipped with the initial condition

$$w(x, 0) = w^0(x), \quad x \in \Omega_0, \tag{10.6}$$

and boundary conditions similar to (9.35)–(9.37). We assume that $\partial\Omega_t = \partial\Omega_i \cup \partial\Omega_o \cup \partial\Omega_{W_t}$ is a disjoint partition of the boundary $\partial\Omega_t$, where the inlet $\partial\Omega_i$ and outlet $\partial\Omega_o$ are independent of time and the set $\partial\Omega_{W_t}$ represents impermeable walls, whose part may move in dependence on time. We prescribe the boundary conditions

$$\rho = \rho_D, \quad \boldsymbol{v} = \boldsymbol{v}_D, \quad \sum_{k=1}^{d}\left(\sum_{l=1}^{d} \tau_{lk}^{V} n_l\right) v_k + \frac{\gamma}{\mathrm{Re\,Pr}} \frac{\partial \theta}{\partial \boldsymbol{n}} = 0 \quad \text{on } \partial\Omega_i, \tag{10.7}$$

$$\sum_{k=1}^{d} \tau_{sk}^{V} n_k = 0, \ s = 1, \ldots, d, \quad \frac{\partial \theta}{\partial \boldsymbol{n}} = 0 \quad \text{on } \partial\Omega_o, \tag{10.8}$$

$$\boldsymbol{v} = \boldsymbol{z}_D, \quad \frac{\partial \theta}{\partial \boldsymbol{n}} = 0 \quad \text{on } \partial\Omega_{W_t}, \tag{10.9}$$

where ρ_D, $\boldsymbol{v}_D$ and $\boldsymbol{z}_D$ are given functions and $\boldsymbol{n} = (n_1, \ldots, n_d)$ is an outer unit normal to $\partial\Omega_t$. The vector function $\boldsymbol{z}_D$ represents the velocity of the moving wall.

10.1.1 Space Discretization of the Flow Problem

For the space semidiscretization we use the discontinuous Galerkin method. We proceed in a similar way as in the previous chapter. We construct a polygonal (or polyhedral) approximation Ω_{ht} of the domain Ω_t. The parts $\partial\Omega_i$, $\partial\Omega_o$ and $\partial\Omega_{W_t}$ of the boundary $\partial\Omega_t$ are approximated by parts $\partial\Omega_{hi}$, $\partial\Omega_{ho}$ and $\partial\Omega_{hW_t}$, respectively, of $\partial\Omega_{ht}$. By $\mathscr{T}_{ht}$ we denote a partition of the closure $\overline{\Omega}_{ht}$ of the domain Ω_{ht} into a finite number of closed simplexes K with mutually disjoint interiors such that $\overline{\Omega}_{ht} = \bigcup_{K \in \mathscr{T}_{ht}} K$. By $\mathscr{F}_{ht}$ we denote the system of all faces of all elements $K \in \mathscr{T}_{ht}$. Further, we introduce the set of boundary faces $\mathscr{F}_{ht}^B = \{\Gamma \in \mathscr{F}_{ht};\ \Gamma \subset \partial\Omega_{ht}\}$, the set of "Dirichlet" boundary faces $\mathscr{F}_{ht}^D = \left\{\Gamma \in \mathscr{F}_{ht}^B;\ \text{a Dirichlet condition is given on } \Gamma\right\}$ and the set of inner faces $\mathscr{F}_{ht}^I = \mathscr{F}_{ht} \setminus \mathscr{F}_{ht}^B$. Moreover, in $\mathscr{F}_{ht}^B$ we distinguish the sets $\mathscr{F}_{ht}^i$, $\mathscr{F}_{ht}^o$ and $\mathscr{F}_{ht}^W$ of all inlet, outlet and wall faces, respectively, and put $\mathscr{F}_{ht}^{io} = \mathscr{F}_{ht}^i \cup \mathscr{F}_{ht}^o$.

Each $\Gamma \in \mathscr{F}_{ht}$ is associated with a unit vector $\boldsymbol{n}_\Gamma$ normal to Γ. For $\Gamma \subset \mathscr{F}_{ht}^B$ the normal $\boldsymbol{n}_\Gamma$ has the same orientation as the outer normal to $\partial\Omega_{ht}$. We set $d(\Gamma) =$ diameter of $\Gamma \in \mathscr{F}_{ht}$. For each $\Gamma \subset \mathscr{F}_{ht}^I$ there exist two neighbouring elements $K_\Gamma^{(L)}, K_\Gamma^{(R)} \in \mathscr{T}_{ht}$ such that $\Gamma \subset \partial K_\Gamma^{(R)} \cap \partial K_\Gamma^{(L)}$. We use the convention that $K_\Gamma^{(R)}$ lies in the direction of $\boldsymbol{n}_\Gamma$ and $K_\Gamma^{(L)}$ lies in the opposite direction to $\boldsymbol{n}_\Gamma$. If $\Gamma \in \mathscr{F}_{ht}^B$, then the element adjacent to Γ will be denoted by $K_\Gamma^{(L)}$.

The approximate solution will be sought in the space of piecewise polynomial functions

$$\boldsymbol{S}_{hpt} = (S_{hpt})^{d+2}, \quad \text{with } S_{hpt} = \{v;\, v|_K \in P_p(K)\ \forall\, K \in \mathscr{T}_{ht}\}, \tag{10.10}$$

where $p > 0$ is an integer and $P_p(K)$ denotes the space of all polynomials on K of degree $\leq p$. For any function $\boldsymbol{\varphi} \in \boldsymbol{S}_{hpt}$ and any face $\Gamma \in \mathscr{F}_{ht}$ we use the standard symbols $\boldsymbol{\varphi}_\Gamma^{(L)}$, and $\boldsymbol{\varphi}_\Gamma^{(R)}$, $\langle \boldsymbol{\varphi} \rangle_\Gamma$ and $[\boldsymbol{\varphi}]_\Gamma$. (See e.g., Sect. 9.2.1.)

The discrete problem is derived in a standard way: We multiply system (10.4) by a test function $\boldsymbol{\varphi}_h \in \boldsymbol{S}_{hpt}$, integrate over $K \in \mathscr{T}_{ht}$, apply Green's theorem, sum over all elements $K \in \mathscr{T}_{ht}$, use the concept of the numerical flux and introduce suitable mutually vanishing terms for a regular exact solution. Moreover, we carry out a linearization of nonlinear terms. Similarly as in Sects. 9.3.2 and 8.4.3, we introduce the partially linearized forms $\hat{\boldsymbol{a}}_h$, $\hat{\boldsymbol{J}}_h^\sigma$, $\boldsymbol{\beta}_h$ and $\boldsymbol{\gamma}_h$, depending now, of course, on time t. The forms $\hat{\boldsymbol{a}}_h$ and $\hat{\boldsymbol{J}}_h^\sigma$ are defined in the same way as in (9.77), (9.78), (9.91) and (9.92), (9.80), (9.81), respectively. For each $t \in I_m$, $m = 1, \ldots, r$, we get the viscous form

$$\begin{aligned}
\hat{\boldsymbol{a}}_h(\bar{\boldsymbol{w}}_h, \boldsymbol{w}_h, \boldsymbol{\varphi}_h, t) = &\sum_{K \in \mathscr{T}_{ht}} \int_K \sum_{s,k=1}^{d} \left(\mathbb{K}_{s,k}(\bar{\boldsymbol{w}}_h) \frac{\partial \boldsymbol{w}_h}{\partial x_k} \right) \cdot \frac{\partial \boldsymbol{\varphi}_h}{\partial x_s} \, \mathrm{d}x \\
&- \sum_{\Gamma \in \mathscr{F}_{ht}^I} \int_\Gamma \sum_{s=1}^{d} \left\langle \sum_{k=1}^{d} \mathbb{K}_{s,k}(\bar{\boldsymbol{w}}_h) \frac{\partial \boldsymbol{w}_h}{\partial x_k} \right\rangle n_s \cdot [\boldsymbol{\varphi}_h] \, \mathrm{d}S \\
&- \sum_{\Gamma \in \mathscr{F}_{ht}^i} \int_\Gamma \sum_{s,k=1}^{d} \mathbb{K}_{s,k}(\bar{\boldsymbol{w}}_h) \frac{\partial \boldsymbol{w}_h}{\partial x_k} n_s \cdot \boldsymbol{\varphi}_h \, \mathrm{d}S \\
&- \sum_{\Gamma \in \mathscr{F}_{ht}^W} \int_\Gamma \sum_{s,k=1}^{d} \mathbb{K}_{s,k}^W(\bar{\boldsymbol{w}}_h) \frac{\partial \boldsymbol{w}_h}{\partial x_k} n_s \cdot \boldsymbol{\varphi}_h \, \mathrm{d}S \\
&- \Theta \Bigg(\sum_{\Gamma \in \mathscr{F}_{ht}^I} \int_\Gamma \sum_{s,k=1}^{d} \left\langle \mathbb{K}_{s,k}^{\mathrm{T}}(\bar{\boldsymbol{w}}_h) \frac{\partial \boldsymbol{\varphi}_h}{\partial x_k} \right\rangle n_s \cdot [\boldsymbol{w}_h] \, \mathrm{d}S \\
&+ \sum_{\Gamma \in \mathscr{F}_{ht}^i} \int_\Gamma \sum_{s,k=1}^{d} \mathbb{K}_{s,k}^{\mathrm{T}}(\bar{\boldsymbol{w}}_h) \frac{\partial \boldsymbol{\varphi}_h}{\partial x_k} n_s \cdot (\boldsymbol{w}_h - \bar{\boldsymbol{w}}_B) \, \mathrm{d}S \\
&+ \sum_{\Gamma \in \mathscr{F}_{ht}^W} \int_\Gamma \sum_{s,k=1}^{d} \left(\mathbb{K}_{s,k}^W(\bar{\boldsymbol{w}}_h) \right)^{\mathrm{T}} \frac{\partial \boldsymbol{\varphi}_h}{\partial x_k} n_s \cdot (\boldsymbol{w}_h - \bar{\boldsymbol{w}}_B) \, \mathrm{d}S \Bigg).
\end{aligned} \tag{10.11}$$

The parameter Θ can attain the values 1, 0 and -1 for the SIPG, IIPG and NIPG version, respectively. Moreover, the penalty form now reads

$$\hat{\boldsymbol{J}}_h^{\sigma}(\bar{\boldsymbol{w}}_h, \boldsymbol{w}_h, \boldsymbol{\varphi}_h, t) = \sum_{\Gamma \in \mathscr{F}_{ht}^I} \int_\Gamma \sigma [\boldsymbol{w}_h] \cdot [\boldsymbol{\varphi}_h] \, \mathrm{d}S + \sum_{\Gamma \in \mathscr{F}_{ht}^i} \int_\Gamma \sigma (\boldsymbol{w}_h - \bar{\boldsymbol{w}}_B) \cdot \boldsymbol{\varphi}_h \, \mathrm{d}S$$
$$+ \sum_{\Gamma \in \mathscr{F}_{ht}^W} \int_\Gamma \sigma (\boldsymbol{w}_h - \bar{\boldsymbol{w}}_B) \cdot \mathscr{V}(\boldsymbol{\varphi}_h) \, \mathrm{d}S \qquad (10.12)$$

where the operator $\mathscr{V}$ is defined by (9.66). The weight σ in $\hat{\boldsymbol{J}}_h^{\sigma}$ is defined by (9.67). The boundary state $\bar{\boldsymbol{w}}_B$ is obtained on the basis of the Dirichlet boundary conditions in (10.7) and (10.9) and extrapolation:

$$\bar{\boldsymbol{w}}_B|_\Gamma = (\rho_D, \rho_D v_{D1}, \rho_D v_{D2}, c_v \rho_D \bar{\theta}_\Gamma^{(L)} + \frac{1}{2}\rho_D |\boldsymbol{v}_D|^2), \quad \Gamma \subset \partial\Omega_i, \qquad (10.13)$$
$$\bar{\boldsymbol{w}}_B|_\Gamma = \bar{\boldsymbol{w}}_\Gamma^{(L)}, \quad \Gamma \subset \partial\Omega_o, \qquad (10.14)$$
$$\bar{\boldsymbol{w}}_B|_\Gamma = (\bar{\rho}_\Gamma^{(L)}, \bar{\rho}_\Gamma^{(L)} z_{D1}, \bar{\rho}_\Gamma^{(L)} z_{D2}, c_v \bar{\rho}_\Gamma^{(L)} \bar{\theta}_\Gamma^{(L)} + \frac{1}{2}\bar{\rho}_\Gamma^{(L)} |\boldsymbol{z}_D|^2), \quad \Gamma \subset \partial\Omega_{W_t}. \qquad (10.15)$$

The quantities $\bar{\rho}, \bar{\theta}$, etc. correspond to the state $\bar{\boldsymbol{w}}_h$. We see that as in (9.63) it is possible to write $\bar{\boldsymbol{w}}_B = BC(\bar{\boldsymbol{w}}_h, \boldsymbol{u}_D)$, where $\boldsymbol{u}_D$ represents the Dirichlet data.

Further, we define the reaction form

$$\boldsymbol{d}_h(\boldsymbol{w}_h, \boldsymbol{\varphi}_h, t) = \sum_{K \in \mathscr{T}_{ht}} \int_K (\boldsymbol{w}_h \cdot \boldsymbol{\varphi}_h) \nabla \cdot \boldsymbol{z} \, \mathrm{d}x. \qquad (10.16)$$

In order to avoid spurious oscillations in the approximate solution in the vicinity of discontinuities or steep gradients, we apply local artificial viscosity forms. They are based on the discontinuity indicator

$$g_t(K) = \int_{\partial K} [\bar{w}_{h,1}^k]^2 \, \mathrm{d}S / (h_K |K|^{3/4}), \quad K \in \mathscr{T}_{ht}, \qquad (10.17)$$

introduced in (8.178). By $[\bar{w}_{h,1}^k]$ we denote the jump of the density function $\bar{w}_{h,1}^k = \bar{\rho}_h$ on the boundary ∂K, corresponding to the state $\bar{w}_h$, and $|K|$ denotes the volume of the element K. Then we define the discrete discontinuity indicator

$$G_t(K) = 0 \ \text{ if } g_t(K) < 1, \quad G_t(K) = 1 \ \text{ if } g_t(K) \geq 1, \quad K \in \mathscr{T}_{ht}, \qquad (10.18)$$

and the artificial viscosity forms defined in analogy to (8.181) and (8.182):

$$\boldsymbol{\beta}_h(\bar{\boldsymbol{w}}_h, \boldsymbol{w}_h, \boldsymbol{\varphi}_h, t) = \nu_1 \sum_{K \in \mathscr{T}_{ht}} h_K G_t(K) \int_K \nabla \boldsymbol{w}_h \cdot \nabla \boldsymbol{\varphi}_h \, \mathrm{d}x, \tag{10.19}$$

$$\boldsymbol{\gamma}_h(\bar{\boldsymbol{w}}_h, \boldsymbol{w}_h, \boldsymbol{\varphi}_h, t) = \nu_2 \sum_{\Gamma \in \mathscr{F}_{ht}^I} \frac{1}{2}\left(G_t(K_\Gamma^{(L)}) + G_t(K_\Gamma^{(R)})\right) \int_\Gamma [\boldsymbol{w}_h] \cdot [\boldsymbol{\varphi}_h] \, \mathrm{d}S,$$

with parameters $\nu_1, \nu_2 = O(1)$. It is also possible to use more sophisticated local artificial viscosity forms defined in an analogous way as in (8.180), (8.183) and (8.184).

Special attention has to be paid to the convection form $\hat{\boldsymbol{b}}_h$. Denoting by $\mathbb{I}$ the unit matrix and taking into account the definition of $\boldsymbol{g}_s$ in (10.5) and notation (8.14), we have

$$\frac{\mathrm{D}\boldsymbol{g}_s(\boldsymbol{w})}{\mathrm{D}\boldsymbol{w}} = \frac{\mathrm{D}\boldsymbol{f}_s(\boldsymbol{w})}{\mathrm{D}\boldsymbol{w}} - z_s \mathbb{I} = \mathbb{A}_s(\boldsymbol{w}) - z_s \mathbb{I}, \tag{10.20}$$

and can write

$$\mathbb{P}_g(\boldsymbol{w}, \boldsymbol{n}) = \sum_{s=1}^d \frac{\mathrm{D}\boldsymbol{g}_s(\boldsymbol{w})}{\mathrm{D}\boldsymbol{w}} n_s = \sum_{s=1}^d \left(\mathbb{A}_s(\boldsymbol{w}) n_s - z_s n_s \mathbb{I}\right).$$

By Lemma 8.6 (namely, relation (8.29)), this matrix is diagonalizable. It means that there exists a nonsingular matrix $\mathbb{T} = \mathbb{T}(\boldsymbol{w}, \boldsymbol{n})$ such that

$$\mathbb{P}_g = \mathbb{T}^{-1} \boldsymbol{\Lambda} \mathbb{T}, \quad \boldsymbol{\Lambda} = \mathrm{diag}(\lambda_1, \ldots, \lambda_{d+2}),$$

where $\lambda_i = \lambda_i(\boldsymbol{w}, \boldsymbol{n})$ are eigenvalues of the matrix $\mathbb{P}_g$. By virtue of (8.29), (8.30) and (10.20),

$$\begin{aligned} \lambda_1(\boldsymbol{w}, \boldsymbol{n}) &= (\boldsymbol{v} - \boldsymbol{z}) \cdot \boldsymbol{n} - a, \\ \lambda_2(\boldsymbol{w}, \boldsymbol{n}) &= \cdots = \lambda_{d+1}(\boldsymbol{w}, \boldsymbol{n}) = (\boldsymbol{v} - \boldsymbol{z}) \cdot \boldsymbol{n}, \\ \lambda_{d+2}(\boldsymbol{w}, \boldsymbol{n}) &= (\boldsymbol{v} - \boldsymbol{z}) \cdot \boldsymbol{n} + a. \end{aligned} \tag{10.21}$$

Now we define the "positive and negative" parts of the matrix $\mathbb{P}_g$ by

$$\mathbb{P}_g^\pm = \mathbb{T}^{-1} \boldsymbol{\Lambda}^\pm \mathbb{T}, \quad \boldsymbol{\Lambda}^\pm = \mathrm{diag}(\lambda_1^\pm, \ldots, \lambda_{d+2}^\pm),$$

where $\lambda^+ = \max(\lambda, 0)$, $\lambda^- = \min(\lambda, 0)$. Using the above concepts, for arbitrary states $\boldsymbol{w}^L$, $\boldsymbol{w}^R$ and a unit 2D vector $\boldsymbol{n}$, we introduce the ALE modified Vijayasundaram numerical flux (cf. Sect. 8.4.3)

$$\mathbf{H}_g(\boldsymbol{w}_L, \boldsymbol{w}_R, \boldsymbol{n}) = \mathbb{P}_g^+\Big(\frac{\boldsymbol{w}_L + \boldsymbol{w}_R}{2}, \boldsymbol{n}\Big)\boldsymbol{w}_L + \mathbb{P}_g^-\Big(\frac{\boldsymbol{w}_L + \boldsymbol{w}_R}{2}, \boldsymbol{n}\Big)\boldsymbol{w}_R. \tag{10.22}$$

On the basis of the above considerations, we can introduce the convection form defined for $\bar{\boldsymbol{w}}_h, \boldsymbol{w}_h, \boldsymbol{\varphi}_h \in \boldsymbol{S}_{hpt}$:

$$\begin{aligned}
\hat{\boldsymbol{b}}_h(\bar{\boldsymbol{w}}_h, \boldsymbol{w}_h, \boldsymbol{\varphi}_h, t) = &- \sum_{K\in\mathscr{T}_{ht}} \int_K \sum_{s=1}^{d} ((\mathbb{A}_s(\bar{\boldsymbol{w}}_h) - z_s\mathbb{I})\boldsymbol{w}_h) \cdot \frac{\partial \boldsymbol{\varphi}_h}{\partial x_s}\,\mathrm{d}x \\
&+ \sum_{\Gamma\in\mathscr{F}_{ht}^I} \int_\Gamma \Big(\mathbb{P}_g^+ (\langle\bar{\boldsymbol{w}}_h\rangle_\Gamma, \boldsymbol{n}_\Gamma)\, \boldsymbol{w}_{h\Gamma}^{(L)} + \mathbb{P}_g^- (\langle\bar{\boldsymbol{w}}_h\rangle_\Gamma, \boldsymbol{n}_\Gamma)\, \boldsymbol{w}_{h\Gamma}^{(R)}\Big) \cdot \boldsymbol{\varphi}_h \mathrm{d}S \\
&+ \sum_{\Gamma\in\mathscr{F}_{ht}^W} \int_\Gamma \boldsymbol{f}_W(\bar{\boldsymbol{w}}_{h\Gamma}^{(L)}, \boldsymbol{n}_\Gamma) \cdot \boldsymbol{\varphi}_h\,\mathrm{d}S \\
&+ \sum_{\Gamma\mathscr{F}_{ht}^{io}} \int_\Gamma \Big(\mathbb{P}_g^+(\bar{\boldsymbol{w}}_{h\Gamma}^{(L)}, \boldsymbol{n}_\Gamma)\boldsymbol{w}_{h\Gamma}^{(L)} + \mathbb{P}_g (\boldsymbol{w}_{h\Gamma}^{(L)}, \boldsymbol{n}_\Gamma)\mathscr{B}(\bar{\boldsymbol{w}}_\Gamma^{(L)}, \boldsymbol{w}_{\mathrm{BC}})\Big) \cdot \boldsymbol{\varphi}_h \mathrm{d}S.
\end{aligned} \tag{10.23}$$

The symbol $\boldsymbol{f}_W$ denotes the boundary flux on the approximation $\partial\Omega_{hW_t}$ of the impermeable moving wall. We proceed here in a different way than in Sect. 8.3.1. On every face $\Gamma \in \mathscr{F}_{ht}^W$ (with the normal $\boldsymbol{n} = \boldsymbol{n}_\Gamma$) we use the relation

$$\sum_{s=1}^{d} \boldsymbol{g}_s(\boldsymbol{w})n_s = \begin{pmatrix} \rho(\boldsymbol{v} - \boldsymbol{z}_D) \cdot \boldsymbol{n} \\ \rho v_1(\boldsymbol{v} - \boldsymbol{z}_D) \cdot \boldsymbol{n} + \mathrm{p}n_1 \\ \vdots \\ \rho v_d(\boldsymbol{v} - \boldsymbol{z}_D) \cdot \boldsymbol{n} + \mathrm{p}n_d \\ E(\boldsymbol{v} - \boldsymbol{z}_D) \cdot \boldsymbol{n} + \mathrm{p}\boldsymbol{z}_D \cdot \boldsymbol{n} \end{pmatrix}, \tag{10.24}$$

which follows from (10.5) and (8.10). In view of (10.9), $\boldsymbol{v} = \boldsymbol{z}_D$ on $\Gamma \in \mathscr{F}_{ht}^W$, and hence

$$\sum_{s=1}^{d} \boldsymbol{g}_s(\boldsymbol{w})n_s = \mathrm{p}(0, n_1, \ldots, n_d, \boldsymbol{z}_D \cdot \boldsymbol{n})^{\mathrm{T}}. \tag{10.25}$$

This leads us to the choice of boundary flux on $\partial\Omega_{hW_t}$ in the form

$$\boldsymbol{f}_W(\bar{\boldsymbol{w}}_{h\Gamma}^{(L)}, \boldsymbol{n}_\Gamma) = \bar{\mathrm{p}}_{h\Gamma}^{(L)}(0, n_1, \ldots, n_d, \boldsymbol{z}_D \cdot \boldsymbol{n})^{\mathrm{T}}, \tag{10.26}$$

where $\bar{\mathrm{p}}_{h\Gamma}^{(L)}$ is the trace of the pressure on $\Gamma \in \mathscr{F}_{ht}^W$, corresponding to the function $\bar{\boldsymbol{w}}_{h\Gamma}^{(L)}$

10.1.2 Time Discretization by the BDF Method

Let us construct a partition $0 = t_0 < t_1 < t_2 < \cdots < t_r = T$ of the time interval $[0, T]$ and define the time step $\tau_n = t_n - t_{n-1}$. We use the approximations $\boldsymbol{w}_h(t_n) \approx \boldsymbol{w}_h^n \in \boldsymbol{S}_{hpt_n}, z(t_n) \approx z^n,\ n = 0, 1, \ldots$ Let us assume that $\boldsymbol{w}_h^n,\ n = 0, \ldots, m-1$, are already known and we want to determine $\boldsymbol{w}_h^m$. We introduce the functions

$$\hat{\boldsymbol{w}}_h^n = \boldsymbol{w}_h^n \circ \mathrm{A}_{t_n} \circ \mathrm{A}_{t_m}^{-1}, \quad n = m, m-1, m-2, \ldots, \tag{10.27}$$

i.e.,

$$\hat{\boldsymbol{w}}_h^n(x) = \boldsymbol{w}_h^n(\mathrm{A}_{t_n}(\mathrm{A}_{t_m}^{-1}(x))), \quad x \in \Omega_{t_m}. \tag{10.28}$$

Obviously, for $n = m$, the definition of $\hat{\boldsymbol{w}}_h^m$ by (10.27) is trivial, since

$$\hat{\boldsymbol{w}}_h^m(x) = \boldsymbol{w}_h^m(x), \quad x \in \Omega_{t_m}. \tag{10.29}$$

The transformation of $\boldsymbol{w}_h^n$ from the domain Ω_{t_n} to Ω_0 reads

$$\tilde{\boldsymbol{w}}_h^n(X) = \boldsymbol{w}_h^n(\mathrm{A}_{t_n}(X)), \quad X \in \Omega_0. \tag{10.30}$$

Then

$$\hat{\boldsymbol{w}}_h^n(x) = \tilde{\boldsymbol{w}}_h^n(\mathrm{A}_{t_m}^{-1}(x)), \quad x \in \Omega_{t_m}. \tag{10.31}$$

In order to define the ALE derivative of the exact solution $\boldsymbol{w}$, by virtue of (10.3), we introduce the function

$$\tilde{\boldsymbol{w}}(X, t) = \boldsymbol{w}(\mathrm{A}_t(X), t), \quad X \in \Omega_0,\ t \in [0, T]. \tag{10.32}$$

We use the approximations

$$\boldsymbol{w}(x, t_n) \approx \boldsymbol{w}_h^n(x), \quad x \in \Omega_{t_n}, \tag{10.33}$$

and thus in view of (10.31),

$$\tilde{\boldsymbol{w}}(X, t_n) \approx \tilde{\boldsymbol{w}}_h^n(X) = \hat{\boldsymbol{w}}_h^n(x), \quad X \in \Omega_0,\ X = \mathrm{A}_{t_m}^{-1}(x),\ x \in \Omega_{t_m}. \tag{10.34}$$

Now, by (10.2) and the above relations, we can obtain the approximation of the ALE derivative of the vector function $\boldsymbol{w}$ at time $t = t_m$ and a point $x \in \Omega_{t_m}$ with the aid of the backward finite difference:

$$\frac{\mathrm{D}^{\mathrm{A}}\boldsymbol{w}}{\mathrm{D}t}(x,t_m) = \frac{\partial \tilde{\boldsymbol{w}}}{\partial t}(X,t_m)|_{X=\mathrm{A}_{t_m}^{-1}(x)}$$
$$\approx \frac{\tilde{\boldsymbol{w}}(X,t_m) - \tilde{\boldsymbol{w}}(X,t_{m-1})}{\tau_m} \approx \frac{\tilde{\boldsymbol{w}}^m(X) - \tilde{\boldsymbol{w}}^{m-1}(X)}{\tau_m}.$$

These relations and (10.34) lead to the first-order *BDF approximation of the ALE derivative* in the form

$$\frac{\mathrm{D}^{\mathrm{A}}\boldsymbol{w}}{\mathrm{D}t}(x,t_m) \approx \frac{\boldsymbol{w}^m(x) - \hat{\boldsymbol{w}}^{m-1}(x)}{\tau_m}, \quad x \in \Omega_{t_m}. \tag{10.35}$$

In a similar way the ALE derivative can be approximated by the *backward difference formula* of order q:

$$\frac{\mathrm{D}^{\mathrm{A}}\boldsymbol{w}_h}{\mathrm{D}t}(x,t_m) \approx \frac{\mathrm{D}^{\mathrm{A}}_{appr}\boldsymbol{w}_h}{\mathrm{D}t}(x,t_m) = \frac{1}{\tau_m}\left(\alpha_0 \boldsymbol{w}_h^m + \sum_{l=1}^{q} \alpha_l \hat{\boldsymbol{w}}_h^{m-l}(x)\right), \quad x \in \Omega_{t_m}, \tag{10.36}$$

with coefficients α_l, $l = 0,\dots,q$, depending on τ_{m-l}, $l = 0,\dots,q-1$, see Sect. 8.4.5. In the beginning of the computation when $m < q$, we approximate the ALE derivative using formulae of the lower order $q := m$.

In nonlinear terms we use the extrapolation for computing the state $\bar{\boldsymbol{w}}_h^m$:

$$\bar{\boldsymbol{w}}_h^m = \sum_{l=1}^{q} \beta_l \hat{\boldsymbol{w}}_h^{m-l}, \tag{10.37}$$

where β_l, $l = 1,\dots,q$, depend on τ_{m-l}, $l = 0,\dots,q-1$. If $m < q$, then we apply the extrapolation of order m. The values of the coefficients $\alpha_l, l = 0,\dots,q$, and $\beta_l, l = 1,\dots,q$, for $q = 1,2,3$ are given in Tables 8.2, 8.3 and 8.4, 8.5, respectively.

By the symbol $(\cdot,\cdot)_{t_m}$ we denote the scalar product in $L^2(\Omega_{ht_m})$, i.e.,

$$(\boldsymbol{w}_h, \boldsymbol{\varphi}_h)_{t_m} = \int_{\Omega_{ht_m}} \boldsymbol{w}_h \cdot \boldsymbol{\varphi}_h \, dx. \tag{10.38}$$

Definition 10.1 The sequence of functions $\boldsymbol{w}_h^m \in \boldsymbol{S}_{hpt_m}$, $m = 1,2,\dots,r$, is called the approximate solution given by the *ALE BDF-DG scheme* if it satisfies

$$\left(\frac{\mathrm{D}^{\mathrm{A}}_{\mathrm{appr}}\boldsymbol{w}_h}{\mathrm{D}t}(t_m), \boldsymbol{\varphi}_h\right)_{t_m} + \hat{\boldsymbol{a}}_h(\bar{\boldsymbol{w}}_h^m, \boldsymbol{w}_h^m, \boldsymbol{\varphi}_h, t_m) \tag{10.39}$$
$$+ \hat{\boldsymbol{b}}_h(\bar{\boldsymbol{w}}_h^m, \boldsymbol{w}_h^m, \boldsymbol{\varphi}_h, t_m) + \hat{\boldsymbol{J}}_h^{\sigma}(\bar{\boldsymbol{w}}_h^m \boldsymbol{w}_h^m, \boldsymbol{\varphi}_h, t_m) + \boldsymbol{d}_h(\boldsymbol{w}_h^m, \boldsymbol{\varphi}_h, t_m)$$
$$+ \boldsymbol{\beta}_h(\bar{\boldsymbol{w}}_h^m, \boldsymbol{w}_h^m, \boldsymbol{\varphi}_h, t_m) + \boldsymbol{\gamma}_h(\bar{\boldsymbol{w}}_h^m, \boldsymbol{w}_h^m, \boldsymbol{\varphi}_h, t_m) = 0 \quad \forall \boldsymbol{\varphi}_h \in \boldsymbol{S}_{hpt_m}.$$

10.1.3 Space-Time DG Discretization

Another technique regarding how to construct a method of high-order accuracy both in space and time is the *space-time discontinuous Galerkin method* (ST-DGM). We again consider a partition $0 = t_0 < t_1 < \cdots < t_r = T$ of the time interval $[0, T]$ and denote $I_m = (t_{m-1}, t_m)$, $\overline{I}_m = [t_{m-1}, t_m]$, $\tau_m = t_m - t_{m-1}$, for $m = 1, \ldots, r$. We define the space $\boldsymbol{S}_{h,\tau}^{p,q} = (S_{h,\tau}^{p,q})^{d+2}$, where

$$S_{h,\tau}^{p,q} = \left\{ \phi\, ;\, \phi|_{I_m} = \sum_{i=0}^{q} \zeta_i \phi_i, \text{ where } \phi_i \in S_{hpt},\ \zeta_i \in P_q(I_m), m = 1, \ldots, r \right\}$$

with integers $p, q \geq 1$, $P_q(I_m)$ denoting the space of all polynomials in t on I_m of degree $\leq q$ and the space S_{hpt} defined in (10.10). For $\boldsymbol{\varphi} \in \boldsymbol{S}_{h,\tau}^{p,q}$ we introduce the following notation:

$$\boldsymbol{\varphi}_m^{\pm} = \boldsymbol{\varphi}(t_m^{\pm}) = \lim_{t \to t_m \pm} \boldsymbol{\varphi}(t), \quad \{\boldsymbol{\varphi}\}_m = \boldsymbol{\varphi}_m^{+} - \boldsymbol{\varphi}_m^{-}. \tag{10.40}$$

Derivation of the discrete problem can be carried out similarly as above. The difference is now that time t is considered continuous, test functions $\boldsymbol{\varphi}_{h\tau} \in \boldsymbol{S}_{h,\tau}^{p,q}$ are used and also the integration over I_m is applied. In order to stick together the solution on intervals I_{m-1} and I_m, we augment the resulting identity by the penalty expression $\left(\{\boldsymbol{w}_{h\tau}\}_{m-1}, \boldsymbol{\varphi}_{h\tau}(t_{m-1}^{+})\right)_{t_{m-1}}$. The initial state $\boldsymbol{w}_{h\tau}(0-) \in \boldsymbol{S}_{h0}^{p}$ is defined as the $L^2(\Omega_{h0})$-projection of $\boldsymbol{w}^0$ on $\boldsymbol{S}_{h0}^{p}$, i.e.,

$$\left(\boldsymbol{w}_{h\tau}(0-), \boldsymbol{\varphi}_h\right)_{t_0} = \left(\boldsymbol{w}^0, \boldsymbol{\varphi}_h\right)_{t_0} \quad \forall \boldsymbol{\varphi}_h \in \boldsymbol{S}_{hp0}. \tag{10.41}$$

Similarly as in Sect. 10.1.2 we introduce a suitable linearization. We can use two possibilities.

(1) We put $\bar{\boldsymbol{w}}_{h\tau}(t) := \boldsymbol{w}_{h\tau}(t_{m-1}^{-})$ for $t \in I_m$. (This represents a simple time extrapolation.)

(2) Each component of the vector-valued function $\boldsymbol{w}_{h\tau}|_{I_{m-1}}$ is a polynomial in t of degree $\leq q$, and we define the function $\bar{\boldsymbol{w}}_{h\tau}|_{I_m}$ by the time prolongation using values of the polynomial vector function $\boldsymbol{w}_{h\tau}|_{I_{m-1}}$ at time instants $t \in I_m$. Thus, we write $\bar{\boldsymbol{w}}_{h\tau}|_{I_m}(t) := \boldsymbol{w}_{h\tau}|_{I_{m-1}}(t)$ for $t \in I_m$.

Definition 10.2 The *ALE space-time DG (ALE ST-DG) approximate solution* is a function $\boldsymbol{w}_{h\tau}$ satisfying (10.41) and the following conditions:

$$\boldsymbol{w}_{h\tau} \in \boldsymbol{S}_{h,\tau}^{p,q}, \tag{10.42a}$$

$$\int_{I_m} \left(\left(\frac{D^A w_{h\tau}}{Dt}, \varphi_{h\tau} \right)_t + \hat{a}_h(\bar{w}_{h\tau}, w_{h\tau}, \varphi_{h\tau}, t) \right) dt \tag{10.42b}$$
$$+ \int_{I_m} \left(\hat{b}_h(\bar{w}_{h\tau}, w_{h\tau}, \varphi_{h\tau}, t) + \hat{J}_h^{\sigma}(\bar{w}_{h\tau}, w_{h\tau}, \varphi_{h\tau}, t) + d_h(w_{h\tau}, \varphi_{h\tau}, t) \right) dt$$
$$+ \int_{I_m} \left(\beta_h(\bar{w}_{h\tau}, w_{h\tau}, \varphi_{h\tau}, t) + \gamma_h(\bar{w}_{h\tau}, w_{h\tau}, \varphi_{h\tau}, t) \right) dt$$
$$+ (\{w_{h\tau}\}_{m-1}, \varphi_{h\tau}(t_{m-1}+))_{t_{m-1}} = 0 \quad \forall \varphi_{h\tau} \in S_{h,\tau}^{p,q}, \quad m = 1, \ldots, r.$$

Remark 10.3 In practical computations, integrals appearing in definitions of the forms $\hat{a}_h$, $\hat{b}_h$, d_h, $\hat{J}_h^{\sigma}$, β_h and γ_h and also the time integrals are evaluated with the aid of quadrature formulae.

The linear algebraic systems equivalent to (10.39) and (10.42) are solved either by the direct solver (e.g., UMFPACK [72]) or by a suitable iteration method (e.g., the GMRES method with block diagonal preconditioning [249].

10.2 Fluid-Structure Interaction

This section is devoted to problems of fluid-structure interaction (FSI). We are concerned with two FSI problems:

- flow-induced airfoil vibrations,
- interaction of compressible flow with elastic structures.

In both cases we deal with 2D models.

10.2.1 Flow-Induced Airfoil Vibrations

In the study of aerodynamical properties of airplane wings or blades of turbines and compressors, the problem of flow-induced airfoil vibrations plays an important role. We consider an *elastically supported airfoil* with two degrees of freedom: the vertical displacement H (positively oriented downwards) and the angle α of rotation around an *elastic axis* EA (positively oriented clockwise), see Fig. 10.1. In this case the boundary of the bounded domain $\Omega_t \subset \mathbb{R}^2$ occupied by gas is formed by three disjoint parts: $\partial\Omega_t = \partial\Omega_i \cup \partial\Omega_o \cup \partial\Omega_{W_t}$, where $\partial\Omega_i$ is the inlet, $\partial\Omega_o$ is the outlet and $\partial\Omega_{W_t}$ denotes the boundary of an airfoil moving in dependence on time.

10.2.1.1 Description of the Airfoil Motion

The motion of the airfoil is described by the system of ordinary differential equations for unknowns H and α:

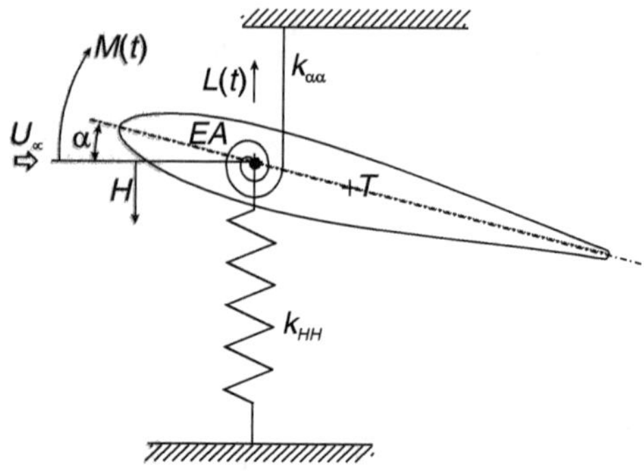

Fig. 10.1 Elastically supported airfoil with two degrees of freedom

$$m\ddot{H} + k_{HH}H + S_\alpha \ddot{\alpha} \cos\alpha - S_\alpha \dot{\alpha}^2 \sin\alpha + d_{HH}\dot{H} = -L(t), \tag{10.43}$$
$$S_\alpha \ddot{H} \cos\alpha + I_\alpha \ddot{\alpha} + k_{\alpha\alpha}\alpha + d_{\alpha\alpha}\dot{\alpha} = M(t).$$

The dot and two dots denote the first-order and second-order time derivative, respectively. This system is derived from the Lagrange equations of the second art (see, e.g., [266]). We use the following notation: $L(t)$—aerodynamic lift force (upwards positive), $M(t)$—aerodynamic torsional moment (clockwise positive), m—airfoil mass, S_α—static moment of the airfoil around the elastic axis EA, I_α—inertia moment of the airfoil around the elastic axis EA, k_{HH}—bending stiffness, $k_{\alpha\alpha}$—torsional stiffness, d_{HH}—structural damping in bending, $d_{\alpha\alpha}$—structural damping in torsion, l—airfoil depth (i.e., the length of an airfoil segment in investigation).

System (10.43) is equipped with the initial conditions prescribing the values $H(0), \alpha(0), \dot{H}(0), \dot{\alpha}(0)$. The aerodynamic lift force L acting in the vertical direction and the torsional moment M are defined by

$$L = -l \int_{\partial\Omega_{Wt}} \sum_{j=1}^{2} \tau_{2j} n_j \, \mathrm{d}S, \quad M = l \int_{\partial\Omega_{Wt}} \sum_{i,j=1}^{2} \tau_{ij} n_j r_i^{\mathrm{ort}} \, \mathrm{d}S, \tag{10.44}$$

where

$$\tau_{ij} = -\mathrm{p}\,\delta_{ij} + \tau_{ij}^V = -\mathrm{p}\,\delta_{ij} + \frac{1}{\mathrm{Re}}\left(\frac{\partial u_i}{\partial x_j} + \frac{\partial u_j}{\partial x_i} - \frac{2}{3}\nabla \cdot \boldsymbol{v}\delta_{ij}\right), \tag{10.45}$$
$$r_1^{\mathrm{ort}} = -(x_2 - x_{EA2}), \; r_2^{\mathrm{ort}} = x_1 - x_{EA1}.$$

By τ_{ij} we denote the components of the aerodynamic stress tensor, δ_{ij} denotes the Kronecker symbol, $\boldsymbol{n} = (n_1, n_2)$ is the unit outer normal to $\partial\Omega_t$ on $\partial\Omega_{Wt}$ (pointing into the airfoil) and $x_{EA} = (x_{EA1}, x_{EA2})$ is the position of the elastic axis (lying in the interior of the airfoil). Relations (10.44) and (10.45) define the coupling of the fluid dynamical model with the structural model.

In contrast to the solution of compressible flow, the numerical solution of the structural problem is not difficult. System (10.43) is transformed to a first-order system and approximated by the Runge-Kutta method. In what follows, we are concerned with the realization of the complete fluid-structure interaction problem.

10.2.1.2 Construction of the ALE Mapping

In the solution of flow-induced airfoil vibrations, for constructing of the ALE mapping the following method can be used. We start from the assumption that we know the airfoil position at time instant t_m, given by the displacement $H(t_m)$ and the rotation angle $\alpha(t_m)$. We want to define the mapping $\mathrm{A}_{t_m} : \overline{\Omega}_{h0} \to \overline{\Omega}_{ht_m}$. We construct two circles K_1, K_2 with center at the elastic axis EA and radii $R_1, R_2,\ 0 < R_1 < R_2$ so that the airfoil is lying inside the circle K_1. The interior of the circle K_1 is moving in vertical direction and rotates around the elastic axis as a solid body together with the airfoil. The exterior of K_2 is not deformed, and in the area between K_1 and K_2 we use an interpolation. First, we define the mapping $\mathbf{H}_{t_m}(X_1, X_2)$, where $\boldsymbol{X} = (X_1, X_2) \in \Omega_{h0}$, describing the vertical motion and rotation:

$$\mathbf{H}_{t_m}(X_1, X_2) = \begin{pmatrix} \cos\alpha(t_m) & \sin\alpha(t_m) \\ -\sin\alpha(t_m) & \cos\alpha(t_m) \end{pmatrix} \begin{pmatrix} X_1 - X_{EA1} \\ X_2 - X_{EA2} \end{pmatrix} + \begin{pmatrix} X_{EA1} \\ X_{EA2} \end{pmatrix} + \begin{pmatrix} 0 \\ -H(t_m) \end{pmatrix},$$

where (X_{EA1}, X_{EA2}) represents the position of the elastic axis at time $t = 0$. If we denote the identical mapping by $\mathbf{Id}(X_1, X_2) = (X_1, X_2)$, we define the mapping $\overline{\mathrm{A}}_{t_m}$ as a combination of $\mathbf{Id}$ and $\mathbf{H}_{t_m}$:

$$\bar{\mathrm{A}}_{t_m}(X_1, X_2) = (1 - \xi)\mathbf{H}_{t_m}(X_1, X_2) + \xi\mathbf{Id}(X_1, X_2), \tag{10.46}$$

where

$$\xi = \xi(\hat{r}) = \min\left(\max\left(0, \frac{\hat{r} - R_1}{R_2 - R_1}\right), 1\right) \tag{10.47}$$

and $\hat{r} = \sqrt{(X_1 - X_{EA1})^2 + (X_2 - X_{EA2})^2}$ is the distance of a point $\boldsymbol{X} \in \Omega_{h0}$ from the elastic axis.

In the case of the space-time DGM it is necessary to construct the ALE mapping for all time instants in the intervals I_m. To this end, we introduce the mapping by the formula

$$\bar{\mathrm{A}}_t(\boldsymbol{X}) = \frac{t_m - t}{\tau_m}\bar{\mathrm{A}}_{t_{m-1}}(\boldsymbol{X}) + \frac{t - t_{m-1}}{\tau_m}\bar{\mathrm{A}}_{t_m}(\boldsymbol{X}), \quad t \in I_m,\ \boldsymbol{X} \in \Omega_{h0}. \tag{10.48}$$

Since the mapping $\bar{\mathrm{A}}_t$ is nonlinear, the elements from the initial triangulation $\mathscr{T}_{h0}$ would be transformed by this mapping to curved elements. Therefore, the ALE mapping A_t is defined as the conforming piecewise linear space interpolation of $\bar{\mathrm{A}}_t$.

The domain velocity is approximated in the BDF method by the formula of order q in the form

$$z^m(\boldsymbol{x}) = \frac{1}{\tau_m}\left(\alpha_0 \boldsymbol{x} + \sum_{l=1}^{q} \alpha_l \mathrm{A}_{t_{m-l}}(\mathrm{A}_{t_m}^{-1}(\boldsymbol{x}))\right) \quad \text{for } \boldsymbol{x} \in \Omega_{ht_m}, \tag{10.49}$$

with coefficients $\alpha_l, l = 0, \ldots, q$, given in Tables 8.2 and 8.3. If $m < q$, then we set $q := m$. In the case of the space-time DGM we use (10.48) and express $z(t)$ in the form

$$z(\boldsymbol{x}, t) = \frac{\mathrm{A}_{t_m}(\mathrm{A}_t^{-1}(\boldsymbol{x})) - \mathrm{A}_{t_{m-1}}(\mathrm{A}_t^{-1}(\boldsymbol{x}))}{\tau_m}, \qquad t \in I_m, \boldsymbol{x} \in \Omega_{ht}. \tag{10.50}$$

10.2.1.3 Coupling Procedure

In solving the complete fluid-structure interaction problem the following coupling algorithm is used.

0. Prescribe $\varepsilon > 0$—the measure of accuracy in the coupling procedure, and an integer $N \geq 0$—the maximal number of iterations in the coupling procedure.
1. Assume that the approximate solution $\boldsymbol{w}_h^k$ of the discrete flow problem (10.39) or the approximate solution $\boldsymbol{w}_{h\tau}|_{I_k}$ of the discrete problem (10.42) and the corresponding lift force L and torsional moment M computed from (10.44) and (10.45) are known.
2. Extrapolate linearly L and M from the interval $[t_{k-1}, t_k]$ to $[t_k, t_{k+1}]$. Set $n := 0$.
3. *Prediction of* H, α: Compute the displacement H and angle α at time t_{k+1} as the solution of system (10.43). Denote it by H^*, α^*.
4. On the basis of H^*, α^* determine the position of the airfoil at time t_{k+1}, the domain $\Omega_{ht_{k+1}}$, the ALE mapping $\mathrm{A}_{ht_{k+1}}$ and the domain velocity z_h^{k+1}.
5. Solve the discrete problem (10.39) at time t_{k+1} or the discrete problem (10.42) in the interval I_{k+1}.
6. *Correction of* H, α: Compute L, M from (10.44) and (10.45) at time t_{k+1} and interpolate L, M in the interval $[t_k, t_{k+1}]$. Compute H, α at time t_{k+1} from (10.43).
7. If $|H^* - H| + |\alpha^* - \alpha| \geq \varepsilon$ and $n < N$, set $H^* = H, \alpha^* = \alpha, n := n + 1$ and go to 4. Otherwise, $k := k + 1$ and go to 2.

If $N = 0$, then the coupling of the flow and structural problems is weak (loose). With increasing N and decreasing ε, the coupling becomes strong.

10.2.1.4 Numerical Examples

In order to demonstrate the applicability and robustness of the developed methods numerical tests were performed. Here we present the results of computations carried out for the flow around the NACA 0012 profile with the following data: $m = 0.086622\,\text{kg}$, $S_a = -0.000779673\,\text{kg m}$, $I_a = 0.000487291\,\text{kg m}^{-2}$, $k_{HH} = 105.109\,\text{N m}^{-1}$, $k_{\alpha\alpha} = 3.696682\,\text{N m rad}^{-1}$, $d_{HH} = 0.0\,\text{N s m}^{-1}$, $d_{\alpha\alpha} = 0.0\,\text{N m s rad}^{-1}$, $\mu = 1.72 \cdot 10^{-5}\,\text{kg m}^{-1}\,\text{s}^{-1}$, far-field pressure $p = 101{,}250\,\text{Pa}$, far-field density $\rho = 1.225\,\text{kg m}^{-3}$, Poisson adiabatic constant $\gamma = 1.4$, specific heat $c_v = 721.428\,\text{m}^2\,\text{s}^{-2}\,\text{K}^{-1}$, heat conduction coefficient $k = 2.428 \cdot 10^{-2}\,\text{kg m s}^{-2}\,\text{K}^{-1}$, airfoil length $c = 0.3\,\text{m}$, airfoil depth $l = 0.05\,\text{m}$, initial conditions for the structural equations $H(0) = -20\,\text{mm}$, $\alpha(0) = 6°$, $\dot{H}(0) = \dot{\alpha}(0) = 0$. The computations started at a time instant $t = -\delta < 0$ with a fixed airfoil. Then at time $t = 0$ the airfoil was released and the FSI process followed.

For the space discretization quadratic polynomials ($p = 2$) were used. In the case of the BDF time discretization, the second-order approximation was used (we denote it by BDF-DGp2q2). In the case of the ST-DG method the quadratic polynomials in space and linear polynomials in time were used (denoted by ST-DGp2q1). For both methods the SIPG variant of the viscous terms was used (i.e., $\Theta = 1$). In the penalty form $\hat{J}_h^\sigma$ the weight σ was defined by (9.67) with the parameter $C_W = 500$ in the interior part of the penalty form, whereas $C_W = 5000$ in the boundary penalty, in order to obtain an accurate approximation of the Dirichlet boundary conditions. The time step is defined as $\tau = 0.003299\,c/v_\infty$ s, where $c = 0.3\,\text{m}$ is the length of the airfoil and v_∞ is the magnitude of the far-field velocity. The constants in the artificial viscosity forms were chosen $\nu_1 = \nu_2 = 0.1$.

With the use of the triangulation at time $t = 0$ shown in Fig. 10.2, low Mach number flow was computed for far-field velocities 20 and 37.5 m/s. The results are shown in Figs. 10.3 and 10.4. We can see that both methods DG-BDF (full lines) and ST-DG (dashed lines) give very similar results. In the case of the far-field velocity 20 m/s the airfoil vibrations are damped. The velocity 37.5 m/s leads already to flutter, when the vibrations are damped no longer. Our results are comparable with computations presented in [179], where the Taylor–Hood finite element method was applied to the model based on the incompressible Navier–Stokes equations.

In the second example, the described methods are applied to the numerical simulation of airfoil vibrations induced by high-speed hypersonic flow with large Reynolds

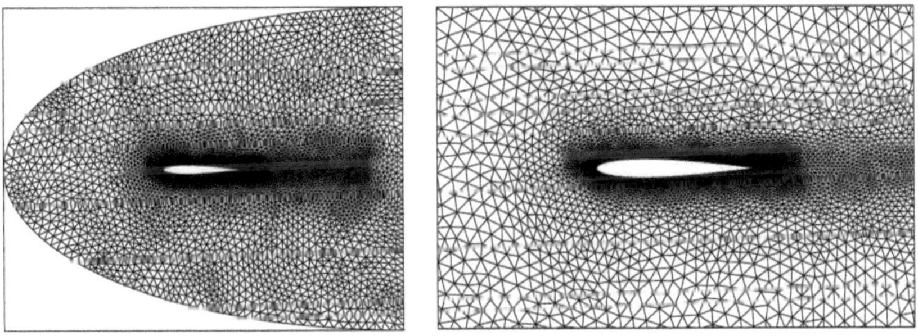

Fig. 10.2 Triangulation at time $t = 0$ with 17,158 elements used for computing subsonic flow and its detail near the airfoil

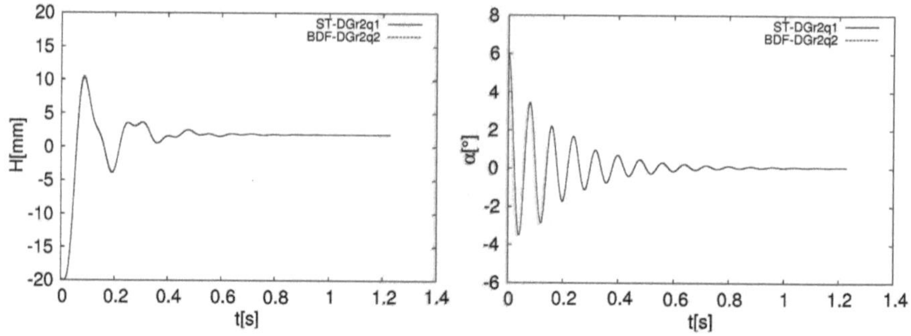

Fig. 10.3 Displacement and rotation angle of the airfoil in dependence on time for far-field velocity 20 m/s and far-field Mach number $M_\infty = 0.0588$

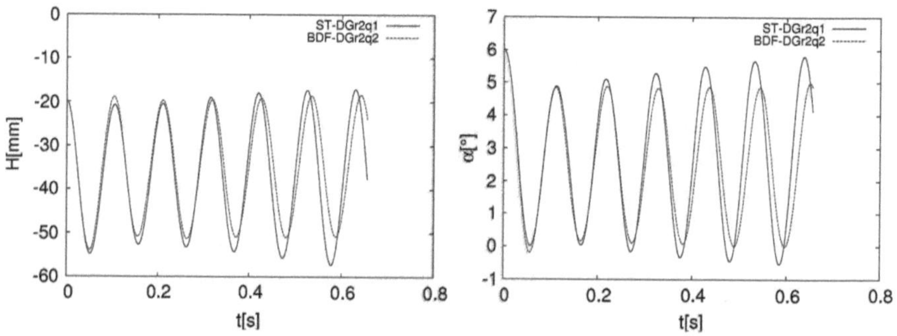

Fig. 10.4 Displacement and rotation angle of the airfoil in dependence on time for far-filed velocity 37.5 m/s and far-field Mach number $M_\infty = 0.1102$

numbers. It appears that in the method combining the DG space discretization with the BDF time discretization some instabilities may appear for flows with far-field Mach numbers higher than 1.5. This is not the case of the ST-DG method, which is very robust and stable for a large range of the Mach and Reynolds numbers.

Here we present the results of the simulation of airfoil vibrations induced by the flow with far-field Mach number $M_\infty = 3$ and Reynolds numbers $\mathrm{Re} = 10^4$ and 10^5 computed using the initial triangulation shown in Fig. 10.5. In this case damped airfoil vibrations were obtained for the same data as above except for bending and torsional stiffnesses, which were now 1000 times higher that before. Figure 10.6 shows the Mach number distribution in the vicinity of the airfoil at several time instants. One can see well resolved oblique shock wave, shock waves leaving the trailing edge and wake. The presented results were computed by the system of programs worked out by J. Česenek [45].

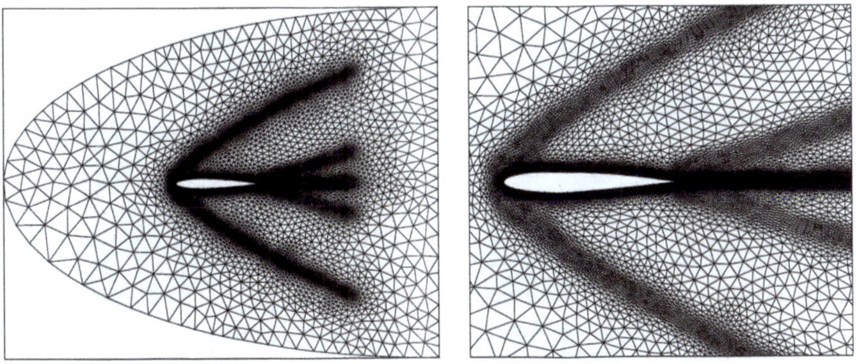

Fig. 10.5 Triangulation at time $t = 0$ with 42,821 elements used for computing hypersonic flow and its detail near the airfoil

10.2.2 Interaction of Compressible Flow and an Elastic Body

In this section, the interaction of compressible flow with an elastic body will be solved. We use the model of dynamical linear elasticity formulated in a bounded open set $\Omega^b \subset \mathbb{R}^2$ representing the elastic body, which has a common boundary with the reference domain Ω_0 occupied by the fluid at the initial time. By $\boldsymbol{u}(X, t) = (u_1(X, t), u_2(X, t))$, $X = (X_1, X_2) \in \overline{\Omega}^b$, $t \in [0, T]$, we denote the displacement of the body.

10.2.2.1 Dynamical Elasticity Equations

The equations describing deformation of the elastic body Ω^b have the form

$$\rho^b \frac{\partial^2 u_i}{\partial t^2} + C\rho^b \frac{\partial u_i}{\partial t} - \sum_{j=1}^{2} \frac{\partial \tau_{ij}^b}{\partial X_j} = 0 \quad \text{in } \Omega^b \times (0, T), \quad i = 1, 2. \tag{10.51}$$

Here ρ^b denotes the material density and τ_{ij}^b, $i, j = 1, 2$, are the components of the stress tensor defined by the generalized Hooke's law for isotropic bodies in the form

$$\tau_{ij}^b = \lambda^b \nabla \cdot \boldsymbol{u}\, \delta_{ij} + 2\mu^b e_{ij}^b(\boldsymbol{u}), \quad i, j = 1, 2. \tag{10.52}$$

By $\boldsymbol{e}^b = (e_{ij}^b)_{i,j=1}^2$ we denote the strain tensor defined by

$$e_{ij}^b(\boldsymbol{u}) = \frac{1}{2}\left(\frac{\partial u_i}{\partial X_j} + \frac{\partial u_j}{\partial X_i}\right), \quad i, j = 1, 2. \tag{10.53}$$

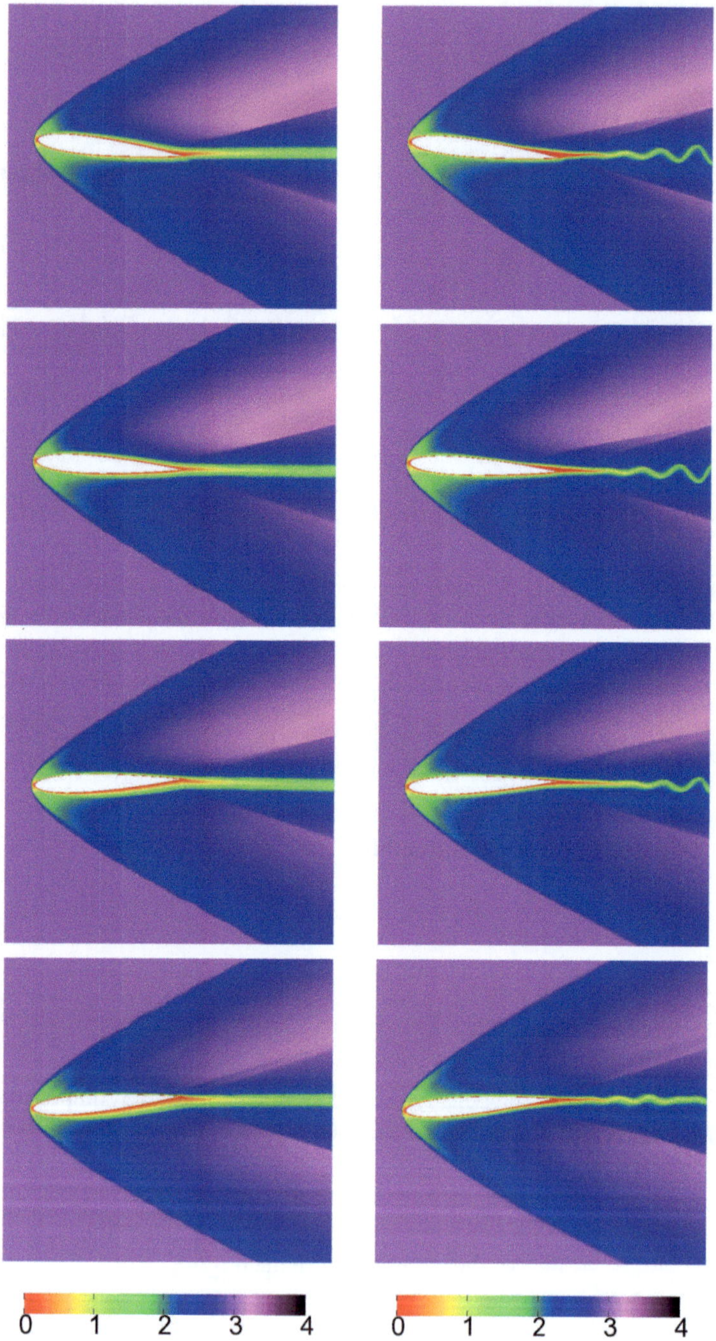

Fig. 10.6 Mach number distribution at time instants $t = 0.0, 0.00039, 0.00078, 0.00117$ s for the far-field velocity 1020 m/s ($M_\infty = 3.0$) and Reynolds numbers $\mathrm{Re} = 10^4$ (*left*) and $\mathrm{Re} = 10^5$ (*right*)

The Lamé coefficients λ^b and μ^b are related to the Young modulus E^b and the Poisson ratio σ^b:

$$\lambda^b = \frac{E^b\sigma^b}{(1+\sigma^b)(1-2\sigma^b)}, \quad \mu^b = \frac{E^b}{2(1+\sigma^b)}. \tag{10.54}$$

The expression $C\rho^b \frac{\partial u_i}{\partial t}$, where $C \geq 0$, is the dissipative structural damping of the system.

We complete the elasticity problem by initial and boundary conditions. The initial conditions read as

$$\boldsymbol{u}(\cdot, 0) = 0, \quad \frac{\partial \boldsymbol{u}}{\partial t}(\cdot, 0) = 0, \quad \text{in } \Omega^b. \tag{10.55}$$

Further, we assume that $\partial\Omega^b = \overline{\Gamma}^b_W \cup \overline{\Gamma}^b_D$, where Γ^b_W and Γ^b_D are two disjoints parts of $\partial\Omega^b$. ($\overline{\Gamma}^b_W$ and $\overline{\Gamma}^b_D$ denote their closures in $\partial\Omega$.) We assume that Γ^b_W is a common part between the fluid and structure at time $t = 0$. This means that $\Gamma^b_W \subset \partial\Omega_{W_0}$. On Γ^b_W we prescribe the normal component of the stress tensor and assume that the part Γ^b_D is fixed. This means that the following boundary conditions are used:

$$\sum_{j=1}^{2} \tau^b_{ij} n_j = T^n_i \quad \text{on } \Gamma^b_W \times (0, T), \quad i = 1, 2, \tag{10.56}$$

$$\boldsymbol{u} = 0 \quad \text{on } \Gamma^b_D \times (0, T). \tag{10.57}$$

By $\boldsymbol{T}^n = (T^n_1, T^n_2)$ we denote the prescribed normal component of the stress tensor.

The structural problem consists in finding the displacement $\boldsymbol{u}$ satisfying equations (10.51) and the initial and boundary conditions (10.55)–(10.57).

10.2.2.2 Formulation of the FSI Problem

Now we come to the formulation of the coupled FSI problem. We denote the common boundary between the fluid and the structure at time t by $\tilde{\Gamma}_{W_t}$. Thus,

$$\tilde{\Gamma}_{W_t} = \left\{ x \in \mathbb{R}^2;\ x = X + \boldsymbol{u}(X, t),\ X \in \Gamma^b_W \right\} \subset \partial\Omega_{W_t}. \tag{10.58}$$

We see that the shape of the domain Ω_t is determined by the displacement $\boldsymbol{u}$ of the part Γ^b_W at time t. The ALE mapping $\mathcal{A}_t$ will be constructed with the aid of a special stationary linear elasticity problem in Sect. 10.2.2.4.

If the domain Ω_t occupied by the fluid at time t is known, we can solve the problem describing the flow, and compute the surface force acting onto the body on the interface $\tilde{\Gamma}_{W_t}$, which can be transformed to the reference configuration, i.e., to the interface Γ_W^b. In case of the linear elasticity model, when only small deformations are considered, we use the transmission condition

$$\sum_{j=1}^{2} \tau_{ij}^b(X) n_j(X) = \sum_{j=1}^{2} \tau_{ij}^f(x) n_j(X), \quad i = 1, 2, \tag{10.59}$$

where τ_{ij}^f are the components of the aerodynamic stress tensor of the fluid, i.e.,

$$\tau_{ij}^f = -\mathrm{p}\delta_{ij} + \tau_{ij}^V, \quad i, j = 1, 2, \tag{10.60}$$

the points x and X satisfy the relation

$$x = X + \boldsymbol{u}(X, t), \quad X \in \Gamma_W^b, \; x \in \tilde{\Gamma}_{W_t}, \tag{10.61}$$

and $\boldsymbol{n}(X) = (n_1(X), n_2(X))$ denotes the outer unit normal to the body Ω^b on Γ_W^b at the point X. Here we consider the dimensional quantities: p is dimensional pressure and τ_{ij}^V is the viscous part of the aerodynamic stress tensor defined by the dimensional velocity in (9.8):

$$\tau_{ij}^V = \mu \left(\frac{\partial v_i}{\partial x_j} + \frac{\partial v_j}{\partial x_i} \right) + \lambda \nabla \cdot \boldsymbol{v}\, \delta_{ij}, \quad i, j = 1, \ldots, 2. \tag{10.62}$$

Further, the fluid velocity is determined on the moving part of the boundary $\tilde{\Gamma}_{W_t}$ by the second transmission condition

$$\boldsymbol{v}(x, t) = \boldsymbol{z}_D(x, t) := \frac{\partial \boldsymbol{u}(X, t)}{\partial t}. \tag{10.63}$$

The points x and X satisfy relation (10.61).

Now we formulate the *continuous FSI problem*: We want to determine the domain Ω_t, $t \in (0, T]$, and functions $\boldsymbol{w} = \boldsymbol{w}(x, t)$, $x \in \overline{\Omega}_t$, $t \in [0, T]$ and $\boldsymbol{u} = \boldsymbol{u}(X, t)$, $X \in \overline{\Omega}^b$, $t \in [0, T]$, satisfying Eqs. (10.4), (10.51), the initial conditions (10.6), (10.55), the boundary conditions (10.7)–(10.9), (10.57) and the transmission conditions (10.59), (10.63).

Theoretical analysis of qualitative properties of this problem, as the existence, uniqueness and regularity of its solution, is open. In what follows, we describe a method for the numerical solution of the elasticity problem.

10.2.2.3 Discrete Structural Problem

The space semidiscretization of the structural problem will be carried out by the conforming finite element method. By Ω_h^b we denote a polygonal approximation of the domain Ω^b. We construct a triangulation $\mathscr{T}_h^b$ of the domain Ω_h^b formed by a finite number of closed triangles with the following properties:

(a) $\overline{\Omega}_h^b = \bigcup_{K \in \mathscr{T}_h^b} K$.
(b) The intersection of two different elements $K, K' \in \mathscr{T}_h^b$ is either empty or a common edge of these elements or their common vertex.
(c) The vertices lying on $\partial\Omega_h^b$ are lying on $\partial\Omega^b$.
(d) The set $\overline{\Gamma}_W^b \cap \overline{\Gamma}_D^b$ is formed by vertices of some elements $K \in \mathscr{T}_h^b$.

Further, by Γ_{Wh}^b and Γ_{Dh}^b we denote the parts of $\partial\Omega_h^b$ approximating Γ_W^b and Γ_D^b.

The approximate solution of the structural problem will be sought in the finite-dimensional space $\boldsymbol{X}_h = X_h \times X_h$, where

$$X_h = \left\{ v_h \in C(\overline{\Omega}_h^b);\ v_h|_K \in P_s(K),\ \forall K \in \mathscr{T}_h^b \right\} \tag{10.64}$$

and $s \geq 1$ is an integer. In $\boldsymbol{X}_h$ we define the subspace $\boldsymbol{V}_h = V_h \times V_h$, where

$$V_h = \left\{ y_h \in X_h;\ y_h|_{\overline{\Gamma}_{Dh}^b} = 0 \right\}. \tag{10.65}$$

Deriving the space semidiscretization can be obtained in a standard way. Multiplying Eq. (10.51) by any test function $y_{hi} \in V_h,\ i = 1, 2$, integrating over Ω_h^b, applying Green's theorem and using the boundary condition (10.56), we obtain an identity containing the form

$$a_h^b(\boldsymbol{u}_h, \boldsymbol{y}_h) = \int_{\Omega_h^b} \lambda^b \nabla \cdot \boldsymbol{u}_h \nabla \cdot \boldsymbol{y}_h \, dX + 2 \int_{\Omega_h^b} \mu^b \sum_{i,j=1}^{2} e_{ij}^b(\boldsymbol{u}_h)\, e_{ij}^b(\boldsymbol{y}_h) \, dX, \tag{10.66}$$

defined for $\boldsymbol{u}_h = (u_{h1}, u_{h2}),\ \boldsymbol{y}_h = (y_{h1}, y_{h2}) \in \boldsymbol{X}_h$. Moreover, we set

$$(\boldsymbol{\varphi}, \boldsymbol{\psi})_{\Omega_h^b} = \int_{\Omega_h^b} \boldsymbol{\varphi} \cdot \boldsymbol{\psi} \, dX, \quad (\boldsymbol{\varphi}, \boldsymbol{\psi})_{\Gamma_{Wh}^b} = \int_{\Gamma_{Wh}^b} \boldsymbol{\varphi} \cdot \boldsymbol{\psi} \, dS. \tag{10.67}$$

We use the approximation $\boldsymbol{T}_h^n \approx \boldsymbol{T}^n$ and the notation $\boldsymbol{u}_h'(t) = \frac{\partial \boldsymbol{u}_h(t)}{\partial t}$ and $\boldsymbol{u}_h''(t) = \frac{\partial^2 \boldsymbol{u}_h(t)}{\partial t^2}$. Then we define the *approximate solution of the structural problem* as a mapping $t \in [0, T] \rightarrow \boldsymbol{u}_h(t) \in \boldsymbol{V}_h$ such that there exist the derivatives $\boldsymbol{u}_h'(t), \boldsymbol{u}_h''(t)$ and the identity

$$(\rho^b \boldsymbol{u}_h''(t), \boldsymbol{y}_h)_{\Omega_h^b} + (C\rho^b \boldsymbol{u}_h'(t), \boldsymbol{y}_h)_{\Omega_h^b} + a_h^b(\boldsymbol{u}_h(t), \boldsymbol{y}_h) = (\boldsymbol{T}_h^n(t), \boldsymbol{y}_h)_{\Gamma_{Wh}^b},$$
$$\forall\, \boldsymbol{y}_h \in \boldsymbol{V}_h, \quad \forall t \in (0, T), \tag{10.68}$$

and the initial conditions

$$\boldsymbol{u}_h(X,0)=0, \quad \boldsymbol{u}'_h(X,0)=0, \quad X\in\Omega_h^b. \tag{10.69}$$

are satisfied.

The discrete problem (10.68), (10.69) is equivalent to the solution of a system of ordinary differential equations. Let functions $\varphi_1,\dots,\varphi_m$ form a basis of the space V_h. Then the system of the $n=2m$ vector functions

$$(\varphi_1,0),\dots,(\varphi_m,0), \quad (0,\varphi_1),\dots,(0,\varphi_m)$$

forms a basis of the space $\boldsymbol{V}_h=V_h\times V_h$. Let us denote them by $\boldsymbol{\varphi}_1,\dots\boldsymbol{\varphi}_n$. Then the approximate solution $\boldsymbol{u}_h$ can be expressed in the form

$$\boldsymbol{u}_h(t)=\sum_{j=1}^{n} p_j(t)\boldsymbol{\varphi}_j, \quad t\in[0,T]. \tag{10.70}$$

Let us set $\boldsymbol{p}(t)=(p_1(t),\dots,p_n(t))^T$. Using $\boldsymbol{\varphi}_i$, $i=1,\dots,n$, as test functions in (10.68), we get the following system of ordinary differential equations

$$\mathbb{M}\boldsymbol{p}''=\boldsymbol{G}-\mathbb{S}\boldsymbol{p}-C\mathbb{M}\boldsymbol{p}', \tag{10.71}$$

where $\mathbb{M}=(m_{ij})_{i,j=1}^n$ is the mass matrix and $\mathbb{S}=(s_{ij})_{i,j=1}^n$ is the stiffness matrix with the elements $m_{ij}=(\rho^b\boldsymbol{\varphi}_j,\boldsymbol{\varphi}_i)_{\Omega_h^b}$ and $s_{ij}=a_h^b(\boldsymbol{\varphi}_j,\boldsymbol{\varphi}_i)$, respectively. The aerodynamic force vector $\boldsymbol{G}=\boldsymbol{G}(t)=(G_1(t),\dots,G_n(t))^T$ has the components $G_i(t)=(\boldsymbol{T}_h^n(t),\boldsymbol{\varphi}_i)_{\Gamma_{Wh}^b}$, $i=1,\dots,n$. System (10.71) is equipped with the initial conditions

$$p_j(0)=0, \quad p'_j(0)=0, \quad j=1,\dots,n. \tag{10.72}$$

One possibility for solving the discrete initial value problem (10.71), (10.72) is the application of the Newmark method [69], which is popular in solving elasticity problems. We consider the partition of the time interval $[0,T]$ formed by the time instants $0=t_0<t_1<\dots<t_r=T$ introduced in Sect. 10.1.2. Let us set $\boldsymbol{p}_0=0$, $z_0=0$, $\boldsymbol{G}_k=\boldsymbol{G}(t_k)$, and introduce the approximations $\boldsymbol{p}_k\approx\boldsymbol{p}(t_k)$ and $\boldsymbol{q}_k\approx\boldsymbol{p}'(t_k)$ for $k=1,2,\dots,r$. The Newmark scheme can be written in the form

$$\begin{aligned}\boldsymbol{p}_{k+1}=\boldsymbol{p}_k+\tau_k\boldsymbol{q}_k+\tau_k^2\Big(&\beta\left(\mathbb{M}^{-1}\boldsymbol{G}_{k+1}-\mathbb{M}^{-1}\mathbb{S}\boldsymbol{p}_{k+1}-C\boldsymbol{q}_{k+1}\right)\\ &+\left(\frac{1}{2}-\beta\right)\left(\mathbb{M}^{-1}\boldsymbol{G}_k-\mathbb{M}^{-1}\mathbb{S}\boldsymbol{p}_k-C\boldsymbol{q}_k\right)\Big),\end{aligned} \tag{10.73}$$

$$q_{k+1} = q_k + \tau_k\Big(\gamma\left(\mathbb{M}^{-1}G_{k+1} - \mathbb{M}^{-1}\mathbb{S}p_{k+1} - Cq_{k+1}\right) + (1-\gamma)\left(\mathbb{M}^{-1}G_k - \mathbb{M}^{-1}\mathbb{S}p_k - Cq_k\right)\Big), \quad (10.74)$$

where $\beta, \gamma \in \mathbb{R}$ are parameters. From Eq. (10.74) we get

$$q_{k+1} = \frac{1}{1+C\gamma\tau_k}\Big(q_k + \tau_k\Big(\gamma\left(\mathbb{M}^{-1}G_{k+1} - \mathbb{M}^{-1}\mathbb{S}p_{k+1}\right) + (1-\gamma)\left(\mathbb{M}^{-1}G_k - \mathbb{M}^{-1}\mathbb{S}p_k - Cq_k\right)\Big)\Big). \quad (10.75)$$

The substitution of (10.75) into (10.73) yields the relation

$$\begin{aligned} p_{k+1} = p_k + \tau_k q_k + \beta\tau_k^2\Big(&\mathbb{M}^{-1}G_{k+1} - \mathbb{M}^{-1}\mathbb{S}p_{k+1} - \frac{C}{1+C\gamma\tau_k}q_k \\ &- \frac{C\gamma\tau_k}{1+C\gamma\tau_k}\left(\mathbb{M}^{-1}G_{k+1} - \mathbb{M}^{-1}\mathbb{S}p_{k+1}\right) \\ &- \frac{C\tau_k}{1+C\gamma\tau_k}(1-\gamma)\left(\mathbb{M}^{-1}G_k - \mathbb{M}^{-1}\mathbb{S}p_k - Cq_k\right)\Big) \\ &+ \left(\frac{1}{2}-\beta\right)\tau_k^2\left(\mathbb{M}^{-1}G_k - \mathbb{M}^{-1}\mathbb{S}p_k - Cq_k.\right. \end{aligned}$$

This implies that

$$\begin{aligned} p_{k+1} = p_k + \tau_k q_k - C\xi_k q_k + \xi_k\left(\mathbb{M}^{-1}G_{k+1} - \mathbb{M}^{-1}\mathbb{S}p_{k+1}\right) \\ + \left(\left(\frac{1}{2}-\beta\right)\tau_k^2 - C(1-\gamma)\xi_k\tau_k\right)\left(\mathbb{M}^{-1}G_k - \mathbb{M}^{-1}\mathbb{S}p_k - Cq_k\right), \end{aligned}$$

which can be written in the form

$$\begin{aligned} \left(\mathbb{I} + \xi_k\mathbb{M}^{-1}\mathbb{S}\right)p_{k+1} = p_k + (\tau_k - C\xi_k)\,q_k + \xi_k\mathbb{M}^{-1}G_{k+1} \\ + \left(C(\gamma-1)\,\xi_k\tau_k + \left(\frac{1}{2}-\beta\right)\tau_k^2\right)\left(\mathbb{M}^{-1}G_k - \mathbb{M}^{-1}\mathbb{S}p_k - Cq_k\right). \end{aligned} \quad (10.76)$$

Here we use the notation

$$\xi_k = \beta\tau_k^2\left(1 - \frac{C\gamma\tau_k}{1+C\gamma\tau_k}\right) = \frac{\beta\tau_k^2}{1+C\gamma\tau_k}. \quad (10.77)$$

If p_k and q_k are known, then p_{k+1} is obtained from system (10.76) and afterwards q_{k+1} is computed from (10.75).

In numerical examples presented in Sect. 10.2.2.6, the parameters $\beta = 1/4$ and $\gamma = 1/2$ were used. This choice yields the second-order Newmark method.

10.2.2.4 Construction of the ALE Mapping for Fluid

The ALE mapping A_t is constructed with the aid of an artificial stationary elasticity problem. We seek $\boldsymbol{d} = (d_1, d_2)$ defined in Ω_0 as a solution of the elastostatic system

$$\sum_{j=1}^{2} \frac{\partial \tau_{ij}^a}{\partial x_j} = 0 \quad \text{in } \Omega_0, \quad i = 1, 2, \tag{10.78}$$

where τ_{ij}^a are the components of the *artificial stress tensor*

$$\tau_{ij}^a = \lambda^a \nabla \cdot \boldsymbol{d}\, \delta_{ij} + 2\mu^a e_{ij}^a, \quad e_{ij}^a(\boldsymbol{d}) = \frac{1}{2}\left(\frac{\partial d_i}{\partial x_j} + \frac{\partial d_j}{\partial x_i}\right), \quad i, j = 1, 2. \tag{10.79}$$

The artificial Lamé coefficients λ^a and μ^a are related to the artificial Young modulus E^a and to the artificial Poisson ratio σ_a in a similar way as in (10.54). On the boundary $\partial\Omega_{t_0}$ conditions for $\boldsymbol{d}$ are prescribed by

$$\boldsymbol{d}|_{\partial\Omega_i \cup \partial\Omega_o} = 0, \ \boldsymbol{d}|_{\Gamma_{W_0} \setminus \Gamma_W} = 0, \ \boldsymbol{d}(X, t) = \boldsymbol{u}(X, t), \ X \in \Gamma_W. \tag{10.80}$$

The solution of problem (10.78)–(10.80) gives us the ALE mapping of $\overline{\Omega}_0$ onto $\overline{\Omega}_t$ in the form

$$\mathrm{A}_t(X) = X + \boldsymbol{d}(X, t), \quad X \in \overline{\Omega}_0, \tag{10.81}$$

for each time t.

Problem (10.78)–(10.80) is discretized by conforming piecewise linear finite elements on the mesh $\mathscr{T}_{h0}$ used for computing the flow field in the beginning of the computational process in the polygonal approximation Ω_{h0} of the domain Ω_0. We introduce the finite element spaces

$$\mathscr{X}_h = \{\boldsymbol{d}_h = (d_{h1}, d_{h2}) \in (C(\overline{\Omega}_{0h}))^2;\ d_{hi}|_K \in P_1(K)\ \forall K \in \mathscr{T}_{h0},\ i = 1, 2\}, \tag{10.82}$$

$$\mathscr{V}_h = \{\boldsymbol{\varphi}_h \in \mathscr{X}_h;\ \boldsymbol{\varphi}_h(Q) = 0 \text{ for all vertices } Q \in \partial\Omega_0\},$$

and the form

$$B_h(\boldsymbol{d}_h, \boldsymbol{\varphi}_h) = \left((\lambda^a + \mu^a)\nabla \cdot \boldsymbol{d}_h, \nabla \cdot \boldsymbol{\varphi}_h\right)_{\Omega_{0h}} + \left(\mu^a \nabla \boldsymbol{d}_h, \nabla \boldsymbol{\varphi}_h\right)_{\Omega_{0h}}. \tag{10.83}$$

Then the approximate solution of problem (10.78), (10.80) is defined as a function $\boldsymbol{d}_h \in \mathscr{X}_h$ satisfying the Dirichlet boundary conditions (10.80) at the vertices on $\overline{\Gamma}_W^b$ and the identity

$$B_h(\boldsymbol{d}_h, \boldsymbol{\varphi}_h) = 0 \quad \forall \boldsymbol{\varphi}_h \in \mathscr{V}_h. \tag{10.84}$$

Using linear finite elements is sufficient, because we need only to know the movement of the points of the mesh.

In our computations we choose the Lamé coefficients λ^a and μ^a as constants corresponding to the Young modulus and Poisson ratio $E^a = 10{,}000$ and $\sigma^a = 0.45$.

If the displacement $\boldsymbol{d}_h^{k+1}$ is computed at time t_{k+1}, then in view of (10.81), the approximation of the ALE mapping is obtained in the form

$$\mathrm{A}_{t_{k+1}h}(X) = X + \boldsymbol{d}_h^{k+1}(X), \quad X \in \Omega_{0h}. \tag{10.85}$$

From the ALE mapping at the time instants t_{k-1}, t_k, t_{k+1} it is possible to approximate the domain velocity with the aid of the second-order backward difference formula

$$z_h^{k+1}(x) = \frac{\alpha_{2,0}x - \alpha_{2,1}\mathrm{A}_{t_kh}(\mathrm{A}_{t_{k+1}h}^{-1}(x)) + \alpha_{2,2}\mathrm{A}_{t_{k-1}h}(\mathrm{A}_{t_{k+1}h}^{-1}(x))}{2}, \quad x \in \Omega_{t_{k+1}h}, \tag{10.86}$$

with coefficients $\alpha_{2,0}, \alpha_{2,1}, \alpha_{2,2}$ from Table 8.3, where we write $k+1$ instead of k.

Remark 10.4 In Sects. 10.2.2.3 and 10.2.2.4, the dynamic elasticity problem was solved with the aid of conforming finite elements used for space discretization and the Newmark method for time discretization. The conforming FEM was also applied to the construction of the ALE mapping. Recently, in works [128, 158, 159, 200–202], all ingredients of the dynamic elasticity discretization and interaction of compressible flow with an elastic body including the construction of the ALE mapping are based on the use of the discontinuous Galerkin method.

10.2.2.5 Coupling Procedure

The realization of the complete fluid-structure interaction problem can be carried out by the following coupling algorithm.

(1) Assume that on time level t_k, the approximate solution $\boldsymbol{w}_h^k$ of the flow problem and the displacement $\boldsymbol{u}_{h,k}$ of the structure are known.
(2) Set $\boldsymbol{u}_{h,k+1}^0 := \boldsymbol{u}_{h,k}$, $l := 1$ and apply the following iterative process:

   (a) Compute the stress tensor τ_{ij}^f and the aerodynamical force acting on the structure and transform it to the interface Γ_{Wh}^b.

(b) Solve the elasticity problem, compute the approximation $\boldsymbol{u}^l_{h,k+1}$ of the displacement at time t_{k+1} and the approximation $\Omega^l_{ht_{k+1}}$ of the flow domain.
(c) Determine the ALE mapping $\mathcal{A}^l_{t_{k+1}h}$ and the approximation $z^l_{h,k+1}$ of the domain velocity.
(d) Solve the flow problem in the domain $\Omega^l_{ht_{k+1}}$.
(e) If the variation of the displacement $|\boldsymbol{u}^l_{h,k+1} - \boldsymbol{u}^{l-1}_{h,k+1}|$ is larger than the prescribed tolerance, go to (a) and $l := l+1$. Else $\boldsymbol{u}_{h,k+1} := \boldsymbol{u}^l_{h,k}$, $k := k+1$ and go to (2).

This algorithm represents the so-called *strong coupling*. If in step (e) we directly set $k := k + 1$ and go to (2) already in the case when $l = 1$, then we get the *weak (loose) coupling*.

10.2.2.6 Numerical Example

In order to demonstrate the applicability of the developed method, we present here results of a numerical experiment carried out for a problem modelling the flow in *vocal folds*.

We consider flow through a channel with two bumps which represent time dependent boundaries between the flow and a simplified model of vocal folds. Figure 10.7 shows the situation at the initial time $t = 0$, the flow computational mesh consisting of 5398 elements and the structure computational mesh with 1998 elements. In Fig. 10.8 we see a detail of the channel near the narrowest part of the channel at the initial time and the positions of sensor points used in the analysis.

The numerical experiments were carried out for the following data: magnitude of the inlet velocity $v_{in} = 4\ \mathrm{m\,s^{-1}}$, the fluid viscosity $\mu = 15 \cdot 10^{-6}\,\mathrm{kg\,m^{-1}\,s^{-1}}$, the inlet density $\rho_{in} = 1.225\,\mathrm{kg\,m^{-3}}$, the outlet pressure $p_{out} = 97611$ Pa, the Reynolds number $\mathrm{Re} = \rho_{in} v_{in} H/\mu = 5227$, heat conduction coefficient $k = 2.428 \cdot 10^{-2}\,\mathrm{kg\,m\,s^{-2}\,K^{-1}}$, the specific heat $c_v = 721.428\,\mathrm{m^2\,s^{-2}\,K^{-1}}$, the Poisson adiabatic constant $\gamma = 1.4$. The inlet Mach number is $M_{in} = 0.012$. The Young modulus and the Poisson ratio have values $E^b = 25{,}000$ Pa and $\sigma^b = 0.4$, respectively, the structural damping coefficient is equal to the constant $C = 100\,s^{-1}$

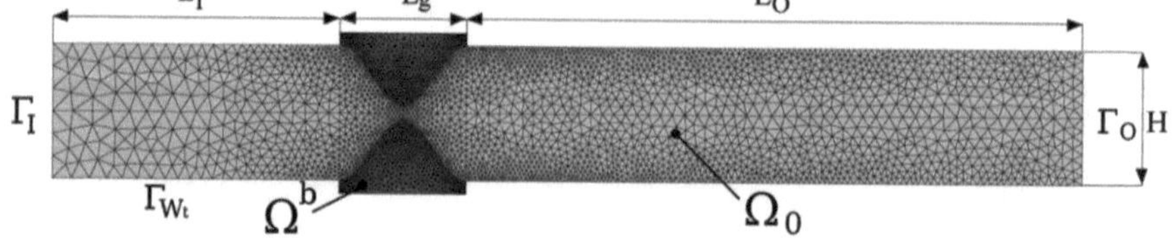

Fig. 10.7 Computational domain at time $t = 0$ with a finite element mesh and the description of its size: $L_I = 50$ mm, $L_g = 15.4$ mm, $L_O = 94.6$ mm, $H = 16$ mm. The width of the channel in the narrowest part is 1.6 mm

Fig. 10.8 Allocation of the sensors

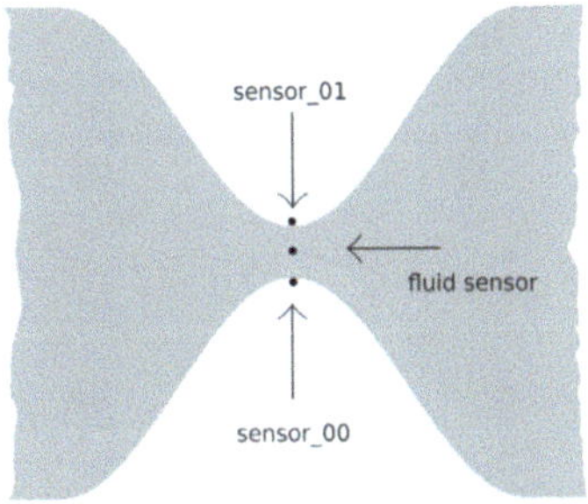

and the material density $\rho^b = 1040\,\mathrm{kg\,m^{-3}}$. The quadratic ($p = 2$) and linear ($s = 1$) elements were used for the DG-BDF approximation of flow and conforming FE-Newmark approximation of the structural problem, respectively.

We present the results obtained by the fluid-structure interaction computation with the strong coupling. Figure 10.9 shows the velocity isolines in the whole channel at several time instants. In Figs. 10.10 and 10.11 we can see the computational mesh and the velocity field near the vocal folds at several time instants. The maximal velocity is $v \approx 54\ \mathrm{m\ s^{-1}}$. We can observe the *Coanda effect* represented by the attachment

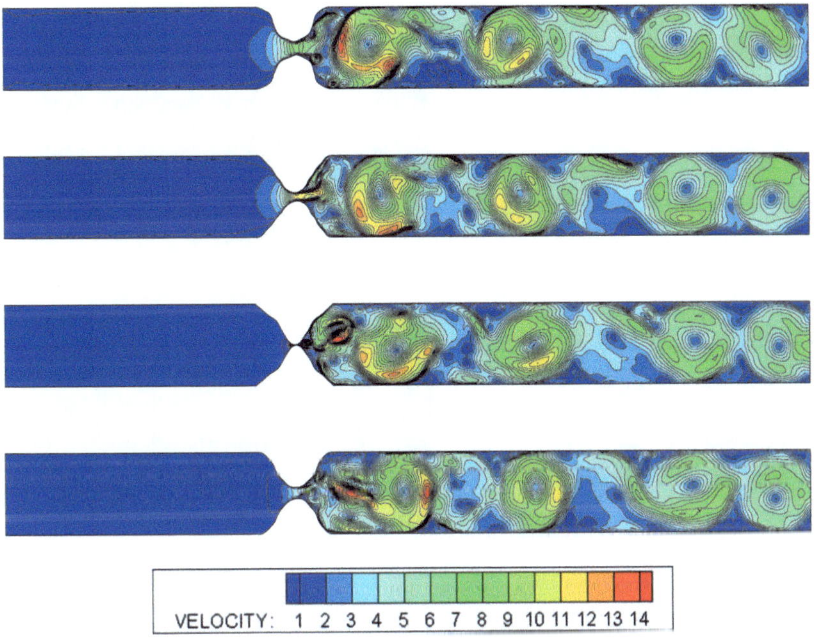

Fig. 10.9 Velocity isolines at time instants $t = 0.1976, 0.1982, 0.1989, 0.1995$ s. The legend shows the dimensionless values of the velocity. For getting the dimensional values multiply by $U^* = 4$

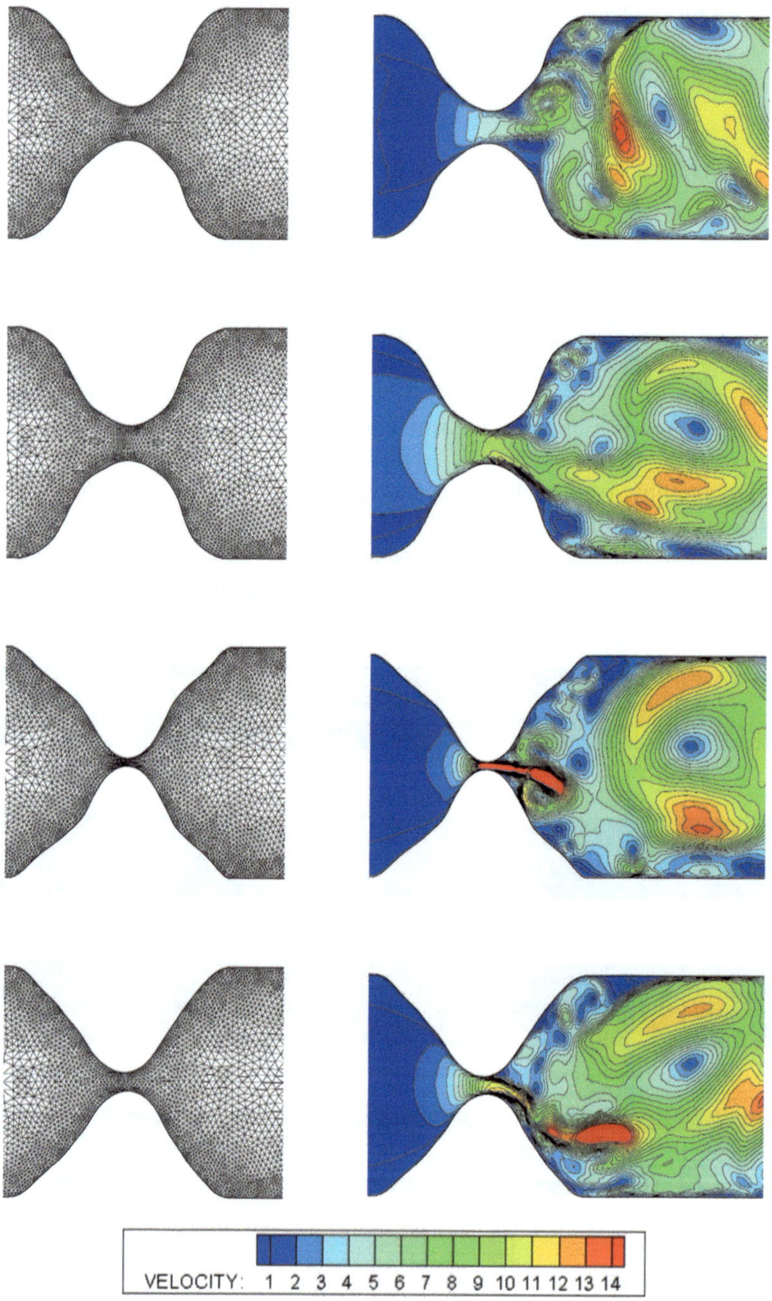

Fig. 10.10 Detail of the mesh and velocity distribution in the vicinity of the narrowest part of the channel at time instants $t = 0.1950, 0.1957, 0.1963, 0.1970$ s. The legend shows the dimensionless values of the velocity. For getting the dimensional values multiply by $U^* = 4$

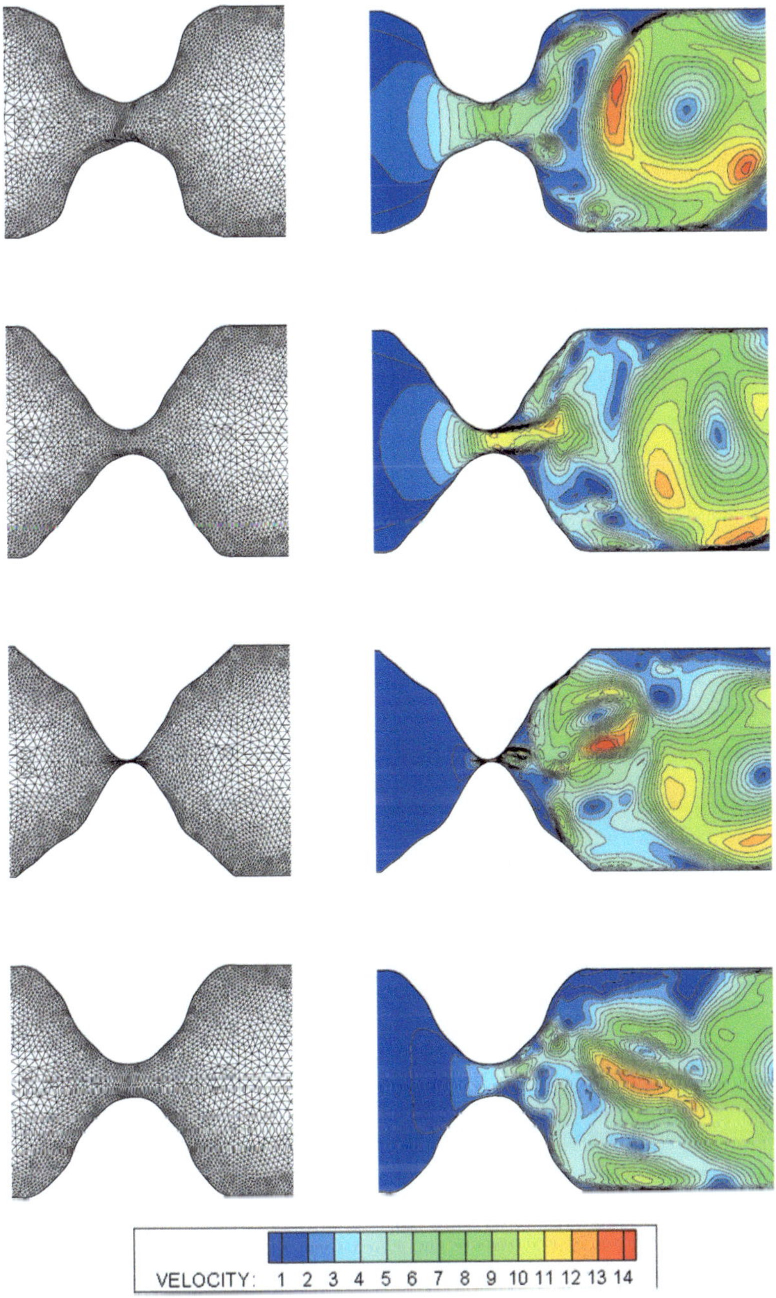

Fig. 10.11 Detail of the mesh and the velocity distribution in the vicinity of the narrowest part of the channel at time instants $t = 0.1976, 0.1982, 0.1989, 0.1995\,\mathrm{s}$. The legend shows the dimensionless values of the velocity. For getting the dimensional values multiply by $U^* = 4$

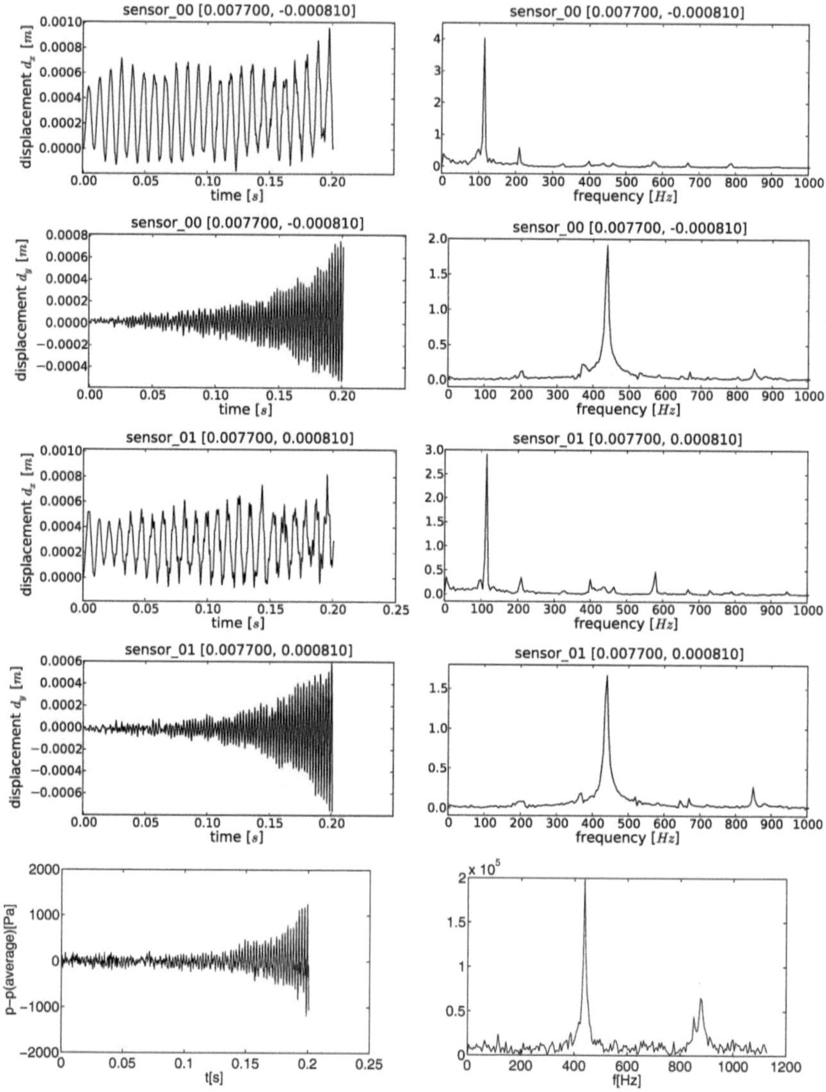

Fig. 10.12 Vibrations of sensor points 00 and 01 on the vocal folds with their Fourier analysis and the fluid pressure fluctuations in the middle of the gap with their Fourier analysis

of the main stream (jet) successively to the upper and lower wall and the formation of large scale vortices behind the vocal folds.

The deformation of vocal folds is tested on two sensor points denoted by 00 and 01 lying on the vocal folds surface shown in Fig. 10.8. The character of the vocal folds vibrations is indicated in Fig. 10.12, which shows the horizontal and vertical displacements d_x and d_y of the sensor points. Moreover, the fluid pressure fluctuations in the middle of the gap as well as the Fourier analysis of the signals are shown. Vocal folds vibrations are not fully symmetric due to the Coanda effect and are composed of the fundamental horizontal mode of vibration with corresponding frequency 113 Hz and by the higher vertical mode with frequency 439 Hz. The increase of vertical vibrations due to the aeroelastic instability of the system results in a fast decrease of the glottal gap. At about $t = 0.2$ s, when the gap was nearly closed, the fluid mesh deformation in this region became too large and the numerical simulation stopped. The dominant peak at 439 Hz in the spectrum of the pressure signal corresponds well to the vertical oscillations of the glottal gap, while the influence of the lower frequency 113 Hz associated with the horizontal vocal folds motion is negligible in the pressure fluctuations. The modeled flow-induced instability of the vocal folds is called *phonation onset* followed in reality by a complete closing of the glottis and consequently by the collisions of vocal folds producing the voice acoustic signal.

References

1. Adams, R.A., Fournier, J.J.F.: Sobolev Spaces. Academic Press, Amsterdam (2003)
2. Akman, T., Karasözen, B.: Variational time discretization methods for optimal control problems governed by diffusion-convection-reaction equations. J. Comput. Appl. Math. **272**, 41–56 (2014)
3. Akman, T., Yücel, H., Karasözen, B.: A priori error analysis of the upwind symmetric interior penalty Galerkin (SIPG) method for the optimal control problems governed by unsteady convection diffusion equations. Comput. Optim. Appl. **57**(3), 703–729 (2014)
4. Akrivis, G., Makridakis, C.: Galerkin time-stepping methods for nonlinear parabolic equations. ESAIM: Math. Model. Numer. Anal. **38**, 261–289 (2004)
5. Alaoui, L.E., Ern, A., Vohralík, M.: Guaranteed and robust a posteriori error estimates and balancing discretization and linearization errors for monotone nonlinear problems. Comput. Methods Appl. Mech. Eng. **200**, 2782–2795 (2011)
6. Appell, J., Zabrejko, P.: Nonlinear Superposition Operators. Cambridge University Press, Cambridge (1990)
7. Arnold, D.N.: An interior penalty finite element method with discontinuous elements. SIAM J. Numer. Anal. **19**(4), 742–760 (1982)
8. Arnold, D.N., Brezzi, F., Cockburn, B., Marini, L.D.: Unified analysis of discontinuous Galerkin methods for elliptic problems. SIAM J. Numer. Anal. **39**(5), 1749–1779 (2002)
9. Atkins, H.L., Shu, C.W.: Quadrature-free implementation of discontinuous Galerkin methods for hyperbolic equations. AIAA J. **36**, 775–782 (1998)
10. Aubin, J.P.: Approximation des problèmes aux limites non homogénes pour des opérateurs nonlinéaires. J. Math. Anal. Appl. **30**, 510–521 (1970)
11. Babuška, I.: The finite element method with penalty. Math. Comp. **27**, 221–228 (1973)
12. Babuška, I., Baumann, C.E., Oden, J.T.: A discontinuous *hp* finite element method for diffusion problems: 1-D analysis. Comput. Math. Appl. **37**, 103–122 (1999)
13. Babuška, I., Strouboulis, T.: The Finite Element Method and Its Reliability. Clarendon Press, Oxford (2001)
14. Babuška, I., Suri, M.: The *hp*-version of the finite element method with quasiuniform meshes. M^2AN Math. Model. Numer. Anal. **21**, 199–238 (1987)
15. Babuška, I., Suri, M.: The *p*- and *hp*-versions of the finite element method. An overview. Comput. Methods Appl. Mech. Eng. **80**, 5–26 (1990)
16. Babuška, I., Suri, M.: The *p*- and *hp*-FEM a survey. SIAM Rev. **36**, 578–632 (1994)
17. Babuška, I., Zlámal, M.: Nonconforming elements in the finite element method with penalty. SIAM J. Numer. Anal. **10**, 863–875 (1973)
18. Baker, G.A.: Finite element methods for elliptic equations using nonconforming elements. Math. Comput. **31**(137), 45–59 (1977)

V. Dolejší and M. Feistauer, *Discontinuous Galerkin Method*,
Springer Series in Computational Mathematics 48,
DOI 10.1007/978-3-319-19267-3

19. Bardos, C., Le Roux, A.Y., Nedelec, J.C.: First order quasilinear equations with boundary conditions. Commun. Partial Differ. Equ. **4**, 1017–1034 (1979)
20. Bassi, F., Bartolo, C.D., Hartmann, R., Nigro, A.: A discontinuous Galerkin method for inviscid low Mach number flows. J. Comput. Phys. **228**, 3996–4011 (2009)
21. Bassi, F., Botti, L., Colombo, A., Rebay, S.: Agglomeration based discontinuous Galerkin discretization of the Euler and Navier-Stokes equations. Comput. Fluids **61**, 77–85 (2012)
22. Bassi, F., Crivellini, A., Rebay, S., Savini, M.: Discontinuous Galerkin solution of the Reynolds averaged Navier-Stokes and k-ω turbulence model equations. Comput. Fluids **34**, 507–540 (2005)
23. Bassi, F., Rebay, S.: A high-order accurate discontinuous finite element method for the numerical solution of the compressible Navier-Stokes equations. J. Comput. Phys. **131**, 267–279 (1997)
24. Bassi, F., Rebay, S.: High-order accurate discontinuous finite element solution of the 2D Euler equations. J. Comput. Phys. **138**, 251–285 (1997)
25. Bassi, F., Rebay, S.: A high order discontinuous Galerkin method for compressible turbulent flow. In: Cockburn, B., Karniadakis, G.E., Shu, C.W. (eds.) Discontinuous Galerkin Method: Theory, Computations and Applications. Lecture Notes in Computational Science and Engineering, vol. 11, pp. 113–123. Springer, Berlin (2000)
26. Bassi, F., Rebay, S.: Numerical evaluation of two discontinuous Galerkin methods for the compressible Navier-Stokes equations. Int. J. Numer. Methods Fluids **40**, 197–207 (2002)
27. Bassi, F., Rebay, S., Savini, M., Mariotti, G., Pedinotti, S.: A high-order accurate discontinuous finite element method for inviscid and viscous turbomachinery flow. In: Proceedings of the Second European Conference on Turbomachinery Fluid Dynamics and Thermodynamics, Antwerpen, Belgium (1997)
28. Baumann, C.E.: An h-p adaptive discontinuous Galerkin finite element method for computational fluid dynamics. Ph.D. thesis, The University of Texas at Austin (1997)
29. Baumann, C.E., Oden, J.T.: A discontinuous hp finite element method for convection-diffusion problems. Comput. Methods Appl. Mech. Eng. **175**(3–4), 311–341 (1999)
30. Baumann, C.E., Oden, J.T.: A discontinuous hp finite element method for the Euler and Navier-Stokes equations. Int. J. Numer. Methods Fluids **31**, 79–95 (1999)
31. Bayliss, A., Turkel, E.: Radiation boundary conditions for wave-like equations. Commun. Pure Appl. Math. **33**, 708–725 (1980)
32. Beam, R.M., Warming, R.F.: An implicit finite-difference algorithm for hyperbolic systems in conservation-law form. J. Comput. Phys. **22**, 87–110 (1976)
33. Beam, R.M., Warming, R.F.: An implicit factored scheme for the compressible Navier-Stokes equations. AIAA J. **16**, 393–402 (1978)
34. Biswas, R., Devine, K.D., Flaherty, J.E.: Parallel, adaptive finite element methods for conservation laws. In: Proceedings of the 3rd ARO Workshop on Adaptive Methods for Partial Differential Equations, Troy, NY, 1992, vol. 14, pp. 255–283 (1994)
35. Blasius, H.: Grenzschichten in Flüssigkeiten mit kleiner Reibung. Z. Math. Phys. **56**, 1–37 (1908)
36. Brenner, S.C.: Poincaré-Friedrichs inequalities for piecewise H-1 functions. SIAM J. Numer. Anal. **41**(1), 306–324 (2003)
37. Brenner, S.C., Scott, R.L.: The Mathematical Theory of Finite Element Methods. Springer, New York (1994)
38. Brezzi, F., Fortin, M.: Mixed and Hybrid Finite Element Methods. Springer, Berlin (1992)
39. Brezzi, F., Manzini, G., Marini, D., Pietra, P., Russo, A.: Discontinuous finite elements for diffusion problems. Technical report, Milano, Istituto Lombardo di Scienze e Lettere (1997)
40. Brezzi, F., Manzini, G., Marini, D., Pietra, P., Russo, A.: Discontinuous Galerkin approximations for elliptic problems. Numer. Methods Partial Differ. Equ. **16**, 365–378 (2000)
41. Castillo, P.: Performance of discontinuous Galerkin methods for elliptic PDEs. SIAM J. Sci. Comput. **24**(2), 524–547 (2002)
42. Castillo, P.: A review of the local discontinuous Galerkin (LDG) method applied to elliptic problems. Appl. Numer. Math. **56**, 1307–1313 (2006)

43. Castillo, P., Cockburn, B., Perugia, I., Schötzau, D.: An a priori error analysis of the local discontinuous Galerkin method for elliptic problems. SIAM J. Numer. Anal. **38**(5), 1676–1706 (2000)
44. Castillo, P., Cockburn, B., Schötzau, D., Schwab, C.: Optimal a priori estimates for the *hp*-version of the local discontinuous Galerkin method for convection-difussion problems. Math. Comp. **71**(238), 455–478 (2002)
45. Česenek, J.: Discontinuous Galerkin method for solving compressible flow (in Czech). Ph.D. thesis, Charles University in Prague (2011)
46. Česenek, J., Feistauer, M.: Theory of the space-time discontinuous Galerkin method for nonstationary parabolic problems with nonlinear convection and diffusion. SIAM J. Numer. Anal. **30**, 1181–1206 (2012)
47. Česenek, J., Feistauer, M., Horáček, J., Kučera, V., Prokopová, J.: Simulation of compressible viscous flow in time-dependent domains. Appl. Math. Comput. **219**(13), 7139–7150 (2013)
48. Česenek, J., Feistauer, M., Kosík, A.: DGFEM for the analysis of airfoil vibrations induced by compressible flow. ZAMM **93**, 387–402 (2013)
49. Chaillou, A., Suri, M.: A posteriori estimation of the linearization error for strongly monotone nonlinear operators. J. Comput. Appl. Math. **205**(1), 72–87 (2007)
50. Chen, H.: Superconvergence properties of discontinuous Galerkin methods for two-point boundary value problems. Int. J. Numer. Anal. Model. **3**(2), 163–185 (2006)
51. Chrysafinos, K., Walkington, N.J.: Error estimates for the discontinuous Galerkin methods for parabolic equations. SIAM J. Numer. Anal. **44**, 349–366 (2006)
52. Ciarlet, P.G.: The Finite Elements Method for Elliptic Problems. North-Holland, Amsterdam (1979)
53. Ciarlet, P.G., Raviart, P.A.: The combined effect of curved boundaries and numerical integration in isoparametric finite element methods. In: Aziz, A. (ed.) The Mathematical Foundations of the Finite Element Method with Applications to Partial Differential Equations, pp. 409–474. Academic Press, New York (1972)
54. Cockburn, B.: An introduction to the discontinuous Galerkin method for convection-dominated problems. In: Quarteroni, A. (ed.) Advanced numerical approximation of non-linear hyperbolic equations. Lecture Notes in Mathematics, vol. 1697, pp. 151–268. Springer, Berlin (1998)
55. Cockburn, B., Dawson, C.: Some extensions of the local discontinuous Galerkin method for convection-diffusion equations in multidimensions. The Mathematics of Finite Elements and Applications. X, MAFELAP 1999 (Uxbridge), pp. 225–238. Elsevier, Oxford (2000)
56. Cockburn, B., Hou, S., Shu, C.W.: TVB Runge-Kutta local projection discontinuous Galerkin finite element method for conservation laws III: one dimensional systems. J. Comput. Phys. **84**, 90–113 (1989)
57. Cockburn, B., Hou, S., Shu, C.W.: TVB Runge-Kutta local projection discontinuous Galerkin finite element for conservation laws IV: the multi-dimensional case. Math. Comput. **54**, 545–581 (1990)
58. Cockburn, B., Kanschat, G., Perugia, I., Schötzau, D.: Superconvergence of the local discontinuous Galerkin method for elliptic problems on Cartesian grids. SIAM J. Numer. Anal. **39**, 264–285 (2001)
59. Cockburn, B., Kanschat, G., Schötzau, D.: The local discontinuous Galerkin method for linearized incompressible fluid flow: a review. Comput. Fluids **34**(4–5), 491–506 (2005)
60. Cockburn, B., Kanschat, G., Schötzau, D.: A locally conservative LDG method for the incompressible Navier-Stokes equations. Math. Comput. **74**(251), 1067–1095 (2005)
61. Cockburn, B., Kanschat, G., Schötzau, D., Schwab, C.: Local discontinuous Galerkin methods for the Stokes system. SIAM J. Numer. Anal. **40**(1), 319–343 (2002)
62. Cockburn, B., Shu, C.W.: TVB Runge-Kutta local projection discontinuous Galerkin finite element for conservation laws II: general framework. Math. Comput. **52**, 411–435 (1989)
63. Cockburn, B., Shu, C.W.: The Runge-Kutta local projection P^1-discontinuous Galerkin method for scalar conservation laws. RAIRO Modél. Math. Anal. Numér. **25**, 337–361 (1991)

64. Cockburn, B., Shu, C.W.: The local discontinuous Galerkin method for time-dependent convection-diffusion systems. SIAM J. Numer. Anal. **35**, 2440–2463 (1998)
65. Cockburn, B., Shu, C.W.: The Runge-Kutta discontinuous Galerkin method for conservation laws. V. Multidimensional systems. J. Comput. Phys. **141**(2), 199–224 (1998)
66. Cockburn, B., Shu, C.W.: Runge-Kutta discontinuous Galerkin methods for convection-dominated problems. J. Sci. Comput. **16**(3), 173–261 (2001)
67. Courant, R.: Variational methods for the solution of problems of equilibrium and vibrations. Bull. Am. Math. Soc. **49**, 1–23 (1943)
68. Crouzeix, M., Raviart, P.A.: Conforming and nonconforming finite element methods for solving the stationary Stokes equations. RAIRO, Anal. Numér. **7**, 33–76 (1973)
69. Curnier, A.: Computational Methods in Solid Mechanics. Kluwer Academic Publishing Group, Dordrecht (1994)
70. Daru, V., Tenaud, C.: High order one-step monotonicity-preserving schemes for unsteady compressible flow calculations. J. Comput. Phys. **193**(2), 563–594 (2004)
71. Darwish, M., Moukalled, F., Sekar, B.: A robust multi-grid pressure-based algorithm for multi-fluid flow at all speeds. Int. J. Numer. Methods Fluids **41**, 1221–1251 (2003)
72. Davis, T.A., Duff, I.S.: A combined unifrontal/multifrontal method for unsymmetric sparse matrices. ACM Trans. Math. Softw. **25**, 1–19 (1999)
73. Dawson, C.: The P^{k+1}-S^k local discontinuous Galerkin method for elliptic equations. SIAM J. Numer. Anal. **4**(6), 2151–2170 (2002)
74. Dawson, C.N., Sun, S., Wheeler, M.F.: Compatible algorithms for coupled flow and transport. Comput. Methods Appl. Mech. Eng. **193**, 2565–2580 (2004)
75. deCougny, H.L., Devine, K.D., Flaherty, J.E., Loy, R.M., Özturan, c, Shephard, M.S.: Load balancing for the parallel adaptive solution of partial differential equations. Appl. Numer. Math. **16**(1–2), 157–182 (1994)
76. Delves, L., Hall, C.: An implicit matching principle for global element calculations. J. Inst. Math. Appl. **23**, 223–234 (1979)
77. Delves, L., Phillips, C.: A fast implementation of the global element methods. J. Inst. Math. Appl. **25**, 179–197 (1980)
78. Demkowicz, L., Rachowicz, W., Devloo, P.: A fully automatic *hp*-adaptivity. J. Sci. Comput. **17**(1–4), 117–142 (2002)
79. Deuflhard, P.: Newton Methods for Nonlinear Problems, Springer Series in Computational Mathematics, vol. 35. Springer, Berlin (2004)
80. Di Pietro, D., Ern, A.: Mathematical Aspects of Discontinuous Galerkin Methods. Mathematiques et Applications, vol. 69. Springer, Berlin (2012)
81. Dick, E.: Second-order formulation of a multigrid method for steady Euler equations through defect-correction. J. Comput. Appl. Math. **35**(1–3), 159–168 (1991)
82. de Dios, B.A., Brezzi, F., Havle, O., Marini, L.D.: L^2-estimates for the DG IIPG-0 scheme. Numer. Methods Partial Differ. Equ. **28**(5), 1440–1465 (2012)
83. Dolejší, V.: Anisotropic mesh adaptation for finite volume and finite element methods on triangular meshes. Comput. Vis. Sci. **1**(3), 165–178 (1998)
84. Dolejší, V.: ANGENER-software package. Charles University Prague, Faculty of Mathematics and Physics (2000). www.karlin.mff.cuni.cz/dolejsi/angen.html
85. Dolejší, V.: Anisotropic mesh adaptation technique for viscous flow simulation. East-West J. Numer. Math. **9**(1), 1–24 (2001)
86. Dolejší, V.: On the discontinuous Galerkin method for the numerical solution of the Navier-Stokes equations. Int. J. Numer. Methods Fluids **45**, 1083–1106 (2004)
87. Dolejší, V.: Analysis and application of IIPG method to quasilinear nonstationary convection-diffusion problems. J. Comp. Appl. Math. **222**, 251–273 (2008)
88. Dolejší, V.: Semi-implicit interior penalty discontinuous Galerkin methods for viscous compressible flows. Commun. Comput. Phys. **4**(2), 231–274 (2008)
89. Dolejší, V.: A design of residual error estimates for a high order BDF-DGFE method applied to compressible flows. Int. J. Numer. Methods Fluids **73**(6), 523–559 (2013)

90. Dolejší, V.: hp-DGFEM for nonlinear convection-diffusion problems. Math. Comput. Simul. **87**, 87–118 (2013)
91. Dolejší, V., Ern, A., Vohralík, M.: A framework for robust a posteriori error control in unsteady nonlinear advection-diffusion problems. SIAM J. Numer. Anal. **51**(2), 773–793 (2013)
92. Dolejší, V., Feistauer, M.: On the discontinuous Galerkin method for the numerical solution of compressible high-speed flow. In: Brezzi, F., Buffa, A., Corsaro, S., Murli, A. (eds.) Numerical Mathematics and Advanced Applications, ENUMATH 2001, pp. 65–84. Springer, Milano (2003)
93. Dolejší, V., Feistauer, M.: Semi-implicit discontinuous Galerkin finite element method for the numerical solution of inviscid compressible flow. J. Comput. Phys. **198**(2), 727–746 (2004)
94. Dolejší, V., Feistauer, M.: Error estimates of the discontinuous Galerkin method for nonlinear nonstationary convection-diffusion problems. Numer. Funct. Anal. Optim. **26**(3), 349–383 (2005)
95. Dolejší, V., Feistauer, M., Hozman, J.: Analysis of semi-implicit DGFEM for nonlinear convection-diffusion problems. Comput. Methods Appl. Mech. Eng. **196**, 2813–2827 (2007)
96. Dolejší, V., Feistauer, M., Kučera, V., Sobotíková, V.: An optimal $L^\infty(L^2)$-error estimate of the discontinuous Galerkin method for a nonlinear nonstationary convection-diffusion problem. IMA J. Numer. Anal. **28**, 496–521 (2008)
97. Dolejší, V., Feistauer, M., Schwab, C.: A finite volume discontinuous Galerkin scheme for nonlinear convection-diffusion problems. Calcolo **39**, 1–40 (2002)
98. Dolejší, V., Feistauer, M., Schwab, C.: On some aspects of the discontinuous Galerkin finite element method for conservation laws. Math. Comput. Simul. **61**, 333–346 (2003)
99. Dolejší, V., Feistauer, M., Sobotíková, V.: Analysis of the discontinuous Galerkin method for nonlinear convection-diffusion problems. Comput. Methods Appl. Mech. Eng. **194**, 2709–2733 (2005)
100. Dolejší, V., Felcman, J.: Anisotropic mesh adaptation and its application for scalar diffusion equations. Numer. Methods Partial Differ. Equ. **20**, 576–608 (2004)
101. Dolejší, V., Havle, O.: The L^2-optimality of the IIPG method for odd degrees of polynomial approximation in 1D. J. Sci. Comput. **42**(1), 122–143 (2010)
102. Dolejší, V., Holík, M., Hozman, J.: Efficient solution strategy for the semi-implicit discontinuous Galerkin discretization of the Navier-Stokes equations. J. Comput. Phys. **230**, 4176–4200 (2011)
103. Dolejší, V.: Kůs, P.: Adaptive backward difference formula—discontinuous Galerkin finite element method for the solution of conservation laws. Int. J. Numer. Methods Eng. **73**(12), 1739–1766 (2008)
104. Dolejší, V., Vlasák, M.: Analysis of a BDF-DGFE scheme for nonlinear convection-diffusion problems. Numer. Math. **110**, 405–447 (2008)
105. Douglas Jr., J., Darlow, B.L., Kendall, R.P., Wheeler, M.F.: Self-adaptive Galerkin methods for one-dimensional, two-phase immiscible flow. In: Proceedings AIME Fifth Symposium on Reservoir Simulation, pp. 65–72. Society of Petroleum Engineers, Denver, Colorado (1978)
106. Douglas Jr, J., Dupont, T.: Interior penalty procedures for elliptic and parabolic Galerkin methods. Computing Methods in Applied Sciences (Second International Symposium, Versailles, 1975), Lecture Notes in Physics, pp. 207–216. Springer, Berlin (1976)
107. Douglas Jr., J., Wheeler, M.F., Darlow, B.L., Kendall, R.P.: Self-adaptive finite element simulation of miscible displacement in porous media. Comput. Methods Appl. Mech. Eng. **47**(1–2), 131–159 (1984) (Special Issue on Oil Reservoir Simulation)
108. Du, Q., Gunzburger, M.D., Ju, L.: Voronoi-based finite volume methods, optimal Voronoi meshes, and PDEs on the sphere. Comput. Methods Appl. Mech. Eng. **192**(35–36), 3933–3957 (2003)
109. Dumbser, M.: Arbitrary high order PNPM schemes on unstructured meshes for the compressible Navier-Stokes equations. Comput. Fluids **39**(1), 60–76 (2010)
110. Dumbser, M., Munz, C.D.: Building blocks for arbitrary high-order discontinuous Galerkin methods. J. Sci. Comput. **27**, 215–230 (2006)

111. Dunavant, D.A.: High degree efficient symmetrical Gaussian quadrature rules for the triangle. Int. J. Numer. Methods Eng. **21**, 1129–1148 (1985)
112. Edwards, R.E.: Functional Analysis, Theory and Applications. Holt, Rinehart and Winston, New York (1965)
113. Eibner, T., Melenk, J.M.: An adaptive strategy for *hp*-FEM based on testing for analyticity. Comput. Mech. **39**(5), 575–595 (2007)
114. Eriksson, K., Estep, D., Hansbo, P., Johnson, C.: Introduction to adaptive methods for differential equations. Acta Numer. **4**, 105–158 (1995)
115. Eriksson, K., Estep, D., Hansbo, P., Johnson, C.: Computational Differential Equations. Cambridge University Press, Cambridge (1996)
116. Eriksson, K., Johnson, C.: Adaptive finite element methods for parabolic problems. I. A linear model problem. SIAM J. Numer. Anal. **28**(1), 43–77 (1991)
117. Eriksson, K., Johnson, C., Thomée, V.: Time discretization of parabolic problems by the discontinuous Galerkin method. RAIRO Modél. Math. Anal. Numér. **19**(4), 611–643 (1985)
118. Ern, A., Vohralík, M.: Aposteriori error estimation based on potential and flux reconstruction for the heat equation. SIAM J. Numer. Anal. **48**, 198–223 (2010)
119. Eymard, R., Gallouët, T., Herbin, R.: Finite volume methods. Handbook of Numerical Analysis, pp. 713–1020. North-Holland, Amsterdam (2000)
120. Fabian, M., Habala, P., Hájek, P., Montesinos, V., Zizler, V.: Banach Space Theory. Springer, New York (2011)
121. Feistauer, M.: On the finite element approximation of functions with noninteger derivatives. Numer. Funct. Anal. Optimiz. **10**, 91–110 (1989)
122. Feistauer, M.: Mathematical Methods in Fluid Dynamics. Longman Scientific & Technical, Harlow (1993)
123. Feistauer, M., Dolejší, V., Kučera, V.: On a semi-implicit discontinuous Galerkin FEM for the nonstationary compressible Euler equations. In: Asukara, F., Aiso, H., Kawashima, S., Matsumura, A., Nishibata, S., Nishihara, K. (eds.) Hyperbolic Problems: Theory, Numerics and Applications, Tenth International Conference in Osaka, September 2004, pp. 391–398. Yokohama Publishers Inc., Yokohama (2006)
124. Feistauer, M., Dolejší, V., Kučera, V.: On the discontinuous Galerkin method for the simulation of compressible flow with wide range of Mach numbers. Comput. Vis. Sci. **10**, 17–27 (2007)
125. Feistauer, M., Dolejší, V., Kučera, V., Sobotíková, V.: $L^\infty(L^2)$-error estimates for the DGFEM applied to convection-diffusion problems on nonconforming meshes. J. Numer. Math. **17**(1), 45–65 (2009)
126. Feistauer, M., Felcman, J., Lukáčová-Medvid'ová, M., Warnecke, G.: Error estimates of a combined finite volume—finite element method for nonlinear convection-diffusion problems. SIAM J. Numer. Anal. **36**(5), 1528–1548 (1999)
127. Feistauer, M., Felcman, J., Straškraba, I.: Mathematical and Computational Methods for Compressible Flow. Clarendon Press, Oxford (2003)
128. Feistauer, M., Hadrava, M., Horáček, J., Kosík, A.: Numerical solution of fluid-structure interaction by the space-time discontinuous Galerkin method. In: Furmann, J., Ohlberger, M., Rohde, C. (eds.) Finite Volumes for Complex Applications VII—Elliptic, Parabolic and Hyperbolic Problems, Proceedings of FVCA 7, Berlin, June 2014, pp. 567–575. Springer, Cham (2014)
129. Feistauer, M., Hájek, J., Švadlenka, K.: Space-time discontinuos Galerkin method for solving nonstationary convection-diffusion-reaction problems. Appl. Math. **52**(3), 197–233 (2007)
130. Feistauer, M., Hasnedlová-Prokopová, J., Horáček, J., Kosík, A., Kučera, V.: DGFEM for dynamical systems describing interaction of compressible fluid and structures. J. Comput. Appl. Math. **254**, 17–30 (2013)
131. Feistauer, M., Horáček, J., Kučera, V., Prokopová, J.: On the numerical solution of compressible flow in time-dependent domains. Math. Bohem. **137**(1), 1–16 (2012)
132. Feistauer, M., Křížek, M., Sobotíková, V.: An analysis of finite element variational crimes for a nonlinear elliptic problem of a nonmonotone type. East-West J. Numer. Math. **1**, 267–285 (1993)

133. Feistauer, M., Kučera, V.: On a robust discontinuous Galerkin technique for the solution of compressible flow. J. Comput. Phys. **224**, 208–221 (2007)
134. Feistauer, M., Kučera, V., Najzar, K., Prokopová, J.: Analysis of space-time discontinuous Galerkin method for nonlinear convection-diffusion problems. Numer. Math. **117**, 251–288 (2011)
135. Feistauer, M., Kučera, V., Prokopová, J.: Discontinuous Galerkin solution of compressible flow in time dependent domains. Math. Comput. Simul. **80**(8), 1612–1623 (2010)
136. Feistauer, M., Najzar, K., Sobotíková, V.: Error estimates for the finite element solution of elliptic problems with nonlinear Newton boundary conditions. Numer. Funct. Anal. Optimiz. **20**, 835–852 (1999)
137. Feistauer, M., Sändig, A.M.: Graded mesh refinement and error estimates of higher order for DGFE solutions of elliptic boundary value problems in polygons. Numer. Methods Partial Differ. Equ. **28**(4), 1124–1151 (2012)
138. Feistauer, M., Sobotíková, V.: Finite element approximation of nonlinear elliptic problems with discontinuous coefficients. RAIRO Modél. Math. Anal. Numér. **24**, 457–500 (1990)
139. Feistauer, M., Švadlenka, K.: Discontinuous Galerkin method of lines for solving nonstationary singularly perturbed linear problems. J. Numer. Math. **12**, 97–118 (2004)
140. Feistauer, M., Ženíšek, A.: Finite element solution of nonlinear elliptic problems. Numer. Math. **50**, 451–475 (1987)
141. Fezoui, L., Stoufflet, B.: A class of implicit upwind schemes for Euler simulations with unstructured meshes. J. Comput. Phys. **84**(1), 174–206 (1989)
142. Frankel, L.: On corner eddies in plane inviscid shear flow. J. Fluid Mech. **11**, 400–406 (1961)
143. Fürst, J.: Modélisation numérique d'écoulements transsoniques avec des schémas TVD et ENO. Ph.D. thesis, Université Mediterranée, Marseille and Czech Technical University Prague (2001)
144. Gajewski, H., Gröger, K., Zacharias, K.: Nichtlineare Operatorgleichungen und Operatordifferentialgleichungen. Akademie-Verlag, Berlin (1994)
145. Gassner, G., Lörcher, F., Munz, C.D.: A discontinuous Galerkin scheme based on a spacetime expansion. I. Inviscid compressible flow in one space dimension. J. Sci. Comput. **32**(2), 175–199 (2007)
146. Gassner, G., Lörcher, F., Munz, C.D.: A discontinuous Galerkin scheme based on a spacetime expansion. II. Viscous flow equations in multi dimensions. J. Sci. Comput. **34**(3), 260–286 (2008)
147. Gautschi, W.: Orthogonal Polynomials: Computational and Approximations. Oxford University Press, New York (2004)
148. Georgoulis, E.H.: hp-version interior penalty discontinuous Galerkin finite element methods on anisotropic meshes. Int. J. Numer. Anal. Model. **3**(1), 52–79 (2006)
149. Georgoulis, E.H., Hall, E., Houston, P.: Discontinuous Galerkin methods on hp-anisotropic meshes I: a priori error analysis. Int. J. Comput. Sci. Math. **1**(2–3), 221–244 (2007)
150. Georgoulis, E.H., Süli, E.: Optimal error estimates for the hp-version interior penalty discontinuous Galerkin finite element method. IMA J. Numer. Anal. **25**(1), 205–220 (2005)
151. Giles, M.B.: Non-reflecting boundary conditions for Euler equation calculations. AIAA J. **42**, 2050–2058 (1990)
152. Godlewski, E., Raviart, P.A.: Numerical Approximation of Hyperbolic Systems of Conservation Laws, Applied Mathematical Sciences, vol. 118. Springer, New York (1996)
153. Grisvard, P.: Singularities in Boundary Value Problems. Springer, Berlin (1992)
154. Gustafsson, B., Fern, L.: Far fields boundary conditions for steady state solutions to hyperbolic systems. In: Nonlinear hyperbolic problems, Proceedings, St Etienne 1986, no. 1270 in Lecture Notes in Mathematics, pp. 238–252. Springer, Berlin (1987)
155. Gustafsson, B., Fern, L.: Far fields boundary conditions for time-dependent hyperbolic systems. SIAM J. Sci. Stat. Comput. **9**, 812–848 (1988)
156. Gustafsson, B., Kreiss, H.O.: Boundary conditions for time-dependent problems with an artificial boundary. J. Comput. Phys. **30**, 333–351 (1979)

157. Guzmán, J., Rivière, B.: Sub-optimal convergence of non-symmetric discontinuous Galerkin method for odd polynomial approximations. J. Sci. Comput. **40**(1–3), 273–280 (2009)
158. Hadrava, M., Feistauer, M., Horáček, J., Kosík, A.: Discontinuous Galerkin method for the problems of linear elasticity with applications to the fluid-structure interaction. In: Simos, T. (ed.) AIP Conference Proceedings 1558, Proceedings of 11th International Conference of Numerical Analysis and Applied Mathematics, pp. 2348–2351. AIP Publishing (2013)
159. Hadrava, M., Feistauer, M., Horáček, J., Kosík, A.: Space-time discontinuous Galerkin method for the problem of linear elasticity. In: Abdulle, A., Deparis, S., Kressner, D., Nobile, F., Picasso, M. (eds.) Numerical Mathematics and Advanced Applications ENUMATH 2013, pp. 391–398, Proceedings of ENUMATH 2013, the 10th European Conference on Numerical Mathematics and Advanced Applications, Lausanne, August 2013. Springer, Cham (2015)
160. Hagstrom, T., Hariharan, S.I.: Accurate boundary conditions for exterior problems in gas dynamics. Math. Comput. **51**, 581–597 (1988)
161. Hairer, E., Norsett, S.P., Wanner, G.: Solving Ordinary Differential Equations I, Nonstiff Problems. No. 8 in Springer Series in Computational Mathematics. Springer, Berlin (2000)
162. Hairer, E., Wanner, G.: Solving Ordinary Differential Equations II. Stiff and Differential-Algebraic Problems. Springer, Berlin (2002)
163. Hartmann, R., Houston, P.: Adaptive discontinuous Galerkin finite element methods for non-linear hyperbolic conservation laws. SIAM J. Sci. Comput. **24**, 979–1004 (2002)
164. Hartmann, R., Houston, P.: Adaptive discontinuous Galerkin finite element methods for the compressible Euler equations. J. Comput. Phys. **183**(2), 508–532 (2002)
165. Hartmann, R., Houston, P.: Symmetric interior penalty DG methods for the compressible Navier-Stokes equations I: method formulation. Int. J. Numer. Anal. Model. **1**, 1–20 (2006)
166. Hartmann, R., Houston, P.: Symmetric interior penalty DG methods for the compressible Navier-Stokes equations II: goal-oriented a posteriori error estimation. Int. J. Numer. Anal. Model. **3**, 141–162 (2006)
167. Hasnedlová, J., Feistauer, M., Horáček, J., Kosík, A., Kučera, V.: Numerical simulation of fluid-structure interaction of compressible flow and elastic structure. Computing **95**, 343–361 (2013)
168. Havle, O., Dolejší, V., Feistauer, M.: DGFEM for nonlinear convection-diffusion problems with mixed Dirichlet-Neumann boundary conditions. Appl. Math. **55**(5), 353–372 (2010)
169. Hecht, F., Pironneau, O., Hyaric, A.L., Ohtsuka, K.: FreeFEM, version 2.22. Université Paris 6, Laboratoire Jacques-Louis Lions, www.freefem.org/ff++
170. Hedstrom, G.W.: Nonreflecting boundary conditions for nonlinear hyperbolic systems. J. Comput. Phys. **30**, 222–237 (1979)
171. Hemker, P.W., Spekreijse, S.: Multiple grid and osher's scheme for the efficient solution of the steady Euler equations. Appl. Numer. Math. **2**, 475–493 (1986)
172. Hendry, J., Delves, L.: The global element method applied to a harmonic mixed boundary value problem. J. Comput. Phys. **33**, 33–44 (1979)
173. Hendry, J., Delves, L., Phillips, C.: Numerical experience with the global element method. Mathematics of Finite Elements and Applications III, pp. 341–348. Academic Press, London (1979)
174. Hesthaven, J., Warburton, T.: Nodal Discontinuous Galerkin Methods: Algorithms, Analysis and Applications. Texts in Applied Mathematics. Springer, New York (2008)
175. van der Heul, D., Vuik, C., Wesseling, P.: A conservative pressure-correction method for flow at all speeds. Comput. Fluids **32**, 1113–1132 (2003)
176. Hindenlang, F., Gassner, G.J., Altmann, C., Beck, A., Staudenmaier, M., Munz, C.D.: Explicit discontinuous Galerkin methods for unsteady problems. Comput. Fluids **61**, 86–93 (2012)
177. Hirch, C.: Numerical computation of internal and external flows. Fundamentals of Numerical Discretization. Wiley Series in Numerical Methods in Engineering, vol. 1. Wiley-Interscience Publication, Chichester (1988)
178. Holík, M.: Discontinuous Galerkin method for convection-diffusion problems. Ph.D. thesis, Charles University Prague (2010)

179. Honzátko, R., Horáček, J., Kozel, K., Sváček, P.: Simulation of free airfoil vibrations in incompressible viscous flow-comparison of FEM and FVM. Acta Polytech. **52**, 42–52 (2012)
180. Houston, P., Robson, J., Süli, E.: Discontinuous Galerkin finite element approximation of quasilinear elliptic boundary value problems I: the scalar case. IMA J. Numer. Anal. **25**, 726–749 (2005)
181. Houston, P., Schötzau, D., Wihler, T.P.: Energy norm a posteriori error estimation of *hp*-adaptive discontinuous Galerkin methods for elliptic problems. Math. Models Methods Appl. Sci. **17**(1), 33–62 (2007)
182. Houston, P., Schwab, C., Süli, E.: Stabilized *hp*-finite element methods for first-order hyperbolic problems. SIAM J. Numer. Anal. **37**(5), 1618–1643 (2000)
183. Houston, P., Schwab, C., Süli, E.: Discontinuous *hp*-finite element methods for advection-diffusion-reaction problems. SIAM J. Numer. Anal. **39**(6), 2133–2163 (2002)
184. Houston, P., Süli, E.: A note on the design of *hp*-adaptive finite element methods for elliptic partial differential equations. Comput. Methods Appl. Mech. Eng. **194**, 229–243 (2005)
185. Houston, P., Süli, E., Wihler, T.P.: A posteriori error analysis of *hp*-version discontinuous Galerkin finite element methods for second-order quasilinear elliptic problems. IMA J. Numer. Anal. **28**, 245–273 (2008)
186. Hozman, J.: Discontinuous Galerkin method for convection-diffusion problems. Ph.D. thesis, Charles University Prague, Faculty of Mathematics and Physics (2009)
187. Ikeda, T.: Maximum principle in finite element models for convection-diffusion phenomena, Lecture Notes in Numerical and Applied Analysis, Mathematics Studies 76, vol. 4. North-Holland, Amsterdam (1983)
188. Jaffre, J., Johnson, C., Szepessy, A.: Convergence of the discontinuous Galerkin finite element method for hyperbolic conservation laws. Math. Models Methods Appl. Sci **5**(3), 367–286 (1995)
189. Jamet, P.: Galerkin-type approximations which are discontinuous in time for parabolic equations in a variable domain. SIAM J. Numer. Anal. **15**(5), 912–928 (1978)
190. Jiránek, P., Strakoš, Z., Vohralík, M.: A posteriori error estimates including algebraic error and stopping criteria for iterative solvers. SIAM J. Sci. Comput. **32**(3), 1567–1590 (2010)
191. John, V., Knobloch, P.: On spurious oscillations at layer diminishing (SOLD) methods for convection-diffusion equations: part I—a review. Comput. Methods Appl. Mech. Eng. **196**, 2197–2215 (2007)
192. Johnson, C., Pitkäranta, J.: An analysis of the discontinuous Galerkin method for a scalar hyperbolic equation. Math. Comput. **46**, 1–26 (1986)
193. Kanschat, G.: Block preconditioners for LDG discretizations of linear incompressible flow problems. J. Sci. Comput. **22–23**, 371–384 (2005)
194. Kanschat, G.: Discontinuous Galerkin Methods for Viscous Incompressible Flow. Advances in Numerical Mathematics. Deutscher Universitäts-Verlag (2007)
195. Kardestuncer, H., Norrie, D.: Finite Element Handbook. McGraw-Hill, New York (1987)
196. Klaij, C.M., van der Vegt, J., der Ven, H.V.: Pseudo-time stepping for space-time discontinuous Galerkin discretizations of the compressible Navier-Stokes equations. J. Comput. Phys. **219**(2), 622–643 (2006)
197. Klaij, C.M., van der Vegt, J., der Ven, H.V.: Space-time discontinuous Galerkin method for the compressible Navier-Stokes equations. J. Comput. Phys. **217**(2), 589–611 (2006)
198. Klein, R.: Semi-implicit extension of a Godunov-type scheme based on low Mach number asymptotics 1: one-dimensional flow. J. Comput. Phys. **121**, 213–237 (1995)
199. Koren, B., Hemker, P.W.: Damped, direction-dependent multigrid for hypersonic flow computations. Appl. Numer. Math. **7**(4), 309–328 (1991)
200. Kosík, A., Feistauer, M., Hadrava, M., Horáček, J.: Discontinuous Galerkin method for coupled problems of compressible flow and elastic structures. In: Simos, T. (ed.) AIP Conference Proceedings 1558, pp. 2344–2347, Proceedings of 11th International Conference of Numerical Analysis and Applied Mathematics. AIP Publishing (2013)
201. Kosík, A., Feistauer, M., Hadrava, M., Horáček, J.: The interaction of compressible flow and an elastic structure using discontinuous Galerkin method. In: Abdulle, A., Deparis, S.,

Kressner, D., Nobile, F., Picasso, M. (eds.) Numerical Mathematics and Advanced Applications ENUMATH 2013, Proceedings of ENUMATH 2013, the 10th European Conference on Numerical Mathematics and Advanced Applications, Lausanne, August 2013, pp. 735–743. Springer, Cham (2015)

202. Kosík, A., Feistauer, M., Hadrava, M., Horáček, J.: Numerical simulation of the interaction between a nonlinear elastic structure and compressible flow by the discontinuous Galerkin method. Appl. Math. Comput. (to appear) (2015)
203. Kroll, N., Bieler, H., Deconinck, H., Couallier, V., van der Ven, H., Sorensen, K. (eds.): ADIGMA—A European Initiative on the Development of Adaptive Higher-Order Variational Methods for Aerospace Applications. Notes on Numerical Fluid Mechanics and Multidisciplinary Design, vol. 113. Springer, Berlin (2010)
204. Kröner, D.: Absorbing boundary conditions for the linearized Euler equations in 2-D. Math. Comput. **57**, 153–167 (1991)
205. Kröner, D.: Numerical Schemes for Conservation Laws. Wiley Teubner, Stuttgart (1997)
206. Kučera, V.: Optimal $L^\infty(L^2)$-error estimates for the DG method applied to nonlinear convection-diffusion problems with nonlinear diffusion. Numer. Func. Anal. Optim. **31**(3), 285–312 (2010)
207. Kučera, V.: On diffusion-uniform error estimates for the DG method applied to singularly perturbed problems. IMA J. Numer. Anal. **34**(2), 820–861 (2013)
208. Kufner, A., John, O., Fučík, S.: Function Spaces. Academia, Prague (1977)
209. Kufner, A., Sändig, A.M.: Some Applications of Weighted Sobolev Spaces. Teubner, Leipzig (1987)
210. Kythe, P.K., Schäferkotter, M.R.: Handbook of Computational Methods for Integration. Chapman & Hall/CRC, Boca Raton (2005)
211. Larson, M.G., Niklasson, A.J.: Analysis of a family of discontinuous Galerkin methods for elliptic problems: the one dimensional case. Numer. Math. **99**(1), 113–130 (2004)
212. Lebedev, N.N.: Special Functions and Their Applications. Dower Publication, New York (1972)
213. Leoni, G.: A First Course in Sobolev Spaces. Graduate Studies in Mathematics, vol. 105. AMS, Providence (2009)
214. Le Saint, P., Raviart, P.A.: On a finite element method for solving the neutron transport equation. In: de Boor, C. (ed.) Mathematical Aspects of Finite Elements in Partial Differential Equations, pp. 89–145. Academic Press, New York (1974)
215. Liesen, J., Strakoš, Z.: Krylov subspace methods (Principles and analysis). Numerical Mathematics and Scientific Computation. Oxford University Press, Oxford (2013)
216. Lions, J.L.: Problèmes aux limites non homogènes à donées irrégulières: Une méthode d'approximation. In: Numerical Analysis of Partial Differential Equations (C.I.M.E. 2 Ciclo, Ispra, 1967), pp. 283–292. Edizioni Cremonese, Rome (1968)
217. Lions, P.L.: Mathematical Topics in Fluid Mechanics. Oxford Science Publications, New York (1996)
218. Lomtev, I., Quillen, C.B., Karniadakis, G.E.: Spectral/hp methods for viscous compressible flows on unstructured 2D meshes. J. Comput. Phys. **144**(2), 325–357 (1998)
219. Makridakis, C.G., Babuška, I.: On the stability of the discontinuous Galerkin method for the heat equation. SIAM J. Numer. Anal. **34**(1), 389–401 (1997)
220. Meister, A.: Comparison of different Krylov subspace methods embedded in an implicit finite volume scheme for the computation of viscous and inviscid flow fields on unstructured grids. J. Comput. Phys. **140**, 311–345 (1998)
221. Meister, A.: Viscous flow fields at all speeds: analysis and numerical simulation. J. Appl. Math. Phys. **54**, 1010–1049 (2003)
222. Meister, A., Struckmeier, J.: Hyperbolic Partial Differential Equations, Theory, Numerics and Applications. Vieweg, Brauschweig/Wiesbaden (2002)
223. Mishev, I.D.: Finite volume methods on Voronoi meshes. Numer. Methods Partial Differ. Equ. **14**(2), 193–212 (1998)

224. Mittal, S.: Finite element computation of unsteady viscous compressible flows. Comput. Methods Appl. Mech. Eng. **157**, 151–175 (1998)
225. More, J., Garbow, B., Hillstro, K.: MINPACK. Argonne National Laboratory. www.netlib.org/minpack
226. Morton, K.W.: Numerical Solution of Convection-Diffusion Problems. Chapman & Hall, London (1996)
227. Nečas, J.: Les Méthodes Directes en Thèorie des Equations Elliptiques. Academia, Prague (1967)
228. Nitsche, J.A.: Über ein Variationsprinzip zur Lösung Dirichlet-Problemen bei Verwendung von Teilräumen, die keinen randbedingungen unteworfen sind. Abh. Math. Sem. Univ. Hamburg **26**, 9–15 (1971)
229. Nomura, T., Hughes, T.: An arbitrary Lagrangian-Eulerian finite element method for interaction of fluid and a rigid body. Comput. Methods Appl. Mech. Eng. **95**(1), 115–138 (1992)
230. Oden, J.T., Babuška, I., Baumann, C.E.: A discontinuous *hp* finite element method for diffusion problems. J. Comput. Phys. **146**, 491–519 (1998)
231. Park, J., Munz, C.D.: Multiple pressure variables methods for fluid flow at all Mach numbers. Int. J. Numer. Methods Fluids **49**, 905–931 (2005)
232. Quarteroni, A., Valli, A.: Numerical approximation of partial differential equations. Springer Series in Computational Mathematics, vol. 23. Springer, Berlin (1994)
233. Quarteroni, A., Valli, A.: Domain Decomposition Methods for Partial Differential Equations. Numerical Mathematics and Scientific Computation. Clarendon Press, Oxford (1999)
234. Reed, W.H., Hill, T.R.: Triangular mesh methods for the neutron transport equation. Technical Report LA-UR-73-479, Los Alamos Scientific Laboratory (1973)
235. Rektorys, K.: The Method of Discretization in Time and Partial Differential Equations. Reidel, Dordrecht (1982)
236. Richter, G.R.: An optimal-order error estimate for the discontinuous Galerkin method. Math. Comput. **50**(181), 75–88 (1988)
237. Richter, G.R.: The discontinuous Galerkin method with diffusion. Math. Comput. **58**(198), 631–643 (1992)
238. Rivière, B.: Discontinuous Galerkin Methods for Solving Elliptic and Parabolic Equations: Theory and Implementation. Frontiers in Applied Mathematics. Society for Industrial and Applied Mathematics, Philadelphia (2008)
239. Rivière, B., Wheeler, M.F., Girault, V.: Improved energy estimates for interior penalty, constrained and discontinuous Galerkin methods for elliptic problems. I. Comput. Geosci. **3**(3–4), 337–360 (1999)
240. Rivière, B., Wheeler, M.F., Girault, V.: A priori error estimates for finite element methods based on discontinuous approximation spaces for elliptic problems. SIAM J. Numer. Anal. **39**(3), 902–931 (2001)
241. Roe, P.L.: Remote boundary conditions for unsteady multidimensional aerodynamic computations. Comput. Fluids **17**, 221–231 (1989)
242. Roller, S., Munz, C.D., Geratz, K., Klein, R.: The multiple pressure variables method for weakly compressible fluids. Z. Angew. Math. Mech. **77**, 481–484 (1997)
243. Romkes, A., Prudhomme, S., Oden, J.T.: A priori error analyses of a stabilized discontinuous Galerkin method. Comput. Math. Appl. **46**(8–9), 1289–1311 (2003)
244. Roos, H.G., Stynes, M., Tobiska, L.: Numerical Methods for Singularly Perturbed Differential Equation. Springer Series in Computational Mathematics, vol. 24. Springer, Berlin (1996)
245. Roos, H.G., Stynes, M., Tobiska, L.: Robust Numerical Methods for Singularly Perturbed Differential Equations. Springer Series in Computational Mathematics. Springer, Berlin (2008)
246. Roubíček, T.: Nonlinear Partial Differential Equations with Applications. Birkhäuser, Basel (2005)
247. Rudin, W.: Real and Complex Analysis, 3rd edn. McGraw-Hill, New York (1987)
248. Runst, T., Sickel, W.: Sobolev Spaces of Fractional Order, Nemytskij Operators, and Nonlinear Differential Equations. Walter de Gruyter, Berlin (1996)

249. Saad, Y., Schultz, M.H.: GMRES: a generalized minimal residual algorithm for solving nonsymmetric linear systems. SIAM J. Sci. Stat. Comput. **7**, 856–869 (1986)
250. Samarskii, A.A., Vabishchevich, P.N.: Numerical Methods for the Solution of Convection-Diffusion Problems. Publisher URSS, Moscow (1999). (in Russian)
251. Schötzau, D., Schwab, C.: An *hp* a priori error analysis of the discontinuous Galerkin time-stepping for initial value problems. Calcolo **37**, 207–232 (2000)
252. Schötzau, D., Wihler, T.P.: Exponential convergence of mixed *hp*-DGFEM for Stokes flow in polygons. Numer. Math. **96**, 339–361 (2003)
253. Schwab, C.: *p*- and *hp*-Finite Element Methods. Clarendon Press, Oxford (1998)
254. Schwab, C.: Discontinuous Galerkin method. Technical report, ETH Zürich (2000)
255. Shu, C.: Essentially non-oscillatory and weighted essentially non-oscillatory schemes for hyperbolic conservation laws. In: Quarteroni, A., et al. (eds.) Advanced Numerical Approximation of Nonlinear Hyperbolic Equations. Lecture Notes Mathematics 1697, pp. 325–432. Springer, Berlin (1998)
256. Sobotíková, V.: Error analysis of a DG method employing ideal elements applied to a nonlinear convection-diffusion problem. J. Numer. Math. **19**(2), 137–163 (2011)
257. Sobotíková, V., Feistauer, M.: On the effect of numerical integration in the DGFEM for nonlinear convection-diffusion problems. Numer. Methods Partial Differ. Equ. **23**, 1368–1395 (2007)
258. Šolín, P.: Partial differential equations and the finite element method. Pure and Applied Mathematics. Wiley-Interscience, New York (2004)
259. Šolín, P., Demkowicz, L.: Goal-oriented *hp*-adaptivity for elliptic problems. Comput. Methods Appl. Mech. Eng. **193**, 449–468 (2004)
260. Šolín, P., Segeth, K., Doležel, I.: Higher-Order Finite Element Methods. Chapman & Hall/CRC Press, London (2003)
261. Stoufflet, B.: Implicit finite element methods for the Euler equations. In: Numerical methods for the Euler Equations of Fluid Dynamics, Proceedings of the INRIA Workshop, Rocquencourt/France 1983, pp. 409–434 (1985)
262. Strang, G.: Variational crimes in the finite element method. In: Aziz, A. (ed.) The Mathematical Foundations of the Finite Element Method with Applications to Partial Differential Equations, pp. 689–710. Academic Press, New York (1972)
263. Sun, S.: Discontinuous Galerkin methods for reactive transport in porous media. Ph.D. thesis, The University of Texas, Austin (2003)
264. Sun, S., Wheeler, M.F.: Discontinuous Galerkin methods for coupled flow and reactive transport problems. Appl. Numer. Math. **52**, 273–298 (2005)
265. Sun, S., Wheeler, M.F.: Symmetric and nonsymmetric discontinuous Galerkin methods for reactive transport in porous media. SIAM J. Numer. Anal. **43**(1), 195–219 (2005)
266. Sváček, P., Feistauer, M., Horáček, J.: Numerical simulation of flow induced airfoil vibrations with large amplitudes. J. Fluids Struct. **23**, 391–411 (2007)
267. Tenaud, C., Garnier, E., Sagaut, P.: Evaluation of some high-order shock capturing schemes for direct numerical simulation of unsteady two-dimensional free flows. Int. J. Numer. Methods Fluids **126**, 202–228 (2000)
268. Thomée, V.: Galerkin Finite Element Methods for Parabolic Problems. 2nd revised and expanded edition. Springer, Berlin (2006)
269. Toro, E.F.: Riemann Solvers and Numerical Methods for Fluid Dynamics. Springer, Berlin (1997)
270. Toselli, A.: *hp* discontinuous Galerkin approximations for the Stokes problem. Math. Models Methods Appl. Sci **12**(11), 1565–1597 (2002)
271. Toselli, A., Schwab, C.: Mixed *hp*-finite element approximations on geometric edge and boundary layer meshes in three dimensions. Numer. Math. **94**(4), 771–801 (2003)
272. van der Vegt, J.J.W., van der Ven, H.: Space-time discontinuous Galerkin finite element method with dynamic grid motion for inviscid compressible flows. I: general formulation. J. Comput. Phys. **182**(2), 546–585 (2002)

273. van der Ven, H., van der Vegt, J.J.W.: Space-time discontinuous Galerkin finite element method with dynamic grid motion for inviscid compressible flows II. Efficient flux quadrature. Comput. Methods Appl. Mech. Eng. **191**, 4747–4780 (2002)
274. Verfürth, R.: A Review of a Posteriori Error Estimation and Adaptive Mesh-Refinement Techniques. Wiley-Teubner Series Advances in Numerical Mathematics. Wiley, Chichester. Stuttgart (1996)
275. Verfürth, R.: A note on polynomial approximation in Sobolev spaces. M^2AN Math. Model. Numer. Anal. **33**, 715–719 (1999)
276. Verfürth, R.: A posteriori error estimation techniques for finite element methods. Numerical Mathematics and Scientific Computation. Oxford University Press, Oxford (2013)
277. Vijayasundaram, G.: Transonic flow simulation using upstream centered scheme of Godunov type in finite elements. J. Comput. Phys. **63**, 416–433 (1986)
278. Vlasák, M.: Numerical solution of convection-diffusion problems by discontinuous Galerkin method. Ph.D. thesis, Charles University Prague (2010)
279. Vlasák, M.: Optimal spatial error estimates for DG time discretizations. J. Numer. Math. **21**(3), 201–230 (2013)
280. Vohralík, M.: A posteriori error estimates, stopping criteria and inexpensive implementation. Habilitation thesis, Université Pierre et Marie Curie—Paris **6**, (2010)
281. Watkins, D.S.: Fundamentals of Matrix Computations. Monographs, and Tracts. Pure and Applied Mathematics, Wiley-Interscience Series of Texts. Wiley, New York (2002)
282. Wesseling, P.: Principles of Computational Fluid Dynamics. Springer, Berlin (2001)
283. Wheeler, M.: An elliptic collocation-finite element method with interior penalties. SIAM J. Numer. Anal **15**(4), 152–161 (1978)
284. Wihler, T.P.: Discontinuous Galerkin FEM for elliptic problems in polygonal domains. Ph.D. thesis, ETH Zürich (2002)
285. Wihler, T.P., Frauenfelder, O., Schwab, C.: Exponential convergence of the *hp*-DGFEM for diffusion problems. Comput. Math. Appl. **46**, 183–205 (2003)
286. Wihler, T.P., Schwab, C.: Robust exponential convergence of the *hp* discontinuous Galerkin FEM for convection-diffusion problems in one space dimension. East-West J. Numer. Math. **8**(1), 57–70 (2000)
287. Ženíšek, A.: Nonlinear Elliptic and Evolution Problems and Their Finite Element Approximations. Academic Press, London (1990)

Index

V. Dolejší and M. Feistauer, *Discontinuous Galerkin Method*,
Springer Series in Computational Mathematics 48,
DOI 10.1007/978-3-319-19267-3

C

D